Handbook of
PESTICIDES
Methods of Pesticide Residues Analysis

T0141184

Handbook of
PESTICIDES

Methods of Pesticide Residues Analysis

Edited by
Leo M.L. Nollet
Hamir Singh Rathore

CRC Press
Taylor & Francis Group
Boca Raton London New York

CRC Press is an imprint of the
Taylor & Francis Group, an **informa** business

CRC Press
Taylor & Francis Group
6000 Broken Sound Parkway NW, Suite 300
Boca Raton, FL 33487-2742

First issued in paperback 2020

© 2010 by Taylor and Francis Group, LLC
CRC Press is an imprint of Taylor & Francis Group, an Informa business

No claim to original U.S. Government works

ISBN-13: 978-0-367-57727-8 (pbk)
ISBN-13: 978-1-4200-8245-6 (hbk)

This book contains information obtained from authentic and highly regarded sources. Reasonable efforts have been made to publish reliable data and information, but the author and publisher cannot assume responsibility for the validity of all materials or the consequences of their use. The authors and publishers have attempted to trace the copyright holders of all material reproduced in this publication and apologize to copyright holders if permission to publish in this form has not been obtained. If any copyright material has not been acknowledged please write and let us know so we may rectify in any future reprint.

Except as permitted under U.S. Copyright Law, no part of this book may be reprinted, reproduced, transmitted, or utilized in any form by any electronic, mechanical, or other means, now known or hereafter invented, including photocopying, microfilming, and recording, or in any information storage or retrieval system, without written permission from the publishers.

For permission to photocopy or use material electronically from this work, please access www.copyright.com (http://www.copyright.com/) or contact the Copyright Clearance Center, Inc. (CCC), 222 Rosewood Drive, Danvers, MA 01923, 978-750-8400. CCC is a not-for-profit organization that provides licenses and registration for a variety of users. For organizations that have been granted a photocopy license by the CCC, a separate system of payment has been arranged.

Trademark Notice: Product or corporate names may be trademarks or registered trademarks, and are used only for identification and explanation without intent to infringe.

Library of Congress Cataloging-in-Publication Data

Handbook of pesticides : methods of pesticide residues analysis / editors, Leo M.L. Nollet and Hamir
   S. Rathore.
    p. cm.
   Includes bibliographical references and index.
   ISBN 978-1-4200-8245-6 (alk. paper)
   1. Pesticide residues in food--Analysis. 2. Pesticides--Analysis. 3. Agricultural chemicals--Analysis.
I. Nollet, Leo M. L., 1948- II. Rathore, Hamir Singh. III. Title.

TX571.P4H36 2010
664'.06--dc22
                                          2009013434

Visit the Taylor & Francis Web site at
http://www.taylorandfrancis.com

and the CRC Press Web site at
http://www.crcpress.com

Dedicated to my teachers
Late Professor Mohsin Qureshi and Professor Shakti Rais Ahmad,
Department of Applied Chemistry, Zakir Husain College of Engineering
and Technology, Aligarh Muslim University, Aligarh, India
and my wife Kamini Rathore.

Hamir Singh Rathore

Contents

PART I   General Aspects

PART II   Techniques and Analysis

PART III   Pesticides and the Environment

Preface

With the introduction of fertilizers and high-yielding varieties of cereals, and other commercial crops, management in agriculture has assumed a new dimension in countries such as India. With the provision of assured irrigation facilities, the intensity of land use has also been stepped up. To avoid risk factors in this high-cost and high-intensity crop management system, farmers need an effective and inexpensive plant protection schedule; this is the reason for the manyfold increase in the use of chemical plant protection in the last two decades.

Chemical plant protection is profit-induced poisoning of the environment. Among the chemicals used, the organochlorine insecticides have been the major cause of anxiety for ecologists, not only because they persist for so long but also because of the ease with which they are taken up into the bodies of living organisms, especially the fatty tissues of both animals and humans.

Our information on the occurrence of residues in various parts of the environment is very uneven and localized. For example, a great deal of data on residues are available in China (29%), the United States (13%), Japan (7%), India (6%), Spain (6%), and Germany (5%), while we know virtually nothing about the extent of pesticide contamination in Africa, South America, and much of Asia (Nepal, Pakistan, Sri Lanka, etc.), although large amounts of organochlorine insecticides have been used in these regions.

Therefore, there are vociferous clashes between those ecologists who believe that all pesticides are bad and should be banned, and agriculturalists and others who believe that continued use of large quantities of pesticides is essential to the survival of humanity. There is thus a need for a balanced approach to this issue, and this can be resolved by collecting information selectively because of the vast literature available on this subject. This book provides simple and inexpensive methods as well as ultrasensitive sophisticated high-priced methods of pesticides residue analysis. It is hoped that it will serve as an important source of knowledge for pesticide users and policy makers as well as a guide for those dealing with pesticide residues analysis.

The editors would like to thank all the contributors for their excellent efforts. Their in-depth knowledge is worthy of appreciation.

Leo M. L. Nollet
Hamir Singh Rathore

Acknowledgment

Hamir Singh Rathore thanks the All India Council for Technical Education, New Delhi, India, for the award of an emeritus fellowship for editing this book.

Editors

Leo M. L. Nollet is a professor of biochemistry, aquatic ecology, and ecotoxicology in the Department of Applied Engineering Sciences at University College Ghent, a member of the Ghent University Association, Ghent, Belgium. His main research interests are in the areas of food analysis, chromatography, and the analysis of environmental parameters.

Dr. Nollet edited for Marcel Dekker, New York—now part of CRC Press of the Taylor & Francis Group—the first and second editions of *Food Analysis by HPLC* and the *Handbook of Food Analysis*. The last edition is a three-volume set. He also edited the third edition of the *Handbook of Water Analysis, Chromatographic Analysis of the Environment* (CRC Press) and the second edition of the *Handbook of Water Analysis* (CRC Press) in 2007. He coedited two books with F. Toldrá that were published in 2006: *Advanced Technologies for Meat Processing* (CRC Press) and *Advances in Food Diagnostics* (Blackwell Publishing). He also coedited *Radionuclide Concentrations in Foods and the Environment* with M. Pöschl in 2006 (CRC Press).

Dr. Nollet has coedited several books with Y. H. Hui and other colleagues: the *Handbook of Food Product Manufacturing* (Wiley, 2007); the *Handbook of Food Science, Technology and Engineering* (CRC Press, 2005); and *Food Biochemistry and Food Processing* (Blackwell Publishing, 2005). Finally, he has also edited the *Handbook of Meat, Poultry and Seafood Quality* (Blackwell Publishing, 2007).

He has worked on the following six books on analysis methodologies with F. Toldrá for foods of animal origin, all to be published by CRC Press:

- *Handbook of Muscle Foods Analysis*
- *Handbook of Processed Meats and Poultry Analysis*
- *Handbook of Seafood and Seafood Products Analysis*
- *Handbook of Dairy Foods Analysis*
- *Handbook of Analysis of Edible Animal By-Products*
- *Handbook of Analysis of Active Compounds in Functional Foods*

He is currently working on *Food Allergens: Analysis, Instrumentation, and Methods* with A. Van Hengel, which is to be published by CRC Press in 2010.

He received his MS (1973) and PhD (1978) in biology from the Katholieke Universiteit Leuven, Leuven, Belgium.

Hamir Singh Rathore is an emeritus fellow (AICTE) in the Department of Applied Chemistry, Zakir Husain College of Engineering and Technology, Aligarh Muslim University, Aligarh, India. Dr. Rathore works in the area of physical and analytical chemistry. More specifically, he works on the synthesis and analysis of inorganic ion exchangers, organic acids, heavy metal ions, and pesticides using chromatography, spectroscopy, spot-test analysis, etc. He has published 125 research/review papers in journals of international repute, has supervised 20 PhD students and an equal number of MPhil students, and has published the following books: *Basic Practical Chemistry* (1982), *Experiments in Applied Chemistry* (1990), both AMU Publications and coauthored with Dr. I. Ali, and the *Handbook of Chromatography: Liquid Chromatography of Polycyclic Aromatic Hydrocarbons* (1993) coedited with Professor Joseph Sherma and published by CRC Press. He has also contributed chapters on pesticide residues analysis in the *Handbook of Food Analysis* (1996), the *Handbook of Water Analysis* (2000), and the *Handbook of Food Analysis* (2004), all published by Marcel Dekker, Inc.

Rathore was awarded his postdoctoral fellowship in 1987 by the Third World Academy of Sciences, Trieste, Italy, and carried out research work on biosensors with Marco Mascini at II University of Rome. He has presented his research work, delivered invited talks, chaired technical sessions, and acted as the sectional president of analytical and environmental chemistry in several conferences in India and abroad (Brazil, Finland, Hungary, Italy, Russia, Spain, the United States, etc.).

Contributors

Fatma U. Afifi
Department of Pharmaceutical Sciences
Faculty of Pharmacy
University of Jordan
Amman, Jordan

Sumit Arora
Dairy Chemistry Division
National Dairy Research Institute
Karnal, India

N. C. Basantia
Avon Food Lab
Delhi, India

Chanbasha Basheer
Department of Chemistry
National University of Singapore
Singapore, Singapore

Abdelkader H. Battah
Department of Pathology and Microbiology
and Forensic Medicine
Faculty of Medicine
University of Jordan
Amman, Jordan

Francesca Bettazzi
Dipartimento di Chimica
Università degli Studi di Firenze
Sesto Fiorentino, Italy

Cristina Blasco
Laboratori de Nutrició i Bromatologia
Facultat de Farmàcia
Universitat de València
Valencia, Spain

Sonia Centi
Dipartimento di Chimica
Università degli Studi di Firenze
Sesto Fiorentino, Italy

Aruna Chhabra
Dairy Cattle Nutrition Division
National Dairy Research Institute
Karnal, India

Claudio De Pasquale
Dipartimento di Ingegneria e Tecnologie Agro
 Forestali
Università degli Studi di Palermo
Palermo, Italy

Mohamed Hamza El-Saeid
Department of Chemistry
Texas Southern University Houston, Texas

and

Soil Science Department
College of Food and Agricultural Sciences
King Saud University
Riyadh, Saudi Arabia

Laura Gámiz-Gracia
Department of Analytical Chemistry
Faculty of Sciences
University of Granada
Granada, Spain

Ana M. García-Campaña
Department of Analytical Chemistry
Faculty of Sciences
University of Granada
Granada, Spain

Pritee Goyal
Department of Chemistry
Faculty of Science
Dayalbagh Educational Institute
Agra, India

Svetlana Grujic
Department of Analytical Chemistry
Faculty of Technology and Metallurgy
University of Belgrade
Belgrade, Serbia

Rima M. Hajjo
Division of Medicinal Chemistry and
    Natural Products
University of North Carolina
Chapel Hill, North Carolina

José F. Huertas-Pérez
Department of Analytical Chemistry
Faculty of Sciences
University of Granada
Granada, Spain

Amjad Mumtaz Khan
Department of Applied Chemistry
Faculty of Engineering and Technology
Aligarh Muslim University
Aligarh, India

Haseeb Ahmad Khan
Department of Biochemistry
College of Sciences
King Saud University
Riyadh, Saudi Arabia

Serena Laschi
Dipartimento di Chimica
Università degli Studi di Firenze
Sesto Fiorentino, Italy

Mila Lausevic
Department of Analytical Chemistry
Faculty of Technology and Metallurgy
University of Belgrade
Belgrade, Serbia

Hian Kee Lee
Department of Chemistry
National University of Singapore
Singapore, Singapore

Jin-Ming Lin
Department of Chemistry
Tsinghua University
Beijing, China

Li-Bin Liu
State Key Laboratory of Environmental
    Chemistry and Ecotoxicology
Research Center for Eco-Environmental
    Sciences
Chinese Academy of Sciences
Beijing, China

Yan Liu
State Key Laboratory of Environmental
    Chemistry and Ecotoxicology
Research Center for Eco-Environmental
    Sciences
Chinese Academy of Sciences
Beijing, China

P. Manisankar
Department of Industrial Chemistry
Alagappa University
Karaikudi, India

Marco Mascini
Dipartimento di Chimica
Università degli Studi di Firenze
Sesto Fiorentino, Italy

Ali Mohammad
Department of Applied Chemistry
Faculty of Engineering and Technology
Aligarh Muslim University
Aligarh, India

Leo M. L. Nollet
Faculty of Applied Engineering Sciences
University College Ghent
Ghent, Belgium

Ilaria Palchetti
Dipartimento di Chimica
Università degli Studi di Firenze
Sesto Fiorentino, Italy

Yolanda Picó
Laboratori de Nutrició i Bromatologia
Facultat de Farmàcia
Universitat de València
Valencia, Spain

Marina Radisic
Department of Analytical Chemistry
Faculty of Technology and Metallurgy
University of Belgrade
Belgrade, Serbia

Hamir Singh Rathore
Department of Applied Chemistry
Faculty of Engineering and Technology
Zakir Husain College of Engineering and
    Technology
Aligarh Muslim University
Aligarh, India

S. K. Saxena
Centre for Analysis, Research and Training
New Delhi, India

Shafiullah
Department of Applied Chemistry
Zakir Husain College of Engineering
    and Technology

and

Chemical Research Unit
A.K. Tibbiya College
Aligarh Muslim University
Aligarh, India

Vivek Sharma
Dairy Chemistry Division
National Dairy Research Institute
Karnal, India

K. K. Singh
Agriculture and Soil Survey
Krishi Bhavan, Bikaner, India

and

Chemical Research Unit
A.K. Tibbiya College
Aligarh Muslim University
Aligarh, India

Pramod Singh
Animal Nutrition Division
Indian Council of Agricultural Research
Research Complex for Northeast
    Hilly Region
Umiam (Barapani), India

Jorge J. Soto-Chinchilla
Department of Analytical Chemistry
Faculty of Sciences
University of Granada
Granada, Spain

Man Mohan Srivastava
Department of Chemistry
Faculty of Science
Dayalbagh Educational Institute
Agra, India

Shalini Srivastava
Department of Chemistry
Faculty of Science
Dayalbagh Educational Institute
Agra, India

Alka Tomar
Centre for Media Studies (Environment)
Research House, Saket Community Centre
New Delhi, India

Tomasz Tuzimski
Department of Physical Chemistry
Faculty of Pharmacy
Medical University of Lublin
Lublin, Poland

Suresh Valiyaveettil
Department of Chemistry
National University of Singapore
Singapore, Singapore

Tatjana Vasiljevic
Department of Analytical Chemistry
Faculty of Technology and Metallurgy
University of Belgrade
Belgrade, Serbia

C. Vedhi
Department of Industrial Chemistry
Alagappa University
Karaikudi, India

S. Viswanathan
Department of Industrial Chemistry
Alagappa University
Karaikudi, India

Balbir K. Wadhwa
Dairy Chemistry Division
National Dairy Research Institute
Karnal, India

1 Introduction

Hamir Singh Rathore

Much of the increase in global agricultural production over the last few decades has come about through the adoption of high-input farming systems. Pesticides are one of the most important components of high-input farming. Now humans have realized the extent to which pests harm crops, cause damage, and transmit diseases to both humans and domestic animals. The use of pesticides to kill pests is not a new concept; about AD 70, Pliny, the elder, recommended that arsenic could be used to kill insects, and the Chinese used arsenic sulfide as an insecticide as early as the late sixteenth century. The use of arsenical compounds has continued and, during the early part of the twentieth century, large quantities of compounds such as lead arsenate were used to control insect pests. Another arsenical compound, Paris green (copper aceto-arsenite), was extensively applied to pools and standing water in the tropics in an attempt to control malaria-transmitting mosquitoes.

Inorganic compounds, which were used as insecticides and fungicides, contained antimony, boron, copper, fluorine, manganese, mercury, selenium, sulfur, thallium, and zinc as their active ingredients and were not found to be very effective as insecticides. However, many such compounds were persistent in the soil. There were instances of crops being damaged by inorganic residues in the soil.

The era of synthetic organic pesticides began about 1940. These chemicals were found to be very effective in controlling pests, so their adoption was extremely rapid. Several new chemicals were developed and used as pesticides. In India, the consumption of pesticides increased from 4,000 MT in 1954–1955 to 70,000 MT in 1985–1986. By the end of seventh five-year plan, it was about 92,000 MT and 1.5 lakhs MT by the year 2000. Presently, India's consumption of pesticides is more than 327 g as compared to 1600 g in the United States, 2000 g in the United Kingdom, and 10 kg in Japan.

Chemical crop protection is profit-induced poisoning of the environment. If on one side pesticides have helped India and other countries in achieving self-sufficiency in food production, on the other hand, their indiscriminate use has considerably polluted the environment. Among the many aspects of pesticide use that have caused environmental damage, the following two are note worthy:

1. Extent of persistence of some soil-applied insecticides under agroclimatic conditions
2. Extent of contamination of environment including food commodities

A literature survey shows that aldrin applied once or repeatedly in soil before sowing of different field crops did not result in accumulation of excessive residues in the soil. About 99% of the initial deposits dissipated in a period of 3–5 months. Groundnut kernels absorbed aldrin/dieldrin residues above the maximum residue levels (MRL). The dissipation of hexachlorocyclohexanes (HCH) has been observed to be quite fast. Its dissipation has been recorded to be 60%–90% under the cover of different field crops, applied once in a crop season or repeatedly in five-year experiments. None of the crops grown in treated soils absorbed HCH residues above the MRL. Chlordane and heptachlor, when applied at the rate of 10 kg a.i./ha before sowing wheat in 1984, dissipated more than 95% in a period of 2 years whereas endosulfan decomposed completely in soil.

Translocation of these toxicants occurred from soils to crops at traces. A systemic pesticide, phorate of organophosphates (OP) group, when applied at sowing/transplanting of paddy, potato, and maize at the rate of 1.4 kg a.i./ha dissipated by more than 95% in 2–4 months. In potato, the pesticide volume of 1.5 kg/ha was found to be unsafe whereas in all other crops the tested doses were found to be safe.

A carbamate oxime compound, aldicarb, dissipated almost completely in 1.5–5 months under the cover of radish, potato, sugarcane, and wheat crops at the level of a dose of 0.5–4 kg a.i./ha. The behavior of carbofuran was found to be similar to that of aldicarb. It dissipated completely in 90 days under the cover of paddy crop. The degradation of carbofuran was equally fast under maize crop. It left negligible residues after 2 months of treatment. Its translocation into crop parts was found to be negligible.

Sporadic and nonsystematic work carried out on monitoring and surveillance of pesticide residues in all over the world in general and in India in particular has revealed the following findings:

- Dichloro-diphenyl-trichloroethane (DDT) residues in soils have been found to be from traces to 7.27 ppm. Its contamination incidences ranged from 85% to 100%. Residues of phorate and methyl parathion were reported as traces in soil. Due to favorable agrometerological conditions in India, the dissipation of pesticides occurred at a much faster rate than that observed under temperate conditions. The time taken for 95% dissipation of organochlorine (OC) insecticides in soil varies from 3 to 10 months in India and from 1 to 30 years in temperate conditions.
- In India, DDT and its metabolites are the main contaminants of water bodies, air, and rain water. It has been reported during 1977–1980 that the Yamuna river water contained 2.5–24 ppt and the Hindon river water contained 217 ppt DDT. Suraj Kund, Damdama lakes, and the Delhi university campus pond contained 195, 377, and 64 ppt of DDT, respectively. Rain water contained 12.5 ppb DDT and 5.29 ppb HCH. Air samples collected in Delhi contained 60.2 ng/m^3 DDT and 438 ng/m^3 HCH.
- Milk samples (48%) and egg samples (100%) contained either DDT or HCH residues varying from traces to 7.0 ppm. In some samples, the concentration of the pesticide residues was above their MRL values. The source of the contamination is probably consumption of contaminated feed by milch animals and poultry.
- Cereal and pulse samples were found to be contaminated with HCH residues at the level of 0.001–1.10 ppm, which is not very alarming. Wheat, rice, and mung grains contained aldrin residues (0.003–16 ppm). In some samples, residues were found to be exceeding the MRL values. The samples collected directly from the field had low level of contamination whereas the samples collected from personal storage had high level (44 ppm) of contamination. In these cereals, HCH was directly mixed during storage, which otherwise is not allowed legally in India.
- About 68% samples of fruits and vegetables had been reported to be contaminated mainly with HCH (2.1 ppm), DDT (6.0 ppm), and endosulfan (1.7 ppm). Grape berries contained carbaryl at the level of 3.28–6.12 ppm. There are indications that the hard pesticides such as DDT and HCH are still used on fruits and vegetables without any consideration of waiting periods by the growers.

It is now well established that the benefits of the pesticide conferred on humankind are great. It has been estimated that pesticides saved millions of lives and prevented hundreds of millions of incidences of serious illnesses due to malaria, typhus, dysentery, and more than 20 other insect-borne diseases. In agriculture, it has been calculated that, even after the effective use of pesticides, pests still cause annual losses of 20% on a global scale. Therefore, these losses would be astronomical in the absence of persistent chemical use.

However, with the laboratory demonstration of the mutagenic, carcinogenic, and teratogenic activities of a number of pesticides, there is an increasing concern that their inadvertent exposure may contribute to cancer incidence and the genetic disease burden in nontarget species. As stated above, due to favorable agroclimatic conditions for the degradation of pesticides, there do not seem to be any fear of accumulation of residues in soil to alarming levels in most of the parts of India. But, pesticides tend to concentrate in the fatty tissue of various organisms. Such substances show poisoning effects even if they are relatively nontoxic. For example, the toxicity of DDT is the same as that of aspirin. However, consumption of 10 tablets of aspirin has no serious effect while an equivalent amount of DDT, if ingested, can prove to be hazardous because the drug (aspirin) is not retained by the body, whereas the insecticide (DDT) is. Therefore, DDT and other OCs act as a commutative poison.

Substances that tend to get stored up in the fatty tissue undergo bioamplification, that is, their concentration progressively increases in those species that occupy a higher tropic level in the food chain. For example, spraying a marsh with DDT to control mosquitoes results in the accumulation of traces of DDT in the cells of microscopic aquatic organisms, such as plankton. The DDT then becomes concentrated in the tissues of the fish and shellfish that feed on the contaminated plankton. The concentration of DDT measured in the fish is 10 times greater than that in plankton. Gulls, which feed upon fish, accumulate still higher concentrations of DDT. For predators, such as ospreys, pelicans, falcons, and eagles, which feed upon gulls, as well fish, the levels of accumulated DDT are often so high that the consequences are disastrous.

The bioaccumulation tendency of pesticide residues has created many critical problems. For example, in a recent study, type 2 diabetes was linked with the presence of OC residues in food including soft drinks and water because of the latter's role in insulin resistance. The residues of chlordane and *trans*-nonachlor were found to be the most effective. The mechanism of action of the bioaccumulative-persistent organic pesticide, however, is not clear. They possibly affect one of the 15-odd enzymes involved in the insulin action pathways.

Birds of prey, which occupy a significantly high position in the food chain, concentrate enormous quantities of pesticides in their tissues. Their body attempts to metabolize the pesticide by altering the normal metabolic patterns. This alteration involves the use of hormones, which normally regulate the calcium metabolism of birds. This is vital to their ability to lay eggs with thick shells. When these hormones are used to metabolize insecticides, they become chemically modified and are no longer available for the construction of egg shells. As a result, the eggs are easily damaged and the survival rate of the offspring declines. The enzymes also modify the bird's sex hormones, leading to abnormally late breeding, prevention of laying of new eggs, or laying of fewer eggs. All these factors result in a decline of bird population.

Some of the OCs are highly toxic to fish. Species like trout and salmon succumb when water contains only traces of DDT, HCH, and other pesticides. Low levels of *bis*(diethylphosphate) anhydride affects the learning ability of goldfish. A low concentration (1.25 ppm) of diazinon results in higher activity in rainbow trout but reduces rate of swimming in tin-plated barbs.

In the light of the above findings, there is a need for regularly undertaking systematic surveillance and monitoring of pesticide residues in different components of environment, i.e., surface and subsurface water, soil, food commodities, etc. to keep pesticidal pollution within safe levels. Nevertheless, we still have not completed establishing the seriousness of the problem. A current literature survey shows that about 1500 papers have been published in the area of pesticides in the year 2006 from countries, namely, Albania, Argentina, Australia, Bangladesh, Belgium, Brazil, Canada, China, Columbia, Croatia, Czech Republic, Denmark, Egypt, Finland, France, Georgia, Germany, Ghana, Greece, Hungary, India, Iran, Israel, Italy, Jamaica, Japan, Jordan, Kazakhstan, Malaysia, Mexico, Morocco, the Netherlands, New Zealand, Nicaragua, North Korea, Norway, Pakistan, Poland, Portugal, Romania, Russia, Saudi Arabia, Singapore, Slovakia, Slovenia, South Africa, South Korea, Spain, Sri Lanka, Sweden, Switzerland, Taiwan, Tanzania, Thailand, Turkey, Uganda, the United Kingdom, the United States, Yemen, and Zimbabwe.

The maximum number of publications is from China (29%) followed by the United States (13.3%), Japan (7%), India (6%), Spain (5.5), and Germany (5%). Canada and Russia account for 2% and 1% of the publications, respectively. Publications from each of the remaining countries are less than 2%. In view of this, there is little doubt about the need for extensive monitoring programs for pesticide residues in the environment. Radioisotopes are already monitored, and there is no reason why this technology should not be extended to pesticides and their possible transport between different areas of the world.

The monitoring of residues in various parts of the physical environment is still inadequate. The rivers in the United States are now being monitored annually for pesticide residues, but in Great Britain, only a few small-scale surveys have been made, and in India, only a few surveys have been made. The monitoring of soils for pesticide residues is much more sporadic and confined to agricultural soils. Moreover, little data on pesticide residues are available in the developing countries, although large quantities of persistent pesticides are regularly used in these areas. This may be due to the lack of costly, sophisticated, and ultrasensitive instruments and their repair facilities in this part of the world.

One of the main aims of this book is to present methods of analysis of pesticide residues using different analytical techniques. The methods are presented by leading researchers in the field, who possess authoritative knowledge. With these techniques, the scientist and the layperson may be able to detect/estimate the pesticide residues in different parts of the physical environment. Then a much more balanced judgment could be made based on the application of techniques compared to knowledge acquired from some of the books and articles that appear in the popular press, which are often emotive and written with a sensational appeal. We hope that the methods presented will be very useful to students, researchers, consultants, and other professionals in environmental science; production, marketing, and pesticide use; and in chemistry, biology, and ecology.

Part I

General Aspects

2 Methods of and Problems in Analyzing Pesticide Residues in the Environment

Hamir Singh Rathore

CONTENTS

2.1 INTRODUCTION

Crop diseases are caused by pests such as insects, bacteria, fungi, and viruses. Pesticides are often classified by the type of pest they control. Another way to think about pesticides is considering their chemical structures, their common source, or their production method. Categories of pesticides include biopesticides, antimicrobials, and pest control devices. Thus pesticides are defined as any substance or a mixture of substances used for preventing, destroying, repelling, or mitigation of any pest.

2.1.1 CHEMICAL PESTICIDES

Pliny proposed the use of arsenic to kill insects as early as about AD 70. The Chinese used arsenic sulfide in the late sixteenth century. Since the early nineteenth century, certain inorganic compounds such as lead arsenate, Paris green (copper aceto-arsenite), sodium fluorosilicate, zinc phosphide, and so on have been used as insecticides. It was not realized at that time that arsenical pesticides could persist in the soil up to 40 years. The era of synthetic organic pesticides began around 1940. The commonly known insecticide, dichloro-diphenyl-trichloroethane (DDT), was originally synthesized in 1874 and rediscovered as an insecticide in 1939. DDT was soon followed by benzene hexachloride (BHC). Since then thousands of compounds have been synthesized and used in agricultural farms. Pesticides can be broadly classified according to their general chemical nature into several principal types as shown in Table 2.1.

TABLE 2.1
Chemical Classification of Pesticides

S. No.	Chemical Type	Example	Structure	Typical Action
A	Organochlorines	*p,p'*-DDT		Insecticide
B	Organophosphates	Malathion		Insecticide
C	Carbamates	Carbaryl		Insecticide
D	Dithiocarbamates	Thiram	$(CH_3)_2 N–CS–S–CNSN (CH_3)_2$	Fungicide
E	Carboxylic acid derivatives	2,4-D		Herbicide
F	Substituted ureas	Diuron		Herbicide

TABLE 2.1 (continued)
Chemical Classification of Pesticides

S. No.	Chemical Type	Example	Structure	Typical Action
G	Triazines	Simazine		Herbicide
H	Pyrethroids	Cypermethrin		Insecticide
I	Neem products	Nimbidin (Azadirachtin)	$C_{35}H_{44}O_{16}$	Insecticide
J	*Others*			
1	Organometallics	Phenylmercury acetate		Fungicide
2	Thiocyanates	Lethane 60		Insecticide
3	Phenols	Dinitrocresol		Insecticide
4	Formamides	Chlordimeform		Insecticide

2.1.2 BIOPESTICIDES

Biological control of pests or biopesticides has been suggested as an effective substitute for chemicals. It is the control of harmful pests by using other pests, plants, or any such living body. The controlling agents include parasites, predators, diseases, protozoa, and nematodes that attack pests. Biopesticides are derived from natural materials such as animals, plants, bacteria, and certain minerals. For example, Canola oil and baking soda have pesticidal applications and are considered biopesticides. Biopesticides have a long history. Even in Neolithic times (about 7000 BC), farmers used biopesticides prepared from seeds of resistant plants. At the end of 2001, there were approximately 195 registered biopesticide active ingredients and 780 products. Biopesticides fall into three major classes:

- Microbial pesticides consist of microorganisms (bacterium, fungus, viruses, or protozoa) as the active ingredient. For example, some fungi control certain weeds, and some other fungi kill specific insects.
- Plant-incorporated protectants (PIPs) are pesticidal substances that plants produce from genetic material, which are added to the plant. For example, scientists can take the gene

from the *Bt* pesticidal protein and introduce the gene into the plant's own genetic material. Then the plant, instead of the *Bt* bacterium, manufactures the substance that destroys the pest. The protein and its genetic material, but not the plant itself, are regulated by the U.S. Environmental Protection Agency (EPA).

• Biochemical pesticides are naturally occurring substances such as insect sex pheromones that interfere with mating as well as various scented plant extracts that attract insect pests to traps.

2.1.3 TYPES OF PESTICIDES

Pesticides are also classified on the basis of their pesticidal actions, namely, algicides (control algae in lakes, canals, swimming pools, etc.), antifouling agents (kill or repel organisms that attach to underwater surfaces such as boat bottoms), antimicrobials (kill organisms such as bacteria and viruses, attractants (attract pests, for example, to lure an insect or rodent into a trap; however, food is considered a pesticide when used as an attractant), biopesticides (pesticides derived from such natural materials as animals, plants, bacteria, and certain minerals), biocides (kill microorganisms), disinfectants and sanitizers (kill or inactivate disease-producing microorganisms on inanimate objects), fungicides (kill fungi including blights, mildews, molds, and rusts), fumigants (produce gas or vapor intended to destroy pests in buildings or soil), herbicides (kill weeds and other plants that grow where they are not wanted), insecticides (kill insects and other arthropods), miticides/acaricides (kill mites that feed on plants and animals), microbial pesticides (microorganisms that kill, inhibit, or outcompete pests, including insects or other microorganisms), nematicides (kill nematodes: microscopic, worm-like organisms that feed on plant roots), ovicides (kill eggs of insect and mites), pheromones (biochemicals used to disrupt the mating behavior of insects), repellants (repel pests, including insects such as mosquitoes and birds), rodenticides (control mice and other rodents), defoliants (cause leaves or other foliage to drop from a plant, usually to facilitate harvest), desiccants (promote drying of living tissues, such as unwanted plant tops), insect growth regulators (disrupt the molting, maturity from pupal stage to adult or other life processes of insects), and plant growth regulators (substances excluding fertilizers or other plant nutrients that alter the expected growth, flowering, or reproduction rate of plants). Insecticides are usually classified into the following three classes according to their mode of action.

2.1.3.1 Stomach or Internal Insecticides

Insecticides, for example BHC, DDT, methoxychlor, lead arsenate, calcium arsenate, Paris green, NaF, fluorosilicates, and compounds of P and Hg, that are taken up by insects, are called stomach poisons or internal insecticides. They control insects such as grass hoppers, caterpillars, and so on.

2.1.3.2 Contact or External Insecticides

Agents or preparations such as toxaphene, chlordane, aldrin, dieldrin, methoxychlor, nicotine, pyrethrins, rotenone, tetraethyl pyrophosphate, malathion, and parathion that destroy insects such as leaf hoppers, thrips, and aphids simply by external body contact belong to this class. BHC and DDT also belong to this class.

2.1.3.3 Fumigants

Chemicals such as HCN, CS_2, nicotine, *p*-dichlorobenzene, methyl bromide, ethylene oxide, and so on, acting on insects through their respiratory system are members of this class. BHC is also a member of this class.

2.1.4   MODE OF PESTICIDE POISONING

Pesticides are divided mainly into four categories depending on their mode of poisoning action.

2.1.4.1   Physical Poisons

Physical pesticides kill living organisms by physical action. For example, endrin penetrates percutaneously through the epidermis of the skin and produces lethal effects. Silica and charcoal dusts also interfere in the inhalation of air through the nasal passage and thereby accumulate in the lungs.

2.1.4.2   Nerve Poisons

Pesticides such as DDT, methyl isocyanide (MIC), malathion, parathion, diazinon, and systox act as nerve poisons. They initiate extreme nervous excitation, cause the release of excessive neuroactive substances, and disrupt nerve activity.

2.1.4.3   Protoplasmic Poisons

Pesticides including endrin, ziram, lead arsenate, and sodium arsenite cause precipitation of protein in the body, resulting in liver damage and ultimately death.

2.1.4.4   Respiratory Poisons

Fumigants such as hydrocyanic gas, methyl bromide, ethylene dichloride, and ethylene dibromide inactivate respiratory enzymes such as oxidases, peroxidases, and reductases, and ultimately, they cause acute suffocation and block the respiratory tract.

Pesticides are invaluable for increasing agricultural production as pests and diseases destroy up to one-third of the crop during growth, harvest, and storage. In spite of the exploitation of alternative methods of pest control, the consumption of pesticides in India rapidly increased from 430 metric tons in 1954 to 80,000 metric tons in 1993. However, the rapidly increasing use of pesticides, often with insufficient technical advice or research, has brought in its wake many environmental problems inimical to the interests of humans. A major concern regarding health hazards is human exposure to chronically low levels of pesticides, and the focus is primarily on carcinogenicity, teratogenicity, allergic reactions, neurotoxicity, and effects on the immune and reproductive systems.

Persistent chemicals such as DDT, dieldrin, and polycyclic hydrocarbons have been reported to alter the level of testosterone and to decrease reproductive ability. Impotency in farm workers on exposure to pesticides, such as DDT, dieldrin, and dichloropropane, has also been reported. Endosulfan has been reported to cause testicular dysfunction, whereas ziram, thiram, and dithane M-45 were found to induce significant increase in the frequency of abnormal sperm. In utero exposure to diethylstilbestrol is reported to be associated with altered male reproductive capacity.

The possibility of exposure to environmental chemicals in pregnant women has aroused a great concern over adverse effects on the developing embryo, and sometimes congenital malformations have been observed. Few pesticides are sensitive to low rate of ovulation. Epidemiological studies have confirmed that women married to men exposed to dibromochloropropane and chloropyrene have a higher rate of abortion. Due to these injurious effects, many of the highly toxic pesticides (chlorinated hydrocarbons) have been banned in the United States, but these chemicals still persist as environmental contaminants and are in widespread use in developing countries. The pesticides banned, under restricted use, unapproved, or under review in India are listed in Table 2.2. New and improved compounds are continually being evaluated to overcome this difficulty.

Most pesticides escape natural degradation processes and persist in most foodstuffs including animal tissues. Various surveys have indicated that the most persistently and highly contaminated

TABLE 2.2
Out-of-Use Pesticides in India

	Name	Chemical Family	Mode of Action
A	*Banned pesticides*		
1	Aldrin	Organochlorine	Insecticide
2	Chlordane	Organochlorine	Insecticide
3	Dibromochloropropame (DBCP)	Organochlorine	Nematioid
4	Endrin	Organochlorine	Insecticide
5	Ethyl parathion	Organophosphorous	Insecticide
6	Heptachlor	Organochlorine	Insecticide
7	Nitrofen	Organochlorine	Preemergence herbicide
8	Paraquat-dimethyl sulfate	Bipyridylium	Contact herbicide and desiccant
9	Pentachlorophenol (PCP)	Organochlorine	Preharvest defoliant, wood preservative, and molluscicide
10	Pentachlornitrobenzene (PCNB)	Organochlorine	Soil fungicide and seed dressing agent
11	Tetradifon	Organochlorine	Acaricide
12	Toxaphene	Organochlorine	Insecticide
B	*Restricted use pesticides*		
1	Aluminum phosphide	Inorganic	Insecticidal fumigant
2	Benzene hexachloride	Organochlorine	Insecticide
3	Chlorbenzilate	Organochlorine	Acaricide
4	Captafol	Organochlorine	Fungicide
5	Dichloro-diphenyl-trichloroethane	Organochlorine	Insecticide
6	Dieldrin	Organochlorine	Insecticide
7	Ethylene dibromide	Organobromine	Fumigant (insecticide and nematicide)
8	Methyl bromide	Organobromine	Fumigant (insecticide)
9	Sodium cyanide	Inorganic	Rodenticide
10	Phenyl mercury acetate	Organomercury	Seed dressing
11	Lindane	Organochlorine	Insecticide
12	Nicotine sulphate	Alkaloid, pyridene, pyrrolidene	Insecticide
C	*Unapproved pesticides*		
1	Azinphos—methyl	Organophosphorous	Insecticide
2	Azinphos—ethyl	Organophosphorous	Insecticide
3	Ammonium sulphamate	Inorganic	Contact and translocated herbicide
4	Binapacry	Nitro compound	Miticide, ovicide and fungicide
5	Calcium arsonate	Inorganic	Insecticide and herbicide
6	Corbophennothion	Organophosphorous	Insecticide and acaricide
7	Chinomethionat	Organophosphorous	Insecticide, acaricide, and fungicide
8	Dicrotophes	Organophosphorous	Contact and systemic fungicide
9	EPN	Organophosphorous	Acaricide and insecticide
10	Fentin acetate	Organotin	Fungicide, algicide, and molluscicide
11	Fentin hydroxide	Organotin	Fungicide
12	Lead acetate	Inorganic	Fungicide
13	Leptophos (phosvel)	Organophosphorous	Insecticide
14	Mephosfolan	Organophosphorous	Systemic insecticide
15	Meviphos (Phosdrin)	Organophophorous	Systemic insecticide and acaricide
16	2,4,5-T	Phenoxy	Selective herbicide
17	Thiodemetera/Dosiptem	Organophosphorous	Systemic insecticide and acaricide
18	Vamidothion	Organophosphorous	Persistent systemic aphicide and miticide

TABLE 2.2 (continued)
Out-of-Use Pesticides in India

	Name	Chemical Family	Mode of Action
D	*Under review pesticides*		
1	Alachlor	Acetamide	Preemergence herbicide
2	Benomyl	Benzimidazole	Systemic fungicide
3	Copper aceto-arsenite (Paris green)	Organocopper	Stomach insecticide
4	Diuron	Urea	Herbicide
5	Ethyl mercury chloride	Organomercry	Fungicide (seed treatment)
6	Fenarimol	Pyrimidine	Propytactic and curative leaf fungicide
7	Menazon	Organophosphorous	Systemic aphicide
8	Methomyl	Carbamate	Insecticide
9	Oxyflourfen	Trifluoromethyl, diphenyl ether	Herbicide (pre- and postemergence)
10	Sodium methane arsonate	Aliphatic	Herbicide
11	Calcium cyanide	Inorganic	Fumigant
12	Phosphamidon/Dimecron	Organophosphorous	Systemic insecticide with strong stomach action
13	Thiometon	Organophosphorous	Systemic insecticide
14	Triazophos	Organophosphons	Insecticide, miticide, and nematicide
15	Tride morph/calixin	Morpholine	Fungicide with both curative and protective properties
16	Manocrotophos	Organophosphorous	Contact and systemic insecticide and acaricide
17	Ziram	Dithiocarbamate, organozinc	Fungicide
18	Zinab	Dithiocarbamate, organozinc	Fungicide

foodstuffs are animal products, followed by leafy vegetables and garden fruits. Among pesticides, DDT, BHC, dieldrin, and lindane appear to be widely distributed. Approximately 50% of all pesticide residues detected in food, as indicated by different surveys, are organochlorines, and 60% of these are found primarily in animal products. It is reported that the propensity of animal tissues to store pesticides, particularly in fat, may be characteristic, irrespective of the amount ingested in food.

A recent survey of the chemical abstracts of the year 2006 shows that a number of books have been written on the subject. Several reviews have been published on the different aspects such as enzyme and pesticide residue buildup; pesticide agents; termite control; assessment of children's exposure to pesticides; quality criteria in pesticide analysis; risk assessment of pesticide residues in food; interfacing geographical information systems and pesticide models; ultrasonic treatment of pesticide; method and system for analysis of organochemicals; advances in the process of pesticides; botanical insecticides and environmental safety; pesticide pathways; pesticide residues in agriculture products; pesticide residues in farm products and trends of their analysis techniques; analysis of traditional Chinese medicines for organochlorine pesticides (OCPs); safety evaluation and risk management for modern chemical pesticides; discovery of new pesticides and development of industrialization in China; multitarget toxicity of organophosphorous pesticides (OPP), and so on.

Considerable work has been carried out on the synthesis of foliar fertilizer for degrading pesticide residue and promoting crop growth; production status quo and opportunity of bromine-containing pesticides; research advances on thiazole insecticides; manufacture of pesticide containing *Piper nigrum* extract; manufacture of composite biological pesticide; countermeasures of microbial pesticide industrialization; preparation of chromones, their use as pesticides, and their applications to plants or soil; phthalamide derivatives and their preparation, agrochemical composition, and use as insecticides; preparation of novel anthranilamides useful for controlling invertebrate pests; iodophenyl-substituted cyclic keto ends and their preparation, agrochemical compositions, and use as pesticides and or herbicides; 5-hetrocyclyl pyrimidines and their preparations; agrochemical compositions and use for controlling unwanted microorganisms; preparations of difluoroalkene derivatives and agricultural pesticides containing them, and so on.

Studies have been performed on emulsifiable pesticide granular formulations and emulsion stability of pesticide microemulsions. Many articles have been published on the adverse effects of pesticides, that is, the mutation effect of pesticides with SOS/UMU chromotest; a physiologically based pharmacokinetic model of organophosphate dermal adsorptions, deconvolution of overlapping pesticides using automass deconvolution and identification system (AMDIS); ecological risks of pesticides; assessment of dermal pesticide exposure with fluorescent tracer; a new empirical approach to estimate the short-range transport and dry deposition of volatilized pesticides; mutagenicity of biodegraded and chlorinated derivations of agricultural chemicals; and toxic properties of pesticides.

Attempts were made to produce new formulations and design new procedures of their applications such as natural pyrethrin–microcapsulated pesticide and its uses; alcohol-based aerosol for pest control; improving the effectiveness of aerial pesticide sprays; and model-based, computer-aided design for controlled release of pesticides.

The following mechanism studies have been reported on the degradation of pesticides: photocatalytic degradation of OPPs comparing glass spring-loaded TiO_2 with a TiO_2 slurry; mechanism and application of the degradation of OPP by TiO_2 photocatalytic process; research on degradation of organophosporous wastewater by TiO_2 nanopowder photocatalysis; heterogeneous photocatalyzed reaction of three selected pesticide derivatives, namely, propham, propachlor, and tebuthiuron in aqueous suspensions of TiO_2; decomposition of pesticides by photocatalyst; degradation of toxaphene by zero-valent iron and bimetallic substrate; detoxification of aqueous solutions of the pesticides "Sevrol" by solar photocatalysis; degradation of pesticides by chlorination according to their basic structure; degradation of three pesticides used in viticulture by electrogenerated Penton's reagents; neutralization of organic pesticides by oxidation; study of processes for pyrolytic decomposition of some classes of pesticides; degradation of OPPs by TiO_2 photocatalytic process; and photocatalytic degradation of two selected pesticide derivatives, dichlorvos and phosphamidon, in aqueous suspension of TiO_2.

Treatment processes to reduce the toxicity and removal of pesticides have been published: treatment of pesticide wastewater-contained cyanogens with chlorine dioxide; treatment of pesticide industry wastes; pesticide solvent system for reducing phytotoxicity; removal of pesticides by a combined ozonation/attached biomass process sequence; evaluation of estrogenic activities of pesticides using an in vitro reporter gene assay; accumulation kinetics of organochlorinated pesticides by triolein-containing semipermeable membrane devices, and so on.

Articles have been published on the effect of high doses of sodium bicarbonate in acute OPP poisoning; field study methods for the determination of bystander exposure to pesticides; quantity–activity relationships of organophosphate compounds and robust analysis; and occupational allergic hazards of agricultural chemicals and their prevention.

2.2   ANALYSIS OF PESTICIDE RESIDUES

The analyses of inorganic pesticides do not differ significantly from standard methods of trace element analyses and are not discussed here. The use of organic pesticides in agriculture is now firmly established and is becoming more extensive. It is relevant to point out that the analyses need to be carried out on a variety of commodities, each with its own extraction and cleanup problems, and the most efficient one must be used, because the quantity and the identity of any residue present are usually required. The analysis is often complicated by chemical changes undergone by the pesticides (Table 2.3) when absorbed into living tissue, adsorbed onto the soil, or exposed to ultraviolet

TABLE 2.3
Pesticide Residues in Different Matrices

S. No.	Matrix	Pesticide	Object	Reference
A	*Pesticide residues in beverages, water, and sediments*			
1	Sediments and fish species	Organochlorine	Monitoring	[1]
2	Sediments and biota	Organochlorine	Levels and distribution	[2]
3	Reservoir	Miscellaneous	Distribution studics	[3]
4	Water hyacinth	Ethion	Rhytoremediation	[4]
5	Raw water	Miscellaneous	Water purification	[5]
6	Alfine glacier	Organophosphorous	Accumulation	[6]
7	Prasae River	Miscellaneous	Statistical studies	[7]
8	Atlantic and Antractic Oceans	DDTs, HCHs, and HCB	Monitoring	[8]
9	Surface water sediments and air precipitation of lake	Organochlorine	Precipitation studies	[9]
10	Surface water	Miscellaneous	Pesticide load prediction	[10]
11	Carbonatic soils	Miscellaneous	Sorption kinetics	[11]
12	Surface water and sediments	Organochlorine	Distribution studies	[12]
13	River water	N-methyl carbonate	Monitoring	[13]
14	Sea water	Organochlorine	Monitoring	[14]
15	Groundwater	Miscellaneous	Quality guidelines	[15]
16	Rain water	Miscellaneous	Monitoring	[16]
17	Fresh water	Carbamate, organophosphorous, and pyrethroid	Toxic effect	[17]
18	Sediments	Miscellaneous	Monitoring	[18]
19	Groundwater	Miscellaneous	Monitoring	[19]
20	Aqueous media	Miscellaneous	Evaluation	[20]
21	Lake water	Organochlorine	Monitoring	[21]
22	Surface sediments	Organochlorine	Monitoring	[22]
23	Sediments	Organochlorine	Monitoring	[23]
24	Wines	Miscellaneous	Survey	[24]
25	Surface water	Miscellaneous	Monitoring	[25]
26	Water and sediments	Miscellaneous	Distribution studies	[26]
27	Surface water	Miscellaneous	Monitoring	[27]
28	Water and surface sediments	Miscellaneous	Screening	[28]
29	Water from industrial and urban sewage	Organonitrogen and organophosphorous	Monitoring	[29]
30	Surface water	Miscellaneous	Transport studies	[30]
31	Lake sediments	DDT and HCH	50 years record	[31]
32	Water	Miscellaneous	Modeling of pesticide transport	[32]
33	Waste water	Miscellaneous	Treatment with Ozone and hydrogen peroxide	[33]
34	Watcr	Miscellaneous	Photodegradation	[34]
35	Sediments	Organochlorine	Persistence	[35]
36	Sediments	Organochlorine	Distribution studies	[36]
37	Water	Miscellaneous	Removal by activated carbon	[37]
38	Estuary sediments	Miscellaneous	Monitoring	[38]
39	Sediments	Organochlorine	Distribution studies	[39]
40	Sediments	Organochlorine	Levels and distribution studies	[40]

(continued)

TABLE 2.3 (continued)
Pesticide Residues in Different Matrices

S. No.	Matrix	Pesticide	Object	Reference
41	Water	Organochlorine	Contamination	[41]
42	Lake water	Organochlorine	Comprehensive assessment	[42]
43	Sea water	Organophosphorous	Estimation of source	[43]
44	Water shed	Miscellaneous	Quality management	[44]
45	Groundwater	Miscellaneous	Leaching	[45]
46	Wine	Miscellaneous	Estimation	[46]
47	Sediments and water	Miscellaneous	Behavior of pesticides	[47]
48	Water	Triazine	Removal by nanofiltration	[48]
49	Drinking water	Miscellaneous	Survey	[49]
50	Surface water	Miscellaneous	Pollution	[50]
51	River water	Miscellaneous	Load of pesticides	[51]
52	Costal water	Miscellaneous	Pollution	[52]
53	Water	Miscellaneous	Indicator	[53]
54	Surface sediments	Organochlorine	Distribution studies	[54]
55	Surface water	Miscellaneous	Losses in surface runoff	[55]
56	River water	Miscellaneous	Pesticides load	[56]
57	Sediments	Miscellaneous	Risk	[57]
58	Fresh water	Paraquat and malathion	Sublethal effects	[58]
B	*Pesticide residues in soils*			
1	Land	Arsenical pesticides	Impact of land disturbance	[59]
2	Soils	Miscellaneous	Chemometric interpretation	[60]
3	Soils	DDT and other selected organochlorines	Pesticides	[61]
4	Sandy loan soils	Cotton pesticides	Degradation and persistence	[62]
5	Soils	Miscellaneous	Effect of metallic cations on soil	[63]
6	Soils	Linuron and 2,4-D	Sorption	[64]
7	Soils	Miscellaneous	Degradation	[65]
8	Soils	Miscellaneous	Pollution	[66]
9	Agriculture soils	Miscellaneous	Residue level and new inputs	[67]
10	Urban soils	Organochlorine	Distribution	[68]
11	Soils	Miscellaneous	Retention	[69]
12	Soils	Organochlorine	Content and composition	[70]
13	Soils	Organochlorine	Analysis	[71]
14	Soils	Miscellaneous	Degradation	[72]
15	Soils	Polytrin-C	Accumulation	[73]
16	Clay loam aggregates	Miscellaneous	Sorption and diffusion	[74]
17	Soils	Permethrin	Handling techniques of samples	[75]
C	*Pesticide residues in agro-ecosystem, fruits, grains, and vegetables*			
1	Cameron high lands	Miscellaneous	Modeling pesticide transport	[76]
2	Nontarget ecosystem	Miscellaneous	Managing pesticide risk	[77]
3	Agricultural products	Miscellaneous	Levels of pesticides	[78]
4	Farm applicators	Miscellaneous	Diagnosed depression	[79]
5	Plant	Systemic xenobiotics	Absorption and translation	[80]
6	Agricultural ecosystem and conservation area	Miscellaneous	Ecogenotoxicology	[81]
7	Agricultural products	Miscellaneous	"Positive list" system	[82]
8	Ecosystem	Miscellaneous	Pesticide exposure data	[83]

TABLE 2.3 (continued)
Pesticide Residues in Different Matrices

S. No.	Matrix	Pesticide	Object	Reference
9	Potato crop	Chlorpyrifos and fenpropimorph	Volatilization of pesticides	[84]
10	Agricultural products	Miscellaneous	Estimation of pesticide residues	[85]
11	Vegetable	Organophosphorous and carbamates	Estimation of pesticide residues	[86]
12	Chille crop	Miscellaneous	Pesticide residues	[87]
13	Agricultural products	Miscellaneous	Systematical analysis	[88]
14	Transgenic cotton	Miscellaneous	Farm-scale evaluation	[89]
15	Cotton	Miscellaneous	Environmental risk	[90]
16	Field ditch	Miscellaneous	Ecological effects	[91]
17	Grasses and sedges	Miscellaneous	Public health risks	[92]
18	Fruits and vegetable	Miscellaneous	Removal efficiency	[93]
19	Cucumber	Organophosphorous	Removal study	[94]
20	Vegetables and fruits	Miscellaneous	Monitoring	[95]
21	Grain, fruits, and vegetable	Miscellaneous	Residue studies	[96]
22	Pulse beetle	Miscellaneous	Management of storage	[97]
23	Grain legumes	Plant-derived products	Action of plant diseases	[98]
24	Chinese cabbage	Miscellaneous	Dynamics of pesticide residues	[99]
25	Wheat	Miscellaneous	Impact of yield quality	[100]
26	Duckweed	Miscellaneous	Effects	[101]
27	Tomatoes	Miscellaneous	Pesticide residues	[102]
28	Herbal preparations and phytomedicines	Miscellaneous	Monitoring	[103]
29	Medicinal plants	Organophosphorous	Monitoring	[104]
30	Fresh apples and products	Miscellaneous	Monitoring	[105]
31	Surface interactions	S-Triazine-type pesticides	Electrochemical impedance study	[106]
32	Apples, lettuce, and potato	Miscellaneous	Monitoring	[107]
33	Potato crops	Miscellaneous	Dynamics of pesticides	[108]
34	Rainbow trout	Fipronil and chiral legacy pesticides	Bioaccumulation, biotransformation, and metabolites formation	[109]
35	Vegetables	Miscellaneous	Contamination	[110]
36	Cockles, Anadara, etc.	Organochlorine	Contamination	[111]
37	Peach, potato, and aphid	Carbaryl and malathion	Characterization of pesticide	[112]
38	Marketed foods	Miscellaneous	Contamination	[113]
39	Apples	Organophosphorous	Estimation	[114]
40	Cucumber	Methyl bromide	Resurgence of pests	[115]
41	Grain of rice	Miscellaneous	Grain growth effect	[116]
42	Barley	Miscellaneous	The fate and efficacy of pesticides	[117]
43	Medicinal materials	Miscellaneous	Removal and detection of pesticide residues	[118]
44	Vegetables	Miscellaneous	Control of pesticide residues	[119]
45	Medicinal material	Organochlorine	Systematic evaluation	[120]
46	Food	Miscellaneous	Risks and remedies	[121]
47	Roots and shoots of rice seedlings	Atrazine	Bioconcentration	[122]
48	Vegetables and fruits	Organophosphorous	Investigation	[123]
49	Barley seedlings	Miscellaneous	Effect of soil pollution	[124]

(continued)

TABLE 2.3 (continued)
Pesticide Residues in Different Matrices

S. No.	Matrix	Pesticide	Object	Reference
D	*Pesticide residues in fish, humans, and other animals*			
1	Tissue of Indo-Pacific humpback dolphins	Organochlorine	Monitoring	[125]
2	Brown front (*Salmo trutta fario*)	Organochlorine	Monitoring	[126]
3	Terrestrial vertebrates	Miscellaneous	Risk assessment	[127]
4	Avian and mammalian wild life	Miscellaneous	Toxicity assessment	[128]
5	Fish tissues	Organophosphorous	Interaction	[129]
6	Salmonid communities	Miscellaneous	Changes	[130]
7	Aquatic macrophytes	Miscellaneous	Interaction	[131]
8	Chicken muscle and eggs	Organochlorine	Pesticide residues	[132]
9	Sediments and fish species	Organochlorine	Residues	[133]
10	Women	Organochlorine	Lifestyle factors	[134]
11	Farmer's group	Miscellaneous	Exposure and genotoxicity correlations	[135]
12	Pregnant women	Miscellaneous	Exposure effect	[136]
13	Patients in hospital	Organophosphorous	Epidemiology	[137]
14	Children and workers	Organophosphorous	Biologic monitoring	[138]
15	Farm family	Miscellaneous	Biomonitoring	[139]
16	Vietnamese and Vietnam food	Miscellaneous	Estimation of pesticide residues	[140]
17	Cattle	Organochlorine	Bioaccumulation	[141]
18	Nile tilapia and Nile perch	Miscellaneous	Estimation	[142]
19	Belgian human plasma	Miscellaneous	Estimation	[143]
20	Human fat and serum	Miscellaneous	Estimation	[144]
21	Parrot feather	Miscellaneous	Uptake	[145]
22	Fish	Organophosphorous	Effect on protein contents	[146]
23	Fish	Organochlorine	Monitoring	[147]
24	Gill tissue of the fish	Butachlor and machet	Histopathological changes	[148]
25	Pest control operators	Chlorfyrifos	Removal of pesticide residues	[149]
26	Women breast adipose tissue	Organochlorine	Pesticide level	[150]
27	Silver carp	Organochlorine	Distribution	[151]
28	Animal feed and meat	Miscellaneous	Contamination	[152]
29	Tripterygium wilfordii	Miscellaneous	Manufacture of pesticide	[153]
30	Crawfish industry	Fipronil	Environmental impact	[154]
31	Human blood	DDT, DDE, and DDD	Pesticide levels	[155]
32	Placentas from male infant	Prenatal Organochlorine	Estimation of pesticide residue	[156]
33	Human placentas	Organochlorine	Estimation of pesticide residue	[157]
34	Common marmosets	Miscellaneous	Effect on sleep, electrocardiogram, and cognitive behavior	[158]
35	Korean human tissues	Organochlorine	Distribution	[159]
36	Human sperm	Miscellaneous	Induction and aneuploidy	[160]
37	Female workers	Carbaryl	Effect of occupational exposure	[161]
38	Daphnia magna	Miscellaneous	Chronic toxicity	[162]
39	Male mice	Kuaishaling	Pathologic observation	[163]

TABLE 2.3 (continued)
Pesticide Residues in Different Matrices

S. No.	Matrix	Pesticide	Object	Reference
40	Occupational workers	Fenvalerate	Exposure of semen quality	[164]
41	Semen quality	Organophosphorous	Impact of exposure	[165]
42	Blood serum	Organochlorine	Estimation of pesticide levels	[166]
43	Green anole lizards	Natural pyrethrins	Effect of temperature on toxicity	[167]
44	Freshwater fish	Miscellaneous	Effect of in vitro pesticide exposures	[168]
45	Breast cancer	Miscellaneous	Proximity effect	[169]
46	Breast tissues	Organochlorine	Occurrence	[170]
47	Edible fish	Organochlorine	Accumulation	[171]
48	United States population	Organophosphorous metabolites	Estimation	[172]
49	Children	Organophosphorous	Risk of brain tumors	[173]
50	Human lymphocyte culture in vitro	Miscellaneous	Induced cytogenic risk assessment	[174]
51	Fisheries products	Miscellaneous	Assessment of consumer's exposure	[175]
52	Tadpoles	Dichlorvos and butachlor	Genotoxicity	[176]
53	Patients	Organophosophorous	Poisoning	[177]
54	Vineyard workers	Miscellaneous	Contamination	[178]
55	Human pregnane and receptor lizards	Miscellaneous	Identification	[179]
56	Patients	Miscellaneous	Toxicoepidemiology	[180]
57	Rat testicular cells	1,2-Dibromo-3-chloropropane	Analysis of DMA damage	[181]
58	Women cotton pickers	Miscellaneous	Effect on reproduction hormones	[182]
59	Human SK-S-SH neuroblastoma cells	Carbamate esters	Neurite outgrowth	[183]
60	Organisms	Organochlorine	Toxic effects	[184]
61	*Aphidoletes aphidimyza* (Rondani)	Miscellaneous	Toxic effects	[185]
62	Honey bees	Miscellaneous	Damage	[186]
63	Microorganisms	Miscellaneous	Degradation	[187]
64	Beeswax	Miscellaneous	Estimation of pesticide levels	[188]
65	Microbes	Organophosphorous	Degradation	[189]
66	*Perna viridis*	Organochlorine	Estimation	[190]
67	Predatory mite	Miscellaneous	Effect of pesticides	[191]
68	Eggs	Organochlorine	Estimation of pesticide levels	[192]
69	Bacteria	Miscellaneous	Identification	[193]
70	Macroinvertebrates	Miscellaneous	Identifications of primary stresses	[194]
71	Birds	Miscellaneous	Assessment of exposure	[195]
72	Yeast strains	Miscellaneous	Effects	[196]
E	*Pesticide residues in air*			
1	Ambient air	Miscellaneous	Estimation of pesticides used in potato cultivation	[197]
2	Air of Mt. Everest region	Miscellaneous	Observation	[198]
3	Ambient air	Organochlorine	Monitoring	[199]

(continued)

TABLE 2.3 (continued)
Pesticide Residues in Different Matrices

S. No.	Matrix	Pesticide	Object	Reference
4	Outdoor air	Organophosphorous	Correlation	[200]
5	Dust sources	Miscellaneous	Monitoring	[201]
6	Air	Organophosphorous	Inhalation exposure to young children	[202]
7	Air	Organochlorine	Monitoring	[203]
F	*Pesticide residues in milk, milk products, oils, and fats*			
1	Commercial yogurt	Organochlorine	Survey	[204]
2	Milk and butter	Organochlorine	Monitoring	[205]
3	Human breast milk	Organochlorine	Monitoring	[206]
4	Edible oils	Organonitrogen	Behavior during refinement of edible oils	[207]

light or sunlight. These changes often produce compounds that are more toxic than the original pesticides, and the analyst has to determine the rate of breakdown, the nature and quantity of these metabolites, and end products, as well as the pesticide residue. The analysis of pesticide residues involves the following steps:

- Detection: Preliminary characterization of pesticide residues.
- Extraction of the residue from the sample matrix using an efficient and selective solvent.
- Removal of interfering substances from the extract: usually referred to as the cleanup or the separation procedure, which often involves either chromatography or solvent partition.
- Concentration of pesticide residue in the cleaned up extract is generally low, below the lower limit of estimation of the available analytical technique; therefore, a suitable precon-centration/enrichment method is coupled with the analytical technique.
- Estimation of the quantity of pesticide residues, together with metabolites and breakdown products, in the cleaned up and enriched extract.
- Confirmation of the presence of the residue by, for example, using a different method or the formation and identification of a derivative.

2.2.1 Detection

Generally, a preliminary characterization of the residue is required before undertaking tedious, time-consuming, sophisticated, and costly instrumental analysis. Spot-test analysis has been found to be inexpensive and simple for on-field detection of the residue. Different characteristics of the analyte have been used to maximize the sensitivity, selectivity, and specificity of the test. Recently, several spot tests, such as the capillary spot test, the thin-layer chromatographic spot test, the paper chromatographic spot test, the ion-exchange spot test, and the enzymatic spot test, have been developed and used for the detection of pesticide residues at trace levels.

2.2.2 Extraction

Several modern extraction techniques have been developed, and they may be divided into the following two groups.

2.2.2.1   Fluid-Phase Partitioning Methods

- Single-drop and liquid microextraction
- Supercritical and pressurized liquid extraction (PLE)
- Microwave-assisted extraction (MAE)
- Simple fluid-phase partitioning extraction

2.2.2.2   Sorptive and Membrane-Based Extraction Methods

- Solid-phase microextraction (SPME)
- Sorptive-phase developments
- Other sorptive techniques
- Hollow-fiber membrane extractions
- Other membrane extraction techniques

The selection of the extraction procedure to be employed is governed by the type of pesticide and the nature of the matrix/sample under examination. The extraction procedure should provide improved recoveries (at least 80% efficient), give higher sample throughput (sufficiently selective), consume less organic solvent, and require minimum cleanup before the determination.

Blasco et al. [210] used PLE for the determination of traces of benzimidazoles and azols, organophosphorous, carbamates, neonicotinoids, and acaricides in oranges and peaches. The extraction was found to be more efficient at a high temperature and pressure (75°C and 7500 psi) using ethyl acetate as the extraction solvent and acidic alumina as the drying agent. Recoveries by PLE were 98%–97% and lower limits of extraction were 0.025–0.25 mg kg^{-1}.

MAE has several similarities with supercritical fluid extraction (SFE) and PLE. There are also significant differences in these procedures. Macan et al. [211] used ultrasonic solvent extraction (USE) for atrazine and fenarimol in soil. They determined the efficiency of USE by using thin-layer chromatography (TLC). Tor et al. [212] used USE for 12 OCPs from soil with petroleum ether–acetone (1/1, V/V). Thus, as newer technologies are developed, they are often compared with other developing or already known techniques. Raymie [208] developed a relatively simple procedure, which they termed "QUECHERS" for the quantitative determination of 229 pesticides in fruits and vegetables. The QUECHERS method compared favorably with traditional methods and is under consideration by regulatory bodies. Hildebrand solubility parameters were used to determine extraction solvents for OCPs in soil and verified via Soxhlet extraction [208].

Stoichev et al. [209] reported a comparative study of extraction procedures for the determination of OCPs in fish. The best results irrespective of fish species were observed by using hexane/methylene chloride in Soxhlet apparatus. PLE was used [303] for carbosulfan and seven of its metabolites in oranges. In SPME the practice of placing a stationary phase coating onto an extraction fiber is derived from gas chromatography (GC). Hence, SPME phases tend to be similar to GC stationary phases, whereas solid-phase extraction (SPE) phases are more similar to liquid chromatography stationary phases. SPME is finding more versatile applications, because it can sample from air as well as liquid. However, SPME is a nonexhaustive technique. SPE was used for a simple and fast extraction of OCPs from avian serum [213].

Pahadia et al. [214] used the matrix SPE technique, which is also known as matrix solid-phase dispersion (MSPD), for organochlorine, organophosphorous, and carbamate pesticides in dairy and fatty foods. In this technique, the biological matrices are mixed with C_{18} (40 µm octadeylsilyl-derivatized silica), the resulting mixture is packed in a column, and the analytes are selectively eluted. MSPD eliminates the need for the tedious homogenization and centrifugation steps found in traditional solvent extraction and also reduces both the analytical time and the amount of solvent used. Pasquale et al. [215] used SPME for OPPs in a range of complex matrices such as water and soil. The applicability of headspace solid-phase microextraction (HS-SPME) to the determination of multiclass pesticides in water was reported by Sakamoto and Tsutsumi [216]. They also studied the effect of temperature on the extraction of 174 pesticides. Anastassiades and Lehotay [217] used single-phase extraction with acetonitrile liquid–liquid partitioning followed by cleanup by dispersive SPE for fast and easy multiresidue analysis of dicholorvos, methamidophos, mevinphos, acephate, o-phenylphenol, omethoate, diazinon, metalyxy, carbaryl, dichlofluanid, captan, thiabendazole, folpet, and imazalil in fruits and vegetables.

The stir bar sorptive extraction (SBSE) was used by Liu et al. [218] to determine OPPs in vegetables. Hydroxy-terminated polydimethylsiloxane (PDMS) prepared by sol–gel method was used as the extraction phase. The detection limit of OPPs was ≤1.2 mg L^{-1} in water. This method was also applied to the analysis of OPPs in vegetable samples, and the matrix effect was studied. The detection limit was 0.05–50 mg g^{-1} in vegetables. They [219] also described an improved sol–gel technology for the preparation of extractive phase on bars used in sorptive microextraction of methamidothion, dichlorvos, acephate, malathion, fenitrothion, fenthion, parathion, and chlorpyrifos at trace levels (0.9–8.0 pg mL^{-1}). The results of the two modes of extraction, that is, SBSE and supersonification sorptive extraction, are compared. It was claimed that sorption extraction (SE) provides a simple, effective, and solvent-free sample preparation technique for selective adsorption and enrichment of analytes in the sample matrix. Liu et al. [220] also demonstrated a new technology for the preparation of SPME containing 3% vinyl group without –OH (having no sol–gel activity) group. This product was thoroughly incorporated into the sol–gel network, and then the vinyl group of PDMS was cross-linked among PDMS chains during the aging process, forming a network. This new, physically incorporated extraction phase of SPME by sol–gel technology was coupled with GC–thermionic specified detector (TSD) and validated for extraction and determination of OPPs in water, orange juice, and red wine. The limits of detection of the method for OPPs were below 10 ng L^{-1} except methidathion. Relative standard deviations (RSDs) were in the range of 1%–20% for the discussed pesticides.

The extracting syringe (ESY), a novel membrane-based technique, was developed by Basu et al. [221] for the analysis of OCPs in raw leachate water. The ESY showed its competency to liquid–solid extraction (LSE) and accelerated solvent extraction (ASE) technologies. The researchers demonstrated its applicability for environmental analysis of organic pollutants toward green techniques for green environment. Basheer et al. [222] developed a novel, multiwalled carbon nanotube (MWCNT)-supported microsolid phase extraction (μ-SPE) procedure. In this technique, 6 mg sample of MWCNTs was packed inside a (2 × 1.5 cm) sheet of porous polypropylene membrane whose edges were heat-sealed to secure the contents. The μ-SPE device was wetted with dichloromethane and was then placed in a stirred, sewage-sludge sample solution to extract OPPs at traces (0.1–50 ug L^{-1}). After extraction, the analytes were desorbed in hexane and analyzed by gas chromatography-mass spectrometry (GC-MS).

Covaci [223] reported the application of solid-phase disk extraction for the extraction of OCPs such as dieldrin, mirex, heptachlor, heptachlorepoxide, hexachlorocyclohexanes (HCH), hexachlorobenzene in serum, cord blood, milk, and follicular and seminal fluid. Leandro et al. [224] described a method based on semiautomated SPE using octadecyl-bonded silica disks for the analysis of 100 pesticides and their transformation products in drinking water. The application of solvent silicone tube extraction (SiSTEx) was developed by Janska et al. [225] for the analysis of 26 pesticides in fruits and vegetables.

2.2.3 CLEANUP OR SEPARATION

Generally the extract must be purified or cleaned up before the determination of the pesticide residue can be carried out. The extract normally contains coextracted matter sufficient to interfere with quantification. For example, a water sample extract from agricultural runoff often contains many biocides, humic materials, and inorganic compounds. The coextractives such as halogen impurities interfere in the electron-capture detection (ECD), fatty materials interfere in TLC, and the coextractives can affect the absorbance in spectrophotometry. There are some exceptional cases also, for example, the extract in hexane from reasonably clean rivers can be analyzed by GC successfully. The coextractives also alter the characteristics of the chromatography column. Thus, the analyst must choose an appropriate cleanup procedure. The methods such as adsorptions methods (column, planar chromatography), solvent partition, distillation, and gel chromatography are generally used. The type of adsorbent and the technique used obviously depend on the nature of the coextractive. A short column of alumina, silica gel, charcoal, and Florisil is sufficient to remove low contents of fat. It is recommended that the extract solution, after reducing to a small volume, be applied to the column and then eluted off by the solvent or solvents and collected in small fractions. The activity of the adsorbent is checked by recovery studies. The adsorbents are activated by strong heating, and some may be activated for a specific purpose by pretreatment with acids or bases (such as alumina) or organic solvents (such as charcoal). The adsorbent can also be deactivated. For example, the deactivation of alumina is carried out by the addition of water and the different grades of alumina thus produced. Cleanup of pesticides is made by a Florisil column; fatty materials can be removed by liquid–liquid partition using dimethyl formamide–hexane and dimethyl sulfoxide–hexane. Planar chromatography can be used for cleanup, provided the extract contains a low concentration of fatty material. The efficiency of the method can be increased by altering the thickness of the adsorbent layer and the mode of development.

Sweep codistillation is a simple, inexpensive, and easy method, because it requires nonspecialized adsorbents, common solvents, and a nonlaborious cleanup procedure. The use of gel filtration is limited in the analysis of pesticides. Recently, a fresh interest in the technique was created by the introduction of Sephadex LH-20, a lipophilic-modified dextran gel, which can achieve separation in nonaqueous media. Its mode of separation is based on molecular size, so it offers a potential cleanup system for OPPs (MW 200–350) from chlorophyll (MW 906) and carotene (MW 536). This technique performs poorly with OCPs. The attractive and useful feature of gel chromatography/gel filtration is that the column can be used repeatedly for long periods without any remarkable change in the elution volumes and the recoveries. Ion-exchange resin column chromatography has also been used in the cleanup, for example in the estimation of diquat in tubers and paraquat in fruit.

2.2.4 Preconcentration/Enrichment

Enrichment of the analyte is required in some cases before cleanup and commonly before using the sophisticated and costly end method of determination. The choice of enrichment method depends on factors such as the volatility and solubility of the pesticide residues, the degree of concentration required, and the nature of the analytical technique to be used. The enrichment methods can be divided into the following two groups.

2.2.4.1 Solvent Removal Methods

In this procedure, the solvent is removed to enrich the dissolved residues by using techniques such as freeze concentration, lyophilization, evaporation, distillation, reverse osmosis, and ultrafiltration. Freeze concentration removes solvent as a solid phase and concentrates the analyte in the unfrozen portion. It is applicable at the level of multiliter volumes of the solution. Carbamate residues in acetonitrile or acetone solution are concentrated by reducing the volume 20-fold and providing 80% recovery. In lyophilization, the solvent is frozen and then removed by sublimation under vacuum. The concentration of the analyte can be enriched several thousand times but recovery is poor. OCPs are preconcentrated successfully. Evaporation is routinely used in the enrichment of nonvolatile pesticide residues of higher melting point. Distillation, particularly steam distillation, is used to enrich nonvolatile pesticide residues. Vacuum distillation is more advantageous but it is more laborious than lyophilization. Reverse osmosis is a recently developed technique. Cellulose acetate membranes can remove 90%–97% solvent of molecular weight less than 200. Ultrafiltration is a filtration under pressure through a membrane, and it gives better yield than lyophilization for 82%–85% enriching of endosulfan residues in solution. It is more selective than reverse osmosis because it concentrates only residues of higher molecular weight (1000).

2.2.4.2 Isolation Methods

In this procedure, the pesticide residues are taken out of solution with the help of techniques such as liquid–liquid extraction, solid–liquid extraction using activated carbon, polymeric adsorbents, in situ polymerized resins, polyurethane foam plugs, inorganic adsorbents (activated alumina, calcium phosphate, Florisil, hydroxylapatite, magnesia, silica gel, etc.), ion-exchange resins, and precipitation. The application of these techniques has already been discussed in extraction and cleanup procedures.

2.2.5 METHODS OF DETERMINATION/ESTIMATION

The methods used in the determination of residues may fall into five groups, namely, biological, spectrophotometric, chromatographic, electrochemical, and radiochemical. The radiochemical methods are extremely limited in use.

2.2.5.1 Bioassay

Bioassay is based in the measurement of growth, death, or some other physiological change in animals, plants, or microorganisms. Any organism that is susceptible to a pesticide may be used for the bioassay of its residues. Among several limitations, the main disadvantages are its lack of specificity and requirement of the isolation of very small quantities of toxicants from the large amount of plant or animal mass. The methods used for bioassaying pesticide residues are the dry method, wet method, diet method, and enzymatic method. A considerable amount of research findings have been published on the inhibition of cholinesterase by pesticides. The test animals are fed with the pesticide residue, and the degree of the inhibition of the enzyme is found to be proportional to the sample toxicity. Direct inhibition measurements are discussed in Chapter 6. A unique advantage of enzyme-linked immunosorbent assay (ELISA), which is based on immunochemistry, is that it requires neither cleanup nor preconcentration. A few recently reported articles are discussed in the following paragraph.

Tschmelak et al. [226–238] reported biosensors as a low-cost tool to determine the steroidal hormone testosterone, pharmaceuticals, antibiotics, hormones, endocrine-disrupting chemicals, and pesticides, such as atrazine, simazine, propazine, metolachlor, 4-chloro-2-methyl phenoxyacetic acid, diuron, 2,4-D, bentazon, 4,4-DDE, dieldrin, carbofuran, and diazinon, at subnanogram levels in real environmental samples of water, milk, and so on. They developed different techniques such as total internal reflectance fluorescence (TIRF), optical automated water analyzer computer-supported system (AWACSS), and affinity chromatographic biosensors. These techniques are rapid and inexpensive and require no sample pretreatments and enrichment unlike commonly used analytical methods. Liu and Lin [239] developed a self-assembling acetylcholinerase on carbon nanotubes for flow injection/amperometric detection of organophosphate pesticides and nerve agents. It has excellent operational lifetime stability without a decrease in the activity of enzyme for more than 20 repeated measurements over a period of 1 week. DNA diffusion [240] was used to estimate apoptosis and necrosis caused by pesticides such as monocrotophos, profenophos, chlorpyrifos, and acephate on human lymphocytes. Wong and Matsumura [241] studied the serum-free BG-1 avarian cell culture model to investigate the ability of arochlor, endosulfan-I, heptachlor, epoxide, cis-permethrin, β-HCH, α-HCH, Kepone, and chlorothalosil to stimulate estrogenic actions. Taba and Hay [242] developed a whole-cell bioreporter for the detection of 2,4-D in soil.

2.2.5.2 Spectrophotometric Determination

Spectrophotometric determination is a simple, inexpensive, and routinely used method, but it does not achieve the sensitivity and selectivity or specificity of chromatographic methods. The spectrophotometric determination is based on ultraviolet and visible light spectrometry/colorimetry, infrared spectroscopy, fluorescence, phosphorescence, and chemiluminescence.

2.2.5.2.1 Ultraviolet and Visible Spectrometry/Colorimetry

As mentioned above, in most of the laboratories, ultraviolet and visible spectrophotometry/ colorimetry is commonly used in the determination of compounds that are difficult to chromatograph, such as 2,4-D, 2,4,5-T and ionic bipyridilium herbicides, namely, diquat and paraquat. Many of the chlorinated pesticides have been determined by colorimetric methods, but with the advent of gas–liquid chromatography with its ultrasensitivity, colorimetric methods have lost favor. However, in Third World countries where the sophisticated, ultrasensitive, and costly instrumentation is either not installed or the repair facilities are not available, colorimetry is still the readily available technique. Feigenbrugel et al. [244] measured the adsorption spectra of 2,4-D, cymoxanil, fenpropidin, isoproturon, and pyrimethanil in aqueous solution in the near-UV region. They [245] also measured near-UV molar absorptivities of alachlor, metolachlor, diazinon, and dichlorvos in aqueous solution by using a new near-UV setup based on a long path length (approximately 1 m). This useful information was used to estimate the potential fate of these pesticides by sunlight photolysis in both aqueous and gas phases. This new UV setup may be considered to eliminate the discussed pesticides in wastewater. It can also be considered for analytical measurements of the pesticides by high-performance liquid chromatography (HPLC) coupled with UV detection.

A colorimetric method was reported by Harikrishna et al. [246] for the quantification of carbofuran in rice, wheat, and water samples. This method is based on the reaction of carbofuran and phenol with p-aminoacitanilide to produce a yellow chromophore with an absorption maxima at 465 nm.

A spectrophotometric method was reported by Deb et al. [247] for the estimation of zinc dimethyl dithiocarbamate (ziram). This procedure involves the complexation reaction of ziram with N-hydroxy-N,N'-diphenyl benzamidine in chloroform at pH 9.5 ± 0.2 and then the reaction of chloroform extract with 4-(2-pyridylazo) naphthol (PAN) to give an intense red color with λ_{max} at 550 nm. Cesnik and Gregorcic [248] validated a method for the determination of dithiocarbamates and thiuram disulfide in apple, lettuce, potatoes, strawberry, and tomato matrix. In this case, two yellow cupric N,N-bis(2-hydroxy ethyl) dithiocarbamate complexes are formed and measured jointly by spectrophotometry. Rathore and Varshney [243] reported a spectrophotometric procedure for the determination of ziram residues in cereals. They have compared the analytical parameters of this method with the other reported spectrophotometric methods. It is specific to the functional group, thiocarbamate, which gives colored complexes with Cu(II). It follows Beer–Lambert's law at 430 nm.

2.2.5.2.2 Infrared Spectroscopy

Infrared spectroscopy has been used for the qualitative identification as well as quantitative determination with one physical measurement. The major disadvantage of this technique is the problem of obtaining a clean extract free from water. Therefore, microinfrared techniques are used to analyze the purified sample either by gas–liquid chromatography or by preparative TLC when multiple residues are present.

2.2.5.2.3 Fluorescence, Phosphorescence, and Chemiluminescence

The selectivity and sensitivity of fluorometric methods are high. The increase in fluorescence with increasing concentrations is usually linear only over a limited range of concentration, and great care has to be taken to ensure that measurements are made in this range. The naturally occurring biological materials also fluoresce, and finding suitable cleanup procedures to remove them is difficult. Therefore, only a few publications exist in this area.

A literature survey shows that the phosphorescence characteristics of certain pesticides are reported and the potential utility of the technique is given. Fluorescence analysis provides a rapid and sensitive method for the determination of carbamate residues, which are not readily determined by GC, and has been used for screening milk samples for carbamates. Perez et al. [249] developed a sensitive and fast chemiluminescence flow-injection method for the direct determination of carbaryl in natural waters and cucumber. The detection limit is 4.9 ng mL^{-1}. They [250] extended this method for the determination of carbofuran at trace levels (0.02 μg mL^{-1}) in lettuce and water samples. Chinchilla et al. [251] devised a new strategy for the chemiluminescent determination of total N-methylcarbamate (NMC) contents in water. It is based on a previous offline alkaline hydrolysis of NMCs to produce methylamine (MA) and the subsequent derivatization of MA with o-phthalaldehyde (OPA). The derivative so obtained is involved in the peroxyoxalate chemiluminescent reaction using imidazole as a catalyst. The chemiluminescence emission was proportional to the total NMC contents. Gracia et al. [252] used this technique for the quantification of carbaryl residues in vegetables.

Hua et al. [253] reported an integrated optical fluorescence multisensor for pesticide residue analysis in water. They demonstrated the procedure for the pollutant estimation with a detection limit below 1 ng L^{-1}. A method based on the combination of a multicommuted flow system and solid-surface fluorescence spectroscopy was developed by Reyes et al. [254]. It was validated for the determination of thiobendazole residues in citrus fruits. It shows remarkably good agreement with liquid chromatography–electron spray/mass spectrometry. Rathore et al. [255] reported a selective fluorescence spot test for the detection of trichloroacetic acid (TCA) at ppm level in soil, water, and vegetation. It is based on the formation of yellow-greenish fluorescent salicylaldazine under UV light. The salicylaldazine is formed by reaction of TCA with salicylic acid in the presence of phenol in sodium hydroxide.

2.2.5.3 Chromatographic Methods

The chromatographic methods used in pesticide analysis may be divided into six groups, for example, paper chromatography, TLC, gas–liquid chromatography, liquid-column chromatography, capillary electrophoresis, and supercritical fluid chromatography (SFC).

2.2.5.3.1 Paper Chromatography

Paper chromatography is a simple, inexpensive, rapid, sensitive, and selective method of identification, estimation, detection, and separation of pesticide residues. Its reproducibility is poor, so it has been replaced by TLC. In the 1960s, fiber glass papers were used for the reverse-phase chromatography of organophosphorous and carbamate pesticides. Acetylated papers were used for the determination of organochlorine herbicides in soil and water, substituted urea herbicides and their trichloroacetates, and for the determination of organophosphorous pesticide residues in foodstuffs. Rathore et al. [256] used chromatography on sodium diethyladithiolarbamate-impregnated paper strips for the separation of metals ions.

2.2.5.3.2 Thin-Layer Chromatography

TLC has grown rapidly in recent years and is now widely accepted as a simple, reproducible, sensitive, quick, and efficient technique for the detection, separation, and determination of most pesticides. Paper chromatography has been replaced by TLC as the latter remains a reliable separation technique. For most pesticide residues, analysis on a 250 μm thick layer of alumina or silica gel gives the best results, but other adsorbents such as kieselguhr, magnesium oxide, polyamide, and microcrystalline cellulose Avicel have also been found to be useful. The applicability of TLC in multiresidue analysis of pesticides in wheat grain was proposed by Tiryaki and Aysal [257]. Triyaki [258] validated some parameters such as recovery, precision, accuracy, calibration, and function of TLC and found the parameters within the required range.

Rathore and Varshney [259] used cellulose-coated glass plates for the separation of dithiocarbamate fungicides. Tuzimski and his collaborators [260–277] published a series of articles on the relationship between R_F values and mobile-phase composition (heptane, chloroform, ethylacetate, terahydrofuran, dioxane, and acetone) for 30 moderately polar pesticides using silica gel layers (octadecyl silica). They also demonstrated TLC as a pilot technique for costly and precise HPLC.

This study includes two-dimensional TLC, reverse-phase TLC (RP-TLC), micropreparative TLC, TLC in combination with diode-array scanning densitometry for identification of fenitrothion in apples, and two-stage fractionation of a mixture of 10 pesticides. Rasmussen and Jacobsen [278] reported TLC for the analysis of triazine herbicides, triazinon herbicides, phenyl urea herbicides, sulfonylurea herbicides, phenoxy acid herbicides, diazinon, and bentazene. They claimed that when compound-specific HPLC and GC methods are to be developed, preliminary use of TLC is a good means of identifying persistent pesticides and their metabolites.

2.2.5.3.3   Gas Chromatography

Gas–liquid chromatography has been proved to be a largely used, most versatile, and sensitive method for pesticide residue analysis. The replacement of glass or metal by columns of Teflon was suggested because of the fact that a metallic column can decompose the analyte. The most commonly used stationary phases are organosilicones SE30, QF1, or DC200; admixture of QF1 and DC200; Apiezon; butane-1,4-diol succinate; versamid 900; Carbowax 20M; GE-X60; phenyl methyl silicone; OV-17; admixture of QF1 and *neo*-pentyl glycol succinate; and so on. Detectors are electron capture (electron affinity), nickel 63, microcoulometric, emission spectrometric, argon plasma at reduced pressure, low-pressure helium, argon–helium, flame ionization, sodium—thermionic, flame photometric, Coulson electrolytic conductivity, flameless ionization (chemi-ionization), alkali flame ionization detectors, and so on.

Aydin et al. [279] determined lindane, mirex, aldrin, heptachlor, methoxychlor, *o,p*-DDE, *p,p′*-DDD, *p,p′*-DDT, and dieldrin in water using ECD. Identification and quantification of organochlorine and OPP residues were reported by Tohboub et al. [280] using full-scan mass spectrometric detection (MSD). OCP residue analysis was made by Krauss et al. [281] in Brazilian human milk using ECD. A capillary column with ^{63}Ni ECD was used by Hajou et al. [282] for determining folpet, HCB, α-HCH, Quintozen, r-HCH, β-HCH, vinclozolin, chlorothalonil, dicofol, penconazole, *trans*-chlordane, procymidene, *o,p*-DDE, *p,p′*-DDE, dieldrin, endrin, and *o,p*-DDT in *Mentha piperita*. Cesnik et al. [283–285] analyzed pesticide residues in fruits and vegetables using GC-MSD. GC-ECD was used by Saxena et al. [286] to determine OCP residues in the milk supplied in Jaipur city in India. This technique [287] was also used to determine OCPs in soil, sediments, vegetables, grains, milk fat, blood serum, and semen. Fast GC with ECD was used [288] to determine pesticide residues at traces in nonfatty food samples.

GC with tandem quadrupole mass spectrometry was used to determine pesticide residues in baby foods [289]. Monitoring of OCP residues in drinking water was done by Lazarov et al. [290]. Sanchez et al. [291,292] developed an online, coupling, reverse-phase liquid chromatography/ GC with nitrogen–phosphorous detection and an automated through oven transfer adsorption–desorption for the determination of organophosphorous and triazine pesticides in olive oil, vegetables, and fruits. Shahi et al. [293] used GC–^{63}Ni ECD for monitoring pesticide residues in market vegetables. GC–MSD was reported by Libin et al. [294] for analyzing multiresidual pesticides in cabbage, carrot, apple, orange, cucumber, and rice. Gracia et al. [295] suggested the use of chemi-luminescence detection as an alternative to UV/VIS, fluorescence, or mass spectrometric detectors. Rubio et al. [296] reported simplified GC–thermionic/EC/MS detection for 32 organochlorine, organophosphorous, and organonitrogen pesticides at µg kg^{-1} levels in virgin oil. A GC–ECD was validated by Dempelou and Liapis [297] for multiresidue analysis of 16 pesticides in fruits.

Patel et al. reported large volume–difficult matrix introduction–GC–time-of-flight–mass spectrometry [298–300] for the determination of pesticide residues in fruit-based baby foods. This procedure eliminates the need for a cleanup step and thus allows rapid determination. They also validated resistive heating–GC–flame photometric detection [299] for the rapid screening of OPPs in fruits and vegetables, GC quadrupole MS [301] for OCPs in fats and oils, and programmable temperature vaporization injection with resistive heating–GC–photometric detection [302] for OPPs. Derivatization of pesticides is carried out to confirm the presence of a particular pesticide, a technique described in Section 2.2.6, and to convert the pesticides into a compound that can be chromatographed, that is, to decrease the pesticide volatility for paper or TLC, increase the volatility (stability) for gas–liquid chromatography, increase the sensitivity of the pesticide to a particular detector, or to avoid non-Gaussian peaks. Derivatization is common for the carbamate pesticides. A carbamate is hydrolyzed to produce amine, which is derivatized to 2,4-dinitrophenyl ethers. This derivative was extracted and injected into a gas–liquid chromatographic column. Carboxylic acid herbicides also have poor gas chromatographic characteristics, but this problem has been overcome by the preparation of methyl derivatives.

2.2.5.3.4 Liquid Column Chromatography

The application of liquid chromatography for pesticide detection depends on the pesticides having sufficiently different partition coefficients in the selected solvent system. As nonvolatile pesticides are analyzed by this technique, an extremely wide range of pesticides can be separated. The problem of the leaching off of the stationary phase has been controlled by introducing new packing materials, on which this stationary phase is chemically bonded to the support, and the use of controlled surface–porosity support. The current advances in this technique are summarized as follows: Soler et al. [303] determined carbosulfan and its metabolites in oranges by liquid chromatography, ion-trap, triple-stage mass spectrometry. A liquid chromatography–mass spectrometry (LC-MS) was established [304] for the purpose of simultaneous determination of carbamate and OPPs in fruits and vegetables. A reversed phase, C_{18} HPLC with UV detection [305,306] was used to separate 2,4,5-T, dichlorprop, dinoseb, carbondazin, and thiabendazol. Lehotay et al. [307,308] used a quick, easy, cheap, effective, rugged, and safe liquid chromatographic method for the determination of 229 pesticide residues in fruits. A coupled-column reversed-phase, liquid chromatography was used for the determination of pesticide residues [309]. Evans et al. [310,311] optimized the ion-trap parameters for the quantification of chlomequat by liquid chromatography–mass spectrometry. Ultraperformance liquid chromatography and HPLC with tandem quadrupole mass spectrometry [312] were used to determine priority pesticides in baby food.

2.2.5.3.5 Capillary Electrophoresis

Capillary electrophoresis (CE) remains a popular technique for the separation of biologically active compounds. Its applications in areas including pharmaceuticals, agrochemicals, carbohydrates, and peptides were reviewed [313]. Kaltsonoudis et al. [314] separated carbendazim and thiabendazole in lemons using capillary electrophoretic method in a low-pH phosphate buffer containing acetonitrile. The potential of CE combined with mass spectrometry was reported by Goodwin et al. [315] for the simultaneous determination of two herbicides, namely, glyphosate and glufosinate, and their metabolites (aminomethylphosphoric acid and methylphosphinicopropionic acid). They [316] also investigated isotachophoresis with conductivity detection for the preconcentration, separation, and determination of glyphosate and glufosinate and their metabolites.

2.2.5.3.6 Supercritical Fluid Chromatography

SFC is a fully mature technique. SFC does not replace HPLC and GC but is used as a complementary technique to both HPLC and GC [317]. This may be because of its high price and sophistication. In this area, much of the research dealt with evolutionary work (new columns, new detectors, better sampling handling, etc.), mechanistic studies of SFC separations, and applications, particularly the resolution of chiral compounds. There is considerable interest in the analysis of food contaminants, particularly antioxidant properties of foods. Several SFC-based assays were developed. A few articles were published on the analysis of phytochemical antioxidants including metabolites of pesticides [318].

2.2.5.4 Electrochemical Techniques

Electrochemical techniques [319] such as polarography, potentiometry, and voltammetry were applied in the analysis of pesticide residues. The technique using electrochemical biosensors is described in biological methods (Chapters 6 and 8). Polarography applied to the analysis of pesticide residues contains an oxidizable or reducible group such as nitro, halogen, carbonyl, and so on. Several workers investigated the use of polarography to detect parathion, malathion, bisdithiocarbamates, triphenyl tin fungicides, and so on. Rafique et al. [320] reported direct current polarographic studies of parathion in micellar medium.

Manisankar et al. [321–326] published several articles on the differential pulse stripping voltametric determination of methyl parathion in soil samples; utilization of sodium montmorillonite clay-modified electrode for the determination of isoproturon and carbendazim in soil and water samples; the cyclic voltametric behavior of dicofol, cypermethrin, monocrotophos, chlorpyrifos, and phosalone; utilization of polypyrrole-modified electrode for the determination of isoproturon and carbendazim in soil and water; and determination of endosulfan, isoproturon, and carbendazim using wall-jet electrode and heteropolyacid montmorillonite clay–modified glass carbon electrode in the presence of the surfactant, cetyl trimethyl ammonium bromide. Perez et al. [327] used micellar electrokinetic chromatography for the determination of herbicides, metribuzin, and its major conversion products in soil.

2.2.5.5   Radiochemical Techniques

Neutron activation analysis was used to determine elements including bromine and chlorine at traces (1 μg kg^{-1}) in milk products and fruits. The technique is of limited use because expensive equipment and skilled technicians are involved. Radioactive isotope was used in analyzing the pesticide metabolites at the level 0.1 μg kg^{-1}. The common isotopes are ^3H, ^{14}C, ^{32}P, ^{35}S, ^{36}Cl, and ^{82}Br, which are introduced in the pesticide molecule. This technique is not available in many laboratories due to lack of equipment to synthesize their own radioactive compounds. An attempt was made to determine dimethoate and phenthoate by isotope dilution analysis.

2.2.5.6   Other Methods

Several other analytical methods have demonstrated a very low detection limit and greater selectivity for pesticide residues. For the first time, Dane et al. [328] used electron monochromator–mass spectrometry (EM-MS) to determine three different dinitroaniline pesticides, flumetralin, pendimethalin, and trifluralin, in traces: 37±9, 10.4±0.6, and 47±17 ng cig^{-1}, respectively. Atomic absorption spectrometry (AAS) is not a common instrument used for determining pesticides. However, Richardson [329] recently developed an AAS method using flow injection and liquid–liquid extraction to determine dimethoxydithiophosphate in water. It was carried out in a continuous mode that enabled the method to be rapid and simple. In this method, only a small volume (250 μL) of the sample was required. The detection limit was found to be 5.0 μg L^{-1}.

2.2.6   Confirmatory Techniques

It is important that the identity and level of pesticide residues determined should be confirmed by another technique, other than the previously employed one, for identification and quantification. If initially TLC is used, the confirmation should be obtained by using alternative developing solvents or visualizing agents. Often, the pesticide can be taken out from the chromatoplate with a suitable solvent and injected directly on to a gas chromatogram. When the residues present are large enough, a simple confirmation technique for gas–liquid chromatography is afforded by TLC.

For a few pesticides, gas–liquid chromatography results using one detector can be confirmed by using a different detector, for example for OCPs, microcoulometric detector, and then electron capture detector can be used. Similarly for OPPs, thermionic and flame photometric detectors can be used.

Nuclear magnetic resonance (NMR) spectroscopy finds an important place in the structural elucidation of pesticides. It is a highly sensitive technique, which is useful for the identification of pesticide residues. Evans et al. [330] performed a detailed study of tandem mass spectrometric analysis of quarternary ammonium pesticides, namely, paraquat, diquat, difenzoquat, mepiquat, and chlormequat, and the study revealed a number of ions that had not been reported previously. This technique along with negative ion electrospray mass spectrometry was used by Goodwin et al. [331,332] for the confirmation of glyphosate, glufosinate, aminomethylphosphonic acid, and methylphosphinicopropionic acid.

Infrared spectra were also used for the confirmation of pesticide residues. They were coupled with TLC or column chromatography to obtain a better resolution because of the cleanup of the analyte. The p-value approach was used to identify or confirm the pesticides at the nanogram level. The p-value is determined by the equilibration of an analyte between equal volumes of two immiscible liquid phases, followed by the analysis of one of the solvents for the analyte. This value can be derived from a single distribution between the phases or from a multiple distributions as in countercurrent distribution.

2.3   AUTOMATION OF PESTICIDE RESIDUES ANALYSIS

Totally automated procedures are available for the determination of pesticide residues in air, foodstuffs, and soil. The automated procedures can be applied right from the sample to the final chart record, and they can be used for the extraction cleanup and enrichment of the sample for injection into a GC. Recently [333] developed systems are SPE followed by elution/in situ derivatization, GC-MS analysis, and sampling using an annular diffusion scrubber/in situ derivatization and HPLC–UV analysis. In an automated system, the sample solutions were transferred to capillary glass tubes and after evaporation of the solvent, they were automatically passed into a heating chamber directly connected to the gas chromatograph. The capacity of the system was 70 samples and was used for the analysis of triazine herbicides and chlorinated hydrocarbons.

One of the automated methods is enzymatic determination of pesticide residues based on the inhibition of cholinesterase. Enzymatic hydrolysis takes place on treating enzyme with thiocholine ester at pH 7.4 to give thiocholine. The latter reacts with 5,5-dithiobis(2-nitrobenzoic acid) to produce yellow anion of 2-thio-2-nitrobenzoic acid, which is determined colorimetrically at 420 nm. The pesticide partially inactivates the enzyme, resulting in decreased hydrolysis of the ester, and the decrease is proportional to the enzyme concentration. Thus, enzyme concentration can be determined in the automated procedure. An automated spectrophotometric method was also used to determine OPPs in crop samples. It is based on the formation of yellow orthophosphomolybdenum complex.

At present, rapid and ultrasensitive, computer-assisted automated procedures are available. Modern gas chromatographic systems can be easily adapted to computers because of the fact that the gas chromatographic signals permit the insertion of digitizers between the chromatography and the computer. Computers are also coupled with spectrophotometry. The recording device receives the analog signal, converts it to a digital signal, and records the impressions on a magnetic tape. The magnetic tape is replayed on the playback system to yield the result through the integrator computer system.

REFERENCES

1. E Markeu and A Nuro. Chlorinated pesticides in the sediments and fish species of the Shkodra Lake. *J Environ Prot Ecol* 6 (3): 538–549, 2005.
2. A Covaci, A Gheorghe, O Huka, and P Schepens. Levels and distribution of organochlorine pesticides, polychlorinated biphenyls and polybrominated diphenyl ethers in sediments and biota from the Danube Delta, Romania. *Environ Pollut* 140 (1): 136–149, 2006.

3. N Xue and X Xu. Composition, distribution and characterization of suspected endocrine-disrupting pesticides in Beijing Guan Ting Reservoir (GTR). *Arch Environ Contam Toxicol* 50 (4): 463–473, 2006.

4. H Xia and X Ma. Phytoremediation of ethion by water hycinth (*Eichhornia crassipes*) from water. *Biores Technol* 97 (8): 1050–1054, 2006.

5. S Haltori, K Tanaka, J Ogura, and S Shiode. The selection of agricultural chemicals for surveillance of parameter based on data of raw water of purification. *Osaku-Shi Suidokyoku Suishitsu Shik keusho Chosa Kenkyu narabini Shikeu Sciseki* 56: 29–34, 2004.

6. S Villa, C Negrelli, V Maggi, A Finizio, and M Vighi. Analysis of a firn core for assessing POP seasonal accumulation on an Alpine glacier. *Ecol Environ Safety* 63 (1): 17–24, 2006.

7. S Sematong, K Zapuang, M Kithumharn, S Issaravanich, and Y Sookkasean. Statistical use of pesticides in Prasal River agricultural area, Rayong province. *Thai J Health Res* 19 (2): 133–144, 2005.

8. RC Montone, S Taniguchi, C Bolain, and RR Weber. PCBs and chlorinated pesticides (DDTs, HCHs and HCB) in the atmosphere of the southwest Atlantic and Antarctic oceans. *Marine Pollut Bull* 50 (7): 778–782, 2005.

9. JT Nyangababo, L Henry, and E Omutange. Organochlorine pesticide contamination in surface water, sediment and air precipitation of Lake Victoria Basin, East Africa. *Bull Environ Contam Toxicol* 75 (5): 960–967, 2005.

10. A Finizio, S Villa, and M Vighi. Predicting pesticide mixtures load in surface waters from a given crop. *Agric Ecosys Environ* 111 (1–4): 111–118, 2005.

11. KP Nkedi, D Shinde, RM Sarabi, Y Ouyang, and L Nieves. Sorption kinetics and equilibria of organic pesticides in carbonatic soils from South Florida. *J Environ Quality* 35 (1): 268–276, 2006.

12. K Fytianos, RJW Meesters, HFR Schoroeder, B Gauliarmen, and N Gantidis. Distribution of organochlorine pesticides in surface water and sediments in Lake Volvi (northern Greece). *Int J Environ Anal Chem* 86 (1–2): 109–118, 2006.

13. K Fytianos, K Pitarakis, and E Bobola. Monitoring of *N*-methyl carbamate pesticides in Pinios River (Central Greece) by HPLC. *Int J Environ Anal Chem* 86 (1–2): 131–145, 2006.

14. O Wurl and JP Obbard. Chlorinated pesticides and PCBs in the sea surface microlayer and sea water samples of Singapore. *Marine Pollut Bull* 50 (11): 1233–1243, 2005.

15. GC Hose. Assessing the need for groundwater quality guidelines for pesticides using the species sensitivity distribution approach. *Human Ecol Risk Assess* 11 (5): 951–966, 2005.

16. A Scheyer, S Morville, P Mirabel, and M Millet. Analysis of trace levels of pesticides in rain water using SOME and GC-tandem mass spectrometry. *Anal Bioanal Chem* 384 (2): 475–487, 2006.

17. TTR Helisoma and K Gabol. Toxic effect of carbamate, organophosphate, pyrethroid and $CuSO_4$ on the fresh water snail *Helisoma trivolvis*. *Biosci Res Bull* 21 (1): 1–10, 2005.

18. Y Sapozhnikova, E Zubcor, N Zubcor, and D Schlenk. Occurrence of pesticides, polychlorinated biphenyls (PCBs) and heavy metals in sediments from the Dniester River, Moldova. *Arch Environ Contam Toxicol* 49 (4): 439–448, 2005.

19. DF Hudak and A Thapinta. Agricultural pesticides in groundwater of Kanchana Buri, Ratcha Buri, and Suphan Buri Provinces, Thailand. *Bull Environ Contam Toxicol* 74 (4): 631–636, 2005.

20. R Sawicki and L Mercier. Evaluation of mesoporous cyclodextrin-silica nanocomposite for the removal of pesticides from aqueous media. *Environ Sci Technol* 40 (6): 1978–1983, 2006.

21. FC Li, YF Shen, HX Wang, and XJ Kong. Monitoring of organochlorine pesticides and polychlorinated biphenyls in Baiyandian Lake using microbial communities. *J Freshwater Ecol* 20 (4): 751–756, 2005.

22. N Xue, D Zhang, and X Xu. Organochlorinated pesticide multiresidues in surface sediments from Beijing Guantig reservoir. *Water Res* 40 (2): 183–194, 2006.

23. H Ding, XG Li, H Liu, J Wang, WR Shen, YC Sun, and XL Shao. Persistent organochlorine residues in sediments of Haihe River and Dagu Drainage River in Tianjin, China. *J Environ Sci* 17 (5): 731–735, 2005.

24. P Edder and D Ortelli. Survey of pesticide residues in Swiss and foreign wines. *Mittei Lebens Chung Hygience* 96 (5): 311–320, 2005.

25. YF Tang, YX Wang, and HS Cai. Transport and face of pesticides in surface water environment—Case study at Hanjiang River catchment. *Wuhan Huag Xue Xuebao* 27 (2): 28–30, 2005.

26. T Kawakami, M Ishizaka, Y Ishi, H Eun, J Miyazaki, K Tamura, and T Higashi. Concentration and distribution of several pesticides applied to paddy fields in water and sediments from Sugao Marsh, Japan. *Bull Environ Contam Toxicol* 74 (5): 954–961, 2005.

27. K Starner, F Spurlok, S Gill, K Goh, H Feng, J Hsu, P Lee, D Tran, and J White. Pesticide residues in surface water from irrigation season monitoring in the San Joaquin Valley, California, USA. *Bull Environ Contam Toxicol* 74 (5): 920–927, 2005.

28. N Xue, X Xu, and Z Jin. Screening 31 endocrine-disrupting pesticides in water and surface sediment samples from Beijing Guanting reservoir. *Chemosphere* 61 (11): 1594–1606, 2005.
29. M Chen, R Ren, ZJ Wang, XT Lin, S Chem, SH Wu, LL Liu, and SF Zhang. Residues of organic nitrogen and organophosphorous pesticide in water from the industrial and urban sewage in Beijing. *Anqu Huanj Gong* 12 (2): 45–48, 2005.
30. YF Tang, YX Wang, and HS Cai. Transport and fate of pesticide in surface water environment. *Wuhan Huag Xue Xuebao* 27 (4): 13–15, 2005.
31. LG Sun, XB Yin, CP Pan, and YH Wang. A 50 years record of dichloro-diphenyl-trichloroethanes and hexachlorocychohexanes in lake sediments and penguin droppings on King George Island. *Maritime Antarctic J Environ Sci* 17 (6): 899–905, 2005.
32. YF Tang, YX Wang, and HS Lai. Numerical modeling of pesticides transport in vadose zone. *Anqu Huanj Gong* 12 (2): 11–14, 2005.
33. A Che, H Sun, and K Li. Device and process for treating pesticide containing wastewater with ozone and hydrogen peroxide. Faming Zhu Shen Gong Shuom CN. 1644, 528 (cl. C02F1/72), 27 Jul 2005, Appl. 10,093,994,22 Dec. 2004, 8 pp.
34. DX Gong, YZ Zou, WX Zhao, RB Yang, and DF Fan. Effect of four pesticides on photodegradation of prochloraz in water. *Nong Huanj Kexue Xuebao* 12 (2): 11–14, 2005.
35. H Zing, XG Li, H Liu, J Wang, WR Shem, Y-C Sun, and XL Shao. Persistent organochlorine residues in sediments of river and Dagu Drainage River in Tianjin China. *J Environ Sci* 17 (5): 731–735, 2005.
36. Q Wu and H Jia. The grain-size destruction of organochlorine pesticides in sediments. *Guang Daxue Xuebao ziram Kexue* 4 (1): 40–45, 2005.
37. A Cougnaud, C Faur, and CP Le. Removal of pesticides from aqueous solution: Quantitative relationship between activated carbon characteristics and adsorption properties. *Environ Technol* 26 (8): 857–866, 2005.
38. M Moazzen, NR Saadat, and AP Gize. Pesticides in Dee Estuary sediments NW England implications for pollution study of costal environment. *J Eng Geol* 1 (1): 67–85, 2003.
39. SJ Chen, XJ Luo, BX Mai, GY Sheng, JM Fu, and EY Zeng. Distribution and mass inventories of polycyclic aromatic hydrocarbons and organochlorine pesticides in sediments of Pearl River Estuary and northern South China Sea. *Environ Sci Toxicol* 40 (3): 709–714, 2006.
40. RQ Yang, AH Lv, JB Shi, and GB Jiang. The levels and distribution of organochlorine pesticides (OCPs) in sediments from the Haihe River China. *Chemosphere* 61 (3): 347–354, 2005.
41. M Jankovska, V Hammer, and L Pasek. Problems of water contamination with organochlorine pesticides. *Vodni Hospo* 55 (3): 51–55, 2006.
42. G Shen, Y Lu, M Wang, and Y Sun. Status and fuzzy comprehensive assessment of combined heavy metal and organochlorine pesticide pollution in the Tahia lake region of China. *J Environ Manage* 76 (4): 355–362, 2005.
43. Y Li, H Heng, X Wang, L Hong, and C Ye. Estimation of the source of organophosphorous pesticides in Xiamen sea area. *Huaj Kexue Xuebao* 25 (8): 1071–1077, 2005.
44. RD Wauchope. Pesticides and watershed-scale modeling solution for water quality management. *J Agric Food Chem* 53 (22): 8834, 2005.
45. AHMF Anwar and MF Bari. Prioritising areas from groundwater monitoring based pesticides leaching. *J Appl Hydrol* 17 (1): 43–52, 2005.
46. H Otteneder and P Majerus. Pesticides residues in wine, transfer grapes. *Bulletin de l'o I.V.* 78 (889–890): 173–181, 2005.
47. R Bromilow, CR De, A Evans, and P Nicholls. Behavior of pesticides in sediment/water systems in outdoor mesocosms. *J Environ Sci Health* 41 (1): 1–16, 2006.
48. Y Zhang, Y Zhang, and GX Chen. Effect of water matrix on the removal of triazine pesticides by nanofiltration. *Guange Daxue Xuebao* 37 (3): 321–324, 2005.
49. B Schellschmidt, HH Dieter, and W Lingh. Pesticides in the drinking water. *Schni Dut Phyto Gesell* 7 (113–128): 158–180, 2004.
50. IK Kontantinon, DG Hela, and TA Albanis. The status of pesticides pollution in surface water of Greece. *Environ Pollut* 141 (3): 555–570, 2006.
51. M Vochlhel, B Freiheit, and D Steffen. The load of pesticides in the River Weser (Germany) after heavy rainfall during the time of application in May 2004. *Hydro Wasser Bewirt* 49 (6): 313–316, 2005.
52. I Kakata. Coastal water pollution is the most urgent environment issue. *Nippon Kaisui Gobk* 59 (6): 414–419, 2005.
53. RS Kookana, RL Correll, and RB Miller. Pesticide impact rating index pesticide risk indicator for water quality. *Water Air Soil Pollut* 5 (3–6): 277–278, 2005.

54. X Luo, S Chen, B Mai, Y Zeng, G Sheng, and J Fu. Distribution of organochlorine pesticides (OCPs) in surface sediments in Pearl River Delta and its adjacent coastal areas of South China Sea. *Huaj Kexue Xuebao* 25 (9): 1272–1279, 2005.

55. A Chinkuyu, T Meixner, J Gish, and C Daughtry. Prediction of pesticide losses in surface runoff from agricultural fields using GLEAMS and RZWQM. *Am Soc Agric Eng* 48 (2): 585–599, 2005.

56. L Comoretto and CS Leatto. Comparing pharmaceutical and pesticides loads into a small Mediterranean River. *Sci Total Environ* 349 (1–3): 201–210, 2005.

57. CN Fung, JG Zheng, DW Connel, X Zhang, HL Wang, JP Giesy, Z Fang, and PKS Lam. Risk posed by trace, organic contaminants in coastal sediments in the Pearl River Delta, China. *Mar Pollut Bull* 50 (10): 1036–1046, 2005.

58. YC Yuan, HC Chen, and YK Yaun. Sublethal effects of paraquate and malathion on the fresh water shrimp, *Macrobrachium nipponense*. *Acta Zool Taiwan* 14 (2): 87–95, 2004.

59. CE Renshaw, CB Bostick, X Feng, KC Wong, SE Wingston, R Karimi, LC Folt, and YC Chen. Impact of land disturbance on the fate of arsenial pesticides. *J Environ Qual* 35 (1): 61–67, 2006.

60. C Goncalves, DS Esteves, CC Joaquim, and MF Alphendurada. Chemometric interpretation of pesticide occurrence in soil samples from an intensive horticulture area in north Portugal. *Anal Chim Acta* 560 (1–2): 164–171, 2006.

61. P Bailey, D Waite, AL Quinnett, and BD Rifley. Residues of DDT and other selected organochlorine pesticides in soils from Saskatchewan, Canada (1999). *Can J Soil Sci* 85 (2): 265–271, 2005.

62. MI Tariq, S Afzal, and I Husain. Degradation and persistence of cotton pesticides in sandy loam soils from Punjab, Pakistan. *Environ Res* 100 (2): 184–196, 2006.

63. K Flogeac, E Guillon, and M Aplin Court. Effect of metallic cations on the sorption of pesticides on soil. *Environ Chem Lett* 3 (2): 86–90, 2005.

64. I Nishiguche, P Moldrup, T Komatsu, Y Kodama, and S Ito. Effect of organic carbon content and pH on sorption of linuron and 2,4-D by three soils. *Hiroshima Daigaku Diaga Kuin Kogaku Kenleyuka Kenleyu Hokokyu* 53 (1): 15–21, 2004.

65. S Beulke, BW Van, CD Brown, M Mitchell, and A Walker. Evaluation of simplifying assumptions on pesticide degradation in soil. *J Environ Qual* 34 (6): 1933–1943, 2005.

66. L Li and XS Zhao. Investigation and assessment on pollution of Ginseng cultivation soil in the cast mountain areas of Jilis Province. *Nongye Huanjing Kexue* 24 (2): 403–406, 2005.

67. HJ Goa, X Jiang, F Wang, YR Bian, DZ Wang, JC Dend, and Dy Yan. Residue levels and new inputs of chlorinated POPs in agriculture soils from Taihu Lake region. *Pedosphere* 15 (3): 301–309, 2005.

68. XH Li, LL Ma, XF Liu, SJ Fu, HX Cheng, and XB Xu. Distribution of organochlorine pesticides in urban soil from Beijing. People's Republic of China. *Bull Environ Contam Toxicol* 74 (5): 938–945, 2005.

69. N Syversen and K Haarstad. Retention of pesticides and nutrients in a vegetated buffer root zone compared of soil with low biological activity. *Int J Environ Anal Chem* 85 (15): 1175–1187, 2005.

70. TB Zhang, Y Rao, HF Wan, GY Gang, and YS Xia. Content and compositions of organochlorinated pesticides in soil of Dongguan City. *Zhongguo Huanjing Kexue* 25 (Suppl.) 89–93, 2005.

71. X Wang, Y Dong, Q An, and H Wang. Difference in analyzing organochlorines from soil with USEPA 8080 and GB/T 14550–93. *Turanj (Nanzing China)* 37 (1): 105–108, 2005.

72. GNJ Saxton. Soil pesticide residue degradation and soil sample management procedures for environmental forensics. Avail UMI, Order No. DA 3166702 from *Diss Abstr Int*, B65 (3): 11652, 2005.

73. MA Qadir, MA Athar, and K Mahmood. Study of accumulation of polytrin-C on different types of soil. *J Pure Appl Sci* 22 (2): 113–117, 2003.

74. BW Van, S Beulke, and CD Brown. Pesticide sorption and diffusion in natural clay loam aggregate. *J Agric Food Chem* 53 (23): 9146–9154, 2005.

75. GN Saxton and B Engel. Permethrin insecticide and soil sample handling techniques of state regulatory agencies. *Environ Forensics* 6 (4): 327–333, 2005.

76. WYW Abdullah, BY Aminuddin, and M Zulkifli. Modelling pesticide and nutrient transport Cameron High lands, Malaysia agro-ecosystems. *Water Air Soil Pollut Focus* 5 (1–2): 115–123, 2005.

77. CM Travisi, P Nijkamp, M Vighi, and P Giacomelli. Managing pesticide risks for non-target ecosystems with pesticides risk indicators: A multicriteria approach. *Int J Environ Technol Manage* 6 (1/2): 141–162, 2006.

78. HB Cesnik, AB Gregorcic, and V Spela. Pesticide residues in agricultural products of Slovence origin in 2005. *Act Chim Slovenia* 53 (1): 95–99, 2006.

79. CL Beseler. Diagnosed depression and low intermediate, and high pesticide exposures in Iowa and North Carolina farm applicators and their spouses enrolled in the agricultural health study. Avail UMI order No DA 3185496, from *Diss Abstr Int B* 66 (8): 4176, 2006.

80. MN Satchivi, EW Stoller, LM Loyd, and DP Briskin. A nonlinear, dynamic, simulation model for transport and whole plant allocation of systemic xenobiotics following foliar application IV: Physiochemical properties requirements for optimum absorption and translocation. *Pesticide Biochem Physiol* 84 (2): 83–97, 2006.

81. FAC Rodrigues, OCLDS Weber, FGDC Eliana, and RGCP Tidon. Pesticide ecogenotoxicology: A comparative evaluation b/w an agricultural ecosystem and a conservation area (national park). *Pesticides* 15: 73–84, 2005.

82. Y Akiyama, N Yoshioka, and K Ichihashi. Study of pesticide residues in agricultural products towards the "positive list" system. *Shokuhin Eiseegaku Zasshi* 46 (6): 305–318, 2005.

83. PJ Van den Brink, CD Brown, and IG Dubus. Using the expert model PERPEST to translate measured and predicated pesticide exposure data into ecological risks. *Ecol Model* 191 (1): 106–117, 2006.

84. M Leistre, JH Smelt, JH Weststrate, F Van den Berg, and R Aaldererik. Volatilization of pesticides chlorpyrifos and fenpropimorph from potato crop. *Environ Sci Technol* 40 (1): 96–102, 2006.

85. M Kojima, K Fukunaga, T Nishiyama, H Harada, A Tokino, and M Tsuji. An investigation of the pesticide residue in agricultural products in Shiga prefecture. *Nippon Shokuhim Kagaku Gakkaishi* 12 (2): 100–106, 2005.

86. L Li, YH Zhao, HZ Liu, and YM Guo. Pesticide of organophosphorus, carbamate and its residue in the vegetable in Enshi city. *Ziran Kexueban* 23 (2): 150–153, 2005.

87. SB Singh, I Mukherjee, M Gopal, and G Kulshrestha. Pesticide residues in chilli crop: Evaluation of chemical module for IPM. *Indian J Agric Chem* 37 (2–3): 69–71, 2004.

88. T Nakamara, K Tonami, F Sokamoto, and C Sosaki. Establishment of systematical analysis for pesticide residue within agricultural products. *Ishikawa-ken Hoken Kankyo Senta Kenya Hokokasho* 41: 46–56, 2004 (Pb-2005).

89. MG Cottaneo, C Yafuso, C Schmidt, CY Huang, M Rahman, C Carl, C Ellers-Kirk, BJ Orr, SE Marsh, C Antilla, P Dutisleul, and Y Carriere. Farm-scale evaluation of the impacts of transgenic cotton on biodiversity, pesticide use, and yield. *Proc Natl Acad Sci USA* 103 (20): 7571–7576, 2006.

90. C Turgut and O Erdogan. The environmental risk of pesticides in cotton production in Aegean region, Turkey. *J Appl Sci* 5 (8): 1391–1393, 2005.

91. D de Zwart. Ecological effects of pesticide use in the Netherlands: Modified and observed effects in the field ditch. *Integr Environ Assess Manage* 1 (2): 123–134, 2005.

92. TM Vekuge, AM John, and MA Kishimba. Concentrations of pesticide residues in grasses and sedges due to point source contamination and the indications for public health risks. *Chemosphere* 61 (9): 1293–1298, 2005.

93. P-T Ku, H-D Kwon, S-H Park, and Y Zee. A study on the removal efficiency of pesticide residues as fruits and vegetables treated by additional materials. *Hun'guk Eungyong Sangmyorg Hwahakhoeji* 48 (4): 388–393, 2005.

94. B Guan, J Liu, and D-X Yuan. Comparative research on removing organophosphorous pesticide residues from cucumber with different Wad water. *Huanjing yu Jiankang Zazhi* 23 (1): 52–54, 2006.

95. AB Gebara CHP Ciscato, M des Ferreira, and SH Moteiro. Pesticide residues in vegetables and fruits monitored in Sao Paulo city, Brazil, 1994–2001. *Bull Environ Contam Toxicol* 75 (1): 163–169, 2005.

96. J Mueller, G Sommerfeld, and R Binner. Result of residue studies on grain, fruits and vegetables. *Schriflenreihe der Deutschen Phytomedizinischen Gesellschaft* 7 (Gesunde Reflanzen-Gesunde Nahrurg): 129–140, 158–180, 2004.

97. RB Jolli, SS Karabhantanal, and TC Jayaprakash. Influence of pesticides and storage methods in the management of pulse beetle (*Callosobruchus* spp.) and their defects on seed viability in mothbeen. *J Entomol Res* 29 (2): 159–162, 2005.

98. RT Gahukar. Plant-derived products against insect pests and plant diseases of tropical grain legumes. *Int Pest Control* 47 (6): 315–318, 2005.

99. Z-Y Zang, C-Z Zhange, X-J Liu, and X-Y Hang. Dynamics of pesticide residues in the autumon chiniso cabbage. *Pest Manage Sci* 62 (4): 354–355, 2006.

100. ZS Szentpetery, Z Hegedus, and M Jolankai. Impact of agrochemicals on yield quantity and pesticides of winter wheat varieties. *Cereal Res Commun* 33 (2–3): 635–640, 2005.

101. C Turgut. The effect of pesticides on duckweed at their predicted environmental concentration in Europe. *Fresenius Environ Bull* 14 (9): 783–787, 2005.

102. E Zerouali, R Salghi, A Hormatallah, B Hammouti, L Bazzi, and M Zaafarani. Pesticide residues in tomatoes grown in green houses in Souss Mass valley in Morocco and dissipation of endosulfan and deltamethrin. *Fresenius Environ Bull* 15 (4): 267–271, 2006.

103. MMT Abd El-Rahman, SM Fahmy, and MY Zaki. Monitoring of pesticide residues in some herbal preparations and phytomedicines. *J Drug Res* 25 (1and 2): 166–169, 2004.

104. SM Fahmy, MY Zabi, and MMT Abd El-Rahman. Monitoring of organophosphorous pesticides residues in certain medicinal plants on the local market. *Egypt J Pharmaceut Sci* 45 (1): 33–40, 2004.

105. R Stefhan, J Ticha, J Hajslova, T Kovalczuk, and V Kocourele. Baby food production chain. Pesticide residues in fresh apples and products. *Food Addit Contam* 22 (12): 1231–1242, 2005.

106. M Hromadova, R Sokolova, L Pospisil, and N Fonelli. Surface interaction of s-triazine type pesticides. An electrochemical impedance study. *J Phys Chem B* 110 (10): 4869–4874, 2006.

107. HB Cesnik, A Gregorcic, SV Bolta, and V Kamecl. Monitoring of pesticide residues in apples, lettuce and potato of the Slovene origin, 2001–04. *Food Addit Contam* 23 (2): 164–173, 2006.

108. GC Lopez-Perz, M Arias-Estevez, E Lopez-Periago, B Soto-Gonzclcz, B Cancho Grande, and J Simal-Crandara. Dynamics of pesticides as potato crops. *J Agric Food Chem* 54 (5): 1797–1803, 2006.

109. BJ Konwick, AW Garrison, MC Black, JK Avants, and AT Risk. Bioaccumulation, biotransformation and metabolite formation of fipironil and chiral legacy pesticides in rainbow trout. *Environ Sci Technol* 40 (9): 2930–2936, 2006.

110. P Amoah, P Drechsel, RL Abaidoo, and WJ Ntow. Pesticide and pathogen contamination of vegetable in Ghana's urban markets. *Arch Environ Contam Toxicol* 50 (1): 1–6, 2006.

111. W Ang, NM Mazlin, LY Heng, BS Ismail, and S Salmijah. Study on organochlorine compounds in cockles, *Anadara granosa* and mussels, *Perna virdis* from agriculture sites in peninsular Malaysia. *Bull Environ Contam Toxicol* 75 (1): 170–174, 2005.

112. W-S Lan, J Cong, H Jiang, S Jiang, and C-L Qiao. Expression and characterization of carboxylesterase. E4 gene from peach potato aphid (*Myzus persicae*) for degradation of carbaryl and malathion. *Biotechnol Lett* 27 (15): 1141–1146, 2005.

113. P Branca, A Longo, FA Grivette, S Dignoni, M Manzo, and A Salzarulo. Pesticide contamination of marketed foods in the Piedment region in 2004. *Industrie Alimentari (Pinerodo, Italy)* 44 (448): 635–646, 2005.

114. DFK Rawn, SC Quade, JB Shields, G Conca, WF Sun, GMA Lacroix, M Smith, A Fouqaet, and A Belanger. Organophosphate levels in apple composites and individual apples from a treated Canadian orchard. *J Agric Food Chem* 54 (5): 1943–1948, 2006.

115. JP Gilreath, TN Motis, BM Santos, JW Noting, JS Locascio, and DO Chellemi. Resurgence of soilborne pests in dousle cropped cucumber after application of methyl bromide chemical alternatives and solarization in tomato. *Hortic Technol* 15 (4): 797–801, 2005.

116. J Wu, B Dong, D Li, H Qui, and G Yang. Effects of four pesticides on grain growth parameters of rice. *Zhongguo Nongye Kexue (Beijing, China)* 37 (3): 376–381, 2004.

117. DM Armutage, DE Baxter, and DA Collins. The fate and efficacy of pesticides applied to malting barley during storage and processing. *Calloques—Institut Nation de La Recherche Agronomique* 101 (Stored Malting Barley): 137–165, 2005.

118. S Li, C Quan, Y Wang, and S Tian. Study on removal and detection of pesticide residues in Chinese medicinal materials. *Zhong Caoyao* 35 (2): 232–234, 2004.

119. S Wei, S He, and M Liang. Supervision and control of pesticide residues in export vegetables in Ningbo. *Nongyo Kexue yu Guanli* 25 (11): 11–14, 2004.

120. KSY Leung, K Chan, CL Chan, and G-H Lu. Systematic evaluation of organochlorine pesticide residues in Chinese materia medica. *Phytother Res* 19 (6): 514–518, 2005.

121. X Wu, J Li, and W Hui. Pesticides in food: Risks and remedies. *Shipin yu Fajiao Gengye* 31 (6): 80–84, 2005.

122. Y-H Su and Y-G Zhu. Bioconcentration of atrazine and chlorophenols into roots and roots of rice seedlings. *Environ Pollut* (the Netherlands) (Pub:2005) 139 (1): 32–39, 2006.

123. Y Bai, L Zhou, J Wang, and X Wang. Investigation of organophosphorous pesticide residues in vegetables and fruits in Shaanxi area. *Xiam Jiaotong Daxue Xueboo, Yixveban* 26 (1): 86–88, 2005.

124. K Przybulbwska. Effects of soil pollution by pesticides and petrochemicals on the growth and development of spring barley seedlings. *Roczniki Gleboznaweze* 56 (42): 129–135, 2005.

125. CCM Leung, TA Jefferson, SK Hung, GJ Zheng, LWY Yeung, BJ Richardson, and PKS Lam. Petroleum hydrocarbon, polycyclic aromatic hydrocarbons, organochlorine pesticides and polychlorinated biphenyls in tissues of Indo-Pacific humfibacks dolphins from South China Waters. *Mar Pollut Bull* 50 (2): 1713–1719, 2005.

126. J Kolarova, Z Svobodova, V Zlabele, T Randak, J Hajsleva, and P Suchan. Organochlorine and PAHs in brown trout (*Salmo trutta fario*) population from Ticha Orlice River due to chemical plant with possible effects to Vitellogenin expression. *Fresenius Environ Bull* 14 (120): 1091–1096, 2005.

127. DL Fischer. Accounting for diffusing exposure patterns between laboratory tests and the field in the assessment of long term risks of pesticides to terrestrial vertebrates. *Ecotoxicology* 14 (8): 855–862, 2005.

128. P Mineau. A review and analysis of study end points relevant to the assessment of long term pesticide toxicity in avian and mammalian wild life. *Ecotoxicology* 14 (8): 775–799, 2005.

129. JD Maul, JL Farris, and MJ Lydy. Interaction of chemical cues from fish tissues and organophosphorous pesticides on ceriodalphnia dubia survival. *Environ Pollut* 141 (1): 90–97, 2006.

130. K Gormley, KL Teather, and DL Daryl. Charges in Salmomid communities associated with pesticide runoff events. *Ecotoxicology* 14 (7): 671–678, 2005.

131. JM Dabrowski, A Bollen, FR Bennett, and R Schulz. Pesticide interception by emergent aquatic macrophyte: Potential to mitigate spray-drift input in agricultural streams. *Agric Ecosyst Environ* 111 (1–4): 340–348, 2005.

132. RS Aulakh, JPS Gill, BS Joia, and HW Ockerman. Organochlorines pesticides residues in poultry feed, chicken muscle and eggs at a poultry farm in Punjab, India. *Trans Sci Food Agric* 86 (5): 741–744, 2006.

133. E Marku and A Nuro. Chlorimated pesticides in the sediments and fish species of the Shkodra Lake. *J Environ Protect Ecol* 6 (3): 538–549, 2005.

134. MF Olea-Serrano, J Iborhizea, J Exposito, P Torne, J Laguna, V Pedraza, and N Olea. Environmental and life style factors for organochem exposure women living in southern Spain. *Chemosphere* 12 (11): 1917–1924, 2006.

135. DP Mattophoulos. Pesticide exposure and genotoxicity correlations within a Greek farmers group. *Int J Environ Anal Chem* 86 (3–4): 215–223, 2006.

136. ZM Gawora, J Jurewicz, and W Hanke. Exposure to pesticides among pregnant women working in agriculture. *Med Pr* 56 (3): 197–204, 2005.

137. C Dharmani and K Joya. Epidemiology of acute organophosphate poisoning in hospital emergency room patients. *Rev Environ Health* 20 (3): 215–232, 2005.

138. RA Fenske, Chensheng, CL Curl, JH Shirai, and JC Kissel. Organophosphorous pesticide exposure among children and workers: An analysis of recent studies in Washington state. *Environ Health Perspect* 113 (11): 1651–1657, 2005.

139. BA Baker, BH Alexander, JS Mandel, JF Acqua Vella, R Honeycutt, and P Chapman. Farm family exposure study: Methods and recruitment practices for a biomonitoring study of pesticides exposure. *J Expo Anal Environ Epidemiol* 15 (6): 491–499, 2005.

140. A Schecter, HT Quynh, O Paepke, R Malesch, JD Censtable, and KC Tung. Halogenated organics in Vietnamese and in Vietnam food; dioxins, dibenzofurans, PCBs, polybrominated diphenyl ethers and selected pesticides. *Organohalogen Compounds (Computer optical disk)* 66 (Dioxin 2004): 3634–3639, 2004.

141. R Sadler, A Seawright, G Show, N Dennison, D Connell, W Barron, and P White. Bioaccumulation of organochlorine pesticides from contaminated soil by cattle. *Toxicol Environ Chem* 87 (4): 575–582, 2005.

142. L Henry and MA Kishimba. Pesticide residues in Nile tilapia (*Oreochromis niloticus*) and Nile perch (*Lates niloticus*) from Southern Lake Victoria, Tanzania. *Environ Pollut* 140 (2): 348–354, 2006.

143. WN Van, A Covaci, KC Kannan, J Gordon, A Chu, G Effe, DE De, and L Goeyens. *Organohalogen Compounds (Computer optical disk)* 66 (Dioxan 2004): 2784–2790, 2004.

144. P Araque, AM Soto, MF Olea-Serrano, SC Sonnen, and N Olea. Pesticides in human fat and serum samples vs total effective xenoestrogen burden. In: JLM Vidal, A G Frenich (Eds.). *Methods Biotechnol, Vol. 19, Pesticide Protocols*. Humana Press Inc., Totowa, NJ, 2006, pp. 207–215.

145. C Turgut. Uptake of modeling of pesticides by roots and shoots of parrot feather (*Myriophyllum aquaticum*). *Environ Sci Pollut Res Int* 12 (6): 342–346, 2005.

146. G Ghanbahadur, S Raut, A More, and SB Wagh. Effect of organophosphate (Nuvan) on protein contents of gills and liver in the fish *Rasbora daniconius*. *Himalayan J Environ Zool* 19 (1): 63–64, 2005.

147. A Mazet, G Keck, and P Berney. Concentration of PCBs, organochlorine pesticides Drome River: Potential effects on others and heavy metals (lead, cadmium, and copper) in fish from the (*Cultra lutra*). *Chemosphere* 61 (6): 810–816, 2005.

148. KS Tilak, K Veeraiah, and PB Thathaji. Histopathological changes in gill tissue of the fish *Channa punctata* exposed to sublethal concentration of butachlor and machete, a herbicide. *J Aquat Biol* 20 (1): 111–115, 2005.

149. RH Fitzgera and HM Manley. Laundring protocols for chlorpyrifos residue removal from pest control operators' overall. *Bull Environ Contam Toxicol* 75 (1): 94–101, 2005.

150. SM Waliszewski, MT Bermudez, RM Infanzon, CS Silva, O Carvajal, P Trujillo, SG Arroyo, RV Dietrimi, VA Soldana, G Melo, S Esquivel, F Castro, H Ocampo, J Torres, and PM Hayward-Jones. Persistent organochlorine pesticide levels in breast adipose tissue in women with malignant and benign breast tumors. *Bull Environ Contam Toxicol* 75 (4): 752–759, 2005.

151. YZ Sun, XT Wang, XH Li, and XB Xu. Distribution of persistent organochlorine pesticides in tissue/ organ of silver carp (*Hypophthalmichthys molitrix*) from Guantung Reservoir China. *J Environ Sci (China)* 17 (5): 722–726, 2005.

152. J Gilbert and H Senyuva. Environmental contaminants and pesticides in animal feed and meat. In: JN Safos (Ed.). *Improving the Safety of Fresh Meat*. Woodhead Publishing Ltd: Cambridge, UK, 2005, pp. 132–155, ISBN: 978-1-85573-95-0.

153. W Hong and C Wu. Manufacture of pesticide from suspension culture cell of tripterygium Wilfordii. Forming Zhuanli Shenqing Gongkai Shumingsshu CN1, 675, 999 (CI. A01N63/00) 5 Oct 2005, Appl. 10059135, 13 Aug 2004, 5 pp.

154. DB Bedient, RD Horsak, D Schlenk, RM Hoving, and JD Pierson. Environmental impact of Fipironil to the Louisiana Crawfish Industry. *Environ Forensics* 6 (3): 289–299, 2005.

155. C La Rocca, V Abate, S Alivernini, N Iacovella, A Montovani, L Silvestotorni, G Spera, and BL Turrio. Levels of PCBs, DDT, DDE, and DDD in Italian human blood samples. *Organohalogen Compounds (Computer optical disk)* 66 (Dioxan 2004): 2755–2760, 2004.

156. H Shem, KM Maiin, M Kaleva, H Virtanen, and KW Schram. Prenatal organochlorine pesticides in placentas from Finland; exposure of male infant born during 1997–2001. *Placenta* 26 (6): 512–514, 2005.

157. MS Souza, GG Megnarelli, MG Revedatti, SS Cruz, and AM Pechen De D'Angelo. Prenatal exposure of pesticides: Analysis of human placental acetyl cholinesterase, glutathione S-transforase and catalase as biomarkers of effect. *Biomarkers* 10 (5): 369–376, 2005.

158. NG Muggleton, AJ Smith EAM Scott, SJ Wilson, and PC Pearce. A long term study of the effects of diazinon on sleep, the electrocorticogram and cognitive behaviour in common marmosets. *J Psychopharmacol* 19 (5): 455–466, 2005.

159. MJ Park, SK Lee, JY Yang, KW Kim, SY Lee, WT Lee, KH Chung, YP Yun, and YE Yoo. Distribution of organochlorines and PCB congeners in Korean human tissues. *Arch Pharmacol Res* 28 (7): 829–838, 2005.

160. K Haerkoenen. Pesticides and the induction of aneuploidy in human sperm. *Cytogenet Genome Res* 111 (3–4): 378–383, 2005.

161. Y Li, L Tan, X Sun, J Ji, Q Wang, L Chen, L Song, and S Wang. Effect of occupational exposure to carbaryl on reproductive and endocrine functions of female workers. *Zhongguo Gongye Yixue Zazhi* 18 (3): 163–165, 2005.

162. Y Tan, S Li, and X Wu. Chronic toxicity of several insecticides to *Daphnia magna*. *Nongyaoxue Xuebo* 6 (3): 62–66, 2004.

163. YP Gan, HG Zhao, LM Diao, and TH She. Pathologic observations of reproductive toxicity of kuaishaling to male mice. *Huanjing Yu Zhiye Yixul* 22 (5): 437–440, 2005.

164. L Tang, S Wang, JM Ji, X Sun, Y Li, Q Wang, and L Chen. Effect of fenvalerate exposure on semen quality among occupational workers. *Contraception* 75 (1): 92–96, 2006.

165. W Li, J Wu, X Zoa, C Xiao, W Zhou, and E Gao. Impact of organophosphorous pesticide exposure on semen quality. *Shengzhi Yu Biyun* 24 (5): 285–290, 2004.

166. SM Waliszewaki, RM Imfanzon, SG Arroyo, RV Pietrini, O Carvajal, P Trujillo, and PM Hayword-Jones. Persistent organochlorine pesticides levels in blood serum in women bearing babies with undescended testis. *Bull Environ Contam Toxicol* 75 (5): 952–959, 2005.

167. LG Talent. Effect of temperature on toxicity of a natural pyrethrin pesticide to green anole lizards (*Anolis carolinensis*). *Environ Toxicol Chem* 24 (12): 3113–3116.

168. AJ Harford, J Andrew, K O' Hallogram, and PFA Weight. The effect of in vitro pesticide exposures on the phagocytic function of four native Australian freshwater fish. *Aquat Toxicol* 75 (4): 330–342, 2005.

169. P Renolds, SE Hurley, RB Gunier, S Yerabati, T Quach, and A Hertz. Residential proximity to agricultural pesticide use and incidence of breast cancer in California 1988–1997. *Environ Health Perspect* 113 (8): 993–1000, 2005.

170. S Burgaz, I Coke, T Cuban, D Bulbul, and M Iscan. Occurrence of organochlorine pesticides in cancerous and non-cancerous breast tissues. *Fresenius Environ Bull* 14 (11): 1024–1030, 2005.

171. AB Munshi, F Preveen, and TH Usmani. Accumulation of organochlorines in edible fishes and mussel from the coastal waters of Karachi, Pakistan. *J Chem Soc Pakistan* 27 (4): 404–408, 2005.

172. DB Barr, R Allen, AO Olsson, R Braro, LM Caltabiano, A Montesano, J Nguyen, S Udurka, D Walder, RD Walker, G Weerasekera, RD Whitehead, SN Schober, and L Larry. Concentration of selective metabolites of organophosphorus pesticides in the United States population. *Environ Res* 99 (3): 314–326, 2005.

173. SS Nielsen, BA Mueller, AJ De Roos, HMA Viernes, FM Fariri, and H Checkoway. Risk of brain tumors in children and susceptibility to organophosphorous insecticides: The potential role of paraoxonase (PONI). *Environ Health Perspect* 113 (7): 909–913, 2005.

174. K Jamil, AP Shaik, and AJ Lakshimi. Pesticide induced cytogenetic risk assessment in human lymphocyte culture in vitro. *Bull Environ Contam Toxicol* 75 (1): 7–14, 2005.

175. SS Wong, GC Li, and SN Chen. A preliminary assessment of consumer's exposure to pesticide residues in fisheries products. *Chemosphere* 62 (4): 674–680, 2006.

176. B Geng, D Yan, and Q Xue. Genotoxicity of pesticide dichlorvos and herbicide, butachlor in Phacoporous megacepalues tadpoles. *Dongwu Xuebao* 51 (3): 447–454, 2005.

177. A Qu, T Wu, H Tian, X Zou, and B Zhang. Dynamic charges of serum TNT α in patients with severe acute organophosphorus poisoning complicated by respiratory failure. *Zhonghue Laodeny Weisheng Zhiyubing Zahi* 23 (2): 150–154, 2005.

178. I Baldi, P Lebailly, S Jean, L Rougetet, S Dulaurent, and P Marquet. Pesticide contamination of workers in vineyards in France. *J Expo Sci Environ Epidemiol* 16 (2): 115–124, 2006.

179. G Lemaire, W Mnif, P Amon, J-M Pascussi, A Pillon, F Robenoelina, H Fenet, E Gomex, C Casellas, J-C Nicolas, V Carailles, M-J Duchesne, and P Balaguer. Identification of new human pregnane α receptor ligands among pesticide using a stable repoter all system. *Toxicol Sci* 91 (2): 501–509, 2006.

180. D Tagwixyi, BD Dexter, and C Nhachi. Toxicoepicdemiology in Zimbabwe: Pesticide poisoning admissions to major hospitals. *Clin Toxicol* 44 (1): 59–66, 2006.

181. J Labaj, D Slamenova, L Hrusovska, and G Brunborg. Analysis of DNA damage induced by pesticide 1,2-dibromo-3-chloropropane(DBCP) in rat testicular cells. *Biologia (Bratislova, Slovakia)* 60 (Suppl. 17): 93–96, 2005.

182. S Rizwan, I Ahmad, M Ashraf, S Aziz, T yashmin, and A Sattar. Advance effect of pesticides on reproduction hormones of women cotton pickers. *Pakistan J Biol Sci* 8 (11): 1588–1591, 2005.

183. P-A Chang, Y-J Wu, W Li, and X-F Leng. Effect of carbamate esters on neurite outgrowth in differentiating human SK-N-SH neuroblastoma cells. *Chem Biol Interact* 159 (1): 65–72, 2006.

184. D Kolankaya. Organochlorine pesticide residues and their toxic effects on the environment and organisms in Turkey. *Int J Environ Anal Chem* 86 (1–2): 147–160, 2006.

185. H Orita and M Kashic. Toxic effect of some pesticides on adults and larvae of *Aphidoletes aphidimyza* (Rodani). *Kyushu Byogaichu Kenkyu Kaiho* 51: 83–88, 2005.

186. F Seefeld. Chemical detection of damage to honey bees caused by pesticides. *Nachrichten blatt des Dutschen Rflanzenschutzdienstes (Braunschweing, Germany)* 58 (2): 59–66, 2006.

187. F Zhu, M Wang, and J Li. Pesticide degradation microorganism. *Weishengwuxue Tongbao* 31 (5): 120–123, 2004.

188. JJ Jimenez, JL Bernal, NM Jesus de, and MT Martis. Residues of organic contaminants in beeswax. *Eur J Liquid Sci Technol* 107 (12): 896–902, 2005.

189. Y Liang, E Zeng, and X Lu. Advances in microbial degradation of organophosphorous pesticides. *Weishengwuxue Zazhi* 24 (6): 51–55, 2004.

190. MK So, X Zhang, JP Grery, CN Fung, HW Fong, J Zheng, MJ Kramer, H yoo, and PKS Lam. Organochlorine and dioxin-like compounds in green-lipped mussels Perna viridis from Hong Kong mariculture zones. *Mar Pollut Bull* 51 (8–12): 677–687, 2005.

191. SS Kim, SO Seo, JC Park, SG Kim, and DI Kim. Effects of selected pesticides on the predatory mite Ambly seius cucumeris. *J Entolol Sci* 40 (2): 107–114, 2005.

192. R Guitart, R Clavero, R Mateo, and M Manez. Levels of persistent organochlorine residues in eggs of greater flamingos from the Guadalquivir Marshes (Doñana), Spain. *J Environ Sci Health B* 40 (5): 753–760, 2005.

193. L Lopez, C Pozo, B Rodelas, C Calvo, B Juarez, MV Martinez-Toledo, and J Gonzalez-Lopes. Identification of bacteria isolated from an aligotrophic lake with pesticide removal capacity. *Ecotoxicology* 14 (3): 299–312, 2005.

194. BS Anderson, BM Phillips, JW Hunt, V Connor, N Richard, and RS Tjeerdema. Identifying primary stressors impacting macro invertebrates in the Salinas River (California USA): Relative effects of pesticides and suspended particles. *Environ Pollut* 141 (3): 402–408, 2006.

195. P Presser and ADM Hart. Assessing potential exposure of birds to pesticides-treated seeds. *Ecotoxicology* 14 (7): 679–691, 2005.

196. J Dela, C Hilan, R Saliba, R Lteif, and P Strehaiano. Effects of some pesticides on two yeast strains. *Saccharomyces cerevisiae* and *Metschinikowia pulcherrima. J International des Science de la Vigne et du vin* 39 (2): 67–74, 2005.

197. LM White, WR Earnest, G Julien, C Garron M, and Leger. Ambient air concentrations of pesticides used in plot cultivation in Prince Edward Island Canada. *Pest Manage Sci* 62 (2): 126–136, 2006.

198. T Li, T Zha, F Wang, XH Qiu, and WL Lin. Observation of organochlorine pesticide in the air of the Mt. Everest region. *Ecotoxicol Environ Safety* 63 (1): 33–41, 2006.

199. AV Konoplev, VA Nikitin, DP Samsonov, GV Chernik, and AM Rychkov. Polychlorinated biphenyl and organochlorine pesticides in the ambient air of Russian Arctic in the Far East. *Meteor Gidrol* (7): 38–42, 2005.

200. M Harnly, R Mc Laughlin, A Bradman, M Anderson, and R Guriev. Correlating agricultural use of organophosphates outdoor air concentrations: A particular concern for children. *Environ Health Perspect* 113 (9): 1184–1189, 2005.

201. KF Mutlaq, AI Rushdi, and BRT Simoneit. Pesticides residues in dust sources in winter and summer seasons in Riyadh city, Kingdom of Saudi Arabia. *Alex Sci Exch* 26 (1): 125–139, 2005.

202. J Kawahara, R Horikoshi, T Yamaguchi, K Kamagai, and Y Yanagisawa. Air pollution and young children's inhalation exposure to organophosphorous pesticide in an agricultural community in Japan. *Environ Int* 31 (8): 1123–1132, 2005.

203. FM Jaward, G Zhang, JJ Nam, AJ Sweetman, JP Obbard, Y Kobara, and KC Jones. Passive air sampling of polychlorinated biphenyls organochlorine compounds and polybrominated diphenyl ethers across Asia. *Environ Sci Technol* 39 (22): 8638–8645, 2005.

204. H Zhang, ZF Chai, HB Sun, and JL Zhang. A survey of extractable persistent organochlorine pollutants in Chinese commercial yogurt. *J Dairy Sci* 89 (5): 1413–1419, 2006.

205. A Kumar, P Dayal, G Singh, FM Prasad, and PE Joseph. Persistent organochlorine residue in milk and butter in Agra city, India. A case study. *Bull Environ Contam Toxicol* 75 (1): 175–179, 2005.

206. A Sudaryato, T Kunisue, H Iwata, S Tanabe, M Niidda, and M Hashmi. Dioxins, PCBs, and organochlorine pesticides in human breast milk from Malaysia. *Organohalogen Compounds (Computer optical disk)* 66 (Dioxin 2004): 2733–2738, 2004.

207. J Fukazawa, Y Suzuki, S Tokairin, K Chimi, T Maruyama, and T Yanagita. Behaviour of 11 organonitrogen pesticides during the refinement of edible oils. *J Oleo Sci* 54 (8): 431–435, 2005.

208. DE Raynie. Modern extraction techniques. *Anal Chem* 78: 3997–4004, 2006.

209. T Stoichev, N Rizor, A Kolarska, F Ribarora, and M Atanassova. Comparison of extraction procedures for determination of organochlorine pesticides in fish. *J Univ Chem Technol Metall* 40 (3): 251–254, 2005.

210. C Blasco, G Font, and Y Pico. Analysis of pesticides in fruits by pressurized liquid extraction and liquid chromatography-ion-trap-triple stage mass spectrometry. *J Chromatogr A* 1098: 37–43, 2005.

211. MK Macan, S Babic, A Zelenika, and J Macan. Determination of atrazine and fenarimol extraction efficiency by thin-layer chromatography. *Agrochimica XLIX* (5–6): 246, 2005.

212. A Tor, ME Aydin, and S Ozcan. Ultrasonic solvent extraction of organochlorine pesticides from soil. *Anal Chim Acta* 459: 173–180, 2006.

213. SE Sundberg, JJ Ellington, and JJ Evans. A simple and fast extraction method for organochlorine pesticides and polychlorinated biphenyls. *J Chromatogr B* 831: 99–104, 2006.

214. S Pahadia, V Sharma, V Sharma, and BK Wadhwa. Evolution of multiresidue analyses of pesticides in dairy and fatty foods: A review. *Indian J Dairy Sci* 58 (2): 75–79, 2005.

215. CD Pasquale, A Jones, A Chartton, and G Alongo. Use of SPME extraction to determine organophosphorus pesticides adsorption phenomena in water and soil matrices. *Int J Environ Anal Chem* 85 (15): 1101–1115, 2005.

216. M Sakamoto and T Tsutsumi. Applicability of headspace solid-phase microextraction to the determination of multi-class pesticides in water. *J Chromatogr A* 1028 63–74, 2004.

217. M Anastassiades and SJ Lehotay. Fast and easy multiresidue method employing acetonitrile extraction/partitioning and "dispersive solid-phase extraction" for the determination. *J AOAC Int* 86 (2): 412–431, 2003.

218. W Liu, Y Hu, J Zhao, Y Xu, and Y Guan. Determination of organophosphorus pesticides in cucumber and potato by stir bar sorptive extraction. *J Chromatogr A* 1095: 1–7, 2005.

219. W Liu, H Wang, and Y Guan. Preparation of stir bar for sorptive extraction using sol-gel technology. *J Chromatogr A* 1045: 15–22, 2004.

220. W Liu, Y Hu, J Zhao, Y Xu, and Y Guan. Physically incorporated extraction phase of solid-phase microextraction by sol-gel technology. *J Chromatogr A* 1102: 37–43, 2006.

221. T Basu, S Bergstrom, A Hussen, J Norberg, and JA Jonsson. Extracting syringe for determinations of organochlorine pesticides in leachate water and soil-water-slurry: A novel technology for environmental analysis. *J Chromatogr A* 111: 11–20, 2006.

222. C Basheer, A A Alnedhary, BSM Rao, S Valliyaveettil, and HK Lee. Development and application of porous membrane protected carbon nanotube micro-solid-phase extraction combined with gas chromatography/mass spectrometry. *Anal Chem* 78: 2853–2858. 2006.

223. A Covaci. Application of solid-phase disk extraction combined with gas chromatographic techniques for determination of organochlorine pesticides in human body fluids. In: JLM Vidal, AG Frenich (Eds.). *Methods in Biotechnology, Vol. 19, Pesticide Protocols*, Humana Press Inc., Totowa NJ, 2005, pp. 49–57.

224. CC Leandro, DA Bishop, RJ Fussill, FD Smith, and BJ Keely. Semiautomatic determination of pesticides in water using solid-phase extraction disks and gas chromatography-mass spectrometry. *J Agric Food Chem* 54: 645–469, 2006.

225. M Janska, SJ Lehotay, K Mastovska, J Hajslova, T Alan, and A Amirav. A simple and expensive "Solvent in silicone tube extraction" approach and its evaluation in the gas chromatographic analysis of pesticides in fluids and vegetables. *J Sep Sci* 29: 66–80, 2006.

226. J Tschmelak, M Kumpf, N Kappel, G Proll, and G Gauglitz. Total internal reflectance fluorescence (TIRF) biosensor for environmental monitoring of testosterone with commercially available immunochemistry: Antibody characterization, assay, development and real sample measurements. *Talanta* 69: 343–350, 2006.

227. J Tschmelak, G Proll, and G Gauglitz. Optical biosensor for pharmaceuticals, antibiotics, hormones, endocrine disrupting chemicals and pesticides in water: Assay optimization process for estrone as example. *Talanta* 65: 313–323, 2005.

228. J Tschmelak, G Proll, J Riedf, J Kaiser, P Kraemmer, L Barzag, JS Wilkinson, P Hua, JP Hole, R Nudd, M Jakson, R Abuknesha, D Barcelo, SR Mozaz, MJL de Alda, F Sacher, J Stress, J Stobodnik, P Oswald, H Zozmenko, E Korenkova, L Tothova, Z Krascsenits, and G Gaulitz. Automated water analyzer computer supported system (AWACSS) Part I: Project objectives basic, technology, immunoassay development, software design and networking. *Biosens Bioelectron* 20: 1499–1508, 2005.

229. J Tschmelak, G Proll, and G Gaulitz. Improved strategy for biosensor-based monitoring of water bodies with diverse organic carbon levels. *Biosens Bioelectron* 21: 979–983, 2005.

230. J Tschmelak, G Proll, J Riedt, J Kaiser, P Kraemmer, L Barzagas, JS Wilkinson, P Hua, JP Hole, R Nudd, M Jakson, R Abuknesha, D Barcelo, SR Muzaz, MJLD Alda, F Sacher, J Stress, J Slobodnik, P Oswald, H Kozmenko, E Korenkova, L Tothova, Z Krascsenits, and G Gauglitz. Biosensors for unattended, cost-effective and continuous monitoring of environmental pollution: Automated water analyser computer supported system (AWACSS) and river analyzer (RIANA). *Int J Environ Anal Chem* 85 (12–13): 837–852, 2005.

231. J Tschmelak, N Kappel, and G Gauglitz. TIRF based biosensor for sensitive detection of progesterone in milk based on Ultra-sensitive progesterone detection in water. *Anal Bioanal Chem* 382: 1895–1903, 2005.

232. G Proll, J Tschmelak, and G Gauglitz. Fully automated biosensors for water analysis. *Anal Bioanal Chem* 381: 61–63, 2005.

233. J Tschmelak, G Proll, and G Gauglitz. Immunosensor for estrone with an equal limit of detection as common analytical methods. *Anal Bioanal Chem* 378: 744–745, 2004.

234. J Tschmelak, G Proll, and G Gauglitz. Sub-nanogram per litre detection of the emerging contaminant progesterone with a fully automated immunosensor based on evanescent field techniques. *Anal Chim Acta* 519: 143–146, 2004.

235. J Tschmelak, M Kumpf, G Proll, and G Gauglitz. Biosensors for seven sulphonamides in drinking, ground and surface water with difficult matrices. *Anal Lett* 37 (8): 1701–1718, 2004.

236. J Tschmelak, G Proll, and G Gauglitz. Ultra-sensitive fully automated immunoassay for detection of propanil in aqueous samples: Steps of progress toward sub-nanogram per liter detection. *Anal Bioanal Chem* 379: 1004–1012, 2004.

237. J Tschmelak, G Proll, and G Gauglitz. Verification of performance with the automated direct optical TIRF immunosensor (river analyser) in single and multi-analyte assays with real water samples. *Biosens Bioelectron* 20: 743–752, 2004.

238. G Proll, M Kumpf, M Mehlmann, J Tschmelak, H Griffith, R Abuknesha, and G Gauglitz. Monitoring an antibody affinity chromatography with label-free optical biosensor technique. *J Immunol Methods* 292: 35–42, 2004.

239. G Liu and Y Lin. Biosensor based on self-assembling acetylcholinesterase on carbon nanotubles for flow injection/amperometric detection of organophosphate pesticides and nerve agents. *Anal Chem* 78: 835–843, 2006.

240. G P Das, AP Shaik, and K Jamil. Estimation of apoptosis and necrosis caused by pesticides in vitro on human lymphocytes using DNA diffusion assay. *Drug Chem Toxicol* 29: 1–10, 2006.

241. PSY Wong and F Matsumura. Serum free BG-1 cell proliferation assay: A sensitive method for determining organochlorine pesticide estrogen receptor activation at the nanomolar range. *Toxicol in Vitro* 20: 382–394, 2006.

242. FA Taba and AG Hay. A simple solid phase assay for the detection of 2,4-D in soil. *J Microbiol Methods* 62: 135–143, 2005.

243. HS Rathore and C Varshney. Simple and selective spectrophotometric method for the routine determination of ziram residues. *J Indian Council Chemists* 22 (2): 71–74, 2005.

244. V Feigenbrugel, SL Calve, and P Mirabel. Molar absorptivities of 2,4-D, cymoxanil, fenpropidin, isoproturon and pyrimethamil in aqueous solution in the near-UV. *Spectrochem Acta A* 63: 103–110, 2006.
245. V Feigenbrugel, C Loew, SL Calve, and P Mirabel. Near-UV molar absorptivities of acetone, alachlor, metolachlor, diazinon and dichlorvos in aqueous solution. *J Photochem Photobiol* 174: 76–81, 2005.
246. V Harikrishna, B Prasad, and NVS Naidu. Methodology for the estimation of carbofuran in rice, wheat and water samples. *J Indian Chem Soc* 82: 183–185, 2005.
247. MK Deb, C Chakravarty, and RK Mishra. Spectrophotometric determination of zinc dimethyl dithiocarbamate (Ziram) with hydroxyamidine and 4-(2-pyridylazo) naphthol. *J Indian Chem Soc* 73: 551–552, 1996.
248. HB Cesnik and A Gregorcic. Validation of the method for the determination of dithiocarbamates and thiuram disulphide on apple, lettuce, potato, strawberry and tomato matrix. *Acta Chim Slov* 53: 100–104, 2006.
249. JFH Perez, AMG Campana, LG Gracia, AG Casado, and Md O Iruela. Sensitive determination of carbaryl in vegetal food and natural waters by flow-injection analysis based on the luminol chemiluminescence reaction. *Anal Chim Acta* 524: 161–166, 2004.
250. JFH Perez, LG Gracia, AMG Campana, AG Casado, and JLM Vidal. Chemiluminescence determination of carbofuran at trace levels in lettuce and water by flow injection analysis. *Talanta* 65: 980–985, 2005.
251. JJS Chinchilla, LG Gracia, AMG Campana, and LC Rodriguez. A new Strategy for the chemiluminescent screening analysis of total *N*-methylcarbamate content in water. *Anal Chem Acta* 541: 113–118, 2005.
252. LG Gracia, LC Rodriguez, JJS Chinchilla, JFH Perez, AG Casads, and AMG Campana. Establishment of signal-recovery functions for calculation of recovery factor: Application to monitoring of contaminant residues in vegetables by chemiluminescence detection. *Anal Bioanal Chem* DOI(10): 1007-10012, 2005.
253. P Hua, JP Hole, JS Wilkinson, G Proll, J Tschmelak, G Gauglitz, MA Jackson, R Nudd, HMT Griffith, RA Abuknesha, J Kaiser, and D Kraemmer. Integrated optical fluorescence multisensor for water pollution. *Opt Exp* 13 (4): 1124–1130, 2005.
254. JFG Reyes, EJL Martinez, PO Barrales, and AM Diaz. Determination of thiobendazole residues in citrus fruits using a multicommuted fluorescence based optosensor. *Anal Chim Acta* 557: 95–100, 2006.
255. HS Rathore, SK Saxena, and T Begum. A selective fluorescence spot test for the detection of TCA in soil and water. *J Indian Chem Soc* 69: 798–799, 1992.
256. HS Rathore, M Kumar, and K Ishratullah. Metal ion chromatography on sodium diethyldithiocarbamate. *Indian J Chem Technol* 13: 84–87, 2006.
257. O Tiryaki and P Aysal. Applicability of TLC in multiresidue methods for the determination of pesticides in wheat grains. *Bull Environ Contam Toxicol* 75: 1143–1149, 2005.
258. O Tiryaki. Methods validation for the analysis of pesticide residues in grain by thin-layer chromatography. *Accred Qual Assur* 11: 506–513, 2006.
259. HS Rathore and C Varshney. Chromatographic behaviour of some dithiocarbamate fungicides on cellulose plates. *J Planar Chromatogr* 20 (4): 287–292, 2007.
260. I Soczewinski, T Tuzimski, and K Pomorska. Chemometric characterization of R_F values of pesticides for thin-layer chromatographic systems of the type silica + weakly polar diluent-polar modifier. *J Planar Chromatogr* 11: 90–93, 1998.
261. E Soczewinski and T Tuzimski. Chemometric characterization of the R_F values of pesticides for thin-layer chromatographic systems of the type silica + weakly polar diluent-polar modifier. Part II. *J Planar Chromatogr* 12: 186–189, 1999.
262. T Tuzimski and E Soczewiniski. Chemometric characterization of the R_F values of pesticides for thin-layer chromatographic system of the type silica–non polar diluent + polar modifier. Part III. *J Planar Chromatogr* 13: 271–275, 2000.
263. T Tuzimski. Chemometric characterization of the R_F values of pesticides in thin-layer chromatography on silica with mobile phases comprising a weakly polar diluent and a polar modifier. Part IV. *J Planar Chromatogr* 15: 124–127, 2002.
264. T Tuzimski and E Soczewinshi. Chemometric characterization of the R_F values of pesticides in thin-layer chromatography on silica with mobile phases comprising a weakly polar diluent and a polar modifier. Part V. *J Planar Chromatogr* 15: 164–168, 2002.
265. T Tuzimski and E Soczewinski. Correlation of retention parameters of pesticides in normal- and reversed phase systems and their utilization for the separation of a mixture of 14 triazines and urea herbicide by means of two-dimensional thin-layer chromatography. *J Chromatogr A* 961: 277–283, 2002.
266. T Tuzimski and E Soczewinski. Use of a database of plots of pesticide retention (R_F) against mobile phase composition, Part I: Correlation of pesticide retention data in normal- and reversed phase systems and their use to separate a mixture of ten pesticides by 2D-TLC. *Chromatographia* 56 (3/4): 219–223, 2002.

267. T Tuzimski. Thin-layer chromatography (TLC) as a pilot technique for HPLC. Utilization of retention database (R_F) vs. eluent composition of pesticides. *Chromatographia* 56 (5/6): 379–381, 2002.

268. T Tuzimski and E Soczewinski. Use of database of plots of pesticide retention (R_F) against mobile-phase composition. Part II. TLC as a pilot technique for transferring retention data to HPLC, and use of the data for preliminary fractionation of a mixture of pesticides by micropreparative column chromatography. *Chromatographia* 26 (3/4): 225–227, 2002.

269. T Tuzimski and E Soczewinski. Correlation of retention data of pesticides in normal and reversed-phase systems and utilization of the data for separation of a mixture of ten urea herbicides by two-dimensional thin-layer chromatography on cyanopropyl-bonded polar stationary phase and on a two-adsorbent-layer multi-K SC 5 plate. *J Planar Chromatogr* 16: 263–267, 2003.

270. T Tuzimski and A Bartosiewicz. Correlation of retention parameters of pesticides in normal and RP systems and their utilization for the separation of a mixture of ten urea herbicides and fungicides by two-dimensional TLC on cyanopropyl bonded-polar stationary phase and two-adsorbent layer multi-K Plate. *Chromatographia* 58 (11/12): 781–787, 2003.

271. T Tuzimski and E Soczewinski. Use of database of plots of pesticide retention (R_F) against mobile-phase compositions for fractionation of a mixture of pesticides by micropreparative thin-layer chromatography. *Chromatographia* 59 (1–2): 121–128, 2004.

272. T Tuzimski. Separation of a mixture of eighteen pesticides by two-dimensional thin-layer chromatography on a cyanopropyl-Bonded polar stationary phase. *J Planar Chromatogr* 17: 328–334, 2004.

273. T Tuzimski. Use of thin-layer chromatography in combination with diode-array scanning densitometry for identification of fenitrothion in apples. *J Planar Chromatogr* 18: 419–422, 2005.

274. T Tuzimski. Two-dimensional TLC with adsorbent gradients of the type silica-ctadecyl silica and silica-cyanopropyl for separation of mixtures of pesticides. *J Planar Chromatogr* 18: 349–357, 2005.

275. T Tuzimski. Two-stage fractionation of a mixture of pesticides by micropreparative TLC and HPLC. *J Planar Chromatogr* 18: 39–44, 2005.

276. T Tuzimski. Two-stage fractionation of a mixture of 10 pesticides by TLC and HPTC. *J Liquid Chromatogr Relat Technol* 28: 463–476, 2005.

277. T Tuzimski and J Wojtowicz. Separation of a mixture of pesticides by 2D-TLC on two-adsorbent-layer multi SC5 Plate. *J Liquid Chromatogr Relat Technol* 28: 277–287, 2005.

278. J Rasmussen and OS Jacobsen. Thin-layer chromatographic methods for the analysis of eighteen different ^{14}C-labeled pesticides. *J Planar Chromatogr* 18: 248–252, 2005.

279. ME Aydin, S Ozcan and S Sari. Organochlorine pesticides in the sewerage system of Knonya Turkey. *Fresenius Environ Bull* 13 (11B): 1303–1308, 2004.

280. YR Tahboub, MF Zaater, and ZA Al-Talla. Determination of the limits of identification and quantification of selected organochlorine and organophosphorous pesticide residues in surface water by full-scan gas chromatography/mass spectrometry. *J Chromatogr A* 1098: 150–155, 2005.

281. T Krauss, AMCB Brag, JM Rosa, K Kypke, and R Malisch. Levels of organochlorine pesticides in Brazilian Human milk. *Organohalogen Compounds* 66: 2739–2744, 2004.

282. R Hajou, Fu Afifi, and A Battach. Determination of multipesticide residues in mentha piperita. *Pharmaceut Biol* 43 (6): 554–562, 2005.

283. HB Cesnik, A Gregorcic, SV Bolta, and V Kmeel. Monitoring of pesticide residues in apples, lettuce and potato of the Slovene origin, 2001–04. *Food Addit Contam* 23 (2): 164–173, 2006.

284. HB Cesnik, A Gregorcic, and SV Bolta. Pesticide residues in agricultural products of Slovene origin in 2005. *Acta Chim Slov* 53: 95–99, 2006.

285. HB Cesnik and A Gregorcic. Validation of the method for the determination of dithiocarbamates and thiuram disulphide on apple, lettuce, potato, strawberry and tomato matrix. *Acta Chim Slov* 53: 100–104, 2006.

286. G Saxena, M Agrawal, A Baroth, and P Bhatnagar. Pesticides residues in the milk of Jaipur city. *J Ecophysiol Occup Health* 5: 77–79, 2005.

287. SM Waliszewski, S Goomez-Arroyo, O Carvajal, R Villalobos-Pietrini, and RM Infanzon. Use del acido sulfurico en las determinaciones de plaguicidas organoclorados. *Rev Int Contam Ambient* 20 (4): 185–192, 2004.

288. M Domotorova, E Matisova, M Kirchner, and J de Zeeuw. MSPD combined with fast GC for ultratrace analysis of pesticide residues in non-fatty food. *Acta Chim Slov* 52: 422–428, 2005.

289. CC Leandro, RJ Fussell, and BJ Keely. Determination of priority pesticides in baby food by gas chromatography tandem quadrupole mass spectrometry. *J Chromatogr A* 1085: 207–212, 2005.

290. B Lazarov, J Manova, Z Bratanova, and N Bonev. Quick method for determination of organochlorine pesticides in drinking waters: Application at accidental contamination. *J Environ Prot Ecol* 6 (3): 521–524, 2005.

291. R Sanchez, JM Cortes, J Villen, and A Vazquez. Determination of organophosphorus and triazine pesticides in olive oil by on-line coupling reversed-phase liquid chromatography/gas chromatography with nitrogen-phosphorus detection and an automated through-oven transfer adsorption-desorption interface. *J AOAC Int* 88 (4): 1255–1260, 2005.

292. JM Cortes, R Sanchez, EMD Plaza, J Villen, and A Vazquez. Large volume GC injection for the analysis of organophosphorus pesticides in vegetables using the through-oven transfer adsorption desorption (TOTAD) interface. *J Agric Food Chem* 54: 1997–2002, 2006.

293. DK Shahi, K Nisha, and A Sharma. Monitoring of pesticide residues in market vegetables at Ranchi, Jharkhand (India). *J Envion Sci Eng* 47 (4): 322–325, 2005.

294. L Libin, Y Hashi, Q Yaping, Z Haixia, and L Jinming. Rapid analysis of multiresidual pesticides in agricultural products by gas chromatography-mass spectrometry. *Chin J Anal Chem* 34 (6): 783–786, 2006.

295. LG Gracia, AMG Compana, JJS Chinchilla, JFH Rerez, and AG Casado. Analysis of pesticides by chemiluminescence detection in the liquid phase. *Trends Anal Chem* 24 (11): 927–942, 2005.

296. MG Rubio, MLFD Cordova, MJA Canada, and AR Medina. Simplified pesticide multiresidue analysis in virgin olive oil by gas chromatography with thermoionic specific, electron-capture and mass spectrophotometric detection. *J Chromatogr A* 1108: 231–239, 2006.

297. ED Dempelou and KS Liapis. Validation of a multi-residue method for the determination of pesticide residues in apples by gas chromatography. *Int J Environ Anal Chem* 86 (1–2): 63–68, 2006.

298. K Patel, RJ Fussell, DM Goodwall, and BJ Keely. Analysis of pesticide residues in lettuce by large volume-difficult matrix introduction gas chromatography-time of flight-mass spectrometry (LV-DMI-GC-TOF-MS). *Analyst* 128: 1228–1231, 2003.

299. K Patel, RJ Fussell, R Macarthur, DM Goodall, and BJ Keely. Method Validation of resistive heating-gas chromatography with flame photometric detection for the rapid screening of organophosphorus pesticides in fruit and vegetables. *J Chromatogr A* 1046: 225–234, 2004.

300. K Patel, RJ Fussell DM Goodall, and BJ Keely. Evaluation of large volume-difficult matrix introduction gas-chromatography-time of flight-mass spectrometry (LV-DMI-GC-TOF-MS) for the determination of pesticides in fruit-based baby foods. *Food Addit Contam* 21 (7): 658–669, 2004.

301. K Patel, RJ Fussell, M Hetmanski, DM Goodall, and BJ Keely. Evaluation of gas chromatography-tandem quadrupole mass spectrometry for the determination of organochlorine pesticides in fats and oils. *J Chromatogr A* 1068: 289–296, 2005.

302. K Patel, RJ Fussell, DM Goodall, and BJ Keely. Application of programmable temperature vaporisation injection with resistive heating-gas chromatography flame photometric detection for the determination of organophosphorus pesticides. *J Sep Sci* 29: 90–95, 2006.

303. C Soler, J Manes, and Y Pico. Determination of carbosulfan and its metabolites in oranges by liquid chromatography ion-trap triple-stage mass spectrometry. *J Chromatogr A* 1109: 228–241, 2006.

304. M Liu, Y Hashi, Y Song, and JM Lin. Simultaneous determination of carbamate and organophosphorus pesticides in fruits and vegetables by liquid chromatography-mass spectrometry. *J Chromatogr A* 1097: 183–187, 2005.

305. KP Prousalis, CK Kaltsonoudis, and T Tsegenidis. A new sample clean-up procedure, based on ion-pairing RP-SPE cartridges, for the determination of ionizable pesticides. *Int J Environ Anal Chem* 86 (1–2): 33–43, 2006.

306. K P Prousalis, DA Polygenis, A syrokou, FN Lamari, and T Tsegenidis. Determination of carbendazim, thiabendazole, and *o*-phenylphenol residues in lemons by HPLC following sample clean-up by ion-pairing. *Anal Bioanal Chem* 379: 458–463, 2004.

307. SJ Lehotay, AD Kok, M Hiemstra, and PV Bodegraven. Validation of a fast and easy method for the determination of residues from 229 pesticides in fruits and vegetables using gas and liquid chromatography and mass spectrometric detection. *J AOAC Int* 88 (2): 595–614, 2005.

308. S J Lehotay. Quick, easy, cheap, effective rugged and safe approach for determining pesticide residues. In: JLM Vidal, AG Frenich (Eds.). *Methods in Biotechnology, Vol. 19, Pesticide Protocols*, Humana Press Inc., Totowa, NJ, 2006, pp. 239–261.

309. E Hogendoorn and E Dijkman. Coupled-column liquid chromatography for the determination of pesticide residues. In: JLM Vidal, A G Frenich (Eds.). *Methods in Biotechnology, Vol. 19, Pesticide Protocols*, Humana Press, Inc, Totowa NJ, 2005, pp. 383–399.

310. CS Evans, JR Startin, DM Goodall, and BJ Keely. Optimization of ion trap parameter for the quantification of chlormequat by liquid chromatography/mass spectrometry and the application in the analysis of pear extracts. *Rapid Commun Mass Spectrom* 14: 112–117, 2000.

311. CS Evans, JR Startin, DM Goodall, and BJ Keely. Improved sensitivity in detection of chlormequat by liquid chromatography-mass spectrometry. *J Chromatogr A* 897: 399–404, 2000.

312. CC Leandro, P Hancock, RJ Fussel, and BJ Keely. Comparison of ultra-performance liquid chromatography and high-performance liquid chromatography for the determination of priority pesticides in baby foods by tandem, quadrupole mass spectrometry. *J Chromatogr A* 1103: 94–101, 2006.

313. TJ Ward. Chiral separations. *Anal Chem* 78 (12): 3947–3956, 2006.

314. CK Kaltsonoudis, FN Lamari, KP Prousalis, NK Karamanos, and T Tsegenidis. Analysis of carbendazim and thiabendazole in lemons by CE-DAD. *Chromatographia* 57 (3/4): 181–184, 2003.

315. L Goodwin, JR Startin, BJ Keely, and DM Goodall. Analysis of glyphosate and glufosinate by capillary electrophoresis-mass spectrometry utilising a sheathless microelectrospray interface. *J Chromatogr A* 1004: 107–119, 2003.

316. L Goodwin, M Hanna, JR Startin, BJ Keely, and DM Goodall. Isotachophoretic separation of glyphosate glufosinate, AMPA and MPP with contactless conductivity detection. *Analyst* 127: 204–206, 2002.

317. MC Henry and CR Yonker. Supercritical fluid chromatography, pressurized liquid extraction and supercritical fluid extraction. *Anal Chem* 78 (12): 3909–3916, 2006.

318. SD Richardson and TA Ternis. Water analysis: Emerging contaminants and current issues. *Anal Chem* 77 (12): 3807–3838, 2005.

319. E Bakker and Y Qim. Electrochemical sensors. *Anal Chem* 78 (12): 3965–3984, 2006.

320. MZA Rafique, HS Rathor, SK Gangwar, and Kabir-ud-Din. Polarographic study on parathion in micellar media. *J Indian Chem Soc* 82: 329–332, 2005.

321. P Manisankar, C Vedhi, and G Selvanathan. Electro chemical determination of methyl parathion using a modified electrode. *Toxicol Environ Chem* 85 (4–6): 233–241, 2003.

322. P Manisankar, G Selvanathan, and G Vedhi. Utilization of sodium montmorillonite clay-modified electrode for the determination of isoproturon and carbendagine in soil and water samples. *Appl Clay Sci* 29: 249–257, 2005.

323. P Manisankar, S Viswanathan, AM Pushalatha, and C Rani. Electrochemical studies and square wave stripping voltametry of five common pesticides on poly 3,4-ethylenedioxythiophene modified wall-jet electrode. *Anal Chim Acta* 528: 157–163, 2005.

324. P Manisankar, G Selvanathan, and C Vedhi. Utilization of polypyrrole modified electrode for the determination of pesticides. *Int J Environ Anal Chem* 85 (6): 409–422, 2005.

325. P Manisankar, G Selvanathan, S Viswanathan, and HG Prabu. Electrochemical determination of some organic pollutants using wall-gel electrode. *Electroanalysis* 14 (24): 1722–1727, 2002.

326. P Manisankar, G Slevanathan, and C Vedhi. Determination of pesticides using heteropolyacid montmorillonite clay-modified electrode with surfactant. *Talanta* 68: 686–692, 2006.

327. JFH Perez, MDO Iruela, AMG Campana, AG Casado, and AS Nararro. Determinations of the herbicide metribuzin and its major conversion products in soil by micellar electrokinetic chromatography. *J Chromatogr A* 1102: 280–286, 2006.

328. AJ Dane, CD Havey, and KJ Voorhees. The detection of nitro pesticides in main stream and sidestream cigarette smoke using electron monochromator-mass spectrometry. *Anal Chem* 78 (10): 3227–3233, 2006.

329. SD Richardson. Water analysis. *Anal Chem* 73 (12): 2719–2734, 2001.

330. CS Evans, JR Startin, DM Goodall, and BJ Keely. Tandem mass spectrometric analysis of quaternary ammonium pesticides. *Rapid Commun Mass Spectrom* 15: 699–707, 2001.

331. L Goodwin, JR Startin, DM Goodall, and BJ Keely. Tandem mass spectrometric analysis of glyphosate, glufosinate, aminomethyl-phosphonic acid and methyl phosphini copropionic acid. *Rapid Commun Mass Spectrom* 17: 963–969, 2003.

332. L Goodwin, JR Startin, DM Goodal, and BJ Keely. Negative ion electrospray mass spectrometry of aminomethyl phosphonic acid and glyphosati: Elucidation of fragmentation mechanisms by multistage mass spectrometry labelling. *Rapid Commun Mass Spectrom* 18: 37–43, 2004.

333. RE Clement, PW Yang, and CJ Koester. Environmental analysis. *Anal Chem* 73 (12): 2761–2790, 2001.

FURTHER READINGS

Books

Chau ASY, Afghan BK, and Robinson JW (Eds.). *Analysis of Pesticide in Water*, Vols. I, II, III. CRC Press, Boca Raton, FL, 1981.

De AK. *Environmental Chemistry*, 4th ed., New Age International (P) Limited, New Delhi, 2003, pp. 167–195.

Dhingra KC. *Handbook of Pesticides: Formulations on Pesticides and Insecticides*, Small Industry Research Institute, Delhi, 1991.

Goss GR. Agricultural granule particle size consideration: ASTM special technical pesticide formulations and delivery systems, STP/460, ASTM Special Technical Publication, IL, pp. 25–30, 2005.

Masoud S and Gregory L (Eds.). *Pesticide Formulations and Delivery Systems*, Vol. 25. ASTM Special Technical Publication, 2006 STP 1470. ASTM International, Conshohocken, PA, p. 214, 2006, ISBN: O-8031-3496-7.

Ming-HO Y. *Environmental Toxicology: Biological and Health Effects of Pollutants*, 2nd ed., CRC Press, Boca Raton, FL, 2005.

Nollet LML (Ed.). *Handbook of Food Analysis*, Vol. 2, Marcel Dekker, New York, pp. 1291–1530, 1996.

Nollet LML (Ed.). *Handbook of Water Analysis*, Marcel Dekker, New York, pp. 487–654, 2000.

Nollet LML (Ed.). *Handbook of Food Analysis: Residues and Other Food Component Analysis*, 2nd ed., revised and expanded. Vol. 2, Marcel Dekker, New York, 2004.

Salyani M (Ed.). *Pesticide Formulations and Delivery Systems*, Vol. 25, ASTM International, Philadelphia, 2006.

Sree Ramulu US. *Chemistry of Insecticides and Fungicides*, 2nd ed., Oxford and IBH Publishing Co. Pvt. Ltd., New Delhi, 1979.

Stenersen J. *Chemical Pesticides: Mode of Action and Toxicology*, CRC Press, Boca Raton, FL, 2004.

Valkenburg WV, Sugavanan B, and Khetan SK. *Pesticide Formulation: Recent Development and Their Application in Developing Countries*, New Age International (P) Limited, New Delhi, 1998.

Yedla S and Dikshit AK. *Abatement of Pesticide Pollution: Removal of Organochlorine Pesticides from Water Environment*, Narosa, New Delhi, 2004.

Reviews

Banasiake U and Hohgardt K. Process for risk assessment of pesticide residues in food. *Schriften Reihe der Dentsehen Phytomedigzinischen Gesellschaft* 7 (Gesunde Pflanzen-Gesnude Nohrung): 99–112, 158–180, 2006.

Bradman A and Whyatt RM. Characterizing exposures to non-persistent pesticides during pregnancy and early childhood in the national children's study: A review of monitoring and measurement methodologies. *Environ Health Perspect* 113 (8): 1092–1099, 2005.

Buncel E, Van Loon GW, Balakrishnan V, Chambers E, Churchill D, Esbata A, Fang W, Kiepek E, Oketuride O, Omakor E, and Shirin S. Graduate student co-workers pesticide pathways. *Canadian Chem News* 57 (4): 18–20, 2005.

Cai X, Wang C, and Zhang Z. Advances in process of pesticides. *Nangyao Kexue Yu Guanli* 25 (6): 35–37, 2004.

Chen Q-B and Ji YL. Safely evaluation and risk management for modern chemical pesticides. *Yunnan Nongye Daxue Xueboo* 20 (1): 99–106, 2005.

Dubois M, Caste C, Despres A-G, Efstathiou T, Nio C, Dumont E, and Parent-Massin D. Safety of oxygreen: A ozone treatment on wheat grain part 2: Is there a substantial equivalence between oxygreen treated wheat grains and untreated wheat grains. *Food Addit Contam* 23 (3): 1–15, 2006.

Ewart DM and Mawson JB. Composition and method for termite control. *Aust Pat Appl Au* 203: 624, 2003 (Cl AO1N65/00).

Feng J. Discovery of new pesticides and development of industrialization in China. *Xiandai Nongyao* 4 (3): 1–9, 2005.

Fenke AR, Bradman A, Whyatt MR, Wolff SM, and Barr BD. Lessons learned for the assessment of children's pesticide exposure, critical sampling and analytical issues for future studies. *Environ Health Perspect* 113 (10): 1455–1462, 2005.

Garrido AF, Martinez LJV, Egea JFG, and Arrebola JFL. Quality criteria in pesticide analysis. *Methods Biotechnol* 19 (Pesticide Protocols): 219–230, 2006.

King GA. Research advances: Enzymes bottles pesticides residue buildup. *J Chem Educ* 83 (3): 346–349, 2006.

Matthews P. A pesticide agent. *Aust Pat Appl AU* 203: 723, 2003 (Cl No A01N S9/20).

Pahadia S, Sharma V, Sharma V, and Wadhwa BK. Evolution of multiresidue analysis of pesticides in dairy and fatty food—A review. *Int J Dairy Sci* 58 (2): 75–78, 2005.

Pamela C-A, Gamal E-DM, Smith DW, and Guest KR. Pesticides and herbicides. *Water Environ Res* (online computer file) 77 (6): 2021–2129, 2005.

Pandey R. Pesticides and sterility. *Everyman's Science* XXXVIII (2): 84–86, 2003.

Sood C and Bhagat RM. Interfacing geographical information systems and pesticide models. *Curr Sci* 89 (8): 1362–1370, 2005.

Tokimoto Y. Method and system for analysis of organochemicals. *Jpn Kokai Tokkyo Koho Jp* 78: 419, 2006 (Cl 90 N 31/00).

Wang J, Wen Z-J, Zhang Z-H, Zhang X-D, Wen F-Y, Wang L, and Xu L. Study progress and application prospect of ultrasonic treatment of pesticide. *Xiandai Nongyo* 4 (5): 22–25, 2005.

Wu YJ and Yang L, Wli. Multitarget toxicity of organophosphorus pesticides. *Huanzing Yu Zhiye Yixue* 22 (4): 367–370, 2005.

Xu T and Li J. Pesticide residues in farm products and trends of their analysis techniques. *Daxue Huaxhe* 18 (6): 5–11, 2003.

Xu X, Hu G, Wang X, and Lee FSC. Progress in analysis of traditional Chinese medicines for organochlorine pesticides residues by gas chromatography. *Fenxi Ceshi Xuebao* 23 (4): 112–117, 2004.

Yanese S, Chatani Y, Kitano R, Nakamura M, Ohfugi M, and Takemae M. Survey of pesticides residues in agricultural products Apr 2003–March 2004. *Kyoto fu Hoken Kankyo-fu Huken Kankyo Kenkyusho Nempo* 49: 67–71, 2004.

Zou X and Jiang Z. On botanical insecticides and environmental safety. *Nongyao Kexue yu Guanli* 25 (9): 22–26, 2004.

3 Pesticides: Past, Present, and Future

Shalini Srivastava, Pritee Goyal, and Man Mohan Srivastava

CONTENTS

3.1 HISTORICAL BACKGROUND

3.1.1 DEFINITION

Generally, the term "pest" refers to any insect, rodent, nematode, fungus, weed, or any other form of terrestrial or aquatic plant, animal, virus, bacteria, or other microorganisms that harm the garden plants, trees, foodstuffs, household articles, or is a vector of diseases. However, for farmers, pests include insects and mites that feed on crops; weeds in the fields; aquatic plants that clog irrigation and damage ditches; agents that cause plant diseases such as fungi, bacteria, viruses, nematodes, snails, slugs, and rodents that consume enormous quantities of plant seedlings and grains. George

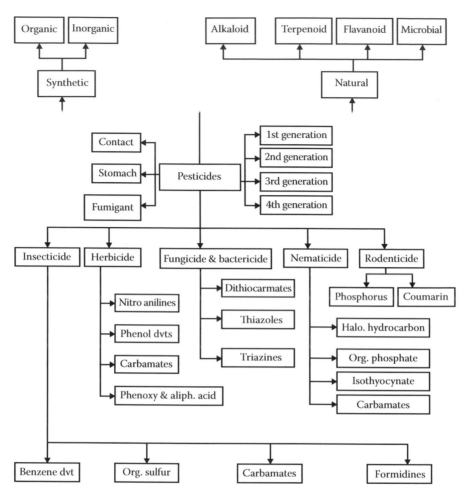

FIGURE 3.1 Classification of pesticides.

Ware, a well-known environmentalist, defined the term "pesticides" as man's intentional additives to improve his environment quality for himself, his animals, and his plants. A pesticide can be used against any form of terrestrial and aquatic plant, animal, or microorganism, which an agency declares as a pest. However, in general, it may be defined as a substance that exerts toxic action on the pests. Owing to the wide diversity, it is really a difficult task to provide a general scheme for the classification of various pesticides. However, pesticides are classified, based on their evolution process, mode of action, chemical nature, and target species. Figure 3.1 presents a schematic view of categorization of pesticides into different classes.

3.1.2 Origin of Pesticides

Dr. Norman E. Borlaug, the Nobel Prize winner for peace (1970), who is known as the Father of Green Revolution, stated in an address to the United Nation's Food and Agriculture Organization that use of pesticides is indispensable for the rise in agricultural production. This is truly reflected in almost all the countries of the world. It is a well-acknowledged fact that pesticides are one of the most important components of modern agricultural technology, and have proved to be important chemical tools for world agricultural production. The activities of harmful organisms result in one-third of the world's agricultural losses. It has been observed that about 13.8% of all the losses are due to insects and mites. Nearly 70,000 species of insects and mites attack all parts of the agricultural plants during their entire period of vegetation and storage. About 10,000 species of them cause substantial economical loss. Farmers, therefore, have no choice but to rely heavily on the use of pesticide chemicals. The enormous economical loss resulting from diverse species of harmful insects, mites, etc., have usurped humankind to develop and use various insecticides, pesticides, and acaricides to protect their crops. The world is now flooded with various types of insecticides and pesticides. The use of these pesticides in agriculture makes it possible to save approximately more than one-third of the crops. The era of chemical plant protection came into being after World War II. This, in fact, provided a great impetus to the development of synthetic pesticides. The DDT was the first synthetic pesticide used and it turned out to be the most powerful and successful chemical pesticide. A German student first synthesized DDT in 1873, while a Swiss entomologist, Dr. Paul Muller, explored its insecticidal activity in 1939. Simultaneously, organophosphate compounds were discovered in Germany during World War II, while studying materials related to the nerve gases, sarin, soman, and tabun. Since then, pesticide research has attracted the attention of the scientific community.

3.1.3 STATUS OF PESTICIDE CONSUMPTION

Today, a wide range of pesticides have been synthesized, formulated, and tested for their bioefficacy against various target species. According to the United States Environmental Protection Agency, there are more than 1180 pesticides, of which 435 are herbicides, 335 are insecticides, and 410 are fungicides. These are sold in more than 32,800 formulations. Worldwide pesticide consumption pattern in 1965 shows that its consumption was 172 T, of which the United States consumes about 24%, Europe 46%, and the remaining 30% is consumed by the developing countries. Pesticide consumption was increased to about twofold in 1980 and threefold in 1990, when compared with that in 1965. This increase led to a similar rise in the agricultural production and farm practices, causing a revolutionary change leading to an incredible possibility that hunger can be eradicated from the world.

India is the third largest consumer of pesticides in the world and the highest among the South Asian countries. In addition, India is the second largest manufacturer of pesticide chemicals in Asia, next to China, and is ranked number 12 globally. About 155 pesticides have been registered under the Insecticide Act 1968, till March 2001, which includes 57 insecticides, 44 fungicides, 33 herbicides and rodenticides, 4 plant-growth regulators, 4 fumigants, 3 acaricides, 1 soil sterilant, 1 molluscicide, and 1 nematicide. The average per-hectare consumption of pesticides in agriculture (calculated based on the total consumption of technical grade pesticides divided by the gross cultivated area) was 1.2 g/ha in 1980 and increased to 431 g/ha in 1992–1993. The leading chemical used in India during 1995–1996 was benzene hexachloride (BHC), which accounted for more than 40% of the total pesticides consumed followed by malathion, methyl parathion, endosulfan, carbaryl, and dimethoate. The states of Uttar Pradesh, Punjab, Andhra Pradesh, Haryana, Gujarat, West Bengal, and Maharashtra account for 79.84% of pesticide consumption. The pesticide consumption remained highest in Andhra Pradesh (12,775 MT), which alone accounted for 15.7% of the total consumption, followed by Uttar Pradesh (11,500 MT), Punjab (7,600 MT), Haryana (5,390 MT), West Bengal (5,338 MT), and Maharashtra (4,898 MT). A study conducted by the Planning Commission, Govt. of India, estimated pesticide consumption in the year 2000 as 1,18,000 tons, out of which 97,000 tons accounted for agriculture and 21,000 tons for public health. Therefore, to fulfill the current requirement, the Indian pesticide industry has installed a capacity of 1,16,000 tons per annum, of which about 70,000 tons is in the organized sector, whereas the rest is in some 500 odd units belonging to the small sector. The imports are currently about 2000 tons only.

3.1.4 IMPACT ON ENVIRONMENT

The increased consumption of pesticides was earlier considered as a good sign of progress in agricultural production. Later man realized the ecological principle that every poison we add into the environment comes right back to us. Pesticides move through air, soil, and water, finding their way into living tissues where they undergo biological magnification. Thus, the deterioration of the ecosystem by the continuous use of fertilizers and pesticides has been observed. How disgusting is it to hear that almost all the surface water in our country, except at the mountains, is unfit for human consumption. The quality of river water deteriorates almost as soon as it enters the plains. At every step, water is extracted from the river and large loads of polluted water are discharged into it. As the river progresses downstream, its quality degrades constantly. The discharges change the physicochemical properties of the water and adversely affect the aquatic life. Chemical fertilizers and pesticides constitute one of the major pollutants of the river. Although the water is treated before supply, the treatment does nothing to remove the pesticide traces and industrial pollutants present in the water; it is really very difficult to get rid of the chemicals and biological contaminants from water. The burden of waterborne disease is about 30.5 million of DALYs (disabled life years). Leaching from agricultural fields has been the most important nonpoint source of pollution to the aquatic environment. Traces of HCH and DDT have been found in rivers of the United States and Europe, where they have been banned for more than two decades. Residues of persistent organochlorines, long banned in the West, have been found at alarming levels in the rivers of India (Figure 3.2).

An Indian parliamentary committee has confirmed allegations by a New Delhi-based nongovernmental organization that soft drinks made by two international cola giants contain excessively high levels of pesticide residues, owing to the use of contaminated groundwater. The parliamentary committee pointed out that all agencies agreed on the presence of pesticides. It stated that the differences in the amounts measured could be attributed to differences in where and when the drinks were manufactured, temperature of storage, and analytical techniques. Perusal of the residue data on pesticides in samples of fruits, vegetables, cereals, pulses, grains, wheat flour, oils, eggs, meat, fish, poultry, bovine milk, butter, and cheese in India indicates their presence in sizable amounts. In India, it is also estimated that 20% of the contamination is above the maximum residue limits

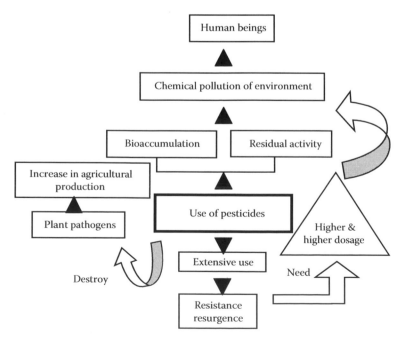

FIGURE 3.2   Today's realization.

(MRLs) fixed. In the European Union (EU), this is estimated to be around 1.4%. Monocrotophos, methyl parathion, and DDVP, all organophosphorus pesticides, are found to be the most prevalent. These are also WHO Class I pesticides. In 2001, 61% of the samples tested were found to be contaminated, of which 11.7% were also above MRLs. However, at present, the fruit samples did not exceed MRLs, but around 15% of the milk samples still exceed MRLs. Hexachlorobenzene (HCB, a fungicide) was identified in water, human milk, and human fat samples collected from Faridabad and Delhi. In a multicentric study to assess the pesticide residues in selected food commodities collected from different states of the country, DDT residues were found in about 82% of the 2205 samples of bovine milk collected from 12 states. About 37% of the samples contained DDT residues above the tolerance limit of 0.05 mg/kg (whole-milk basis). The highest level of DDT residues found was 2.2 mg/kg. The proportion of the samples with residues above the tolerance limit was maximum in Maharashtra (74%), followed by Gujarat (70%), Andhra Pradesh (57%), Himachal Pradesh (56%), and Punjab (51%). In the remaining states, this proportion was less than 10%. Data on 186 samples of 20 commercial brands of infants' milk formulae showed the presence of residues of DDT and HCH isomers in about 70% and 94% of the samples with a maximum level of 4.3 and 5.7 mg/kg (fat basis), respectively. The average total DDT and BHC consumed by an adult were reported to be 19.24 and 77.15 mg/day, respectively. Fatty food was the main source of these contaminants. In another study, the average daily intake of HCH and DDT by Indians were reported to be 115 and 48 mg per person, respectively, which were higher than those observed in most of the developed countries.

In one study, the tested samples were found to be 100% contaminated with low but measurable amounts of pesticide residues. Among the four major chemical groups, residue levels of organophosphorous insecticides were the highest, followed by carbamates, synthetic pyrethroids, and organochlorines. About 32% of the samples showed contamination with organophosphorous and carbamate insecticides above their respective MRL values. The monitoring study indicated that though all the vegetable samples were contaminated with pesticides, only 31% of the samples contained pesticides above the prescribed tolerance limit. Samples of vegetables collected at beginning, middle, and end of seasons were analyzed for organochlorine levels. Maximum pesticide residues were detected in cabbage (21.24 ppm), cauliflower (16.85 ppm), and tomato (17.046 ppm), collected at the end of season, and okra (17.84 ppm) and potato (20.60 ppm), collected at the middle of season. The organo chlorine pesticide (OCP) residue levels in majority of the samples were above the maximum acceptable daily intake (ADI) prescribed by WHO, 1973. Twelve most-commonly used pesticides were selected to study the residual effects on 24 samples of freshly collected vegetables. Most of the samples showed the presence of high levels of malathion. Furthermore, DDE, a metabolite of DDT, BHC, dimethoate, endosulfan, and ethion were also detected in few samples. However, leafy vegetables like spinach, fenugreek, and mustard seem to be the most affected. Radish also showed high levels of contamination. Vegetable samples collected at harvest from farmer's fields around Hyderabad and Guntur recorded HCH residues above the MRL (0.25 ppm). However, residues of DDT and cypermethrin were found to be below the MRL (3.5 and 0.2 ppm, respectively), and mancozeb residues were above the MRL (2 ppm) only in bitter gourd. Furthermore, residues of HCH, DDT, aldrin (including dieldrin), endosulfan, and methyl parathion in vegetables of Srikakulam were below the MRL. Detectable levels of residues of commonly used pesticides were observed in tomato (33.3%), brinjal (73.3%), okra (14.3%), cabbage (88.9%), and cauliflower (100%) samples. However, the levels of concentrated pesticide residues were lower than the MRLs prescribed. In a study to estimate various OCPs in different food items collected from 10 localities in Lucknow city, wheat flour and eggs were found to contain maximum concentration of OCP residues. Furthermore, the estimates of dietary intake of total HCH (1.3 g) and lindane (0.2 mg) was about one-and-a-half times higher than that of the ADI, and 100 times the values reported from the United Kingdom and the United States. Out of 400 food stuffs tested, 23.7% were positive for pesticide residues. Higher rates were found in animal products (30%), cereals and pulses (26.3%), and vegetables (24%).

Pesticides are hazardous because of their potential toxicity and environmental contamination. In India, high residual levels of BHC, lindane, heptachlor, endosulfan, and dieldrin have been found in just about everything necessary for life from food to water. The list includes tea in which DDT residues are so high that its export has been refused. Also, significantly high levels of food contamination with HCH, DDT, aldrin, and dieldrin were evident throughout India. Delhites have one of the highest levels of DDT bio-accumulated in their body fat. The average daily intake of HCH and DDT in India is higher than those observed in most developed nations. In India, 1–2 kg of pesticides are used each year for every man, woman, and child. For a country like India, where human exposure to pesticides is reportedly one of the highest in the world, this is a serious concern.

The pesticides are highly persistent and found in the aquatic systems for decades. The continued use of synthetic broad-spectrum insecticides on crops resulted in the resurgence of insect species.

The use of organochlorine pesticides is completely banned in many western countries but in India, some of these are still being used on farms for the control of insect pests in forage and other crops (Figure 3.3).

Recently, pesticides and related compounds have been correlated with cancer-causing agents. These pesticides slowly enter our bodies and after years, cause cancer, immune system disorders, hormonal or reproductive system disorders, affecting even the fetus. Cancers of lymphatic and hematopoietic systems and brain are also associated with the exposure to pesticides. Studies have linked the rising incidence of non-Hodgkin's lymphoma (NHL), a form of cancer, to the increased use of organophosphate pesticides and phenoxy herbicides and the cumulative effects of these pollutants on the human system. Frequent use of herbicides, particularly 2,4-dichlorophenoxyacetic acid (2,4-D), has been associated with a 200%–800% increased risk of NHL. In addition to people working in agriculture, the general population is also at a high risk of NHL because of the use of these pesticides in homes. The use of pesticides, particularly 2,4-D, 2,4,5-trichlorophenoxy acetic acid, 2-methyl-4-chlorophenoxy acetic acid, and organophosphate pesticides have increased over the last 40 years. This increase has played a significant role in contributing to the rising incidence of NHL. Pesticides are capable of altering the processes of tumor genesis, and are reported to cause a variety of cancers through an immunological mechanism (Figure 3.4).

Species	Insecticides	Resistance level
Singhara beetle	DDT	1–8 times
Galeracella	HCH	1–10 times
Cotton Ball Worm	Pyrethroid	1–10 times
Spodoptera	Cl. Hydrocarbons	1–10 times
S. litara	BHC, endosulphan, mono chlorophos	1–30 times
T. Casterneum	Cl. Hydrocarbons, DDT	1–15 times

FIGURE 3.3 Development of resistance in insects/pest: A challenge.

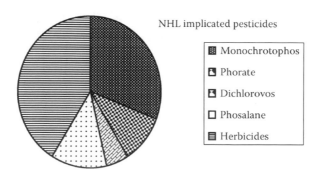

NHL implicated pesticides

- Monochrotophos
- Phorate
- Dichlorovos
- Phosalane
- Herbicides

FIGURE 3.4 Pesticides and cancer.

Adding to the concern about the carcinogenic effects of pesticides, the latest finding is a new discipline of science called the immunotoxicity, which examines substances producing a negative impact on the immune system. Reduced immunity influences incidence of cancer. A weak or devastated immune system allows cancerous cells to escape and form a tumor. Farmers have significant higher risks of Hodgkin's disease, melanoma, multiple myeloma, leukemia, and cancers of lip, stomach, and prostrate. One can only imagine the kind of havoc pesticides can play in a country where a large percentage of population is malnourished and, consequently suffers from immunodeficiency. Thus, it becomes mandatory to screen the agricultural products for such contaminants. To allay the public concern and to protect the environment, regulatory restrictions are imposed on these groups of pesticides. Accordingly, consumption and sale of some pesticides have been banned, withdrawn, or severely restricted.

3.1.5 Impact of Pesticides on Nontarget Organism

When a pesticide encounters a surface or an organism, that contact is called a pesticide exposure. For humans, a pesticide exposure means acquiring pesticides in or on the body. The toxic effect of a pesticide exposure depends on the quantity of the pesticide involved and how long it remains there. Our body gets exposed to pesticides in four main ways:

- *Oral exposures* are often caused by either not washing hands before eating, drinking, smoking or chewing, mistaking the pesticide for food or drink, accidentally applying pesticides to food, or splashing pesticide into the mouth owing to carelessness or accident.
- *Inhalation exposures* are often caused by prolonged contact with the pesticides in closed or poorly ventilated spaces, breathing vapors from fumigants and other toxic pesticides, breathing vapors, dust or mist while handling pesticides without appropriate protective equipment, inhaling vapors present immediately after a pesticide is applied, and using a respirator that fits poorly or using an old or inadequate filter, cartridge, or canister.
- *Eye exposures* are caused by splashing or spraying pesticides in eyes, applying pesticides in windy weather without eye protection, rubbing eyes or forehead with contaminated gloves or hands, and pouring dust, granule, or powder formulations without eye protection.
- *Dermal exposures* are often caused by not washing hands after handling pesticides or their containers, splashing or spraying pesticides on unprotected skin or eyes, wearing pesticide-contaminated clothing (including boots and gloves), applying pesticides in a windy weather, wearing inadequate personal protective equipment while handling pesticides, and touching pesticide-treated surfaces.

Pesticides cause four types of acute effects:

Acute oral effects: Some pesticides can burn our mouth, throat, and stomach severely. Other pesticides when swallowed will not burn our digestive system, but will be absorbed and carried in blood throughout the body and may harm in various ways.

Acute inhalation effects: Some pesticides, making it difficult to breathe, can burn the entire respiratory system. Other pesticides that are inhaled may not harm the respiratory system, but are carried quickly in blood throughout the body where they can harm in various ways.

Acute dermal effects: Contact with some pesticides may harm the skin. These pesticides may cause skin to itch, blister, crack, or change color. Other pesticides can pass through skin and eyes and get into the body.

Acute eye effects: Some pesticides that get into the eyes can cause temporary or permanent blindness or severe irritation. Other pesticides may not irritate the eyes, but may pass through the eyes and into the body.

Allergic effects: These are the harmful effects that some people develop reacting to substances that do not cause the same reaction in most other people. The allergic effects include

- Systemic effects, such as asthma or even life-threatening shock
- Skin irritation, such as rash, blisters, or open sores
- Eye and nose irritation, such as itchy, watery eyes, and sneezing

3.1.6 Fate of Pesticides and Transfer Processes

When a pesticide is released into the environment, many changes occur, which sometimes may be beneficial. The leaching of some herbicides into the root zone can give better weed control. Occasionally, releasing pesticides into the environment can be harmful, as the entire applied chemical may not reach the target site. Runoff can displace a herbicide away from the target weeds. The chemical is wasted, weed control is reduced, and the chances of crop damage and pollution of soil and water are more. Or, some of the pesticide may drift downwind and outside of the intended application site.

Many processes affect pesticides in the environment. These processes include adsorption, transfer, breakdown, and degradation. Transfer includes processes that displace the pesticide away from the target site. These include volatilization, spray drift, runoff, leaching, absorption, and crop removal.

Adsorption is the binding of pesticides to soil particles. The amount of a pesticide adsorbed to the soil varies with the type of pesticide, soil type, moisture content, soil pH, and soil texture. Pesticides are strongly adsorbed to soils that are high in clay or organic matter. They are not as strongly adsorbed to sandy soils. Most soil-bound pesticides are less likely to give off vapors or leach through the soil. Plants also find it difficult to absorb them. Therefore, one may require the higher rate listed on the pesticide label for soils high in clay or organic matter.

Volatilization is the process of solids or liquids changing into gas, which can move away from the initial application site. This movement is called vapor drift. Vapor drift from some herbicides can damage nearby crops. Pesticides volatize most readily from sandy and wet soils. Hot, dry, or windy weather and small spray drops increase volatilization. Where recommended, incorporating the pesticide into the soil can help reduce volatilization.

Spray drift is the airborne movement of spray droplets away from a treatment site during application. Spray drift is affected by spray droplet size—the smaller the droplets, the more likely they will drift, wind speed—the stronger the wind, the more the pesticide spray will drift, and distance between the nozzle and target plant or ground—the greater the distance, the more the wind can affect the spray.

Runoff is the movement of pesticides in water over a sloping surface. The pesticides are either mixed in the water or bound to eroding soil. Runoff can also occur when water is added to a field faster than it can be absorbed into the soil. Pesticides may move with runoff as compounds dissolved in the water or attached to the soil particles. The amount of pesticide runoff depends on the slope, the texture of the soil, the soil moisture content, the amount and timing of a rain-event (irrigation or rainfall), and the type of pesticide used.

Pesticide runoff can be reduced by using minimum tillage techniques to reduce soil erosion, grading surface to reduce slopes, diking to contain runoff, and leaving border vegetation and plant cover to contain runoff.

Leaching is the movement of pesticides in water through the soil. Leaching occurs downward, upward, or sideways. The factors that influence pesticides leaching into the groundwater include characteristics of the soil as well as the pesticide used, and their interaction with water from a rain-event, such as irrigation or rainfall. Leaching can be increased when the pesticide is water soluble, the soil is sandy, a rain-event occurs shortly after spraying, and the pesticide is not strongly adsorbed to the soil.

3.1.7 BREAKDOWN PROCESSES

Degradation is the process of pesticide breakdown after application. Microbes, chemical reactions, and light or photodegradation break down pesticides. This process may take anywhere from hours or days to years, depending on the environmental conditions and the chemical characteristics of the pesticide. Pesticides that break down quickly generally do not persist in the environment or on the crop. However, pesticides that break down too rapidly may only provide short-term control.

Microbial breakdown is the breakdown of chemicals by microorganisms, such as fungi and bacteria. Microbial breakdown tends to increase when temperatures are warm, soil pH is favorable, soil moisture and oxygen are adequate, and soil fertility is good.

Chemical breakdown is the breakdown of pesticides by chemical reactions in the soil. The rate and type of chemical reactions that occur are influenced by the binding of pesticides to the soil, soil temperatures, and pH levels. Many pesticides, especially the organophosphate insecticides, break down more rapidly in alkaline soils.

Photodegradation is the breakdown of pesticides by sunlight. All pesticides are susceptible to photodegradation to some extent. The intensity and spectrum of sunlight, length of exposure, and the properties of the pesticide influence the rate of breakdown. Pesticides applied to foliage are more exposed to sunlight than pesticides that are incorporated into the soil. Pesticides may break down faster inside plastic-covered greenhouses than inside glass greenhouses, since glass filters out much of the ultraviolet light that degrades pesticides.

3.2 EMERGING TRENDS OF PESTICIDES AND REGULATIONS

There is no doubt in the fact that the extensive use of chemical biocides has helped in increasing the agricultural production and destroying a variety of pathogens, but at the same time, their consumption is raising the level of chemicals in the environment and thus, resulting in chemical pollution of the environment, which is now becoming a threat to the survival of nontarget species, including human beings. The adverse effects of the chemicals in the environment are not only limited to us, but is also passed on to the future generation by the way of genetic mutation, birth defects, inherited diseases, and so on. The pests have developed resistance to conventional pesticides and hence, may need higher doses. See also Figure 3.5.

3.2.1 CHANGING PATTERN OF CONVENTIONAL PESTICIDES CONSUMPTION

Considering the demerits of conventional pesticides and related disease management, India had shifted steadily toward environmentally safe products. In India, the average per hectare consumption of pesticides in agriculture was 1.2 g/ha in 1953–1954, which increased to 431 g/ha in 1992–1993. However, thereafter, consumption gradually declined to 288 g/ha in 1999–2000. There is a definite

Pesticides	Reduction in consumption (%)
Organochlorine	40–14.5
Carbamates	15–4.5
Synthetic pyrethroids	10–5
Organophosphate	30–74
Natural pesticides	0–2

FIGURE 3.5   Journey of pesticides (1995–2006).

trend of decline in the consumption of pesticides in India in all the states, particularly in those states that were earlier using high amounts of pesticides like Andhra Pradesh, Tamil Nadu, Karnataka, and Maharashtra. The typical example is Andhra Pradesh where pesticide consumption declined from 12,775 to 7,000 MT. In general, the decline is more conspicuous in the case of pesticides. Consumption of many organochlorine and organophosphorous pesticides and exports of food commodities having undesirable residues has been banned. The tendency is to discard the use of persistent and toxic pesticides and replace them with natural insecticides. Natural compounds have increasingly become the focus of those interested in developing new insecticides.

3.2.2 ECOFRIENDLY PESTICIDES

The various new strategies for plant protection and insect control in the development of nonconventional, ecofriendly, nonpollutant, and biodegradable pesticides are as follows:

- Insect repellents
- Insect attractants
- Juvenile hormones
- Pheromones
- Synergists
- Pesticides of plant origin

3.2.2.1 Insect Repellents

Control of insects can be achieved by ways other than causing rapid deaths. Plants produce many compounds that are insect repellents or act to alter insect feeding behavior, growth, development (molting), and behavior during mating and oviposition. Insect repellents are the groups of chemical agents that prevent damage to plants or animals by making the conditions unattractive or offensive to the pests. The first chemical repellent was discovered soon after man became acquainted with fire. In addition, plant-derived insect repellents have also been explored. Most insect repellents are volatile in nature. In some cases, the same terpenoids can repel certain undesirable insects while attracting insects that are more beneficial. Insect repellents include dimethylphthalate or pyrethrum (ingredient of mosquito repellent cream/coils), naphthalene or *p*-dichlorobenzene (ingredient of mothballs or flakes), and mercury (an ingredient of tablets for protecting stored grain). Extracts of citronella plant (*Andropogon nardus*) contain gerniol, cintronellol, borneol, and terpenes, which are considered as the principal mosquito repellents. Repellents are used in the form of ointments, pastes, solutions, emulsions, and aerosols.

Some important insect repellents commonly used are benzyl benzoate, indalone, Rutgers 612, dimethylphthalate, and butoxy polypropylene glycol.

3.2.2.2 Insect Attractants

These are the chemical agents that attract insects by olfactory stimulation. They attract the insects into traps or to poison baits. Attractant can be defined as the chemical substances whose vapors attract insects. They can be further classified into two groups: food attractants and sexual attractants. Usually, food attractants are components of food products, proteins, enzymes, and molasses. The important synthetic attractant is eugenol, which attracts flies. The 9:1 mixture of geraniol and eugenol forms an attractant for Japanese's beetle adults.

Some important insect attractants are geraniol, eugenol, trimedlure, and isoamyl salicylate.

Fermenting sugars serve as attractants for moth and butterflies. They may be supplemented by essential oils for specific requirements. Thus, anethole is used for cooling moth while isoamyl salicylate is used for tomato and tobacco hornware moths. Metaldehyde, $[CH_3CHO]_4$, is used as an attractant in poison baits for snails and slugs. Trimedlure is an attractant for melon fly.

3.2.2.3 Juvenile Hormones (JH)

Insect hormones are a group of physiologically active organic compounds, which regulate the specific development of insects from egg to the adult insect. Juvenile hormones are one of the important members of insect hormones. These are secreted by corpora allata (a part of the brain) and control complicated postembryonic developments. Juvenile hormones cause developmental disturbances during ontogenetic stage, which prevent insect reproduction. Such an arrest of reproduction is a better and timely approach. They do not have direct toxicity to insects. Normally, they do not kill the target but prevent them from reproduction. Moreover, they do not cause any danger to man and warm-blooded animals. The JH-like compounds occur in human organs and in our daily edibles like milk, cream, and bread.

The JH-like compounds form a new class of compounds characterized as insect sterilants, which cause all developmental disturbances like female sterility, ovicidal effects, adult emergence failure, and finally inability of mating. The JH along with mounting hormone (MH) maintain larval character of growing insects. They also regulate the intensity of motor activity by affecting the central nervous system. The JH-like compounds have also been reported to cause reversal of adult in larval stage in some insect species and are capable of preventing the metamorphosis of insects. In addition to their JH activity, their chemical stability and retention of activity under field conditions make them suitable as the fourth generation of insecticides. Various synthetic analogues of aliphatic and aromatic natural JH have been synthesized. A good amount of attention has been paid on structure–relativity relationship of these compounds. Both acyclic and cyclic juvabione derivatives with several structural modifications have been synthesized and explored for their bioefficacy. The JH-like compounds are commercially available in the market; juvenoid has almost double the activity of pyrethrum.

3.2.2.4 Pheromones

These are the chemicals that are secreted into the external environment by an animal, which elicit a specific reaction in a receiving individual of the same species. These are released by one sex only and trigger behavior patterns in the other sex that facilitates mating. They are produced by certain male insect species. Mostly, they serve as short-range mating stimulants after the individuals have come into proximity.

Pheromone enables pests to be controlled specifically; their application in small quantities avoid contamination of food and fodder as well as environment. They can be used to attract insects to a site where they are destroyed by treatment with insecticides.

3.2.2.5 Synergists

The term synergist may be defined as a chemical which by itself is not toxic to insects at doses used, but when combined with an insecticide, greatly enhances the toxicity of the insecticide.

Some important synergists are sulfoxide, safrole, and tropitol. They reduce the amount of pesticides required for killing insects and simultaneously reduce the risk of environmental pollution. Thus, they have reduced the use of pesticides. Of the several different groups of chemicals able to synergize insecticides, the most widely studied are methylenedioxyphenyl derivatives. Some common examples of synergists are piperonyl butoxide, safrole, sulfoxide, tropitol, and thanite.

3.2.2.6 Pesticides of Plant Origin

Use of natural plant products is now emerging as one of the prime means to protect crop and the environment from pesticidal pollution (Figure 3.6). Plant pesticides possess an array of properties, including insecticidal activity, repellency to pests and insect growth regulation, and toxicity to nematodes, mites, and other agricultural pests. Some of these indigenous resources have been in use

Acanthaceae	Cucurbitaceae	Papaveraceae
Agavaceae	Dioscoreaceae	Piperaceae
Annonaceae	Ebenaceae	Poaceae
Apocynaceae	Euphorbiaceae	Polygonaceae
Araceae	Flacourtiaceae	Polypodiaceae
Aristolochiaceae	Guttiferae	Ranunculaceae
Asclepiadaceae	Helleboraceae	Rosaceae
Balanitaceae	Hippocastanaceae	Rubiaceae
Berberidacea	Hypericaceae	Rutaceae
Boraginaceae	Labialae	Sapindaceae
Brassicacea	Lamiaceae	Sapotaceae
Burseraceae	Lauraceae	Simaroubaceae
Capparaceae	Leguminosae	Solanaceae
Capparidaceae	Liliaceae	Stemonaceae
Celastraceae	Loganiaceae	Taxaceae
Chenopodiaceae	Lycopodiaceae	Theaceae
Compositae	Magnoliaceae	Umbeliferae

FIGURE 3.6   Some plant families with insecticidal activity.

for over a century to minimize losses caused by pests and disease in agricultural production. Herbal pesticides have many advantages over synthetic pesticides. Some of them are as follows:

- Plant pesticides, in general, possess low mammalian toxicity and have least or no health hazards, and are environment-friendly.
- There is practically no risk of developing pest resistance to these products, when used in natural forms.
- Less hazards exist for nontarget organisms.
- Pest resurgence has not been reported except in synthetic pyrethrin.
- No adverse effects on plant growth, seed viability, and cooking quality of grains.
- Less expensive and easily available because of their natural occurrence, especially in oriental countries.

Owing to the prohibitive cost of synthetic pesticides and the problems of environmental pollution caused by continuous use of these chemicals, there is a renewed interest in the use of plants for crop protection. A number of agricultural entomologists, nematologists, and pathologists all over the world are now actively engaged in research to use plant products against agricultural pests and disease to minimize losses caused by them to public health, food materials, etc.

The secondary metabolic compounds of plants are a vast repository of compounds with a wide range of biological activities. This diversity is largely the result of coevolution of hundreds of thousands of plant species with each other and with an even greater number of species of microorganisms and animals. Thus, unlike compounds synthesized in the laboratory, secondary metabolites from plants are virtually guaranteed to have biological activity that is highly likely to function in protecting the plants from pathogen, herbivores, and competitors. The knowledge of the pests to which the producing plant is resistant may provide useful clues in predicting what pests may be controlled using a particular species. Empirical knowledge of plants useful for combating insect

1. Achook	16. Neemite
2. Azadit	17. Neemolin
3. Bioneem	18. Neemrich I and II
4. Echo Neem	19. Neemshield
5. Fortune AZA	20. Nimba
6. Margocide CK20 EC	21. Nimbecidine
7. Navneem 50 EC	22. Nimbosol
8. Neem A1 and A2	23. Neemgreen
9. Neemacitin	24. Neemlin
10. Neemacin	25. Pestoneem
11. NeemAzal	26. Rakshak
12. NeemPlus	27. RD-9 Repelin
13. Neemark	28. Soluneem
14. NeemGold	29. Unim
15. Neemguard	30. Vepacide I and II

FIGURE 3.7 Commercially available neem-based pesticides.

pests has accumulated over the millennia in different cultures in different parts of the world. For example, Indian neem tree (*Azadirachta indica* A. Juss; family: Meliaceae) leaves have been used in India and the neighboring countries for the control of insect infestation in food grain, clothing, etc., since Vedic times (1500 BC) (Figure 3.7). The dried seed powder of custard apple (*Annona squamosa* L; family: Annonaceae) has been used as an insecticide in India and many other tropical countries since ancient times.

Nicotine from tobacco plant (*Nicotiana tabacum* L; family: Solanaceae) was first isolated in 1828, and its aqueous extract for plant protection has been used in Europe since the sixteenth century. Nicotine and nicotine phosphate have been marketed since 1901. It is effective against a wide range of insects and kills them through feeding, contact, and fumigation. In the early nineteenth century, 2200 tons of the alkaloid was used as a crop protection agent.

Another important insecticide from plant is pyrethrum (*Chrysanthemum cinerariaefolium* Vis., *C. coccineum* Wild; family: Compositae). It is the most important insecticide and is commercially cultivated on a large scale in several countries, including Kenya, Tanzania, Ecuador, Brazil, erstwhile USSR, Japan, and India. The insecticidal properties of the several *Chrysanthemum* species were known for centuries in Asia. Even today, powders of the dried flowers of these plants are sold as insecticides. Products based on pyrethrum are essentially used in indoor application and are especially valued for their rapid knockdown (of flying insects) effect, photoability, and low mammalian toxicity. Formulated preparation contains synergists, such as piperonyl butoxide, which enhance their insecticidal activity.

Various plants of genus *Derris, Lonchocarpus, Milletia, Mundulea, Tephrosia* (family: Leguminosae) possess rotenone, a potent insecticide. Rotenone was first isolated in 1902 from *Derris chinensis* Benth. In 1930s, preparations of roots from the genera *Derris, Lonchocarpus,* and *Tephrosia* containing rotenone were used as commercial insecticides.

Ryania (*Ryania speciosa* Vahl; family: Flacourtiaceae) consists of powdered roots and stems of the South American plant. It is used as a commercial insecticide against European corn borer. Sabadilla consists of powdered barley-like seeds of the South and Central American plant (*Schoenocaulon officinale* Gray; family: Liliaceae), which the local people have been using for a long time to control insect infestation.

There are many other plant species identified that contain insecticidal principles and are used at one time or the other by different people in different regions of the world. With the introduction of modern scientific methods of research, the knowledge of insecticidal plants has expanded vastly. Natural compounds have increasingly become the focus of those interested in developing new pesticides. Over 2000 plant species belonging to several families are now known to possess insecticidal properties.

Tens of thousands of secondary products of plants have been identified and there are estimates that hundreds of thousands of these compounds may exist. Plants are known to produce a wide range of secondary metabolites such as alkaloids, terpenoids, polyacetylenes, flavonoids, quinones, phenylpropanoids, and amino acids. There is growing evidence that most of these compounds are involved in the interaction of plants with other species, primarily in the defense of the plant from plant pests. Thus, these secondary compounds represent a large reservoir of chemical structures with biological activity. This resource is largely untapped for use as pesticides.

Most commercially successful pesticides have been discovered by screening compounds synthesized in the laboratory for pesticidal properties. The average number of compounds that must be screened to discover a commercially viable pesticide has increased dramatically.

Being environmentally benign, phytochemicals are ideally suited for incorporation in integrated pest management's programs, particularly in developing countries, which are rich in plant biodiversity and many phytochemical biopesticides can be developed from indigenous plants. This approach has led to the discovery of several commercial pesticides derived from plants.

3.2.3 Combinatorial Approach—A Bright Future

There are problems associated with plant-produced phytotoxins as potential insecticides. Various bioactive principles have been isolated from various plant species, but only a few have reached to the level of commercialization. The most common problem is that the amount of purification initially conducted is a variable for which there is no general rule. The isolation of bioactive principle depends on the particular growth stage of a plant. The methods used for isolation of an active molecule are lengthy, time-consuming, and dependent on the source. A small amount of the compound is generally isolated and is generally weak in comparison with the synthetic chemicals. Collection of natural sources, storage, production, formulation, and application (spray) is a tedious work. Plant-based pesticides show slow rate of efficacy when compared with the conventional chemical pesticides. The biopesticides are to be highly specific because certain stages of the life cycle of the pest may almost be immune. Most of the pesticides require damp climate, initial spray distribution, and movement of their hosts for dissemination. In case of few natural products, structural identification might prove to be a difficult task. The major drawback of biopesticides is limited phytochemical stability, i.e., temperature, UV or sunlight, rainfall, humidity, and other environmental factors readily affect them. They need various stabilizers like antioxidants, ascorbic acid, retinoic acid, and flavonoids to obtain the desired persistence. When compared with synthetic pesticides, molecular complexity, limited environmental stability, low activity of plant-based pesticides are discouraging. It appears that none of the approaches is suitable in a larger sense. The best choice is a combination, i.e., the use of semisynthetics. Plant-derived chemicals are recently used as model compounds for development of semisynthetic insecticides. Plants contain a virtually untapped reservoir of pesticides that can be template for synthetic pesticides. Structural manipulations in the bioactive principles derived from plant-based pesticides are supposed to improve the activity and toxicological properties with no or less deleterious environmental effects. Considering the demerits of synthetic and plant-base pesticides, it might be better to prepare synthetic analog of the biologically active principle derived from the plant. Semisynthetic pesticides can prove to be an excellent alternative approach to synthetic pesticides having favorable toxicological and environmental expectations.

Physostigmine, a carbamate alkaloid, isolated from calabar beans (*Physostigma venenosum* Balf.) has been recognized for its insecticidal (anticholinesterase) activity. Structural modifications have been made on the bioactive principle of this plant-derived carbamate, which resulted in synthetic analog isolan. Isolan was found to be nonpersistent and nontoxic to human beings. At present, 25 different products are in the market and their worldwide production is about 35,000 tons per year.

Nereistoxin is a biocide metabolite of the annelid (shell fish) *Lumbriconereis heteropoda*. Marenzeller has also served as the starting skeleton for developing synthetic insecticide. Bis-thiocarbamate (cartap) is a synthetically tailored molecule of nereistoxin. It is found to be a highly effective insecticidal agent against rice-leaf beetle and nontoxic to human beings.

Pyrethrin, a plant-derived insecticide, is found to be photolabile. Its use was prohibited because of its undesirable half-life in the open environments. To make it photostable, safe, and nonpersistent in the environment, synthetic variations have been carried out and two classes of pyrethrin analogs have been synthesized. One class includes bioallethrin, bioresmethrin, and biophenothrin. In this class of compounds, the acid component, i.e., chrysanthemic acid of pyrethrin, has been retained and only the alcohol moiety has been changed. These are recommended for household, public health, industrial, and certain agricultural uses. In a second class of compounds, both the acid and alcohol moieties differ from those present in natural compounds. In deltamethrin and cypermethrin, halogen atoms have replaced two vinyl methyls. They are registered for agricultural applications and about 18 compounds of this class are in the market. Cypermethrin is a mixture of eight isomers and active component constitutes only 18%–25%. Deltamethrin is a single chemical entity and is the most important pyrethroid to date.

This combinational approach seems to be a time-tested strategy for the development of biologically active synthetic analogs from a naturally occurring biologically active molecule. In this endeavor, natural products having biological activities have served well. However, sincere efforts toward the research on the above lines are required.

3.3 PEST MANAGEMENT REGULATORY AGENCY

3.3.1 AN OVERVIEW

Pesticides are carefully regulated through a program of premarket scientific assessment, enforcement, education, and information dissemination. These activities are shared among federal, provincial, territorial, or municipal governments, and are governed by various acts, regulations, guidelines, directives, and bylaws. Although it is a complex process, regulators at all levels work together toward the common goal—helping to protect citizens from any risks posed by pesticides and ensuring that pest control products do what they claim to do on the label. Fungicides, herbicides, insecticides, and antimicrobials are to be identified for possible human health effects of pesticides and established for the levels at which humans can be exposed to the products without any harm. Studies assessed include short- and long-term toxicity, carcinogenicity (the capacity to cause cancer), genotoxicity (the capacity to cause damage to chromosomes), and teratogenicity (the capacity to produce fetal malformations). The toxicology sections are responsible for setting ADIs, the amount of a compound that can be consumed daily for a lifetime with no adverse effects. ADIs always have safety factors built in, ranging from 100 to 1000. These safety factors are designed to take into account the potential differences in response, both within the same species (adults vs. children) and between species (animals vs. humans). These assessments take into account the different exposures that people could have to pesticides, such as those who work with the pesticides (formulators, applicators, and farmers) and bystanders (people working or living near where a pesticide is used). They also take into consideration the differing exposures that adults and children would have. Exposure data considered include residues found in air and on surfaces indoors and outdoors following application in domestic, commercial, and agricultural situations. Along with the exposure estimates, the end points identified in toxicity studies are also considered during the risk assessment. Routes and duration of exposure, and the species tested in toxicity studies are also considered. Assessments of the effectiveness of personal protective equipment are often performed, and wearing such equipment during handling of the product may be required as a condition of registration.

The food residue exposure assessment is conducted to set the MRLs for pesticides on food, both domestic and imported. Dietary risk assessments are also carried out to assess the potential daily intake of pesticide residues from all possible food sources. Dietary risk assessments take into account the different eating patterns of infants, toddlers, children, adolescents, and adults and so include a detailed evaluation of the foods and drinks that infants and children consume in quantity, such as fruits and fruit juices, milk and soy products. To address the environmental concerns that may arise from the intended use of a product, authorities also make recommendations for restrictions on use that would lessen the risk. This could include label statements outlining buffer zones, timing and frequency of the applications, and the rate at which the product can be applied. Only if there is sufficient scientific evidence to show that a product does not pose unacceptable health or environmental risks and that it serves a useful purpose, will a decision to register be made. A registration is normally granted for a term of 5 years, subjected to renewal.

Environment Protection Agency (EPA) regulates pesticides under three laws:

Federal Food, Drug, and Cosmetic Act (FFDCA): The FFDCA is the law under which EPA sets tolerances for pesticides. The EPA can essentially ban a pesticide by setting a tolerance—the amount of pesticide residue that is allowed to legally remain on food. The U.S. Department of Agriculture's (USDA) Agricultural Marketing Service is responsible for monitoring residue levels in or on food.

Federal Insecticide, Fungicide, and Rodenticide Act (FIFRA): To sell a pesticide, a company must also register it with EPA under FIFRA. For pesticides used on food, EPA can only register uses for pesticides that have a tolerance. Pesticide registrants must register and gain EPA approval of their products as well as for each specific use (i.e., use as a bug spray indoors is one registration and use for a specific crop outdoors is another). The EPA must review registered pesticides on a 15 year cycle.

Food Quality Protection Act: The FQPA amended the first two laws. According to one National Research Council report, "The great majority of individual naturally occurring and synthetic chemicals in the diet appear to be present at levels below which any significant adverse biological effect is likely, and so low that they are unlikely to pose any appreciable cancer risk."

3.3.2 Pesticides Regulation in India

In India, the production and use of pesticides are regulated by a few laws, which mainly lay down the institutional mechanisms. In addition to procedures for registration, licensing, quality regulation, etc., these laws also try to lay down standards in the form of MRLs, and average daily intake levels. Through these mechanisms, chemicals are sought to be introduced into farmers' fields and agricultural crop production without jeopardizing the environment or consumer health. These legislations are governed and administered by different ministries—the regulatory regime and its enforcement have several lacunae stemming from such an arrangement. An added dimension is that administration of the legislations includes both state governments and the central government.

The Central Insecticides Act 1968 is meant to regulate the import, manufacture, storage, transport, distribution, and use of pesticides with a view to prevent risk to human beings, animals, and the environment. Through this Act, a Central Insecticide Board has been set up to advice the state and central governments on technical matters and for including insecticides into the Schedule of the Act. Around 625 pesticides have been included in the Schedule thus far. The Board is supposed to specify the classification of insecticides on the basis of their toxicity, their suitability for aerial application, to advise the tolerance limits for insecticide residues, to establish minimum intervals between applications of insecticides, specify the shelf-life of various insecticides. Then, there is a Registration Committee, which registers each pesticide in the country after scrutinizing their formulae and claims made by the applicant as regards its efficacy and safety to human beings and animals. The Registration Committee is also expected to specify the precautions to be taken against poisoning through the use or handling of insecticides. Around 181 pesticides have been registered by the Committee so far in India. Then, there are other institutions like Central Insecticides Laboratory and Insecticides Inspectors to ensure that the quality of insecticides sold in the market is as per norms. The Central Insecticides Laboratory is also meant to analyze samples of materials for pesticide residues as well as to determine the efficacy and toxicity of insecticides. This laboratory is also responsible for ensuring the conditions of registration. As per this legislation, the central government will register the pesticides, whereas the marketing licenses are allowed by state governments. Pesticides and their contamination of food products are sought to be regulated through some concepts like MRLs, average daily intake (ADIs), and good agricultural practices (GAPs).

MRL is the maximum concentration of a pesticide residue resulting from the use of a pesticide according to GAP. It is the limit that is legally permitted or recognized as acceptable in or on a food, agricultural commodity, or animal food. The concentration is expressed in milligrams of pesticide residue per kilogram of the commodity. These data are evaluated and the no-observed-adverse-effect level (NOAEL) is calculated from the chronic toxicity studies. In case of toxic pesticides, acute reference dose is also taken into consideration. This NOAEL and acute reference dose are supposed to be taken as the starting information for prescribing the tolerance limits of pesticides in food commodities. The NOAEL is usually referred to in terms of milligrams of that particular pesticide per kilogram of body weight. From this NOAEL, the ADI is calculated by dividing the figure normally with a safety factor of 100. The figure 100 is taken into consideration as a multiple of 10 (10×10), where the first 10 provides for interspecies variation, while the second 10 provides for intraspecies variation. Therefore, ADI, which is expressed in terms of milligrams per kilogram of body weight, is an indication of the fact that if a human being consumes that amount of pesticide everyday, throughout his lifetime, it will not cause appreciable health risk on the basis of well-known facts at the time of the evaluation of that particular pesticide.

3.3.3 GENUINE CONCERNS: INDIAN PERSPECTIVES

Ministry of Agriculture (MOA) regulates the manufacture, sale, import, export, and use of pesticides through the "Insecticide Act, 1968." There is a clear conflict of interest in this arrangement. The MOA, which is supposed to promote pesticides to increase food production, has also been assigned the task of regulating pesticides. Agricultural scientists are generally not health specialists, because the health impacts of pesticides are more invisible than visible. Acts and Rules are primarily geared toward regulating the acute health effects of pesticides. The focus on the chronic health effects is highly inadequate—the result is poor scrutiny of pesticides from chronic toxicity point of view. Terms like chronic toxicity or ADI are missing from the entire act. New pesticides can be registered and used for 2 years without considering any health and safety consideration. No data are required for the neurotoxicity, teratogenicity, effect on reproduction, carcinogenicity, metabolism, mutagenicity, and health records of industrial workers. However, neither fixing of ADI nor setting of MRLs on food commodities is a part of the registration process. Prevention of Food Adulteration Act, Ministry of Health, monitors and regulates pesticide contamination in food commodities and sets MRLs of pesticides on food commodities. Until 2004, 181 pesticides registered, MRLs for only 71 were notified under the PFA. Even today, out of the 194 pesticides registered, MRLs for only 121 have been notified under the PFA. Pesticides are registered for use on "Y" number of crops, but MRLs are set only for "$Y-X$" number of crops.

3.4 CONCLUSION

There is a necessity to rework on our systems. In India and other countries, we will have to completely rework on the existing regulatory mechanism for registering and using pesticides. The FAO's International Code of Conduct on distribution and use of pesticide is a good starting point and is the least that must done. We will have to revise our standards to make sure ADI is not exceeded. Therefore, we need to make sure that the standards are enforced. Information is made available to the public and the entire process is transparent. Slowly, the world is moving beyond finding linkages between pesticides and the diseases they cause, as it is not important any more. It is understood that these toxins will have implications, even if we cannot prove it by scientific means. What is more important is to know how much and how many of these chemicals are trespassing human bodies. The new idea in regulation is to use biomonitoring studies to regulate chemicals. Can the following be introduced in our country?

- Safe and wise use of policy for pesticides
- Scientific standard setting—including ADI
- Harmonization between registration and MRLs
- Reregistration to consider new scientific data in decision making
- Comparative risk assessment methodology before introducing new pesticides
- Transparency and accountability in registration
- Better surveillance and enforcement—not only for food but also for body burden
- Public disclosure of monitoring data and use of data for regulation—ban problematic pesticide
- Global product assessment and liability convention

FURTHER READINGS

Agnihotri NP. Pesticide consumption in agriculture in India—An update. *Pest Res J* 12(1): 150–155, 2000.
Ames BN and Gold LS. Environmental pollution, pesticides, and the prevention of cancer: Misconceptions. *FASEB J* 11: 1041–52, 1997.
Arnason JT, Philogene BJR, and Morand P (eds.). *Insecticides of Plant Origin*, American Chemical Society, Washington, DC, pp. 11–24, 1989.
Balandrin MF, Klocke JA, Wurtele S, and Bollinger WH. Natural plant chemicals: Sources of Indus. *Med Mater Sci* 228: 1154–1160, 1985.
Bhatnagar VK. Pesticide pollution—Trends and perspective, *ICMR Bull* 31(9), 85, 2001.
Brandson C and Hommannn K (eds.). *The Cost of Invocation: Valuing the Economy-Wide Cost of Environmental Degradation in India*, Asia Environment division, World Bank, Mimeo, 1995.
Chandra B (ed.). *Regulation of Pesticides in India*, Centre for Science and Environment, New Delhi, 2006.
Chin HB (ed.). *The Effect of Processing on Residues in Foods. Pesticide Residues and Food Safety: A Harvest of Viewpoints*, American Chemical Society, Washington, DC, 1991.
Dethe MD, Kale VD, and Rane SD. Pesticide residues in/on farmgate samples of vegetables. *Pest Manage Hort Ecosyst* 1(1): 49–53, 1995.
Duke SO. Naturally occurring chemical compounds as herbicides. *Rev Weed Sci* 2: 15–44, 1986.
Grainage M and Ahmed S. *Handbook of Plants with Pest-Control Properties*, Wiley-Interscience, New York, 1988.
Green MB, Hartley GB, and TF West (eds.). *Chemicals for Crop Improvement and Pest Management*, Pergamon, New York, 1987.
Griessbuhler H (ed.). *Advances in Pesticide Science*, Pergamon, Oxford, 1, 45–53, 1979.
Jacobson M and Crosby DG (eds.). *Naturally Occurring Insecticides*, Marcel Dekker, New York, 1971.
Kaphalia BS, Farida S, Siddiqui S, and Seth TD. Contamination levels in different food items & dietary intake of organ chlorine pesticide residues in India. *Ind J Med Res* 81: 71–78, 1985.
Kashyap R, Iyer LR, and Singh MM. Evaluation of daily dietary intake of dichlorodiphenyltrichloroethane (DDT) and benzenehexachloride (BHC) in India. *Arch Environ Health* 49: 63, 1994.
Klocke JA (ed.). Natural plant compounds useful in insect control. *Am Chem Soc* 330: 396–415, 1987.

Kranthi KR, Jadhav DR, Wanjari RR, Ali SS, and Russell D. Carbamate and organophosphate resistance in cotton pests. *Bull Entom Res* 91: 37–46, 2001.

Kranthi KR, Jadhav DR, Kranthi S, Wanjari RR, Ali SS, and Russell DA. Insecticide resistance in five major insect pests of cotton in India. *Crop Protec* 21: 449–460, 2002.

Krishnamoorthy SV and Regupathy A. Monitoring of HCH and DDT residues in groundnut and sesamum oils. *Pest Res J* 2: 145, 1990.

Kudesia VP and Charya MU (eds.). *Pesticide Pollution*, Pragati Prakashan, India, 1993.

Kumari B, Kumar R, Madan VK, Singh R, Singh J, and Kathpal TS. Magnitude of pesticidal contamination in winter vegetables from Hisar, Haryana. *Environ Monit Assess* 87(3): 311–318, 2003.

Lederberg J, Shope RE, and Oaks SC (eds.). *Emerging Infections: Microbial Threats to Health in the United States*, National Academy Press, Washington, DC, 1992.

Lydon J and Duke SO. Herbicide from natural compounds. *Weed Tech* 1: 122–128, 1987.

Lydon J and Duke SO. Pesticide effects on secondary metabolism of higher Plants. *Pestic Sci* 25: 361–373, 1989.

McLaren JS. Biologically active substances from higher plants: status and future potential. *Pestic Sci* 17: 559–578, 1986.

Metcalf RL and Mckelvey JJ (eds.). *The Future for Insecticides: Needs and Prospects*, Wiley Interscience, New York, 1976.

Miyamoto J and Kearney PC (eds.). *Pesticide Chemistry: Human Welfare and the Environment*, Pergamon, Oxford, 253, 1983.

Mohapatra SP, Gabhiye VT, Agnihotri NP, and Raina M. Insecticide pollution of Indian rivers. *Environmentalist* 15(14): 41–44, 1995.

Mukherjee D, Roy BR, Chakraborty J, and Ghosh BN. Pesticide residues in human foods. *Ind J Med Res* 72: 577–582, 1980.

Nair A and Pillai MKK. Monitoring of hexachlorobenzene residues in Delhi and Faridabad, India. *Bull Environ Contam Toxicol* 42: 682, 1989.

Rao GMVP, Rao NH, and Raju K. Insecticide resistance in field populations of American bollworm, *Helicoverpa armigera* Hub. (Lepidoptera: Noctuidae). *Resist Pest Manag Newslett* 15(1): 15–17, 2005.

Repetto R and Baliga S (eds.). *Pesticides and the Immune System: The Public Health Risks*, World Resources Institute, Washington DC, 1996.

Robert R. Pesticides and public health: Integrated methods of mosquito management. *Emerg Infect Dis* 7(1): 17–23, 2001.

Sasi KS and Sanghi R. Analyzing pesticide residues in winter vegetables from Kanpur. *Ind J Environ Health* 43: 154–158, 2001.

Swaminathan MS and Kochhar SL (eds.). *Plants and Society*, Macmillan, London, 350–366, 1989.

Yang RZ and Tang CS. Plants used for pest control in china: A literature review. *Econ Bot* 42: 376–406, 1988.

Zahm SH and Blair A (eds.). *Cancer Research*, World Resources Institute, Washington DC, 1992.

4 Scope and Limitations of Neem Products and Other Botanicals in Plant Protection: A Perspective

K. K. Singh, Alka Tomar, and Hamir Singh Rathore

CONTENTS

4.1   INTRODUCTION

It is estimated that one-third of the world's agricultural production is destroyed by 20,000 species of field and storage pests. Synthetic pesticides are widely used now due to their effectiveness, relatively long shelf life, and ease with which they can be transported, stored, and sprayed. However, they cause serious problems. The first problem is toxicity. It is estimated that three million cases of pesticide poisoning happen each year, 20,000 of which prove to be fatal. Pollution of soil, water, and air, health hazards to human beings, livestock, and in fact pollution of the environment constitute the second major problem. Development of pest resistance and resurgence, necessitating the use larger and repeated doses, escalating the cost to farmers, and environmental hazards, is the third biggest problem [1].

The control of pests and diseases is, thus, a problem as old as agriculture. Until 1945, the weapons used against pests were mainly based on metallic salts, mercury or sulfur, a few synthetic substances, and a limited number of natural products, such as nicotine, rotenones, rianoids, quassinoids, and pyrethroids. Not all these substances were particularly efficient in protecting the cultivated plants from the attack of fungi and insects [2].

Prior to the discovery of organochlorine and organophosphate insecticides in the late 1930s and early 1940s, botanical insecticides were important products for pest management in industrialized countries. In spite of the wide recognition that many plants possess insecticidal properties, only a handful of pest control products directly obtained from plants, that is, botanical insecticides, are in use in developed countries [3].

Ideally, insecticides should reduce pest populations, be target-specific (kill the pest but not other organisms), break down quickly, and have low toxicity to humans and other mammals. Although synthetic insecticides (e.g., chlorinated hydrocarbons, organophosphates, and pyrethroids) have been an important part of pest management for many years, the disadvantages and risks in using them have become apparent. Some synthetic insecticides leave unwanted residues in food, water, and the environment. Some are suspected carcinogens, and low doses of many insecticides are toxic to mammals. Natural products are preferred because of their innate biodegradability [4]. As a result, many people are looking for less hazardous alternatives to conventional insecticides [5].

Some alternatives include less-toxic or natural products, such as botanicals. Botanicals are insect toxins that are derived or extracted from plants or plants parts. Many botanical insecticides have been known and used for hundreds of years, but they were displaced from the marketplace by synthetic insecticides in the 1950s. These old products, and some newer, plant-derived products, deserve consideration for use in pest control. Botanical insecticides have different chemical components and modes of action.

Plants have long been a source of insecticides and continue to be evaluated. The plants that have shown potential in pest control during the past few years are neem (*Azadirachta indica* A. Juss), chinaberry (*Melia toosendn* L. and *M. azedarach* L.), West Indian Mahogany (*Swietenia mahagoni* Jacq.), custard apple (*Annona squamosa* L.), French marigold (*Tagetes patula* L.), and thunder-god-vine (*Tripterygium wilfordii* Hook). Currently two plants in particular provide botanical products that can be useful for control of certain landscape plant pests—pyrethrum (from *Chrysanthemum cinerariaefolium*) and neem (from *A. indica* A Juss.). Others include nicotine (from *Nicotiana tabacum*), quassin (from *Quassia amara* and *Picrasma excelsa*), ryania (from *Ryania speciosa*), and rotenone (from *Derris elliptica* and *Lonchocarpus* sp.) [6].

In this chapter, botanical plant materials for plant protection are described in detail. The situation for India is depicted and may differ for other countries.

4.2   HISTORY OF BOTANICAL PESTICIDES

At least 2000 plants are known to have pesticidal activity. Neem is not the first botanical pesticide. Pyrethrins, which are naturally derived from daisy-like flowers of certain species of *Chrysanthemum*, have been used for centuries. Almost 2000 years ago the Chinese knew that *Chrysanthemum* plants had insecticidal value; some 2400 years ago the Persians used them. Not until recent centuries, however, were the potentials of the pyrethrums, extracted from the flowers, fully appreciated. Supposedly, an American trader, who had learned the secret while traveling in the Caucasus, introduced the insecticide into Europe early in the nineteenth century. Last century, Yugoslavia became the center of the world's pyrethrum industry, but after World War I, Japan became the main producer. With supplies cut off during World War II, the Allies began producing the flowers in Kenya. Since the 1960s, pyrethrum production has been established in the New Guinea highlands as well [7].

Like neem products, pyrethrins are valued for their low toxicity to mammals and birds. However, the ingredients in these insecticidal chrysanthemums are lethal to insects in a different way from those in neem. They are nerve poisons and contact insecticides. Pyrethrum has quick knockdown properties and is the active ingredient in millions of aerosol spray cans people use against flies and mosquitoes (Table 4.1).

TABLE 4.1
Summary of Botanical Pesticides and Toxicity to Mammals and Insect Pests

SN	Chemical Name/ Trade Name	Plant Botanical Name	Oral LD_{50}[a]	Formulation Used	Mode of Action	Target Organisms/Insect or Disease Pathogen
1	Pyrethrins	*Chrysanthemum cinerariaefolium* (flowers)	1,200–1,500	Dust, spray and aerosol "bombs"	Contact activity, paralytic, and knockdown	Broad range of pests including ants, aphids, roaches, fleas, flies, ticks. Exposed caterpillars, sawfly larvae, leaf beetles, leaf hoppers, white fly, exposed thrips, mites, powdery mildew, rust, leaf roller
2	Neem	*Azadirachta indica*	10,000–13,000	Oil or extract products (azadirachtin), rose defense (neem oil)	Insect growth regulators (IGRs), repellent	Sucking and chewing insects, immature insects, powdery mildew, aphids, leaf beetles: adults, black spot, caterpillars, lace bug, leaf miner, mealy bugs, spider mites, thrips, and whitefly
3	Rotenone	*Derris* sp. and *Lonchocarpus* sp. (root)	60–1,500	1% dust or a 5% powder for spraying	Inhibits cellular respiration (broad-spectrum contact and stomach poison)	Aphids, certain beetles (asparagus beetle, bean leaf beetle, Colorado potato beetle, cucumber beetle, flea beetle, strawberry leaf beetle, and others) and caterpillars and predatory mites, slugs
4	Sabadilla (Red Devil)®	*Schoenocaulon officinale* (Tropical lily)	4,000	Baits, dusts or sprays, wettable powder	Contact and stomach activity	Bugs (Hemiptera: squash bugs, harlequin bugs, thrips, caterpillars, leaf hoppers and stink bugs, etc.), adults, and immatures (nymphs), beetles: larvae adults (bean, Japanese, tortoise, etc.), leaf miners, aphids, leaf beetles, adults
5	Ryania	*Ryania speciosa*	750–1,200	—	Slow-acting stomach poison	Cooling moths, leaf hoppers, caterpillars, true bugs (hemiptera—including stink bugs, lygus bugs, etc.), thrips
6	Nicotine (Black-Leaf 40)®	*Nicotiana tabacum*	50–60	40% liquid concentrate of nicotine sulfate	Broad-spectrum contact and stomach poison, fast-acting nerve toxin	Insects with piercing–sucking mouth parts, aphids, whiteflies, leaf hoppers, thrips, and mites

Sources: Modified from Henn, T. and Weinzierl, W., *Alternatives in Insect Management: Botanical Insecticides and Insecticidal Soaps*, University of Illinois CES, Circular 1296, p. 6, 1989; Ellis, B. and Bradley, F., *The Organic Gardener's Handbook of Natural Insect and Disease Control*, Rodale Press, Emmaus, PA, 1996; Stoll, G., *Natural Protection in the Tropics*, Margraf Verlag, Weikersheim, Germany, 2000.

[a] LD_{50} is the median lethal dose (mg) of toxicant per kilogram of body weight of the test animal, which kills 50% of the test animals. A low LD_{50} indicates a more toxic substance.

Despite the development of many synthetic insecticides, this chemical from chrysanthemums has maintained its position as a major commercial product. Its production is more than 10,000 ton world over. Although powerful synthetic analogues have been developed, demand for the natural material has remained high, and for the past several years it has been in short supply.

The increasing concern of man for the environment seems likely to result in a rising demand for pesticides from plants rather than from petroleum. Such "soft" pesticides represent the hope that agricultural pests can be controlled while maintaining environmental stability.

Now neem, another botanical pesticide, can perhaps step up to take an equally important, but complementary, role in the rising soft market.

4.3 TYPES OF BOTANICAL PESTICIDES

4.3.1 PYRETHRINS

Pyrethrin is an extract of the dried flowers of the pyrethrum, *C. cinerariaefolium*, commercially grown in Kenya. The active ingredients contained in the plant are various compounds known as pyrethrins. The word "pyrethrum" is the name for the crude flower dust itself, and the term "pyrethrins" refers to the six related insecticidal compounds that occur naturally in the crude material. Pyrethrins have a rapid "knockdown" effect on many insects and are irritating, which has caused them to be used for such purposes as wasp and hornet sprays, household aerosols, or for flushing cockroaches.

Most insects are highly susceptible to low concentrations of pyrethrins. The toxins cause immediate knockdown or paralysis on contact, but insects often metabolize them and recover. Pyrethrins break down quickly and have a short residual and low mammalian toxicity, making them among the safest insecticides in use. Pyrethrins may be used against a broad range of pests including ants, aphids, roaches, fleas, flies, and ticks. They are available in dusts, sprays, and aerosol "bombs." Pyrethrins have very low toxicity to mammals and are rapidly broken down when exposed to light. As a result, certain pyrethrins are the only insecticides registered for use in food handling areas. Pyrethrins are widely labeled for use on most food crops as well.

Labels for pyrethrins list many insects. However, with respect to their use on shade trees and shrubs, they are probably most useful for control of exposed caterpillars, sawfly larvae, leaf beetles, and leafhoppers. Their short persistence can limit effectiveness, yet it also helps minimize impacts on natural enemies.

4.3.2 NEEM

Neem insecticides are extracted from the seeds of the neem tree, *A. indica*, that grows in arid tropical and sub-tropical regions on several continents. This plant has long been used in Africa and Southern Asia as a source of pharmaceuticals, such as wound dressings and toothpaste. The active ingredient is used both as feeding deterrent and a growth regulator. The treated insect usually cannot molt to its next life stage and dies. It acts as a repellent when applied to a plant and does not produce a quick knockdown and kill. It has low mammalian toxicity and does not cause skin irritation in most formulations. The neem tree supplies at least two compounds with insecticidal activity (azadirachtin and salanin) and other unknown compounds with fungicidal activity. More recently its ability to control insects has been developed. Products include ornazin, AZA-Direct, and Azatin.

Neem seed extracts contain oils and a variety of compounds that can affect insect development. Most important is azadirachtin, which has various effects from inhibition of feeding, interference with molting or egg production, and disruption of hormones important in growth. Treated insects rarely show immediate symptoms, and death may be delayed a week or longer, usually occurring during a molt. Affected insects are often sluggish and feed little.

The low toxicity and broad labeling of neem insecticides recommend its use. Furthermore, effects on beneficial species are minimal. Slow action and a limited range of susceptible insect species are the primary limitations of neem insecticides.

4.3.3 ROTENONE

Rotenone occurs in the roots of two tropical legume *Lonchocarpus* species in South America, *Derris* species in Asia, and several other related tropical legumes. Insects quickly stop feeding, and death occurs several hours to a few days after exposure. Rotenone degrades rapidly when exposed to air and sunlight. It is not phytotoxic, but it is extremely toxic to fish and moderately toxic to mammals. It may be mixed with pyrethrins or piperonyl butoxide to improve its effectiveness.

Rotenone is a broad-spectrum contact and stomach poison that is effective against leaf-feeding insects, such as aphids, certain beetles (asparagus beetle, bean leaf beetle, Colorado potato beetle, cucumber beetle, strawberry leaf beetle, and others), and caterpillars, as well as fleas and lice on animals. It is commonly sold as a 1% dust or a 5% powder for spraying.

4.3.4 SABADILLA

Sabadilla is derived from the ripe seeds of *Schoenocaulon officinale*, a tropical lily plant that grows in Central and South America. The alkaloids in Sabadilla affect insect nerve cells, causing loss of nerve function, paralysis, and death. The dust formulation of Sabadilla is the least toxic of all registered botanical insecticides. However, pure extracts are very toxic if swallowed or absorbed through the skin and mucous membranes. It breaks down rapidly in sunlight and air, leaving no harmful residues.

Sabadilla is a broad-spectrum contact poison but has some activity as a stomach poison. It is commonly used in organic fruit and vegetable production against squash bugs, harlequin bugs, thrips, caterpillars, leafhoppers, and stink bugs. It is highly toxic to honey bees, however, and should only be used in the evening, after bees have returned to their hives. Formulations include baits, dusts, or sprays [8].

4.3.5 RYANIA

Ryania is extracted from stems of a woody South American plant, *R. speciosa*, and causes insects to stop feeding soon after ingestion. It works well in hot weather. Ryania is moderate in acute or chronic oral toxicity in mammals. It is generally not harmful to most natural enemies but may be toxic to certain predatory mites. Ryania has longer residual activity than most other botanicals.

It is used commercially in fruit and vegetable production against caterpillars (European corn borer, corn earworm, and others) and thrips. Its effectiveness may be enhanced if mixed with rotenone and pyrethrin.

4.3.6 NICOTINE

It is a simple alkaloid derived from tobacco, *N. tabacum*, and other *Nicotiana* species. It is a fast-acting nerve toxin and is highly toxic to mammals. Insecticidal formulations generally contain nicotine in the form of 40% nicotine sulfate, which is diluted in water and applied as a spray. Dusts can irritate skin and are normally not available for garden use. Nicotine is used primarily for insects with piercing–sucking mouth parts such as aphids, whiteflies, leafhoppers, thrips, and mites. Nicotine is more effective when applied during warm weather. It was registered for use on a wide range of vegetables and fruit crops but is no longer registered commercially [11].

Yet there are many plant materials in a country such as India whose pesticidal properties are not well exploited. Although many pure chemical compounds were isolated from these plants, entomological and field studies are lacking. The list of such plants include *A. squamosa* (seeds, leaves), *Tephrosia purpurea* (pods, roots), *T. villosa* (pods, roots), *Pongamia pinnata* (leaves and seeds), *Lantana camara* (leaves, stems, and flowers), *Ocimum sanctum* (holy basil, leaves, whole plant), *Vitex negundo* (leaves), *Zingiber officinale* (rhizome), *Curcuma longa* (rhizome), *Allium sativum* (garlic bulbs), and *A. cepa* (onion bulbs). In addition, some botanical insecticides that had enjoyed use in North America and Western Europe have lost their regulatory status as approved products. These include nicotine (from *N. tabacum*), quassin (from *Q. amara* and *P. excelsa*), and ryania (from *R. speciosa*). As a consequence, the only botanicals in wide use in North America and Europe are pyrethrum (from *C. cinerariaefolium*) and rotenone (from *Derris* spp. and *Lonchocarpus* spp.), although neem (*A. indica* A. Juss) is approved for use in the United States and regulatory approval is pending in Canada and Germany.

Botanical pesticides do not produce knockdown effect on insects. The insect population get reduced after spraying. Promising results will be obtained if botanical pesticides are sprayed as prophylactic agents.

4.4 OILSEEDS INSECT-PESTS MANAGEMENT

Though India has become self-sufficient in the production of food grains, the performance of oil-seeds is dismal. One of the major causes for the low productivity of these crops is the losses from insect-pests viz., mustard aphid (*Lipaphis erysimi* Kalt.) in rapeseed mustard, white grub (*Holotrichia consanguineum* Blanch) in safflower, sesame leaf webber (*Antigastra catalaunalis* Dup.) in some sesame are the most serious [12].

Spraying of neem seed kernel extract (NSKE) or neem cost extract (NCE) or neem leaf extract (NLE) at 10% or neem oil (NO) at 5% on the 35th and 45th day after sowing proved effective against leaf miner (*Aproaerema modicella*) or groundnut [13]. NO also gave effective control of groundnut jassid (*Balclutha hortensis*) [14]. Koshiya and Ghelan [15] observed that NLE and neem seed extract (NSE) at 15% concentration effectively controlled the tobacco caterpillar infesting groundnut. Similarly, neem cake mulch effectively reduced the termite (*Odontotermes* spp.) damage on groundnut [16]. Udaiyan and Ramarathinam [17] observed that NO and neem-based formulations could control red hairy caterpillar (*Amsacta albistriga*) to the extent of 92.5% against 100% for quinalphos in groundnut. NO at lower concentration of 0.025% gave 85.4%–90.5% reduction in safflower aphid (*Dactynotus carthami*) [18]. Higher concentrations of NO resulted in 100% reduction in aphid population [19]. Singh et al. [20] reported NO to be effective against mustard aphid (*L. erysimi*). Neem extracts can also effectively control the mustard sawfly (*Athalia proxima*). The extract of neem reduced the mustard aphid (*L. erysimi*) infestation and increased the yields of mustard in Bangladesh [21]. Neem oil, its ethanolic extract, and some of its constituents were tested for contact toxicity against the mustard aphid [22]. The LC_{50} values for oil, ethanolic extract of oil, salanin, a derivative of salanin, and a nonterpenoid fraction were found to be 0.674%, 0.328%, 0.05%, 0.09%, and 0.104%, respectively [23] (Table 4.2).

Petroleum ether extracts of 10 plant species at 0.5%, 1.0%, and 1.5% concentration were found to be effective against mustard sawfly (*A. proxima* Klug), of which 15% extract of *A. indica* leaves caused higher larval mortality [20]. Three neem-based formulations at 0.05%–12.0% against flea beetle (*Phyllotreta cruciferae*) on rape (Canola) were tested [24]. The effects of neem products, that is, neem oil, neem leaf extract (500 g leaves extracted with 500 mL of distilled water), and neem decoction (500 g of leaves boiled in 500 mL of distilled water), were tested on the groundnut jassid, *B. hortensis* [14]. The neem oil at 25% caused highest mortality (97.5%) of the cicadellid species, and the neem leaf extract at 5% caused 83.4% mortality. RD-9 Repelin at 3000 ppm was also effective for managing the aphids and leafhopper in groundnut [25].

Powdered seeds of neem sprayed at 0.2%–0.8% suspension six times at weekly intervals acted as a feeding inhibitor against sesame leaf webber and pod borer, *A. catalaunalis* [26].

TABLE 4.2
LC$_{50}$ Values of Neem Seed Oil and Its Constituents against *L. erysimi* (Kalt.)

Test Material	Appropriate Yield (g kg^{-1} Oil)	LC$_{50}$ (%)
Hexane extract (oil)	—	0.674
Ethanol-soluble fraction	160	0.328
Ethanol-insoluble fraction	—	Nontoxic
Nonterpenoid	1	0.104
Epinimbin	5	No toxicity at 0.3%
Nimbin	5	No toxicity at 0.3%
Salannin	16	0.3%
Salannin derivative	0.5	0.096

Source: Modified from Kumar, V. et al., Oilseeds insect-pests management, in *Neem in Sustainable Agriculture*, Narwal, S.S., Tauro, P., and Bisla, S.S. (eds.), Scientific Publishers, Jodhpur, India, 1997, Chapter 14, pp. 207–213.

Aphicide activity of neem products (Azadirachtin) have been reported in safflower aphid, *D. carthami* [12,27]. In laboratory studies, the lipid-associated limonoids derived from fresh neem seeds and limonoidal residue from expeller grade neem oil proved toxic to safflower aphid [28].

4.5 INSECT-PESTS MANAGEMENT IN PULSES

Among the pulses, chickpea (*Cicer arietinum* L.) and pigeonpea (*Cajanus cajan* L.) are major crops in India, accounting for 55% of the total grain legume production. Most of the Indian population being vegetarian depends on pulses for dietary protein; however, the gap between their demand and availability is widening owing to ever increasing human population. This gap is mainly due to low productivity of grain legumes, which are subjected to various biotic and abiotic stresses. Among the biotic stresses, pests (insects, nematodes, and pathogens) are the most important. Various parts and products of a neem tree (*A. indica* A. Juss) suppress these pests.

Replin, Neemark, Neem 25 EC, and Neem 25 WDP provided up to 80% protection from *H. armigera* in pigeonpea [29]. However, neem products in combination with endosulphan (0.7%) gave better control. The efficacy of various neem products such as NSKE, NLE, NO, Neemark, and Repelin against gram pod borer has been observed by many workers [30–35]. NO (3%) and NSKE (5%) were as effective as endosulphan 0.05% against gram pod borer on green gram [36]. NO (5%) also recorded low damage of this pest on bengalgram [37]. Nimbecidine (0.2%) and Neemgold (0.5%) as second spray, after the first spray with conventional insecticides such as endosulphan and monocrotophos, can reduce the overall use of insecticides, without any significant reduction in yield and satisfactory control of the gram pod borer on chickpea [38]. Effective control of *Maruca testulalis* on laboratory is suggested with the use of NO (5%) and the parasite *Bracon hebetor* [39]. The efficacy of *Helicoverpa* nuclear polyhedrosis virus (HNPV) increases when applied in combination with neem products [40,41]. The combination treatment of HNPV at 500 larval equivalent (LE) ha^{-1} + NSKE (5%) recorded minimum pod and grain damage and the maximum grain yield [42].

4.6 INSECT-PESTS MANAGEMENT IN COTTON

Pest problems on cotton have increased during the last decade due to the cultivation of high-yielding susceptible varieties, excessive use of fertilizers and irrigation, change in the cropping pattern, constraints in the adoption of integrated pest management (IPM) strategies, and reliance on insecticides for pest management. The excessive and indiscriminate use of pesticides for pest management for higher yield has caused extensive damage to the cotton ecosystem, which has become so fragile that any little disturbance may lead to crop failure despite the best pest management strategies. The failure of crop in many parts of India in the recent past has been the outcome of excessive and indiscriminate use of insecticide, which has adversely affected the socioeconomic status of farmers, leading to many suicidal deaths

Research work has shown that neem-based products are effective over a wide range of pests. They may not kill the pest instantaneously but incapacitate it in several ways. The precise effect of various neem extracts on an insect is often difficult to pinpoint. Neem acts as a contact poison, particularly against soft-bodied insects and larvae. It acts systematically because of absorption by roots and translocation to plant parts when applied into the soil or sprayed on the plant. Neem products affect differently physiological processes in insects, for example, metamorphosis including insect growth regulation, adult fertility, toxicity, and they also affect behavior, having antifeedant and oviposition deterrent effects [43]. These effects are presented in Table 4.3 for important insect-pests of

TABLE 4.3
List of Neem Products/Pesticides Used against Major Pests of Cotton

Insect-Pests/Neem Products	Mode of Action	References
Amrasca biguttala		
Neemolin 0.2%–1.0%	Insecticidal toxicity	[45]
Navneem 95EC @750 mL/ha	Insecticidal toxicity	[49]
Navneem 20EC @1.25 L/ha		
Neemark 500–700 mL/ha	Insecticidal toxicity	[50]
Inde-Ne 20EC @250–2500 mL/ha		
Neem oil 2%–3%	Ovipositional deterrent	[51,52]
NSKE @5%	Insecticidal toxicity	[52]
Margocide 30EC @10 mL/ha	Insecticidal toxicity	[53]
Margocide 20EC @5 mL/ha		
Aphis gossypii		
NSKE @5%	Insecticidal toxicity	[52]
Neem oil 1%–3%	Insecticidal toxicity	[52,54]
Nimbecidine @0.5%–2%	Insecticidal toxicity	[17]
Bemisia tabaci		
Neem oil 1%–5%	Insecticidal toxicity, Ovipositional deterrent	[45,54–56]
Neem oil 1%–5%	Insect growth regulator/inhibitor (IGR)	[57,58]
Neem oil 1%–5%	Ovipositional deterrent	[59]
Neemark @500–700 mL/ha	Insecticidal toxicity	[50]
Ind-Ne @250–500 mL/ha		
Neemolin @150 L/ha	IGR	[57,58]
Neemolin @150 L/ha	Antibeedant	[45]
Margocide 30EC @10 mL/L	Insecticidal toxicity	[53]
NSKE aqueous 0.2%–2%	IGR, Ovipositional deterrent	[60]
NSKE in water @2%	Insecticidal toxicity	[54,61]
Thrips tabaci		
Margosan @0.2%	IGR	[62]
NSKE @5%, Neemark @1%	Insecticidal toxicity	[52]
Neem oil @2%–3%	Insecticidal toxicity	[63]

TABLE 4.3 (continued)
List of Neem Products/Pesticides Used against Major Pests of Cotton

Insect-Pests/Neem Products	Mode of Action	References
Earias insulana		
Neem oil @0.5%–3%	Insecticidal toxicity	[54,64]
Margosan, Salanin @0.02%–2%	Antifeedant	[62,65]
NSKE @0.075%–1%	Insecticidal toxicity	[66]
NSKE @5%	IGR	[66,67]
Margosan, Neemix, Repelin, Azatin, Neemzal, Achook (1%)	Insecticidal toxicity	[62]
Helicoverpa armigera		
NSKE @5%	Insecticidal toxicity, IGR/inhibitor, ovipositional deterrent	[54,68–70]
Margocide @0.1%, NSKE @3%–5%, Replin @1%	Insecticidal toxicity	[70]
Neemrich @0.75%, Replin @2%, Achook @1%, Neem Azal @0.04%	Insecticidal toxicity, IGR/inhibitor	[71]
Azadirachtin @10ppm	Antifeedant	[72]
Pectinophora gossypiella		
Neem oil @2%–5%	Insecticidal toxicity	[61,64]
Neemrich 20EC @1%	Insecticidal toxicity	[50,73]
Spodoptera litura		
Neemrich @0.15%	IGR/inhibitor, Antifeedant	[74–76]
Replin @2%		
Neemark @0.75%		
Neemoil @0.5%–3%		
Azadirechin @10ppm	Ovicidal sterility	[77,78]
Neemolin @0.2%–1%	IGR/inhibitor	[45]

cotton. The important biological effects of neem-based products on different insect pests are shown in Figure 4.1.

Simmonds et al. [44] studied the antifeedant activity of azadirachtin, and its analogues are most active against the Egyptian cotton leaf worm (*Spodoptera littoralis*). Leaves treated with Neemolin at 1% or extracts at 0.5% were not preferred for feeding by larvae, which subsequently consumed

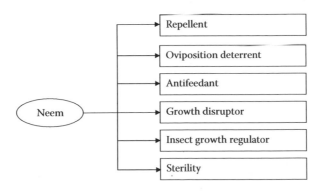

FIGURE 4.1   Biological effects of neem on insect pests.

only 2%–8% of leaf area compared with 72% of untreated leaves [45]. When azadirachtin-treated final instar larvae of *S. litura* were allowed to pupate, the females had low ovarian weight due to reduction in proteins and nucleic acid content. Further, azadirachtin reduced yolk deposition, affected the ovarial sheath, and disrupted interfollicular tissues. The follicular epithelium was intensely damaged as the ovary developed [46]. Finally, azadirachtin reduced fecundity up to 90% in the next generations, and the proportion of eggs hatching was only 54% [47].

Singh [45] reported strong repellence with the use of Neemolin. The odor of the neem leaves and other plant products did not, however, affect adult longevity in the spotted bollworm, *Earias vittella*, although it reduced fecundity and egg hatching [48]. Azadirachtin reduced pupation and adult eclosion in *S. litura* by 47% and 42%, respectively [47]. Rearing of neonate larvae of *S. litura* treated with Neemolin (1%) led to a considerable reduction in larval population, pupal weight, and malformation in adults [45].

Aqueous and solvent extracts of neem leaves and seed, neem oil, and many commercial formulations have been evaluated against pests of cotton, which have exhibited antifeedant, ovipositional and growth retardant, and even toxic effect against different pests. Neem products along with their mode of action in insect-pests management in cotton are described in Table 4.3. Neem-based formulations have multiaction principles, which act together on behavior and physiological process of the pest; hence, there is little chance of development of resistance in insect pests to them. Moreover, neem can be effectively used in combination with insecticides, biocontrol agents, and other pest management strategies with maximum possible stability to environment.

4.7   NEEM AS NEMATICIDE

The management of nematodes through nonchemical methods and the search for eco-friendly pesticides from plants are gaining importance [79,80]. Among plants, neem is the most important, because all its parts and derivatives have been extensively explored and reviewed for nematode management [81–86]. The importance of neem for plant protection has been known since ancient times, but studies against plant parasitic nematodes started only in the late 1960s in India. Earlier workers mainly used neem cake as soil amendment for its manurial value and control of soilborne pests and pathogens. The use of neem in nematological research has been reviewed [79,83,87,88].

Neem is used as soil amendment. The most common methods for application of neem are as fresh-dry leaves, cake, seed kernel, and seed coat against root-knot nematodes (Table 4.4). The application of neem products as soil amendments helps in nematode control in three ways: (1) it changes the physicochemical properties of the soil to favor the growth of useful microflora and inhibits that of harmful nematodes, (2) toxins produced during decomposition are directly toxic to nematodes, and (3) manurial effects of neem products boost the crop root and shoot development and induce resistance/tolerance to plant parasitic nematodes.

Neem leaf is used in both fresh and dry form for the amendment. Root-knot incidence due to *M. javanica* was reduced by the application of 5%–10% (w/w) of fresh leaves to infected soil [85], while in chickpea 400 g ha^{-1} was found effective [89]. The fresh leaves were effective in controlling root-knot nematode singly or in combination with nematicides and fertilizers [90,91]. Neem seed, seed kernel, and seed coat were also found to be effective at 1%–2% w/w of soil or more, in recent studies against the root-knot nematode in mungbean and chickpea [83,84,96,111].

Neem cake is found to be an excellent organic fertilizer rich in plant nutrients because of its high N, P, K, Ca, Mg, and S contents. It is effective against phytonematodes and soilborne plant pathogens. Extensive work has been done on the application of neem cake as a soil amendment. Neem cake was either applied on the basis of nitrogen (content of the cake 5%–7%) requirement of the crop or on the basis of weight per unit area (ha) or unit weight of soil (kg). Incorporation of neem cake into *M. javanica*-infested soil, 3 weeks before planting okra, tomato, or potato, reduced the number of galls [85]. Gowda and Setty [112] studied the comparative efficacy of neem cake and methomyl

TABLE 4.4
Neem Products as Soil Amendments against Root-Knot Nematodes

Neem Products	Host Crop	Nematode Species	Doses	References
Leaf (fresh)	Tomato	*M. javanica*	5%–10%	[85]
	Chickpea	*M. javanica*	500 g ha^{-1}	[89–91]
	Tomato	*M. incognita*	1%–5%	[88,92]
			1%–3%	[93]
			0.5%–2%	[94]
Seed	Chickpea	*M. incognita*	1%–2%	[95,96]
Seed kernel	Chickpea	*M. incognita*	1%	[84,96]
Seed coat	Chickpea	*M. incognita*	1%–2%[a]	[84,91,96]
Cake	Tomato	*M. incognita*	1%–10%[a]	[97]
			1429–1714[a]	[98]
	Tomato and okra	*M. incognita*	1225[b]	[92]
	Tomato, okra, and potato	*M. incognita*	1800[b]	[85,99]
	Mungbean	*M. incognita*	1714[b]	[100]
	Tobacco	*M. incognita*	2435[b]	[101]
	Betelvine	*M. incognita*	100[b]	[102,103]
Cake	Sunflower	*M. incognita*	2.5 tons ha^{-1}	[104]
Cake	Brinjal	*M. incognita*	500 kg ha^{-1}	[105]
Cake	Mulberry	*M. incognita*	1 kg ha^{-1}	[106]
Cake	Tomato	*M incognita*	0.5%	[107]
Cake	Japanese mint	*M. incognita*	2%	[108]
Cake	Mushroom	*Aphelenchoides composticola*	20 g	[109]
Cake	Brinjal	*M. incognita*	0.5 ton ha^{-1}	[110]

[a] w/w of soil.
[b] kg ha^{-1}.

on the growth of tomato and root-knot development by *Meloidogyne incognita*. Mohammed et al. [113] observed that the addition of neem, sesame, soybean, or cotton seed cake to the soil improved root and shoot growth of potted *Citrus reticulata* seedlings and considerably reduced *Tylenchulus semipenetrans* populations. Nematodes associated with the fodder crop berseem (*Trifolium alexandrinum*) were suppressed by the application of neem cake. In the subsequent season, fresh fodder yield was increased and nematode populations were decreased by the residual effects of the oilcakes [114]. In betel vine, soil application of neem oil cake at 1 ton ha^{-1} in trenches near the root zone at the time of flowering was the most effective treatment in controlling *M. incognita* and reducing the number of root galls 10–12 months after application [115].

Azmi [116] found that the application of neem oil cake performed the best in reducing the nematode population and increasing the seedling growth of subabul, *Leucaena leucocephala*. Spot application of neem and karanj (*P. pinnata*) oilcakes at 400 kg ha^{-1} was most effective in reducing the root-knot nematode population in tomato [117]. Jonathan et al. [118] studied the effect of organic amendments on the control of sugarcane nematodes and found that neem cake (2 ton ha^{-1}) and press mud (25 ton ha^{-1}) were most effective in reducing the populations of *M. incognita*, *Pratylenchus coffeae*, and *Helicotylenchus dihystera*. In citrus, the application of neem cake (20 g/plant) resulted in maximum shoot length and shoot weight and maximum reduction in soil nematode population [119]. Root dipping of okra seedlings in water-soluble extracts of neem oil cake for 50 min gave beneficial results in terms of reduction in damage due to *M. incognita* [120]. Alam [121] found that oilseed cakes of neem, castor, mustard, and groundnut each at 110 kg N ha^{-1}, singly and in different combinations, significantly controlled populations of plant parasitic nematodes on tomato, egg plant, chilli, okra, cabbage, and cauliflower. Acid extract of neem cake at 1:10 dilution was most effective in controlling *M. incognita* population on *Vigna unguiculata* [122]. The application of neem cake at 35 g per plant and karanj cake at 44 g per plant reduced *M. incognita* population and increased the growth parameters and yield of brinjal [123]. Integrated application of neem cake (2.5 ton ha^{-1}) + carbofuran 1 kg a.i. ha^{-1} at 45 days after planting was effective in reducing *M. incognita* population in ginger [124].

In mulberry, application of neem and pongamia oil cakes at 2 ton ha^{-1} was equally effective as sebuphos (1 kg a.i. ha^{-1}) in increasing the leaf yield and reducing *M. incognita* population [106].

Jain and Gupta [125] observed that neem cake at 80 q ha^{-1} was more effective than carbofuran at 2 kg a.i. ha^{-1} in reducing the root-knot nematode population and improved tomato growth. Neem cake was more effective than neem oil and *Calotropis procera* extract in cowpea plants [126]. Neem cake @2% w/w was effective in reducing the population of *M. incognita* as well as increasing the growth parameters of Japanese mint (*Mentha arvensis*) [108]. Jonathan et al. [127] reported significant reduction in *M. incognita* and *H. multicinctus* population in banana plants treated with neem cake (1.5 ton ha^{-1}) or press mud (15 ton ha^{-1}) coupled with enhanced fruit yield. Neem cake at 20 g dose treatment of compost beds effectively increased the number of fruiting bodies as well as the yield of *Agaricus bisporus* by reducing the population of *Aphelenchoides composticola* [109]. Srivastava [128] proved that neem cake amendment (2.5 ton ha^{-1}) caused maximum reduction of *M. incognita* population in papaya coupled with enhanced yield (51.8 kg per tree). Integration of Vesicular-arbuscular mycorrhizal fungus (VAM) and neem cake (0.5 ton ha^{-1}) improved the plant growth parameters and yield and reduced *M. incognita* population in brinjal [110]. Tariq and Siddiqui [129] evaluated that bare root dip of tomato seedlings in extracts of neem oil cake + carbofuran for 720 min caused the highest inhibition of root-knot larvae penetration and root-knot development.

4.8   BOTANICALS IN PLANT DISEASE MANAGEMENT

Plant products have assumed special significance in the present-day strategy of developing environmentally sound methods of plant disease control, especially for vegetables. The use of pesticides results in the problem of residual toxicity and pollution in the environment. However, the use of botanicals helps in avoiding the risk of phytotoxicity and the accumulation of harmful residues on plants and in soil. Since the first record of the effect of plant extract of *Acorus calamus* L. on *Alternaria* sp. and *Helminthosporium* sp. [130], several reports have been published on various fungal, bacterial, and viral pathogens of crop plants [131]. With the increase in the awareness of the toxic hazards of pesticides to crops and environment, the importance of indigenous botanicals in plant disease control has been emphasized [132].

Plant derivatives were found to contain an array of chemicals, and these chemicals were reported to have various types of influences on the pathogens as well as on the host plant. Kamalakkannan [133] observed that the extracts of plant species were effective in reducing the rice blast disease caused by *Pyricularia oryzae*. Jayalakshmi [134] analyzed the components of the plant species that were found to be effective in reducing the chili mosaic virus disease.

Earlier reports on vegetables indicate that initial screening studies were conducted with several parts of plants against *Alternaria* species. The efficacy of leaf extracts of *Acacia loculate*, *Ficus religiosa*, *Amaranthus viridis*, *M. azedarach* on *A. tenuis*, *N. glutinosa* on *A. brassicola* [135], flower extract of *Lawsonia alba* on *A. alternata* [136], and root extract of *Achyranthes aspera* on *A. alternata* [136] has been reported.

Preliminary screening of 44 plant extracts for testing their antifungal property against *A. solani* (Ell. and Mart; Jones and Grout) causing blight disease of tomato was carried out by the poisoned food technique and spore inhibition method. The results revealed that the bulb extract of garlic (*A. sativum*), leaf extract of vilvam (*Aegle marmelos*), and flower extract of red periwinkle

TABLE 4.5
Effect of Plant Extracts/Products on the Spore Germination and Mycelial Growth of *A. solani*

Common Name	Botanical Name	Spore Germination[a]		Mycelial Growth	
		Germination (%)	Inhibition (%)	Germination (mm)	Inhibition (%)
Onion bulb	*Allium cepa*	17.8	90.3	12.7	85.9
Neem seed	*Azadirachta indica*	24.8	81.6	21.7	75.9
Neem cake	*Azadirachta indica*	42.5	52.3	64.0	28.9
Neem leaf	*Azadirachta indica*	22.4	84.8	21.3	76.3
Vilvam leaf	*Aegle marmelos*	19.3	88.6	14.7	83.7
Betel leaf	*Piper betle*	20.2	87.5	19.0	78.9
Red periwinkle	*Catharanthus roseus*	18.8	89.1	15.3	83.0
Allitin	Synthetic product (based on the active principle in *A. sativum*)	14.7	93.3	10.0	88.9
Control	Sterile water	77.6	0	90.0	0
CD (*P* = 0.05) (mean of three replications)		3.7	0	13.3	0

Source: Modified from Narasimhan, V. et al., Efficacy of botanicals for the management of blight disease of two vegetable crops, in *Neem for the Management of Crop Diseases*, Mariappan, V. (ed.), Associated Publishing Co., New Delhi, India, 1995, pp. 69–76.

[a] Data after angular transformation.

(*Catharanthus roseus*) at 10% concentration, prepared in water, was most effective in inhibiting spore germination and mycelial growth of pathogen [137] (Table 4.5). Among the eight plant species screened against *A. tenuissima* (Kunze) wilt, shere causing blight disease of onion, the leaf extracts of *A. marmelos* and *Prosopis juliflora* were effective in inhibiting spore germination and mycelial growth (Table 4.6).

The efficacy of plant extracts in inhibiting virus diseases was studied by preinoculation application and the symptoms and incubation were recorded [138]. The results of screening of different plant extracts for pathogenic fungi in postharvest diseases *in vitro* and *in vivo* are presented in Tables 4.7 and 4.8.

All the plant extracts had considerably (from 86% to 100%) inhibited the pathogenic fungi on banana studied *in vitro*.

Of the 10 plant extracts tested *in vivo*, those of *Aloe vera*, *C. gigantea*, *Cassia auriculata*, and *Delonix regia* had considerable effect, whereas those of neem, *Mangifera indica*, *Thevetia peruviana*, and *L. leucocephala* had moderate effect and the rest had very low effect in reducing different diseases of onion.

The results of *in vitro* efficacy of plant extracts on the phytopathogenic bacterium *Xanthomonas campestris* pv. *citri* are presented in Table 4.9.

Of the 16 plant extracts tested *in vitro* against *X. campestris citri*, the cause of acid lime canker, crude and centrifuged extracts of neem cake, *A. sativum*, *Polyalthia longifolia*, and *L. camara* had superior efficacy (60–137 mm² inhibition area), whereas those of *Croton sparsiflorus*, *Bougainvillea spectabilis*, *Ficus bengalensis*, *Datura stramonium*, *Parthenium hysterophorus*, neem leaf, *C. auriculata*, *Vinca rosea*, and *Eucalyptus* sp. had moderate effect (25–56 mm² inhibition area) and the rest had the least inhibitory effect. Ether extracts were found to be more inhibitory than crude or centrifuged extracts (80–305 mm² inhibitory area).

TABLE 4.6
Effect of Plant Extract/Products on the Spore Germination of Mycelial Growth of *A. tenuissima*

Common Name	Botanical Name	Spore Germination[a] (%)	Mycelial Growth (%)
Bulb extract			
1. Garlic	*Allium sativum* L.	52.7	52.3
Leaf extract			
2. Pomegranate	*Punica granatum* L.	50.2	48.3
3. Gundumani	*Adenanthera pavonia* L.	48.0	47.7
4. Neem	*Azadirachta indica* A. Juss	46.1	43.3
5. Betel vine	*Piper betle* L.	50.4	49.7
6. Vilvam	*Aegle marmelos* Corr.	22.1	10.3
7. Prosopis	*Prosopis juliflora* L.	25.5	14.7
8. Red gulmohar	*Delonix regia* Ref.	46.0	40.0
9. Control	Sterile water	80.6	78.7
CD (*P* = 0.05) (mean of three replications)		2.1	3.2

Source: Modified from Narasimhan, V. et al., Efficacy of botanicals for the management of blight disease of two vegetable crops, in *Neem for the Management of Crop Diseases*, Mariappan, V. (ed.), Associated Publishing Co., New Delhi, India, 1995, pp. 69–76.

[a] Data after angular transformation.

TABLE 4.7
Percentage Inhibition of Pathogenic Fungi on Banana Fruits *in Vitro*

Host Extract (0.8%)	Colletotrichum musae	Botryodiplodia theobromae	Gloeosporium musarum	Fusarium moniliforme	Aspergillus niger	Rhizopus stolonifer
1. Neem	98.68	98.27	96.61	100.00	54.61	98.88
2. *Bougainivillea spetabilis*	96.05	99.08	87.84	98.82	54.61	97.00
3. *Parthenium hysterophorus*	98.81	95.40	86.59	97.35	75.64	33.66
4. *Croton sparsiflorus*	96.74	84.24	77.56	86.76	86.84	14.88
5. Control	0.00	0.00	0.00	0.00	0.00	0.00

Source: Modified from Seethalakshmi, V., Studies on post-harvest diseases of banana, MSc (Ag) thesis, AP Agricultural University, Hyderabad, India, 1991.

Of the nine plant extracts used against brinjal mosaic, the extracts of *Mirabilis jalapa*, *B. spectabilis*, and *P. chilensis* had superior efficacy, whereas neem and eucalyptus extracts had moderate efficacy (Table 4.10).

Several workers showed the efficacy of plant extracts in inhibiting the growth of pathogenic fungi in vitro as well as in reducing the diseases *in vivo*. Mishra and Dixit [143] observed that extracts of *A. sativum* stopped the growth of *Absidia spinosa*, *A. tenuis*, and *Cephalosporium graminium*. Bhowmik and Vardhan [144] showed that extracts of *Cinnamomum camphora* and *C. roseus* inhibited growth, sporulation, and spore germination of *Curvularia lunata*.

TABLE 4.8
Effect of Plant Extracts on Pathogenic Fungi on Onion Bulbs *in Vivo*

Plant Extract	Percentage Reduction Over Control		
	Aspergillus flavus	*A. niger*	*Rhizopus* sp.
1. *Aloe vera*	40.10	44.29	39.33
2. *Calotropis gigantia*	17.70	16.53	42.01
3. *Cassia auriculata*	27.02	31.60	31.37
4. *Delonix regia*	0.00	37.60	36.11
5. *Mangifera indica*	5.26	22.85	22.07
6. *Moringa oleifera*	4.27	8.37	6.47
7. Neem	2.71	28.03	27.23
8. *Thevetia peruviana*	0.00	27.22	24.65
9. *Prosopis specifera*	0.00	12.04	10.95
10. *Leucaena leucocephala*	17.32	22.90	23.20

Source: Modified from Sudhakara Rao, P. Storage diseases of onion (*Allium cepa* L.) and garlic (*Allium sativum* L.), MSc (Ag) thesis, AP Agricultural University, Hyderabad, India, 1987.

TABLE 4.9
Effect of Plant Extracts on *X. campestris citri in Vitro*

SN	Plant Extract	Mean Area of Inhibition (mm^2)		
		Crude Extract	Centrifuged Extract	Ether Extract
1	*Cassia auriculata*	25	31	53
2	*Ficus bengalensis*	46	42	104
3	Neem leaf	28	35	131
4	Neem cake	137	137	—[a]
5	*Bougainvillea spectabilis*	54	56	80
6	*Calotropis gigantia*	13	16	—
7	*Parthenium hysterophorus*	31	31	84
8	*Tridax procumbens*	10	12	—
9	*Taphrosia purpurea*	89	89	—
10	*Croton sparsiflorus*	56	58	134
11	*Lantana camara*	62	60	—
12	*Allium sativum*	134	135	—
13	*Vinca rosea*	25	28	—
14	*Datura stramonium*	31	28	—
15	*Eucalyptus* sp.	25	20	—
16	*Polyalthia longifolia*	137	131	305
17	Control	0	0	0

Source: Modified from Mohan, C. and Moses, G.J., Effect of extracts of neem and other plants of *Xanthomonas campestris citri*, the incident of citrus canker. Abstract. Source See Ref. No. 3, pp. 41, 1993.

[a] Not tested.

TABLE 4.10
Effect of Plant Extracts on Inhibition of Brinjal Mosaic Disease

SN	Host Extract	Percentage Inhibition	Incubation Period (Days)
1	Achras zapota	41.7	15.1
2	Neem	58.4	15.8
3	*Bougainvillea spectabilis*	75.0	18.5
4	*Catharanthus roseus*	16.7	15.4
5	*Eucalyptus citrodera*	58.4	15.0
6	*Mirabilis jalapa*	100.0	0.0
7	*Polyalthia longifolia*	0.0	15.2
8	*Prosopis chilensis*	83.4	19.8
9	*Tamarindus indica*	25.0	15.2
10	Control	0.0	15.20

Source: Modified from Bharati, M., Studies on mosaic disease of brinjal (*Solanum melongena* L.) caused by cucumber mosaic virus, MSc (Ag) thesis, AP Agricultural University, Hyderabad, India, 1992.

Singh et al. [145] observed that aqueous plant extracts of *D. stramonium, Cannabis sativus, Eucalyptus* sp., *Thuja sinensis*, and tobacco were effective at 100% concentration against hill bunt of wheat when inoculated seed was dipped in the plant extract.

Grainge et al. [146] reported that extracts of *A. sativum* were inhibitory to the phytopathogenic bacterium, *X. campestris oryzae.* Hatagalung [147] observed that garlic suppressed the incidence of bacterial wilt of tomato. Dhaliwal and Dhaliwal [148] reported the inhibition of tobacco mosaic virus (TMV) on *Phaseolus vulgaris* by extracts of *A. cepa* and *A. sativum.* Chaudhary and Saha [149] showed the inhibition of black gram leaf crinkle by different plant extracts.

4.9 PROSPECTS AND CONSTRAINTS

Neem and other botanicals are still not relied upon as protectants even in the countryside, because there is no industry to provide readymade and efficient products. Hence, farmers use their own preparations, which are not quite effective and stable. Thus, industries may be encouraged to manufacture efficient products. More stable and effective ingredients in neem should be identified and formulated as insecticides for stored grains. The use of neem products needs to be popularized in urban areas for short-term grains storage, and the effects of neem on the quality of food grains is to be studied.

Several toxic substances from various parts of neem have been identified. However, the efficacy of a particular toxic principle against a specific pathogen has not been worked out so far. There is also an urgent need to develop an effective delivery system of neem products against a particular pathogen or groups of pathogen to get their maximum benefits. The work on these lines may provide an alternative to the method of chemical control and keep the environment free from pollution by hazardous chemicals.

The limited residual life of neem derivatives may be considered as an economic disadvantage in cases of severe onslaught of major pests. Secondly, the temperature, rainfall, pH of treated surface, and other environmental factors also exert a negative impact on the efficacy of neem-based pesticides.

Neem cake and foliage as soil amendments suppress parasitic nematodes and enhance the growth and yields of these crops. However, mycotoxic substances in neem for controlling fungal diseases are comparatively less studied in India.

There are certain constraints in the adoption of neem for cotton pest management. Cotton growers habitually adopt chemicals that are of wide spectrum and can give immediate knockdown effect. Due to high input cost, farmers do not want to take any risk and wish to keep the crop completely free from pests. It is certain that neem formulations as such cannot provide quick knockdown but can play an important role in reducing the negative effect of insecticides. Moreover, farmers are not aware of the long-term benefits of the eco-friendly approach to cotton pest management. They need to be educated and encouraged to adopt neem formulations and other botanicals as component of IPM.

4.10 CONCLUSION

Estimates of losses to crops and crop products due to pests and diseases usually vary between 10% and 30%, which amounts to thousands of crores of rupees every year. According to an estimate, pests and diseases in various crops in the field and in stored grains cause an annual loss of about Rs. 20,000 crores in India. The use of pesticides has become an integral part of the present-day improved agricultural technology. The indiscriminate use of these chemicals has not only induced resistance in pests but also affected nontarget organisms and led to the contamination of ground water and soil ecosystem. The problem is further compounded by natural processes of bioaccumulation and biomagnification along the food chain, resulting in health hazards to consumers.

With the modern techniques now available and the attention being given to the area of botanicals, we look forward to interesting developments on the biological activity of neem and its products so as to exploit them as pesticides, fungicides, and nematicides in a world where one is in constant fear of rupture of ecosystems and of undesirable consequences for human health from excessive use of nonbiodegradable pesticides. Conservation of neem and other plants with biorational activity needs quick survey and screening before they are destroyed by deforestation. In India, we are favorably placed with regard to the immense availability of neem for the manufacture of effective pesticidal products for use in agriculture. Extension activists should simultaneously educate the farmers about neem products available in their surroundings so that these can be used effectively.

There is an urgent need to develop newer and safer chemicals and formulations to combat the menace of pests; it is equally important to adopt IPM to increase their efficiency. The promotion and use of neem derivatives and other botanicals are thus imperative for sustainable agriculture.

REFERENCES

1. KK Singh and Gayatri Verma. Neem and its pesticidal characteristics. *Science and Culture* 67(1–2): 30–41, 2001.
2. CS Logfren, DW Anthony, and GA Mount. Size of aerosol droplets impinging on mosquitoes as determined by SEM. *Journal of Economic Entomology* 66: 1085–1088, 1973.
3. MB Isman. Neem and other botanical insecticides: Barriers to commercialization. *Phytoparasitica* 25(4): 339–344, 1997.
4. TN Ananthakrishnan. Insect plant interactions problems and perspectives. In: *Dynamics of Insect Plant Interaction* (Eds. TN Ananthakrishnan and A Raman), pp. 1–11. Oxford & IBH Publishing Company, New Delhi, India, 1988.
5. PL Pedigo. *Entomology and Pest Management*, pp. 742, Prentice Hall, Upper Saddle River, NJ, 1999.
6. M Wink. Production and application of phytochemicals from an agricultural perspective. In: *Phytochemistry and Agriculture* (Eds. TA Van Beek and H Breteler), pp. 171–213. Clarendon Press, Oxford, U.K., 1993.
7. M Jacobson. Insecticides from plants: A review of literature 1994–1953. *U. S. Department of Agriculture Handbook*, p. 154, Washington DC, 1958.
8. T Henn and W Weinzierl. *Alternatives in Insect Management: Botanical Insecticides and Insecticidal Soaps*, University of Illinois CES, Circular 1296, p. 6, 1989.
9. B Ellis and F Bradley. *The Organic Gardener's Handbook of Natural Insect and Disease Control*, Rodale Press, Emmaus, PA, 1996.
10. G Stoll. *Natural Protection in the Tropics*, Margraf Verlag, Weikersheim, Germany, 2000.

11. A Carr, M Smith, L Gilkeson, J Smillie, B Wolf, and F Rodales. *Chemical Free Yard & Garden*, pp. 456, Rodale Press Inc., New York, 1991.

12. RS Goyal. Studies on better utilization of neem seed cake biological activity of alcohol extract and its isolates. MSc thesis, Post Graduate School, IARI, New Delhi, India, 1967.

13. N Chandramohan and P Sivasubramanian. Evaluation of neem products against leaf miner, *Aproaerema modicella* D. *Neem Newsletter* 4(4): 44, 1987.

14. V Nandagopal, RK Jaroli, R Kumar, and PS Reddy. Neem products: Possible insecticides on the groundnut jassid, *Balclutha hortensis. International Arachis Newsletter* 8: 22, 1990.

15. DJ Koshiya and AB Ghelan. Antifeedant activity of different plant derivatives against *Spodoptera litura* F. on groundnut. Source as per Reference No. 1, pp. 270–275, 1990.

16. DVS Rao, KN Singh, JA Wightman, and GVR Rao. Economic status of neem cake mulch for termite control in groundnut. *International Arachis Newsletter* 9(1): 12–13, 1991.

17. K Udaiyan and S Ramarethinan. Bioefficacy of neem derivatives on some major pests of rice, jowar, cotton, groundnut, vegetable and tea. *Pestology* 14(1): 40–52, 1994.

18. C Devakumar, VS Saxena, and SK Mukerjee. Evaluation of neem (*Azadirachta indica* A. Juss) limonoids and azadirachtin against safflower aphid (*Dactynotus carthami* H R L). *Indian Journal of Entomology* 48: 467–470, 1986.

19. A Mani, K Kumudanathan, and CA Jagdish. Anti-feedant property of certain neem products against red hairy caterpillar, *Amsacta albistriga* Walk. *Neem Newstalk* 7: 16, 1990.

20. YP Singh, S Pandey, MB Guddewar, A Shukla, and ML Saini. Efficacy of some plant extracts against mustard sawfly (*Athalia proxima* Klug). *Plant Protection Bulletin* 43: 26–30, 1991.

21. KH Kabir and MD Mia. Effectiveness of some indigenous materials as repellent against the mustard aphid. *Bangladesh Journal of Zoology* 15: 87–88, 1987.

22. RP Singh, C Devakumar, and S Dhingra. Activity of neem (*Azadirachta indica* A. Juss) seed kernel extracts against the mustard aphid, *Lipaphis erysimi. Phytoparasitica* 16: 225–229, 1988.

23. V Kumar, H Singh, and R Kumar. Oilseeds insect-pests management, Chapter 14. In: *Neem in Sustainable Agriculture* (Eds. SS Narwal, Patric Tauro, and SS Bisla), pp. 207–213, Scientific Publishers, Jodhpur, India, 1997.

24. P Palaniswamy and I Wise. Effects of neem-based products on the number of feeding activity of a crucifer flea beetle, *Phyllotreta cruciferae* (Goeze) (Coleoptera: Chrysomellidae) on Canola. *Journal of Agricultural Entomology* 11: 49–60, 1994.

25. T S Subramaniam, B Sivaprasad, and VVLN Prasad. Neem oil 93EC (RD-9 Replin)—An effective botanical for pest management. In: *World Neem Conference*, Bangalore, India, pp. 17, India Tobacco Company, Calcutta, 1993 (Abstract).

26. SS Chadha. Use of neem (*Azadirachta indica* A Juss) seed as feeding inhibitor against *Antigastra catalaunalis* Dupon. (Lepidoptera: Pyrallidae) a sesame (*Sesamum indicum* L.) pest in Nigeria. *East African Agriculture and Forestry Journal* 42: 257–272, 1977.

27. KP Srivastava and BS Parmar. Evaluation of neem oil emulsifiable concentrate against sorghum aphids. *Neem Newletter* 2: 7, 1985.

28. C Devakumar and SS Riar. Identification of the regulators in Neem: Results of ten year studies. In: *World Neem Conference*, Bangalore, India, pp. 52, India Tobacco Company, Calcutta, 1993 (Abstract).

29. JN Sachan and SS Lal. Role of botanical insecticide in *Helicoverpa armigera* management in pulses. Source as per reference No. 12, pp. 16, 1990.

30. HCL Gupta, A Kumar, and AR Khoja. Utility of neem and its derivative in the management of *Helicoverpa armigera* infesting chickpea and tomato. Source as per reference No. 12, pp. 31, 1993.

31. B Jhansi Rani. Studies on Biological Effects of Neem (*Azadirachta indica*, A. Juss) Seed Derivatives on *Helicoverpa armigera*. PhD thesis, Indian Agricultural Research Institute, New Delhi, India, 1988.

32. SS Sharma and B Daniya. Efficacy of some plant oil against pod borer, *Heliothis armigera* on chickpea. *Neem Newsletter* 3(2): 15–17, 1986.

33. RP Singh, Y Singh, and SP Singh. Field evaluation of neem (*Azadirachta indica* A. Juss) seed kernel extracts against the pod borers of pigeon pea, *Cajanus cajan* L. Millsp. *Indian Journal of Entomology* 47: 111–112, 1985.

34. SN Sinha. Neem in integrated management of *Helicoverpa armigera* infesting chickpea. Source as per reference No. 12, pp. 16, 1993.

35. SN Sinha and KN Mehrotra. Diflubenzuron and neem oil in the control of *Heliothis armigera* infesting chickpea. *Indian Journal of Agricultural Science* 58: 348–239, 1988.

36. PC Sundra Babu, S Kuppuswamy, and PV Subha Rao. Management of pod borer on lab. *Tamil Nadu Agricultural University Newsletter* 13(5): 2, 1983.

37. PC Sundra Babu and B Rajasekaran. Evaluation of certain synthetic pyrethroids and vegetable products for the control of Bengal gram pod borer, *Heliothis armigera*. *Pesticides* 18: 58–59, 1989.
38. CP Srivastava, OP Singh, and KN Singh. Testing of neem derivatives with some commonly used insecticides to control pod borer in chickpea. *Pestology* 18(8): 36–39, 1994.
39. K Jhansi and PC Sundara Babu. Evaluation of the effect of parasites, fungus and neem oil individually and in combination against spotted pod borer, *Maruca testulalis*. *Andhra Agriculture Journal* 34: 340–342, 1987.
40. S Jayaraj. Neem in pest control: Progress and prospective. (Souvenir). *World Neem Conference*, Bangalore, pp. 37–43, Indian Tobacco Co., Calcutta, 1993.
41. NR Supra, DW Deshmukh, and SU Satpute. Efficacy of endosulphan and herbal products alone and in combination with nuclear polyhedrosis virus against pod borer, *H. armigera* (Hb.) on gram and effect of intercropping on the incidence of this pest. *Pestology* 15(5): 5–9, 1991.
42. SV Sarode, YS Jumde, RO Deotale, and HS Thakare. Neem seed kernel extract for management of *Helicoverpa armigera* (Hubner) in chickpea. *International Chickpea and Pigeonpea Newsletter* 1: 21–23, 1994.
43. H Schmutterer. Properties and potential of natural pesticides from the neem tree, *Azadirachta indica*. *Annual Review of Entomology* 35: 271–297, 1990.
44. MSJ Simmonds, WM Blaney, SV Ley, JC Anderson, R Banteli, AA Denholm, PCW Green, RB Grossman, C Giuitteridge, L Jenneans, SC Smith, PL Toogood, and A Wood. Behavioural and neurophysiological responses of *Spodoptera littoralis* to azadirachtin and a range of synthetic analogues. *Entomologia Experimentals et Applicata* 77: 69–80, 1995.
45. KN Singh. Potential of neem for insect pest management. *Pestology* 20(3): 29–33, 1996.
46. Shashi Gupta. Effect of Azadirachtin on Proteins and Nucleic Acid of *Spodoptera litura*. PhD thesis, Indian Agricultural Research Institute, New Delhi, India, 1988.
47. GSG Ayyangar and PJ Rao. Carry over effects of Azadirachtin on development and reproduction of *Spodoptera litura* (Fabr.). *Annals of Entomology* 9: 55–57, 1991.
48. A Shukla, SC Pathak, and RK Agrawal. Effect of some plant odours in the breeding environment on fecundity and hatching of okra shoot and fruit borer, *Earias vittella* under laboratory conditions. *Crop Research* 13(1): 157–161, 1997.
49. AK Dhawan, GS Srimannarayana, GS Simwat, and K Nagalah. Management of cotton pests in upland cotton with Navneem 95 EC. In: *1st National Symposium on Allelopathy in Agroecosystem (Agriculture and Forestry)* (Eds. P Tauro and SS Narwal), pp. 157–158, Indian Society of Allelopathy, Chaudhary Charan Singh, Haryana Agricultural University, Hisar, India, 1992.
50. AK Dhawan and GS Simwat. Management of sucking pests of cotton with natural plant products. In: *Proceedings of the 1st National Symposium on Allelopathy in Agroecosystem (Agriculture and Forestry)* (Eds. P Tauro and SS Narwal), pp. 154–155, Indian Society of Allelopathy, Chaudhary Charan Singh Haryana Agricultural University, Hisar, India, 1992.
51. KN Saxena and A Basit. Inhibition of oviposition by voferrles of certain plants and chemicals in the leaf hopper, *Amrasca devastans*. *Journal of Chemical Ecology* 8: 329–338, 1982.
52. PL Tadas, HK Kene, and SD Deshmukh. Effect of raw cotton seed oil against sucking pests of cotton. *Punjab Rao Krishi Vidyapeeth Research Journal* 18: 142–143, 1994.
53. JP Singh and JP Gupta. Impact of various insecticides on intermittent population of jassid and whitefly infesting American cotton during different spray schedules followed for the control of bollworm complex. *Journal of Entomological Research* 17: 297–303, 1993.
54. S Uthasamy and P Gajendran. Efficacy of neem products for the management of pests of cotton. *Neem Newsletter* 9(1): 18, 1992.
55. NV Rao, AS Reddy, and PS Reddy. Relative efficacy of some new insecticides on insect pests of cotton. *Indian Journal of Plant Protection* 18: 53–58, 1990.
56. SN Puri, BB Bhosle, M Ilyas, GD Butler Jr., and TJ Henneberry. Detergents and plant derived oils for control of the sweet potato whitefly on cotton. *Crop Protection* 13: 45–48, 1994.
57. K Natrajan and VT Sundaramurthy. Effect of neem oil on cotton whitefly (*Bemisia tabaci*). *Indian Journal of Agricultural Science* 60: 290–291, 1990.
58. BS Nandihaili, P Hugar, and BV Patil. Evaluation of neem and neem products against cotton whitefly, *Bemisia tabaci*. *Karnataka Journal of Agricultural Science* 3: 58–61, 1990.
59. K Natrajan, VT Sundaramurthy, and P Chidambaram. Usefulness of fish oil rosin soap in the management of whitefly and other sap feeding insects on cotton. *Entomology* 16: 229–232, 1991.
60. DL Coudriet, AN Kishaba, and DE Meyercmrk. Laboratory evaluation of neem seed extract against larvae of the cabbage looper and beet armyworm. *Journal of Economical Entomology* 79: 39–41, 1986.

61. SA Nimbalkar, SM Khode, YM Taley, and KJ Patil. Bioefficiency of neem and karanj seed extracts along with other commercial plant products against bollworms on cotton. In: *Botanical Pesticides in Integrated Pest Management* pp. 252–255, Indian Society of Tobacco Science, Rajahmundry, India, 1993.
62. KRS Ascher, M Eliahu, NE Nemny, and J Meisner. Neem seed kernel extract as an inhibitor of growth and fecundity in *Spodoptera littoralis*. *Proceedings of the 2nd International Neem Conference*, Rauischholzhausen, Germany, pp. 331–344, 1984.
63. F Ahmad, FR Khan, and MR Khan. Comparative efficacy of some traditional and non traditional insecticides against sucking insect-pests of cotton. *Sarhad Journal of Agriculture* 11(6): 733–739, 1995.
64. SA Salem. Response of cotton bollworms, *Pectinophora gossypiella* and *Earias insulana* to seed neem kernel pure oil. *Annals of Agricultural Science Moshtohor* 19: 597–607, 1991.
65. J Meisner, KRS Ascher, R Aly, and JD Warthen. Response of *Spodoptera littoralis* and *Earias insulana* larvae to *azadirachtin* and *salanin*. *Phytoparasitica* 9: 27–32, 1981.
66. J Meisner, M Kehat, M Zur, and C Eizick. Response of *Earias insulana* Boisd. larvae to neem (*Azadirachta indica*) kernel extract. *Phytoparasitica* 6: 85–88, 1978.
67. J Meisner, M Kehat, M Zur, and F Chava. Response of *Earias insulana* Boisd. larvae to neem (*Azadirachta indica* A. Juss) kernel extract. *Phytoparasitica* 6: 85–86, 1991.
68. BR Jhansi and RP Singh. Identification of effective and inexpensive neem (*Azadirachta indica* A. Juss) seed kernel extract for management of *Helicoverpa armigera* Hubner, Paper presented at *4th International Neem Conference*, Bangalore, India, February 24–28, 1993.
69. NC Patel, VM Valand, and SN Patel. Ovicidal action of some new insecticides against eggs of *Helicoverpa armigera* (Hubner). *Indian Journal of Plant Protection* 23: 99–100, 1995.
70. DM Mehta, JR Patel, and NC Patel. Ovicidal and oviposition deterrent effect of botanicals individually and in combination with endosulfan on *H. armigera*. *Indian Journal of Plant Protection* 22: 214–215, 1994.
71. K Raman, S Ganesa, and BN Vyas. Studies on the effect of neemrich insecticide on feeding development and control of cotton and vegetable pests. In: *National Symposium on Recent Advances in Integrated Pest Management*, Abstract pp. 107. Indian Society for the Advancement of Insect Science, Punjab Agricultural University, Ludhiana, pp. 154–155, 1992.
72. R Babu, K Murugan, and G Vanithakumari. Interference of azadirachtin on the food, utilization efficiency and midgut enzymatic profiles of *Helicoverpa armigera* Hubner. *International Journal of Environmental Toxicology* 6(2): 81–84, 1996.
73. AD Khurana. Evaluation of gossyplure and neem oil formulation along with insecticidal spray schedule for the control of pink bollworm on cotton. *Journal of Insect Science* 6: 303–304, 1993.
74. GR Rao, G Raghavaiah, and B Nagalingam. Effect of botanicals on certain behavioural responses and on the growth inhibition of *Spodoptera litura*. In: *Botanical Pesticides in Integrated Pest Management* pp. 175–182, Indian Society for Tobacco Science, Rajahmundry, India, 1993.
75. AK Dhawan and GS Simwat. Evaluation of some insecticides alone and in combination with other insecticides against boll worm complex and tobacco caterpillar infesting cotton. In: *World Neem Conference*, Bangalore, India, pp. 19–20, Indian Tobacco Company, Calcutta, 1993 (Abstract).
76. O Koul. Anifeedant and growth inhibiting effects of calamus oil and neem oil on *Spodoptera litura* under laboratory conditions. *Phytoparasitica* 15: 169–180, 1987.
77. O Koul. Azadirachtin interaction with development of *Spodoptera litura* Fab. *Indian Journal of Experimental Biology* 23: 160–163, 1985.
78. D Singh and SS Bhathal. Survival, growth and development effects of some neem based products on *Spodoptera litura*. In: *Proceedings of the 1st National Symposium on Allelopathy in Agroecosystems (Agriculture and Forestry)*, Hisar, India (Eds. P Tauro and SS Narwal), pp. 163–165, February 12–14, 1992.
79. V Mojumder. Nematodes. In: *The Neem Tree—A Source of Unique Products for Pest Management and Other Purposes* (Ed. H Schmutterer), pp. 129–150, VCH Postfach, Weinheim, Germany, 1995.
80. JJ Thomason. Challenges facing nematology. Environmental risks with nematicides and the need for new approaches. See Ref. 60, pp. 469–476, 1987.
81. MM Alam. Neem in nematode control. In: *Nematode Biocontrol: Aspects and Prospects* (Eds. MS Jairajpuri, MM Alam, and I Ahmad), pp. 51–55, CBS Publishers, New Delhi, India, 1990.
82. MM Alam. Bioactivity against phytonematodes. In: *Neem Research and Development* (Eds. NS Randhawa and BS Parmar), pp. 123–143, Society of Pesticide Science, Indian Agricultural Research Institute, New Delhi, India, 1993.

83. SD Mishra and V Mojumder. Management of plant-parasitic nematodes through soil amendments of soil. In: *Nematodes Pest Management in Crops* (Eds. DS Bhatti and RK Walia), pp. 132–147, CBS Publishers, New Delhi, India, 1994a.

84. V. Mojumder and SD Mishra. Management of nematodes pests. In: *Neem in Agriculture* (Eds. BS Parmar and RP Singh), pp. 40–48, IARI Research Bulletin No. 40, Indian Agricultural Research Institute, New Delhi, India, 1993a.

85. RS Singh and K Sitaramaiah. Incidence of root-knot of okra and tomatoes in oilcake amended soil. *Plant Disease Reporter* 50: 668–672, 1967.

86. K Vijayalakshmi. Studies on the effect of some agricultural chemicals and soil amendments on some plant parasitic nematodes. PhD thesis, IARI, New Delhi, India, 1976.

87. Indian Agricultural Research Institute. *Neem in Agriculture*, Research Bulletin No. 40, 63 pp., Indian Agricultural Research Institute, New Delhi, India, 1983.

88. K Vijayalakshmi, HS Gaur, and BK Goswami. Neem for the control of plant parasitic nematodes—mini review. *Neem Newsletter* 2: 35–42, 1985.

89. DC Gupta and K Ram. Neem and Dhatura leaves along with chemicals used against *M. javanica* infecting chickpea. *Indian Journal of Nematology* 17: 84–86, 1981.

90. K Ram and DC Gupta. Studies on the control of *Meloidogyne javanica* infecting chickpea (*Cicer arietinum*). *Haryana Agricultural University Journal of Research* 11: 77–81, 1981.

91. K Ram and DC Gupta. Efficacy of plant leaves nematicides and fertilizers alone and in combination against *Meloidogyne javanica* infected chickpea (*Cicer arietinum* L.). *Indian Journal of Nematology* 12: 221–222, 1982.

92. AS Srivastava, RC Pandey, and S Ram. Application of organic amendments for the control of root-knot nematode, *Meloidogyne javanica* (Treub.). *Labdev Journal of Science and Technology* 9B: 203–205, 1972.

93. BK Goswami and K Vijaylakshmi. Effect of some indigenous plant materials and oilcake amended soil on the growth of tomato and root-knot nematode population (Abstract). *Indian Journal of Nematology* 11: 121, 1981.

94. CGP Rao and R Bajaj. Effect of decomposing organic matter on plant growth and the incidence of root-knot in *Lycopersicon esculentum*. *Indian Phytopathology* 37: 160–164, 1984.

95. SD Mishra and V Mojumder. Toxic behaviour of neem products on soil and plant nematodes. *International Symposium on "Allelopathy in Sustainable Agriculture, Forestry and Environment"* Abstract (Eds. SS Narwal, P Tauro, GS Dhaliwal, and J Prakash), pp. 130, Indian Society Allelopathy, Haryana Agricultural University, Hisar, India, 1994b.

96. V Mojumder and SD Mishra. Management of root-knot nematode, *Meloidogyne incognita* through neem products (Abstract). In: *Allelopathy in Sustainable Agriculture, Forestry and Environment* (Eds. SS Narwal et al.), p. 115, Indian Society of Allelopathy. Haryana Agricultural University, Hisar, India, 1994 c.

97. D Bhattacharya and BK Goswami. A study on the comparative efficacy of neem and groundnut oil cakes against root-knot nematode, *Meloidogyne incognita* as influenced by microorganisms on sterilized and unsterilized soil. *Indian Journal of Nematology* 17: 81–83, 1987.

98. SD Mishra and SK Prasad. Effect of soil amendments on nematodes and crop yields. II. Oilseed cakes, organic matter and inorganic fertilizers at different levels of *Meloidogyne incognita*. *Indian Journal of Entomology* 40: 120–135, 1978.

99. K Sitaramaiah. Mechanism of reduction of plant parasitic nematodes in soils amended with organic materials. In: *Progress in Plant Nematology* (Eds. SK Saxena, MW Khan, A Rashid, and RM Khan), pp. 263–295, CBS Publishers and Distributors, New Delhi, India, 1990.

100. K Vijayalakshmi and SK Prasad. Effect of some nematicides, oilseeds, cakes and inorganic fertilizers on nematodes and crop growth. *Annals of Agriculture Research* 3: 133–139, 1982.

101. MV Desai, MH Shah, SN Pillai, and AS Patel. Oilseeds in control of root-knot nematodes. *Tobacco Research* 5: 105–108, 1973.

102. GB Jagdale, AB Pawar, and K S Darekar. Effect of organic amendments on root-knot nematodes infecting betelvine. *International Nematology Network Newsletter* 2: 7–10, 1985a.

103. GB Jagdale, AB Panwar, and KS Darekar. Studies on control of root-knot nematode on betelvine. *International Nematology Network Newsletter* 2: 10–13, 1985b.

104. V Devappa, K Krishnappa, and BMR Reddy. Management of root-knot nematode *Meloidogyne incognita* on sunflower. *Mysore Journal of Agricultural Science* 31(2): 155–158, 1997.

105. S Kumar and S Valivadu. Evaluation of organic amendments for the management of root-knot and reniform nematodes infesting brinjal as compared with carbofuran. *Management in Horticultural Ecosystems Pest* 2(2): 71–74, 1996.

106. SDD Govindaiah, MT Himanthraj, and AK Bajpai. Nematicidal efficacy of organic manures, inter-crops, mulches and nematicide against root-knot nematodes in mulberry. *Indian Journal of Nematology* 19(1): 25–28, 1997.

107. JA Jacob and MM Haque. Effect of neem products on the suppression of root-knot nematode *Meloidogyne incognita* in tomato. In: *Nematology: Challenges and opportunities* in 21st century. *Proceedings of the 3rd International Symposium of Afro-Asian Society of Nematologists.* Sugarcane Breeding Institute, Coimbatore, India, April 16–19, 1998.

108. R Singh and V Kumar. Management of *Meloidogyne incognita* infecting *Mentha arvensis*, by neem cake and carbofuran. *Annals of Plant Protection Sciences* 8(2): 141–143, 2000.

109. Gitanjali and SN Nandal. Effect of neem products and dazomet for the management of *Aphelenchoides compositicola* on white button mushroom (*Agaricus bisporus*) under semi-commercial conditions. *Indian Journal of Nematology* 31(1): 52–57, 2001.

110. A Borah and PN Phukan. Comparative efficacy of *Glomus fasciculatum* with neem cake and carbo-furan for the management of *Meloidogyne incognita* on brinjal. *Indian Journal of Nematology* 34(2): 129–132, 2004.

111. V Mojumder and SD Mishra. Management of root-knot nematode *Meloidogyne incognita* infecting chickpea with neem seed coat. *Annals Agricultural Research* 13: 388–390, 1992b.

112. DN Gowda and KGH Setty. Comparative efficacy of neem cake and methomyl on the growth of tomato and root-knot and development. *Current Research* 7(7): 118–120, 1978.

113. HJ Mohammed, SI Husain, and AJ Al Zarari. Effect of oil cakes on the growth of *Citrus reticulatae* L. and the control of citrus nematodes in Iraq. *Phytopathologica Mediterranea* 19(2–3): 153–154, 1980.

114. N Hasan and RK Jain. Effect of soil amendments on fodder production, photosynthetic pigments and nematodes associated with berseem (*Trifolium alexandrinum* L.) followed by bajra (*Penniseum typhoides* L.). *Agricultural Science Digest* 4(1): 12–14, 1984.

115. A Acharya and NN Padhi. Effect of neem oil cake and sawdust against root-knot nematodes, *Meloid-ogyne incognita* on betelvine (*Piper betle* L.). *Indian Journal of Nematology* 18(1): 105–106, 1988.

116. MI Azmi. Control of plant parasitic nematodes by incorporation of deoiled cakes of tree seeds of subabul, *Leucaena leucocephala* (Lam.) de wit. *Indian Forester* 116(10): 812–814, 1990.

117. KS Darekar, NL Mhase, and SS Shekle. Effect of placement of oilseed cakes on the control of root-knot nematodes on tomato. *International Nematology Network Newsletter* 7(1): 5–7, 1990.

118. EI Jonathan, SV Krishnamoorthy, ML Mandharan, and K Muthukrishanan. Effect of organic amend-ments on the control of sugarcane nematodes. *Bharatiya Sugar* 16(6): 37–49, 1991.

119. PP Reddy, MS Rao, and M Nagesh. Management of the citrus nematodes on acid lime by integration of *Trichoderma harzianum* with oil cakes. *Nematologia Mediterranea* 24(2): 265–67, 1996.

120. TA Khan, S Nisar, and SI Husain. Effect of water soluble extracts of certain oilcakes on the develop-ment of root-knot disease of okra (*Abelmoschus esculentus* var. Sevendhari). *Current Nematology* 2, 167–170.

121. MM Alam. Control of plant parasitic nematodes with oilseed cakes on some vegetables in field. *Pakistan Journal of Nematology* 9(1): 21–30, 1991.

122. K Alagumalai, M Trivalluvan, N Nagendra, and Panjaly Ramraj. Effect of acid extract of neem cake on the population build up of *Meloidogyne incognita*. *Environment and Ecology*, 13(2): 275–277, 1995.

123. SG Thakur and KS Darekar. Effect of some non-edible oilseed cakes against root-knot nematode, *Meloidogyne incognita* and growth parameters of brinjal. *Current Nematology* 6(1): 21–26, 1995.

124. MS Sheela, H Bai, T Jiji, and KJ Juriyan. Nematodes associated with ginger rhizosphere and their management in Kerala. *Pest Management in Horticulture Ecosystems* 1(1): 43–48, 1995.

125. RK Jain and DC Gupta. Efficacy of neem (*Azadirachta indica*) cake as nursery bed treatment in the management of root knot nematode (*Meloidogyne javanica*) infecting tomato. *Indian Journal of Nematology* 27(2): 249–251, 1998.

126. ZH Latif, R Ahmad, and M Inam-ul-Haq. Effect of seed treatments with neem cake, neem oil and latex of oak on the germination of cowpea and its vulnerability to root-knot nematode (*Meloidogyne incognita*). *Pakistan Journal of Nematology* 11(1): 52–55, 1999.

127. EI Jonathan, G Gajendran, and WW Manuel. Management of *Meloidogyne incognita* and *Helicoty-lenchus multicinctus* in banana with organic amendments. *Nematologia Mediterranea* 28: 103–105, 2000.

128. SS Srivastava. Efficacy of organic amendments in the management of root knot nematodes (*Meloidogyne incognita*) infecting papaya. *Indian Journal of Nematology* 32(2): 183–184, 2002.

129. I Tariq and MA Siddiqui. Evaluation of nematicidal properties of neem for the management of *Meloidogyne incognita* on tomato. *Indian Journal of Nematology* 35(1): 56–58, 2005.

130. Mann and Burns. Pharmology and chemotherapy of plant products. In: *The Materia Medica*, pp. 63–67, School of Tropical Medicine, Calcutta, 1927.

131. RJ Shragia. Chemical control of plant diseases: An exciting future. *Annual Review of Phytopathology* 13: 257–267, 1975.

132. Krishna Kumar and YL Nene. Antifungal properties of *Cleome isocandra* L. extracts. *Indian Phytopathology* 21: 445–446, 1968.

133. A Kamalakkanan. Management of Rice Blast (*Pyricularia oryzae*) Disease by Using Chemicals and Plant Extracts. MSc (Ag) thesis, TNAU, Coimbatore, pp. 201, 1994.

134. V Jayalakshmi. Management of Chilli Mosaic Virus by Plant Products. MSc (Ag) thesis, pp. 142, TNAU, Coimbatore, India, 1994.

135. PS Shekhawat and R Prasada. Antifungal properties of some plant extracts. II. Growth inhibition studies. *Science and Culture* 37: 40–41, 1971.

136. DK Pandey, RN Tripathi, and MN Tripathi. Antifungal activity in some seed extracts. *Environment India* 4: 83–85, 1995.

137. V Narasimhan, M Vijayan, and V Senthilnathan. Efficacy of botanicals for the management of blight disease of two vegetable crops. In: *Neem for the Management of Crop Diseases* (Ed. V Mariappan), pp. 69–76, Associated Publishing Co., New Delhi, India, 1995.

138. GJ Moses. Potential of neem and other plant extracts for management of crop diseases. In: *Neem for the Management of Crop Diseases* (Ed. V Mariappan), pp. 91–97, Associated Publishing Co., New Delhi, India.

139. V Seethalakshmi. Studies on Post-Harvest Diseases of Banana. MSc (Ag) thesis, AP Agricultural University, Hyderabad, India, 1991.

140. P Sudhakara Rao. Storage Diseases of Onion (*Allium cepa* L.) and garlic (*Allium sativum* L.) MSc (Ag) thesis, AP Agricultural University, Hyderabad, India, 1987.

141. C Mohan and GJ Moses. Effect of extracts of neem and other plants of *Xanthomonas campestris citri*, the incident of citrus canker. Abstract. Source See Ref No. 3, pp. 41, 1993.

142. M Bharati. Studies on Mosaic Disease of Brinjal (*Solanum melongena* L.) caused by Cucumber Mosaic Virus. MSc (Ag) thesis, AP Agricultural University, Hyderabad, India, 1992.

143. SB Misra and DN Dixit. Fungicidal spectrum of the leaf extract of *Allium sativum*. *Indian Phytopathology* 29: 448–449, 1976.

144. BN Bhowmick and V Vardhan. Antifungal activity of some leaf extracts of medicinal plants on *Curvularia lunata*. *Indian Phytopathology* 34: 385–386, 1981.

145. S Singh, LB Goel, SK Sharma, and SK Nayar. Fungitoxicants and plant extracts in the control of hill bunt of wheat. *Indian Phytopathology* 25: 297–298, 1972.

146. M Grainge, L Berger, and S Ahmad. Effects of extracts of *Artabortrys uncinatus* and *Allium sativum* on *Xanthomonas campestris oryzae*. *Current Science* 54: 90, 1995.

147. L Hatagalung. Garlic bulbs as a material for suppressing the incidence of bacterial wilt (*Pseudomonas solanacearum*) on tomato. *Bulletin Penelitian Horticulture* 16: 184–193, 1988.

148. AS Dhaliwal and GK Dhaliwal. Inhibition of TMV multiplication by extracts from *Allium cepa* and *A. sativum*. *Advancing Frontiers of Plant Sciences* 28: 305–310, 1971.

149. AK Chaudhary and NK Saha. Inhibition of urdbean leaf crinkle virus by different plant extracts. *Indian Phytopathology* 38: 366–368, 1988.

Part II

Techniques and Analysis

5 Analysis of Pesticides in Food Samples by Supercritical Fluid Chromatography

Mohamed Hamza El-Saeid and Haseeb Ahmad Khan

CONTENTS

5.1 INTRODUCTION

5.1.1 SIGNIFICANCE OF PESTICIDE ANALYSIS IN FOOD SAMPLES

Pesticides are the chemicals that control insects, weeds, fungi, and other pests that destroy almost half of the world's food crops each year. Pesticides are regarded as the most economical and effective tools for maintaining the demand/supply ratio of agricultural products stable for saving the increasing world population. The prudent use of pesticides adhering to the guidelines, including the appropriate concentration and volume of pesticide solution as well as correct timings of spray, can significantly minimize pesticide contamination of food. However, their indiscriminate use, often by illiterate or untrained workers, leads to contamination of foods, the consumption of which poses severe health hazards to humans. It has been estimated that about 90% of the human exposure to pesticide residues is caused by eating contaminated foods. Vegetables are essential constituents of human diet. Currently there is a growing trend toward adapting a vegetarian diet to minimize the risks of major diseases, such as cancer, diabetes, and atherosclerosis. Factually, the intake of raw or partially cooked vegetables is more beneficial than overcooked vegetables due to heat-induced losses of biomolecules (e.g., vitamins) and denaturation of fibrous components. However, the consumption of contaminated raw vegetables is more hazardous than cooked vegetables as cooking can break down the pesticides before their entry into the human body.

A wide range of pesticide formulations, including insecticides, herbicides, and fungicides, is applied during various stages of vegetable production, starting from sowing to reaping for proper crop protection. Estimation of pesticide residues in vegetables is therefore important to ensure hygiene in dietary products and to save our population from pesticide-related health effects. The methodology of pesticide analysis mainly comprises two protocols; one for the extraction of targeted pesticide(s) from specific matrices followed by appropriate preconcentration (if necessary, depending on the sensitivity of the method) and the second protocol deals with the determination of extracted pesticides. This chapter highlights the efficiencies and advantages of two advanced extraction techniques, supercritical fluid extraction (SFE) and microwave-assisted extraction (MAE), for the extraction of pesticides from vegetable samples followed by their determination using supercritical fluid chromatography (SFC).

5.1.2 SUPERCRITICAL FLUIDS

For every substance there is a temperature above which it can no longer exist as a liquid, no matter how much pressure is applied. Similarly, there is a pressure above which the substance can no longer exist as a gas no matter how high the temperature is raised. These points are called the supercritical temperature and supercritical pressure, respectively, and are the defining boundaries on a phase diagram (Figure 5.1) for a pure substance, beyond which the substance has properties intermediate between a liquid and a gas and is called a supercritical fluid. In this region the fluid has good solvating power and high diffusivity, which make it a good choice as a mobile phase in chromatography.

Supercritical fluids are produced by heating a gas above its critical temperature or compressing a liquid above its critical pressure. A substance such as CO_2 can exist in solid, liquid, and gaseous phases under various combinations of temperature and pressure, and there is a point, the critical temperature, where the liquid and vapor have the same density. In this physical state,

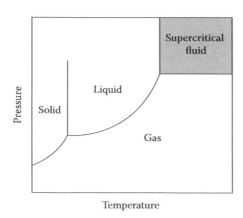

FIGURE 5.1   Pressure–temperature phase diagram.

CO_2 is a good solvent for many organic substances. It also has valuable properties as a mobile phase in chromatography as its viscosity remains similar to that of a gas, permitting high flow rates. The benefits of using supercritical fluids are their liquid-like densities offering higher solubility and increased column loading.

5.1.3 SUPERCRITICAL FLUID EXTRACTION

Although analytical extraction approaches have been improved, most of them rely on time-consuming procedures such as Soxhlet extraction, which requires large volumes of expensive and toxic solvents [1,2]. SFE is rather a new technique for the extraction of analytes from sample matrices. The main advantages of using supercritical fluids for extractions are that they are inexpensive, contaminant free, and less costly to dispose safely than organic solvents. Because of distinct properties, such as low viscosity, high diffusivity, and adjustable density, SFE has gained increased attention as a better alternative to conventional extraction methods [3]. Carbon dioxide is the commonly used extraction solvent due to its suitable critical temperature (31.2°C) and pressure (72.8 atm) and its nontoxic and nonflammable properties [2,3]. Moreover, CO_2 can easily be removed by reducing the pressure. Supercritical fluids can have solvating powers similar to those of organic solvents, but with higher diffusivities, lower viscosity, and lower surface tension. The solvating power can be adjusted by changing the pressure, temperature, or addition of a modifier to the supercritical fluid. A common modifier is methanol (typically 1%–10%), which increases the polarity of supercritical CO_2. Water must be removed before performing SFE, because a highly water-soluble analyte will prefer to partition into the aqueous phase, resulting in a poor SFE recovery. Proper choice of the extraction fluid will also allow the analyst to conduct the extraction at low temperatures, a key feature to extend the application of SFE for the extraction of thermally liable substances [4,5].

The instrumentation of SFE consists of a source of fluid in a pressurized cylinder, a fluid delivery module, an extracting vessel, a backpressure regulating device, modifier pumps, and a collector for procuring the extract after SFE. Analytes are trapped by letting the solute-containing supercritical fluid decompress into an empty vial, through a solvent, or onto a solid sorbent material. Extractions are done in dynamic, static, or combination modes. In dynamic extraction, the supercritical fluid continuously flows through the sample in the extraction vessel and out the restrictor to the trapping vessel. In static mode, the supercritical fluid circulates in a loop containing the extraction vessel for some period of time before being released through the restrictor to the trapping vessel. In the combination mode, a static extraction is performed for some period of time followed by a dynamic extraction. Coextraction of unwanted solutes along with the target analyte may occur in SFE. These interferences can be removed either by conventional sample cleanup methods or by using a sorbent column downstream the SFE device.

Rissato et al. [6] used SFE for the extraction of various pesticides, including organochlorines, organophosphates, organonitrogens, and pyrethroids, from soil samples before their estimation by gas chromatography (GC). SFE has been used for the extraction of 17 organochlorine and organophosphate pesticides from a table-ready food containing crude vegetables, white bread, and vegetable oil [7]. El-Saeid [8] has applied SFE for the extraction of various pesticides from canned food, fruits, and vegetables. Aquilera et al. [9] have extracted 22 pesticides from rice samples using SFE, with recoveries ranging 74%–98%. They used 15 mL CO_2 volume, 50°C temperature, 200 atm pressure, and methanol as modifier [9]. SFE has been applied for the extraction of different pesticides from honey; the optimal conditions were reported to be 400 bar pressure, 90°C temperature, and acetonitrile as modifier [10]. Zhao et al. [11] used SFE for the extraction of 12 organochlorine pesticides from *Angelicae sinensis*. The optimized extraction conditions were found to be pure CO_2 (without modifier), 15 MPa pressure, 60°C temperature, and an extraction time of 20 min. Several organophosphate pesticides, including fenitrothion, chlorpyrifos, diazinon, methamidophos, edifenphos, mevinphos, fenthion, and acephate, have been extracted from agro wastewater. This procedure was performed at 90°C and 325 atm for 20 min for static extraction followed by 40 min of dynamic extraction [12]. King and Zhang [13] have reported a rapid SFE methodology for the extraction of carbamate pesticides, including carbaryl, 3-hydroxycarbofuran, carbofuran, aldicarb, and methiocarb. Sun and Lee [14] have extracted four carbamates (propoxur, propham, chlorpropham, and methiocarb) from soils using supercritical CO_2 at 300 kg/cm^3 modified with 10% methanol. The extraction was performed at 60°C and lasted for 30 min, resulting in up to 92% recoveries.

5.1.4 Microwave-Assisted Extraction

The extraction of organic compounds by microwave irradiation is a recent technique, first used in 1986 by Ganzler and coworkers [15]. For MAE, the sample is suspended in a suitable extraction solvent and irradiated in a microwave oven (2450 MHz) for an appropriate time, without allowing the suspension to boil. This frequency of microwaves corresponds to a wavelength of 12.2 cm and energy of 0.23 cal/mol. This frequency can cause only molecular rotations without altering the molecular structure. The electrical component of the wave changes 4.9×10^9 times per second, which generates a disorganized movement of polar molecules, causing rapid heating [16]. Nonpolar solvents such as hexane and toluene are not affected by microwave energy and therefore require polar additives for use with MAE [17]. The main advantages of MAE are reductions in extraction time and solvent quantity; hence, the technique is environment friendly.

MAE has earlier been reported to be superior to ultrasonic as well as Soxhlet methods for the extraction of chlorinated pesticides from animal feed [18], whereas Barriada-Pereira et al. [19] have found comparable recoveries of organochlorine pesticides from vegetable samples using MAE (81.5%–108.4%) and Soxhlet (75.5%–132.7%). Diagne et al. [20] also observed comparative recoveries of fenitrothion residues in beans using MAE (89.8%) and Soxhlet (88.4%). A comparative study of five extraction techniques revealed the ability of MAE to retrieve a high concentration of pesticides from dietary composites [21]. Dichlorvos has been extracted from vegetables using 10% aqueous ethylene glycol as extractant and medium microwave power for 10 min [22]. Singh et al. [23] used MAE for simultaneous extraction of thiamethoxam, imidacloprid, and carbendazim in fresh and cooked vegetable samples, with recoveries ranging from 68% to 106%. There was no breakdown of pesticides in cooked vegetables, and the parent compounds were extracted intact [23]. Bouaid et al. [24] have reported a simple and rapid (9 min) MAE method for extraction of organophosphate pesticides from 1.5 to 2.5 g of orange peel using 10 mL of hexane/acetone mixture at 90°C and microwave power set at 50% (475 W). The recovery of pesticides ranged between 93% and 101%. Vryzas et al. [25] applied MAE for extraction of carbamates from tobacco and peaches, with a recovery range of 80%–100%. Sun and Lee [14] have achieved 85%–105% recoveries of carbamates in soil using 30 mL methanol as extractant and 6 min microwave heating at 80°C. Although more than 85% recoveries were obtained by both MAE and SFE, slightly higher recoveries of propoxur, propham, and methiocarb were achieved using MAE, whereas SFE showed slightly higher recovery for chlorpropham [14]. Various classes of pesticides, including parathion, methyl parathion, DDE, HCB, simazine, and paraquat, have been extracted from soil samples using MAE, with recoveries more than 80% [26]. MAE has been used for the extraction of organic contaminants, including pesticides, polychlorinated biphenyls, and polycyclic hydrocarbons, from marine sediments and tissues [27,28]. The herbicide methabenzthiazuron has been extracted from soil using MAE; the recoveries of herbicides are dependent on soil type [29]. Triazine herbicides have been extracted from soil samples using MAE with water containing 1% methanol as extractant [30]. MAE was performed at 105°C for 3 min using 80% output of maximum power (1200 W), resulting in 76.1%–87.2% recoveries for triazine [30]. MAE has been used for the extraction of multiresidue phenoxyalkanoic acid herbicides and their phenolic conversion products from soil using aqueous methanol as extraction medium; the recoveries were found to be above 80% [31]. El-Saeid et al. [32] have performed a comparative evaluation of SFE and MAE for extraction of atrazine from food samples.

5.1.5 SUPERCRITICAL FLUID CHROMATOGRAPHY

In recent years, SFC has emerged as a powerful green technology in industries such as pharmaceutical, agricultural, food and environmental sciences. The main difference between SFC and other chromatographic techniques (GC and high-performance liquid chromatography [HPLC]) is the use of a supercritical fluid as the mobile phase in SFC. The supercritical fluid demonstrates unique characteristics that can greatly enhance the efficiency of the process. SFC has several advantages over other conventional chromatographic techniques; compared with HPLC, SFC provides rapid separations without the use of organic solvents. With the desire for environmentally conscious technology, the use of organic chemicals as those used in HPLC could be reduced with the use of SFC. Because SFC generally uses CO_2, which is collected as a byproduct of other chemical reactions or is collected directly from the atmosphere, it contributes no new chemicals to the environment. In addition, SFC separations can be done faster than HPLC separations, because the diffusion of solutes in supercritical fluids is about 10 times greater than that of liquids and about three times less than that of gases. SFC has been shown to be superior than conventional HPLC in terms of high resolution, lower analysis time, and better separation potential for drugs and their metabolites [33,34]. SFC has gained wide recognition as "the Future of HPLC." The higher efficiency, economical impact, safety benefit, and ease of use have made SFC a highly promising technology.

Part of the theory of separation in SFC is based on the density of the supercritical fluid, which corresponds to the solvating power. As the pressure in the system is increased, the supercritical fluid density increases and correspondingly its solvating power increases. Therefore, as the density of the supercritical fluid mobile phase is increased, components retained in the column can be made to elute. This is similar to temperature programming in GC or using a solvent gradient in HPLC. In SFC, the mobile phase is initially pumped as a liquid and is brought into the supercritical region by heating it above its supercritical temperature before it enters the analytical column. It passes through an injection valve, where the sample is introduced into the supercritical stream and then into the analytical column. It is maintained supercritical as it passes through the column and into the detector by a pressure restrictor placed either after the detector or at the end of the column. The restrictor is a vital component as it keeps the mobile phase supercritical throughout the separation and often must be heated to prevent clogging. Although there are a number of possible fluids that may be used in SFC as the mobile phase, CO_2 is widely adapted due to its low cost, low interference with chromatographic detectors, and good physical properties (nontoxic, nonflammable, suitable critical temperature). However, the main disadvantage of CO_2 is its inability to elute very polar or ionic compounds. This drawback can be overcome by adding a small portion of a second fluid called a modifier fluid. The modifier is generally an organic fluid that is completely miscible with CO_2. The addition of the modifier fluid improves the solvating ability of the supercritical fluid and sometimes enhances the selectivity of the separation. It can also improve the separation efficiency by blocking some of the highly active sites on the stationary phase.

Toribio et al. [35] have performed chiral separation of six triazole pesticides, including cyproconazole, propiconazole, diniconazole, hexaconazole, tebuconazole, and tetraconazole, using SFC equipped with a Chiralpak AD column. The same column type was also used for the separation of triadimefon and triadimenol enantiomers and diastereoisomers using SFC [36]. Dost et al. [37] have developed a sensitive SFC-mass spectrometry method for separation and quantification of three classes of pesticides, including triazines, carbamates, and sulfonylureas, in soil samples. A multiresidue method based on SFC and solid-phase extraction has been reported for the analysis of 35 common contaminants including pesticides in water; the detection limit ranged from 0.4 to 2.6 μg/L [38]. SFC coupled with SFE has been used for the analysis of sulfonylureas, their precursors, and metabolites in complex matrices such as soil, vegetation, and cell culture medium [39]. Wheeler and McNalley [40] carried out a comparative evaluation of packed-column versus capillary column SFC for the analysis of herbicides and insecticides.

5.2 MATERIALS AND METHODS

5.2.1 PESTICIDE STANDARDS

The common/chemical names, formulas, and structures of the pesticides used in this study are given in Table 5.1. All the standards of pesticides including eight pyrethroid insecticides (allethrin, resmethrin, phenothrin, permethrin, tetramethrin, cypermethrin, deltamethrin, and one metabolite, phenoxybenzyl alcohol), eight herbicides (trifuralin, tillam, chlorthal, alachlor, propazin, terbuthylazin, atrazine, and simazine), and seven fungicides (PCNB, CDEC, dichlon, captan, captafol, thiram, and carboxin) were obtained from ChemService Inc., West Chester, PA. The standard solutions of pesticides were prepared in HPLC-grade methanol to give a concentration range of 0.001–10 ppm. All the standards as well as samples were filtered with 0.2 μm filter disks before analysis [41].

5.2.2 FOOD SAMPLES

The food products, including the raw whole potatoes, peeled potatoes, and frozen mixed vegetables (carrots, potatoes, green beans, peas, lima beans, okra, corn, onion, and celery), were purchased from a local market. Prior to analysis, all the food products were cleaned, washed, cut, homogenized, and dehydrated at 50°C overnight in an electric oven under vacuum. The powdered food samples were stored in glass containers and kept at −5°C until analyzed.

5.2.3 SPIKING OF FOOD SAMPLES WITH PESTICIDES

Dried food samples were spiked with known amounts (0.16–1.60 μg/g) of pesticides. The spiked food samples were extracted by using SFE or microwave solvent extraction for pesticide residue analysis using SFC.

5.2.4 EXTRACTION OF PESTICIDES

5.2.4.1 Supercritical Fluid Extraction

An SFE apparatus (Model 7680T, Hewlett Packard, USA) was used. The system comprised of an automated restrictor and a solid-phase sorbent trap prepacked with 30 μm Hypersil ODS into which the CO_2 extraction solvent was decompressed during collection. The pesticide extraction method of Khan [42] was used after modifications for optimal conditions for pyrethroids extraction (Table 5.2). A known amount of food sample (5 g) was transferred into the extraction thimble. The extraction process was carried out in three steps. The first step was performed to eliminate hydrocarbons and nonpolar compounds; the second step was performed to extract the pesticides; the third step was performed to wash the thimble and ODS trap as well as to ensure a complete extraction of target pesticides. Appropriate ranges of values were tested for each step to optimize the extraction of pesticides (Table 5.2). The extracted sample was eluted from the trap with 1.5 mL of methanol at a flow rate of 0.4 mL/min and a trap temperature of 40°C and collected in an auto sampler vial. The ODS trap was regenerated between extractions by rinsing with 2.0 mL of methylene chloride followed by 2.0 mL of methanol at 1 mL/min to waste [41]. The entire extraction procedure was automated and controlled by Hewlett Packard Chemstation software.

5.2.4.2 Microwave-Assisted Extraction

A microwave solvent extraction system (Model MES-1000, CEM Corporation, Matthews, NC) consisting of lined extraction vessels (LEV) was used. This 950 W microwave instrument is specifically designed for organic solvents. The safety features of this system prevent the ignition of

TABLE 5.1
List of Pesticides Used in This Study

Common/Chemical Name	Formula	Action	Structure
Allethrin/2-methyl-4-oxo-3-(2-propenyl)-2-cyclopenten-1-yl 2,2-dimethyl-3-(2-methyl-1-propenyl) cyclopropanecarboxylate	$C_{19}H_{26}O_3$	Insecticide	
Resmethrin/[5-(phenylmethyl)-3-furanyl]methyl 2,2-dimethyl-3-(2-methyl-1-propenyl) cyclopropanecarboxylate	$C_{22}H_{26}O_3$	Insecticide	
Phenothrin/(3-phenoxyphenyl)methyl 2,2-dimethyl-3-(2-methyl-1-propenyl) cyclopropanecarboxylate	$C_{23}H_{26}O_3$	Insecticide	

(continued)

TABLE 5.1 (continued)
List of Pesticides Used in This Study

Common/Chemical Name	Formula	Action	Structure
Permethrin/(3-phenoxyphenyl)methyl 3-(2,2-dichloroethenyl)-2,2-dimethylcyclopropanecarboxylate	$C_{21}H_{20}Cl_2O_3$	Insecticide Acaricide	
Tetramethrin/((1,3,4,5,6,7-hexahydro-1,3-dioxo-2H-isoindol-2-yl) methyl 2,2-dimethyl-3-(2-methyl-1-propenyl) cyclopropanecarboxylate	$C_{19}H_{25}NO_4$	Insecticide	
Cypermethrin/cyano(3-phenoxyphenyl)methyl 3-(2,2-dichloroethenyl)-2,2-dimethylcyclopropanecarboxylate	$C_{22}H_{19}Cl_2NO_3$	Insecticide Acaricide	

Deltamethrin/(S)-cyano(3-phenoxyphenyl) methyl (1R,3R)-3-(2,2-dibromoethenyl)-2,2-dimethylcyclopropanecarboxylate — $C_{22}H_{19}Br_2NO_3$ — Insecticide

(S)-alcohol (1R)-cis-acid

Treflan (trifluralin)/2,6-dinitro-N,N-dipropyl-4-(trifluoromethyl) benzenamine — $C_{13}H_{16}F_3N_3O_4$ — Herbicide

Tillam (pebulate)/S-propyl butylethylcarbamothioate — $C_{10}H_{21}NOS$ — Herbicide

Chlorthal/2,3,5,6-tetrachloro-1,4-benzenedicarboxylic acid — $C_8H_2Cl_4O_4$ — Herbicide

Alachlor/2-chloro-N-(2,6-diethylphenyl)-N-(methoxymethyl) acetamide — $C_{14}H_{20}ClNO_2$ — Herbicide

(continued)

TABLE 5.1 (continued)
List of Pesticides Used in This Study

Common/Chemical Name	Formula	Action	Structure
Propazine/6-chloro-N,N′-bis(1-methylethyl)-1,3,5-triazine-2,4-diamine	$C_9H_{16}ClN_5$	Herbicide	
Terbythylazine/6-chloro-N-(1,1-dimethylethyl)-N′-ethyl-1,3,5-triazine-2,4-diamine	$C_9H_{16}ClN_5$	Herbicide	
Atrazine/6-chloro-N-ethyl-N′-(1-methylethyl)-1,3,5-triazine-2,4-diamine	$C_8H_{14}ClN_5$	Herbicide	

Name	Formula	Use
Simazine/6-chloro-N,N'-diethyl-1,3,5-triazine-2,4-diamine	$C_7H_{12}ClN_5$	Herbicide Algicide
PCNB (quintozene)/pentachloronitrobenzene	$C_6Cl_5NO_2$	Fungicide
CDEC (sulfallate)/2-chloro-2-propenyl diethylcarbamodithioate	$C_8H_{14}ClNS_2$	Fungicide Herbcide
Dichlone/2,3-dichloro-1,4-naphthalenedione	$C_{10}H_4Cl_2O_2$	Fungicide Algicide
Captan/3a,4,7,7a-tetrahydro-2-[(trichloromethyl)thio]-1H-isoindole-1,3(2H)-dione	$C_9H_8Cl_3NO_2S$	Fungicide

(continued)

TABLE 5.1 (continued)
List of Pesticides Used in This Study

Common/Chemical Name	Formula	Action	Structure
Captafol/3a,4,7,7a-tetrahydro-2-[(1,1,2,2-tetrachloroethyl)thio]-1H-isoindole-1,3(2H)-dione	$C_{10}H_9Cl_4NO_2S$	Fungicide	
Thiram/tetramethylthioperoxydicarbonic diamide	$C_6H_{12}N_2S_4$	Fungicide	
Carboxin/5,6-dihydro-2-methyl-N-phenyl-1,4-oxathiin-3-carboxamide	$C_{12}H_{13}NO_2S$	Fungicide	

TABLE 5.2
Optimized Conditions for SFE of Pesticides

Parameter/Condition	Step 1		Step 2		Step 3	
	Tested Values	Optimized Values	Tested Values	Optimized Values	Tested Values	Optimized Values
Modifier (%)	0	0	0–2	0	20–35	30
CO_2 density (g/mL)	0.20–0.50	0.25	0.60–0.75	0.67	0.60–0.75	0.67
Chamber temperature (°C)	30–50	40	70–95	80	70–95	80
Flow rate (mL/min)	0.5–2.0	1.0	2.0–3.0	2.5	2.0–3.0	2.5
Pressure (psi)	900–1200	1117	3000–3800	3469	3000–3800	3469
Nozzle temperature (°C)	40–60	45	40–60	45	40–60	45
Time (min)	5–10	5	30–50	40	5–15	10

TABLE 5.3
Optimized Conditions for MAE of Pesticides

Parameter/Condition	Tested Values	Optimized Values
Microwave power (%)	60–80	75
Temperature (°C)	100–135	125
Solvent volume (mL)	40–65	60
Pressure (psi)	70–95	85
Time (min)	20–35	30

flammable or explosive solvents. Additionally, a solvent vapor detector system automatically turns off the microwave magnetron if solvent vapors are detected in the microwave cavity. Double-walled extraction vessels, made of inner a Teflon PFA liner and Ultem polyetherimide outer bode, suitable for use with organic solvents were used.

Preweighed food samples (5 g) were extracted with 60 mL of solvent (acetone–hexane, 3:2). The extraction conditions are given in Table 5.3. After extraction, the food samples were filtered and concentrated using a rotary evaporator. The residue on the filter paper was reextracted thrice with 10 mL of methanol to ensure the complete extraction of pesticides. The filtrates were concentrated in a rotary evaporator [41].

5.2.5 DETERMINATION OF PESTICIDES

A Hewlett Packard Supercritical Fluid Chromatograph (SFC, Model G 1205A) attached to an HP 1050 diode array detector, modifier pump, and a silica column (Alltec Hypersil APS, 25 micron, length 205 mm, ID 4.6 mm) was used. Earlier reported methods [43,44] were modified for the analysis of different groups of pesticides. Chromatographic conditions were optimized, and the analysis of pyrethroids was performed at an oven temperature of 60°C, pressure 130–200 bar, flow rate 1–3 mL/min, and 2% methanol as modifier; pyrethroid peaks were detected at a wavelength of 220 nm (Table 5.4). Herbicides and fungicides were run at an oven temperature of 30°C, pressure 80–150 bar, flow rate 1–2 mL/min, and 2%–3% modifier. The herbicides and fungicides were detected at 220 and 210 nm, respectively (Table 5.4).

TABLE 5.4
Protocol for Chromatographic Determination of Various Pesticides Using SFC

| | | Optimized Values for | | |
Parameter/Condition	Tested Values	Insecticides	Herbicides	Fungicides
Concentration (ppm)	0.2–10.0	0.30–1.0	0.25–1.0	0.45–1.0
Temperature (°C)	20–80	60	30	30
Pressure (bar)	50–250	130–200	80–150	80–150
Flow rate (mL/min)	0.5–4	1–2	1–2	1–2
Modifier (%)	1–4	2	2	2
Wavelength (nm)	200–250	220	220	210
Run time (min)	8–40	25	10	12

5.3 RESULTS

5.3.1 EXTRACTION OF PESTICIDES FROM FOOD SAMPLES

5.3.1.1 Optimized Conditions for Pesticide Extraction Using SFE

The optimized conditions for extraction of pesticides from vegetables using SFE are shown in Table 5.2. These conditions apply to all the pesticides used in this study and therefore constitute a single protocol for extraction of all these pesticides together in a multiresidue analysis. The entire procedure of SFE comprising three sequential steps was completed in 55 min. Comparatively milder conditions were applied for short time (5 min) in the beginning to isolate interfering components, including hydrocarbons and various nonpolar species. All the parameters including CO_2 density, chamber temperature, flow rate, and pressure have to be increased while switching from step 1 to step 2. The optimal settings for step 2 and step 3 have been found to be the same except for the use of 30% modifier and a shorter time in step 3 (Table 5.2). There is no requirement of modifier in the first two steps, whereas the use of modifier in the last step is important to ensure the complete extraction of pesticides from vegetable samples.

5.3.1.2 Optimized Conditions for Pesticide Extraction Using MAE

The optimized conditions for MAE of pesticides are shown in Table 5.3. Of the various ranges tested, a microwave power of 75%, temperature of 125°C, and pressure of 85 psi were found to be appropriate for MAE. The extraction was completed in 30 min using 60 mL of solvent (Table 5.3). After extraction, the food samples were filtered by a buchner funnel, and the filtrates were concentrated using a rotary evaporator. The residue that remained on the filter paper was reextracted thrice with 10 mL of methanol, and the whole filtrate was rotary concentrated.

5.3.2 DETERMINATION OF PESTICIDES USING SFC

5.3.2.1 Optimized Conditions for Pesticide Analysis Using SFC

The optimum measuring temperature is 60°C for pyrethroid insecticides and 30°C for herbicides and fungicides (Table 5.4). The optimal pressure range is 130–200 bar for pyrethroids and 80–150 bar for both herbicides and fungicides. A flow rate of 1–2 mL/min and the use of 2% modifier have been found to be suitable for analysis of pesticides using SFC. A comparatively longer run time (25 min) is required for pyrethroid insecticides than the shorter run times for herbicides (10 min) and fungicides (12 min). The appropriate wavelength for UV detection of insecticides and herbicides has been found to be 220 nm and for fungicides, 210 nm (Table 5.4).

5.3.2.2   Retention Times and Limits of Detection

The retention times (RTs) and lower limits of detection (LOD) of various insecticides, herbicides, and fungicides are given in Tables 5.5 through 5.7, respectively. These values have been obtained by using the standard solutions of individual pesticides and performing the SFC with optimized conditions, as mentioned before. The RTs of insecticides ranged from 8.4 to 22.9 min while all of the peaks are distinctly identified (Figure 5.2). The LOD of various insecticides ranged between 0.31 and 0.62 ppm (Table 5.5). The peaks for all the eight herbicides are well resolved on chromatograms (Figure 5.3), with their RTs ranging 3.7–9.2 min and LOD ranging 0.25–0.69 ppm (Table 5.6). The RTs for seven fungicides (Figure 5.4) ranged between 4.1 and 19.6 min and their LODs, between 0.45 and 0.78 ppm (Table 5.7).

TABLE 5.5
RTs and Lower LOD of Different Pyrethroid Insecticides

Pyrethroids	RT (MIN)	LOD (ppm)
Allethrin	8.4	0.35
Resmethrin	9.2	0.52
Phenothrin	10.5	0.38
Permethrin (*cis* and *trans*)	12.8, 13.5	0.31
Tetramethrin	14.5	0.46
Cypermethrin (mixed isomers)	16.6, 17.1, 17.8	0.38
Deltamethrin	19.9	0.54
Phenoxybenzyl alcohol	22.9	0.62

TABLE 5.6
RTs and Lower LOD of Different Herbicides

Herbicides	RT (MIN)	LOD (ppm)
Trifuralin	3.7	0.56
Tillam	4.0	0.69
Chlorthal	4.7	0.29
Alachlor	5.5	0.38
Propazin	7.4	0.44
Terbuthylazin	7.7	0.51
Atrazin	8.2	0.45
Simazin	9.2	0.25

TABLE 5.7
RTs and Lower LOD of Different Fungicides

Fungicides	RT (MIN)	LOD (ppm)
PCNB	4.1	0.58
CDEC	4.8	0.45
Dichlon	5.5	0.52
Captan	7.2	0.59
Captafol	8.3	0.66
Thiram	9.2	0.78
Carboxin	10.6	0.53

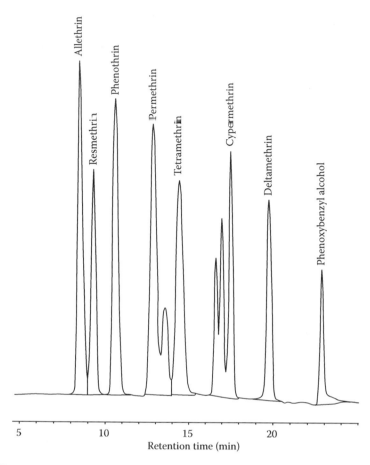

FIGURE 5.2   Chromatogram showing the separation of pyrethroid insecticides using SFC.

5.3.2.3   Recoveries of Pesticides from Food Samples

The recoveries of pyrethroid insecticides from whole potatoes, peeled potatoes, and mixed vegetables ranged 93.8%–99.8%, 92.3%–105.8%, and 93.6%–102.7%, respectively, with the use of SFE. The corresponding recovery ranges while using MAE were found to be 94.2%–102%, 96.6%–101.2%, and 96%–103% (Table 5.8). The extraction of the common metabolite of pyrethroids, phenoxyben-zyl alcohol, was highly efficient using either SFE or MAE, resulting in high recoveries (Table 5.8). The recoveries of herbicides using SFE ranged 78.2%–110.8% (whole potatoes), 66.7%–112.9% (peeled potatoes), and 97.4%–114.3% (mixed vegetables) (Table 5.8). The recovery patterns of her-bicides using MAE appeared to be 71.8%–110.6% (whole potatoes), 79.3%–111.7% (peeled pota-toes), and 91.1%–109.2% (mixed vegetables). Only one herbicide, propazine, showed poor recovery (<80%) from potatoes, whereas its recovery from vegetables was satisfactory using SFE (104%) or MAE (97.3%) (Table 5.9). One of the fungicides, thiram, could not be extracted by MAE, whereas its recovery with the use of SFC was 95.2% (whole potatoes), 57.4% (peeled potatoes), and 78.6% (mixed vegetables) (Table 5.10). The recovery ranges for fungicides using SFE were found to be 70.7%–109% (whole potatoes), 57.4%–110.4% (peeled potatoes), and 78.6%–113.4% (mixed vegetables). The recoveries of various fungicides (excluding thiram) using MAE were in the range of 60.8%–99.4% (whole potatoes), 62.6%–112% (peeled potatoes), and 57.5%–91.1% (mixed vegetables) (Table 5.10).

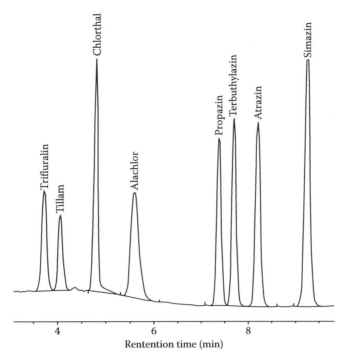

FIGURE 5.3   Chromatogram showing the separation of herbicides using SFC.

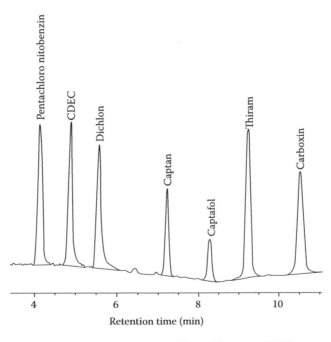

FIGURE 5.4   Chromatogram showing the separation of fungicides using SFC.

TABLE 5.8
Percent Recovery of Pyrethroid Insecticides from Food Samples Using SFE or MAE

Pyrethroids	Whole Potatoes		Peeled Potatoes		Mixed Vegetables	
	SFE	MAE	SFE	MAE	SFE	MAE
Allethrin	95.70 + 1.79	97.33 ± 1.86	98.23 ± 1.16	101.23 ± 2.00	98.40 ± 1.07	99.40 ± 2.20
Resmethrin	99.20 ± 0.85	96.53 ± 2.23	105.87 ± 2.93	100.70 ± 3.18	93.67 + 0.68	101.83 ± 5.99
Phenothrin	93.83 ± 2.76	99.47 ± 0.33	96.80 ± 1.66	98.37 ± 1.00	94.70 ± 2.49	99.10 ± 5.05
Permethrin	99.80 ± 7.47	102.00 ± 0.50	100.37 ± 1.39	99.20 ± 1.30	102.03 ± 3.59	103.03 ± 2.26
Tetramethrin	96.23 ± 0.86	94.20 ± 2.55	93.07 ± 0.65	97.97 ± 4.28	95.83 ± 0.52	96.00 ± 2.71
Cypermethrin	99.20 ± 0.85	97.83 ± 0.39	92.30 ± 2.05	98.97 ± 0.66	97.07 ± 1.88	96.87 ± 1.30
Deltamethrin	98.03 ± 0.87	97.33 ± 1.92	93.83 ± 1.61	96.60 ± 1.63	102.77 ± 1.35	103.23 ± 2.82
Phenoxybenzyl alcohol	113.83 ± 4.69	110.90 ± 1.53	113.30 ± 2.69	106.00 ± 2.45	109.70 ± 0.78	108.97 ± 2.83

Note: Values are the mean of three replicates ± standard deviation.

TABLE 5.9
Percent Recovery of Herbicides from Food Samples Using SFE or MAE

Herbicides	Whole Potatoes		Peeled Potatoes		Mixed Vegetables	
	SFE	MAE	SFE	MAE	SFE	MAE
Trifuralin	110.80 ± 0.90	110.60 ± 0.80	110.40 ± 1.10	106.30 ± 1.10	109.40 ± 0.50	104.90 ± 2.80
Tillam	95.10 ± 1.60	96.80 ± 3.20	109.60 ± 1.00	104.10 ± 0.40	97.40 ± 2.20	97.50 ± 2.20
Chlorthal	104.50 ± 0.70	109.40 ± 2.20	110.00 ± 0.80	108.70 ± 3.50	113.70 ± 1.30	91.10 ± 0.90
Alachlor	94.90 ± 2.50	90.70 ± 1.30	112.90 ± 0.50	111.70 ± 1.50	109.60 ± 1.00	92.20 ± 1.10
Propazin	78.20 ± 1.90	71.80 ± 4.30	66.70 ± 2.10	79.30 ± 2.30	104.00 ± 0.40	97.30 ± 2.90
Terbuthylazin	85.70 ± 2.00	83.50 ± 1.90	85.00 ± 2.70	87.10 ± 2.70	114.30 ± 4.40	109.20 ± 2.20
Atrazin	96.60 ± 1.20	89.20 ± 2.00	82.50 ± 1.90	94.00 ± 2.60	110.50 ± 3.00	98.40 ± 0.70
Simazin	95.20 ± 0.90	86.70 ± 2.90	108.00 ± 0.50	99.20 ± 0.40	111.50 ± 0.70	102.20 ± 1.20

TABLE 5.10
Percent Recovery of Fungicides from Food Samples Using SFE or MAE

Fungicides	Whole Potatoes		Peeled Potatoes		Mixed Vegetables	
	SFE	MAE	SFE	MAE	SFE	MAE
PCNB	70.70 ± 1.00	60.90 ± 0.90	76.00 ± 2.00	62.60 ± 0.70	83.50 ± 1.90	57.50 ± 2.10
CDEC	109.00 ± 3.70	99.40 ± 1.30	109.70 ± 0.40	106.20 ± 3.10	113.40 ± 0.90	77.30 ± 1.30
Dichlon	88.10 ± 1.30	60.80 ± 0.50	77.30 ± 1.20	112.00 ± 1.50	80.10 ± 2.40	62.50 ± 0.60
Captan	107.20 ± 2.90	97.70 ± 1.00	96.70 ± 1.20	87.70 ± 1.10	97.60 ± 0.50	88.60 ± 1.80
Captafol	102.77 ± 1.35	97.00 ± 1.20	110.40 ± 0.50	92.20 ± 1.00	96.50 ± 1.90	91.10 ± 0.90
Thiram	95.20 ± 0.90	00.00 ± 0.00	57.40 ± 2.00	00.00 ± 0.00	78.60 ± 2.50	00.00 ± 0.00
Carboxin	89.30 ± 0.70	77.10 ± 1.20	91.60 ± 0.90	89.60 ± 0.90	105.00 ± 2.60	66.80 ± 2.10

5.4 CONCLUSIONS

Since conventional extraction techniques such as Soxhlet extraction, sonication, and mechanical shaking are laborious, time consuming, and require large volumes of toxic organic solvents, much attention is being paid to develop more efficient and environment friendly techniques for rapid extraction of pesticides from complex matrices. Both SFE and MAE have emerged as efficient and rapid techniques for extraction of pesticides from a wide range of matrices. Our findings clearly indicate the suitability of SFE and MAE for rapid and efficient extraction of various groups of pesticides from food samples. The chromatographic separation and determination of pesticides using SFC offer the advantages of high sensitivity, greater resolution, minimal use of organic solvents, and versatility for analyzing different groups of pesticides in food samples.

REFERENCES

1. Antunes, P., Gil, O., and Gil, M. G. 2003. Supercritical fluid extraction of organochlorines from fish muscle with different sample preparation. *J Supercrit Fluids* 25:135–142.
2. Luque de Castro, M. D. and Garcia Ayuso, L. E. 1998. Soxhlet extraction of solid materials: an outdated technique with a promising future. *Anal Chim Acta* 369:1–10.
3. Hawthorne, S. B. 1990. Analytical-scale supercritical fluid extraction. *Anal Chem* 62:633A–642A.
4. Mchugh, M. A. and Krukonis, V. J. 1986. *Supercritical Fluid Extraction: Principles and Practice.* Butterworths: Boston, MA.
5. Stahle, E., Quirin, K. W., and Gerard, D. 1988. *Dense Gases for Extraction and Refining.* Verlag: S. Berlin.
6. Rissato, S. R., Galhiane, M. S., Apon, B. M., and Arruda, M. S. 2005. Multiresidue analysis of pesticides in soil by supercritical fluid extraction/gas chromatography with electron-capture detection and confirmation by gas chromatography-mass spectrometry. *J Agric Food Chem* 53:62–69.
7. Aguilera, A., Brotons, M., Rodriguez, M., and Valverde, A. 2003. Supercritical fluid extraction of pesticides from a table-ready food composite of plant origin (gazpacho). *J Agric Food Chem* 51:5616–5621.
8. El Saeid, M. H. 2003. Pesticide residues in canned foods, fruits and vegetables: The application of supercritical fluid extraction and chromatographic techniques in the analysis. *Sci World J* 3:1314–1326.
9. Aguilera, A., Rodriguez, M., Brotons, M., Boulaid, M., and Valverde, A. 2005. Evaluation of supercritical fluid extraction/aminopropyl solid-phase "in-line" cleanup for analysis of pesticide residues in rice. *J Agric Food Chem* 53:9374–9382.
10. Rissato, S. R., Galhiane, M. S., Knoll, F. R., and Apon, B. M. 2004. Supercritical fluid extraction for pesticide multiresidue analysis in honey: Determination by gas chromatography with electron-capture and mass spectrometry detection. *J Chromatogr A* 1048:153–159.
11. Zhao, C., Hao, G., Li, H., and Chen, Y. 2002. Supercritical fluid extraction for the separation of organochlorine pesticides residue in *Angelica sinensis*. *Biomed Chromatogr* 16:441–445.
12. Yu, J. J. 2002. Removal of organophosphate pesticides from wastewater by supercritical carbon dioxide extraction. *Water Res* 36:1095–1101.
13. King, J. W. and Zhang, Z. 2002. Derivatization reactions of carbamate pesticides in supercritical carbon dioxide. *Anal Bioanal Chem* 374:88–92.
14. Sun, L. and Lee, H. K. 2003. Optimization of microwave-assisted extraction and supercritical fluid extraction of carbamate pesticides in soil by experimental design methodology. *J Chromatogr A* 1014:165–177.
15. Ganzler, K., Salgo, A., and Valko, K. 1986. Microwave extraction. A novel sample preparation method for chromatography. *J Chromatogr* 371:299–306.
16. Letellier, M. and Budzinski, H. 1999. Microwave assisted extraction of organic compounds. *Analusis* 27:259 271.
17. Barnabus, I. J., Dean, J. R., Fowlis, I. A., and Owen, S. P. 1995. Extraction of polycyclic aromatic hydrocarbons from highly contaminated soils using microwave energy. *Analyst* 120:1897–1903.
18. Gfrerer, M., Chen, S., Lankmayr, E. P., Quan, X., and Yang, F. 2004. Comparison of different extraction techniques for the determination of chlorinated pesticides in animal feed. *Anal Bioanal Chem* 378:1861–1867.
19. Barriada-Pereira, M., Concha-Grana, E., Gonzalez-Castro, M. J., Muniategui-Lorenzo, S., Lopez-Mahia, P., Prada-Rodriguez, D., and Fernandez-Fernandez, E. 2003. Microwave-assisted extraction versus Soxhlet extraction in the analysis of 21 organochlorine pesticides in plants. *J Chromatogr A* 1008:115–122.
20. Diagne, R. G., Foster, G. D., and Khan, S. U. 2002. Comparison of soxhlet and microwave-assisted extractions for the determination of fenitrothion residues in beans. *J Agric Food Chem* 50:3204–3207.
21. Rosenblum, L., Garris, S. T., and Morgan, J. N. 2002. Comparison of five extraction methods for determination of incurred and added pesticides in dietary composites. *JAOAC Int* 85:1167–1176.

22. Chen, Y. I., Su, Y. S., and Jen, J. F. 2002. Determination of dichlorvos by on-line microwave-assisted extraction coupled to headspace solid-phase microextraction and gas chromatography-electron-capture detection. *J Chromatogr A* 976:349–355.

23. Singh, S. B., Foster, G. D., and Khan, S. U. 2004. Microwave-assisted extraction for the simultaneous determination of thiamethoxam, imidacloprid, and carbendazim residues in fresh and cooked vegetable samples. *J Agric Food Chem* 52:105–109.

24. Bouaid, A., Martin-Esteban, A., Fernandez, P., and Camara, C. 2000. Microwave-assisted extraction method for the determination of atrazine and four organophosphorus pesticides in oranges by gas chromatography (GC). *Fresenius J Anal Chem* 367:291–294.

25. Vryzas, Z., Papadakis, E. N., and Papadopoulou-Mourkidou, E. 2002. Microwave-assisted extraction (MAE)-acid hydrolysis of dithiocarbamates for trace analysis in tobacco and peaches. *J Agric Food Chem* 50:2220–2226.

26. de Andrea, M. M., Papini, S., and Nakagawa, L. E. 2001. Optimizing microwave-assisted solvent extraction (MASE) of pesticides from soil. *J Environ Sci Health B* 36:87–93.

27. Jayaraman, S., Pruell, R. J., and McKinney, R. 2001. Extraction of organic contaminants from marine sediments and tissues using microwave energy. *Chemosphere* 44:181–191.

28. Numata, M., Yarita, T., Aoyagi, Y., and Takatsu, A. 2004. Evaluation of a microwave-assisted extraction technique for the determination of polychlorinated biphenyls and organochlorine pesticides in sediments. *Anal Sci* 20:793–798.

29. Baez, M. E., Aponte, A., and Sanchez-Rasero, F. 2003. Microwave-assisted solvent extraction of the herbicide methabenzthiazuron from soils and some soil natural organic and inorganic constituents. Influence of environmental factors on its extractability. *Analyst* 128:1478–1484.

30. Shen, G. and Lee, H. K. 2003. Determination of triazines in soil by microwave-assisted extraction followed by solid-phase microextraction and gas chromatography-mass spectrometry. *J Chromatogr A* 985:167–174.

31. Patsias, J., Papadakis, E. N., and Papadopoulou-Mourkidou, E. 2002. Analysis of phenoxyalkanoic acid herbicides and their phenolic conversion products in soil by microwave assisted solvent extraction and subsequent analysis of extracts by on-line solid-phase extraction-liquid chromatography. *J Chromatogr A* 959:153–161.

32. El Saeid, M. H., Kanu, I., Anyanwu, E. C., and Saleh, M. A. 2005. Impact of extraction methods in the rapid determination of atrazine residues in foods using supercritical fluid chromatography and enzyme linked immunosorbent assay: Microwave solvent vs supercritical fluid extractions. *Sci World J* 5:11–19.

33. Toribio, L., del Nozal, M. J., Bernal, J. L., Alonso, C., and Jimenez, J. J. 2005. Comparative study of the enantioselective separation of several antiulcer drugs by high-performance liquid chromatography and supercritical fluid chromatography. *J Chromatogr A* 1091:118–123.

34. Wang, Z., Li, S., Jonca, M., Lambros, T., Ferguson, S., Goodnow, R., and Ho, C. T. 2006. Comparison of supercritical fluid chromatography and liquid chromatography for the separation of urinary metabolites of nobiletin with chiral and non-chiral stationary phases. *Biomed Chromatogr* 20:1206–1215.

35. Toribio, L., del Nozal, M. J., Bernal, J. L., Jimenez, J. J., and Alonso, C. 2004. Chiral separation of some triazole pesticides by supercritical fluid chromatography. *J Chromatogr A* 1046:249–253.

36. del Nozal, M. J., Toribio, L., Bernal, J. L., and Castano, N. 2003. Separation of triadimefon and triadimenol enantiomers and diastereoisomers by supercritical fluid chromatography. *J Chromatogr A* 986:135–141.

37. Dost, K., Jones, D. C., Auerbach, R., and Davidson, G. 2000. Determination of pesticides in soil samples by supercritical fluid chromatography-atmospheric pressure chemical ionisation mass spectrometric detection. *Analyst* 125:1751–1755.

38. Toribio, L., del Nozal, M. J., Bernal, J. L., Jimenez, J. J., and Serna, M. L. 1998. Packed-column supercritical fluid chromatography coupled with solid-phase extraction for the determination of organic microcontaminants in water. *J Chromatogr A* 823:163–170.

39. McNally, M. E. and Wheeler, J. R. 1988. Supercritical fluid extraction coupled with supercritical fluid chromatography for the separation of sulfonylurea herbicides and their metabolites from complex matrices. *J Chromatogr* 435:63–71.

40. Wheeler, J. R. and McNally, M. E. 1987. Comparison of packed column and capillary column supercritical fluid chromatography and high-performance liquid chromatography using representative herbicides and pesticides as typical moderate polarity and molecular weight range molecules. *J Chromatogr* 410:343–353.

41. El Saeid, M. H. 1999. New techniques for residue analysis of pesticides in foods. PhD dissertation, Texas Southern University, Houston, TX and Al Azhar University, Cairo, Egypt.
42. Khan, S. U. 1995. Supercritical Fluid Extraction of bound pesticide residues from soil and food commodities. *J Agric Food Chem* 43:1718–1723.
43. Nishikawa, Y. 1993. Enantiomer separation of synthetic pyrethroids by subcritical and supercritical fluid chromatography with chiral stationary phases. *Anal Sci* 9:33–37.
44. France, J. E. and Voorhees, K. J. 1988. Capillary supercritical fluid chromatography with ultraviolet multichannel detection of some pesticides and herbicides. *J High Resol Chrom* 11:692–696.

6 Disposable Electrochemical Biosensors for Environmental Analysis

*Serena Laschi, Sonia Centi, Francesca Bettazzi,
Ilaria Palchetti, and Marco Mascini*

CONTENTS

6.1   INTRODUCTION

In recent years electrochemical sensors and biosensors have become an accepted part of analytical chemistry, since they satisfy the expanding need for rapid and reliable measurements.

Like many other technologies, electrochemical sensors and biosensors have benefited from the growing power of new materials, design, and processing tools; thus many technologies are available to fabricate miniaturized, simple-to-operate, and low-cost devices. Among these thick-film technology is one of the most used, since the equipment needed is less complex and costly than those of others; moreover, thick-film electrochemical transducers can be easily mass produced at low cost and thus used as disposable; in electrochemistry a disposable sensor offers the advantage of not suffering from the electrode fouling that can result in loss of sensitivity and reproducibility [1]. Nowadays disposable thick-film electrochemical transducers are produced mainly by the screen-printing technique. Screen-printed electrodes are planar devices, based on different layers of inks printed on a plastic or ceramic substrate. Many articles [2–8] in the recent years report the use of these devices for environmental as well for clinical or food analysis, and many of these articles are related to the use of these electrodes in the field of electrochemical biosensors. One of the most used strategies in screen-printed electrode production is the use of carbon inks for the fabrication of the working electrode surface, since this material is relatively inexpensive, shows a wide working potential range, is inert, has good electrical conductivity, and has a relatively high hydrogen overpotential. Moreover, a carbon screen-printed electrode surface can be easily modified using biomolecules, redox compounds, or catalytic particles, thus increasing the range of compounds detectable.

In this chapter we present some protocols for the detection of environmental organic pollutants, namely, organophosphoric and carbamic pesticide, and herbicides, based on disposable screen-printed biosensors.

6.2   DISPOSABLE CARBON-MODIFIED ELECTROCHEMICAL BIOSENSORS FOR ORGANOPHOSPHORUS AND CARBAMATE PESTICIDES

Organophosphorus and carbamate compounds have come into widespread use in agriculture, because they show low environmental persistence; nevertheless, they exert a high acute toxicity. The principal effect of these compounds is the inhibition of the enzyme acetylcholinesterase (AChE), which is essential for terminating the action of the neurotransmitter acetylcholine (ACh). Actually, the intoxication by these compounds results in an accumulation of endogenous ACh and continual stimulation of the nervous system. Due to their toxicological activity, some of these compounds have also been used as chemical warfare agents (CWAs) [1,2].

The most frequently used methods for the unambiguous identification of organophosphorus and carbamate compounds are based on gas chromatography (GC) in combination with mass spectrometry (GC–MS) and/or tandem mass spectrometry (GC–MS/MS), liquid chromatography (LC) coupled with MS, and nuclear magnetic resonance (NMR) spectrometry [3].

An alternative to the chromatographic determination of these compounds is the use of biosensor-based techniques. Biosensors result in rapid, simple, and selective methods for the fast analysis of these compounds, because they combine the selectivity of the enzymatic reactions with operational simplicity. The bioanalytical detection of organophosphate and carbamate pesticides using cholinesterases (ChEs), either free in solution or immobilized as a biorecognition element in biosensors, has a long tradition. The promising results obtained in this research field have allowed the use of ChE in combination with a variety of transducers, such as potentiometric [4,5], amperometric [6,9–13], or optical transducers [14].

If a salt of acetyl- or butyrylthiocholine (ATCh and BTCh, respectively) is used as a substrate for the ChE enzymes, thiocholine (TCh) is produced during the enzymatic reaction. Thiol-containing compounds are known as oxidable at the surface of solid electrodes, but the oxidation generally requires high potential values on a suitable electrode [5,15]. This can be overcome using chemically modified carbon electrodes [16,18–22].

In this work, screen-printed carbon electrodes (SPCEs) were modified by incorporating in the ink an optimized percentage of cobalt(II)-phthalocyanine (CoPC) [23]. As reported in the literature [24], among the electrochemical mediators, CoPC was indicated as one of the most suitable for the detection of thiol-containing molecules [16,17], and the resulting oxidation signals occur at lower voltages, thus limiting the electrochemical interference of other oxidable compounds. Using these modified SPCEs, under optimized chronoamperometric conditions, it is possible to detect pesticides, such as carbofuran, through the study of the AChE activity. Actually, the AChE free in solution is incubated with the pesticide. The inhibitory effect of the pesticide determines the decrease in the catalytic activity of AChE. As a consequence, less thiocholine is produced, and thus a current value lower than that recorded in a blank solution is obtained. This current decrease is correlated with the pesticide concentration. A detection limit (DL) of 2.0×10^{-10} M for carbofuran was found in an analysis time of 15 min.

In order to increase the versatility of the device, a reproducible and reliable immobilization strategy of AChE onto the SPCE surface was studied. The AChE was immobilized by cross-linking with glutaraldehyde, bovine serum albumin (BSA), and Nafion onto the surface of the modified SPCE. The composition of the surface protein layer (enzyme units, glutaraldehyde, BSA, and Nafion amounts) was optimized to obtain a high and reliable response toward the substrate and AChE inhibitors.

In the optimized conditions, the dynamic range for carbofuran detection was 10^{-10} to 10^{-7} M with a DL of 4.9×10^{-10} M, for an analysis time of 15 min. This is an important feature, considering that the immobilization can determine a loss of the activity of the enzyme, which influences the sensitivity as well as dynamic range of the pesticide detection [4]. Moreover, the proposed method was less prone to electrochemical interferences since the incubation and measurement were performed in two separate steps.

6.2.1 OBJECTIVES

1. To detect organophosphorus and carbamate pesticides using an acetylcholinesterase (AChE) based biosensor. The inhibitory effect of the pesticide determines the decrease in the catalytic activity of AChE; as a consequence, less thiocholine (TCh) was produced from acetylthiocholine (ATCh), the enzymatic substrate.
   Therefore, the current value, due to the oxidation of TCh at the modified carbon screen-printed electrodes (SPCEs), is lower than that recorded in a blank solution. This current decrease is correlated with the pesticide concentration.
2. To test a cobalt(II)-phthalocyanine (CoPC)-modified SPCEs as transducers of an acetylcholinesterase (AChE)-based biosensor.
3. To test a biosensor in a standard pesticide solution.

6.2.2 Materials and Instrumentation

- AChE from Electric Eel (EC 3.1.1.7), ATCh chloride, BSA, Nafion (perfluorinated ion-exchange resin) 5 WT%, carbofuran (2,3-dihydro-2,2-dimethyl-7-benzo-furanol N-methylcarbamate), glycine, glutaraldehyde 25% v/v, acetonitrile HPLC grade, sodium dihydrogenophosphate, disodium hydrogenophosphate, KCl, and NaClO 14%.
- CoPC modified screen-printed electrodes (Ecobioservices, http://www.ebsr.it).
- All the electrochemical measurements are performed in phosphate buffer 0.05 M at pH 8.0 with 0.1 M KCl (measuring buffer). ATCh, AChE, and other dilutions are prepared in the measuring buffer.

6.2.3 Procedure

6.2.3.1 Modified SPCEs Preparation

CoPC-modified, carbon screen-printed electrodes are prepared by mixing the carbon ink with CoPC powder in an amount equivalent to 5% m/m of the total carbon in the printing ink. The mixture is then homogenized. The produced sensors are stored in the dark at room temperature. The scheme of an electrochemical cell and electrode design is reported in Ref. [9] and can be obtained from Ecobioservices (Florence, Italy). Before use, the pseudo Ag reference electrode is oxidized using NaClO 14% solution, to avoid the oxidation of the Ag pseudo-reference by thiols during measurements.

6.2.3.2 Immobilization of AChE onto CoPC-Modified SPCEs

AChE is immobilized onto the electrode surface by cross-linking with glutaraldehyde, BSA, and Nafion. A first enzyme solution is prepared by mixing 3.3 mL of the measuring buffer with 50 μL of AChE solution and 132 mg of BSA. Then, to 300 μL of this mixture are added 6 μL of glutaraldehyde 25% v/v and 90 μL of Nafion 5% m/m. The final reagent concentrations are AChE 7.5 U mL^{-1}, BSA 3% m/m, glutaraldehyde 0.25%, and Nafion 0.25% m/m, respectively.

Finally, 7 μL of this mixture is casted onto the working area of a CoPC-modified electrode. When the enzymatic layer is dried, electrodes are dipped in a 0.1 M glycine solution for 30 min. This is a blocking treatment, necessary to saturate the surface sites not involved in the enzymatic immobilization. Biosensors are then stored at +4°C until use.

6.2.3.3 Blank Measurements

The enzyme-modified working electrode is covered with 10 μL of buffer; after 5 min, the solution is removed. Then, 200 μL of enzymatic substrate solution (1 mM) is casted onto the cell; after 10 min, the potential is applied, and the current response at 30 s is evaluated. Chronoamperometric measurements are performed at the applied potential of +0.1 V vs. pseudo Ag/AgCl reference electrode.

6.2.3.4 Inhibition Measurements

The enzyme-modified working electrode is covered with 10 μL of buffer with inhibitor; after 5 min, the solution is removed, and the biosensor is washed with the buffer. Then, 200 μL of enzymatic substrate solution (1 mM) is casted onto the cell; after 10 min, the potential is applied, and the current response at 30 s is evaluated. Chronoamperometric measurements are performed at the applied potential of +0.1 V vs. pseudo Ag/AgCl reference electrode. All potentials are referred to the screen-printed Ag/AgCl pseudo-reference electrode; the experiments are carried out at room temperature (25°C).

6.2.4 DISCUSSION

As reported in the literature [9,10], among the electrochemical mediators, CoPC was indicated as one of the most suitable for the detection of thiol-containing molecules, and the resulting oxidation signals occur at lower voltages, thus limiting the electrochemical interference of other oxidable compounds. Using CoPC-modified SPCEs, under optimized chronoamperometric conditions, it is possible to detect pesticides, such as carbofuran, through the study of the AChE activity.

In Figure 6.1 a typical inhibition curve is reported. This was obtained by plotting the inhibition percentage ($I\%$) vs. the carbofuran concentration.

$$I\% = 100 \left[(I_1 - I_2)/I_1 \right]$$

where I_2 is the oxidation current obtained for the sample—carbofuran solution, and I_1 the oxidation current obtained for a blank solution (incubation without pesticide).

The investigated pesticide concentration range is 10^{-11} to 10^{-6} M. For concentrations higher than 10^{-6} M the signal is leveled off, and an $I\%$ of 100 is generally obtained. The DL can be calculated by substituting the blank minus 3 SD (standard deviation) in the equation of the linear portion of the inhibition curve (Figure 6.2).

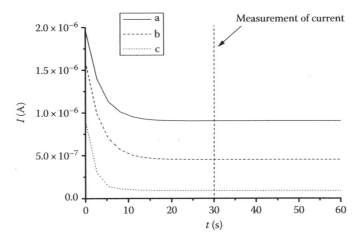

FIGURE 6.1 Typical chronoamperograms obtained after incubation with different concentrations of carbofuran: (a) 0 M, (b) 10^{-9} M, and (c) 10^{-7} M.

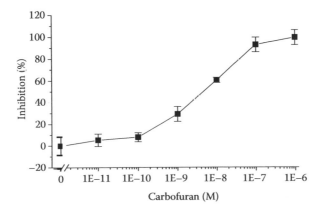

FIGURE 6.2 Inhibition plot of carbofuran onto AChE-based biosensor. (From Laschi, S. et al., *Enzyme Microb. Technol.*, 40, 485, 2007. With permission.)

An important feature of the proposed method is that the sample incubation and the electrochemical measurement are performed in two separate steps, and among them a washing procedure is also included. This guarantees that the proposed method is less prone to electrochemical interferences, since oxidable substances, eventually present in the sample, are washed off before the electrochemical measurement.

For further information on this topic, please refer to Ref. [23].

All materials and reagents for this experiment are available at Ecobioservices and Researches s.r.l. (Florence, Italy, http://www.ebsr.it).

6.3 PSII-BASED BIOSENSOR FOR THE DETECTION OF PHOTOSYNTHETIC INHIBITORS

In this section two different approaches in the detection of inhibitors of photosynthesis are described. The first is based on the use of an amperometric photosystem II (PSII)-based biosensor and the second on the use of commercial kits, coupled with a dedicated handheld instrument for the measurement of chlorophyll fluorescence in photosynthetic material.

6.3.1 INTRODUCTION

6.3.1.1 Photosynthesis

Photosynthesis is the process that converts the energy of light to chemical forms of energy that can be used by biological systems [25,26]. Photosynthesis is carried out by many different organisms, ranging from plants to bacteria. The best-known form of photosynthesis is the one performed by higher plants and algae as well as by cyanobacteria and other similar organisms, which are responsible for a major part of photosynthesis in oceans. All these organisms convert CO_2 (carbon dioxide) to organic material by reducing this gas to carbohydrates in a rather complex set of reactions. Electrons for this reaction ultimately come from water, which is then converted to oxygen and protons. The energy for this process is provided by light, which is absorbed by pigments (primarily chlorophylls and carotenoids).

Light energy is absorbed by individual pigments and transferred to chlorophylls that are in a special protein environment where the actual energy conversion event occurs [27]. Pigments together with proteins, involved with this actual primary electron transfer event, are called "reaction centers." All photosynthetic organisms that produce oxygen have two types of reaction centers, named photosystem II and photosystem I (PSII and PSI), both of which are pigment/protein complexes that are located in specialized membranes called thylakoids [28].

The entire process can also be energetically described by the so-called z-scheme (Figure 6.3). The PSII complex is a light-driven, water plastoquinone oxidoreductase; light energy is absorbed by light-harvesting complexes that contain most of the pigment associated with PSII [29]. The excitation energy is transferred from antenna to the "core" of the PSII complex where the primary photochemistry takes place. The photochemical reactions result in the accumulation of oxidizing equivalents in the oxygen-evolving complex (OEC): four oxidizing equivalents are used to convert two molecules of water into oxygen [30].

The PSII core is the minimal unit that is capable of catalyzing full PSII functions. It is composed of a reaction center, which consists of the D_1 and D_2 polypeptides, cytochrome b_{559}, the PsbI protein, six chlorophyll and two pheophytin molecules, an inner antenna of chlorophyll-binding protein termed CP43 and CP47, and the extrinsic lumenally bound protein of the OEC [28].

Some compounds can interfere with the overall process, blocking the photosynthesis through the replacement of Q_B and the interaction with the D_1 subunit. Among these compounds some classes of herbicides can be found.

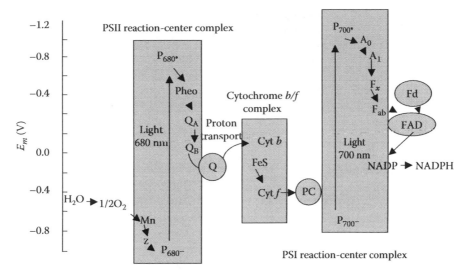

FIGURE 6.3   The z-scheme for photosynthesis.

6.3.2   Environmental Pollution: The Problem of Herbicides

6.3.2.1   Herbicides in the Environment

The massive use of soil for agricultural activities and the increase in food production achieved during the last decades caused the widespread use of fertilizers and pesticides [31].

The persistence of herbicides in soil and water allows the contamination of the entire food chain, causing health and ecological consequences [32]. The health effects of pesticides depend on the type of pesticide and the dose and frequency of exposure. Consequently, their content in surface and drinking waters has been regulated by specific laws, and their levels should be monitored frequently [33,34].

6.3.2.2   Methods of Analysis

Herbicide analysis in surface, ground, and drinking water is generally performed by GC, HPLC often coupled with MS, and recently, capillary electrophoresis (CE). These techniques offer the possibility of performing analysis with high sensitivity, also limiting matrix effects. Generally, a preconcentration step of the water samples is necessary (100- to 1000-folds) by solid-phase extraction [35,36] or liquid–liquid extraction [37]; in the case of solid matrices an extraction is needed [38]. The DLs depend on the molecular structure of the herbicide and on the techniques used to extract and to analyze the sample.

The disadvantages in using of these methods are related to the requirements of expensive equipments, associated with long and expensive procedures, which are not suitable to perform rapid screening analysis on a large number of samples.

For that reason, attention has been focused on new approaches in herbicide analysis. Among them immunochemical methods give high sensitivity and selectivity, but the disadvantages are related to the complex procedures necessary for the production of the antibodies. Moreover, the high specificity of the antigen–antibody reaction can be a problem, because only one compound or few analogues can be bound and measured [39,40].

6.3.2.3   Herbicides as Photosynthetic Inhibitors

The PSII complex is essential to the regulation of photosynthesis, because it catalyzes the oxidation of water into oxygen and supports electron transport. The above property confers to photosynthetic materials and particularly to PSII a great potential for various biotechnological applications.

Among all the pesticides used, photosynthetic inhibitor herbicides represent 30% of the herbicide currently used in agriculture, with the world consumption amounting to many thousand tons. This group consists of several classes of chemicals such as triazines (atrazine), phenylureas (e.g., diuron), or phenols (ioxynil). These compounds all belong to different families but have a common mode of action: binding specifically to the chloroplast D1 protein, with subsequent interruption of the electron and proton flow through PSII. The secondary plastoquinone acceptor Q_B is reversibly associated with D1 protein to perform an important role in PSII electron chain. This secondary acceptor acts to connect the single electron transfer event of the reaction center with the pool of free plastoquinones in the membrane by operating as a two-electron gate. Several classes of photosynthesis-inhibiting herbicides are able to compete with Q_B for the binding site within D1 protein, thereby inhibiting the electron transfer thought PSII and leading to plant damage and death. The natural functions of D1 protein as mediator and herbicide binding agent can be the basis for application of PSII, chloroplasts, or thylakoid membranes, in biosensors for environmental monitoring.

In addition, it has been reported that it is also possible to detect herbicides that have other modes of action, such as alachlor and glyphosate, and even some insecticides [41–43]. Besides, toxic metal cations, such as Pb, Cd, Cr, Cu, Ni, Zn, and Hg, may also cause a wide variety of deleterious effects when taken up by plants, because the photosynthetic apparatus seems to be particularly sensitive to their effect [44–47].

6.3.2.4   Methods for the Measurement of Photosynthetic Activity

The evaluation of the activity of photosynthetic material can be performed using electrochemical and spectrophotometric techniques. A series of electron transfer reactions between donors and acceptor substances, immobilized in the membrane or dissolved in aqueous phase, are involved in the conversion of solar energy in photosynthetic systems or in the cell respiration process [48]. The redox character of the components of the most relevant biological apparatus suggests the use of electrochemical methods to follow the reactions that occur during the metabolic process.

The most used electrochemical strategies to measure PSII activity are based on the detection of oxygen evolution by using voltammetric [49] or amperometric [47–54] techniques. Several types of artificial electron acceptors can be used as electrochemical mediators to maximize the photosynthetic response [55]. Among various possible compounds, the use of potassium ferricyanide [31,48,53] or substituted benzoquinones as 2,3,5,6-tetramethylbenzo-1,4-quinone (duroquinone), or 2,5 dichlorobenzoquinone, can be used [31,53,54]. Regarding spectrophotometric measurements, chlorophyll fluorescence is another widespread method to measure photosynthetic activity [55,56]. The light energy absorbed by chlorophyll molecules in a leaf can undergo one of three fates: it can be used to drive photosynthesis (photochemistry); it can be dissipated as heat; or it can be reemitted as chlorophyll fluorescence [57]. These three processes occur in competition. Thus, an increase in the efficiency of one will result in the decrease of the other two. In vivo fluorescence increases when photosynthesis declines or when it is inhibited. This method was often employed to detect the photosynthetic activity of immobilized photosynthetic material.

Recently, a portable fluorimeter, especially designed to perform measurements on extracted photosynthetic material, was proposed on the market (LuminoTox; Lab-Bell, Canada [58]). Different kits necessary to assess global toxicity or to detect herbicides in water samples through the inhibition effect of toxicants, photosynthetic apparatus, and thylakoid membranes of algae are available [59].

Among the new approaches in herbicide analysis, great interest has been devoted to the development of photosynthetic material-based biosensors [31,51]; different physical or chemical immobilization procedures of the biological material are described. Problems related to the operational stability of photosynthetic material, such as the storage stability and their effect on sensitivity toward inhibition effects of toxicants, appear relevant.

Glass microfibers were used to immobilize whole algae cells for the construction of a biosensor based on fluorescence measurements, resulting in a tool with good sensitivity toward herbicide inhibition effect and a stability of 2 weeks [41]. Green algae and cyanobacteria entrapment on filter paper disks was used to develop a biosensor to detect warfare agents [43] and environmental pollutants [60] by fluorescence measurements.

The inclusion of photosynthetic material in a polymeric network was often used. Natural polymers such as agarose or alginate were employed in the development of an electrochemical biosensor through the immobilization of PSII particles from cyanobacteria on graphite-based electrodes [31]. Problems related to stability and poor adhesion of the immobilized material on electrode surface were thereby observed. Nevertheless, synthetic polymers such as polyacrylamide gel, poly(vinylalcohol) polymers, and poly(vinylalcohol) bearing styrylpyridinium (PVA–SbQ) polymers were also used [54,61]. In particular the use of PVA–SbQ polymers represents an interesting tool for the stabilization of entrapped thylakoid membranes and herbicide detection through photocurrent measurements [28,62]. These polymers were also used in the development of a colorimetric assay for herbicide detection [63].

Covalent fixing of the photosynthetic material on a support was a technique less used due to the intrinsic denaturing effect of the immobilizing agents. Nevertheless, glutaraldehyde was shown to preserve the activity in chloroplasts, and it was commonly used with the addition of some protein, such as collagen or BSA [31,42]. It is known that glutaraldehyde creates a network of bonds with the free $-NH_2$ groups of both added protein and photosynthetic material, protecting the latter from the formation of too many bonds and preventing the loss of biological functions. An electrochemical biosensor was developed using PSII particles extracted from cyanobacteria immobilized with BSA/glutaraldehyde mixture on screen-printed electrodes [31]. With this device, limits of detection in the nanomolar range were found for triazine standard solution analysis and a half-life stability of sensors of 24 h was evaluated.

In conclusion electrochemical biosensors based on photosynthetic material seem to offer an interesting tool for herbicide detection. The necessary simple instrumental setup can be easily modified to produce handheld instruments, with low costs. Besides the possible use of disposable sensors, they can offer the possibility of producing biosensors for in situ screening tests. Nevertheless, problems related to biological material instability were highlighted in previous works and a characterization of biosensor shelf life would be necessary in the development of a simple but efficient device.

6.3.3 EXPERIMENTAL PART

6.3.3.1 PSII-Based Biosensor for the Detection of Photosynthetic Inhibitors

Screen-printing technology offers the possibility to produce sensors, in a large number with low costs; thus, these electrodes are suitable for the development of screening assays. The flexibility of the technique allows the creation of different electrochemical cell designs with the possibility of being adapted to the most diverse applications. Moreover, as SPE are miniaturized and planar, they can be used as drop-on sensors, using a few microliters of the sample solution.

An example of one application of an amperometric biosensor, based on the activity of PSII, is described in the following paragraphs. In the described work the surfaces of carbon-based, electrochemical, screen-printed electrodes were modified with thylakoid membranes extracted from spinach leaves. Thus, thylakoid membranes and carbon-based, screen-printed electrodes were coupled to build up an amperometric biosensor for the detection of photosynthetic inhibitors in a "one-shot" measuring set-up.

6.3.3.1.1 Objectives

1. To detect herbicides through the inhibition of the photosynthetic process using a PSII-based biosensor. The inhibitory effect of herbicides on PSII activity was evaluated through the recording, by amperometric measurements, of the current due to the reoxidation of the reduced form of the DQ, which was formed during the photosynthesis.

   Once the system had been illuminated, the photosynthetic reaction took place, releasing oxygen and reducing DQ. This was then reoxidized at the electrode surface by applying a suitable potential. Because the formation of reduced DQ occurred only under illumination conditions, the oxidation step involved a transient electron flow; in this way a current peak was recorded, which was proportional to the PSII activity. When the inhibitor is present in the solution the photosynthetic activity decreases, resulting in a lower current peak.
2. To test the sample solution.

6.3.3.1.2 Materials and Instrumentation

- PSII-based, screen-printed biosensors (Ecobioservices and Researches s.r.l., http://www.ebsr.it)
- Measurement buffer (Required composition: MES 0.015 mol L^{-1}, mannitol 0.05 mol L^{-1}, NaCl 0.1 mol L^{-1}, $MgCl_2$ 0.005 mol L^{-1}, chloramphenicol 0.00005 mol L^{-1}, and DQ 0.0002 mol L^{-1})
- Methacrylate cell box
- Temporized illumination unit
- Potentiostat
- Connector for screen-printed electrodes

Preparation of testing solutions

Prepare the buffer 10X concentrated without DQ (MES 0.15 mol L^{-1}, mannitol 0.5 mol L^{-1}, NaCl 1 mol L^{-1}, $MgCl_2$ 0.05 mol L^{-1}, and chloramphenicol 0.0005 mol L^{-1}).

Concentrated solutions of DQ can be prepared in MeOH and they can be added to the 1X buffer with a final concentration of 0.0002 mol L^{-1}.

- Preparation of measurement buffer (blank, 1X buffer): dilute the 10X buffer 10-fold (i.e., 100 μL of 10X buffer + 900 μL of MilliQ water). Add DQ to the concentration mentioned above.
- Preparation of solution of standards: prepare a concentrated solution of standard in MeOH (i.e., atrazine 0.05 mol L^{-1}). Dilute this solution in the measurement buffer prepared as described above, to the desired concentration (useful range between 10^{-8} and 10^{-6} mol L^{-1}).
- Preparation of real samples: filter an aliquot of the sample through a 0.45 μm syringe filter. Buffer the solution using 10X buffer (i.e., add 100 μL of 10X buffer + 900 μL real sample). Finally, add DQ to the required concentration of 0.0002 mol L^{-1}.

Experimental procedure (Figure 6.4)

1. Locate a screen-printed biosensor in the slot created on the bottom part of the cell.
2. Fix the top part with screws.
3. Check, through the well opening, that the position of the electrode surface is centered.
4. Drop 50 μL of the testing solution (containing the standard solution or real sample) in the well and close it with the cap that houses the LED.

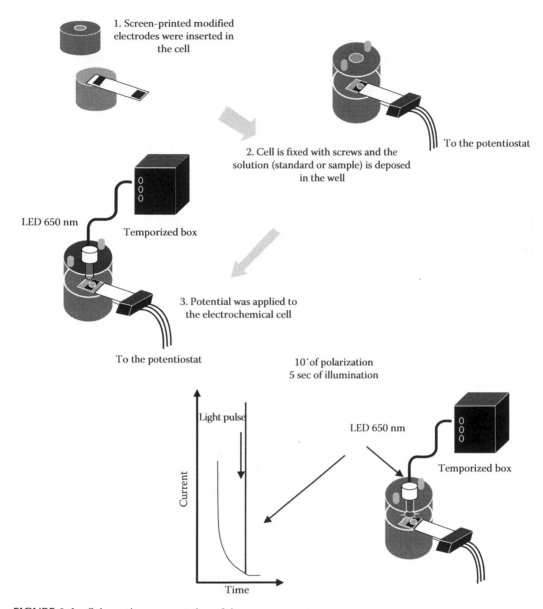

FIGURE 6.4 Schematic representation of the measurement.

5. Apply the working potential at the electrode (+620 mV) and, at the same time, activate the light-pulsing temporization unit by pressing the button.
6. After 10 min the light pulse will be applied; DQ reduction will then be achieved at the electrode surface, and the peak current will be obtained (Figure 6.5).
7. After the measurement, wash the cell with MilliQ water and replace the biosensor for a new measurement.

6.3.3.1.3 Discussion

The inhibition values can be calculated using the following formula

$$[(I \text{ blank} - I \text{ sample})/I \text{ blank}]*100$$

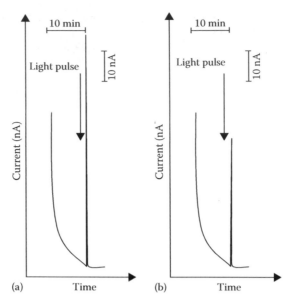

FIGURE 6.5 Amperometric signal obtained on illumination of SPEs modified with thylakoid membranes, before (a) and after (b) incubation with the herbicide (diuron 5×10^{-2} mg L^{-1}).

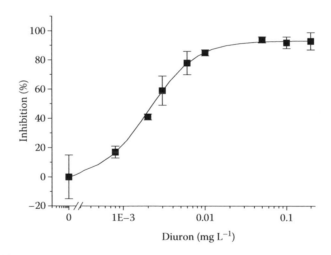

FIGURE 6.6 Inhibition plot of herbicide (diuron) onto SPEs modified with thylakoid membranes.

where I blank is the current value obtained for a blank (test solution containing only buffer) and I sample the value obtained for different inhibitor concentrations. An example of the calibration plot obtained for diuron is reported in Figure 6.6.

For further information on this topic, please refer to Ref. [53].

All materials and reagents for this experiment are available at Ecobioservices and Researches s.r.l. (Florence, Italy, http://www.ebsr.it).

6.3.3.2 Objectives

1. To detect herbicides through the inhibition of photosynthetic process using chlorophyll fluorescence determination. The LuminoTox test is based on chlorophyll fluorescence emission by a Stabilized Aqueous Photosynthetic System (SAPS) or Photosynthetic Enzyme complex (PECs) contained in the kits. This material is obtained with a procedure protected

by a patent. The LuminoTox measurement uses a fluorescence dissipation process emission by PSII complexes [64].

To evaluate the photochemical quantum yield (Φp photosynthetic efficiency) by fluorescence measurements, the algae photosystems must be incubated in dark. To achieve the fluorescence measurements, two levels of light illumination are needed. The first level (F1) is obtained after the application of a low-light intensity to determine the fluorescence of chlorophyll molecules that absorb light in SAPS or PECs photosystems. If this light intensity is too weak to drive the photosynthesis process, all the PECs will be oxidized or in "open state". The second level (F2) is reached following the application of a high-intensity pulse. Saturation pulse will induce the closure of all enzyme complexes (reduced state).

In the following section an experimental application of chlorophyll measurements to the detection of photosynthetic inhibitor is proposed. The brochure describing the complete experimental procedure provided for the test kits LuminoTox (Figure 6.7) is available on the Lab-Bell Web site (www.lab-bell.com).

2. To test sample solution

6.3.3.2.1 Materials and Instrumentation
- Test kits: SAPS or PECs
- Glass four-side cuvette
- Timer
- Pipette

Preparation of testing solutions

Prepare the buffer 10X concentrated without DQ (MES 0.15 mol L^{-1}, mannitol 0.5 mol L^{-1}, NaCl 0.1 mol L^{-1}, MgCl$_2$ 0.05 mol L^{-1}, and chloramphenicol 0.0005 mol L^{-1}).

- Reconstitute photosynthetic material using the supplied buffers and store as suggested by producer.
- Preparation of solution of standards: prepare a concentrated solution of the standard in MeOH (i.e., atrazine 0.05 mol L^{-1}). Dilute this solution in the measurement buffer prepared as described above, to the desired concentration (useful range between 10^{-8} and 10^{-6} mol L^{-1}).

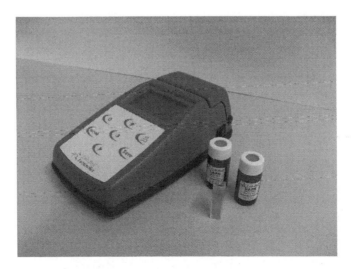

FIGURE 6.7 LuminoTox instrument. (Picture taken by the author.)

- Preparation of real samples: filter an aliquot of the sample through a 0.45 μm syringe filter. Buffer the solution using 10X buffer (i.e., add 200 μL of 10X buffer + 1800 μL real sample).

Experimental procedure

1. Add 2 mL of blank solution (buffer solution)
2. Add 100 μL of reconstituted photosynthetic material.
3. Mix the solution by inverting the cuvette several times.
4. Locate a cuvette in the cuvette-bearing unit, close the lid and wait for 10 min.
5. Press "0" button.
6. Read the results; save the measurement using the "save" option. Blank values will be stored in the internal memory of the instrument until the end of the experimental session.
7. Replace the cuvette if disposable or wash it.
8. Add 2 mL of the solution sample or the standard solution.
9. Repeat points 3 and 4.
10. Press " ▶ " button.
11. Read the results on the screen; save the measurement using the "save" option.

6.3.3.2.2 Discussion

In the LuminoTox analyzer, light intensities were chosen to provide partial oxidation and reduction of SAPS and PECs photosystems.

Φp is evaluated as follows:

$$\Phi p = (F2 - F1)/F2$$

In the LuminoTox, the relative efficiency (Eff.) is evaluated to obtain a better sensitivity:

$$\text{Eff.} = (F2(\text{sample}) - F1(\text{sample}))/F2(\text{blank})$$

$F2$ blank is referred to as the fluorescence value obtained in the control solution.

These values are directly displayed on the screen of the instrument together with the inhibition calculated for each sample respect to the blank. Measurement can also be saved and stored in the internal memory of the instrument.

Table 6.1 shows results obtained for the investigated standard solutions of atrazine:

All materials and reagents for this experiment are available at Lab-Bell (Shawinigan, Quebec, Canada, http://www.lab-bell.com).

TABLE 6.1
Inhibition % Obtained Using PECs for Testing Standard Atrazine Solutions (Data Available on LuminoTox Data Sheet)

Atrazine Solution (μg mL^{-1})	Volume (mL)	Inhibition (%)
0	2.0	0 ± 5
0.006	2.0	20 ± 5
0.2	2.0	62 ± 5

6.4 PCB ANALYSIS USING IMMUNOSENSORS BASED ON MAGNETIC BEADS AND CARBON-SCREEN-PRINTED ELECTRODES IN MARINE SEDIMENT AND SOIL SAMPLES

6.4.1 INTRODUCTION

Polychlorinated biphenyls (PCBs) are a group of 209 structurally related chemical compounds, consisting of two connected benzene rings and 1–10 chlorine atoms. The positions of the chlorine substituents on the rings are denoted by numbers assigned to each of the carbon atoms, with the carbons supporting the bond between the rings designated 1 and 1′ (Figure 6.8).

Any single chemical compound in the PCB category is called a "congener." The name of a congener specifies the total number of chlorine substituents and the position of each chlorine. Although the physical and chemical properties vary widely across the congeners, PCBs have low water solubilities and low vapor pressure. The molecular weight of PCBs varies from 188.7 for $C_{12}H_9Cl$ to 498.7 g·mol^{-1} for $C_{12}Cl_{12}$.

PCBs were manufactured and sold under many names, even if the most common are the "Aroclor" series. Aroclor refers to a mixture of individual chlorinated biphenyl compounds with different degrees of chlorination. The most common mixtures in the Aroclor series are Aroclor 1016, Aroclor 1242, Aroclor 1248, Aroclor 1254, and Aroclor 1260.

Some of the 209 congeners of PCBs, the so-called coplanar congeners, in addition to being stereochemically similar to the planar 2,3,7,8-tetrachlorodibenzo-*p*-dioxin (2,3,7,8-TCDD), also show a biochemical activity and toxicity similar to 2,3,7,8-TCDD. In particular, it has been experimentally shown that PCB77, PCB126, and PCB169 produce toxic effects in terrestrial mammals similar to those produced by 2,3,7,8-TCDD, and for this reason they are indicated as "dioxin-like" molecules.

PCBs are among the 16 chemicals designated as persistent organic pollutants (POPs) that are the subject of negotiations on a global agreement for their control. POPs are highly stable compounds that persist in the environment, accumulate in the fatty tissues of most living organisms since they are lipophilic, and are toxic to humans and wildlife [62].

PCBs were once used in industrial applications, particularly as electrical insulating fluids and as heat-exchange fluids, until concern over possible adverse effects on the environment and on human health resulted in the cessation of PCB production and an ultimate ban on manufacture in most countries. Because of their remarkable electrical insulating properties and their flame resistance, they soon gained widespread use as insulators and coolants in transformers and other electrical equipment where these properties are essential. For several decades, PCBs were also routinely used in the manufacture of a wide variety of common products, such as plastics, adhesives, paints, varnishes, and carbonless copying paper. Despite their ban almost a quarter of a century ago, these pollutants are largely diffused in the environment. Their presence is mainly due to their physical and chemical properties, such as low inflammability, chemical stability, and solubility in most organic solvents. Because of their characteristics, all degradation mechanisms are difficult and environmental and metabolic degradation is generally very slow.

Because of their bioaccumulation in the food web, an important source of exposure for humans to these contaminants (>90%) is via food consumption, particularly fish, meat, and dairy products. As they are nonpolar and highly lipophilic, the highest concentrations are found in fatty foods

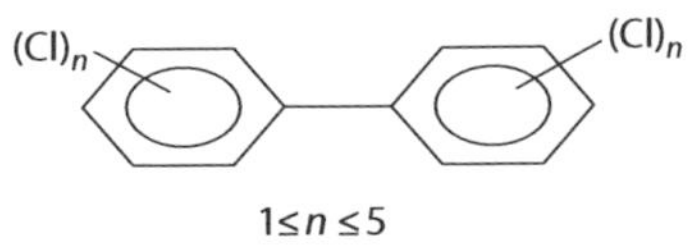

FIGURE 6.8 Chemical structure of a PCB molecule.

[63,64], and several articles report their presence in human milk. It is, therefore, very important to know the distribution of contaminants such as PCBs in the environment and in food.

Moreover, PCBs can be unintentionally produced as by-products in a wide variety of chemical processes that contain chlorine and hydrocarbon sources, during water chlorination and by thermal degradation of other chlorinated organics.

PCB levels in the polluted environmental samples change according to the matrices considered; for example in soil their concentrations can be in the range of several ng g^{-1}–mg g^{-1} (this last value for highly contaminated sites), whereas in sediments the upper limit is generally hundreds of μg g^{-1}. Very high levels can also be reached in living organisms, as a result of bioaccumulation processes.

6.4.2 Analytical Methods for PCB Detection

Conventional methods used for the analysis of PCBs are largely dependent on the use of chromatography and spectrometry or a combination of both as well as GC coupled to an electronic capture detector (GC–ECD) [65–67]. Despite the high degree of specificity, selectivity, and accuracy, there are a number of disadvantages associated to these techniques. For example, the equipment is expensive, sample preparation and analysis can be complicated and time consuming, and the requirement for trained personnel does not permit in situ monitoring [68]. Alternative methods such as immunoassays, which could provide inexpensive and rapid screening techniques for sample monitoring for laboratory and field analysis, are more and more requested.

Immunoassay methods are simple, sensitive, and selective for PCB testing, allow rapid measurements, and have also been applied to PCB detection [69–72].

Among the high number of immunoassay techniques, enzyme-linked immunosorbent assays (ELISAs) combined with a colorimetric end point measurement are the most widely used. These techniques have also been introduced on the market as PCB ELISA kits by many companies.

Immunosensors are affinity biosensors and are defined as analytical devices that detect the binding of an antigen to its specific antibody by coupling the immunochemical reaction to the surface of a device (transducer). Among the different kinds of transducers (piezoelectric, electrochemical, and optical), electrochemical immunosensors, based on the electrochemical detection of immunoreaction, have been widely reported. They have been the subject of increasing interest, mainly because of their potential application as alternative immunoassay techniques in areas such as clinical diagnostics [73] and environmental and food control [74,75].

Most electrochemical immunosensors use antibodies or antigens labeled with an enzyme that generates an electroactive product, which can be detected at the electrochemical transducer surface. The combination of high enzyme activity and selectivity with the sensitive methods of electrochemical detection provides a basis for the development of immunosensors. Horseradish peroxidase (HRP) and alkaline phosphatase (AP) are popular enzyme labels and can be used with a variety of substrates.

Most electrochemical immunosensors use screen-printed electrodes produced by thick-film technology as transducers: the importance of screen-printed electrodes in analytical chemistry is related to the interest in the development of disposable and inexpensive immunosensors. Moreover, screen printing allows the fast mass production of highly reproducible electrodes at low cost for disposable use.

Electrochemical immunosensors based on screen-printed electrodes have recently been applied to the detection of environmental pollutants such as PCBs. In this case the screen-printed electrodes are both the solid-phase for the immunoassay and the electrochemical transducers: antibody or antigen molecules are directly immobilized at the sensor surface (transducer), and one of these species is enzyme-labeled to generate an electroactive product that can be detected at the screen-printed electrodes surface.

The use of the electrode surface as the solid phase in immunoassay can present some problems: a shielding of the surface by antibody or antigen molecules can cause hindrance in the electron transfer, resulting in a reduced signal and so a loss of sensitivity. There are different ways that can be used to improve the sensitivity of the system; an interesting approach could involve the screen-printed electrodes use only for the transduction step, whereas the affinity reaction could be performed using another physical support [76].

The use of microparticles as the solid phase is a relatively new possibility, widely documented in the literature [77–80]. Various kinds of beads can be used in an electrochemical biosystem. The beads range from nonconducting (glass, etc.), conducting (graphite particles) to magnetic materials. Particles are available with a wide variety of surface functional groups and size and have the possibility of reaction kinetics similar to those found in free solution. Graphite or magnetic particles represent the most commonly used beads in bioelectroanalytical systems.

At present various coated magnetic particles are offered on the market. Proteins have been coupled covalently to the surface of superparamagnetic particles, with stable and selective groups to bind protein-specific ligands with minimal interference and nonspecific binding. These particles respond to an applied magnetic field and redisperse on removal of the magnet. They consist of 36%–40% magnetite dispersed within a copolymer matrix consisting of styrene and divinyl-benzene. Their binding capacity varies with the bead size, composition, and the size of the binding ligand.

The use of magnetic beads as solid support to perform the immunoassay allows obtaining a high yield of solid-phase-antibody binding and of the affinity reaction: thus, the probability that antibodies meet the magnetic beads or the antibody-coated magnetic beads meet the analyte is very high, when the solution is kept under stirring. After molecular recognition on the magnetic beads, it is possible to build up the immunosensing surface by localizing the immunomagnetic beads on the working electrode area of a screen-printed electrode with the aid of a magnet and performing the electrochemical measurement.

This approach separates the steps relative to the immunoreaction from the step of electrochemical detection, and for this reason, the working electrode surface is easily accessible by the enzymatic product, which diffuses onto the bare electrode surface. Using this strategy, finding the optimum conditions for the immunoassay on the magnetic beads and for electrochemical detection on the transducer (carbon screen-printed electrodes) is much easier than in the usual one (electrode) surface systems, because optimum conditions for immunoassay do not conform with those for electrochemical detection and vice versa [64,81].

This configuration based on the use of two surfaces, magnetic beads for immunoassay and screen-printed electrodes for electrochemical detection, allows to obtain a faster and a more sensitive detection of the immunoreaction than using a unique surface (screen-printed electrode): in this case it is possible to perform the electrochemical measurement faster (less then 30 min) and improve the sensitivity (around two magnitude orders) [80]. For this reason this approach is advised in the development of an electrochemical immunosensor specific to any analyte.

6.4.3 Immunoassay Scheme for PCBs Detection

The choice of the immunoassay scheme is related to the dimensions of the target molecule: when the antigen is a small molecule such as a PCB molecule, it has only one epitope, and therefore only one antibody can recognize it. For this reason, only a competitive immunoassay scheme can be developed for PCB determination. The competitive immunoassays, as the name indicates, are based on a competition reaction between two reagents for a third one. A competitive assay can be carried out in two different ways:

- The specific antibodies can be immobilized on a solid phase and then competition between an antigen and an antigen derivate labeled by enzyme can be performed. This scheme is

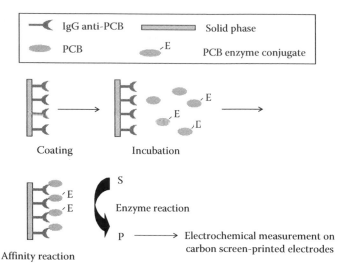

FIGURE 6.9   Representative scheme of a direct competitive assay format.

called direct competitive assay (Figure 6.9). In Section 6.4.4, an example of this kind of assay is reported.

- A derivate of the antigen is immobilized on a solid phase, and then the molecular recognition between the antigen and the specific antibody occurs in a vial. Then this solution is added to the solid phase. If the specific antibody is not labeled by an enzyme, the extent of the affinity reaction can be evaluated by adding a secondary labeled antibody able to bind the first antibody.

In both cases, a label is used to estimate the extent of the affinity reaction. The signal is inversely proportional to the amount of analyte present in the sample.

It is necessary to optimize the following parameters to perform a competitive assay:

- Antibody concentration: the sensitivity of a competitive assay is related to the concentration of the antibody.
- Antigen-labeled concentration: this has to be in a limited amount to saturate the antibodies immobilized on the solid phase.
- Incubation time of the competition solution: it is important that the affinity reaction has completely occurred.

6.4.4   PCB Pollution in Environment and Food Samples

Soren Jenson, a Swedish researcher at the University of Stockholm, first identified PCBs as an environmental problem in 1966: he identified the presence of PCBs in human blood.

The evidence of acute PCB toxicity came from two industrial incidents—one in Japan in 1968 and the other in Taiwan in 1979: cooking oil was contaminated by large quantities of PCBs and PCDFs. Adults, who ingested the contaminated oil, exhibited chloracne and dark brown pigmentation of the skin and lips.

Due to their lipophilic nature, PCBs tend to accumulate or reside in those environmental compartments that are nonpolar and are amenable to lipid accumulation, such as the organic components of sediments. PCB presence in polar substances, such as water, is minimal. PCBs are not volatile and thus do not persist in air in any appreciable concentration. Therefore, the major sources of environment exposure to environmental species remain soils and sediments.

Analysis of environmental or food samples containing PCBs includes matrix preparation, extraction, and determination [63]. In this section some extraction methods, which can be easily coupled to an electrochemical immunosensor, are described. Such extraction techniques have to be simple, fast, and useful for in situ measurements and must not require trained personnel.

Matrix preparation: Samples collected in the field are usually preserved by freezing after dissecting into small pieces. Homogenization and grinding to rupture cell membranes appear to be the most commonly used pretreatment procedures for tissue matrices (e.g., fish muscle tissue).

Extraction: Extractions generally rely on a favorable partition of PCBs from the sample matrix into the extraction matrix. The more favorable the partition coefficient, the higher the extraction efficiency. Since PCBs are lipophilic, the extraction methods are based on the isolation of lipids from the sample matrix. It should be noted that the concentration of planar or non-ortho-substituted PCBs, which are considered the most toxic PCB congeners, is generally $\approx$1000-fold lower (ng kg^{-1}) than those of nonplanar or ortho-substituted PCBs (μg kg^{-1}). In addition, other compounds, such as pesticides, lipid, biological material, or chlorophyll from plants, are also extracted and can interfere with the analysis. Sample-recovery measurements can be made by using the method of standard addition.

Many techniques are used for PCB extraction; among them the most used are the following:

- Liquid–liquid extraction (LLE): the most common LLE configuration uses about 100 mL of solvent per 5–50 g of sample. Samples are generally frozen before extraction to disintegrate the tissues, which are then homogenized by sodium sulfate to bind water present in the sample, followed by overnight drying. The dried powder is then extracted using a suitable solvent. It is essential to match the solvent polarity, and generally, a combination of nonpolar, water-immiscible solvents with solvents of various polarities are used.
- Solid-phase extraction (SPE): it is widely accepted as an alternative extraction/cleanup method. SPE has been used for the extraction of PCBs in various types of human milk. Milk powder and evaporated milk were constituted with water prior to extraction. The milk sample was mixed with 5 mL of water and 10 mL of methanol and sonicated, followed by passing the sample through the column.
- Ultrasonic extraction (USE): it is a simple extraction technique, in which the sample is immersed in an appropriate organic solvent in a vessel and placed in an ultrasonic bath. The efficiency of extraction depends on the polarity of the solvent, the homogeneity of the matrix, and the ultrasonic time. The mixture of the sample and organic solvent is separated by filtration.
- Microwave-assisted extraction (MAE): it has attracted growing interest, as it allows rapid extraction of solutes from solid matrices by employing microwave energy as a source of heat. The portioning of the analytes from the sample matrix to the extractant depends upon the temperature and the nature of the extractant.

An extraction procedure performed by sonication method for dried marine sediments and soil followed by the analysis of the extracts using an electrochemical immunosensor based on magnetic beads and carbon screen-printed electrodes is described in Section 6.4.5.

6.4.5 Objectives

1. To build up an electrochemical immunosensor using magnetic beads as the solid phase and carbon screen-printed electrodes (SPCEs) as transducers
2. To test the electrochemical immunosensor in PCB standard solutions
3. To do the PCB analysis in marine sediment and soil extracts

6.4.6 MATERIALS AND INSTRUMENTS

- Magnetic beads coupled with protein G (Dynal Biotech). Sheep polyclonal antibodies against the congener PCB28 (IgG anti-PCB28) and the corresponding tracer PCB28-AP (4,4'-dichlorobiphenyl-alkaline phosphatase conjugate). PCB standard solutions (AccuStandard Inc., New Haven). α-naphthyl phosphate (analytical grade). Nitrocellulose filter 0.45 μm.
- The saline solutions used are as follows:
  - 0.1 M sodium-phosphate solution pH 5 for washing and coating of the magnetic beads (Solution A)
  - 0.3 mM sodium-phosphate buffer pH 7.2 containing 5 mM NaCl and methanol 1% v/v as a working assay buffer for the competition (solution B)
  - Diethanolamine buffer 0.1 M pH 9.6 containing 1 mM $MgCl_2$ and 100 mM KCl (solution C) for the electrochemical measurements
- Sample mixer with 12 tube mixing wheel and a magnet (Dynal Biotech).
- The electrochemical cells used are three electrode strips, based on a carbon working electrode, a carbon counter electrode and a silver pseudo-reference electrode [82]. (Ecobioservices, http://www.ebsr.it).
- Potentiostat.
- Sonicator for sample extraction (model VC 100) from Vibracell Sonics and Materials.

6.4.7 DIRECT COMPETITIVE IMMUNOASSAY USING MAGNETIC BEADS AS SOLID PHASE

6.4.7.1 Antibody-Coated Bead Preparation

- Perform all reactions at room temperature and under a delicate stirring in the sample mixer.
- Treat 10 μL of magnetic beads coupled with protein G with 500 μL of solution A to remove the NaN_3 preservative as advised by the manufacturer.
- Add beads to 500 μL of IgG anti-PCB28 solution 100 μg mL^{-1} prepared in solution A.
- After 20 min of incubation time, place the tube on a magnet holding block to allow the precipitation of the beads on the bottom of the test tube; remove the supernatant and wash the beads twice with 500 μL of solution A.

6.4.7.2 Affinity Reaction

Preparation of blank solution (containing the tracer only):

- Mix 50 μL of suspension containing antibody-coated beads with 940 μL of solution B, 10 μL of the PCB28-AP conjugate solution diluted 1:10 with respect to the stock solution.

Preparation of competition solutions:

- Mix 50 μL of suspension containing antibody-coated beads with 930 μL of solution B, 10 μL of the PCB28-AP conjugate solution diluted 1:10 with respect to the stock solution in solution B and 10 μL of sample (standard solution or extract).
- After 20 min of incubation time, magnetically separate the beads and remove the supernatant.
- After two washing steps with 500 μL of solution B, resuspend the beads in 100 μL of solution B.

6.4.7.3 Electrochemical Measurement

- Transfer 10 μL of bead suspension onto the surface of the working electrode.
- Localize the beads onto the electrode, and place the magnet holding block on the bottom part of the electrode.

- Deposit 60 μL of α-naphthyl phosphate 1 mg mL^{-1} in solution C on the screen-printed electrode, taking care to close the electrochemical cell.
- Choose the differential pulse voltammetry (DPV) mode in the μAutolab software program. The parameters are as follows: range potential 0/+600 mV, step potential 7 mV, modulation amplitude 70 mV, standby potential 200 mV, and interval time 0.1 s.

The analytical signal is the height of the peak due to the oxidation of the enzymatic product (α-naphthol).

- After 5 min of incubation time, perform the electrochemical measurement of the enzymatic product by DPV (Figure 6.10).

6.4.8 ANALYSIS OF PCB MIXTURES

Prepare standard solutions in the concentration range 0.1–2000 μg mL^{-1} in solution B and follow the procedure described.

Figure 6.11 shows DPV signals recorded for different Aroclor 1248 solutions and the corresponding dose-response curve, which is estimated by nonlinear regression using the following logistic equation by GraphPad Prism 4 program:

$$Y = \frac{\text{bottom} + (\text{top} - \text{bottom})}{1 + 10^{[\log EC_{50} - X] \text{Hillslope}}}$$

where
   top is the Y value at the top plateau of the curve
   bottom is the Y value at the bottom plateau of the curve
   EC_{50} is the antigen concentration necessary to halve the current signal
   X is the $\log_{10}$[free antigen]
   Hillslope is the slope of the linear part of the curve.

The signal is reported as B_x/B_o percentage units, that is, the percentage of the signal decrease with respect to the blank value (solution containing the tracer only), taken as 100% of the response vs. the logarithm of the congener concentration. The curve exhibits the sigmoidal shape typical of a competitive immunoassay; a signal decrease is observed for concentrations greater than 0.001 μg mL^{-1}, whereas the lowest current is measured at Aroclor 1248 concentration of 20 μg mL^{-1}.

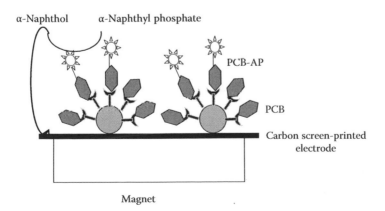

FIGURE 6.10 Functionalized magnetic particle blocked on a carbon-screen-printed electrode by a magnet.

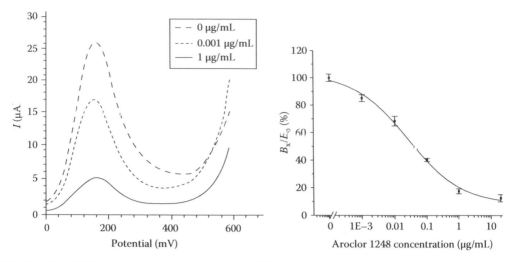

FIGURE 6.11 DPV voltammograms recorded for different Aroclor 1248 standard solutions, and on the right, the corresponding calibration curve.

The DL of the electrochemical immunosensor based on magnetic beads has been estimated to be equal to 0.4 ng mL^{-1}. This depends on the affinity of the antibodies for the antigen and is defined as the lowest analyte concentration that can be distinguished and is calculated by evaluation of the mean of the blank solution (containing the tracer only) response minus two times the SDs.

The EC$_{50}$, which corresponds to the analyte concentration necessary to displace 50% of the label, has been evaluated to be 20 ng mL^{-1}.

The immunoassay test has been replicated to evaluate its reproducibility; for this purpose, three repetitions of the calibration curves in the concentration range 0–20 µg mL^{-1} are carried out, and the average of coefficient of variation (CV) is 7%.

6.4.9 PCB ANALYSIS IN MARINE SEDIMENT AND SOIL EXTRACTS

Marine sediments and soil samples are analyzed by the following procedure:

Marine sediments/soil are dried in the oven at 70°C for 5 h. 0.5 g of dried marine sediments/soil is added to 10 mL of methanol and after a short mixing time (2 min) the mixture is sonicated for 2 min and filtered using a nitrocellulose 0.45 µm filter; then 10 µL of the extract is mixed with 50 µL of suspension containing antibody-coated beads, 930 µL of solution B, and 10 µL of the PCB28-AP conjugate solution diluted 1:10 with respect to the stock solution. After 20 min of incubation time, the beads are washed twice and resuspended in 100 µL of solution B. The electrochemical measurement is performed following the described procedure [64].

6.4.10 DISCUSSION

Table 6.2 shows the results obtained following the described procedure.

One gram of dried marine sediments/soil is artificially spiked with 1 mL of PCB mixture in the concentration range 5–500 µg kg^{-1} (e.g., Aroclor 1248 mixture) and kept in contact for 16 h.

The measured signals are reported as signal percentage (B_x/B_o (%)), giving 100% value to the blank solution. Data obtained from the experiments can be used to indicate the level of PCB pollution in the samples: in fact the theoretical dose–response curve can be divided into three bands corresponding to different pollution levels.

If the sample signal is included in the B_x/B_o (%) range 100–80, 80–40, or 40–0, the sample is respectively considered not polluted, medium polluted, or highly polluted.

TABLE 6.2
Results Obtained from the Analysis of Marine Sediment and Soil Samples, Extracted by the Sonication Assisted Method

Sample Code	Spiked Concentration of Aroclor 1248 ($\mu g\ kg^{-1}$)	B_x/B_o (%)	Classification
	Marine Sediment		
A	5	68 ± 5	Medium polluted
B	50	53 ± 6	Medium polluted
C	500	26 ± 4	Highly polluted
	Soil		
D	5	85 ± 2	Non polluted
E	50	52 ± 6	Medium polluted
F	100	44 ± 5	Medium polluted

The possibility of performing a measurement that allows us to know within a short time whether a sample is contaminated or not is important when many samples have to be analyzed; therefore, these experiments are a valid analytical tool to carry out a fast screening of many samples.

REFERENCES

1. Sanchez-Santed F., Canadas F., Flores P., Lopez-Grancha M., and Cardona D. Long-term neurotoxicity of Paraoxon and chlorpyrifos: Behavioural and pharmacological evidence. *Neurotoxicol. Terathol.*, 2004, 26, 35–317.
2. Noort D., Benschop H. P., and Black R. M. Biomonitoring of exposure to chemical warfare agents: A review. *Toxicol. Appl. Pharmacol.*, 2002, 184, 116–126.
3. Hooijschur E. W. J., Hulst A. G., De Jong A. L., De Reuver L. P., Van Krimpen S. H., Van Baar B. L. M., Wils E. R. J., Kientz C. E., and Brinkman U. A. Th. Identification of chemicals related to the chemical weapons convention during an interlaboratory proficiency test. *TrAC*, 2002, 21, 116–130.
4. Palleschi G., Bernabei M., Cremisini C., and Mascini M. Determination of organophosphorus insecticides with a choline electrochemical biosensor. *Sens. Actuators. B*, 1992, 7, 513–517.
5. Tran-Minh C., Pandey P. C., and Kumaran S. Studies on acetylcholine sensor and its analytical application based on the inhibition of cholinesterase. *Biosens. Bioelectron.*, 1990, 5, 461–471.
6. Cagnini A., Palchetti I., Lionti I., Mascini M., and Turner A. P. F. Disposable ruthenized screen-printed biosensors for pesticides monitoring. *Sens. Actuators. B Chem.*, 1995, 24, 85–89.
7. Arduini F., Ricci F., Tuta C. S., Moscone D., Amine A., and Palleschi G. Detection of carbamic and organophosphorous pesticides in water samples using a cholinesterase biosensor based on Prussian Blue-modified screen-printed electrode. *Anal. Chim. Acta*, 2006, 580, 155–162.
8. Shi M., Xu J., Zhang S., Liu B., and Kong J. A mediator-free screen-printed amperometric biosensor for screening of organophosphorus pesticides with flow-injection analysis (FIA) system. *Talanta*, 2006, 68, 1089–1095.
9. Palchetti I., Cagnini A., Del Carlo M., Coppi C., Mascini M., and Turner A. P. F. Determination of anticholinesterase pesticides in real samples using a disposable biosensor. *Anal. Chim. Acta*, 1997, 337, 315–321.
10. Hernandez S., Palchetti I., and Mascini M. Determination of anticholinesterase activity for pesticides monitoring using a thiocholine sensor. *Intern. J. Environ. Anal. Chem.*, 2000, 78, 263–278.
11. Silva Nunes G., Jeanty G., and Marty J.-L. Enzyme immobilization procedures on screen-printed electrodes used for the detection of anticholinesterase pesticides: Comparative study. *Anal. Chim. Acta*, 2004, 523, 107–115.
12. Bucur B., Fournier D., Danet A., and Marty J.-L. Biosensors based on highly sensitive acetylcholinesterases for enhanced carbamate insecticides detection. *Anal. Chim. Acta*, 2006, 562, 115–121.

13. Vakurov A., Simpson C. E., Daly C. L., Gibson T. D., and Millner P. A. Acetylcholinesterase-based biosensor electrodes for organophosphate pesticide detection: I. Modification of carbon surface for immobilization of acetylcholinesterase. *Biosens. Bioelectron.*, 2004, 20, 1118–1125.
14. Marty J.-L., Mionetto N., Lacorte S., and Barceló D. Validation of an enzymatic biosensor with various liquid chromatographic techniques for determining organophosphorus pesticides and carbaryl in freeze-dried waters. *Anal. Chim. Acta*, 1995, 311, 265–271.
15. Barceló D., Lacorte S., and Marty J.-L. Validation of an enzymatic biosensor with liquid chromatography for pesticide monitoring, *TrAC*, 1995, 14, 334–340.
16. Hart J. P. and Hartley I. C. Voltammetric and amperometric studies of thiocholine at a screen-printed carbon electrode chemically modified with cobalt phthalocyanine: Studies towards a pesticide sensor. *Analyst*, 1994, 119, 259–263.
17. Pereira-Rodrigues N., Cofré R., Zagal J. H., and Bedioui F. Electrocatalytic activity of cobalt phthalocyanine CoPc adsorbed on a graphite electrode for the oxidation of reduced l-glutathione (GSH) and the reduction of its disulfide (GSSG) at physiological pH. *Bioelectrochemistry*, 2007, 70, 147–154.
18. Martorell D., Céspedes F., Martínez-Fàbregas E., and Alegret S. Determination of organophosphorus and carbamate pesticides using a biosensor based on a polishable, 7,7,8,8-tetracyanoquino-dimethane-modified, graphite—Epoxy biocomposite. *Anal. Chim. Acta*, 1997, 337, 305–313.
19. Silva Nunes G., Skládal P., Yamanaka Y., and Barceló D. Determination of carbamate residues in crop samples by cholinesterase-based biosensors and chromatographic techniques. *Anal. Chim. Acta*, 1998, 362, 59–68.
20. Ricci F., Arduini F., Amine A., Moscone D., and Palleschi G. Characterisation of prussian blue modified screen-printed electrodes for thiol detection. *J. Electroanal. Chem.*, 2004, 563, 229–237.
21. Bonnet C., Andreescu S., and Marty J.-L. Adsorption: An easy and efficient immobilisation of acetylcholinesterase on screen-printed electrodes. *Anal. Chim. Acta*, 2003, 481, 209–211.
22. Vakurov A., Simpson C. E., Daly C. L., Gibson T. D., and Millner P. A. Acetylcholinesterase-based biosensor electrodes for organophosphate pesticide detection: I. Modification of carbon surface for immobilization of acetylcholinesterase. *Biosens. Bioelectron.*, 2004, 20, 1118–1125.
23. Laschi S., Ogończyk D., Palchetti I., and Mascini M. Evaluation of pesticide-induced acetylcholinesterase inhibition by means of disposable carbon-modified electrochemical biosensors. *Enzyme Microb. Technol.*, 2007, 40, 485–489.
24. Shahrokhian S. and Yazdani J. Electrocatalytic oxidation of thioglycolic acid at carbon paste electrode modified with cobalt phthalocyanine: Application as a potentiometric sensor. *Electrochim. Acta*, 2003, 48, 4143–4148.
25. http://photoscience.la.asu
26. Tommos C. and Babcock G. T. Proton and hydrogen currents in photosynthetic water oxidation. *Biochim. Biophys. Acta Bioenerget.*, 2000, 1458, 199–219.
27. Blankenship, R. E. Photosynthetic antennas and reaction centers: Current understanding and prospects for improvement. In Nozik, A.J., Ed. *Research Opportunities in Photochemical Sciences*. 1996 online at http://photoscience.la.asu.edu/photosyn/education/antenna.html.
28. Dekker J. P. and Boekema E. J. Supramolecular organization of thylakoid membrane proteins in green plants. *Biochim. Biophys. Acta Bioenerget.*, 2005, 1706, 12–39.
29. Giardi M. T. and Pace E. Photosynthetic proteins for technological applications. *Trends Biotechnol.*, 2005, 23, 257–263.
30. Barber J. Water, water everywhere, and its remarkable chemistry. *Biochim. Biophys. Acta Bioenerget.*, 2004, 1655, 123–132.
31. Koblížek M., Malý J., Masojídek J., Komenda J., Kučera T., Giardi M. T., Mattoo A. K., and Pilloton R. A biosensor for the detection of triazine and phenylurea herbicides designed using PSII coupled with screen-printed electrode. *Biotechnol. Bioeng.*, 2002, 78, 110–116.
32. Muñoz de la Peña, M., Mahedero C., and Bautista-Sánchez A. Monitoring of phenylurea and propanil herbicides in river water by solid-phase-extraction high performance liquid chromatography with photoinduced-fluorimetric detection. *Talanta*, 2003, 60, 279–285.
33. Giardi M. T., Esposito D., Leonardi C., Matoo A. K., Margonelli A., and Angelini G. Patent Upica, Italy, 112. European Patent 2003, EP 1134585, 2000.
34. Brewster J. D., Lightfield A. R., and Bermel P. L. Storage and immobilisation of photosystem II reaction centers used in a assay for herbicides. *Anal. Chem.*, 1995, 67, 1296–1299.
35. Porazzi E., Pardo Martinez M., Fanelli R., and Benfenati E. GC–MS analysis of dichlobenil and its metabolites in groundwater. *Talanta*, 2005, 68, 146–154.

36. Moret S., Sánchez J. M., Salvadó V., and Hidalgo M. The evaluation of different sorbents for the preconcentration of phenoxyacetic acid herbicides and their metabolites from soils. *J. Chromatogr. A*, 2005, 1099, 55–63.

37. Escuderos-Morenas M. L., Santos-Delgado M. J., Rubio-Barroso S., and Polo-Díez L. M. Direct determination of monolinuron, linuron, and chlorbromuron residues in potato samples by gas chromatography with nitrogen–phosphorus detection. *J. Chromatogr. A*, 2003, 1011, 143–153.

38. Baranowska I., Barchańska H., and Pyrsz A. Distribution of pesticides and heavy metals in trophic chain. *Chemosphere*, 2005, 60, 1590–1599.

39. González-Martínez M. A., Puchades R., Maquieira A., Ferrer I., Marco M. P., and Barceló D. Reversible immunosensor for the automatic determination of atrazine. Selection of performance of three polyclonal antisera. *Anal. Chim. Acta*, 1999, 386, 201–210.

40. Cao Y., Lu Y., Long S., Hong J., and Sheng G. Development of an ELISA for the detection of bromoxynil in water. *Environ. Inter.*, 2005, 31, 33–42.

41. Naessens M., Tran Minh C., and Leclerc J. C. Fiber optic biosensor using *Chlorella vulgaris* for determination of toxic compounds. *Ecotoxicol. Environ. Safety*, 2000, 46, 181–185.

42. Labergè D., Chartrand J., Rouillon R., and Carpentier R. In vitro phytotoxicity screening test using immobilized spinach thylakoid. *Environ. Toxicol. Chem.*, 1999, 18, 2851–2858.

43. Sanders C. A., Rodriguez M., and Greenbaum E. Stand-off tissue-based biosensors for the detection of chemical warfare agents using photosynthetic fluorescence induction. *Biosens. Bioelectron.*, 2001, 16, 439–446.

44. Rouillon R., Tocabens M., and Carpentier R. A photoelectrochemical cell for detecting pollutant-induced effects on the activity of immobilized cyanobacterium *Synechoccuc* sp PCC7942. *Enzyme Microb. Technol.*, 1999, 25, 230–235.

45. Rouillon R., Boucher N., Gingras Y., and Carpentier R. Potential for the use of photosystem II submembrane fractions immobilized in poly(vinylalcohol) to detect heavy metals in solution or in sewage sludge. *J. Chem. Technol. Biotechnol.*, 2000, 75, 1003–1007.

46. Rouillon R., Piletsky S. A., Breton F., Piletska E. V., and Carpentier R. Photosystem II biosensors for heavy metals monitoring. In Giardi, M.T. and Piletska, E.V., *Biotechnology Intelligence Unit Biotechnological Applications of Photosynthetic Proteins: Biochips, Biosensors and Biodevices*, 2007, New York, SpringerLink.

47. Rodriguez-Mozaz S., Marco M.-P., Lopez de Alda M. J., and Barceló D. Biosensors for environmental monitoring of endocrine disruptors: A review article. *Anal. Bioanal. Chem.*, 2004, 378, 588–598.

48. Lu Y., Xu J., Liu B., and Kong J. Photosynthetic reaction center functionalized nano-composite films: Effective strategies for probing and exploiting the photo-induced electron transfer of photosensitive membrane protein. *Biosens. Bioelectron.*, 2007, 22, 1173–1185.

49. Rizzuto M., Polcaro C., Desiderio C., Koblížek M., Pilloton R., and Giardi M. T. Herbicide monitoring in surface water samples with a PSII-based biosensor. II Workshop on Chemical Sensors and Biosensors (Roma, Italy), 1999, 348–356.

50. Touloupakis E., Giannoudi L., Piletsky S. A., Guzzella L., Pozzoni F., and Giardi M. T. A multi-biosensor based on immobilized photosystem II on screen-printed electrodes for the detection of herbicides in river water. *Biosens. Bioelectron.*, 2005, 20, 1984–1992.

51. Rouillon R., Mestres J.-J., and Marty J.-L. Entrapment chloroplasts and thylakoids in polyvinylalcohol–SbQ. Optimisation of membrane preparation and storage conditions. *Anal. Chim. Acta*, 1995, 311, 437–442.

52. Giardi M. T., Guzzella L., Euzet P., Rouillon R., and Esposito D. Detection of herbicide subclasses by an optical multibiosensor based on an array of photosystem II mutants. *Environ. Sci. Technol.*, 2005, 39, 5378–5384 B.

53. Bettazzi F., Laschi S., and Mascini M. One-shot screen-printed thylakoid membrane-based biosensor for the detection of photosynthetic inhibitors in discrete samples. *Anal. Chim. Acta*, 2007, 589, 14–21.

54. Bettazzi F., Laschi S., and Mascini M. Disposable biosensors assembled with thylakoid membranes from spinach leaves for herbicide detection. *Chem. Anal.*, 2005, 50, 117–128.

55. Maly J., Di Meo C., De Francesco M., Masci A., Masojidek J., Sugiurad M., Volpe A., and Pilloton R. Reversible immobilization of engineered molecules by Ni–NTA chelators. *Bioelectrochemistry*, 2004, 63, 271–275.

56. Boucher N., Lorrain L., Rouette M.-E., Perron E., Déziel N., Tessier L., and Bellemare F. Rapid testing of toxic chemicals. American Laboratory, October 2004.

57. Frense, D., Muller A., and Beckmann D. Detection of environmental pollutants using an optical biosensor with immobilized algae cells. *Sens. Actuators. B*, 1998, 51, 256–260.

58. Park I. H., Seo S. H., and Lee H. J. Photosynthetic characteristics of polyvinylalcohol-immobilised spinach chloroplasts. *Korean J. Bot.*, 1991, 34, 215–221.

59. Rouillon R., Sole M., Carpentier R., and Marty J. L. Immobilization of thylakoids in polyvinyl alcohol for the detection of herbicides. *Sens. Actuators. B*, 1995, 26–27, 477–479.

60. Piletskaya E. V., Piletsky S. A., Sergeyeva T. A., El'skaya A. V., Sozinov A. A., Marty J.-L., and Rouillon R. Thylakoid membranes based test system for detecting trace quantities of the photosynthesis-inhibiting herbicide in drinking water. *Anal. Chim. Acta*, 1999, 391, 1–7.

61. World Intellectual Property Organization International Publication Number WO 2004/046717 A1. Biosensors, and method and kits for using same.

62. Fránek M., Deng A., Kolář V., and Socha J. Direct competitive immunoassays for the coplanar polychlorinated biphenyls. *Anal. Chim. Acta*, 2001, 444, 131–142.

63. Ahmed F. E. Analysis of polychlorinated biphenyls in food products. *TrAC Trend. Anal. Chem.*, 2003, 22, 170–185.

64. Centi S., Laschi S., Fránek M., and Mascini M. A disposable immunomagnetic electrochemical sensor based on functionalised magnetic beads and carbon-based screen-printed electrodes (SPCEs) for the detection of polychlorinated biphenyls (PCBs). *Anal. Chim. Acta*, 2005, 538, 205–212.

65. Udai S. G., Schwartz H. M., and Wheatley B. Congener specific analysis of polychlorinated biphenyls (PCBs) in serum using GC/MSD. *Anal. Chim. Acta*, 1995, 30, 1969–1977.

66. Berset J. D. and Holzer R. Determination of coplanar and ortho substituted PCBs in some sewage sludges of Switzerland using HRGC/ECD and HRCG/MSD. *Chemosphere*, 1996, 32, 2317–2333.

67. Luthy R. G., Dzombak D. A., Shannon M. J. R., Unterman R., and Smith J. R. Dissolution of PCB congeners from an Aroclor and an Aroclor/hydraulic oil mixture. *Wat. Res.*, 1997, 31, 561–573.

68. Fillmann G., Galloway T. S., Sanger R. C., Depledge M. H., and Readman J. W. Relative performance of immunochemical (enzyme-linked immunosorbent assay) and gas chromatography-electron-capture detection techniques to quantify polychlorinated biphenyls in mussel tissues. *Anal. Chim. Acta*, 2002, 461, 75–84.

69. Šišak M., Fránek M., and Hruška K. Application of radioimmunoassay in the screening of polychlorinated biphenyls. *Anal. Chim. Acta*, 1995, 311, 415–422.

70. Fránek M., Pouzar M. V., and Kolar V. Enzyme-immunoassays for polychlorinated biphenyls. Structural aspects of hapten-antibody binding. *Anal. Chim. Acta*, 1997, 347, 163–176.

71. Lambert N., Fan T. S., and Pilette J. F. Analysis of PCBs in waste oil by enzyme immunoassay. *Sci. Total Environ.*, 1997, 196, 57–61.

72. Zajicek J. L., Tillitt D. E., Schwartz T. R., Schmitt C. J., and Harrison R. O. Comparison of an enzyme-linked immunosorbent assay (ELISA) to gas chromatography (GC) measurement of polychlorinated biphenyls (PCBs) in selected US fish extracts. *Chemosphere*, 2000, 40, 539–548.

73. Luppa P. B., Sokoll L. J., and Chan D. W. Immunosensors—Principles and applications to clinical chemistry. *Clin. Chim. Acta*, 2001, 314, 1–26.

74. Sharma S. K., Sehgal N., and Kumar A. Biomolecules for development of biosensors and their applications. *Curr. Appl. Phys.*, 2003, 3, 307–316.

75. Kröger S., Piletsky S., and Turner A. P. F. Biosensors for marine pollution research, monitoring and control. *Marine Pollution Bull.*, 2002, 45, 42–34.

76. Thévenot D. R., Toth K., Durst R. A., and Wilson F. S. Electrochemical biosensors: Recommended definitions and classifications. *Biosens. Bioelectron.*, 2001, 16, 121–131.

77. Solé S., Merkoçi A., and Alegret S. New materials for electrochemical sensing III. Beads. *Trends in Anal. Chem.*, 2001, 20, 102–110.

78. E. Paleček E., Kizek R., Havran L., Billova S., and Fojta M. Electrochemical enzyme-linked immunoassay in a DNA hybridization sensor. *Anal. Chim. Acta*, 2002, 469, 73–83.

79. Centi S., Tombelli S., Minunni M., and Mascini M. Aptamer-based detection of plasma proteins by an electrochemical assay coupled to magnetic beads. *Anal. Chem.*, 2007, 79, 1466–1473.

80. Centi S., Silva E., Laschi S., Palchetti I., and Mascini M. Polychlorinated biphenyls (PCBs) detection in milk samples by an electrochemical magneto-immunosensor (EMI) coupled to solid-phase extraction (SPE) and disposable low-density arrays. *Anal. Chim. Acta*, 2007, 594, 9–16.

81. Centi S., Laschi S., and Mascini M. Improvement of analytical performances of a disposable electrochemical immunosensor by using magnetic beads. *Talanta*, 2007, 73, 394–399.

82. Ecobioservices, http://ebsr.it.

7 Determination of Pesticides by Matrix Solid-Phase Dispersion and Liquid Chromatography–Tandem Mass Spectrometry

Svetlana Grujic, Tatjana Vasiljevic, Marina Radisic, and Mila Lausevic

CONTENTS

7.1 LIQUID CHROMATOGRAPHY–TANDEM MASS SPECTROMETRY

Although gas chromatography (GC) has traditionally been applied for pesticide residue analysis, the use of liquid chromatography (LC) has grown rapidly in the last decade. Modern pesticides, together with their degradation products, can be considered as typical candidates for LC separation,

141

because of their medium to high polarity, and their thermolability and/or low volatility [1]. In general, optimization of LC separation is tedious and time-consuming, even with the support of the computer-assisted retention modeling [2,3]. Most LC-based methods use common ultraviolet (UV), fluorescence, or electrochemical detection occasionally combined with postcolumn treatment, for example, derivatization. Mass spectrometry (MS) has the advantage over conventional detectors, because it can provide information for unambiguous analyte identification even with poor LC separation. Tandem mass spectrometry (MS/MS) uses two stages of mass analysis—one to preselect an ion and the second to analyze fragments induced by collision of an ion with an inert gas, such as argon or helium. LC coupled with MS/MS (LC–MS/MS) is capable of differentiation between analyte and matrix signal, as well as between the analytes that coelute, thus permitting quantification of pesticide traces in very complex matrices.

Numerous papers on the LC–MS analysis of pesticides and related compounds in various sample matrices have been published [4–10]. LC–MS is the method of choice for carbamates, as their thermal lability prohibits GC analysis [11]. While triazines are readily amenable to GC–MS, this is not true for their hydroxy- and des-alkyl degradation products [12]. Owing to the thermal lability of the urea group, phenylureas are frequently analyzed using LC–MS [13,14]. As chlorinated phenoxy acid (CPA) herbicides can be analyzed by GC–MS only after derivatization [15], LC–MS has often been employed in their analysis [16]. Sulfonylureas are thermally labile and cannot be readily derivatized. Therefore, LC–MS is the technique of choice for their analysis [17]. In a recent review paper, GC–MS versus LC–MS/MS have been evaluated for determination of 500 high-priority pesticides [18]. For each of the selected pesticides, the applicability and sensitivity of both the methods were compared. Only for one class of pesticides, the organochlorine compounds, GC–MS achieved better performance. For all other classes, higher sensitivity was attained using LC–MS with multiple-stage analysis.

7.1.1 LC–MS/MS INSTRUMENTATION

7.1.1.1 Ionization Sources

Unlike GC–MS, where electron impact and chemical ionization are most commonly used [19–21], the soft ionization techniques applied for LC–MS analyses are electrospray ionization (ESI) and atmospheric pressure chemical ionization (APCI). ESI is the ionization technique recommended for polar, ionized, and high molecular weight compounds, and hence, it is frequently used for the pesticide analysis. Liquid-phase chemistry plays a key role in the ion formation in ESI, which is an interface that works well on any compound that is ionized in solution, such as ammonium quaternary compounds and acidic herbicides [22–24]. APCI is a more energetic source than electrospray, as both pneumatic nebulization and high temperatures (350°C–500°C) are applied to evaporate sample solution. These high temperatures must be taken into account when working with thermally degradable compounds. Thurman et al. [25] evaluated the performance of APCI and ESI in both positive-ion and negative-ion modes for the analysis of 75 pesticides from various compound classes (Table 7.1). They also proposed that ionization-continuum diagram could be useful for selecting APCI or ESI. Although the employment of ionization-continuum diagram can be helpful for choosing between APCI and ESI source, there are a great number of pesticides that can be analyzed by both APCI and ESI sources with a satisfactory sensitivity. Many times, the interface selected for a particular pesticide is a matter of individual preference, derived from experience and available techniques as well as matrix properties.

7.1.1.2 Mass Analyzers

The most commonly used mass spectrometers that allow MS/MS experiments are triple quadrupole (TQ) and quadrupole ion trap (QIT). This is mainly due to their easier operating performance, their

TABLE 7.1
Sensitivity of Different LC–MS Ionization Modes for the Determination of Various Classes of Pesticides

Pesticide Class	APCI+	APCI–	ESI+	ESI–
Phenylurea herbicides	++[a]	+/0	++	+/0
Triazine herbicides	++	0[c]	++	0
Sulfonylurea herbicides	++	+[b]	++	+
Organophosphate insecticides	++	+	++	0
Carbamate insecticides	++	0	++	0
Acetanilide herbicides	+	+	+	0
Bipyridylium herbicides	0	0	++	0
Chlorophenoxy acid herbicides	0	+	0	++
Phenolic compounds	0	0/+	0	0/+
Organochlorine insecticides	0	0	0	0

Source: Adapted from Thurman, E.M. et al., *Anal. Chem.*, 73, 5441, 2001.

[a] Very sensitive.

[b] Moderately sensitive.

[c] Low response.

better robustness for routine analysis, and their relatively low cost, compared with time-of-flight (TOF) or Fourier transform-ion cyclotron resonance (FT-ICR) instruments. Dual MS analysis can be either tandem in space or tandem in time (Figure 7.1). Tandem in space means that two mass spectrometers are in series; therefore, precursor ions and product ions are created and analyzed separately. The various steps of the process take place simultaneously, but are separated in space (Figure 7.1a). Among the various combinations, the TQ and hybrid quadrupole time-of-flight

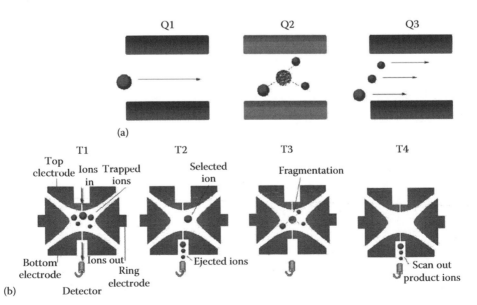

FIGURE 7.1   Tandem mass spectrometry: (a) in space, TQ, where Q1 and Q3 are quadrupole mass filters and Q2 is the collision chamber; (b) in time, QIT, where T1, T2, T3, and T4 are the time segments.

(Q-TOF) mass spectrometers are the most successful ones in pesticide analysis [26]. Dual analysis can also be tandem in time (Figure 7.1b), as achieved in QITs in which the sequence of events takes place in the same space, but are separated in time.

Mass separation in a quadrupole mass filter is based on achieving a stable trajectory for ions of specific m/z values in a hyperbolic electrostatic field. In TQ instruments (Figure 7.1a), an ion of interest is preselected with the first mass filter Q1, collisionally activated with energies up to 300 eV with argon in the pressurized collision chamber Q2, and the fragmentation products are analyzed with the third quadrupole Q3 [26]. The TQ was the most applied mass analyzer in pesticide analysis, because it was the first commercially available instrument. TQ analyzers display high sensitivity when working in multiple reaction monitoring (MRM) mode, and are therefore best suited for achieving the strict maximum residue levels (MRLs) regulated for various toxic compounds in different matrices.

The QIT is a small, low-cost, easy-to-use, fast, sensitive, and versatile mass analyzer. Its unique feature is its ability to perform multiple stages of MS [27,28]. Also, it can simultaneously store positive and negative ions at once for extended periods of time. These characteristics make QIT an attractive option for detection of pesticides, as shown in many studies (Table 7.2). When performing MS/MS, QIT instruments are generally less sensitive than TQ analyzers, but they have the advantage of working in product-ion scan without losses in sensitivity and the possibility of performing multiple-stage fragmentation (MS^n). Such advantages are important tools for unambiguous

TABLE 7.2
Selected Ion-Trap Applications in Pesticide Residue Analysis

Compound	Ionization Mode	Technique	MS^n	LOD	Application	Reference
Insecticides						
Organophosphorus	NCI	GC	MS	0.002–0.04 mg kg^{-1}	Fruit analysis	[29]
	APCI+	LC	MS^2		Fruit analysis	[30]
			MS^3			
Carbamate	EI	GC	MS^2		Fragmentations studies	[31]
	ESI+	LC	MS	0.001–0.01 mg kg^{-1}	Fruit juice analysis	[32]
			MS^2			
	ESI+	LC	MS^2	0.1–0.3 µg kg^{-1}	Fruit analysis	[33]
	APCI+	LC	MS^2		Fruit analysis	[34]
			MS^3			
Pyrethroid	ESI+	LC	MS	0.1–0.4 mg kg^{-1}	Vegetable analysis	[35]
	ESI+	LC	MS^2		Fruit analysis	[36]
Urea	ESI+	LC	MS^2	3–10 µg kg^{-1}	Fruit analysis	[33]
Herbicides						
Phenoxy acid	EI	GC	MS	0.1–0.04 µg g^{-1}	Cereal analysis	[37]
Phenylurea	ESI+	LC	MS	0.005–0.05 mg kg^{-1}	Fruit juice analysis	[32]
			MS^2			
Organophosphorus	ESI–	LC	MS^3		Fragmentations study	[38]
Fungicides						
Conazole	ESI+	LC	MS^2		Fragmentations pathway	[39]
			MS^3		in fruits	
Benzamide	EI	GC	MS^2		Degradation products	[40]
	CI				in fruits and drinks	
Benzimidazole	ESI+	LC	MS^2	0.1–3 µg kg^{-1}	Fruit analysis	[33]
	ESI+	LC	MS^2	0.03 ng mL^{-1}	Fruit juice analysis	[41]

identification of trace compounds, such as pesticides in complex food matrices. The disadvantages of QIT are low resolution, interfering side-reactions (because all reactions occur in the same space), and a limited dynamic range.

7.2 MATRIX EFFECT

The high selectivity of LC–MS/MS methods results in chromatograms without any noticeable interference in the form of extra chromatographic peaks or peak shoulders. Nevertheless, matrix components coeluting with the analytes from the LC column can interfere with the ionization process in the electrospray, causing ionization suppression or enhancement [42]. This effect is very common when working with complex matrices and is known as the matrix effect [43]. Comprehensive studies on the matrix effect demonstrated that it is frequently accompanied by significant deterioration of the analytical method precision [44]. The constituents of the sample can extensively influence the MS detector response and consequently the recovery rate [45,46]. The suppression of the analyte signal has been typically discussed in the literature [36,47,48]. The enhancement of the analyte signal by matrix components has been less common [45]. The extent to which the analyte signal is affected can be simply determined from the signal difference of analyte in standard solution and spiked blank extract, i.e., matrix-matched standard [10].

To overcome the matrix effect when quantifying, several approaches are available. Isotopic dilution is the best option if the labeled target compounds are available [49]. This approach allows signal suppression (or enhancement) to be corrected, as both labeled and native compounds will undergo the same suppression effect. However, the availability of isotopically labeled analogues as internal standards is frequently limited, so this calibration method is usually used when only one analyte is to be determined. Quantification by standard addition is another possibility to correct matrix suppression, but this method is not convenient when a high number of samples is to be analyzed. Matrix-matched calibration is another way to compensate the matrix effect and it was successfully used in numerous recently published papers [36,50–52]. This calibration method will be explained in Examples 1 and 2.

7.3 SAMPLE PRECONCENTRATION METHODS

Although there are thousands of methods for determining pesticides in fruits, vegetables, and other complex food matrices, the pesticide residue analysis still represents an analytical challenge. An adequate analytical method for residue analysis should be sensitive, selective, accurate, automated, cheap, applicable to a wide range of pesticides and matrices, and capable of providing unambiguous structural information. Also, in recent times, special attention is paid to a development of environmentally safe analytical methods. However, such perfect methods are not encountered in practice.

An important aspect in every analytical method for detection of analytes at trace levels in a variety of complex matrices is the efficient sample preparation. This step is the least evolved part of most analytical procedures. Sample preparation protocols for pesticide analysis adopted in many standardized analytical methods include repeated extractions, purification, and concentration steps, making them time-consuming and expensive to perform, especially when a large number of samples must be analyzed. Relatively large volumes of toxic, expensive, and flammable solvents are used and, therefore, subsequent evaporation and disposal of the solvent is needed. In many cases, during sample preparation emulsions can be formed, decreasing the extraction efficiency and prolonging the time of the procedure. As the most commonly employed organic solvents do not selectively extract the targeted analytes, clean-up procedures are often needed for the isolation of analytes from the matrix components [53].

The conventional methods used for the determination of pesticide residues in various matrices are usually based on liquid–liquid extraction (LLE) followed by a clean-up step. Analyte determination is then performed by GC or high-performance liquid chromatography (HPLC) with selective detectors. Such procedures are lengthy and tedious, and generally do not keep pace with the advances in the analytical technology. Solvent extraction methods have the following disadvantages [54]:

1. Environmental contamination by large volumes of toxic and expensive solvents. The amount of solvent (ethyl acetate, acetonitrile, hexane, dichloromethane, etc.) is usually greater by a factor of 10^8–10^{10} than that of the pesticide residues to be determined.
2. Their inefficiency as screening methods. These methods are too complex and do not allow gathering of relevant data in time to prevent contaminated food from entering the market.
3. The difficulty of automation.
4. The formation of emulsions.
5. False-positive results owing to the lack of specificity.
6. The newly developed groups of pesticides that are more polar and/or thermodegradable.

7.3.1 MATRIX SOLID-PHASE DISPERSION

To overcome the drawbacks of the classical LLE methods, in recent years, major attention has been paid to simplification, miniaturization, and improvement of the sample extraction and clean-up methods to replace LLE. The ideal sample preparation method should be fast, accurate, precise, and should employ small volumes of organic solvents. Moreover, the sample preparation should be adaptable to field work and should make use of less costly materials [53,55]. With these current trends, in the last 15 years, alternative solid-phase-based extraction techniques have been developed. These include solid-phase extraction (SPE), solid-phase microextraction (SPME) [56], matrix solid-phase dispersion (MSPD) [57,58], stir-bar sorptive extraction (SBSE) [59–61], microwave-assisted extraction [62], and supercritical-fluid extraction [63,64].

For the analysis of solid and semisolid environmental, food, or biological matrices, MSPD method was introduced by Barker et al. [57]. This simple pretreatment technique, based on the blending of the sample with an abrasive solid support material, combines several analytical techniques. It performs simultaneous disruption and homogenization of the sample, as well as the extraction, fractionation, and clean-up of analytes in a single step. The sample is fractionated and homogenized with sorbent (e.g., SiO_2, Al_2O_3, Florisil, or C_{18}), using mortar and pestle, to obtain a fine semidry mixture. Then, a small column or cartridge is filled with this mixture. In this way, a unique chromatographic column is prepared and the sample is dispersed throughout the entire column packing [58]. The obtained extracts are ready for analysis, but can be subjected to further purification. In recent years, the MSPD procedure has become very popular in the pesticide analysis and has been introduced in many multiresidual methods for the analysis of various matrices, and some of them are presented in Table 7.3.

The MSPD method has been proven to be a good alternative to the classical LLE method [54,76]. When compared with LLE, in which analytes to be extracted are partitioned between two immiscible liquids, in the MSPD method, the analytes are partitioned between a solid sorbent and a semisolid sample matrix, and the analytes are required to have a greater affinity for the sorbent than for the sample matrix. The target compounds are retained on the solid phase and then removed by eluting with a solvent of great affinity for the analytes. For example, when the extraction efficiency of MSPD method for determination of carbendazim residues in fruit juices is compared with that obtained by classical LLE technique, the results show that both the procedures work well at all concentration levels and the values of the relative standard deviations (RSDs) are almost identical. However, the average recoveries for MSPD are higher. Furthermore, MSPD typically uses amounts of less than 1 g of the sample (versus 50 g, common for LLE) and, as a consequence, the amount of matrix constituents in the final extract is lower than that of the solvent extraction. This results in cleaner chromatograms. Moreover, the MSPD method is more rapid and requires lower consumption of toxic organic solvents (i.e., 10 mL versus 250 mL, common for LLE).

TABLE 7.3
Selected MSPD Methods for Analysis of Pesticides in Various Matrices

Matrix	Sorbent	Eluent	Analytical Method	Reference
Animal fat	Al_2O_3	Heptane	LC–UV	[65]
Apple juice	Diatomaceous earth	Hexane/CH_2Cl_2 (1:1, v/v)	GC–MS	[66]
Citrus fruits	C_{18}	CH_2Cl_2/CH_3OH (80:20, v/v)	LC–MS/MS	[67]
Citrus fruits	C_8	CH_2Cl_2	LC–UV	[68]
Fish tissue	C_{18}	CH_3CN	GC–ECD[a]	[69]
Fruit juices	Florisil	Ethyl acetate	GC–NPD[b]	[70]
Fruits, vegetables	C_8	CH_2Cl_2/CH_3CN (60:40, v/v)	LC–MS	[54]
Fruits, vegetables	C_{18}	Ethyl acetate	GC–NPD; GC–ECD; GC–MS	[71]
Honeybee	C_{18}	CH_2Cl_2/CH_3OH (85:15, v/v)	LC–MS	[72]
Milk	Sand	H_2O at 90°C	LC–MS	[73]
Olives	Aminopropyl	CH_3CN	GC–MS; LC–MS/MS	[74]
Oranges	C_8	CH_2Cl_2	LC–MS	[75]
Plants	Acidic silica	CH_2Cl_2/CH_3OH (5:1, v/v)	LC–UV	[76]
Tea	Florisil	n-Hexane/CH_2Cl_2 (1:1, v/v)	GC–ECD; GC–MS	[77]
Tobacco	Florisil	n-Hexane	GC–ECD	[78]
Vegetables	Florisil	CH_2Cl_2	GC–ECD; GC–MS	[79]

[a] ECD, electron-capture detection.
[b] NPD, nitrogen–phosphorus detection.

The main advantages of MSPD method are [53]

1. It requires small amounts of sample and the consumption of toxic, flammable, and expensive solvents is substantially reduced when compared with the classical extraction methods. In this way, it decreases the environmental contamination and increases the analyst's safety.
2. The analytical protocol is drastically simplified and shortened, enhancing the access to timely data on residue levels present in the sample.
3. It is suitable for use with many different types of matrices that may contain residues of chemical contaminants.
4. It eliminates the possibility of emulsion formation.
5. It enhances the extraction efficiency of the analytes as the entire sample is exposed to the extractant.

The principles of MSPD procedure have been described in detail [58]. It includes the following steps:

1. Solid or semisolid, viscous sample is placed in a glass mortar and blended with a sorbent using a glass pestle. The amount of sorbent depends on the sample type. Sample/sorbent ratio typically ranges from 1:1 to 1:4. The shearing forces generated by the blending process disrupt the sample structure and provide a more finely divided material for extraction. Some procedures use abrasives that also possess the properties of a drying agent, such as anhydrous Na_2SO_4 or silica, producing the material that is quite dry for subsequent extraction.
2. The material is transferred to a column, often an empty syringe barrel or a cartridge with a stainless steel or polypropylene frit, cellulose filter, or a plug of silanized glass wool at the bottom that retains the sample. A second frit or plug is then placed on top of the material. The material is compressed using a modified syringe plunger to form a column packing.

3. The packing of the column is eluted using pure solvent or solvent mixtures. There are two possibilities regarding elution. The first is that prior to elution, the washing step is performed. In this way, interfering compounds of the sample matrix are eluted, while the analytes are retained on the column. Then, the analytes are eluted using a different solvent and the extract is collected. The second possibility is omission of the washing step. The target analytes are directly eluted, while interfering matrix components are retained on the column.

4. Finally, the obtained extract is either directly analyzed or additional clean-up is performed. That is, MSPD-based methods use organic solvents that are not selective in extracting the target compounds from complex matrices, making it difficult to obtain clear extracts, free of interferences. This is regarded as the greatest disadvantage of MSPD when analyzing extracts by LC coupled to common LC detectors, such as ultraviolet or diode array detectors (DADs). Therefore, clean-up of the extract is often included in MSPD procedures. However, the clean-up step in not needed when using mass detector, as in LC–MS or GC–MS, and MSPD method is in fact adapted to these advanced analytical techniques.

Reversed-phase materials, such as C_8- and C_{18}-bonded silica, are often used as the solid support. Silica, alumina, carbon materials [80,81], and chemically modified sorbents [82] are less frequently used. The elution solvents used for efficient desorption of the target analytes from the column range from alkanes through toluene, dichloromethane, and alcohols to water at high temperatures. Pesticides are often eluted with hexane [78], ethyl acetate [70,71,83], dichloromethane [41,68,75,79], or their mixtures. The preferred sorbent/eluent combination is primarily determined by the polarity of the target analytes and the nature of the sample matrix. Therefore, in the optimization of the MSPD procedure, several sorbent/solvent combinations should be tested.

7.4 DETERMINATION OF CARBENDAZIM RESIDUES IN FRUIT JUICES BY MSPD FOLLOWED BY LC–MS/MS ANALYSIS (EXAMPLE 1)

Carbendazim is one of the most frequently detected pesticides in fruits and fruit products [83–87]. It is registered for use on fruits, sunflower, sugar beet, and wheat. The related fungicides, benomyl and thiophanate-methyl, are usually degraded to carbendazim directly on fruits as well as during the analysis [76,84]. The extensive use of these pesticides enforces the need for the development of a rapid, selective, and sensitive analytical method for the determination of carbendazim residues in fruit juices. The European Union (EU) has established the MRLs for carbendazim in fruit from 0.1 mg kg^{-1} (for berries and small fruit) to 0.5 mg kg^{-1} (for citrus fruit) [88].

7.4.1 MSPD PROCEDURE

The fruit juice sample, with pH value adjusted to 6, was sonified in ultrasonic bath for 15 min. Then, 1 mL of fruit juice was blended with 1 g of diatomaceous earth for 5 min using mortar and pestle. Homogeneous powdery mixture was transferred into a 6 mL cartridge, i.e., SPE tube. Two Teflon frits were used to retain the packing. The SPE tube was placed on vacuum manifold and dried by vacuum suction for 5 min. The packing was soaked with ca. 2 mL of dichloromethane. After 1 min, by applying a slight vacuum (1 mL min^{-1} flow), the column was eluted with 10 mL of dichloromethane. Extracts were collected in graduated conical tubes (15 mL) and concentrated under gentle nitrogen stream, in a water bath at 30°C, to ca. 1 mL. To avoid evaporation to dryness, 1 mL of methanol was added and evaporation was continued till the final volume of 0.4 mL. The obtained extracts were filtered into autosampler vial through 0.45 μm nylon syringe filter and analyzed.

For recovery studies, fortified fruit juice samples were prepared by spiking with standard carbendazim solution (1 µg mL^{-1}) to produce samples containing pesticide at concentrations ranging from 10 to 500 ng mL^{-1}. The samples were kept overnight at room temperature. Unspiked juice samples were used as blanks.

7.4.2 Calibration

Matrix-matched calibration was used. The standards were blank juice sample extracts fortified at 10, 50, 100, 250, and 500 ng mL^{-1}. The matrix-matched standards were prepared for each sample type (apple, peach, cherry, raspberry, and orange juice) by adding the aliquots of standard carbendazim solution (1 µg mL^{-1}) to the blank extracts obtained following the MSPD procedure. The MS detector response was linear in the studied concentration range, with correlation coefficient of 0.9995. Each analysis was performed in the following sequence: first, matrix-matched standards were analyzed to form a 5-point external calibration curve; then, the blanks were analyzed; finally, the spiked or real samples.

7.4.3 LC–MS Analysis

Surveyor LC system (Thermo Fisher Scientific, Waltham, MA, USA) was used. The column was reverse-phase Zorbax Eclipse® XDB-C18 (Agilent Technologies, Santa Clara, CA, USA), 4.6 mm × 75 mm i.d. and 3.5 µm particle size. Separation was isocratic. Mobile phase consisted of methanol and 0.1% acetic acid (60:40, v/v). Mobile phase flow rate was 0.5 mL min^{-1}. An aliquot of 10 µL of the final extract was injected into the LC system.

The MS used was LCQ Advantage ion trap (Thermo Fisher Scientific, USA). The ESI technique was used and positive ions were analyzed. The instrument tuning parameters were optimized for m/z 192.0 (protonated carbendazim molecule, Figure 7.2a) as well as MS/MS fragmentation: m/z 192.0 → m/z 160.2 (Figure 7.2b), by injection of standard carbendazim solution (1 µg mL^{-1}) with a syringe pump to the mobile-phase flow. The following parameters were determined: capillary temperature (290°C), sheath gas flow (N_2, 38 au, i.e., 38 arbitrary units), source voltage (4.5 kV), capillary voltage (4 V), and collision energy with helium atoms in trap (30%). Results were processed using Xcalibur® v. 1.3 (Thermo Fisher Scientific, USA) software package.

Typical mass chromatograms of the final extract from the fruit juice are shown in Figure 7.3. For every sample, the total ion chromatogram (TIC, m/z 150.0–350.0, Figure 7.3a), selected ion monitoring chromatogram (SIM, m/z 192.0, Figure 7.3b), as well as the selected reaction monitoring chromatogram (SRM, m/z 192.0 → m/z 160.2, Figure 7.3c) were obtained. Carbendazim concentration was calculated by external calibration, using the peak area in SRM mode.

7.4.4 Optimization of MSPD Method

In the MSPD method development, the ratio of fruit juice and sorbent was initially optimized. Diatomaceous earth was used as sorbent, following the procedure of Perret et al. [83]. A sorbent should completely absorb the juice; however, the mixture should be powdery enough to be easily transferred into a column and continually eluted. The optimal ratio of juice and sorbent was found to be 1 mL of juice and 1 g of sorbent. However, the increase in the sorbent quantity did not improve the results.

The solvent used for elution of carbendazim from the column should be selective and efficient. Several frequently used solvents were tested: ethyl acetate [48,83,85–87], methanol–dichloromethane mixture [76], and dichloromethane [84]. The major drawback of using ethyl acetate as well as methanol was slow solvent evaporation, i.e., methanol has lower volatility than, for example, dichloromethane, and with ethyl acetate, a persistent double layer was formed. Dichloromethane, as a single-component solvent, provided the best results and was chosen as eluent. It was determined that 10 mL of dichloromethane was enough for complete elution of carbendazim from the column.

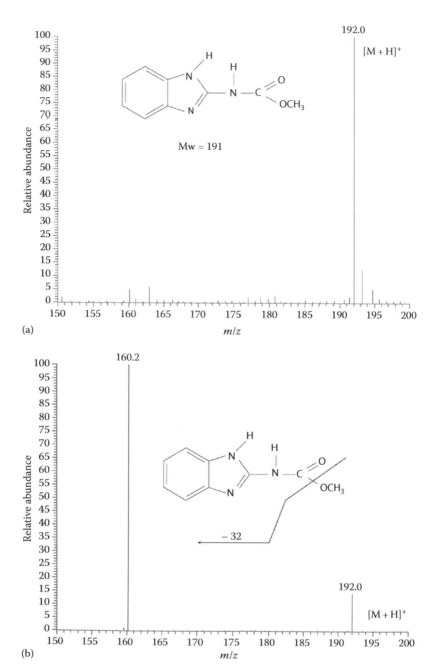

FIGURE 7.2 Mass spectra of carbendazim: (a) protonated molecule $[M + H]^+$, *m/z* 192.0; (b) fragment ion, *m/z* 160.2 (*m/z* 192.0 → *m/z* 160.2). (From Grujic, S. et al., *Food Addit. Contam.*, 22, 1132, 2005. With permission.)

7.4.5 RECOVERY STUDIES

The subsequent step in the optimization of MSPD method was to determine the pH value that will result in the highest recoveries. Carbendazim is a basic pesticide ($pK_a = 4.48$ [89]) and it was assumed that pH value would influence the efficiency of carbendazim extraction from fruit

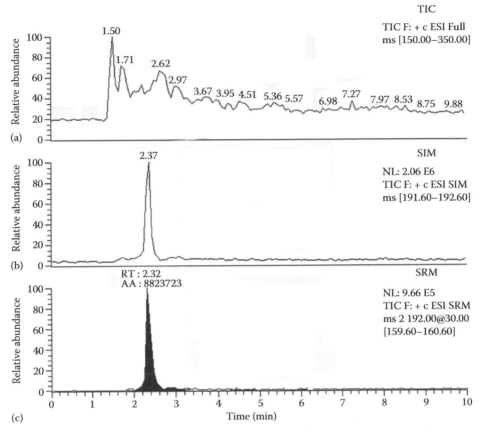

FIGURE 7.3 Mass chromatograms of the final extract from fruit juice: (a) total ion chromatogram; (b) chromatogram of protonated molecule; (c) chromatogram of selected fragmentation product. (From Grujic, S. et al., *Food Addit. Contam.*, 22, 1132, 2005. With permission.)

juice. Prior to recovery studies, the absence of carbendazim in fruit juice matrix was confirmed by analyses of the blanks under reported experimental conditions. The effect of pH was studied over the range 2–10. With increase in the pH value from 2 to 6, the recoveries also increased. In the range 6–10, a slight decrease in the recoveries was observed. The highest recovery of 90% was obtained at pH ~6; therefore, this pH value was chosen as the optimal. The pH values of the investigated fruit juices were in the range 2.57–3.87. It was determined whether the amount of the extracted carbendazim would be much lower than the real one if the pH value was not adjusted to 6.

To study the matrix effect, methanol-based standards as well as the matrix-matched standards of carbendazim at 10, 50, 100, 250, and 500 ng mL^{-1} were analyzed by LC–MS/MS method. The matrix effect is defined as the ratio of analyte response in the matrix-matched standard to its response in the solvent-based standard. It was calculated as carbendazim response in matrix-matched standard divided by carbendazim response in solvent standard multiplied by 100 [10]. As shown in Figure 7.4, matrix components contribute to the signal enhancement (34%) as well as the signal suppression (5%). The matrix influence seems to decrease with the increase in the analyte concentration. However, at higher concentration levels it seems to become constant. This indicates that to get accurate results, matrix-matched standards need to be used.

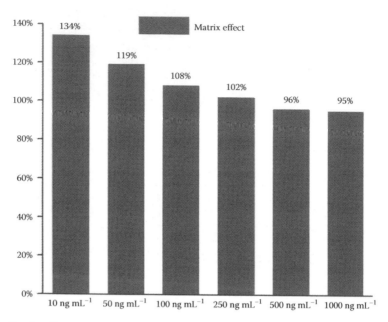

FIGURE 7.4 Matrix effect of apple juice over carbendazim concentration range of 10–1000 ng mL^{-1}. (From Grujic, S. et al., *Food Addit. Contam.*, 22, 1132, 2005. With permission.)

7.4.6 Validation of the Analytical Method

The linearity of the developed analytical method was tested using apple juice spiked at concentrations ranging from 10 to 500 ng mL^{-1}. It was determined that the method was linear with a correlation coefficient (*r*) of 0.99976.

The repeatability of the developed analytical procedure was tested by analysis of three replicate samples spiked at each concentration level in the range 10–500 ng mL^{-1}. The repeatability of the method expressed as RSD was less than 12%. For comparable fruit analysis using MSPD followed by LC–MS/MS, the RSDs were less than 9% [83] and less than 16% [36].

The limit of detection (LOD) of the analytical method was 0.03 ng mL^{-1}. Chromatograms obtained for juice samples spiked at 10 ng mL^{-1} were used. The LOD was calculated as the concentration giving the value of signal-to-noise ratio of 3 (S/N = 3) in SRM mode. The limit of quantification (LOQ) was 0.1 ng mL^{-1}, calculated as the concentration giving the value of ratio S/N = 10. The LOQ was 1000 times lower than MRLs set by the EU, indicating that the developed method is appropriate for quantification of carbendazim residues in fruit juices. This was validated by analyses of apple, peach, cherry, raspberry, and orange juice, spiked at 10, 50, 250, and 500 ng mL^{-1}. The results presented in Table 7.4 show that recoveries were over 82% and RSDs were ≤12%, regardless of the sample matrix or the spiking level.

7.4.7 Real Samples

A survey on carbendazim residues in commercially available fruit juices was performed. Apple, peach, cherry, raspberry, and orange juices of four different brands, produced by four different domestic companies, were purchased from local supermarkets and analyzed by following the developed method. The obtained results are presented in Table 7.5. Many of the investigated fruit juice samples contain carbendazim residues, corroborating that this pesticide is extensively used in fruit production. However, the detected levels were always below the MRLs set by the EU.

TABLE 7.4
**Mean Recoveries and Repeatability (in Brackets) (n = 3) of the Developed
Method for Carbendazim in Different Fruit Juices at Four Fortification Levels**

	Recovery, % (RSD, %) Spiking Level			
	10 ng mL^{-1}	50 ng mL^{-1}	250 ng mL^{-1}	500 ng mL^{-1}
Apple juice	96 (2)	102 (4)	87 (7)	87 (3)
Peach juice	102 (4)	88 (2)	92 (5)	98 (4)
Cherry juice	93 (11)	98 (12)	98 (5)	96 (3)
Raspberry juice	89 (2)	82 (5)	96 (5)	93 (4)
Orange juice	101 (12)	100 (4)	94 (9)	92 (3)

Source: Adapted from Grujic, S. et al., *Food Addit. Contam.*, 22, 1132, 2005.

TABLE 7.5
**Carbendazim Residues in Commercially Available Fruit Juices of
Four Different Brands**

	Carbendazim Concentration (ng mL^{-1}) (RSD, %)[a]			
	A	B	C	D
Apple juice	4.9 (8)	—	—	3.1 (10)
Peach juice	55 (11)	4.2 (7)	17 (6)	18 (11)
Cherry juice	19 (10)	10.9 (3)	21 (10)	11.6 (5)
Raspberry juice	24 (8)	2.0 (10)	13 (8)	4.4 (9)
Orange juice	18 (11)	26 (12)	—	12 (8)

Source: Adapted from Grujic, S. et al., *Food Addit. Contam.*, 22, 1132, 2005.
[a] $n = 2$.

7.5 MULTIRESIDUE LC–MS/MS DETERMINATION OF NINE PESTICIDES IN FRUIT JUICES (EXAMPLE 2)

The use of multiresidue method of analysis is generally preferred for the reduced analysis time and cost, especially when the pesticide application history is not known [90]. Owing to the low detection levels required by the regulatory bodies and the complex nature of matrices in which the target compounds are present, trace level detection and identification with prior efficient sample preparation are important aspects in the analytical method. In the previous example, a rapid and reliable method based on MSPD with diatomaceous earth for determining carbendazim residues in various fruit juice matrices was shown. Our further efforts are aimed at the development of a multiresidue pH-dependent method that will include different chemical classes of pesticides, which are among the most commonly used pesticides in fruit production and preservation in Serbia. The objective is to determine the presence of nine pesticides (acephate, carbendazim, monocrotophos, acetamiprid, dimethoate, simazine, carbofuran, atrazine, diuron), belonging to several chemical classes, in fruit juices, with MSPD sample preparation followed by LC–ESI-MS/MS analysis. Structures of selected pesticides are presented in Table 7.6, with their activity, chemical class, and molecular weight.

TABLE 7.6
Common Name, Activity, Structure, Chemical Class, and Molecular Weight of the Pesticides under Investigation

Common Name	Activity	Structure	Chemical Class	M_w
Acephate	Insecticide		Organophosphorus	183
Carbendazim	Fungicide		Benzimidazole	191
Monocrotophos	Insecticide		Organophosphorus	223
Acetamiprid	Insecticide		Neonicotinoid	222
Dimethoate	Insecticide		Organophosphorus	229
Simazine	Herbicide		Triazine	201
Carbofuran	Insecticide		Carbamate	221
Atrazine	Herbicide		Triazine	215
Diuron	Herbicide		Phenylurea	232

7.5.1 LC–MS Analysis

As reported in Example 1, Surveyor LC system was used for separation of analytes on the reverse-phase C18 column, 4.6 × 75 mm i.d., and 3.5 μm particle size. Before the separation column, a precolumn of 4.6 × 12.5 mm i.d. and 5 μm particle size (Agilent, USA) was installed. The mobile phase consisted of water (A), methanol (B), and 10% acetic acid (v/v) (C). The gradient change was as follows: 0 min, B 33%, C 1%; 7.5 min, B 58%, C 0.6%; 16 min, B 76%, C 1%; 20 min, B 100%, C 0%. The initial conditions were re-established and held for 15 min to ensure minimal carryover between injections. The flow rate of the mobile phase was observed to be 0.5 mL min^{-1}. An aliquot of 10 μL of the final extract was injected into the LC system.

Mass spectra were obtained by LCQ Advantage ion trap (Thermo Fisher Scientific, USA), using ESI technique. All the pesticides were analyzed in the positive ionization (PI) mode. Detection of all analytes was based on the isolation of the protonated molecule, [M + H]$^+$. However, in some instances, adduct ions such as [M + Na]$^+$, [M + K]$^+$, [M + CH$_3$OH]$^+$, or other adducts with solvents, can be used for this purpose [35,91]. In general, the formation of adduct ions other than protonated molecules is an undesired process as their fragmentation does not produce structurally significant fragment ions necessary for analyte identification. However, it was shown earlier [91] that they can be used for the purpose of identification and quantification of the analytes. Subsequent MS/MS fragmentations of the isolated ions were carried out using parameters listed in Table 7.7. The SRM mode was used for the quantification of all pesticides. Acquisition was conducted in five time segments, as shown in Table 7.7.

7.5.2 Sample Preparation

Previously developed MSPD procedure for the determination of carbendazim in fruit juices [41] was employed as the preparation method for nine selected pesticides. When employing a single-residue sample preparation method for the multiresidue analysis, some method parameters must be revised and optimized. The effect of pH on pesticide recoveries was studied over the range 2–8. The pH value was observed to have a decisive influence on the carbendazim recovery. In other words, the highest recoveries for carbendazim were obtained with a pH value adjusted to 6. For other tested pesticides, the pH influence was not that prominent; however, a slight increase in the recoveries was observed at pH 6. Therefore, this pH value was used in the sample preparation. In the previous procedure, dichloromethane was employed as a single extraction solvent. For the purpose of multiresidue analysis, other extraction solvents, such as ethyl acetate and methanol, were also tested. For all the tested pesticides, unsatisfactory results were obtained with both ethyl acetate and methanol. Therefore, dichloromethane was chosen as the extraction solvent in the sample preparation method.

TABLE 7.7
LC–MS and MS/MS Parameters for the Analysis of Selected Pesticides

Pesticide	Time Segment	Time (min)	Parent Ion (*m/z*)	Isolation Width	Daughter Ion (*m/z*)	Collision Energy (%)
Acephate	I	0.0–2.8	184	2	143	34
Carbendazim	II	2.8–5.8	192	2	160	32
Monocrotophos	II		224	2	193	38
Acetamiprid	III	5.8–9.0	223	2	126	36
Dimethoate	III		230	2	199	26
Simazine	IV	9.0–13.8	202	2	124	36
Carbofuran	IV		222	2	165	30
Atrazine	V	13.5–18.0	216	2	174	34
Diuron	V		233	2	72	30

7.5.3 OPTIMIZATION OF LC–MS/MS ANALYSIS

To determine the instrumental conditions that would allow unambiguous identification of the analytes at trace levels, optimization of LC–MS/MS analysis was carried out. The tuning of the instrument was performed for each pesticide using single-pesticide standard solution as well as mixed-analyte standard solution, both prepared at 10 µg mL^{-1}. The detector response for a specific analyte can be different, depending on whether it is a single-analyte solution or a mixture with other analytes. All standard solutions were infused with the syringe pump at 5 µL min^{-1} to the mobile phase consisting of water, methanol, and 10% acetic acid mixed at a ratio of 66:33:1, with the flow rate of 0.5 mL min^{-1}. However, the detector responses for each pesticide in single-pesticide solution and pesticide mixture were the same. Identification of the parent ion as well as the choice of the ionization mode for each analyte were performed in the full-scan mode by recording the mass spectra from m/z 50 to 500 in both positive and negative ionization modes. Acephate, carbendazim, and diuron gave signals in both the ionization modes, but their responses were much higher in the positive mode. As all the other tested pesticides gave signals only in the positive ionization mode, it was selected as the ionization mode for the analysis.

For each analyte, optimization of the isolation width of the parent ion, selection of the optimal collision energy, and identification of the most abundant daughter ion were also carried out, in the selected reaction mode. SRM detection was separated in five time segments, each acquiring data for one or two substances, as shown in Table 7.7. If more analytes are included in the method, then more time segments should be used to perform detection with sufficient instrument sensitivity. In other words, the sensitivity of MS detector decreases as the number of simultaneously recorded transitions increases. In general, the use of MS/MS allows analysis without complete chromatographic separation of the analytes, as it is rare to find molecules that elute at the same retention time and share the same MS/MS transition [92]. However, a certain degree of separation is necessary to enable programming of various SRM transitions into different time segments along the chromatogram. Figure 7.5 illustrates a typical chromatogram obtained under selected time-scheduled conditions for apple juice spiked at the concentration of 10 ng mL^{-1}.

7.5.4 MATRIX-MATCHED CALIBRATION AND MATRIX EFFECT

As described in Example 1, matrix-matched calibration was used to compensate the matrix effect. Matrix-matched standards were prepared for each sample type (apple, peach, and raspberry juice) at 10, 50, 100, 250, and 500 ng mL^{-1}. The linearity of the calibration curves was calculated for all the tested pesticides in all the investigated matrices. Curves displayed good linearity over selected concentration range with regression correlation coefficients ranging from 0.9908 for monocrotophos in peach juice to 0.9997 obtained for dimethoate in apple juice.

The matrix effect was calculated for all the juice matrices, as explained earlier. Figure 7.6 shows the matrix effect of apple juice at concentrations of 10, 50, and 100 ng mL^{-1}, for every tested pesticide. Diuron showed the highest signal enhancement (up to 90%) and acephate displayed the highest signal suppression (up to 30%), both at the concentration level of 10 ng mL^{-1}. Acephate, carbendazim, dimethoate, and diuron showed decrease in the matrix effect with increase in the concentration. In the case of monocrotophos, the signal was enhanced at lower concentration, and suppressed at higher concentration. For other pesticides, the general pattern could not be determined. For instance, simazine showed signal enhancement for 10 and 100 ng mL^{-1}, but for 50 ng mL^{-1}, no matrix effect was observed. Evidently, the matrix influence was very much variable and dependent on the tested pesticide, concentration of the pesticide, as well as the analyzed matrix. Also, for the specific combination of pesticides and matrix, the matrix effect was observed to vary from one set of measurements to the other. This indicates that it is not possible to test the matrix effect only once and consider it to be constant [8]. Therefore, for an accurate quantification, the use of matrix-matched calibration is necessary. However, in the case when detector response is variable

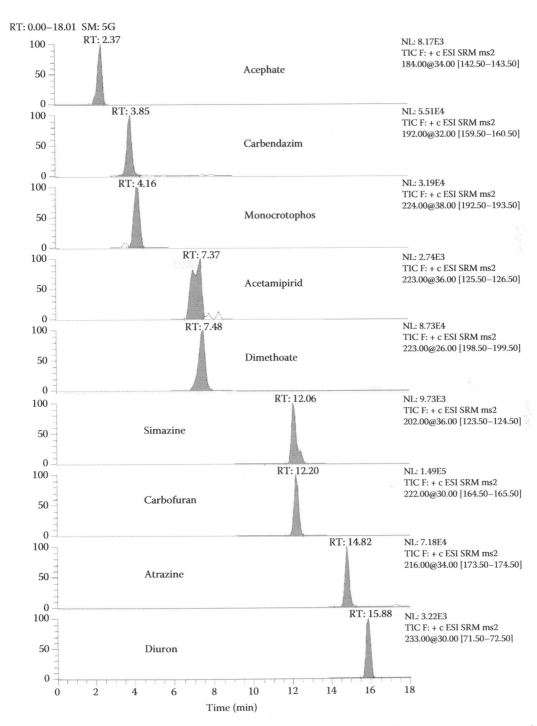

FIGURE 7.5 Mass chromatograms of the final extract from apple juice spiked at the concentration of 10 ng mL^{-1}.

with time, a single-level calibration may be used, as it may provide more accurate results than multilevel calibration. When single-level calibration is employed, the sample response should be within ±20% of the calibration standard response if the MRL is exceeded. On the other hand, if the MRL is not exceeded, then the sample response should be within ±50% of the calibration response [93].

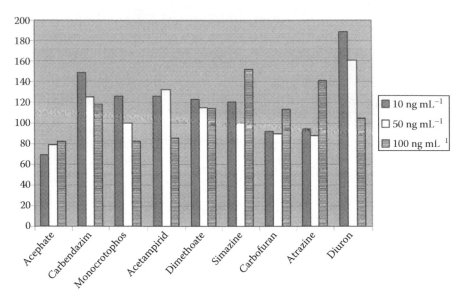

FIGURE 7.6 Matrix effect of apple juice at pesticide concentration of 10, 50, and 100 ng mL^{-1}.

7.5.5 VALIDATION OF THE ANALYTICAL METHOD

To validate the developed analytical procedure, the recoveries, repeatability, and limits of detection and quantification were determined for nine investigated pesticides in all the juice matrices. Recovery studies were carried out by spiking the pesticide-free juice samples with the appropriate volumes of mixed-analyte standard solution (2 μg mL^{-1}) to produce concentrations at 10 and 100 ng mL^{-1}. Recoveries and repeatability of the method were determined by the analysis of three replicate samples, and the results are presented in Table 7.8. Recoveries were satisfactory, ranging from 75% to 110%. The repeatability of the method, expressed as RSD, was between 2% and 28%. However, a change in the MS sensitivity during an analytical run resulted in a few high RSD values (>20%, Table 7.8) [94].

TABLE 7.8
Mean Recoveries and Repeatability (in Brackets) (n = 3) of the Developed Method for Different Juice Samples Spiked at Two Concentration Levels

	Recovery, % (RSD, %)					
	Apple Juice		**Peach Juice**		**Raspberry Juice**	
Pesticide	**10 ng mL^{-1}**	**100 ng mL^{-1}**	**10 ng mL^{-1}**	**100 ng mL^{-1}**	**10 ng mL^{-1}**	**100 ng mL^{-1}**
Acephate	110 (6)	97 (13)	95 (7)	107 (5)	92 (8)	83 (4)
Carbendazim	86 (12)	83 (15)	91 (9)	88 (10)	99 (19)	104 (12)
Monocrotophos	88 (7)	86 (5)	94 (12)	93 (9)	101 (8)	93 (11)
Acetamiprid	75 (2)	83 (2)	97 (14)	90 (28)	85 (5)	87 (6)
Dimethoate	107 (13)	90 (5)	100 (4)	93 (7)	98 (2)	91 (4)
Simazine	79 (4)	91 (5)	81 (15)	86 (13)	94 (12)	83 (4)
Carbofuran	91 (9)	92 (8)	102 (4)	91 (5)	91 (8)	92 (14)
Atrazine	102 (12)	96 (13)	98 (8)	102 (15)	78 (5)	93 (6)
Diuron	91 (5)	92 (6)	104 (17)	98 (8)	102 (6)	83 (21)

Analytical method detection limits were calculated for S/N = 3 in SRM mode for all pesticides. Method quantification limits, corresponding to the concentrations giving the value of ratio S/N = 10, were also calculated. For calculations, chromatograms obtained for the juice samples spiked at the concentration of 10 ng mL^{-1} were used. The estimated values of LODs were in the range of 0.02–0.59 ng mL^{-1}, whereas the LOQ values were in the range of 0.07–2.37 ng mL^{-1}. The highest LOD and LOQ values were observed for acetamipirid, indicating that the sensitivity of the MS detector was the lowest for this pesticide. The MS detector was the most sensitive for carbendazim, as the LOD and LOQ values were the lowest. The MRLs for fruits set by the EU include all tested pesticides, except monocrotophos, simazine, and diuron. The MRLs range from 0.01 mg kg^{-1} (acetamiprid in raspberry) to 0.2 mg kg^{-1} (carbendazim in pome fruit) [88,95–100]. To our knowledge, there are no regulations available for pesticide residues in fruit juices. However, the calculated LODs and LOQs were at least two orders of magnitude lower than the established MRLs, indicating that the proposed method is suitable for quantification of selected pesticides in fruit juices.

7.5.6 REAL SAMPLES

A survey on the residues of nine selected pesticides in commercially available fruit juices was performed. Apple, peach, and raspberry juices of different domestic brands were purchased from local supermarkets and analyzed following the developed method. The study showed that almost 90% of the investigated fruit juices contained pesticide residues. Carbendazim was the most frequently detected pesticide, as it was found in almost 80% of the investigated samples. The highest value of carbendazim residue (74.5 ng mL^{-1}) was found in peach juice. The lowest detected carbendazim concentration was 1 ng mL^{-1} in apple juice. Dimethoate and monocrotophos were also frequently detected, in 23%–30% of the tested juice samples. Dimethoate residues ranged from 0.32 ng mL^{-1} for apple juice to 1.21 ng mL^{-1} for raspberry juice. Concentration of monocrotophos was in the range 0.58–5.83 ng mL^{-1} in raspberry juice. Almost 60% of the analyzed juice samples contained more than one pesticide residue. However, the detected levels were always below the MRLs set by the EU.

A positive result on the pesticide residues in tested sample usually requires additional confirmation, i.e., if only one transition is acquired, false-positive results might be reported [43]. Confirmation is usually made by a second analysis of the positive sample extract using analyte-identification method with two or more transitions [101]. However, confirmation is not mandatory if positive results are below the MRLs [93].

REFERENCES

1. F Hernández, ÓJ Pozo, JV Sancho, L Bijlsma, M Barreda, and E Pitarch. Multiresidue liquid chromatography tandem mass spectrometry determination of 52 non gas chromatography-amenable pesticides and metabolites in different food commodities. *J. Chromatogr. A* 1109:242–252, 2006.
2. A Onjia, T Vasiljevic, D Cokesa, and M Lausevic. Factorial design in isocratic high-performance liquid chromatography of phenolic compounds. *J. Serb. Chem. Soc.* 67:745–751, 2002.
3. T Vasiljevic, A Onjia, D Cokesa, and M Lausevic. Optimization of artificial neural network for retention modeling in high-performance liquid chromatography. *Talanta* 64:785–790, 2004.
4. C Blasco, G Font, J Mañes, and Y Picó. Solid-phase microextraction liquid chromatography/tandem mass spectrometry to determine postharvest fungicides in fruits. *Anal. Chem.* 75:3606–3615, 2003.
5. JV Sancho, ÓJ Pozo, T Zamora, S Grimalt, and F Hernández. Direct determination of paclobutrazol residues in pear samples by liquid chromatography–electrospray tandem mass spectrometry. *J. Agric. Food Chem.* 51:4202–4206, 2003.
6. K Yu, J Krol, M Balogh, and I Monks. A fully automated LC/MS method development and quantification protocol targeting 52 carbamates, thiocarbamates, and phenylureas. *Anal. Chem.* 75:4103–4112, 2003.
7. GL Hall, J Engebretson, MJ Hengel, and T Shibamoto. Analysis of methoxyfenozide residues in fruits, vegetables, and mint by liquid chromatography–tandem mass spectrometry (LC–MS/MS). *J. Agric. Food Chem.* 52:672–676, 2004.

8. C Jansson, T Pihlström, BG Österdahl, and KE Markides. A new multi-residue method for analysis of pesticide residues in fruit and vegetables using liquid chromatography with tandem mass spectrometric detection. *J. Chromatogr. A* 1023:93–104, 2004.

9. D Ortelli, P Edder, and C Corvi. Multiresidue analysis of 74 pesticides in fruits and vegetables by liquid chromatography–electrospray–tandem mass spectrometry. *Anal. Chim. Acta* 520:33–45, 2004.

10. A Sannino, L Bolzoni, and M Bandini. Application of liquid chromatography with electrospray tandem mass spectrometry to the determination of a new generation of pesticides in processed fruits and vegetables. *J. Chromatogr. A* 1036:161–169, 2004.

11. I Liška and J Slobodník. Comparison of gas and liquid chromatography for analysing polar pesticides in water samples. *J. Chromatogr. A* 733:235–258, 1996.

12. WMA Niessen. *Liquid Chromatography–Mass Spectrometry*, 3rd edn., Boca Raton: Taylor & Francis Group, 2006.

13. Y Li, JE George, CL McCarty, and SC Wendelken. Compliance analysis of phenylurea and related compounds in drinking water by liquid chromatography/electrospray ionization/mass spectrometry coupled with solid-phase extraction. *J. Chromatogr. A* 1134:170–176, 2006.

14. I Losito, A Amorisco, T Carbonara, S Lofiego, and F Palmisano. Simultaneous determination of phenyl- and sulfonyl-urea herbicides in river water at sub-parts-per-billion level by on-line preconcentration and liquid chromatography–tandem mass spectrometry. *Anal. Chim. Acta* 575:89–96, 2006.

15. MI Catalina, J Dallüge, RJJ Vreuls, and UATh Brinkman. Determination of chlorophenoxy acid herbicides in water by in situ esterification followed by in-vial liquid–liquid extraction combined with large-volume on-column injection and gas chromatography–mass spectrometry. *J. Chromatogr. A* 877:153–166, 2000.

16. M Takino, S Daishima, and T Nakahara. Automated on-line in-tube solid-phase microextraction followed by liquid chromatography/electrospray ionization–mass spectrometry for the determination of chlorinated phenoxy acid herbicides in environmental waters. *Analyst* 126:602–608, 2001.

17. E Ayano, H Kanazawa, M Ando, and T Nishimura. Determination and quantitation of sulfonylurea and urea herbicides in water samples using liquid chromatography with electrospray ionization mass spectrometric detection. *Anal. Chim. Acta* 507:211–218, 2004.

18. L Alder, K Greulich, G Kempe, and B Vieth. Residue analysis of 500 high priority pesticides: Better by GC–MS or LC–MS/MS? *Mass Spectrom. Rev.* 25:838–865, 2006.

19. TM Vasiljevic, MD Lausevic, and RE March. Mass spectrometry analysis of polychlorinated biphenyls: Chemical ionization and selected ion chemical ionization using methane as a reagent gas. *J. Serb. Chem. Soc.* 65:431–438, 2000.

20. M Lausevic, X Jiang, RE March, and CD Metcalf. Analysis of polychlorinated biphenyls by quadrupole ion trap mass spectrometry. Part II: Comparison of mass spectrometric detection using electron impact and selected-ion chemical ionization with electron capture detection. *Rapid Commun. Mass Spectrom.* 9:927–936, 1995.

21. M Lausevic and RE March. Modulated resonant excitation of selected polychlorobiphenyl molecular ions in an ion trap mass spectrometer. *J. Mass Spectrom.* 31:1244–1252, 1996.

22. E Moyano, DE Games, and MT Galceran. Determination of quaternary ammonium herbicides by capillary electrophoresis/mass spectrometry. *Rapid Commun. Mass Spectrom.* 10:1379–1385, 1996.

23. R Jeannot, H Sabik, E Sauvard, and E Genin. Application of liquid chromatography with mass spectrometry combined with photodiode array detection and tandem mass spectrometry for monitoring pesticides in surface waters. *J. Chromatogr. A* 879:51–71, 2000.

24. CS Evans, JR Startin, DM Goodall, and BJ Keely. Tandem mass spectrometric analysis of quaternary ammonium pesticides. *Rapid Commun. Mass Spectrom.* 15:699–707, 2001.

25. EM Thurman, I Ferrer, and D Barceló. Choosing between atmospheric pressure chemical ionization and electrospray ionization interfaces for the HPLC/MS analysis of pesticides. *Anal. Chem.* 73:5441–5449, 2001.

26. Y Picó, C Blasco, and G Font. Environmental and food applications of LC–tandem mass spectrometry in pesticide-residue analysis: An overview. *Mass Spectrom. Rev.* 23:45–85, 2004.

27. JB Plomley, M Lausevic, and RE March. Determination of dioxins/furans and PCBs by quadrupole ion-trap gas chromatography–mass spectrometry. *Mass Spectrom. Rev.* 19:305–365, 2000.

28. JB Plomley, M Lausevic, and RE March. Analysis of dioxins and polychlorinated biphenyls by quadrupole ion-trap gas chromatography–mass spectrometry. In *Current Practice of Gas Chromatography–Mass Spectrometry*, ed. WMA Niessen, pp. 95–116. New York: Marcel Dekker, Inc, 2001.

29. KS Liapis, P Aplada-Sarlis, and NV Kyriakidis. Rapid multi-residue method for determination of azinphos methyl, bromopropylate, chlorpyrifos, dimethoate, parathion methyl and phosalone in apricots and peaches by using negative chemical ionization ion trap technology. *J. Chromatogr. A* 996:181–187, 2003.

30. C Blasco, G Font, and Y Picó. Analysis of pesticides in fruits by pressurized liquid extraction and liquid chromatography–ion trap–triple stage mass spectrometry. *J. Chromatogr. A* 1098:37–43, 2005.
31. J Yinon and A Vincze. Collision-induced dissociation (CID) processes in some carbamate and phenylurea pesticides studied by ion-trap MS/MS. *Int. J. Mass Spectrom.* 167/168:21–33, 1997.
32. G Sagratini, J Mañes, D Giardiná, P Damiani, and Y Picó. Analysis of carbamate and phenylurea pesticide residues in fruit juices by solid-phase microextraction and liquid chromatography–mass spectrometry. *J. Chromatogr. A* 1147:135–143, 2007.
33. J Zrostlíková, J Hajšlová, T Kovalczuk, R Štěpán, and J Poustka. Determination of seventeen polar/thermolabile pesticides in apples and apricots by liquid chromatography/mass spectrometry. *J. AOAC Int.* 86:612–622, 2003.
34. C Blasco, G Font, and Y Picó. Multiple-stage mass spectrometric analysis of six pesticides in oranges by liquid chromatography–atmospheric pressure chemical ionization–ion trap mass spectrometry. *J. Chromatogr. A* 1043:231–238, 2004.
35. T Chen and G Chen. Identification and quantitation of pyrethroid pesticide residues in vegetables by solid-phase extraction and liquid chromatography/electrospray ionization ion trap mass spectrometry. *Rapid Commun. Mass Spectrom.* 21:1848–1854, 2007.
36. C Soler, J Mañes, and Y Picó. Liquid chromatography–electrospray quadrupole ion-trap mass spectrometry of nine pesticides in fruits. *J. Chromatogr. A* 1048:41–49, 2004.
37. C Sánchez-Brunete, AI García-Valcárcel, and JL Tadeo. Determination of residues of phenoxy acid herbicides in soil and cerals by gas chromatography–ion trap detection. *J. Chromatogr. A* 675:213–218, 1994.
38. L Goodwin, RJ Startin, MD Goodall, and JB Keely. Tandem mass spectrometric analysis of glyphosate, glufosinate, aminomethylphosphonic acid and methylphosphinicopropionic acid. *Rapid Commun. Mass Spectrom.* 17:963–969, 2003.
39. EM Thurman, I Ferrer, JA Zweigenbaum, JF García-Reyes, M Woodman, and AR Fernández-Alba. Discovering metabolites of post-harvest fungicides in citrus with liquid chromatography/time-of-flight mass spectrometry and ion trap tandem mass spectrometry. *J. Chromatogr. A* 1082:71–80, 2005.
40. A Angioni, A Garau, P Caboni, et al. Gas chromatographic ion trap mass spectrometry determination of zoxamide residues in grape, grape processing, and in the fermentation process. *J. Chromatogr. A* 1097:165–170, 2005.
41. S Grujic, M Radisic, T Vasiljevic, and M Lausevic. Determination of carbendazim residues in fruit juices by liquid chromatography–tandem mass spectrometry. *Food Addit. Contam.* 22:1132–1137, 2005.
42. DL Buhrman, PI Price, and PJ Rudewicz. Quantitation of SR 27417 in human plasma using electrospray liquid chromatography/tandem mass spectrometry: A study of ion suppression. *J. Am. Soc. Mass Spectrom.* 7:1099–1105, 1996.
43. F Hernández, JV Sancho, and OJ Pozo. Critical review of the application of liquid chromatography/mass spectrometry to the determination of pesticide residues in biological samples. *Anal. Bioanal. Chem.* 382:934–946, 2005.
44. BK Matuszewski, ML Constanzer, and CM Chavez-Eng. Strategies for the assessment of matrix effect in quantitative bioanalytical methods based on HPLC–MS/MS. *Anal. Chem.* 75:3019–3030, 2003.
45. C Blasco, Y Picó, J Mañes, and G Font. Determination of fungicide residues in fruits and vegetables by liquid chromatography–atmospheric pressure chemical ionization mass spectrometry. *J. Chromatogr. A* 947:227–235, 2002.
46. B Albero, C Sánchez-Brunete, A Donoso, and JL Tadeo. Determination of herbicide residues in juice by matrix solid-phase dispersion and gas chromatography–mass spectrometry. *J. Chromatogr. A* 1043:127–133, 2004.
47. A Di Corcia, C Crescenzi, A Laganà, and E Sebastiani. Evaluation of a method based on liquid chromatography/electrospray/mass spectrometry for analyzing carbamate insecticides in fruits and vegetables. *J. Agric. Food Chem.* 44:1930–1938, 1996.
48. MJ Taylor, K Hunter, KB Hunter, D Lindsay, and S Le Bouhellec. Multi-residue method for rapid screening and confirmation of pesticides in crude extracts of fruits and vegetables using isocratic liquid chromatography with electrospray tandem mass spectrometry. *J. Chromatogr. A* 982:225–236, 2002.
49. O Núñez, E Moyano, and MT Galceran. LC–MS/MS analysis of organic toxics in food. *TrAC Trend. Anal. Chem.* 24:683–703, 2005.
50. S Grimalt, JV Sancho, ÓJ Pozo, JM García-Baudin, ML Fernández-Cruz, and F Hernández. Analytical study of trichlorfon residues in kaki fruit and cauliflower samples by liquid chromatography–electrospray tandem mass spectrometry. *J. Agric. Food Chem.* 54:1188–1195, 2006.
51. JA Wang and W Cheung. Determination of pesticides in soy-based infant formula using liquid chromatography with electrospray ionization tandem mass spectrometry. *J. AOAC Int.* 89:214–224, 2006.

52. A Sannino. Determination of three natural pesticides in processed fruit and vegetables using high-performance liquid chromatography/tandem mass spectrometry. *Rapid Commun. Mass Spectrom.* 21:2079–2086, 2007.

53. S Bogialli and A Di Corcia. Matrix solid-phase dispersion as a valuable tool for extracting contaminants from foodstuffs. *J. Biochem. Biophys. Meth.* 70:163–179, 2007.

54. M Fernández, Y Picó, and J Mañes. Determination of carbamate residues in fruits and vegetables by matrix solid-phase dispersion and liquid chromatography–mass spectrometry. *J. Chromatogr. A* 871:43–56, 2000.

55. Y Picó, M Fernández, MJ Ruiz, and G Font. Current trends in solid-phase-based extraction techniques for the determination of pesticides in food and environment. *J. Biochem. Biophys. Meth,* 70:117–131, 2007.

56. CG Zambonin, M Quinto, N De Vietro, and F Palmisano. Solid-phase microextraction–gas chromatography mass spectrometry: A fast and simple screening method for the assessment of organophosphorus pesticides residues in wine and fruit juices. *Food Chem.* 86:269–274, 2004.

57. SA Barker, AR Long, and CR Short. Isolation of drug residues from tissues by solid phase dispersion. *J. Chromatogr. A* 475:353–361, 1989.

58. SA Barker. Matrix solid-phase dispersion. *J. Chromatogr. A* 885:115–127, 2000.

59. SA Barker. Applications of matrix solid-phase dispersion in food analysis. *J. Chromatogr. A* 880:63–68, 2000.

60. J Beltran, FJ López, and F Hernández. Solid-phase microextraction in pesticide residue analysis. *J. Chromatogr. A* 885:389–404, 2000.

61. EM Kristenson, L Ramos, and UATh Brinkman. Recent advances in matrix solid-phase dispersion. *TrAC Trends Anal. Chem.* 25:96–111, 2006.

62. V Camel. Microwave-assisted solvent extraction of environmental samples. *TrAC Trend. Anal. Chem.* 19:229–248, 2000.

63. S Bøwadt and SB Hawthorne. Supercritical fluid extraction in environmental analysis. *J. Chromatogr. A* 703:549–571, 1995.

64. N Motohashi, H Nagashima, and C Párkányi. Supercritical fluid extraction for the analysis of pesticide residues in miscellaneous samples. *J. Biochem. Biophys. Meth.* 43:313–328, 2000.

65. N Furusawa. A toxic reagent-free method for normal-phase matrix solid-phase dispersion extraction and reversed-phase liquid chromatographic determination of aldrin, dieldrin, and DDTs in animal fats. *Anal. Bioanal. Chem.* 378:2004–2007, 2004.

66. XG Chu, XZ Hu, and HY Yao. Determination of 266 pesticide residues in apple juice by matrix solid-phase dispersion and gas chromatography–mass selective detection. *J. Chromatogr. A* 1063:201–210, 2005.

67. C Soler, J Mañes, and Y Picó. Routine application using single quadrupole liquid chromatography–mass spectrometry to pesticides analysis in citrus fruits. *J. Chromatogr. A* 1088:224–233, 2005.

68. AI Valenzuela, R Lorenzini, MJ Redondo, and G Font. Matrix solid-phase dispersion microextraction and determination by high-performance liquid chromatography with UV detection of pesticide residues in citrus fruit. *J. Chromatogr. A* 839:101–107, 1999.

69. HM Lott and SA Barker. Comparison of a matrix solid phase dispersion and a classical extraction method for the determination of chlorinated pesticides in fish muscle. *Environ. Monit. Assess.* 28:109–116, 1993.

70. B Albero, C Sánchez-Brunete, and JL Tadeo. Determination of organophosphorus pesticides in fruit juices by matrix solid-phase dispersion and gas chromatography. *J. Agric. Food Chem.* 51:6915–6921, 2003.

71. M Navarro, Y Picó, R Marín, and J Mañes. Application of matrix solid-phase dispersion to the determination of a new generation of fungicides in fruits and vegetables. *J. Chromatogr. A* 968:201–209, 2002.

72. M Fernández, Y Picó, and J Mañes. Rapid screening of organophosphorus pesticides in honey and bees by liquid chromatography–mass spectrometry. *Chromatographia* 56:577–583, 2002.

73. S Bogialli, R Curini, A Di Corcia, A Laganà, M Nazzari, and M Tonci. Simple and rapid assay for analyzing residues of carbamate insecticides in bovine milk: Hot water extraction followed by liquid chromatography–mass spectrometry. *J. Chromatogr. A* 1054:351–357, 2004.

74. C Ferrer, MJ Gómez, JF García-Reyes, I Ferrer, EM Thurman, and AR Fernández-Alba. Determination of pesticide residues in olives and olive oil by matrix solid-phase dispersion followed by gas chromatography/mass spectrometry and liquid chromatography/tandem mass spectrometry. *J. Chromatogr. A* 1069:183–194, 2005.

75. C Blasco, G Font, and Y Picó. Comparison of microextraction procedures to determine pesticides in oranges by liquid chromatography–mass spectrometry. *J. Chromatogr. A* 970:201–212, 2002.
76. M Michel and B Buszewski. Optimization of a matrix solid-phase dispersion method for the determination analysis of carbendazim residue in plant material. *J. Chromatogr. B* 800:309–314, 2004.
77. YY Hu, P Zheng, YZ He, and GP Sheng. Response surface optimization for determination of pesticide multiresidues by matrix solid-phase dispersion and gas chromatography. *J. Chromatogr. A* 1098:188–193, 2005.
78. J Cai, Y Gao, X Zhu, and Q Su. Matrix solid phase dispersion–Soxhlet simultaneous extraction clean-up for determination of organochlorine pesticide residues in tobacco. *Anal. Bioanal. Chem.* 383:869–874, 2005.
79. E Viana, JC Moltó, and G Font. Optimization of a matrix solid-phase dispersion method for the analysis of pesticide residues in vegetables. *J. Chromatogr. A* 754:437–444, 1996.
80. G Kyriakopoulos and D Doulia. Adsorption of pesticides on carbonaceous and polymeric materials from aqueous solutions: A review. *Sep. Purif. Rev.* 35:97–191, 2006.
81. T Vasiljevic, J Spasojevic, M Bacic, A Onjia, and M Lausevic. Adsorption of phenol and 2,4-dinitrophenol on activated carbon cloth: The influence of sorbent surface acidity and pH. *Sep. Sci. Technol.* 41:1061–1075, 2006.
82. T Đurkić, A Perić, M Laušević et al. Boron and phosphorus doped glassy carbon: I. Surface properties. *Carbon* 35:1567–1572, 1997.
83. D Perret, A Gentili, S Marchese, M Sergi, and G D'Ascenzo. Validation of a method for the determination of multiclass pesticide residues in fruit juices by liquid chromatography/tandem mass spectrometry after extraction by matrix solid-phase dispersion. *J. AOAC Int.* 85:724–730, 2002.
84. A Sannino. Investigation into contamination of processed fruit products by carbendazim, methyl thiophanate and thiabendazole. *Food Chem.* 52:57–61, 1995.
85. M Fernández, Y Picó, and J Mañes. Pesticide residues in oranges from Valencia (Spain). *Food Addit. Contam.* 18:615–624, 2001.
86. C Blasco, M Fernández, Y Picó, G Font, and J Mañes. Simultaneous determination of imidacloprid, carbendazim, methiocarb and hexythiazox in peaches and nectarines by liquid chromatography–mass spectrometry. *Anal. Chim. Acta* 461:109–116, 2002.
87. Y Su, SH Mitchell, and S Mac AntSaoir. Carbendazim and metalaxyl residues in post-harvest treated apples. *Food Addit. Contam.* 20:720–727, 2003.
88. Commission Directive 2007/12/EC of 26 February 2007 amending certain Annexes to Council Directive 90/642/EEC as regards maximum residue levels of penconazole, benomyl and carbendazim. *Off. J. Eur. Union* L59:75–83.
89. S Takeda, K Fukushi, K Chayama, Y Nakayama, Y Tanaka, and S Wakida. Simultaneous separation and on-line concentration of amitrole and benzimidazole pesticides by capillary electrophoresis with a volatile migration buffer applicable to mass spectrometric detection. *J. Chromatogr. A* 1051:297–301, 2004.
90. CM Torres, Y Picó, MJ Redondo, and J Mañes. Matrix solid-phase dispersion extraction procedure for multiresidue pesticide analysis in oranges. *J. Chromatogr. A* 719:95–103, 1996.
91. S Grujic, T Vasiljevic, M Lausevic, and T Ast. Study on the formation of an amoxicillin adduct with methanol using electrospray ion trap tandem mass spectrometry. *Rapid Commun. Mass Spectrom.* 22:67–74, 2008.
92. B Køppen and NH Spliid. Determination of acidic herbicides using liquid chromatography with pneumatically assisted electrospray ionization mass spectrometric and tandem mass spectrometric detection. *J. Chromatogr. A* 803:157–168, 1998.
93. SANCO/2007/3131. 2007. Method validation and quality control procedures for pesticide residues analysis in food and feed. http://www.cc.europa.eu/food/plant/protection/resources/qualcontrol_en.pdf (accessed February 14, 2008).
94. K Granby, JH Andersen, and HB Christensen. Analysis of pesticides in fruit, vegetables and cereals using methanolic extraction and detection by liquid chromatography–tandem mass spectrometry. *Anal. Chim. Acta* 520:165–176, 2004.
95. Commission Directive 2002/71/EC of 19 August 2002 amending the Annexes to Council Directives 76/895/EEC, 86/362/EEC, 86/363/EEC and 90/642/EEC as regards the fixing of maximum levels for pesticide residues (formothion, dimethoate and oxydemeton-methyl) in and on cereals, foodstuffs of animal origin and certain products of plant origin, including fruit and vegetables. *Off. J. Eur. Union* L225:21–28.

96. Commission Directive 2003/118/EC of 5 December 2003 amending the Annexes to Council Directives 76/895/EEC, 86/362/EEC, 86/363/EEC and 90/642/EEC as regards maximum residue levels of acephate, 2,4-D and parathion-methyl. *O. J. Eur. Union* L327:25–32.

97. Commission Directive 2006/4/EC of 26 January 2006 amending the Annexes to Council Directives 86/362/EEC and 90/642/EEC as regards maximum residue levels for carbofuran. *Off. J. Eur. Union* L23:69–77.

98. Commission Directive 2006/61/EC of 7 July 2006 amending the Annexes to Council Directives 86/362/EEC, 86/363/EEC and 90/642/EEC as regards maximum residue levels for atrazine, azinphos-ethyl, cyfluthrin, ethephon, fenthion, methamidophos, methomyl, paraquat and triazophos. *Off. J. Eur. Union* L206:12–26.

99. Commission Directive 2007/7/EC of 14 February 2007 amending certain Annexes to Council Directives 86/362/EEC and 90/642/EEC as regards the maximum residue levels of atrazine, lambda-cyhalothrin, phenmedipham, methomyl, linuron, penconazole, pymetrozine, bifenthrin and abamectin. *Off. J. Eur. Union* L43:19–31.

100. Commission Directive 2007/11/EC of 21 February 2007 amending certain Annexes to Council Directives 86/362/EEC, 86/363/EEC and 90/642/EEC as regards maximum residue levels of acetampirid, thiacloprid, amazosulfuron, methoxyfenozide, *S*-metholachlor, milbemectin and tribenuron. *Off. J. Eur. Union* L63:26–37.

101. Commission Decision 2002/657/EC of 12 August 2002 implementing Council Directive 96/23/EC concerning the performance of analytical methods and the interpretation of results. *Off. J. Eur. Communities* L221:8–36.

8 Analysis of Pesticide Residue Using Electroanalytical Techniques

P. Manisankar, S. Viswanathan, and C. Vedhi

CONTENTS

8.1 INTRODUCTION

The increasing world population with growing demands has led to a situation where protection of the environment has become a major issue and a crucial factor for the future development of the industrial and agricultural processes. Pollution of water and the environment by toxic and nonbiodegradable organic materials of industrial or agricultural origin poses serious health hazards to all the living organisms. Harmful toxic organics can be classified into two groups: (1) toxic organic chemicals discharged into the environment as "wastes" due to industrial activity and (2) various "useful" toxic chemicals, such as pesticides and other agrochemicals, which are necessary for improving the yield. Agricultural development, especially involving the use of artificial fertilizers, pesticides, growth factors, etc., causes major pollution problems [1]. Pesticides, in general, are chemically or biologically active substances, which are of anthropogenic origin, used for killing or controlling unwanted organisms. They are good slaves but bad masters owing to their toxic side effects. The pervasive use of organic chemicals like pesticides in the agricultural fields has created the need for fast, easier, and affordable methods of toxicity assessment, as these chemicals cause adverse effects in humans and other higher-order animals. It is evident that the concentration level of pesticides in water samples should be very low, and, in fact, the European community has stated that the maximum admissible concentration level for each individual pesticide should be in the range of 0.1–0.5 ng mL^{-1} in drinking water. Owing to their high toxicity, an important task for environmental analytical chemistry is to monitor the concentration of pesticides in waters, soil leaches, and plant and animal tissues [2].

8.2   ROLE OF ELECTROANALYSIS IN POLLUTION CONTROL

Analytical chemistry plays an important role in the protection of the environment. This branch of chemistry finds application in the determination of pollutant concentrations—both quantitative and qualitative—in the biosphere, in determining the pollutant pathway from the source to man, as well as in elucidating further transformations into other substances along this pathway, e.g., as a result of the interaction among various pollutants, metabolism, etc. Analytical chemistry also helps in the evaluation of the effectiveness of various processes that prevent the formation of pollutants or eliminate those already formed. Most applications of environmental analysis involve trace determinations, often at a parts per billion (ppb) level or lower. However, the high sensitivity of the methods must be accompanied further by sufficient selectivity, precision, and accuracy. Furthermore, easy, simple treatment and rapidity of the analytical procedure are also desirable. As a series of analyses is often required, methods that are easy to automate are advantageous. In the selection of the method, the cost of instrumentation that must be available in numerous laboratories is also important. Measurements must often be carried out in the field and thus large apparatuses are excluded even if the method fulfills all the other criteria. It need not be emphasized that microanalytical instruments should be applicable to a wide range of substances (provided that they are not single-purpose analyzers or monitors) and that it is advantageous if several components can be determined simultaneously.

Electrochemistry offers promising approaches for the determination and destruction of pollutants [3]. Modern voltammetric methods have a scope of applicability beyond almost all the other modern instrumental methods in quantitatively determining the inorganic, organometallic, and organic pollutants in trace and ultratrace levels [4]. The sensitive limits of electroanalytical techniques are presented in Table 8.1.

TABLE 8.1
Sensitivity Limits of Electroanalytical Techniques

z	Techniques
10^{-4} to 10^{-5}	AC polarography, thin-layer coulometry
10^{-5} to 10^{-6}	Chronocoulometry, classical polarography
10^{-6} to 10^{-7}	Derivative polarography, square wave polarography, linear sweep voltammetry, and chemical stripping analysis
10^{-7} to 10^{-8}	Pulse polarography, amperometry, and conductivity (aqueous)
10^{-8} to 10^{-9}	Anodic stripping with hanging mercury drop electrodes
10^{-9} to 10^{-10}	Anodic stripping with thin-film electrodes or solid electrodes

Note: z, sensitivity limits (m dm^{-3}).

8.3   ELECTROANALYTICAL METHODS

A number of electrochemical methods, such as polarography, voltammetry, potentiometry, amperometry, and impedance techniques are now available [4–8] to chemists for investigations. To some extent, all the commonly employed electroanalytical methods can be used in environmental analyses: the choice of the method depends on the character of the compound to be determined and the matrix in which it occurs, as well as on the sensitivity and selectivity requirements. Electroanalytical methods for monitoring pesticides may be categorized into those based on (a) polarography/voltammetric, (b) potentiometric, (c) conductometric, and (d) coulometric approaches.

Electroanalytical techniques are classified as dynamic or passive, depending on whether the process of measurement itself forces concentration changes at the electrolyte interface. Thus, techniques such as voltammetry belong to the former category, whereas potentiometry is an example of a passive technique. Techniques such as voltammetry are useful primarily for defining the electrochemical behavior of the targeted pollutants. A variety of voltammetric techniques have been developed in which an external potential is applied to the electrochemical cell and the resulting current is measured. Subsequently, a voltammogram, a plot of the current vs. the applied potential, is recorded. The various ways of implementation of voltammetry are recognized as follows:

- The solution may be moving or quiescent with respect to the electrodes.
- The waveform for the applied potential may be varied.
- The timing sequence of the current measurement with respect to the potential waveform may be varied.
- The electrode and cell geometry, which affect the current response, may be varied.

8.3.1   POLAROGRAPHY

The invention of polarography in 1922 by Professor J. Heyrovsky represented a qualitative change in the field of electroanalytical chemistry, which at this stage is restricted to potentiometry and controlled-current electrolysis. The main contributions of Professor Heyrovsky were the recognition of the importance of potential and its control, the opportunities offered by the measurement of limiting currents, and the possibility to extend electrochemical studies to irreversible systems. The sensitivity of polarographic methods of analysis, enabling the determination of electroactive species up to approximately 10^{-5} M solutions, is superior to or at least comparable with most of the other contemporary techniques.

8.3.2   VOLTAMMETRY IN PESTICIDE RESIDUE ANALYSIS

8.3.2.1   Cyclic Voltammetry

This is perhaps the most versatile electroanalytical technique. The effectiveness of cyclic voltammetry (CV) results from its capability for rapidly observing redox behavior over a wide potential range. Indeed, CV has been termed as electrochemical spectroscopy. CV allows one to scan the potential of the working electrode in the anodic/cathodic direction and then reverse the scan in the opposite direction [9]. The electrode system used in CV is dictated by the nature of the medium as well as by the process being studied. The most common electrodes used are planar platinum discs, platinum wires, hanging mercury drops, and carbon paste electrodes. This technique is readily applied and various systems can be extensively studied owing to its experimental simplicity. CV is like polarography, a relatively simple technique that needs relatively little experimental effort, and provides a great deal of useful information about the electrochemical behavior. This technique is one of the most powerful and popular electrochemical diagnostic tools. One of the most useful features of CV is its ability to generate a potentially reactive species and then to examine it by reversal [10]. Once a mechanism is defined by CV, one can carry out a quantitative study by step

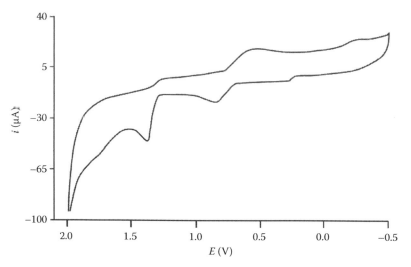

FIGURE 8.1   Cyclic voltammetric behavior of 0.99 mM dm^{-3} isoproturon at pH 1.0 on glassy carbon electrode (GCE). (From Manisankar, P. et al., *Int. J. Environ. Anal. Chem.*, 85, 409, 2005. With permission.)

techniques or by hydrodynamic voltammetry [11]. CV is hardly a technique of choice for quantifying the analyte concentration, especially at trace levels. This is because of its limited sensitivity owing to the capacitative current flowing at the electrode/electrolyte interface.

A typical cyclic voltammogram of the compound isoproturon studied at pH 1.0 is presented in Figure 8.1. From this behavior, two anodic peaks and one cathodic peak can be observed. The anodic peak around +1.3 V is well shaped with a higher current, and the remaining anodic peak around +0.85 V and cathodic peak around +0.58 V are of lesser current with little sharpness. The peak current increases with an increase in the sweep rate for all the peaks. No linearity is observed in the correlation between the peak current and the sweep rate, but linearity is seen when i_p is correlated with the square root of the sweep rate, with good linear correlation ($r^2 = 0.998$), indicating a diffusion-controlled reaction. The plot of log i_p vs. log(sweep rate) was also linear with a slope of 0.3437, confirming the diffusion-controlled reaction.

8.3.2.2 Differential Pulse Voltammetry

The application of conventional polarography is impossible and advanced modes have to be applied, such as the differential pulse mode. This is one of the most important and versatile achievements for electrochemical trace analysis, achieved through the pioneering work of Barker [12], which is now incorporated in every voltammetric device as the most significant function for analytical purposes. Recording of the response from the differential pulse mode applied in the voltammetry is carried out according to the principle introduced by Parry and Osteryoung [13]. In a differential pulse, the excitation waveform consists of small amplitude pulses superimposed on a staircase waveform. The major component of the current difference is the Faradaic current, which flows owing to an oxidation or reduction at the electrode. The capacitive current component owing to the electrical charging of the double layer is mostly removed. Because of this, the differential pulse voltammetry (DPV) gives higher signal-to-noise (S/N) ratios than other DC methods for quantitative analysis. The current is sampled both just before the application of the pulse and at the end of the pulse. The output is the current difference plotted vs. the base potential, and the pulse amplitude remains constant with respect to the base potential. However, the base potential is not constant but is scanned in small steps. The important parameters in this voltammogram are the peak potential and the peak current. Many heavy metals and organics have been determined by this pulse technique up to the range of

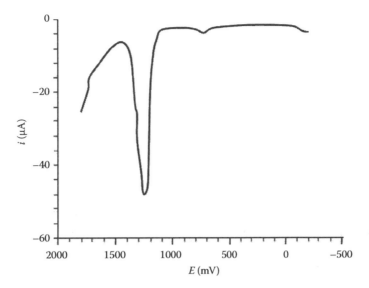

FIGURE 8.2 Differential pulse stripping behavior of 0.99 mM dm^{-3} isoproturon at pH 1.0 on GCE.

10^{-7}–10^{-9} M dm^{-3}. Figure 8.2 shows a representative DPV of isoproturon at an acid pH of 1.0, where one peak can be observed. The characteristic of this oxidation peak is similar to that in CV. The peak current increases with respect to the sweep rate and concentration. The log i_p vs. log(sweep rate) plot leads to a straight line with good correlation ($r^2 = 0.997$). The slope, 0.2546, suggests a diffusion-controlled reaction.

8.3.2.3 Square Wave Voltammetry

Square wave voltammetry (SWV) is one of the major voltammetric techniques provided by modern computer-controlled electroanalytical instruments. SWV is a large-amplitude differential technique, in which a waveform composed of symmetrical square waves is superimposed on a base staircase potential. The current is sampled twice during each square wave cycle, once at the end of the forward pulse and once at the end of the reverse pulse. The reverse pulse causes the reverse reaction of the product of the forward pulse. Sensitivity increases owing to the fact that the net current is larger than either the forward or the reverse components (since it is the difference between them). The total current response depends on both the reduction and the reoxidation currents. The major advantages of SWV are its sensitivity, speed, fine shape and position of the peak, and easy repetitive monitoring. As a result, the analysis time is drastically reduced and sensitivity is highly increased. To compare the results found in the CV studies, the voltammetric experiments were also carried out with the square wave technique. An illustration of the square wave voltammogram of isoproturon at pH 1.0 is presented in Figure 8.3. The peak current increases with an increase in the frequency and concentration. The log i_p is correlated with log(frequency) and a straight line with good correlation is obtained. Here, also, the diffusion-controlled reaction is confirmed from the slope around 0.5.

8.3.2.4 Stripping Voltammetry

In the 1970s and 1980s, even sensitivity of the pulse method was often insufficient, particularly for analyses of samples investigated in environmental chemistry and for analyses of biological materials. Research on increasing the sensitivity of electroanalytical methods has led to the development of the technique of stripping voltammetry, in which the analyte is first accumulated on the surface of a mercury drop of the mercury-covered solid electrode or other electrode materials, either

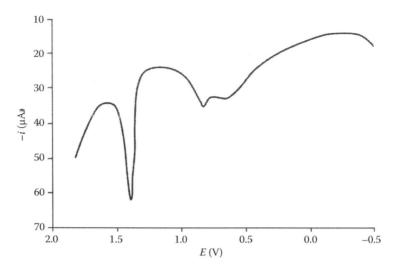

FIGURE 8.3   Square wave stripping behavior of 0.99 mM dm^{-3} isoproturon at pH 1.0 on GCE.

by electrolysis or by adsorption. After a chosen period of time, the accumulated species is electrolyzed and the resulting current–voltage curve is recorded and measured. The resulting "stripping voltammogram" shows peaks, the heights of which are generally proportional to the concentration of the corresponding electroactive species and the potentials of which have the same qualitative significance as their half-wave potentials in polarography. Such a combination of an effective accumulation step with an advanced measurement procedure results in a very low detection limit, and makes the stripping analysis one of the most important techniques in trace analysis. Using such techniques, ultratrace analysis can be carried out with solutions containing as little as 10^{-10} or 10^{-11} M dm^{-3} of analyte [14,15]. Stripping voltammetry thus extends the range of classical polarography by three or four orders of magnitude, making the analyses in the ppb range possible.

The original stripping analysis method involved the cathodic electrodeposition of amalgam-forming metals onto a hanging mercury drop-working electrode, followed by the anodic voltammetric determination of the accumulated metal during a positive-going potential scan [16]. Numerous advances during the 1980s and 1990s, however, have led to the development of alternative preconcentration schemes and advanced measurement procedures that further enhance the scope and power of stripping analysis [17,18]. Consequently, numerous variants of stripping analysis exist currently, differing in their method of accumulation and measurement. A report on stripping analysis was presented by Brainina and Neyman [19], while reviews on adsorptive stripping voltammetry were presented by Kalvoda and Kopanica [20], van den Berg [21], and Paneli and Voulgaropoulos [22]. When the compound contains an electrochemically reducible or oxidizable group, the peak current on the voltammetric curve recorded after completion of the accumulation period corresponds practically only to the reduction or oxidation of the whole amount of the adsorbed electroactive species. Stripping voltammetry enables recurrent determinations of the organic compounds in the concentration range from 1×10^{-6} to 1×10^{-9} M dm^{-3}. Electroactive organic compounds (pesticides, growth simulators, drugs, dyes, etc.) are determined in ppb or lower concentration ranges [23]. Organic compounds in the form of gases and vapors in the atmosphere are determined after absorption in a suitable solution or solvent [24]. Stripping voltammetric analyses have been used for the determination of pesticides pollution in a wide variety of environmental matrices. The challenge in the analyses of these complex matrices is to circumvent the interference from the matrix components, for example, organic matter in food and geological samples, proteins in biological samples, etc. A relatively recent capability of stripping analyses is the determination of pesticides. This has been rendered possible largely by the advent of adsorptive interfacial accumulation. The same electrode is used in both the concentration and the stripping process. Deposition and stripping are also made with different solutions, which are suitable for the stripping process with deep sea and remote analyses. Compared with other highly sensitive analytical methods, such as GC-MS and high-performance liquid chromatography (HPLC), stripping voltammetry gives the same performance at a lower cost.

8.3.2.4.1　Mercury Electrode
The positive and negative features linked with the utilization of mercury as the electrode material have been reviewed in many literatures. Despite its toxicity and limited positive potential window, mercury is still the electrode material of interest, especially in stripping analysis. A mercury thin-film electrode (MTFE), constructed by electrodepositing a thin mercury film onto a suitable solid electrode, has become popular in environmental electroanalysis. A hanging mercury drop electrode is often used for adsorptive stripping analyses of pesticides.

8.3.2.4.2　Solid Electrodes
In recent years, the use of solid electrodes for analyses has gained popularity, and one of the primary reasons is their applicability to anodic oxidations. However, the effective utility of solid electrodes for voltammetric analysis is often hampered by a gradual fouling of the surface. Therefore, appropriate protection of the solid electrodes or periodic in situ regeneration of their activity is highly desirable.

8.3.2.4.3　Carbon-Based Electrodes
Glassy carbon is a popular choice of electrode material, and a review of its physical and electrochemical properties is available in the literature [25]. With well-polished surfaces, fast electron-transfer kinetics can be achieved for pesticide analyses. Carbon-paste electrodes are prepared by mixing finely powdered graphite or other carbonaceous material with a liquid such as Nujol, paraffin oil, silicone grease, or bromonaphthalene. These electrodes have the virtues of easy preparation, low cost, surface renewability, amenability to chemical modification, and very low background currents. A major disadvantage is their poor stability in organic solvents. In general, carbon-paste electrode works best in aqueous solutions. An electrochemical stripping voltammetric method for analyzing organophosphate (OP) compounds was developed using a carbon-paste electrochemical (CPE) transducer [26]. In this report, OPs were observed to strongly adsorb onto a CPE surface and provide facile electrochemical quantitative methods for electroactive OP compounds. Operational parameters were optimized, and the stripping voltammetric performance was studied using SWV. The adsorptive stripping voltammetric response was highly linear over the 1–60 µM range of methyl parathion examined (2 min adsorption), with a detection limit of 0.05 µM dm^{-3} (10 min adsorption) and good precision (RSD = 3.2%, n = 10). These findings could lead to the widespread use of electrochemical sensors to detect OP contaminates.

　　Carbon fiber electrodes are increasingly being used in electroanalysis. Pyrolytic carbon films offer rates of electron transfers comparable with or even better than those attainable with glassy carbon without electrode pretreatment [27]. Furthermore, doped diamond is an intriguing electrode material for electroanalysis [28]. Chemical inertness and low electrode capacitance are positive attributes of this material. Recent electrochemical studies have shown the ability of carbon nanotubes (CNTs) to promote certain types of electron-transfer reactions [29], minimize electrode-surface fouling, and enhance electrocatalytic activity [30]. Electrochemical determination of some organic pollutants using wall-jet GCE has also been reported [31]. A square-wave stripping voltammogram of some pesticides on a GCE is given in Figure 8.4.

8.3.2.4.4　Role of Modified Electrodes in Pesticide Analysis
Numerous studies have been done on modified electrodes. Chemical modification of the electrode surface is an essential key to increase the sensitivity and specificity. Numerous electrode-modification

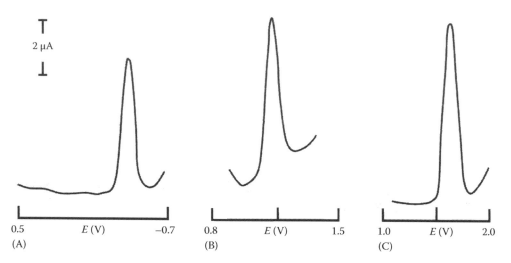

FIGURE 8.4   Square wave stripping voltammogram of (A) endosulfan; (B) isoproturon; and (C) carbendazim on GCE.

methods have been investigated for applications in electrochemical sensors, e.g., clay electrodes, solid polymer electrolytes, membrane-modified electrodes, conducting polymers, sol–gel films, self-assembled monolayers, ceramic materials, enzyme-modified electrodes, etc. Owing to the fact that there are a great number of published studies involving electrode modification for electroanalytical applications, only a brief overview is presented here, with representative examples. Advances in materials science and engineering have made it possible and advantageous to employ numerous chemical detection and analysis techniques in the design and fabrication of electrochemical sensors, and work continues fervently in this research field. The conductive or semiconductive layer on the electrode surface, which demonstrates the desired chemical properties, is an extremely useful technique for sensor design. Solid electrolytes have been used widely for the detection and monitoring of gaseous analytes, and they are especially appealing for this sensing application, as no solution is required in which the analyte must be dissolved prior to measurement. Rather, the solid-state sensor is simply exposed to the target analyte gas(es) or vapor(s), and the analyte(s) is (are) selectively intercalated into or adsorbed on to the electrolyte layer. This interaction serves to alter the chemical potential of the electrolyte film, which results in an electrical signal that can be detected, transduced, and amplified with high sensitivity. Solid-state sensors employing this design have been developed for numerous analyte species of interest, in occupational and environmental health.

8.3.2.4.5   Clay-Modified Electrode

Cheap and naturally occurring, readily available clay minerals are widely employed as modifiers. Their well-defined layered structures, flexible adsorptive properties, and potential as catalysts or catalytic supports make them interesting materials when compared with other modifications. Clay modifications are also made with conducting surfaces. Clays are heterogeneous materials, and each individual clay has a different compositions and particle sizes. Furthermore, clay films are imperfect stacks of clay layers and contain many defects, such as holes and pores of various sizes. This provides different adsorptive sites for the clay-modified electrodes. Adsorption can also be on the external surfaces of the clay or at the edges of the clay sheets. Thus, this is one of the promising areas in the development of sensors.

Natural clay minerals have been particularly useful for the fabrication of solid electrolyte films in sensors for pesticides. Manisankar et al. reported the detection of pesticides like endosulfan, o-chlorophenol, isoproturan, methylparathion, carbendazim, and malathion using montmorillonite-clay-modified GCE with high sensitivity [32]. Squarewave stripping voltammogram of the said

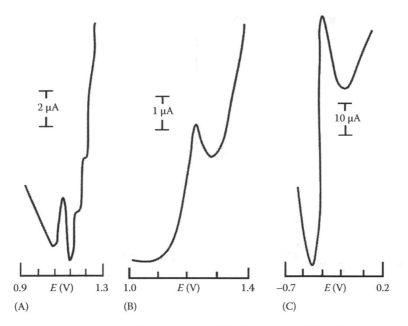

FIGURE 8.5 Square wave stripping voltammogram of (A) isoproturon; (B) carbendazim; and (C) methyl parathion at montmorillonite clay-modified GCE. (From Manisankar, P. et al., *Talanta*, 68, 686, 2006. With permission.)

pesticides on montmorillonite clay-modified GCE is presented in Figure 8.5. Furthermore, the electrochemical detection coupled with liquid chromatographic (LCECD, liquid chromatography with electrochemical detector) systems has been reported for amitrole detection [33,34]. Zen et al. reported a Nafion/lead–ruthenium oxide pyrochlore chemically modified electrode to improve the detection limit to 0.15 ng (20 μL sample loop) [35]. They also reported another sensitive detection scheme for amitrole using a disposable nontronite-coated screen-printed carbon electrode by flow injection analysis (FIA). The SWa-1 nontronite clay is of particular interest owing to its high iron content and demonstrated electrochemical activity towards a variety of compounds [36–41]. Its chemical stability and layer structure make nontronite suitable for use as a matrix in electroanalysis [36–39]. As disposable screen-printed electrodes (SPEs) can be mass produced at low cost and used in diverse fields [42,43], a systematic investigation was carried out in this study to couple the SPE technique with nontronite clay for amitrole. Recently, Liu and Lin developed an electrochemical sensor for OP pesticides and nerve agents using zirconia nanoparticles [44].

8.3.2.4.6 Conducting Polymer-Coated Electrodes
Conducting polymer-modified electrodes offer a powerful route to assembling, at a molecular level, multicomponent systems with complementary functions. Electrodes modified with conducting organic polymers are employed to improve the sensitivity and selectivity, as well as the ability to detect electroactive species [45,46]. Such improvements have been achieved by producing modified surfaces that provide more efficient preconcentration, excluded interferants, enhance the rate of electron transfer, or produce unique, non-Faradaic signals [47]. The significance of dynamic polymer coatings is its multilayered nature, and the fact that it provides three-dimensional reaction zones at the electrode surface. This gives rise to an increase in the flux of reactions that occur there, which in turn increases the sensitivity. The possibilities of using conducting polymers, such as polypyrrole, polyphenylene, polyaniline, and polythiophene, in electrochemical sensors have been recognized for many years [48,49]. Organic conducting polymers demonstrate tremendous versatility in terms of chemical properties and range of conductivities and therefore offer considerable promise for

many commercial applications, including polymer-modified electrochemical sensors [50–53]. However, the short longevities of organic conducting polymer films attached to metal electrodes have restricted the advances in sensor technology for many years. Nevertheless, great strides have been made recently in materials chemistry that have reintroduced organic conducting polymers and oligomers to the scientific limelight [54]. Novel organic polymers and copolymers are being produced and investigated, which offer electronic and electrochromic properties that have long been sought but are now demonstrating longer-term stability that was, for well over a decade, not achievable. It is predicted that recent successes in organic conducting polymer research and development will soon pay dividends, especially in terms of the availability of new electrochemical sensors that will be more robust and rugged. Electrochemical sensors based on conducting polymers hold promise for an expanded array of applicable airborne, environmental, and biological analytes. Square wave stripping voltammetric analysis of pesticides, such as isoproturon, carbendazim, and methyl parathion (Figure 8.6), on polypyrrole-modified GCEs was reported by Manisankar et al. [55]. The range of determination was 0.5–300 ng mL^{-1} for isoproturon, 5–500 ng mL^{-1} for carbendazim, and 20–500 ng mL^{-1} for methyl parathion. The limit of detection (LOD) was 0.5 ng mL^{-1} for isoproturon, 5 ng mL^{-1} for carbendazim, and 15 ng mL^{-1} for methyl parathion. The relative standard deviations found for five identical measurements of the stripping current at 100 ng mL^{-1} analyte concentration was 2.81% for isoproturon, 3.33% for carbendazim, and 2.96% for methyl parathion.

There are numerous advances that have resulted in the manufacturing of portable electroanalytical devices, which are more rugged and user-friendly for making on-site measurements. These include the following examples: (a) disposable SPEs for ease of use, enhanced sensitivity, reduced contamination, and less interference; (b) membrane electrodes for optimized specificity, increased sensitivity, and minimization of interferences; (c) lightweight, robust materials for the fabrication of rugged, light instruments; and (d) advances in battery technology for size minimization and longer device lifetime.

The good response with poly 2,3-ethylenedioxy thiophene modified (PEDOT/GCEs) may be attributed to the availability of electroactive surfaces capable of accommodating the pesticide molecules. Dicofol (DCF) exhibited one well-defined reduction peak owing to the reduction of hydrolyzed product 4,4-dichlorobenzophenone [22] around −1.375 V at pH 13.0. In a similar manner, cypermethrin (CYP) exhibited one well-defined reduction peak at −1.527 V at pH 13.0, owing to the

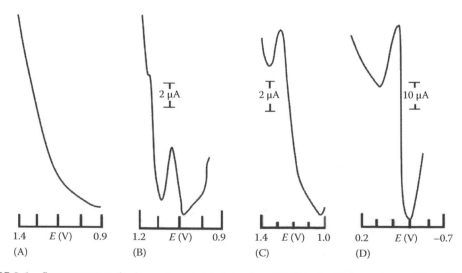

FIGURE 8.6 Square wave stripping voltammetric behavior of (A) blank; (B) isoproturon; (C) carbendazim; and (D) methyl parathion under optimum conditions. (From Manisankar, P. et al., *Int. J. Environ. Anal. Chem.*, 85, 409, 2005; Manisankar, P. et al., *Toxicological. Environ. Chem.*, 85(4–6), 233, 2003. With permission.)

reduction of 3-phenoxybenzaldehyde, which was formed by the hydrolysis of the ester group [56]. Monocrotophos (MCP) showed a reduction peak at −1.742 V in neutral medium, which may be a result of the electroreduction of the carbon–carbon double bond present in the MCP [57]. Chlorpyrifos (CPF) underwent two-step reductions at around −1.6 and −1.7 V. The chlorine atoms and oxygen attached to pyridine in CPF inductively exert an electron-withdrawing effect on the pyridine moiety and increase the electrophilicity of the C–N bond, which in turn enhances the reducibility. Hence, in the first step, the C–N of pyridine is reduced and, in the second step, the formed intermediate undergoes $2e^-$ reduction followed by dehalogenation to produce the product [58]. Phosalone (PAS) exhibited three oxidation peaks at around 0.0, 0.3, and 0.6 V and a small reversible couple centered at −0.65 V at pH of 13.0. The third peaks around 0.6 V, showed irregular behavior, because of its closeness to medium discharge potential. PAS was observed to undergo a base-catalyzed hydrolysis and the amide ring was opened. The secondary amine and phenoxide formed underwent one electron oxidation separately, to produce the corresponding cation radical and quinone. The well-defined oxidation peaks at 0.0 and 0.3 V can be considered for further studies. The stripping voltammograms (Figure 8.7) were recorded with increasing amounts of the corresponding organic pollutants. The LOD is the lowest concentration that can be distinguished from the noise level. In this study, the concentration of the pollutants giving an S/N ratio of 3:1 was 0.09 µg dm^{-3} for DCF.

8.3.2.4.7 Screen-Printed Electrodes

The advent of screen-printing techniques for the fabrication of inexpensive, disposable electrodes has been a boon to electroanalytical chemistry for various applications [59]. SPEs can be manufactured in bulk at a relatively low cost, and their effective performance has been demonstrated for environmental, biomedical, and occupational hygiene monitoring [60]. For instance, such electrodes have recently been employed for on-site monitoring of airborne lead at trace levels [61], and they have been used for the determination of this toxic metal in other matrices as well [62–66]. Disposable SPEs have also been employed for measuring lead in human blood samples. An advantage offered

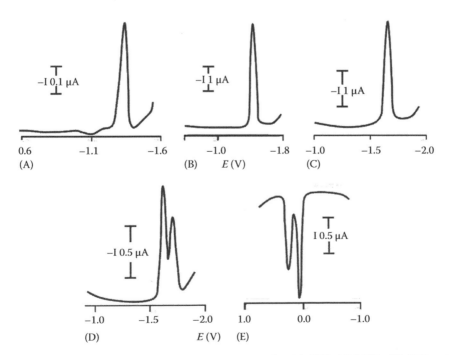

FIGURE 8.7 Square wave stripping voltammograms of (A) DCF; (B) CYP; (C) MCP; (D) CPF; and (E) PAS at optimum conditions. (From Manisankar, P. et al., *Anal. Chim. Acta*, 528, 157, 2005. With permission.)

by SPEs is associated with their single-use application, which avoids problems from electrode fouling owing to repeated analyses using the same electrode surface. Hence, there is no threat of electrode poisoning from reusing the same sensor surface for successive analyses. By employing these
electrodes as sensors in field-portable, battery-powered instrumentation, it is now possible to obtain
excellent sensitivity and selectivity for the on-site measurement of numerous analytes of interest.
It is anticipated that more applications of SPEs in environmental and biological monitoring will
appear in the near future.

8.3.3 Amperometric Methods for Pesticide Analysis

An electrochemical detector is often a component of a chromatography system for separating the
constituents of a complex mixture. The pesticides and their residues in the environment can be
detected by both oxidative and reductive amperometries [2]. The limits for reductive detection
are usually less favorable than for oxidative detection, owing to the background resulting from
the reduction of traces of O_2, hydrogen ions, and trace metals. However, many types of pesticides
have been successfully determined via reductive amperometry, and pesticides containing reducible groups (e.g., nitrocompounds) can be reductively determined. Furthermore, pesticides of the
carbamate class and amine-derived compounds can be oxidatively determined, while those with a
thiocarbonyl group can be oxidized at modest potentials and thus determined directly or through
complexation of the thiocarbonyl group with mercury ions [58,67]:

$$Hg + 2TU \rightarrow Hg(TU)_2^{2+} + 2e^-$$

where TU is thiourea.

The limiting current of this oxidation is proportional to the ligand concentration. Although the
detection at +0.19 V is not very sensitive (the detection limit is ~10 ng dm^{-3}), the analysis has good
selectivity. Therefore, pesticides with a thiocarbonyl group can be determined in the environmental
samples without pretreatment. The FIA system, in conjunction with electrochemical detection, is a
popular method for analyzing large numbers of samples.

8.3.4 Flow Analysis of Pesticides

8.3.4.1 Flow Injection Analysis

The manual handling of solutions (known as "beaker chemistry") remains the Achilles heel of modern analytical instrumentation. It is currently being replaced by FIA, which is computer compatible
and allows automated handling of the sample and reagent solutions with a strict control over the
reaction conditions. FIA was first described in Denmark by Ruzicka and Hansen in 1975. Since then,
the technique has grown into a discipline covered by six monographs and more than 15,000 research
papers. The scope of the method grew from a serial assay of the samples to a tool for enhancing the
performance of spectroscopic and electrochemical instruments. In its simplest form, the sample zone
is injected into a flowing carrier stream of reagent. As the injected zone moves downstream, the sample solution disperses into the reagent, forming the product. A flow-through the detector placed downstream records the desired physical parameter, such as colorimetric absorbance or fluorescence.

The modern FIA system usually consists of a high-quality multichannel peristaltic pump, an
injection valve, a coiled reactor, a detector such as a photometric flow cell, and an autosampler.
Additional components may include a flow-through heater to increase the speed of the chemical
reactions, columns for sample reduction, debubblers, and filters for particulate removal. The typical FIA flow rate is 1 mL/min, the typical sample volume consumption is 100 µL per sample, and
the typical sampling frequency is two samples per minute. The FIA assays usually result in sample
concentration accuracies of a few percent.

8.3.4.2 Sequential Injection Analysis

Sequential injection analysis (SIA) is the second-generation approach for FIA-compatible assays. The SIA usually consists of a single-channel high-precision bi-directional pump, a holding coil, a multiposition valve, and a flow-through detector. The system is initially filled with a carrier stream into which a zone of sample and a zone of reagent(s) are sequentially aspirated into a holding coil, forming a linear stack. These zones become overlapped owing to the parabolic profile induced by the differences between the flow velocities of the adjacent streamlines. Flow reversals and flow accelerations further promote mixing. The multiposition valve is then switched to the detector position and the flow direction is reversed, propelling the sample/reagent zones through the flow cell.

The advantage of SIA over the more traditional FIA is that SIA typically consumes less than one-tenth of the reagent and produces far less waste—an important feature when dealing with expensive chemicals, hazardous reagents, or online/remote site applications. One disadvantage of SIA is that it tends to run slower than FIA.

Measurements in flowing liquids are becoming progressively more important in all branches of analytical chemistry, including environmental analysis. These involve not only continuous monitoring of substances, e.g., in natural waters, waste waters, process streams, etc., but also laboratory analyses of discrete samples using the methods of continuous flow analysis (CFA), FIA, and especially HPLC. Analyses in flow systems generally permit an increase in the sample throughput, save manual work, and lend themselves readily to extensive automation. As a large proportion of analyses involves determinations of traces of substances in complex matrices (this is especially important in environmental and clinical analyses), the methods employed must simultaneously be highly sensitive, reproducible, and selective.

8.3.4.3 Wall-Jet Flow Analysis

Electrochemical methods of analysis of flowing liquids have traditionally been used in the process of stream monitoring and recently have also been increasingly employed in the methods of CFA, FIA, and especially HPLC. Among the many available electrochemical methods, only a few are important in analytical practice. Low-frequency conductometry and high-frequency impedance measurements have limited applications and will not be discussed here. The most important methods, based on charge-transfer reactions at an electrode–solution interface, are ion-selective electrode (ISE), potentiometry, polarography, voltammetry, and coulometry. A wall-jet flow cell with an experimental setup is given in Figure 8.8 [31].

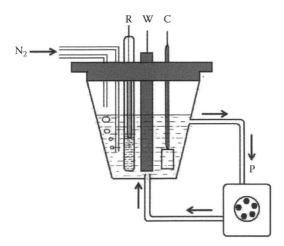

FIGURE 8.8   Flow cell: W, wall-jet working electrode; R, reference electrode; C, counterelectrode; P, peristaltic pump. (From Manisankar, P. et al., *Anal. Chim. Acta.*, 528, 157, 2005. With permission.)

8.4 SENSORS IN PESTICIDE ANALYSIS

8.4.1 MOLECULARLY IMPRINTED POLYMER SENSORS

The concept of molecularly imprinted polymers (MIPs) continues to fascinate researchers because of its intuitive beauty, ruggedness, and versatility. While most research in this direction is targeted to the design of chromatographic stationary phases, their use in electrochemical sensors is expanding for electroactive analytes. In an earlier work, Mandler et al. explored the sol–gel polymers imprinted with the OP pesticide parathion and performed gas- and liquid-phase partitioning experiments as well as CV studies [68]. The imprinted films showed more than a 10-fold increase in equilibrium binding over the nonimprinted polymers, and discriminated well against a range of other structurally similar OPs. Furthermore, determinations of degradation products of nerve agents in human serum by solid-phase extraction using MIP [69,70] and MIP-containing imidazoles and bivalent metal complexes for the detection and degradation of organophosphotriester pesticides [71] were also reported.

8.4.2 MICROCHIPS AND ULTRAMICROSENSORS

In addition to ultramicroelectrodes, chemistry in miniature has also been realized through chemical analyses on microchips [72]. Microdevices for electrochemical analysis on a micrometer scale have been fabricated using centimeter-sized chips comprising glass, silicon, or inert polymeric materials. Microfluidic circuits have been fabricated, which provide a "total analysis" system including sample introduction and pretreatment, chemical reaction, detection, and separation or isolation of reaction products [73]. Analytical performance on a small scale is improved by means of speed and efficiency, as reactions can be completed effectively and rapidly through the implementation of the lab-on-a-chip concept. In an application of "microelectrochemistry," potentiometric detection on a chip has demonstrated electrochemical behavior similar to conventional electrochemical cells and microelectrodes [74]. For example, a microscale capillary electrophoresis system with amperometric detection has been fabricated [75]; the device has been employed in the assays of mixtures of nitroaromatic explosives and catecholamines. Another exciting related development is the fabrication of disposable microchips for blood-chemistry biosensors [76]. Microscale electrochemical detection technology offers tremendous potential for many other analytical applications, especially for on-site screening measurements. It is only a matter of time before this "chemistry-on-a-chip" technology is employed in the manufacturing of electrochemical sensors for occupational hygiene, environmental pollution, contamination, and related applications.

Developments in the design and fabrication of ultramicroelectrodes [77] offer considerable promise in the advancements of electrochemical sensors. Ultramicroelectrodes have proven to be particularly useful for bio-monitoring purposes [78,79], and they have also been employed for environmental and industrial hygiene measurements [80,81]. These extremely minute electrodes provide fantastic sensitivity, as their size is diminished; the S/N ratio increases even though the magnitude of the detected signal is smaller [82]. Although the magnitude of the current signal is decreased as the electrode dimensions are reduced, with modern electronics, it is possible to measure extremely small signals with low-noise operational amplifiers and associated instrumentation. In this way, the favorable mass-transport characteristics offered by the electrodes of minute size can be used for analytical advantage. As the electrode size is reduced, the rate of analyte diffusion to the electrode surface is increased dramatically, thereby enhancing the sensitivity. Extremely low detection limits may be achieved with ultramicroelectrodes, and many hazardous substances demand that detection limits be as low as possible.

Another advantage of ultramicroelectrodes is that often no supporting electrolyte is necessary, owing to the favorable mass-transfer characteristics of tiny electrodes [77]. Hence, it may be possible to use ultramicroelectrodes to measure analytes having very high redox potentials [83]. In this way, it may become possible to monitor toxic airborne species, e.g., polyaromatic hydrocarbons, which have previously been unattainable for measurement by electroanalysis. Arrays of ultramicroelectrodes can be employed to give increased sensitivity and selectivity for environmental monitoring [84].

8.4.3 Sensor Arrays

As the sensor size decreases with the corresponding increase in the S/N ratios, it has become possible to use arrays of sensors to further increase the sensitivity as well as improve the selectivity. For example, new sensor arrays based on amperometric detection, coupled with chemical modification and pattern recognition techniques, have been fabricated, which significantly enhance the analytical performance [85,86]. Filho et al. reported that the sensor array is efficient for the detection of atrazine, imazaquin, metribuzin, and paraquat in contaminated waters, once the limit of detection for each sensor is laid well below the values established by the Environmental Protection Agency (EPA) [87].

8.5 APPLICATION OF ELECTROANALYTICAL METHODS ON PESTICIDES MONITORING IN ENVIRONMENTAL SAMPLES

The wide use of pesticides in modern agricultural practices has led to improvement in the levels of residues present in soil, water, and food. The soil receives large quantities of pesticides as an inevitable result of their application to crops. Most lipophilic pesticide residues reside mainly in the soil. The soil constitutes an integral part of the human exposure pathway, whereby the pollutant ultimately enters the food chain. As is the case with many other triazines, ametryn is a selective herbicide used on corn and potatoes for general weed control. The main problem of this pesticide is its persistence in groundwater. It moves both vertically and laterally in soil owing to its high-water solubility. Manisankar et al. developed an electroanalysis technique for pesticides such as isoproturon, carbendazim, methyl parathion, endosulfan, and o-chlorophenolin in soil samples using square wave stripping voltammetry and PEDOT [poly(3,4-ethylenedioxythiophene)]-coated GCE (Figure 8.9).

Table 8.2 summarizes the electroanalytical monitoring of pesticides pollution in various environmental matrices.

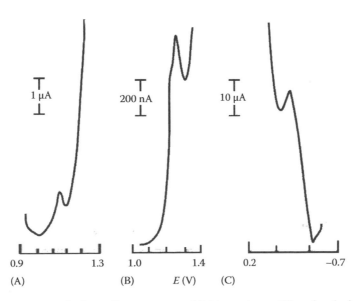

FIGURE 8.9 Square wave stripping voltammogram, of (A) isoproturon; (B) carbendazim; and (C) methyl parathion in soil samples.

TABLE 8.2
Electroanalytical Monitoring of Pesticides in Various Environmental Matrices

S. No.	Analyte	Class	Methods	Matrices	Remarks	Ref.
1	Guthion	Organophosphorous	FIA amperometric	Commercial formulations	LOD 4.1×10^{-7} M dm^{-3}	[88]
2	Organophosphorous and cabamate	Organophosphorous carbamate	Amperometric with cholinesterase	Commercial formulations	Cobalt phthalocyanine-modified carbon-paste electrode and cholinesterase enzyme LOD 0.3 mg dm^{-3}	[89]
3	Tetramethrin Monocrotophos Phosphamidon Dichlorvos	Insecticide Organophosphorous	AdSWSV Polarography Cyclic voltammetry	Soil and water samples Commercial formulations	Hg electrode LOD 8.5×10^{-9} M dm^{-3} LOD 1×10^{-8} M dm^{-3}	[90] [91]
4	Formetanate Chlordimeform	Insecticide Acaricide	DPP	Commercial formulations	Hg, LOD 1×10^{-9} M dm^{-3}	[92]
5	Carbendazim	Fungicide	DPV	Wine, apple, and soil samples	Graphite electrode modified with silicone OV-17 LOD 9.1 ng mL^{-1}	[93]
6	Allethrin	Pyrethroid	DPV	Commercial formulations	GCE, LOD 1.5×10^{-5} M dm^{-3}	[94]
7	Chlorpyrifos	Insecticide	DPP	Irrigation water Waste water	Hg electrode LOD 8.7×10^{-7} M dm^{-3}	[95]
8	S-Triazines, atrazine-desethyl, and aziprotryne	Herbicide	Amperometric	Environmental and tap water samples	Metallic copper electrode and GCE, LOD 0.2 μg mL^{-1}	[96]
9	Thiram		SWV	Grapes and pharmaceutical preparations	Cylindrical carbon fiber microelectrode LOD 4.3×10^{-7} M dm^{-3}	[97]
10	Imidacloprid	Insecticide	AdSWSV	River water samples	HMDE, LOD 1.6×10^{-8} M dm^{-3}	[98]
11	Metribuzin	Herbicide	AdDPSV	River water samples		[99]
12	Buprofezin	Insecticide	AdDPSV	Soil, tap water, and treated water	HMDE, LOD 2.2 μg dm^{-3}	[100]
13	Cypermethrin Deltamethrin	Pyrethroid Insecticide	Cyclic voltammetry DPV	Agrochemical products Natural water	HMDE	[101]
14	Endosulfan	Insecticide	AdSWSV and AdDPSV	Soil samples	GCE, LOD 9×10^{-9} M L^{-1}	[102]
15	Amitrole	Insecticide	SWV and FIA		Nontronite-coated screen-printed electrode LOD 0.33 μM in SWV and 0.07 ng in a 20 μL sample loop by FIA	[103]

No.	Pesticide	Type	Technique	Sample	Electrode/Details	Ref.
16	Isoproturon	Herbicide	AdSWSV	Natural water and soil	Wall-jet electrode system, LOD 10^{-4} mg mL^{-1}	[31]
17	Carbendazim	Fungicide	AdSWSV	Natural water and soil	Wall-jet electrode system, LOD 10^{-4} mg mL^{-1}	[31]
18	Endosulfan	Insecticide	AdSWSV	Natural water and soil	Wall-jet electrode system, LOD 10^{-4} mg mL^{-1}	[31]
19	Endosulfan	Insecticides	AdDPSV	—	Sodium montmorillonite clay-modified GCE, LOD 5 ppb	[104]
20	Molinate	Herbicide	SWV		HMDE LOD 3.5×10^{-8} M dm^{-3}	[105]
21	Dicofol, cypermethrin, monocrotophos, chlorpyrifos, and phosalone	Insecticides	AdSWSV	Soil extract	Wall-jet poly 3,4-ethylenedioxythiophene-modified GCE LOD < 0.09 and < 1.0 µg dm^{-3}	[106]
22	Atrazine and ametryne	Triazine herbicides	SWV	Natural water samples	Copper solid amalgam electrode LOD 3.06 µg dm^{-3} for atrazine and 3.78 µg dm^{-3} for ametryne	[107]
23	Paraquat	Herbicide	SWV	Natural water samples	Pyrolitic graphite electrode modified by metallophthalocyanine LOD 26.53 µg dm^{-3} and LOQ 88.23 µg dm^{-3}	[108]

Note: AdSWSV, adsorptive square wave stripping voltammetry; AdDPSV, adsorptive differential pulse stripping voltammetry; SWV, square wave voltammetry; DPV, differential pulse voltammetry; DPP, differential pulse polarography; FIA, flow injection analysis; HMDE, hanging mercury drop electrode; LOQ, limit of quantification.

8.6 CONCLUSION

Electrochemical sensor research is a diverse, healthy field in which a significant level of activity is carried out in countries all over the world. A new approach to modify the electrochemical waveform has also been proposed to effectively deal with charging currents for voltammetric analysis, and this may have important implications on the design of low-detection limit sensors of this type. Several electrochemical approaches developed for the analyses of pesticides in environmental matrices were discussed in this chapter. It is encouraging that electroanalyses for environmental monitoring are conceptually very similar to those used for clinical or food analyses. One of the difficulties in natural environments is the fouling of the electrode surface, which can also occur in a clinical or food analysis. Thus, analytical method developments in various application areas are closely linked. Apart from this advantage, the real challenges for the future are those of good electrode materials, miniaturization, and measurements made as close to real time as possible. Although monitoring of an industrial process where the sample matrix is usually known can give accurate results in the environmental analysis in the field, where the matrix is probably to some extent unknown, it is unlikely that the results would be very accurate. The need for sample pretreatment should be minimized to reduce analysis time and allow the probing of natural speciation; thus, excellent and easy-to-use electrode protection strategies need to be developed. Field analyses should probably be used in the first instance, in a diagnostic sense, as an alarm sensor for environmental agencies, companies, etc., because closer control must be maintained, after which specific sensors can be installed if necessary. In this way, multispecies sensors of toxicity, as described earlier, can be extremely valuable. This also reduces the problem of accurate calibration of the electrochemical sensor for field use, particularly if disposable sensors are being employed. Immuno-biochips, which are currently under study and development, are not only miniaturizing the microtiter plate. Apart from the parallelization of the measurements already performed on the microtiter plate, the immuno-biochips show a new functionality owing to the integration of a part of or the entire detection system. This approach enables, on the one hand, reduction of the operating times of the different test steps and, on the other hand, a real integration in a complete system, particularly incorporating microfluidic parts.

REFERENCES

1. M Farré, C Gonçalves, S Lacorte, and D Barceló, MF Alpendurada. Pesticide toxicity assessment using an electrochemical biosensor with *Pseudomonas putida* and a bioluminescence inhibition assay with *Vibrio fischeri. Anal. Bioanal. Chem.* 373: 696–703, 2002.
2. K Rajeshwar and JG Ibanez. *Environmental Electrochemistry: Fundamentals and Applications in Pollution Sensors and Pollutant Treatment*, Academic Press, New York, 1997.
3. PM Bersier and J Bersier. *Contemporary Electroanalytical Chemistry* (Ed. A Ivaska), Plenum Press, New York, 1990.
4. R Kalvoda. *Electroanalytical Methods in Chemical and Environmental Analysis*, Plenum Press, New York, 1987.
5. BB Damaskin. *The Principles of Current Methods for the Study of Electrochemical Reactions*, McGraw Hill, New York, 1967.
6. R Adams. *Electrochemistry at Solid Electrodes*, Marcel Dekker, New York, 1969.
7. A Welssberger and BW Rosstter. *Physical Methods of Chemistry—Electrochemical Methods Part–II*, Wiley InterScience, New York, 1971.
8. DD Macdonald. *Transient Technique in Electrochemistry*, Plenum Press, New York, 1977.
9. AJ Bard and LR Faulkner. *Electrochemical Methods, Fundamentals and Applications*, John Wiley & Sons, New York, 1980.
10. AJ Fry. *Synthetic Organic Electrochemistry*, Marcel Dekker, New York, 1975.
11. PT Kissinger and WR Heineman. *Laboratory Techniques in Electroanalytical Chemistry*, 1st edition, Marcel Dekker Inc., New York, 1984.
12. D Pletcher and FC Walsh. *Industrial Electrochemistry*, 2nd edition, Chapman & Hall, London, 1990.
13. EP Parry and RA Osteryoung. Evaluation of analytical pulse polarography. *Anal. Chem.* 37: 1634–163, 1965.

14. F Vydra, K Stulik, and E Julakova. *Electrochemical Stripping Analysis*, Halsted Press, New York, 1976.
15. J Wang. *Stripping Analysis*, VCH Publishers, Deerfield Beach, FL, 1985.
16. TR Copeland and RK Skogerboe. Anodic stripping voltammetry. *Anal. Chem.* 46: 1257A–1268A, 1974.
17. J Wang. Recent advances in stripping analysis. *Fres. J. Anal. Chem.* 337: 508–511, 1990.
18. M Esteban and E Casassas. Stripping electroanalytical techniques in environmental analysis. *Trends Anal. Chem.* 13: 110–117, 1994.
19. K Brainina and E Neyman. *Electrochemical Stripping Methods*, American Chemical Society, Washington, DC, 1993.
20. R Kalvoda and M Kopanica. Adsorptive stripping voltammetry in trace analysis. *Pure Appl. Chem.* 61: 97–112, 1989.
21. CMG van den Berg. Potentials and potentialities of cathodic stripping voltammetry of trace elements in natural waters. *Anal. Chim. Acta* 250: 265–276, 1991.
22. MG Paneli and A Voulgaropoulos. Applications of adsorptive stripping voltammetry in the determination of trace and ultratrace metals. *Electroanalysis* 5: 355–373, 1993.
23. R Kalvoda. *Contemporary Electroanalytical Chemistry* (Ed. A Ivaska), Plenum Press, New York, 1990.
24. MD Manita, RM Salikhzhdanova, and CF Javorskaja. *Sovremennyie Metody Opredeleniya Atmosfernkh Zagryaznenii*, Naselenykh Mest, Medicine, Moscow, 1980.
25. WE van der Linden and JW Dieker. Glassy carbon as electrode material in electroanalytical chemistry. *Anal. Chim. Acta* 119: 1–24, 1980.
26. G Lin and Y Lin. Electrochemical stripping analysis of organophosphate pesticides and nerve agents. *Electrochem. Comm.* 7(4): 339–343, 2005.
27. CF McFadden, PR Melaragno, and JA Davis. Fabrication of pyrolytic carbon film electrodes by pyrolysis of methane on a machinable glass ceramic. *Anal. Chem.* 64: 742–746, 1990.
28. VA Pedrosa, L Codognoto, SAS Machado, and LA Avaca. Is the boron-doped diamond electrode a suitable substitute for mercury in pesticide analyses? A comparative study of 4-nitrophenol quantification in pure and natural waters. *J. Electroanal. Chem.* 573: 11–18, 2004.
29. RH Baughman, C Cui, AA Zakhidov, Z Iqbal, JN Barisci, GM Spinks, GG Wallace, A Mazzoldi, DD Rossi, AG Rinzler, O Jaschinski, S Roth, and M Kertesz. Carbon Nanotube Actuators. *Science* 284: 1340–1344, 1999.
30. CE Banks, RR Moore, TJ Davies, and RG Compton. Investigation of modified basal plane pyrolytic graphite electrodes: Definitive evidence for the electrocatalytic properties of the ends of carbon nanotubes. *Chem. Commun.* 16: 1804–1805, 2004.
31. P Manisankar, G Selvanathan, S Viswanathan, and HG Prabu. Electrochemical determination of some organic pollutants using wall-jet electrode. *Electroanalysis* 14: 1722–1727, 2002.
32. P Manisankar, G Selvanathan, and C Vedhi. Determination of pesticides using heteropolyacid montmorillonite clay-modified electrode with surfactant. *Talanta* 68: 686–692, 2006.
33. JM van der Poll, M Vink, and JK Quirijins. Determination of amitrole in plant tissues and sandy soils by capillary gas chromatography with alkali flame ionization detection. *Chromatographia* 30: 155–158, 1990.
34. V Pichon and M-C Hennion. Comparison of on-line enrichment based on ion-pair and cation-exchange liquid chromatography for the trace-level determination of 3-amino-1,2,4-triazole (aminotriazole) in water. *Anal. Chim. Acta* 284: 317–326, 1993.
35. J-M Zen, A Senthil Kumar, and M-R Chang. Electrocatalytic oxidation and trace detection of amitrole using a nafion/lead-ruthenium oxide pyrochlore chemically modified electrode. *Electrochim. Acta* 45: 1691–1699, 2000.
36. A Fitch. Clay-modified electrodes: A review. *Clays Clay Miner.* 38: 391–400, 1990.
37. MT Carter and AJ Bard. Clay modified electrodes: Part VII. The electrochemical behavior of tetrathiafulvalcnium-montmorillonite modified electrodes. *J. Electroanal. Chem.* 229: 191–214, 1987.
38. J-M Zen and C-W Lo. A glucose sensor made of an enzymatic clay-modified electrode and methyl viologen mediator. *Anal. Chem.* 68: 2635–2640, 1996.
39. J-M Zen, S-H Jeng, and H-J Chen. Catalysis of the electroreduction of hydrogen peroxide by nontronite clay coatings on glassy carbon electrodes. *J. Electroanal. Chem.* 408: 157–163, 1996.
40. J-M Zen, S-H Jeng, and H-J Chen. Determination of paraquat by square-wave voltammetry at a perfluorosulfonated ionomer/clay-modified electrode. *Anal. Chem.* 68: 498–502, 1996.

41. J-M Zen and P-J Chen. A selective voltammetric method for uric acid and dopamine detection using clay-modified electrodes. *Anal. Chem.* 69: 5087–5093, 1997; J-M Zen, W-M Wang, G Ilangovan. Adsorptive potentiometric stripping analysis of dopamine on clay-modified electrode. *Anal. Chim. Acta* 372: 315–321, 1998.

42. J-M Zen, H-H Chung, and A Senthil Kumar. Determination of lead(II) on a copper/mercury-plated screen-printed electrode. *Anal. Chim. Acta* 421: 189–197, 2000.

43. J-M Zen, H-H Chung, and A Senthil Kumar. Electrochemical impedance study and sensitive voltammetric determination of Pb(II) at electrochemically activated glassy carbon electrodes. *Analyst* 125: 1139–1146, 2000.

44. G Liu and Y Lin. Electrochemical sensor for organophosphate pesticides and nerve agents using zirconia nanoparticles as selective sorbents. *Anal. Chem.* 77: 5894–5901, 2005.

45. A Ivaska. Analytical applications of conducting polymers. *Electroanalysis* 3: 247–254, 1991.

46. MD Imisides, R John, PJ Riley, and GG Wallace. The use of electropolymerization to produce new sensing surfaces: A review emphasizing electrode position of heteroaromatic compounds. *Electroanalysis* 3: 879–889, 1991.

47. RW Murray. Chemically modified electrodes. In: *Electroanalytical Chemistry* (Ed. A J Bard), Vol. 13, p. 191, Marcel Dekker Inc., New York, 1984.

48. J Janata. *Principles of Chemical Sensors*, Plenum Press, New York, 1989.

49. K Ashley. Electrochemical studies of strained molecules and conducting polymers. PhD dissertation, University of Utah, 1987.

50. M Josowicz, J Janata, K Ashley, and S Pons. Electrochemical and ultraviolet-visible spectroelectrochemical investigation of selectivity of potentiometric gas sensors based on polypyrrole. *Anal. Chem.* 59: 253–258, 1987.

51. TA Skotheim (Ed.). *Electroresponsive Molecular and Polymeric Systems*, Vol. 2, Marcel Dekker, New York, 1991.

52. RW Murray (Ed.). *Molecular Design of Electrode Surfaces, Techniques of Chemistry Series*, XXII, Wiley, New York, 1992.

53. MD Ryan, EF Bowden, and JQ Chambers. Dynamic electrochemistry: Methodology and application. *Anal. Chem.* 66: 360R–427R, 1994.

54. M Freemantle. Organic materials get more active. *Chem. Eng. News* 79: 49–56, 2001.

55. (a) P Manisankar, G Selvanathan, and C Vedhi. Utilization of polypyrrole modified electrode for the determination of pesticides. *Int. J. Environ. Anal. Chem.* 85: 409–422, 2005. (b) P Manisankar, C Vedhi, and G Selvanathan. Electrochemical determination of methyl parathion using modified electrodes. *Toxicological. Environ. Chem.* 85(4–6), 233–241, 2003.

56. JL Andersen and DJ Chesney. Liquid chromatographic determination of selected carbamate pesticides in water with electrochemical detection. *Anal. Chem.* 52: 2156–2161, 1980.

57. JL Andersen, KK Whiten, JD Brewster, T-Y Ou, and WK Nonidez. Microarray electrochemical flow detectors at high applied potentials and liquid chromatography with electrochemical detection of carbamate pesticides in river water. *Anal. Chem.* 57: 1366–1373, 1985.

58. HB Hanekamp, P Bos, and RW Frei. Design and selective application of a dropping mercury electrode amperometric detector in column liquid chromatography. *J. Chromatogr. A* 186: 489–496, 1979.

59. JP Hart and SA Wring. Recent developments in the design and application of screen-printed electrochemical sensors for biomedical, environmental and industrial analyses. *Trends Anal. Chem.* 16: 89–103, 1997.

60. J Wang. Decentralized electrochemical monitoring of trace metals: From disposable strips to remote electrodes. *Analyst* 119: 763–766, 1994.

61. K Ashley, KJ Mapp, and M Millson. Ultrasonic extraction and field-portable anodic stripping voltammetry for the determination of lead in workplace air samples. *Am. Ind. Hyg. Assoc. J.* 59: 671–679, 1998.

62. J Wang and B Tian. Screen-printed stripping voltammetric/potentiometric electrodes for decentralized testing of trace lead. *Anal. Chem.* 64: 1706–1709, 1992.

63. K Ashley, M Hunter, LH Tait, J Dozier, JL Seaman, and PF Berry. Field investigation of on-site techniques for the measurement of lead in paint films. *Field Anal. Chem. Technol.* 2: 39–50, 1998.

64. K Ashley, R Song, CA Esche, PC Schlecht, PA Baron, and TJ Wise. Ultrasonic extraction and portable anodic stripping voltammetric measurement of lead in paint, dust wipes, soil, and air: An interlaboratory evaluation. *J. Environ. Monit.* 1: 459–464, 1999.

65. I Palchetti, C Upjohn, APF Turner, and M Mascini. Disposable screen-printing electrodes (SPE) mercury-free for the lead detection. *Anal. Lett.* 33(7), 1231–1246, 2000.

66. C Yarnitsky, J Wang, and B Tian. Hand-held lead analyzer. *Talanta* 51: 333–338, 2000.
67. JW Lawrence, F Iverson, HB Hanekamp, P Bos, and RW Frei. Liquid chromatography with UV absorbance and polarographic detection of ethylenethiourea and related sulfur compounds: Application to rat urine analysis. *J. Chromatogr. A* 212: 245–250, 1981.
68. S Marx, A Zaltsman, I Turyan, and D Mandler. Parathion sensor based on molecularly imprinted sol–gel films. *Anal. Chem.* 76: 120–126, 2004.
69. T Yamazaki, E Yilmaz, K Mosbach, and K Sode. Towards the use of molecularly imprinted polymers containing imidazoles and bivalent metal complexes for the detection and degradation of organophosphotriester pesticides. *Anal. Chim. Acta* 435: 209–214, 2001.
70. MZ Hui and L Qin. Determination of degradation products of nerve agents in human serum by solid phase extraction using molecularly imprinted polymer. *Anal. Chim. Acta* 435: 121–127, 2001.
71. MCB Lopez, MJL Castanon, AJM Ordieres, and PT Blanco. Electrochemical sensors based on molecularly imprinted polymers. *Trends Anal. Chem.* 23: 36–48, 2004.
72. M Freemantle. Downsizing chemistry: Chemical analysis and synthesis on microchips promise a variety of potential benefits. *Chem. Eng. News* 77(7): 27–36, 1999.
73. SB Cheng, CD Skinner, J Taylor, S Attiya, WE Lee, G Picelli, and DJ Harrison. Development of a multichannel microfluidic analysis system employing affinity capillary electrophoresis for immunoassay. *Anal. Chem.* 73: 1472–1479, 2001.
74. R Tantra and A Manz. Integrated potentiometric detector for use in chip-based flow cells. *Anal. Chem.* 72: 2875–2878, 2000.
75. J Wang, BM Tian, and E Sahlin. Micromachined electrophoresis chips with thick-film electrochemical detectors. *Anal. Chem.* 71: 5436–5440, 1999.
76. JW Choi, CH Ahn, S Bhansali, and HT Henderson. A new magnetic bead-based, filterless bio-separator with planar electromagnet surfaces for integrated bio-detection systems. *Sens. Actuat. B* 68: 34–39, 2000.
77. M Fleischmann, S Pons, DR Rolison, and PP Schmidt. *Ultramicroelectrodes*, Datatech Science, Morganton, NC, 1987.
78. RM Wightman, LJ May, J Bauer, D Leszczyszyn, and E Kristensen. In: *Chemical Sensors and Microinstrumentation* (Eds. R W Murray, R E Dessy, W R Heineman, J Janata, W R Seitz), pp. 114–128. ACS Symposium Series, No. 403, American Chemical Society, Washington, DC, 1989.
79. Kevin Ashley. Developments in electrochemical sensors for occupational and environmental health application. *J. Hazard. Mater.* 102: 1–12, 2003.
80. K Ashley. Electroanalytical applications in occupational and environmental health. *Electroanalysis* 6: 805–820, 1994.
81. WM Draper, K Ashley, CR Glowacki, and PR Michael. Industrial hygiene chemistry: Keeping pace with rapid change in the workplace. *Anal. Chem.* 71: 33R–60R, 1999.
82. S Pons and M Fleischmann. The behavior of microelectrodes. *Anal. Chem.* 59: 1391A–1399A, 1987.
83. MH Kowalski, JW Pons, PJ Stang, S Pons, NH Werstiuk, and K Ashley. Electrochemical and photoelectronic spectral study of compounds with high ionization potentials: Anodic oxidation of vinyl triflates in aprotic solvents. *J. Phys. Org. Chem.* 3: 670–676, 1990.
84. R Feeney and SP Kounaves. Microfabricated ultramicroelectrode arrays: Developments, advances, and applications in environmental analysis. *Electroanalysis* 12: 677–684, 2000.
85. R Solná, S Sapelnikova, P Skládal, M Winther-Nielsenc, C Carlssond, J Emnéusb, and T Ruzgasb. Multienzyme electrochemical array sensor for determination of phenols and pesticides. *Talanta* 65: 349–357, 2005.
86. E Bakker and Y Qin. Electrochemical sensors. *Anal. Chem.* (Review) 78(12): 3965–3984, 2006.
87. NC Filho, ES Medeiros, ST Tanimoto, LHC Mattoso. Electrets, 2005. ISE-12. 2005 12th International Symposium. Sept. 11–14, 2005, pp. 424–427.
88. H Méndez, R Carabias Martínez, E Rodríguez Gonzalo, and J Pérez Trancon. Determination of the pesticide guthion by flow-injection analysis with amperometric detection. *Electroanalysis* 2: 487–491, 1990.
89. P Skládal. Determination of organophosphate and carbamate pesticides using a cobalt phthalocyanine-modified carbon paste electrode and a cholinesterase enzyme membrane. *Anal. Chim. Acta* 252: 11–15, 1991.
90. P Hernández, F Galán-Estella, and L Hernández. Determination of tetramethrin by adsorptive stripping with square wave voltammetry. *Electroanalysis* 4: 45–49, 1992.
91. M Subbalakshmamma and S Jayarama Reddy. Electrochemical reduction behavior and analysis of some organophosphorous pesticides. *Electroanalysis* 6: 521–526, 1994.

92. M Subbalakshmamma and S Jayarama Reddy. Electrochemical behavior of formetanate and chlordime-form pesticides. *Electroanalysis* 6: 612–615, 1994.
93. P Hernandez, Y Ballesteros, F Galan, and L Hernandez. Determination of carbendazim with a graphite electrode modified with silicone OV-17. *Electroanalysis* 8: 941–946, 1996.
94. DC Coomber, DJ Tucker, and AM Bond. Electrochemical reduction of pyrethroid insecticides in non-aqueous solvents. *J. Electroanal. Chem.* 426: 63–73, 1997.
95. ASR Al-Meqbali, MS El-Shahawi, and MM Kamal. Differential pulse polarographic analysis of chlorpyrifos insecticide. *Electroanalysis* 10: 784–786, 1998.
96. A Zapardiel, E Bermejo, JA Pérez, and M Chicharro. Determination of s-triazines with copper and glassy carbon electrodes. Flow injection analysis of aziprotryne in water samples. *Fresen. J. Anal. Chem.* 367(5): 461–466, 2000.
97. MA Hernández-Olmos, L Agui, P Yanez-Sedeno, and JM Pingarron. Analytical voltammetry in low-permitivity organic solvents using disk and cylindrical microelectrodes. Determination of thiram in ethyl acetate. *Electrochim. Acta* 46: 289–296, 2000.
98. A Guiberteau, T Galeano, N Mora, P Parrilla, and F Salinas. Study and determination of the pesticide imidacloprid by square wave adsorptive stripping voltammetry. *Talanta* 53: 943–949, 2001.
99. J Skopalová, K Lemr, M Kotoucek, L Cáp, and P Barták. Electrochemical behavior and voltammetric determination of the herbicide metribuzin at mercury electrodes. *Fresen. J. Anal. Chem.* 370(7): 963–969, 2001.
100. MS Ibrahim, KM Al-Magboul, and MM Kamal. Voltammetric determination of the insecticide buprofezin in soil and water. *Anal. Chim. Acta* 432: 21–26, 2001.
101. H Chaaieri Oudou, Rosa M Alonso, and Rosa M Jiménez. Voltammetric study of the synthetic pyrethroid insecticides cypermethrin and deltamethrin and their determination in environmental samples. *Electroanalysis* 13: 72–77, 2001.
102. C Yardmc and N Özaltn. Electrochemical studies and differential pulse polarographic analysis of lansoprazole in pharmaceuticals. *Analyst* 126: 361–366, 2001.
103. J-M Zen, H-P Chen, and A Senthil Kumar. Disposable clay-coated screen-printed electrode for amitrole analysis. *Anal. Chim. Acta* 449: 95–102, 2001.
104. P Manisankar, C Vedhi, S Viswanathan, and HG Prabu. Investigation on the usage of clay modified electrode for the electrochemical determination of some pollutants. *J. Environ. Sci. Health Part B* 39: 89–100, 2004.
105. MF Barroso, OC Nunes, MC Vaz, and C Delerue-Matos. Square-wave voltametric method for determination of molinate concentration in a biological process using a hanging mercury drop electrode. *Anal. Bioanal. Chem.* 381(4): 879–83, 2005.
106. P Manisankar, S Viswanathan, AM Pusphalatha, and C Rani. Electrochemical studies and square wave stripping voltammetry of five common pesticides on poly 3,4-ethylenedioxythiophene modified wall-jet electrode. *Anal. Chim. Acta* 528: 157–163, 2005.
107. D De Souza, RA De Toledo, A Galli, GR Salazar-Banda, MR Silva, GS Garbellini, LH Mazo, LA Avaca, and SA Machado. Determination of triazine herbicides: Development of an electroanalytical method utilizing a solid amalgam electrode that minimizes toxic waste residues, and a comparative study between voltammetric and chromatographic techniques. *Anal. Bioanal. Chem.* 387(6): 2245–2253, 2007.
108. IC Lopes, D De Souza, SAS Machado, and AA Tanaka. Voltammetric detection of paraquat pesticide on a phthalocyanine-based pyrolitic graphite electrode. *Anal. Bioanal. Chem.* 388: 1907–1914, 2007.

9 Use of Planar Chromatography in Pesticide Residue Analysis

Tomasz Tuzimski

CONTENTS

9.1 INTRODUCTION: ADVANTAGES OF PLANAR CHROMATOGRAPHY

Pesticides are widespread throughout the world. The composition of pesticide mixtures occurring in environmental samples depends on geographical area, season of the year, number of farms, and quantity and intensity of use of plant-protection agents. The variety of their mixtures in different matrices, for example, rivers, is very large. Many sample-preparation techniques are used in pesticide residue analysis; the method selected depends on the complexity of the sample, the nature of the matrix and the analytes, and the analytical techniques available. Planar chromatography is an important analytical method, with other chromatographic techniques such as gas chromatography (GC), high-performance liquid chromatography (HPLC), and supercritical fluid chromatography (SFC). Thin-layer chromatography (TLC), although less sensitive and efficient than some other separation methods, has many advantages. Planar chromatography is most effective for the low-cost analysis of samples requiring minimal sample cleanup. Planar chromatography is especially suitable for field analysis at sites where the concentration of compounds (e.g., pesticides) might be high (e.g., study and liquidation of dumping grounds of toxic substances; in the chemical industry and the transport, storage, and distribution of pesticides). Planar chromatography is also selected for screening step analysis, because

- Single use of a stationary phase minimizes sample preparation requirements.
- Parallel separation of numerous samples enhances high throughput.
- Ease of postchromatographic derivatization improves method selectivity and specifity.
- Detection and/or quantitation steps can easily be repeated under different conditions.
- All chromatographic information is stored on the plate and can be (re-) evaluated if required.
- Several screening protocols for different analytes can be carried out simultaneously.
- Selective derivatizing reagents can be used for individual or group identification of the analytes.
- Detection of the separated spots with specific and sensitive color reagents.
- Visual detection of ultraviolet (UV)-absorbing compounds is possible in field analyses using a UV lamp.
- Detection by contact with x-ray film, digital bio- and autoradiography, and even quantitative assay using enzymes is possible.
- TLC plates can be documented by videoscans or photographs.
- Planar chromatography combined with modern videoscanning and densitometry enables quantitative analysis.
- Planar chromatography coupled with densitometry enables detection of the spots or zones through scanning of the chromatograms with UV-Vis light in the transmission, reflectance, or fluorescence mode.
- With multiwavelength scanning of the chromatograms, spectral data of the analytes can be directly acquired from the TLC plates and can further be compared with the spectra of the analytes from the software library.
- Additional information for structural elucidation can be obtained by planar chromatography combined with MS (fast atom bombardment [FAB] and liquid secondary ion mass spectrometry [SIMS]).
- The whole procedure of chromatographic development can be followed visually, so any distortion of the solvent front, and so on, can be observed directly.
- The chromatogram can be developed simply by dipping the plate into a mobile phase.
- Two-dimensional (2D) development with a single adsorbent is possible.
- 2D development on, for example, silica–octadecyl silica coupled layers (Multi-K SC5 and CS5 dual phase) is possible.
- Planar chromatography is also the easiest technique that performs multidimensional separation (e.g., by graft chromatography or multidimensional chromatography).

Summing up, planar chromatography is one of the principal separation techniques, which also plays a key role in pesticide residue analysis. In the first part of the chapter the reader will gain useful information to avoid some problems in performing planar chromatography experiments, and in the second part he or she will find useful information for application of planar chromatography for separation, detection, and qualitative and quantitative determination of pesticides belonging to different chemical classes of samples (water, soil, food, and other samples).

9.2 CHARACTERISTIC OF TLC METHOD: CHAMBERS, SAMPLE APPLICATION, AND CHROMATOGRAM DEVELOPMENT

The stages of planar chromatography procedure, such as sample application, chromatogram development, registration of chromatogram, and its evaluation, cannot be presently performed in one run using commercially available devices. It means that planar chromatography analysis cannot be completely automated in contemporary laboratory practice. Chromatographers have to separately optimize each of the mentioned stages using more or less sophisticated devices. There are various equipments for semi- or fully automatic operations at the mentioned stages of planar chromatography procedures. At each stage of the TLC procedure the chromatographer should possess basic skills that substantially help in accomplishing TLC experiments correctly, to obtain reliable, repeatable, and reproducible results. He/she might meet many pitfalls during work with the TLC mode. Some fundamental books help in overcoming these problems [1–5]. The present chapter gives some information that can draw the reader's attention to the procedures and equipments mentioned, which have been often applied and proven in contemporary planar chromatography practice.

9.2.1 MODERN CHAMBERS FOR TLC

Various chambers have been used for the development of thin-layer chromatograms. The classification of chromatographic chambers can be performed taking into account the following:

- Volume of vapor atmosphere inside the chamber—unsaturation or saturation with the vapor mobile-phase system.
- Direction of mobile-phase migration—linear development in which solvent migrates through a rectangle or square chromatographic plate from one of its edges to the opposite edge with constant width of the front of the mobile phase (Figure 9.1a) or radial development (including circular (Figure 9.1b) and anticircular (Figure 9.1c) types of radial developments); in the circular type the mobile phase is delivered at the center of the chromatographic plate and its front migrates toward the periphery of the adsorbent layer, meanwhile in the anticircular type the mobile phase migrates in the opposite direction and its front again is circular.
- Configuration of the chromatographic plate in the chamber (horizontal—then the chamber is named as horizontal chamber—or vertical—then the chamber is named as vertical one).
- Degree of automation of chromatogram development (including temperature and humidity control, eluent and vapor-phase delivery to the chromatographic chamber, and drying the chromatographic plate).

Regarding the volume of vapor atmosphere two main types of the chamber can be distinguished: normal (conventional) chambers (N-chambers) and sandwich chambers (S-chambers). The above classification is not unequivocal because a chamber can belong to more than one chamber type.

9.2.1.1 Conventional Chambers (N-Chambers)

The N-chambers are typically made of glass as a vessel possessing cuboid or cylindrical form. Their dimensions are about 230 × 230 × 80 mm for the respective development of 200 × 200 mm

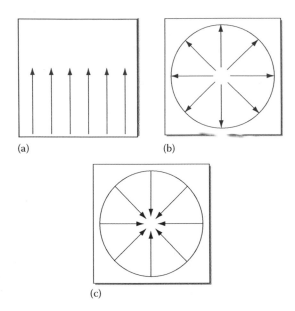

FIGURE 9.1   Modes of development in planar chromatography. (Adapted from Dzido, T.H. and Tuzimski, T., in *Thin Layer Chromatography in Phytochemistry*, Eds. Waksmundzka-Hajnos, M., Sherma, J., and Kowalska, T., Taylor & Francis, Boca Raton, FL, 2008.)

TLC plates or 130 × 130 × 50 mm for the respective development of 100 × 100 mm TLC plates. This type of chamber can be very easily applied for conditioning (with vapor phase by use saturation pads or an adequate size of filter paper [blotting paper]) in the chromatographic chamber, which is very important, especially when mixed mobile phase is used for chromatogram development. Then the repeatability of retention values is higher in comparison with development without vapor saturation.

Another type of N-chamber is the cuboid twin-trough chamber, which can be conveniently used for chromatography under different conditions of vapor saturation [1]. A schematic view of the chamber is presented in Figure 9.2. The bottom of the chamber is divided by a ridge into two parallel troughs. This construction of the chamber enables us to perform chromatogram development in three modes: without chamber saturation (Figure 9.2a), with chamber saturation (Figure 9.2b), and chamber saturation with one solvent followed by development with another one (Figure 9.2c) [5].

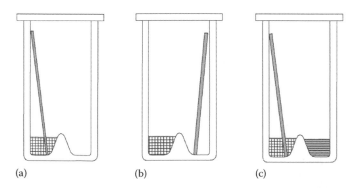

FIGURE 9.2   The twin-trough chamber with various variants of chromatogram development. (Adapted from Dzido, T.H. and Tuzimski, T., in *Thin Layer Chromatography in Phytochemistry*, Eds. Waksmundzka-Hajnos, M., Sherma, J., and Kowalska, T., Taylor & Francis, Boca Raton, FL, 2008.)

9.2.1.2 Horizontal Chambers for Linear Development

As was mentioned here the chromatographic plate is positioned horizontally in this chamber type. TLC horizontal chambers, which are most often applied in experiments, are produced by Camag (Muttenz, Switzerland), Desaga (Heidelberg, Germany), and Chromdes (Lublin, Poland). All have similar construction. The elements of the three types of horizontal chambers described below (horizontal developing chamber, horizontal DS [Dzido-Soczewiński] chamber, and H-separating chamber) are made of Teflon and glass, so they are very resistant to all solvents applied for chromatographic separations. The differences consist in the eluent delivery system. One example of a horizontal chamber (horizontal DS chamber manufactured by Chromdes, Lublin, Poland) is presented as a cross-section before and during chromatogram development in Figure 9.3a and b, respectively [8,9]. The main feature of the chamber is the formation of a vertical meniscus of the solvent (dark area) between the slanted bottom of the mobile-phase reservoir (2) and the glass strip (1). Chromatogram development is started by shifting the glass strip to the edge of the chromatographic plate (3) with the adsorbent layer face down, which brings the solvent in contact with the chromatographic plate. During development the meniscus of the solvent moves in the direction of the chromatographic plate, which makes the chamber very economical (the solvent can be exhausted from the reservoir almost completely). Conditioning of the chamber atmosphere can be performed by pouring some drops of solvent onto the bottom of the chamber (lined with blotting paper) (7) after removing glass plates (6). All kinds of plates (foil and glass backed, of dimension from 5 × 10 to 20 × 20 cm) can be developed in these chambers depending on the chamber type and size. The maximum distance of chromatogram development is equal to 20 cm. The consumption of solvents is very low, for example, 3–5 mL for 100 × 200 mm plates.

Another example of a horizontal chamber (horizontal developing chamber manufactured by Camag, Muttenz, Switzerland) is presented in Figure 9.4 as a cross-section [6,7]. The chromatographic plate (1) is positioned with the adsorbent layer face down and is fed with the solvent from the reservoir (3). Chromatogram development is started by tilting the glass strip (4) to the edge of the chromatographic plate. Then a planar capillary is formed between the glass strip and the wall of

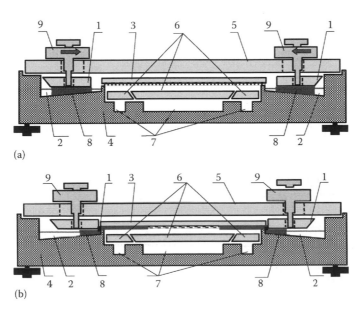

FIGURE 9.3 Horizontal DS-II chamber (Chromdes): (a) before development, (b) during development. 1, cover plate of the mobile-phase reservoir; 2, mobile-phase reservoir; 3, chromatographic plate with layer face down; 4, body of the chamber; 5, main cover plate; 6, cover plates (removable) of the troughs for vapor saturation; 7, troughs for saturation solvent; 8, mobile phase; 9, mobile-phase distributor/injector. (Courtesy of Chromdes, Lublin, Poland, www.chromdes.com)

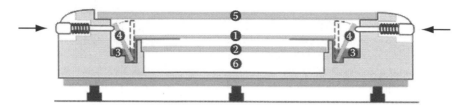

FIGURE 9.4 Horizontal developing chamber (Camag): 1, chromatographic plate with layer face down; 2, counter plate (removable); 3, troughs for solution of the mobile phase; 4, glass strip for transfer of the mobile phase by capillary action to the chromatographic plate; 5, cover glass plate. (Courtesy of Camag, Muttenz, Switzerland, www.camag.com)

the solvent reservoir in which the solvent instantaneously rises, feeding the chromatographic plate. The maximum distance of development with this chamber is 10 cm. The chambers are offered for 10×10 and 20×10 mm plates.

The two horizontal chambers types described above possess the following methodological possibilities [5]:

- Double number of samples in comparison with conventional chambers can be separated on one plate (due to two solvent reservoirs on both sides of the plate, which enable simultaneous development of two chromatograms from two opposite edges) [10].
- Saturation of the adsorbent layer with vapors of the mobile phase or another solvent [10,12].
- 2D separation of four samples on one plate simultaneously [11].
- Multiple development [13,14].
- Stepwise gradient elution [15,16].
- Zonal sample application for preparative separation (in horizontal DS chamber) [17,18].
- Continuous development (in SB/CD and horizontal DS chambers)* [18].
- Short bed-continuous development (in SB/CD and horizontal DS chambers)* [18,19].
- Development of six different chromatograms on one plate simultaneously (high-performance thin-layer chromatography [HPTLC] vario chamber from Camag or horizontal DS-M chamber from Chromdes) [18].

More detailed description of these methodical possibilities can be found by the reader in the following references [1,2,18,19].

Another horizontal chamber (H-separating chamber) for TLC is manufactured by Desaga [20]. Its principle of action is based on the Brenner–Niedervieser (BN) chamber [21]. The chromatographic plate is fed with solvent from the reservoir using a wick made of porous glass. The chambers are manufactured for 5×5 cm and 10×10 cm plates.

Other horizontal chambers such as vario-KS-chamber [22,23], SB/CD-chamber [24], sequence-TLC developing chamber [25], ES-chamber [19], and ES-chamber modified by Rumiński [26] and by Wang et al. [27,28] have been described in the literature and applied in some laboratories; however, these are not commercially offered at present.

9.2.1.3 Horizontal Chambers for Radial (Circular and Anticircular) Development

The mode of radial development of planar chromatograms is rarely applied in laboratory practice. Radial development using circular mode can be easily performed with a Petri dish [29]. The so-called U-chamber was used for circular and anticircular developments. This mode, including

---

* SB/CD chamber is not manufactured at present.

both types of development, can be carried out using commercial U-chambers (Camag) [30]. However, at present these chambers are not commercially offered by this firm probably due to low interest of the customers. In spite of some advantages regarding separation efficiency, the practitioners prefer to apply linear development rather than radial development. Two main reasons explain this status: at first highly sophisticated chamber construction and its maintenance to develop chromatograms and the second reason is about the shortage of equipment and software for chromatogram evaluation. Subsequently, methods have been developed by a number of researchers to control the mobile-phase movement [31].

9.2.1.4 Automatic Chambers

Repeated development in planar chromatography is based on the fact that a single development does not always result in satisfactory separation. Automatic developing chambers (ADCs) are especially suitable for routine analysis due to repeatable conditions provided by the instrumental control of the chromatographic process—so all chromatograms are repeatable and reproducible. The automatic chambers often applied in TLC experiments are produced by Desaga (Heidelberg, Germany) [32] and Lothar Baron Laborgeräte [33]. A new ADC 2 is demonstrated in Figure 9.5 [6]. The main advantages of this chamber according to the Camag manufacturer specification are [6] as follows:

- Fully automatic development of 10×10 cm and 20×10 cm chromatographic plates.
- Twin trough chamber is applied for chromatogram development.
- Manual methods previously applied with twin trough chamber can be conveniently adapted for automatic development with ADC 2 chamber.
- Development under conditions of controlled humidity.
- All operations necessary to run the separation process can be introduced from the keypad of the chamber.

FIGURE 9.5 Automatic developing chamber, ADC 2. (Courtesy of Camag, Muttenz, Switzerland, www. camag.com)

- All those operations can be programmed earlier and can be introduced from a computer using the manufacturer's software.
- Complete separation process proceeds without any influence of manual operations.
- Data relevant to separation procedures can be stored in the computer memory and can be kept and applied if required.
- Can be applied with the requirements of good laboratory practice (GLP) and good manufacturing practice (GMP).

Chromatogram development can be performed under conditions of temperature control using a device named TLC Thermo Box (Desaga [32] and Lothar Baron Laborgeräte [33]). The separation process can be carried out at temperatures in the range below 10°C to 20°C above room temperature with a precision equal to ±0.5°C.

A laboratory-made, temperature-controlled, removable, horizontal micro-TLC chamber unit is presented in Figure 9.6a and b [34]. The chamber unit (made of chromium-coated brass or Teflon) works inside a foam-insulated metal oven connected to an external liquid circulating thermostat Neslab RTE7 (Thermo Electron Corporation, Newington, NH, USA). The system provides a constant TLC plate temperature ranging from −20°C to +80°C with an accuracy of ±0.5°C.

A series of solvents are used as the mobile phase for the development of the TLC plate in a special device for an automated multiple development (AMD) of a chromatogram described by Perry et al. [35] and a programmable setup constructed by Burger [36] and produced by Camag. The apparatus for AMD of a chromatogram (AMD 2) is presented in Figure 9.7 [6]. Five different solvents (in five bottles) are used for preparation of eluent solutions, so gradient development can be accomplished with a similar number of the mobile-phase components. A full separation process comprising 20–25 steps takes a long time. However, this is compensated by simultaneous separation of many samples on one chromatographic plate and using the system outside working hours without inspection. Therefore, the final analysis is characterized by a relatively high throughput. This throughput can be increased by reduction of the number of steps of the AMD procedure. Application of special software for the simulation of the planar chromatography process can additionally enhance this procedure [37,38].

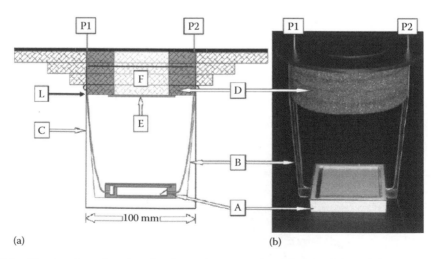

(a)   (b)

FIGURE 9.6 Construction of a microchamber unit support placed inside of the stainless steel submersible container working in a thermostated water bath. Section drawing (a) and perspective view (b) show a microchamber module. A, aluminium tape support for the chamber unit; B, stainless steel submersible container; C, foam-made container lid; D, glass window; E, removable insulation window; F, injection and saturation liquid pipes; P1 and P2, level of the external heating/cooling liquid (L). (From Zarzycki, P.K., *J. Chromatogr. A*, 1187, 250, 2008. With permission.)

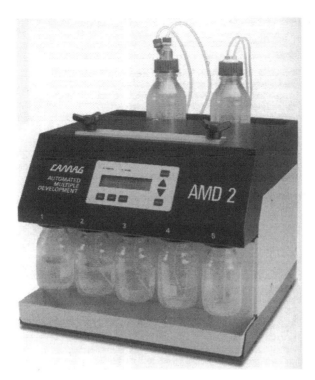

FIGURE 9.7 Device for AMD of chromatograms, AMD 2. (Courtesy of Camag, Muttenz, Switzerland, www.camag.com)

Advantages of this device are the following:

- Chromatographic plate (usually HPTLC plate) is developed repeatedly in the same direction.
- Each step of the chromatogram development follows complete evaporation of the mobile phase from the chromatographic plate and is performed over longer migration distance of the solvent front than the one before.
- Each step of the chromatogram development uses a solvent of lower elution strength than the one used in the preceding run; it means that a complete separation process proceeds under conditions of gradient elution.
- The focusing effect of the solute bands takes place during the separation process, which leads to very narrow component zones and high efficiency of the chromatographic system comparable to HPLC.

Poole and Belay [39] reviewed the essential methods and parameters of multiple development techniques in planar chromatography (including also AMD). Evaluation of parameters such as change in the zone width vs. number of developments, zone separation vs. number of developments through AMD, and several typical applications of AMD are described.

9.2.2 SAMPLE APPLICATION

Resolution of the chromatographic system is dependent on the size of the starting zone (spot) of the solute. The sample shape as streaks or bands is advantageous with regard to resolution and quantitative analysis. The sample can be applied to the stationary phase by spotting, dipping, spraying, or sampling through a syringe. Conventional application of the sample mixture on the chromatographic plate can be performed with a calibrated capillary or a microsyringe. More advantageous modes of sample application can be performed with a semiautomatic applicator or fully automated device. All these modes can be applied for analytical and preparative separations as well.

9.2.2.1 Sample Application in Analytical Thin-Layer Chromatography

The sample spotting can be performed by hand operation using a disposable micropipette, calibrated capillary, or microsyringe. It can also be performed by using a special device, for example, Nanomat (Camag) [6], in which the capillary is held by a dispenser. Application of the chromatographic plates with a preconcentration zone or an aerosol applicator can be done by, for example, that manufactured by Camag (Linomat 5 and Automatic TLC Sampler 4) [6] and Desaga (HPTLC-Applicator AS 30) [32]. In the former case the sample spot is focused in the preconcentration zone when the solvent migrates through it during the chromatographic process.

Automatic aerosol applicators have gained higher popularity in laboratory practice in spite of relatively high price due to some important features [5]:

- Starting sample spot is very small; typical diameter is about 1 mm.
- Dimension of the spot is not dependent on the solvent type of the sample solution.
- High repeatability of sample volume applied on the layer—very important for quantitative analysis.
- Various sample shapes can be obtained—dot (spot), streak, band, or rectangle.
- Various sample volumes can be applied.

9.2.2.2 Sample Application in Preparative Layer Chromatography

Adsorbent layers used for preparative separations are thicker than those for analytical separations. Sample application for preparative separations in planar chromatography usually requires spotting larger volumes of the sample solution on the plate—its solution is usually deposited on the almost whole width of the chromatographic plate in the shape of a band, streak, or rectangle. This procedure of sample application can be performed manually using a capillary or microsyringe, but this mode is tedious and needs many manual operations; the shape of the starting band is often not appropriate, leading to lower resolution of the zones on the final chromatogram. More experience is necessary when sample application is performed by moving the tip of pipette or syringe needle over a start line without touching the layer surface [40].

When the sample mixture is more complex, then the starting zone should be formed as a very narrow band, which leads to a higher resolution of bands on the chromatogram. Very good results can be obtained using automatic aerosol applicators (e.g., Linomat 5, Camag) as mentioned above. The starting sample zone can be formed in a desired shape.

A convenient mode of sample application in the shape of a narrow band (1 mm wide) to a plate up to 40 cm wide can be performed using a TLC sample streaker from Alltech. In this case the volume of sample solution depends on syringe capacity.

Especially large quantities of sample can be applied on the chromatographic plate as described in Ref. [41]. This mode was adapted by Nyiredy and Benkö [42] for extraction and separation of components from plant materials. The sample solution is mixed with the specified quantity of the adsorbent. The solvent is evaporated and the residue (bulky adsorbent with deposited sample on it) is introduced to the start line of the chromatographic plate.

A horizontal ES chamber [19,43] or horizontal DS chamber (Chromdes) [5,18] can be very easily used for band sample application. In the first stage of this procedure the adsorbent layer of the chromatographic plate is fed with a sample solution instead of the solvent (the mobile phase). When desired the sample volume is introduced and then the chromatographic plate is supplied with a solvent to proceed with the chromatographic process. This procedure possesses two advantages: no sophisticated equipment is necessary to perform the sample application for preparative separation and during sample application frontal chromatography is performed, which leads to preliminary separation of the components of the sample mixture.

9.2.3 Chromatogram Development

As mentioned above chromatogram development in TLC can be performed applying linear or radial modes. Both modes can be performed in a simple way using a conventional chamber and applying very complicated procedures, including sophisticated devices. Involved operations and procedures depend on various variables about properties of the sample, adsorbent layer, solvent, mode of detection, and evaluation of the chromatogram. Some aspects, especially about the mobile phase, are discussed in this chapter.

9.2.3.1 Mobile Phases Applied in TLC

Mobile phases used for TLC have to fulfill various requirements. They must not chemically affect and/or dissolve the stationary phase, because this leads to modification of properties of the chromatographic system. They must not produce chemical transformations of the separated compounds. The multicomponent mobile phase applied in TLC must be used only once, not repeatedly, because the volatility of solvents produces a continuous modification of quantitative composition of the mobile phase, which negatively affects the chromatographic repeatability. The mobile phase must be easily eliminated from the adsorbent layer and must be compatible with detection methods. Reproducibility can be greatly affected by the conditions and the time of preservation of the mobile-phase solution.

Chemical information about mobile-phase properties is essential to the initial selection of the chromatographic system and detection properties. Choice of the mobile phase (and also the stationary phase) is dependent on many factors about the properties of the compounds to be separated (Table 9.1) [44]. When the properties of the mobile phase and stationary phase of TLC systems are considerably different, then the separation selectivity is expected to be high. In general, if the stationary phase is polar the mobile phase should be apolar or slightly polar; such a system is named a normal-phase (NP) system. If stationary phase is nonpolar then the mobile phase should be polar, and such a chromatographic system is named as a reversed-phase (RP) one. The choice of the mobile phase is dependent not only on the properties of the adsorbent and its activity but also on the structure and the type of separated analytes. Various solvents can be used as the components of the mobile phase in planar chromatography. and their choice of the chromatographic process is based on eluotropic and isoelutropic series. The mobile phase applied in planar chromatography can be composed of one, two, or more solvents.

9.2.3.1.1 NP Planar Chromatography

The retention of solutes on inorganic polar adsorbents (silica, alumina) or moderately polar adsorbents (cyanopropyl, diol, or aminopropyl) originates in the interactions of the polar adsorption sites on the surface with polar functional groups of the compounds. This mode was previously called as adsorption or liquid–solid chromatography. Generally, the strength of molecular interactions of the stationary phase with polar molecules of analytes increases in the order: cyanopropyl < diol < aminopropyl « silica ≈ alumina stationary phases. Basic compounds are very strongly retained by silanol groups of silica gel, and acidic compounds show increased affinity to aminopropyl stationary phase. Aminopropyl and diol stationary phases show affinity to compounds with proton-acceptor and/or proton-donor functional groups (e.g., alcohols, esters, ethers, ketones). Other polar compounds are usually more strongly retained on cyanopropyl than aminopropyl chemically modified stationary phases. The alumina surface comprises hydroxyl groups, aluminium cations, and oxide anions and is more complex than silica gel. Alumina favors interactions with π electrons of solute molecules and often yields better separation selectivity than silica for analytes with different

TABLE 9.1
Hints for Stationary and Mobile-Phase Selection with Respect to the Type of Sample Components

Polarity of Compound			Sample Information
			1. Polarity of Compound
Compound	Stationary Phase	Mobile Phase	Comments

NP chromatography with nonaqueous mobile phases

Compound	Stationary Phase	Mobile Phase	Comments
Low (hydrophobic)	Polar adsorbents (silica or, less often, alumina)	Nonpolar mobile phase (nonaqueous)	It is difficult to separate the compounds of low polarity on silica gel due to their weak retention. Selection of solvents for the mobile phase is limited because most solvents demonstrate too high elution strength (compounds show very high values of R_F or migrate to the front of the mobile phase)
High (hydrophilic)			Compounds of high polarity are difficult to separate on silica gel because of strong retention. Selection of solvents for the mobile phase is limited because most solvents are of too lower elution strength for these solutes (compounds have very low values of R_F stay on the start line of chromatogram)

NP chromatography with aqueous mobile phases (pseudo-RP system)

Compound	Stationary Phase	Mobile Phase	Comments
Very high (very hydrophilic)	Polar adsorbents (silica or, less often, alumina)	Polar mobile phase (aqueous)	Compounds of very high polarity (very high hydrophilic, e.g., alkaloids which additionally show strong interactions with silanol groups of silica-based stationary phases) are difficult to separate on silica gel with nonpolar mobile phases because of strong retention (stay on the start line of chromatogram even when 100% methanol is applied as the mobile phase) and should be chromatographed with more polar eluents containing water

NP chromatography with nonaqueous mobile phases and RP chromatography with aqueous mobile phases

Compound	Stationary Phase	Mobile Phase	Comments
Intermediate polarity	Moderately polar-bonded phases, chemically bonded on silica support: cyanopropyl— $(CH_2)_3$-CN, diol —$(CH_2)_3$-O-CH_2-CHOH-CH_2-OH, or aminopropyl —$(CH_2)_3$-NH_2	Nonpolar mobile phase (nonaqueous) and polar mobile phase (aqueous)	Compounds of intermediate polarity are separated on polar chemically bonded stationary phases, because their molecules can interact with silanol groups of silica gel. They demonstrate good separation selectivity in both NP (nonaqueous) and RP (aqueous) systems. The moderately polar stationary phases are compatible with the water mobile phase within the whole concentration range Many solvents can be selected to prepare the mobile phase. The only limitation is concerned with the miscibility of the mobile-phase components. In addition, the migration velocity front of the mobile phase varies less with the solvent composition in comparison to typical RP systems

RP chromatography with aqueous mobile phases

Low (hydrophobic)	Nonpolar adsorbents (chemical modification is based on reactions of the silanol (≡ Si-OH) groups on the silica surface with organosilanes to obtain stationary phases of the type ≡ Si-R, where R is an aliphatic chain of the type —C_1, C_2, C_8, C_{18})	Polar mobile phase (aqueous)	Compounds of low polarity are difficult to separate in systems with nonpolar adsorbents because of very strong retention. Mobile-phase selection is limited because most solvents show too low elution strength for these separations
			Compounds of high polarity are difficult to separate with nonpolar adsorbents because of weak retention. Appropriate mobile-phase selection is restricted because most solvents show too strong elution strength for these solutes
High (hydrophilic)			

RP chromatography with nonaqueous mobile phases

Low (very hydrophobic)	Nonpolar adsorbents: The stationary phase of the type ≡ Si-R (R is the aliphatic chain of various length, e.g., — C_1, C_2, C_8, C_{18}) is formed after reaction of the silanol (≡ Si-OH) groups of the silica surface with organosilanes	Polar mobile phase (nonaqueous)	The separation of compounds of low polarity (very hydrophobic samples) is difficult with aqueous mobile phases on nonpolar adsorbents because of very strong retention (stay on the start line of chromatogram even when acetonitrile–water (99:1, v/v) mobile phase is applied. The solutes (e.g., lipids) can be chromatographed using the mobile phase composed of more polar (acetonitrile, methanol) and less polar (tetrahydrofuran, chloroform, methylene chloride, acetone, methyl-t-butyl ether) organic solvents or various mixtures of these solvents. The retention decreases with increasing concentration of the less-polar solvent in the mobile phase (multicomponent eluent may contain even hexane or heptane)

2. Molecular Mass (MW)

Compound	MW <1000	Organic soluble	Nonionic	NP systems with silica or with chemically bonded stationary phase (aminopropyl, cyanopropyl, diol)
				RP systems with chemically bonded stationary phase (aminopropyl, cyanopropyl, diol) and alkylsiloxane-bonded stationary phases (C_2, C_6, C_8, C_{18})
		Water soluble		NP systems on silica and on chemically bonded stationary phases (aminopropyl, cyanopropyl, diol)

TABLE 9.1 (continued)
Hints for Stationary and Mobile-Phase Selection with Respect to the Type of Sample Components

Compound	Polarity of — Stationary Phase	Polarity of — Mobile Phase	Sample Information — Comments
MW >1000		Ionic	RP systems on chemically bonded stationary phases (aminopropyl, cyanopropyl, diol) and alkylsiloxane-bonded stationary phases with ligands of the type C_2, C_6, C_8, C_{18} IPC in RP systems with chemically bonded stationary phases (aminopropyl, cyanopropyl, diol) and alkylsiloxane-bonded stationary phases with ligands of the type C_2, C_6, C_8, C_{18}
		Organic soluble	Precipitation chromatography
		Water soluble	Cellulose

3. pK_a Value of Compound

pK_a Value of Compound	Acid–Base Behavior	Stationary Phase	Mobile Phase	Comments
Low values of pK_a	Strong acid or weak base	Nonpolar adsorbents: alkylsiloxane-bonded stationary phases (with ligands of the type C_2, C_6, C_8, C_{18}, phenyl) or chemically bonded stationary phases of the type aminopropyl, cyanopropyl, diol	RP systems: buffered polar mobile phase with controlled pH	When an acidic or basic molecule undergoes ionization (i.e., is converted from an uncharged species into a charged one) it becomes much less hydrophobic (more hydrophilic). When pH value of the mobile phase is equal to pK_a of the compounds of interest, then values of concentration of its ionized and unionized forms are identical (i.e., the values of concentration of B and BH^+ or HA and A^- in the mobile phases are equal). It means that retention changes of these solutes in principle take place in the pH range from the value pK_a −1.5 to the value pK_a +1.5. The relationship between retention of the solute and mobile-phase pH in RP systems-is more complicated for compounds with two or more acid and/or basic groups
High values of pK_a	Strong base or weak acid			

Note: See Refs. [44,45].

number or spacing of double bonds. The stationary phase in an NP system is more polar than that in the mobile phase. The mobile phase in this chromatographic mode is usually a binary (or more component) mixture of organic solvents of different polarity, for example, ethanol + chloroform + heptane. In principle, the elution strength of solvents applied in NP systems increases according to their polarity, for example,: hexane $\approx$ heptane $\leq$ octane < methylene chloride < methyl-t-butyl ether < ethyl acetate < dioxane < acetonitrile $\approx$ tetrahydrofuran < 1-propanol $\approx$ 2-propanol < methanol.

The retention of compounds in NP systems generally increases in the order: alkanes < alkenes < aromatic hydrocarbons $\approx$ chloroalkanes < sulfides < ethers < ketones $\approx$ aldehydes $\approx$ esters < alcohols < amides « phenols, amines, and carboxylic acids.

The sample retention is enhanced when the polarity of the stationary phase increases and the polarity of the mobile phase decreases.

9.2.3.1.2 RP Planar Chromatography

Silica gel chemically modified with various ligands, for example, C_2, C_8, C_{18} alkyl chains, or amino-propyl, cyanopropyl, diol, is the most popular stationary phase in RP planar chromatography (RP TLC). The mobile phase used in RP TLC is more polar than the adsorbent and is usually composed of two (or more) solvents, for example, water + water-soluble organic solvent (methanol, acetonitrile, tetrahydrofuran, acetone). The organic solvent in the mobile-phase solution is often named as modifier. The sample retention increases when its polarity decreases and when the polarity of the mobile phase increases. In general, the polarity decrease (increase of elution strength) of solvents applied in RP TLC can be presented according to the order: methanol, acetonitrile, dioxane, tetrahydrofuran, 1- and 2-propanol.

Samples containing ionized or ionizable organic analytes are often separated in RP chromatography with buffers as the components of the mobile phase. The pH value of the buffer solution should be in the range of 2–8 due to lower stability of the stationary phases beyond this range. However, this requirement is often neglected because application of the chromatographic plate in a typical experiment is performed only once. The addition of a buffer to the mobile phase can be applied to suppress the ionization of acidic or basic solutes and to eliminate undesirable chromatographic behavior of ionic species.

Ionic analytes can also be chromatographed in RP systems with additives to the mobile phase. The example is ion-pair chromatography (IPC) performed in RP systems—ionogenic surface-active reagent (containing a strongly acidic or strongly basic group and a hydrophobic moiety in the molecule) is added to the mobile phase. The retention of solutes in IPC systems can be controlled by changing the type and/or concentration of the ion-pair reagent and of the organic solvent in the mobile phase. A very important parameter of the mobile phase of the IPC system is its pH, which should be adjusted to an appropriate value. Acidic substances can be separated with tetrabutylammonium or cetyltrimethyloammonium salts, whereas basic analytes can be separated by using C_6-C_8-alkanesulfonic acids or their salts in the mobile phases. The retention generally rises with the concentration increase of the ion-pair reagent in the mobile phase (higher concentration of this reagent in the mobile phase leads to enhancement of its uptake by the nonpolar stationary phase). However, it should be mentioned that a very high concentration of the ion-pair reagent in the mobile phase does not significantly affect the retention. Generally, the retention of ionogenic solutes also increases with an increase in the number and size of alkyl substituent in the molecule of ion-pair reagent.

9.2.3.2 Solvent Properties and Classification

The solvents used as components of the mobile phases in TLC should be of appropriate purity and of low viscosity, inexpensive, and compatible with the stationary phase and binder being used. The solvent of the sample mixture should be of the lowest elution strength possible (in the case of sample application on the chromatographic plate using a capillary or microsyringe). Some basic physicochemical parameters (viscosity, dipole moment, refractive index, dielectric constant, etc.) are used for the characterization of the solvent ability for molecular interactions, which are of great importance for chromatographic retention, selectivity, and performance. The physical constants mentioned above for some common solvents used in chromatography are collected in a few books and articles [1,2,4,45,46].

Solvent strength (eluent strength, elution strength) refers to the ability of the solvent or solvent mixture to elute the solutes from the stationary phase. This strength rises with an increase in solvent polarity in NP systems. A reversed order of elution strength takes place for RP systems. Solvent polarity is connected with molecular interactions of solute–solvent, including dispersion (London), dipole–dipole (Keesom), induction (Debye), and hydrogen-bonding interactions [47].

The first attempts of solvent classifications were performed for characterization of liquid phases applied in GC (Rohrschneider and McReynolds) [48,49]. Another solvent classification was by Hildebrand [50–52]. In this classification, the solubility parameter was derived based on values of cohesion energy of pure solvents.

Snyder's polarity scale has gained significance for solvent classification in liquid chromatography practice in which the parameter P' is used for characterization of solvent polarity [44]. This parameter was calculated based on the distribution constant, K, of test solutes (ethanol, dioxane, nitromethane) in gas–liquid (solvent) systems. Ethanol was chosen for characterization of the solvent with regard to its basic properties (proton-acceptor properties), dioxane to characterize its acidic properties (proton-donor properties), and nitromethane to describe dipolar properties of the solvent. The sum of the log K values of these three test compounds is equal to the parameter P' of the solvent. In addition, each value of log K of the test solutes was divided by parameter P'; then the relative values of three types of polar interaction were calculated for each solvent: x_d for dioxane (acidic), x_e for ethanol (basic), and x_n for nitromethane (dipolar). These x_i values were corrected for nonpolar (dispersive) interactions and were demonstrated in a three-component coordinate plot, on equilateral triangle, Figure 9.8. Snyder characterized more than 80 solvents and obtained 8 groups of solvents on the triangle [53,54]. The triangle was named as Snyder-Rohrschneider solvent selectivity triangle (SST). This classification of solvents is useful for selectivity optimization in liquid

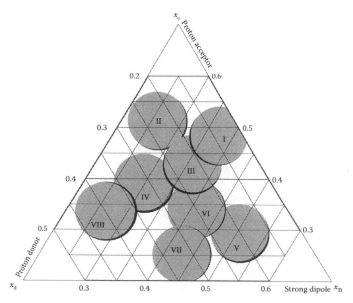

FIGURE 9.8 The solvent selectivity triangle (SST). (Adapted from Snyder, L.R., *J. Chromatogr. Sci.*, 16, 223, 1978.)

chromatography. Solvents belonging to the groups should demonstrate differentiated separation selectivity. Especially the solvents located close to various triangle corners should demonstrate the most different separation selectivity. Another advantageous feature of the SST is that the number of solvents applied in the optimization procedure can be reduced to the members representing each group from the SST. This approach was tested with success for normal [55] and RP systems [56]. However, for some chromatographic systems the prediction of selectivity changes failed [57].

Some empirical scales of solvent polarity based on kinetic or spectroscopic measurements have been described [58] to present their ability for molecular interactions.

There are several solvatochromic classifications of solvents, which are based on spectroscopic measurement of their different solvatochromic parameters [58–63]. The ET(30) scale [58] is based on the charge-transfer absorption of 2,6-diphenyl-4-(2,4,6-triphenyl-N-pyridino) phenolate molecule (know as Dimroth and Reichardt's betaine scale). The Z scale [59,60] is based on the charge-transfer absorption of N-ethyl-4-methocycarbonyl) pyridinium iodine molecule (developed by Kosower and Mohammad). The scale based on Kamlet–Taft solvatochromic parameters has gained growing popularity in the literature and laboratory practice [61–64]. The following parameters can be distinguished in this scale: dipolarity/polarizability (π^*), hydrogen-bond acidity (α) and basicity (β) (see Table 9.2). The solvatochromic parameters are average values for a number of selected solutes and somewhat independent of solute identity. Some representative values for solvatochromic parameters of common solvents used in TLC are summarized in Table 9.2.

These parameters were normalized in a similar way as x_d, x_e, x_n parameters of Snyder. The values of α, β, and π^* for each solvent were summed up and divided by the resulted sum. Then fractional parameters were obtained (fractional interaction coefficients): α/Σ (acidity), β/Σ (basicity), and π^*/Σ (dipolarity). These values were plotted on a triangle diagram similarly as in Snyder–Rohrschneider SST. In Figure 9.8 the SST based on normalized solvatochromic parameters is plotted for some common solvents applied in liquid chromatography [52].

More comprehensive representation of parameters characterizing solvent properties can be expressed based on Abraham's model in which the following equation is used [62–65]:

$$\log K_L = c + l \log L^{16} + rR_2 + s\pi_2^H + a\sum \alpha_2^H + b\sum \beta_2^H, \tag{9.1}$$

where
   $\log K_L$ is the gas–liquid distribution constant
   $\log L^{16}$ is the distribution constant for the solute between a gas and n-hexadecane at 298 K
   R_2 is the excess molar refraction (in $cm^3/10$)
   π_2^H is the ability of the solute to stabilize a neighboring dipole by virtue of its capacity for orientation and induction interactions
   $\sum \alpha_2^H$ is the effective hydrogen-bond acidity of the solute
   $\sum \beta_2^H$ is the hydrogen-bond basicity of the solute

All these parameters with the exception of $\log K_L$ are the solute descriptors. As can be seen, the parameters s, a, and b represent polar interactions of the solvent molecule with a solute as dipole–dipole, hydrogen-bond basicity, and hydrogen-bond acidity, respectively; the parameter r represents the ability of the solvent molecule to interact with n- or π-electrons of the solute molecule. In addition to previous classifications of solvents, this model takes into account molecular interactions with cavity formation in the solvent for solute molecule and dispersion interactions between solvent and solute. These effects are presented by constants c and l.

The values of the discussed parameters are given in Table 9.2. The chromatographer can compare these data and others in this table that can be helpful for optimization of retention and separation selectivity.

TABLE 9.2
Parameters Applied for Characterization of Solvents for Liquid Chromatography

Solvent	Selectivity Group	Snyder's Classification Based on Selectivity Triangle ($S_{S RP}$ is an Empirical Solvent Strength Parameter Used in RP System)					Kamlet–Taft and Coworkers' Classification						Abraham's Model Classification					
		Solvent Strength		Solvent Selectivity					Solvatochromic Parameters				System Constant for Distribution between Gas Phase and Solvent (Abraham's Model)					
		(P')	$(S_{S RP})$	x_e	x_d	x_n	$E_{T(30)}$	E_T^N	π_1^*	α_1	β_1	r	s	a	b	l	c
n-Butyl ether	I	2.1		0.44	0.18	0.38	33.0	0.071	0.27	0	0.46						
Diisopropyl ether		2.4		0.48	0.14	0.38	34.1	0.105	0.27	0	0.49						
Methyl tert.-butyl ether		2.7					34.7	0.124									
Diethyl ether		2.8		0.53	0.13	0.34	34.6	0.117	0.27	0	0.47						
n-Butanol	II	3.9		0.59	0.19	0.25	49.7	0.586	0.47	0.79	0.88						
2-Propanol		3.9		0.55	0.19	0.27	48.4	0.546	0.48	0.76	0.95						
1-Propanol		4.0		0.54	0.19	0.27	50.7	0.617	0.52	0.78							
Ethanol		4.3	3.6	0.52	0.19	0.29	51.9	0.654	0.54	0.83	0.77	−0.21	0.79	3.63	1.31	0.85	0.01
Methanol		5.1	3.0	0.48	0.22	0.31	55.4	0.762	0.60	0.93	0.62	−0.22	1.17	3.70	1.43	0.77	0
Tetrahydrofuran	III	4.0	4.4	0.38	0.20	0.42	37.4	0.207	0.58	0	0.55						
Pyridine		5.3		0.41	0.22	0.36	40.5	0.302	0.87	0	0.64						
Methoxyethanol		5.5		0.38	0.24	0.38											
Dimethylformamide		6.4		0.39	0.21	0.40	43.2	0.386	0.88	0	0.69						
Acetic acid	IV	6.0		0.39	0.31	0.30	51.7	0.648	0.64	1.12							
Formamide		9.6		0.38	0.33	0.30	55.8	0.775	0.97	0.71							

Dichloromethane	V	4.3		0.27	0.33	0.40	40.7	0.309	0.82	0.30	0						
1,1-Dichloroethane		4.5		0.30	0.21	0.49	41.3	0.327	0.81	0	0						
Ethyl acetate	VI	4.4		0.34	0.23	0.43	38.1	0.228	0.55	0	0.45						
Methyl ethyl ketone		4.7		0.35	0.22	0.43			0.67	0.06	0.48						
Dioxane		4.8	3.5	0.36	0.24	0.40	36	0.164	0.55	0	0.37						
Acetone		5.1	3.4	0.35	0.23	0.42	42.2	0.355	0.71	0.08	0.48						
Acetonitrile		5.8	3.1	0.31	0.27	0.42	45.6	0.460	0.75	0.19	0.31	−0.22	2.19	2.38	0.41	0.73	0
Toluene	VII	2.4		0.25	0.28	0.47	33.9	0.099	0.54	0	0.11	−0.22	0.94	0.47	0.10	1.01	0.12
Benzene		2.7		0.23	0.32	0.45	34.3	0.111	0.59	0	0.10	−0.31	1.05	0.47	0.17	1.02	0.11
Nitrobenzene		4.4		0.26	0.30	0.44	41.2	0.324	1.01	0	0.39						
Nitromethane		6.0		0.28	0.31	0.40	46.3	0.481									
Chloroform	VIII	4.3		0.31	0.35	0.34	39.1	0.259	0.58	0.44	0	−0.60	1.26	0.28	1.37	0.98	0.17
Dodecafluoroheptanol		8.8		0.33	0.40	0.27											
Water		10.2	0	0.37	0.37	0.25	63.01	1.000	1.09	1.17	0.18	0.82	2.74	3.90	4.80	−2.13	−1.27

Note: See Refs. [44,45,53–68,114–120].

One of the first attempts of the solvent systematization with regard to their elution properties was formulated by Trappe as the eluotropic series [69]. Pure solvents were ordered according to their chromatographic elution strength for various polar adsorbents in terms of the solvent strength parameter ε^0 defined according to Snyder [70,71] and expressed by Equation 9.2:

$$\varepsilon^0 = \Delta G_S^0 / 2.3RTA_S, \tag{9.2}$$

where

ΔG_S^0 is the adsorption free energy of solute molecules
R is the universal gas constant
T is the absolute temperature
A_S is the area occupied by the solvent molecule on the adsorbent surface

The parameter ε^0 represents adsorption energy of the solvent per unit area on the standard activity surface. Solvent strength is the sum of many types of intermolecular interactions.

Neher [72] proposed an equieluotropic series, which gives the possibility of replacing one solvent mixture by another one: composition scales (approximately logarithmic) for solvent pairs are subordinated to give constant elution strengths for vertical scales. Equieluotropic series of mixtures are approximately characterized by constant retention, but these can often show different selectivity. The scales, devised originally for planar chromatography on alumina layers, were later adapted to silica by Saunders [73] who determined accurate retention data by HPLC and subordinated the composition scale to Snyder's elution strength parameter [74].

Snyder [75] proposed the calculation of the elution strength of multicomponent mixtures. The solvent strength ε_{AB} of the binary solvent mobile phase is given by the relationship

$$\varepsilon_{AB} = \varepsilon_A^0 + \frac{\log\left(X_B 10^{\alpha n_b(\varepsilon_B^0 - \varepsilon_A^0)} + 1 - X_B\right)}{\alpha n_b}, \tag{9.3}$$

where

ε_A^0 and ε_B^0 are the solvent strength of two pure solvents A and B, respectively
X_B is the mole fraction of the stronger solvent B in the mixture
α is the adsorbent activity parameter
n_b is the adsorbent surface area occupied by a molecule of the solvent B

The equation for solvent strength for a ternary mixture was also derived [75]. These equations were tested for a series of mobile phases on alumina [75–77] and silica [78] demonstrating good agreement with experimental data especially for the last adsorbent. Some discrepancies were observed for alumina when different classes of solutes were investigated [57].

9.2.3.3 Optimization of the Mobile-Phase Composition

Identification and quantitation of analytes are the objective of each analysis. Reliable results of this analysis can be obtained with TLC mode when the resolution, R_S, of sample components is satisfactory, at least greater than 1.0. The resolution can be expressed according to the equation:

$$R_S = 0.25 \cdot \underbrace{\left(\frac{K_2}{K_1} - 1\right)}_{i} \cdot \underbrace{\sqrt{K_f N}}_{ii} \cdot \underbrace{\left(1 - \overline{R_F}\right)}_{iii}, \tag{9.4}$$

where
  N is the plate number of the chromatographic system
  K is the distribution constant of the solute 1 or 2

The distribution constant is related to the retention factor, k, according to the following equation:

$$K = k V_s / V_m,$$ (9.5)

where V_s/V_m is the ratio of the stationary and the mobile-phase volumes. Relationships between k and R_F (retardation factor) is as follows:

$$k = \frac{1 - R_F}{R_F}.$$ (9.6)

As seen, the resolution in TLC can be optimized by adjusting the three variables mentioned above: (1) selectivity, (2) performance, and (3) retention. If the distribution constant of two solutes is the same, then separation is impossible. The resolution increases when the plate number is higher. In planar chromatography, the performance of the chromatographic system is dependent on R_F—higher R_F leads to higher performance. On the other hand, retention increase (decrease of R_F) is responsible for increased resolution. It means that both variables, performance and retention, should be characterized by optimal value of R_F for which the resolution reaches the maximum value. This value is close to 0.3; compare Figure 9.9 where resolution is plotted vs. retardation factor. Typical mixtures

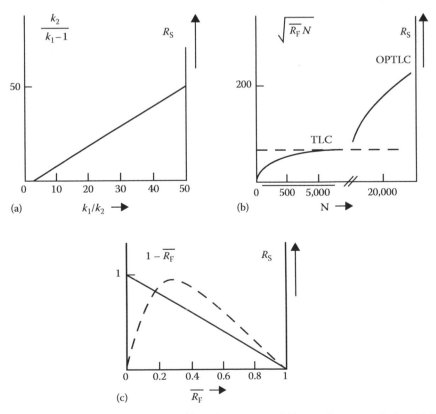

FIGURE 9.9 The influence of (a) selectivity, (b) performance, and (c) retention on resolution. (Adapted from Snyder, L.R., *Principles of Adsorption Chromatography*, Marcel Dekker, New York, 1968.)

are more complicated (multiple component), and it is not possible to separate all components with R_F values close to 0.3. The optimal R_F range of separated solutes in the chromatogram practically is 0.2–0.8 or thereabouts, and if of the correct selectivity, will distribute the sample components evenly throughout this R_F range [74].

Typical selection of the solvent is based on eluotropic series, for example, for most popular silica and less often used alumina and/or ε^0 parameter. A simple choice of the mobile phase is possible by microcircular technique on the basis of eluotropic series or for binary and/or more component mobile phases [79,80]. In the microcircular technique, after spotting the sample mixture in a few places on the chromatographic plate the selected solvents are applied in the center of each spot by means of a capillary. Then the sample bands migrate radially, and different chromatographic behavior of spotted mixture can be observed: nonsuitable solvent when spotted in a mixture gives a clench spot (too weak solvent strength) or periphery fringe (too strong solvent strength) and suitable solvent when the spotted mixture forms zones that are spread over the entire surface of circular development.

A single solvent rarely provides suitable separation selectivity and retention in chromatographic systems. A typical solution of the mobile phase is selected by adjusting an appropriate qualitative and quantitative composition of a two (binary) or more component mixture. The dependence of retention on the composition of the mobile phase can be predicted using a few popular approaches reported in the literature and used in laboratory practice.

The semiempirical model of adsorption chromatography (for NP systems) was independently created and published some time ago by Snyder [74] and Soczewiński [81]. This approach has been called the Snyder–Soczewiński model [82,83]. With some simplification, both authors' models lead to an identical equation describing the retention as a function of the concentration of the more polar modifier in binary mobile phase comprising less polar diluent (e.g., in NP system of the type: silica-polar modifier (ethyl acetate) + nonpolar diluent (n-heptane)).

$$R_M = \log k = \text{const} - m \log C_{\text{mod}}, \tag{9.7}$$

where

C_{mod} is the mole fraction (or volume fraction) of the polar component (modifier) in the mobile phase

m is the constant

k is the retention factor

$R_M = \log((1 - R_F)/R_F)$

The value of m is interpreted as the number of solvent molecules displaced by the solute molecule from the adsorbent surface (or the ratio of the area occupied by the solute molecule and by the modifier solvent).

The typical experimental relationships between R_M and eluent concentration expressed as logarithmic scale are straight lines and usually not parallel. The distance between lines and their slopes give information about variations of selectivity. The slope is dependent on the eluent strength and number of polar groups in the solute molecule. For some examples the lines cross (changes in spot sequence on chromatogram). Moreover, for some diluent-modifier pairs, the vertical distances, ΔR_M, between the lines are differentiated, showing individual selectivities of the systems relative to various pairs of solutes.

For the RP systems an analogous semilogarithmic equation was reported by Snyder [84] and is presented bellow:

$$R_M = \log k = \log k_w - S\varphi_{\text{mod}}, \tag{9.8}$$

where

$\log k_w$ is the retention factor of the solute for pure water as the mobile phase

φ_{mod} is the volume fraction of the modifier (e.g., methanol)

A similar equation was reported for partition systems of paper chromatography by Soczewiński and Wachtmeister [85]. For $\varphi_{mod} = 1$ (pure modifier), $S = \log k_w - \log k_{mod}$, $S = \log (k_w/k_{mod})$—the logarithm of hypothetical partition coefficient of solute between water and modifier (actually miscible) [82]. The constant S increases with decreasing polarity of the organic solvent and is a measure of its elution strength. On the other hand, S rises with an increase in size of the solute molecule. The above equation can be used for prediction of retention and selectivity for a reasonable concentration range. However, for a broad concentration range this equation does not predict solute retention with good precision. In cases of broad concentration range of the mobile phase, the following equation was reported [86]:

$$\log k = \log k_w + a\varphi + b\varphi^2, \tag{9.9}$$

where a and b are constants that are dependent on the solute and the mobile-phase type. Deviations from this equation occur especially beyond the concentration range $0.1 < \varphi < 0.9$, that is, for high and low concentrations of water. These deviations are explained by several reasons. Conformational changes in the alkyl chain structure of the stationary phase at a high water concentration in the mobile phase can influence this effect. When the concentration of water is low, then its participation in the hydrophobic mechanism of retention is eliminated, and additionally, molecular interactions of the solute and unreacted silanols can occur.

The significance of the relationships between retention and composition of the mobile phase for prediction of separation of sample components inspired many authors to investigate the problem of prediction of retention more deeply. One example is finding the dependence of $\log k$ vs. $E_{T(30)}$ solvatochromic parameter [87]. This relationship shows a very good linearity. Another approach is based on the methodology for linear salvation energy relationships. In this mode the solvatochromic parameters described above were applied for formulation of equations, which were used for retention prediction in various chromatographic systems. An important advantage of this mode is that sample descriptors were determined from other experiments as the chromatographic ones. The disadvantage of this approach is that system constants applied in the equation should be individually determined for each chromatographic system, including various qualitative and quantitative compositions of the mobile phase.

The dependence of retention on the composition of the mobile phase can also be described using different theoretical models:

- Martin–Synge model of partition chromatography [88,89]
- Scott–Kucera model of adsorption chromatography [90,91]
- Kowalska model of adsorption and partition chromatography [92–94]
- Ościk thermodynamic model [95,96]

It is purposeful to discuss in more detail about the modes of retention and selectivity optimization that can be applied to obtain appropriate chromatographic resolution. Various strategies were described in the scientific literature, for example, overlapping resolution (ORM) mapping scheme [56,96–98], window diagram method [99–101], computer-assisted method [102–107], and chemometric methods [108–113]. However, it seems that the strategy of separation optimization based on classification of solvents by Snyder (or solvatochromic parameters) and the PRISMA method described by Nyiredy [114–117] is the most suitable in laboratory practice for planar chromatography separations of sample mixtures. This opinion is expressed taking into account the simplicity of this procedure and low costs of operations involved (no sophisticated equipment and expensive software are necessary).

As mentioned above, solvents from each group of SST show different selectivity, which can lead to changes in the separation order. When the average solvent strengths and selectivity values are calculated for each solvent group of SST, then linear correlations of these quantities are found for solvent groups I, II, III, IV, and VIII and for solvent groups I, V, and VII [118]. Solvents of group VI do not belong to either correlations due to different ability for molecular interactions in comparison with solvents of the remaining groups. It was mentioned here that the solvents belonging to the groups in the corners of SST triangle (groups I, VII, VIII) and from its middle part (group VI) are the most often applied in NP systems of planar chromatography. Nyiredy et al. [115] suggested the selection and testing of 10 solvents with various strengths from 8 selectivity groups of SST (diethyl ether (I); 2-propanol, and ethanol (II); tetrahydrofuran (III); acetic acid (IV); dichloromethane (V); ethyl acetate, dioxane (VI); toluene (VII); chloroform (VIII)). All these solvents are miscible with hexane (or heptane) the solvent strength of which is about 0. Experiments were performed in unsaturated chambers.

For separation of nonpolar compounds the solvent strength of the mobile phase can be controlled by change of hexane (heptane) concentration. The separation of polar compounds can be varied (optimized) by adding a polar solvent to the mobile phase (e.g., low concentration of water). Thereby, the R_F values of the sample compounds should be brought within the range 0.2–0.8. The next step of the mobile-phase optimization system is to construct a tripartite PRISMA model, which is used for correlation of the solvent strength (S_T) and selectivity of the mobile phase (Figure 9.10).

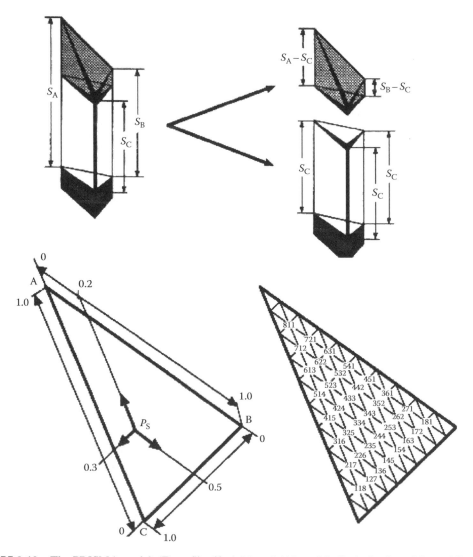

FIGURE 9.10 The PRISMA model. (From Siouffi, A.M. and Abbou, M., Optimization of the mobile phase, Chapter 3, in *Planar Chromatography, A Retrospective View for the Third Millennium*, Ed. Nyiredy, Sz., Springer Scientific Publisher, Budapest, 2001. With permission.)

The upper portion of the frustum serves to optimize polar compounds, the center part does so for nonpolar compounds, while the lower part symbolizes the modifiers. It enables to choose the number of the mobile-phase components in the range from two to five. The optimization process is detailed in the literature [114–117]. The PRISMA method represents a useful approach for the optimization of mobile phase, especially in cases of complex samples from plants containing a great number of unknown components [119]. The PRISMA model works well also for RP systems [120]. The procedure was used for selection of the mobile phases to separate synthetic red pigments in cosmetics and medicines [121], cyanobacterial hepatotoxins [122], drugs [123], and pesticides [124].

9.2.4 Classification of the Modes of Chromatogram Development

As mentioned above, chromatogram development can be performed applying linear or radial modes. In this section various methodological possibilities of these modes will be discussed with special attention to linear development.

9.2.4.1 Linear Development

9.2.4.1.1 Isocratic Linear Development
Isocratic linear development is the most popular mode of chromatogram development in analytical and preparative planar chromatography, and also in phytochemistry analysis. It can be easily performed in conventional chambers and horizontal chambers of all types. The mobile phase in the reservoir is brought in contact with the adsorbent layer, and then the movement of the eluent front takes place. The chromatogram development is stopped when the mobile-phase front reaches the desired position. In the isocratic mode of chromatogram development plates of different sizes are applied (usually 5×5 cm, 10×10 cm, and 10×20 cm and this makes the migration distance equal to about 4, 9, or 18 cm, respectively). The eluent can be supplied to the chromatographic plate simultaneously from its opposite edges (in horizontal developing chamber from Camag or horizontal DS chamber from Chromdes) so that the number of separated samples can be doubled in comparison with development in the vertical chamber or with development in the horizontal chamber when performed from one edge of the plate.

9.2.4.1.2 Continuous Isocratic Development
In the conventional mode of chromatogram development, the chromatographic plate is placed in the developing chamber. The development is finished when the eluent front reaches the end of the chromatographic plate or the desired position on the plate. However, the development can proceed further if some part of the plate extends out of the chamber, allowing the mobile phase to evaporate and ensuring that solvent migration is continuous and development is performed over the entire length of the plate, with evaporation proceeding with constant efficiency. To enhance the efficiency of evaporation a blower or heating block can be applied to the exposed part of the chromatographic plate. To ensure continuous development the mobile phase can be evaporated at the end of the glass cover plate by use of nitrogen stream also [125]. In Figure 9.11a the cross-section of the DS chamber is presented during continuous development (also compare Figure 9.3). Under these conditions the planar chromatogram development is more similar to the column chromatography mode than to the conventional development. In case of incomplete separation of the components of lower R_F values, some increase in separation can be obtained when applying this mode. In Figure 9.11b and c this procedure is schematically demonstrated. As presented, the chromatogram development has proceeded to the end (front of the mobile phase reached the end of the chromatographic plate) and mixture components of higher R_F value are well separated as opposed to these of lower values (Figure 9.11b). In this situation the continuous development should be performed. The end part of the chromatographic plate, which comprises the bands of good resolution, needs to be exposed as

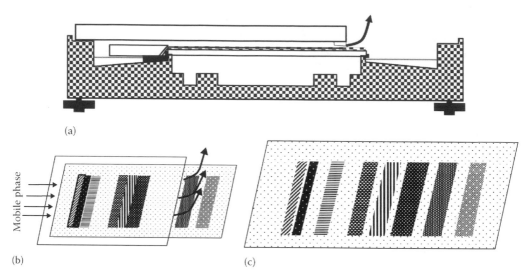

FIGURE 9.11 Schematic demonstration of horizontal DS chamber applied for continuous development: (a) cross-section of DS chamber during continuous development, (b) part of the plate with bands of lower retention exposed but with bands of higher retention covered to enable further development, (c) final chromatogram. (Adapted from Dzido, T.H. and Tuzimski, T., in *Thin Layer Chromatography in Phytochemistry*, Eds. Waksmundzka-Hajnos, M., Sherma, J., and Kowalska, T., Taylor & Francis, Boca Raton, FL, 2008.)

demonstrated in Figure 9.11b. The components of lower R_F values can migrate through a longer distance, which usually leads to improvement of their separation (Figure 9.11c). If necessary a larger part of the chromatographic plate can be exposed in the next stage of continuous development to obtain an improvement of separation of components of even higher retention than those located on the exposed part of the chromatographic plate.

9.2.4.1.3 Short Bed-Continuous Development

The migration distance varies with time according to the equation

$$Z_t = \kappa t^{1/2}, \tag{9.10}$$

where Z_t, κ, and t are the distance of the solvent front traveled, constant, and migration time, respectively. The development of planar chromatograms on long distance (e.g., 18 cm) usually takes a lot of time. The development of planar chromatograms is more and more time consuming with gradual decrease of mobile-phase velocity, which takes place in the planar chromatography process. Therefore, initially high flow of the mobile phase was used to accelerate the chromatographic analysis in the SB/CD. In the SB/CD this path is very short, typically equal to several centimeters [19,126–128]. The eluent strength should then be much weaker than that in the conventional development, because several void volumes of eluent migrate through the layer. This is the reason why this mode is preferentially applied for analytical separations. The development of a chromatogram on a short distance with simultaneous evaporation of the mobile phase from the exposed part of the chromatographic plate can be very conveniently performed by means of horizontal chambers. The SB/CD mode was introduced by Perry [126] and further popularized by Soczewiński et al. [19,127] using a horizontal equilibrium sandwich chamber.

The principle of the SB/CD technique is demonstrated in Figure 9.12. Instead of chromatogram development over a distance of 18 cm (Figure 9.12a), continuous elution over a short distance, for example, 5 cm, with simultaneous evaporation of the mobile phase from the exposed part of the

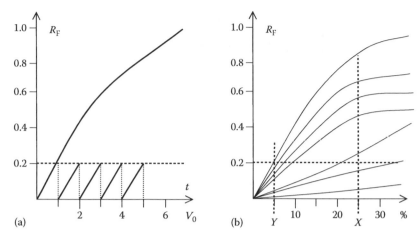

FIGURE 9.12 (a) Principle of SBCD, elution with five interstitial volumes on 4 cm distance (5 × 4 cm) is faster than single development on 20 cm distance, (b) R_F values of sample components plotted as a function of modifier concentration. Optimal concentration (Y) for SBCD (5 × 4 cm) is lower than that for development on full distance of 20 cm (X). (Adapted from Soczewiński, E., *Chromatographic Methods Planar Chromatography*, Vol. 1, Eds. Kaiser, R.E., and Dr. Alfred Huetig, Verlag, Heidelberg, Basel, New York, 1986, pp. 79–117.)

chromatographic plate (Figure 9.12b) can be performed. Several void volumes pass throughout the short chromatographic plate bed. However, the flow rate of the eluent depends on the efficiency of solvent evaporation, and the flow rate of the mobile phase can be higher if a more volatile solvent is used. It is often necessary to increase the efficiency of evaporation of the mobile phase from the end of the short plate by the application of a heater and/or blower.

In the SB/CD mode, a better resolution relative to the conventional development can be obtained for a similar migration distance of solutes. It is well known that the best resolution of the mixture components can be obtained in conventional development if average R_F value is equal to 0.3. However, in the continuous development the applied mobile phase is of lower eluent strength, for example, eluent strength that enables to reach the average value $R_F = 0.05$. Under such conditions, several void volumes of the mobile phase should pass through the chromatographic system. If the average migration distance of the component mixture is similar to that of the conventional development, then the resolution obtained with continuous development is better. This effect is explained by higher selectivity of the chromatographic system with a mobile phase of lower eluent strength and by better kinetic properties of the chromatographic system. At lower eluent strength the molecules of the components spend more time in the stationary phase, and the flow rate of the mobile phase is higher (closer to optimal value) under the condition determined by the efficiency of solvent evaporation from the exposed part of the plate.

The SB/CD is especially used in a marked increase of detection sensitivity of solutes, for example, to the analysis of trace polyaromatic hydrocarbons in river water samples. The SB/CD technique can be used to preconcentrate the sample solution directly on the thin layer. The results of experiments are similar to these when precoated plates with a narrow weakly adsorbing zone are used. In the first step, the dilute samples are spotted along the layer in a series 2–3 cm long. In the second step, the solutes are then eluted with a volatile solvent under a narrow cover plate forming sharp starting zones, and if necessary, evaporation of the eluent can be accelerated by a stream of nitrogen. Next, the cover plate is removed to completely evaporate the solvent. After drying the chamber is covered and the plate with the concentrated starting zones is developed with a suitable eluent. The resolution obtained by the SB/CD mode is better than that by the continuous mode, and the development time is also shorter. Additionally, the spot diameter is very small, which leads to a better detection level.

9.2.4.1.4　*Two-Dimensional Separations*

One of the most attractive features of planar chromatography is the ability to operate in the 2D mode. Two-dimensional TLC (2D-TLC) is performed by spotting the sample in the corner of a square chromatographic plate and by development in the first direction with the first eluent. After the development is completed the chromatographic plate is then removed from the developing chamber, and the solvent is allowed to evaporate from the layer. The plate is rotated through 90° and then developed with the second solvent in the second direction, which is perpendicular to the direction of the first development. In 2D-TLC the layer is usually of continuous composition, but two different mobile phases must be applied to obtain a better separation of the components. If these two solvent systems are of approximately the same strength but of optimally different selectivity, then the spots will be distributed over the entire plate area and in the ideal case the spot capacity of the 2D system will be the product of the spot capacity of the two constituent 1D systems. If the two constituent solvent systems are of the same selectivity but of different strengths, the spots will lie along a straight line; if both strength and selectivity are identical, the spots will lie along the diagonal.

Computer-aided techniques enable identification and selection of the optimum mobile phases for separation of different groups of compounds. The first report on this approach was by Guiochon and coworkers, who evaluated 10 solvents of fixed composition in 2D separation of 19 dinitrophenyl amino acids chromatographed on polyamide layers [129]. The authors introduced two equations for calculation of the separation quality—the sum of the squared distances between all the spots, D_A, and the inverse of the sum of the squared distances between all the spots, D_B. Streinbrunner et al. [130] proposed other functions for identification of the most appropriate mobile phases—the distance function DF and the inverse distance function IDF, which are the same form as D_A and D_B, respectively, but which use distances rather than the squares of distances. The planar response function PRF has been used as optimization criterion by Nurok et al. [131]. Strategies for optimizing the mobile phase in planar chromatography (including 2D separation) [132] and overpressured layer chromatography (OPLC) (including 2D OPLC) [133] have also been described. Another powerful tool is the use of graphical correlation plots of retention data for two chromatographic systems that differ with regard to modifiers and/or adsorbents [134].

The largest differences were obtained by combination of NP systems and RP system with the same chromatographic layer, for example, cyanopropyl [135,136]. Nyiredy [2,137] described the technique of joining two different adsorbent layers to form a single plate. In addition, the largest differences were obtained by combination of NP systems of the type silica/nonaqueous eluent and RP systems of the type octadecyl silica/water + organic modifier (methanol, acetonitrile, dioxane) on multiphase plates with a narrow zone of SiO_2 and a wide zone of RP-18 (or vice versa), which were commercially available from Whatman (Multi-K SC5 or CS5 plates) [138–141].

In 2D development the mixtures can be simultaneously spotted at each corner of the chromatographic plate so that the number of separated samples can be higher in comparison with the "classical 2D development" [9]. An example of this type of 2D development is illustrated in Figure 9.13a through d. Figure 9.13d, which shows a videoscan of the plate that shows separation of three fractions of the mixture of nine pesticides by 2D planar chromatography with NP/RP systems on a chemically bonded cyanopropyl stationary phase.

The multidimensional separation can be performed using different mobile phases in systems with single-layer or bilayer plates. Graft TLC is a multiple system in which chromatographic plates with similar or different stationary phases are used. Compounds from the first chromatographic plate after chromatogram development can be transferred to the second plate, without scraping, extraction, or respotting the bands by use of a strong mobile phase [2]. Graft TLC, a novel multiplate system with layers of the same or different adsorbents for isolation of the components of natural and synthetic mixtures on preparative scale, was first described by Pandey et al. [142]. Separation of alkaloids by graft TLC on different, connected, adsorbent layers (diol and octadecyl silica) has also been reported [143]. Graft TLC separation (2D planar chromatography on connected layers) of mixture of phenolic acids [144], saponins [145], and three mixtures of pesticides was also described [146]. An example of this technique is demonstrated in Figure 9.14 [147].

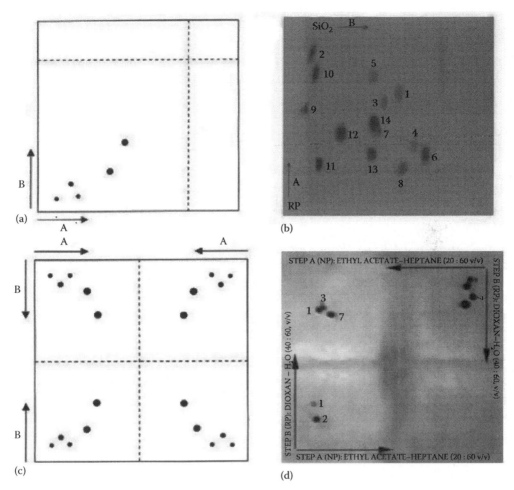

FIGURE 9.13 Two-dimensional development, (a) schematic presentation of 2D-chromatogram. (Adapted from Dzido, T.H., *Planar Chromatography, A Retrospective View for the Third Millennium*, Ed. Nyiredy, Sz., Springer Scientific Publisher, Budapest, 2001.) (b) 2D-chromatogram of the 14-component mixture of pesticides presented as videoscan of dual-phase Multi-K CS5 plate in systems. A (first direction): methanol–water (60:40, v/v) on octadecyl silica adsorbent, B (second direction): tetrahydrofuran–*n*-heptane (20:80, v/v) on silica gel. (From Tuzimski, T. and Soczewiński, E., *J. Chromatogr. A*, 961, 277, 2002. With permission.) (c) Schematic presentation of 2D-chromatogram of four samples simultaneously separated on the plate. (Adapted from Dzido, T.H., Chapter 4 in *Planar Chromatography, A Retrospective View for the Third Millennium*, Ed. Nyiredy, Sz., Springer Scientific Publisher, Budapest, 2001.) (d) 2D-chromatograms of three fractions of the mixture of nine pesticides presented as videoscan of the HPTLC plate (cyanopropyl) in systems with A (first direction): ethyl acetate–*n*-heptane (20:80, v/v), B (second direction): dioxane–water (40:60, v/v). (From Tuzimski, T. and Soczewiński, E., *Chromatographia*, 59, 121, 2004. With permission.)

Horizontal chambers can be easily used for 2D separations. The only problem seems to be the sample size. In a conventional 2D separation used for analytical purposes, the sample size is small. The quantity of the sample can be considerably increased when using a spray-on technique with an automatic applicator. Soczewiński and Wawrzynowicz have proposed a simple mode to enhance the size of the sample mixture with the ES horizontal chamber [43].

9.2.4.1.5 Multiple Development

Multiple development is the mode in which the direction of development is identical for each development step but the development distance and mobile-phase composition can be varied in each step.

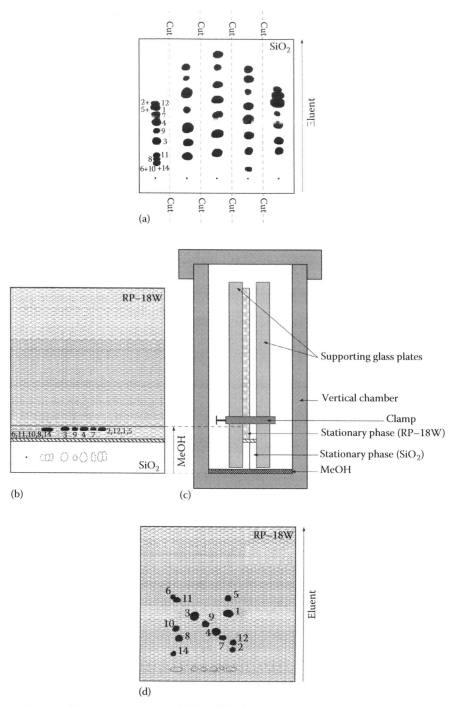

FIGURE 9.14 Transfer of the mixture of pesticides from the first plate to the second one. (a) First development with partly separated mixtures of pesticides on silica plate. After development the silica plate was dried and cut along the dashed lines into 2 × 10 cm strips. (b) A narrow strip (2 × 10 cm) was connected (2 mm overlap—hatched area) to 10 × 10 cm HPTLC RP-18W plate along the longer (10 cm) side of the strip. The partly separated mixture of pesticides was transferred in a vertical chamber to the second plate using methanol as a strong eluent to a distance about of 1 cm. (c) Schematic diagram of cross-section of the two connected adsorbent layers. (d) The HPTLC RP-18W plate was developed in the second dimension with organic water eluent in the DS chamber. (From Tuzimski, T., *J. Planar Chromatogr.*, 20, 13, 2007. With permission.)

The chromatogram is developed several times on the same plate and each step of the development follows the complete evaporation of the mobile phase from the chromatographic plate of the previous development. On the basis of the development distance and the composition of the mobile phase used for consecutive development steps, multiple development techniques are classified into four categories [137]:

- Unidimensional multiple development (UMD), in which each step of chromatogram development is performed with the same mobile phase and the same migration distance of eluent front.
- Incremental multiple development (IMD), in which the same mobile phase but an increasing development distance in each subsequent step is applied.
- Gradient multiple development (GMD), in which the same development distance but a different composition of the mobile phase in each step is applied.
- Bivariant multiple development (BMD), in which the composition and development distance is varied in each step of the chromatogram development.

These modes of chromatogram development are mainly applied for analytical separations due to the good efficiency, which is comparable to HPLC.

The sophisticated device used in this mode is manufactured by Camag and is known as AMD or AMD 2 system. AMD mode enables both isocratic and GMD. In a typical isocratic AMD mode, the development distance is increased during consecutive development steps, whereas the mobile-phase strength is constant. In the initial stage of the AMD gradient procedure, the solvent of the highest strength is used (e.g., methanol, acetonitrile, or acetone); in the next stages—an intermediate or base solvent of medium strength (e.g., chlorinated hydrocarbons, ethers, esters, or ketones); and in the final stage—a nonpolar solvent (e.g., heptane, hexane) [148].

Several parameters must be considered to obtain the best separation in AMD mode: choice of solvents, gradient profile of solvents, and number of steps. All modes of multiple development can be easily performed using chambers for automatic development, which are manufactured by some firms. However, these devices are relatively expensive. Typical horizontal chambers for planar chromatography should be considered for application in multiple development in spite of more manual operations in comparison with the automatic chromatogram development. Especially horizontal DS chamber could be considered for separations with multiple development. This chamber can be easily operated due to its convenient maintenance, including cleaning the eluent reservoir. For the separation of a more complicated sample mixture, a computer simulation could be used to enhance the efficiency of the optimization procedure [37,38,149–152].

9.2.4.1.6 Gradient (Stepwise and Continuous) Development

The separation efficiency is much better than that in isocratic development due to the elimination of the "general elution problem" (especially when investigated sample mixtures comprise components of various polarity with a wide range of k values) and the presence of the compressing effect of the gradient and enhanced mutual displacement of the solutes especially effective for moderate k values [153].

A typical isocratic planar chromatogram of the mixtures containing compounds of various polarity is composed of bands of medium retention (R_F values in the range from 0.1 to 0.8), of lower retention ($0.8 < R_F < 1$), and of higher retention ($0.0 < R_F < 0.1$). For such a mixture the bands of lower and higher retention are not well separated and are located close to the mobile-phase front and start lines, respectively. All the components can be separated only if a suitable continuous or stepwise mobile-phase gradient is chosen for chromatogram development.

Gradient elution, both continuous and stepwise, can be performed in a sandwich chamber with a glass distributor (horizontal DS and ES chambers). The principle of the mode is then based on the introduction of mobile-phase fractions of increasing strength following one after another in a series

The device for continuous gradient elution in the horizontal chamber described by Nyiredy [41] and presented above (Figure 9.15) seems to be a very interesting solution both to analytical and preparative applications.

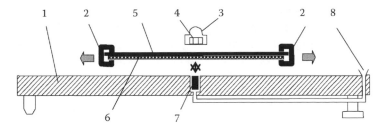

FIGURE 9.16 "Sequence TLC" developing chamber (Scilab). 1, support with solvent source (reservoir); 2, holding frame; 3, magnet holder; 4, magnet; 5, cover plate; 6, TLC plate; 7, wick with iron core; 8, solvent entry. (Adapted from Bunčak, P., GIT, *Suppl., Chromatographie*, 3, 1982.)

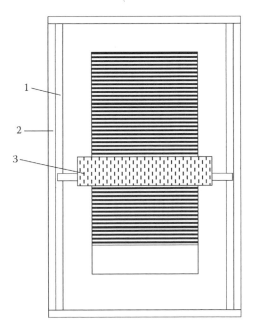

FIGURE 9.17 Top view of ES chamber modified by Wang et al. 1, supporting plate; 2, spacing plate; 3, distributor. (Adapted from Su, P., Wang, D., and Lan, M., *J. Planar Chromatogr.*, 14, 203, 2001.)

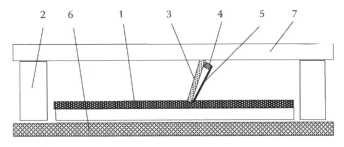

FIGURE 9.18 Cross-section of ES chamber with funnel distributor (modified by Wang et al.), 1, spacing plate; 2, base plate; 3, distributor; 4, glue; 5, slide; 6, thin-layer plate; 7, cover plate. (Adapted from Lan, M. et al., *J. Planar Chromatogr.*, 16, 402, 2003.)

9.2.4.1.8 Temperature Control

Temperature control in planar chromatography is rare. Most planar chromatography analyses are usually performed at room temperature in nonthermostated developing chambers. The optimum chromatographic separation is a compromise between maximum resolution and minimum analysis time. In classical planar chromatography the total analysis time is the same for all solutes and the solutes' mobility is driven by nonforced flow of the mobile phase—capillary action. The efficiency and selectivity of a chromatographic process and the precision and reproducibility of analysis are temperature dependent. The running time is strongly affected by the developing distance, the degree of saturation of vapor of the mobile phase, and the viscosity and particle size of the stationary phase. The mobile-phase viscosity depends on the mobile-phase composition and is decreased with temperature increase. The last effect leads to an increase in the mobile-phase flow rate and eventually to a shortening of chromatogram development. The relationship between the retention parameter of solutes (R_M) and the reciprocal of absolute temperature ($1/T$) is often linear (van't Hoff plot). Zarzycki has described some technical problems associated with temperature-controlled planar chromatography [159]. The author has also described a construction of a simple developing device designed for temperature control of TLC plates [160].

Dzido also described an adaptation of the horizontal DS chamber to planar chromatography with temperature control [161]. The author has also observed the change in development time at RP TLC systems at a temperature of 58°C in comparison with that at 15°C [161]. This chamber enables precise temperature control of the chromatographic system, because the chromatographic plate is located between two heating coils connected to a circulating thermostat.

The influence of temperature and mobile-phase composition on the retention of different cyclodextrins and two macrocyclic antibiotics has been examined by RP TLC using wide-range (0%–100%) binary mixtures of methanol–water [162]. Using a thermostated chamber for planar chromatography, the interactions between cyclodextrins and n-alcohols were investigated [163]. The influence of temperature on retention and separation of cholesterol and bile acids in RP TLC systems was also reported [164].

9.2.4.2 Radial Development

Radial development of planar chromatograms can be performed circularly and anticircularly. Capillary action is the driven force for the mobile-phase movement in these modes. Otherwise, in rotation planar chromatography, which is another mode of radial development, the centrifugal force is responsible for the mobile-phase movement. This mode was described by Hopf [165] for the first time. Different modifications of this technique have been also reported [166–174].

9.2.4.2.1 Circular Development

In the circular mode of chromatogram development, the samples are applied in a circle close to the center of the plate, and the eluent enters the plate at the center. The mobile phase is moved through the stationary phase from the center to the periphery of the chromatographic plate, and the sample components form zones like rings. In the first report of circular development by Izmailov and Schraiber, a chamber was not used [175]. Circular or anticircular mode of chromatogram development in a closed system was first carried out in a Petri dish containing eluent and a wick that touches the layer, supported on top of the dish, at its central point. An example of such a chromatogram is presented in Figure 9.19 [176]. The chromatogram was obtained with the circular U-chamber from Camag (Figure 9.20), which can be used for preparative and analytical separations.

The chamber for circular development described by Botz et al. [177] and modified by Nyiredy [178] is especially suitable for preparative planar chromatogram development, in which various sample mixtures (solid or liquid) can be applied on the chromatographic plate.

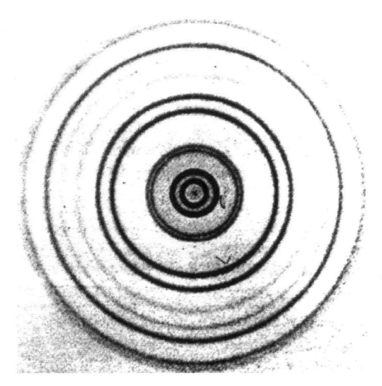

FIGURE 9.19 Circular chromatography of dyes on precoated silica gel high-performance TLC plate; lipophilic dyes, mobile-phase: hexane-chloroform-NH$_3$, 70:30. (From Ripphahn, J. and Halpaap, H., *HPTLC High Performance Thin-Layer Chromatography*, Eds. Zlatkis, A. and Kaiser, R.E., Elsevier, Institute of Chromatography, Amsterdam, Bad Dürkheim, Germany, 1977, pp. 189–221. With permission.)

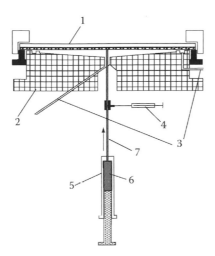

FIGURE 9.20 Cross-section view of the U-chamber (Camag, Muttenz). 1, chromatographic plate; 2 body of the chamber; 3, inlet or outlet for parallel or counter gas flow, to remove vaporized mobile phase, to dry or moisten (impregnate) the plate; 4, syringe for sample injection; 5, dosage syringe to maintain the flow of the mobile phase; 6, eluent; 7, capillary. (Adapted from Kaiser, R.E., *HPTLC High Performance Thin-Layer Chromatography*, Eds. Zlatkis, A. and Kaiser, R.E., Elsevier, Institute of Chromatography, Amsterdam, Bad Dürkheim, Germany, 1977, pp. 73–84.)

The obtained separation quality, using the circular mode, was considerably higher than that when a linear ascending development was performed in a twin-trough chamber. Even linear development using plates with a preconcentration zone produced lower separation quality in comparison with that in circular development [177]. The authors advise that separation using circular development with the chamber described has advantages relative to the separation efficiency obtained in linear development.

However, the advantages are referred to chromatogram development in a vertical N-chamber (twin-trough chamber). The separation quality was not compared with that using the horizontal mode of linear development. The authors have reported one disadvantage of the mode: recovering of bands of interest can be performed only by scraping the adsorbent from the plate. This is more complicated than in the case of linear chromatogram development using rectangular plates.

In spite of the advantages mentioned above the circular development is rarely used in contemporary practice of planar chromatography.

9.2.4.2.2 Anticircular Development

In anticircular mode of chromatogram development, the sample mixture is spotted at the circumference of the plate, and the mobile phase is moved from the circumference to the center of the plate. The sample application capacity is larger than that in the circular mode because of the long start line. In this mode of chromatogram development, especially good separation is observed for high R_F values. Anticircular development is very rarely applied in planar chromatographic practice for analytical separation.

This mode of separation was introduced by Kaiser [179]. Studer and Traitler adapted an anticircular U-chamber from Camag for preparative separations on 20×20 cm plates [180]. However, the mode has not gained much popularity in laboratory practice probably because a more sophisticated equipment is necessary to perform the separation.

Issaq [181] has proposed the application of conventional chambers to perform anticircular development. In this mode a commercially available chromatographic plate is divided into triangular plates, and the sample (or samples) is spotted along the base of the triangular plate. Wetting of the mobile phase is started when the base of the triangular chromatographic plate contacts the solvent. It means that all kinds of developing chambers (N-chambers and S-chambers) can be easily used in this mode of chromatogram development. The bands on the plate after preparative chromatogram development are narrower than the original bands on the start line of the plate, depending on their migration distance (Figure 9.21) [181]. It means that the bands are more concentrated and require less solvent for development and less solvent to elute from the plate as well.

9.2.5 Combinations of Different Modes of Chromatogram Development

The application of multidimensional planar chromatography (MD-PC) combined with different separation systems and modes of chromatogram development is often necessary for performing the separation of more complicated multicomponent mixtures. High separation efficiency can be obtained using modern planar chromatographic techniques, which comprise 2D development, chromatographic plates with different properties, a variety of solvent combinations for mobile-phase preparation, various forced-flow techniques, and multiple development modes. By combining these possibilities, MD-PC can be performed in various ways. Giddings defined multidimensional chromatography, as a technique that includes two criteria [182]:

- The components of the mixture are subjected to two or more separation steps in which their migration depends on different factors.
- When two components are separated in any single step, they always remain separated until completion of the separation.

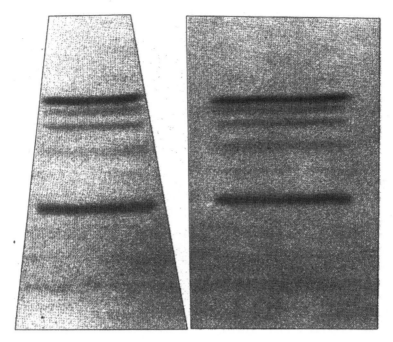

FIGURE 9.21 Comparison of the separation of streaks of dyes on triangular and rectangular 5 × 20 cm plates. (From Issaq, H.J., *J. Liq. Chromatogr.*, 3, 789, 1980. With permission.)

Nyiredy divided MD-PC techniques as follows [2,183,184]:

- Comprehensive 2D planar chromatography (PC × PC)—multidimensional development on the same monolayer stationary phase, two developments with different mobile phases or using a bilayer stationary phase, and two developments with the same or different mobile phases.
- Targeted or selective 2D planar chromatography (PC + PC)—technique, in which following the first development from the stationary phase a heart-cut spot is applied to a second stationary phase for subsequent analysis to separate the compounds of interest.
- Targeted or selective 2D planar chromatography (PC + PC)—second mode technique, in which following the first development, which is finished and the plate dried, two lines must be scraped into the layer perpendicular to the first development and the plate developed with another mobile phase, to separate the compounds that are between the two lines. For the analysis of multicomponent mixtures containing more than one fraction, the separation of components of the next fractions should be performed with suitable mobile phases.
- Modulated 2D planar chromatography (^{n}PC)—technique, in which on the same stationary phase the mobile phases of decreasing solvent strength and different selectivity are used.
- Coupled-layer planar chromatography (PC–PC)—technique, in which two plates with different stationary phases are turned face to face (one stationary phase to second stationary phase) and pressed together so that a narrow zone of the layers overlaps and the compounds from the first stationary phase are transferred to the second plate and separated with a different mobile phase.
- Combination of MD-PC methods—technique, in which the best separation of a multicomponent mixture is realized by parallel combination of stationary and mobile phases, which are changed simultaneously. By the use of this technique, for example, after separation of compounds in the first dimension with changed mobile phases, the plate is dried and the separation process is continued in perpendicular direction by use of the grafted technique with a changed mobile phase (based on the idea of coupled TLC plates, denoted as graft TLC in 1979 [142]).

A new procedure for separation of complex mixtures by combination of different modes of MD-PC was described [124,185]. By the help of this new procedure 14 or 22 compounds from a complex mixtures were separated on 10 × 10 cm TLC and HPTLC plates [124,185]. In Figure 9.22 an example of this procedure is presented step by step for the separation of 14 compounds from complex mixtures on a TLC plate [124].

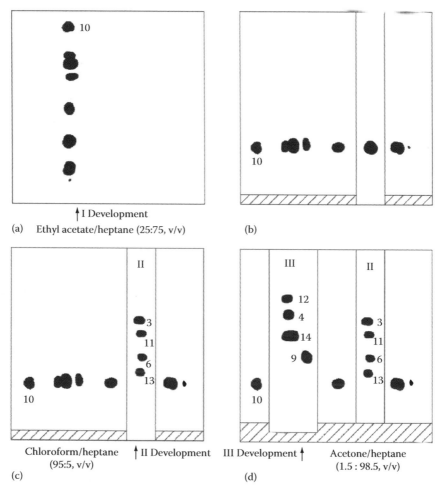

FIGURE 9.22 Illustration of step-by-step selective MD-PC separation. (a) The dried plate after first separation (first development); (b) The plate prepared for the separation of the second group of compounds: two lines (about 1 mm thick) are scraped in the stationary phase perpendicular to the first development in such a way that the spot(s) of target compounds are between these lines. In addition, the strip of adsorbent layer of 5 mm width is removed from the plate along its lower edge to prevent wetting the layer outside the area fixed by these lines during the second development (hatched lines indicate the removed part of the stationary phase). Therefore, the mobile phase wets a narrow strip of the layer only during the second run; (c) The dried plate after separation of the second group of pesticides (3, 6, 11, 13) by use of double development with chloroform–*n*-heptane (95:5, v/v) as the mobile phase at the same distance (UMD); (d) The plate after separation of the components of the fourth group of pesticides with acetone–*n*-heptane (1.5:88.5, v/v) as the mobile phase in the third development; (e) The plate after separation of two components of the third group (2, 7) of pesticides with toluene as the mobile phase in the fourth development; (f) The plate after separation of three components of the first group (1, 5, 8) of pesticides with ethyl acetate–dichloromethane (10:90, v/v) as the mobile phase in the fifth development. (From Tuzimski, T., *J. Sep. Sci.*, 30, 964, 2007. With permission.)

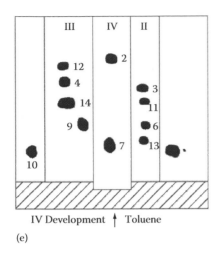

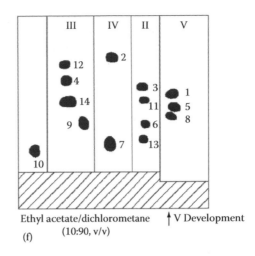

IV Development ↑ Toluene

(e)

Ethyl acetate/dichlorometane ↑ V Development
(10:90, v/v)

(f)

FIGURE 9.22 (continued)

9.2.6 RECAPITULATION: STATIONARY AND MOBILE PHASES FOR IDENTIFICATION OF PESTICIDES AND APPLICATION OF TLC IN THE STUDY OF LIPOPHILICITY OF PESTICIDES

Modes of chromatogram development, sample application, and application of appropriate stationary and mobile phases are the key elements that influence the resolution of the mixture compo nents and efficiency of quantitative and qualitative analysis. Optimization of these elements can be effectively performed on the basis of a good understanding of the theoretical fundamentals and practical knowledge of planar chromatography. Sophisticated equipments, methods, and software are inherent elements of today's planar chromatography and can effectively facilitate optimization of chromatographic separation. Thanks to these features, this method is a powerful analytical and separation technique in a contemporary analysis, which has gained growing popularity in laboratory practice. The literature on planar chromatography regarding the problems discussed in this chapter is very broad. Only part of this literature is cited in the references of this chapter, which is an additional evidence of the meaning and interest in this technique, especially for separation and analysis of pesticides in environmental samples.

The correct identification of pesticides and their quantitative analysis in samples of natural origin are possible after optimization of the mobile phase and suitable selection of the stationary phase for their standards. The retention vs. eluent composition relationships of nearly 100 pesticides was presented for various eluent/adsorbent systems of TLC, including both nonaqueous NP systems (e.g., heptane + ethyl acetate/silica) and RP systems (e.g., water + methanol/octadecyl silica) [186]. For NP systems, the following polar adsorbents were used: silica, alumina, florisil, bonded-phase adsorbents –CN–, DIOL- and NH_2-silica; and several polar modifiers (ethyl acetate, tetrahydrofuran, dioxane, diisopropyl ether); and for aqueous RP systems methanol, acetonitrile, and tetrahydrofuran. The retention–eluent composition relationships are represented as R_F vs. volume percent concentration plots; group selectivities were compared by $R_{F II}$ vs. $R_{F I}$ correlation plots, which indicate how chosen sets of pesticides are separated in the pair of eluent/adsorbent systems. The correlation plots also inform on the separation of these sets of compounds in 2D-TLC using a pair of eluent–adsorbent systems, directly giving the distribution of spots on the plate area for chosen suitable eluent compositions [187–190].

An example of the separation of a mixture of 18 pesticides by 2D-TLC on a cyanopropyl-bonded polar stationary phase is shown in Figure 9.23 [136]. Correlation of retention data of pesticides in NP and RP systems and utilization of the data for separation of a mixture of 10 urea herbicides by 2D-TLC on a cyanopropyl-bonded stationary phase and on a two-adsorbent-layer Multi-K SC5 plate

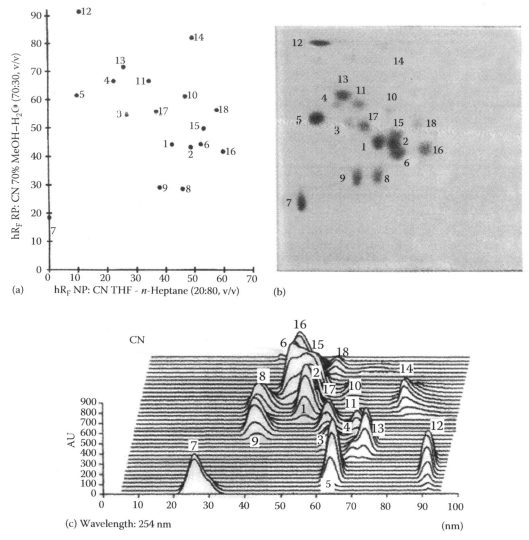

FIGURE 9.23 (a) Correlation between NP hR_F values obtained with tetrahydrofuran–n-heptane, 20 + 80 (v/v), as mobile phase and reversed-phase hR_F values obtained with methanol–water, 70 + 30 (v/v), as the mobile phase on cyanopropyl-bonded silica gel. This pair of NP and RP systems was chosen for 2D-TLC. The videoscan (b) and densitogram (c) of the plate show the separation achieved for the 18-component pesticide mixture. (From Tuzimski, T., *J. Planar Chromatogr.*, 17, 328, 2004. With permission.)

were reported [191]. The separation of a mixture of the new pesticides admitted for marketing in Poland has been separated by use of 2D-TLC [192]. *N*-nitroso-triazine herbicides (cyanazine and terbuthylazine and their reaction products) using different phases for each dimension on silica plates separated by 2D-TLC were described [193].

The R_F vs. eluent composition plots also provide information on preconcentration of pesticides by solid-phase extraction: eluent compositions corresponding to low R_F values indicate strong adsorption and concentration of solutes even from large sample volumes, whereas high R_F values indicate the possibility of elution from the SPE column in a low volume of eluent. The result of a chromatographic study presented in Figure 9.24 indicates that micro TLC performed on RP18 plates wetted with water and organic/water mobile phases can be a useful method for the estimation of SPE behavior of analytes separated on the octadecylsilica-packed cartridges [34]. A linear relationship

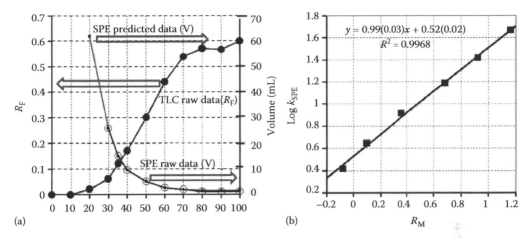

FIGURE 9.24 Methanol/water elution profiles of the estetrol observed on the C18 cartridges (circles; graph a) and RP-18W plates (black dots; graphs a) as well as the relationship between SPE and micro-TLC data based on the R_M and log k_{SPE} values (graph b). Volume parameter (right Y-axis; graph a) corresponds to the experimentally measured breakthrough volumes of the estetrol. Small black diamonds on graph a represent the predicted trajectory of the estetrol SPE elution using micro-TLC data via log $k_{SPE} = R_M$ equation. Slope and intercept coefficients of the linear regression equation (log $k_{SPE} = aR_M + b$) were calculated by the use of the retention data obtained from methanol/water mobile phases ranging from 30% to 70% (v/v) and for the V_0 parameter of SPE tubes = 0.55 mL (measured for methanol). (From Zarzycki, P.K., *J. Chromatogr. A*, 1187, 250, 2008. With permission.)

was observed between R_M values of, for example, estetrol plotted against the logarithmic form of the SPE retention factor (k_{SPE}) [34].

TLC study of the synthesis of some unsaturated chlorotriazine derivatives with herbicidal activity was described [194]. Chromatographic behavior of dithiocarbamate fungicides on cellulose plates was reported [195]. TLC as a pilot technique for the choice of suitable HPLC systems for the analysis of pesticide samples was reported [113,196]. TLC as a pilot technique for transferring retention data to HPLC and use of the data for preliminary fractionation of a mixture of pesticides by micropreparative column chromatography was described [197]. Two-stage fractionation of a mixture of pesticides by micropreparative TLC and HPLC was described [198,199]. Use of a database of plots of pesticide retention (R_F) against mobile-phase compositions for fractionation of a mixture of pesticides by micropreparative TLC was reported [200].

Lipophilicity is an important characteristic of organic compounds with interesting environmental activity. The quantitative structure–activity relationship (QSAR) and quantitative structure–retention relationship (QSRR) studies have found growing acceptance and application in agrochemical research. The retention parameter from TLC, which is normally used in QSRR, is the R_M value, defined as $\log(1/R_f - 1)$, where R_f is the ratio of a distance passed by the analyte to that attained on the plate by the solvent front. In the case of RP TLC extrapolation is usually performed to pure water (buffer) as a hypothetical eluent, and then the extrapolated R_M value is usually denoted by R_M^0. Various determination methods are used to determine lipophilicity. Hansch et al. proposed the use of log P [201], the ratio of concentrations of components in two phases, more exactly the octanol–water partition coefficient, measured with the shake-flask method. If the analyte has lower values of log P (higher values of energy of hydration), then the substance has stronger affinity to the water phase, and the analyte is hydrophilic. Biagi et al. [202] found a good correlation between log P, log k, and R_M values, determined by TLC. The relationship between the hydrophobicity and specific hydrophobic surface area of 12 fungicides and herbicides determined by HPLC compared with RP TLC was described [203]. A TLC study of the lipophilicity of triazine herbicides was also reported [204]. A study was reported [205] of the relationship between the R_f values for a group of organophosphorus (OPs) insecticides obtained by TLC and a series of topological descriptors. The retention of eight cyanophenyl herbicides on water-insoluble β-cyclodextrin polymer beads (BCDP) was determined by TLC. The effect of pH, salt concentration in the mobile phase, and various physicochemical parameters of herbicides on the herbicides–BCDP interaction were calculated by using 2D nonlinear maps of principal component parameters and variables [206]. Interaction of pesticides with a β-cyclodextrin derivative studied by RP TLC and principal component analysis (PCA) was described [207]. Comparison of different properties—log P, log k_w, and φ_0 values—as descriptors of hydrophobicity of some fungicides was also studied [208]. Planar chromatography in studies of the hydrophobic properties of some herbicides was described [209]. RP TLC behavior of some s-triazine derivatives was reported [210]. A partial least-squares study of the effects of organic modifier and physicochemical properties on the retention of some thiazoles was also described [211]. TLC and OPLC for evaluation of the hydrophobicity of s-triazine derivatives was reported [212]. TLC and OPLC were used to study the hydrophobicity of homologous s-triazines [213].

9.3 SAMPLE PREPARATION

Analysis of environmental samples requires a good extraction method for sample preparation. The great variety of samples and pesticides to be analyzed requires numerous sample preparation methods. Sample preparation methods such as liquid–liquid and Soxhlet extraction together with large column chromatographic cleanup on adsorbents such as, for example, florisil are still widely used before planar analysis [214]. Liquid–liquid extraction (LLE) has a very long history and does not require special instrumentation and materials. These are the main reasons for which it is employed in most methods for the determination of pesticides, including official ones, rather than the much more recent methods such as SFE and SPE [215]. LLE has an advantage in a great variety of extraction systems available, which have been thoroughly studied theoretically, optimized, and widely tested in practice. LLC offers many possibilities for fine-tuning of extraction efficiency and selectivity by variation of the experimental conditions. However, it also has several serious drawbacks and therefore, it is gradually being replaced by SFE and SPE. These drawbacks are [216]

- The extraction yield is often insufficient, and another preconcentration step must be included before the actual analysis.
- Phase separation is often complicated by the formation of emulsions.
- Large volumes of solvents are required.
- The extraction procedure tends to be tedious, slow, and difficult to automate.

SPE is one of various techniques available to the analyst to bridge the gap that exists between the sample collection and analysis step. SPE is a very important alternative to LLE in analysis of pesticides. The concentration of pesticides in the original samples is frequently very low, so a preconcentration method should be applied. Advantages of SPE include faster analyses, lower cost, and less organic solvent consumption. SPE can be used in water analysis, owing to the fact that it provides a high enrichment ratio. It also enables satisfactory cleanup of dirty samples. Large volumes of water can be extracted with a barely small effort and can be eluted with small quantities of organic solvent. Examples of SPE for the recovery of analytes before planar chromatography, including the separation and determination of different classes of pesticides in water, were described [217–226].

Table 9.3 includes some classical extraction methods [217–235], procedures (step by step), and recoveries for different classes of pesticides on different types of SPE cartridges.

Occasionally, extraction on SPE columns is time consuming because of the limited water flow rate through the adsorptive bed. To shorten the sample-preparation time, the SPE column can be replaced by a thin adsorptive disk made from Teflon fibers and containing an adsorbent [218]; these are known as Empore disks. Disk extraction has been reported to use 90% less solvent than LLE

TABLE 9.3
Examples of Sample Preparation for TLC Analysis of Pesticides

Examples of Sample Preparation for TLC Analysis of Pesticides in Water Samples

Pesticides	Type of Sample	Extraction Method	Procedure (Step by Step)	Recovery (%)	Reference
Sulfonylurea herbicides (metsulfuron-methyl, chlorsulfuron, bensulfuron-methyl, tribenuron, chlorimuron-ethyl) and bensulfuron-methyl residues	Tapwater	SPE 3 mL C18 SPE columns (Bellefonte, USA)	Conditioning: 5 mL methanol followed by 10 mL distilled water; flow rate 2–5 mL min^{-1}; sample: pH 3.0–3.5 by addition of dilute (1:10); washing: distilled water (5 mL); elution: glacial acetic acid in ethyl acetate (0.1% (v/v), 10 mL)	Recoveries of bensulfuron-methyl from water spiked at three levels: 5 µg kg^{-1}, 96% 10 µg kg^{-1}, 92% 20 µg kg^{-1}, 96%	[217]
Alachlor, atrazine, α-cypermethrin	Water	SPE C18 47 mm Empore disk (Varian, Harbor City, USA)	Samples were prefiltered through 0.45 µm filters and were acidified with hydrochloride acid to pH 2–2.5; extraction disks were washed with 2C mL eluent and were conditioned by passage of 20 mL methanol through the disk under vacuum, spiked water sample, dried, and eluted with 20 mL eluent (10 mL hexane then 10 mL dichloromethane)	Optimum recovery of atrazine and alachlor was achieved by use of dichloromethane, whereas the best recovery of α-cypermethrin was achieved by use of hexane Optimum recoveries were 94.7 ± 4.4 for atrazine, 104.2 ± 3.7 for alachlor, and 100.2 ± 4.2 for α-cypermethrin	[218]
7 phenylurea herbicides (chlortoluron, diuron, fluometuron, isoproturon, linuron, methabenzthiazuron, neburon)	Drinking water	SPE 3 mL, 500 mg, Bakerbond SPE C18 cartridges (J.T. Baker, Germany)	Washed: 10 mL methanol, conditioned 5 mL water, sample application—flow rate 2–3 mL min^{-1}, washed: 3 mL water, dried, eluted: 5 mL acetonitrile, evaporated to dryness and redissolved in 50 µL acetonitrile	Recovery rates were between 91% and 102% except for methabenzthiazuron (87%)	[219]
7 OPs pesticides: azinophos methyl, chlorpyriphos, diazinon, fenthion, methamidophos, methidathion, omethoate	Water samples	SPE 6 mL, 200 mg, Bakerbond SPE SDB-1 cartridges (J.T. Baker, Germany)	Washed: 5 mL ethyl acetate, conditioned 2 × 5 mL methanol, next 2 × 5 mL water, sample application, washed: 2 × 5 mL water, dried 30 min^{-1}, eluted: 2 mL n-hexane–acetone (1:1), next 3 mL ethyl acetate, evaporated to dryness and redissolved in 100 µL ethyl acetate	Recovery rates were between 94% and 102% for both types of cartridges except for chlorpyriphos on C18	[220]

(continued)

TABLE 9.3 (continued)
Examples of Sample Preparation for TLC Analysis of Pesticides

Pesticides	Type of Sample	Extraction Method	Procedure (Step by Step)	Recovery (%)	Reference
Diflubenzuron residues	Water (deionized and river)	SPE 3-mL, 200 mg, Bakerbond SPE C18 cartridges (J.T. Baker, Germany) SPE cartridges connected to a 75 mL reservoir	Washed: 2 × 3 mL ethyl acetate, conditioned 1 × 3 mL methanol, next 1 × 3 mL water, sample application—flow rate 10 mL min⁻¹, washed: 2 × 3 mL water, dried 30 min⁻¹, eluted: 2 mL ethyl acetate, evaporated to dryness and redissolved in 100 µL ethyl acetate. Washed: 5 mL portions of acetonitrile, methanol, and water, sample application—pressure 15 mmHg, washed: 35 mL acetonitrile–water (3:7), eluted: 2 mL acetonitrile, evaporated to dryness and redissolved in 1.0 mL ethyl acetate	Recoveries of diflubenzuron from water spiked at 50 µg L⁻¹ were 95%–97% and RSD of 2%–3%	[221]
265 pesticides	Water	SPE 1 g Bakerbond SPE C18 cartridges (J.T. Baker, Frankfurt, Germany)	For neutral compounds: washing: methanol, ethyl acetate, water—each twice the volume of cartridge; 1 L water samples, flow rate 8 mL min⁻¹, dried 3 h, eluted: 6 mL of methanol, evaporated to dryness and redissolved in 100 µL toluene. For phenoxycarbocylic acids: Washing: acetone, methanol, water (pH < 2)—each twice the volume of cartridge; 1 L acidified water samples (pH < 2 with HCl), flow rate 8 mL min⁻¹, dried 3 h, eluted: 2.5 mL of methanol, evaporated to dryness and redissolved in 100 µL of methanol	The recoveries of 184 pesticides were estimated from spiked tap water samples at the 100 ng L⁻¹ concentration level. The great majority of them result in recovery rates better than 50%	[222]
Pesticides	Drinking water	SPE 1 g Bakerbond SPE C18 and SDB cartridges (J.T. Baker, Frankfurt, Germany)	Conditioning: n-hexane, dichloromethane, methanol, water (pH 2); 500 mL acidified water samples (pH 2), dried, washing: n-hexane, eluted: methanol and/or dichloromethane		[223]

Analytes	Sample	SPE	Procedure	Results	Ref.
12 pesticides and their metabolites	Water	SPE 6 mL Bakerbond SPE C18 cartridges, 3 mL Bakerbond SPE silica cartridges (J.T. Baker)	Washing: 10 mL of acetonitrile, water (pH 2 with HCl); sample application: flow rate 1–6 mL min⁻¹, 1 L acidified water sample (pH 2 with HCl) was percolated, then the cartridge was connected to a column packed with activated carbon and eluted with 3 mL of acetonitrile. The eluate direct from cartridge or purified through silica cartridge was eluted with 3 mL of acetonitrile, evaporated to dryness, and redissolved in 0.2 mL of acetonitrile-n-heptane (95:5)		[224]
Atrazine, clofentezine, chlorfenvinphos, hexaflumuron, terbuthylazine, lenacyl, neburon, bitertanol, and metamitron	Water samples from Wieprz-Krzna Canal from Łęczyńsko-Włodawskie Lake District (South-East Poland)	SPE C18/SDB-1 (C18 500 mg on top + SDB 200 mg on bottom/6 mL), C18 (2000 mg/6 mL), C18 Polar Plus (3000 mg/6 mL), and CN (1000 mg/6 mL) Bakerbond SPE cartridges (J.T. Baker, Deventer, Holland)	For SPE assays, each cartridge was conditioned with 3 × 2 mL CH_2Cl_2, 3 × 2 mL methanol and 3 × 2 mL water. After being loaded with the water samples (1 L; flow rate 10 mL min⁻¹; pressure 75 mmHg), the SPE column was washed with methanol–H_2O (5:95, v/v), followed by vacuum drying for 1 min, and eluted with 5 mL methanol and next 5 ml CH_2Cl_2	Recovery rates were high for all extraction materials except for all pesticides on CN, which had lower values	[225]
21 pesticides	Water samples from eight lakes from Łęczyńsko-Włodawskie Lake District (South-East Poland)	SPE C18/SDB-1 (C18 500 mg on top + SDB 200 mg on bottom/6 mL), C18 (2000 mg/6 mL), C18 Polar Plus (3000 mg/6 mL), and CN (1000 mg/6 mL) Bakerbond SPE cartridges (J.T. Baker, Deventer, Holland)	For SPE assays, each catridge was conditioned with 3 × 2 mL CH_2Cl_2, 3 × 2 mL methanol and 3 × 2 mL water. After being loaded with the water samples (1 L; flow rate 10 mL min⁻¹; pressure 75 mmHg), the SPE column was washed with methanol–H_2O (5:95, v/v), followed by vacuum drying for 1 min, and eluted with 5 mL methanol and next 5 mL CH_2Cl_2	Recovery rates were high for all extraction materials except for all pesticides on CN, which had lower values	[226]

(continued)

TABLE 9.3 (continued)
Examples of Sample Preparation for TLC Analysis of Pesticides

Pesticides	Type of Sample	Extraction Method	Procedure (Step by Step)	Recovery (%	Reference
			Examples of Sample Preparation for TLC Analysis of Pesticides in Soil Samples		
8 insecticides (oxamyl, pirimicarb, carbaryl, malathion, phosalone, fenitrothion, tetradifon, methoxychlor) from three different groups of compounds (organophosphorus, organochlorine, and carbamate pesticides)	Soil	Soxhlet apparatus	25 g sample was extracted with 150 mL of dichloromethane for 8 h. The extract was transferred into a separator to separate the water phases. Organic extracts were dried with anhydrous sodium sulfate (20 g) during 24 h and (after filtration) were concentrated to approximately 10 mL volume. The solid components were separated from the suspensions obtained, and then the clear solutions were concentrated to 2 mL volume	The recovery rates were between 71.6% and 81.6% for C 8 Polar Plus cartridges. SD (%) were between 7.5 and 9.9	[227]
		SPE C18 Polar Plus and others (C8, C18) (J.T. Baker, Deventer, Holland)	1 mL volume of the extract was dissolved in methanol and then it was mixed with 800 mL water. The emulsion was extracted, retained in the column, dried for 20 min, washed by 3 × 2 mL of n-hexane, eluted with 2 × 2 mL of methanol, and concentrated to 1 mL	Mixture of pesticides was spiked into 25 g of soil samples: 10 µg of pirimicarb, 50 µg of oxamyl, 60 µg of phosalone, 100 µg of malathion, and 20 µg of other pesticides	
107 pesticides	Soil	SFE Suprex model (Suprex Inc., Pittsburg, USA) with Prepmaster, Accutrap, and modifier pump	The extraction cell was filled with 2 g of the soil sample and extracted under conditions of fluid: temperature: 40°C, pressure: 400 atm, flow rate: 2.5 ml min⁻¹; cryogenic trap temperature: −40°C, static extraction: 10 min, dynamic extraction: 30 mL CO_2, modifier addition: 300 µL methanol into extraction vessel, continuous addition of 6% methanol to the fluid, restrictor temperature: 40°C	Recovery: samples (2 g) were spiked with 2 mL of standard solution of the respective pesticides (500 µg mL⁻¹ in methanol). After evaporation of the solvent (18 h), the soil was transferred to the extraction vessel. The sample was dried under a gentle stream of nitrogen and finally dissolved in 500 µL of methanol. Recovery rates for 20 pesticides were between 8.4% and 82.3%	[228]

Examples of Sample Preparation for TLC Analysis of Pesticides in Food

Tricyclazole, thiram, and folpet	Tomatoes	LLE	The sample (10 g) was weighed into a 150 mL conical flask with stopper and a known amount of spiking solution was added. After equilibration, the fortified samples were mixed with 50 mL acetone–dichloromethane (1:1) and 3 g anhydrous sodium sulfate and extracted by mechanical vibration at room temperature for 30 min. The extract was filtered through a glass funnel containing 5 g anhydrous sodium sulfate into a 150 mL round-bottom flask. Extraction of the sample was repeated with another 50 mL extraction solution and the filtrate was collected in the same flask, evaporated to dryness at 35°C–40°C, and the residue was eluted by 3 × 1 mL of methanol and concentrated to 2 mL	Recoveries of the pesticides from tomatoes were between 67.66% and 98.02%, and RSD values were between 0.13% and 22.06%	[229]
Atrazine and simazine	Honey	Ultrasonic solvent extraction (USE): 20 l ultrasonic cleaning bath (model UZ-20R, Iskra, Šentjernej, Slovenia) at 30 kHz working frequency and 400 W power at 25°C ± 2°C	Homogenized honey samples pretreated at temperatures 33°C ± 2°C were placed in ultrasonic bath and half bath height in 100 mL glass tubes. Temperature in honey samples did not exceed 35°C in all experiments. The extracts were passed through Whatman 40 filter, the filtrates were transferred into 100 mL separating funnel, and allowed to separate. Next the extracts were evaporated at 35°C to dryness, and the residues were dissolved in 1 mL of methanol	The best recovery of pesticides from spiked honey samples was achieved after extraction by ultrasonic solvent extraction, which was carried out in three steps for 20 min using 20 mL of benzene: water 1:1 (v/v) Recoveries obtained by USE were 92.3% ± 2.8% and 94.2% ± 3.1% for atrazine and simazine, respectively	[230]
		Shake-flask extraction method	Five gram of spiked honey sample was suspended in 20 mL of benzene: water 1:1 (v/v) and shaken mechanically in an Erlenmeyer flask for 4 h. The extract was filtered and evaporated to dryness as described above	Recoveries obtained by shake-flask extraction method were 71.9% ± 5.1% and 75.7% ± 4.5% for atrazine and simazine, respectively	

(continued)

TABLE 9.3 (continued)
Examples of Sample Preparation for TLC Analysis of Pesticides

Pesticides	Type of Sample	Extraction Method	Procedure (Step by Step)	Recovery (%)	Reference
Fenitrothion	Apples and fresh apple juice	Shake-flask extraction method: Sample preparation (juice) Ultrasonic solvent extraction (USE): Sample preparation (extraction on apples). Ultrasonic bath (Bandelin electronic RK 100 H, Berlin, Germany)	Apples (1 kg) were collected and the juice squeezed out. 150 mL dichloromethane was added to 150 mL of apple juice and shaken in a separatory funnel; next, the bottom fraction was separated and filtered. The obtained extract was evaporated to dryness and the dried residue dissolved in methanol and transferred to 5 mL volumetric flask and made up to volume with methanol. Apples were homogenized and 15 g of sample was placed in a 100 mL glass flask with 60 mL acetone–dichloromethane (1:1, v/v) mixture and 2 g of anhydrous Na_2SO_4 and hermetically closed. The flask was introduced into an ultrasonic bath with distilled water for 15 min at 20°C. The liquid was passed through filter and washed into a flask with 10 mL of the solvent mixture. The fractions were combined in a flask and concentrated to dryness by rotary vacuum evaporation. The dry extract was dissolved in 5 mL acetone or methanol	Method of recovery was studied by analyzing five replicates of samples spiked at 120 and 260 µg mL^{-1} levels. Recoveries were in the range 71.9–85.1% for 120 µg mL^{-1} level and 73.4–85.4 for 260 µg mL^{-1} level. The average recoveries from the spiked samples and SD were 79.5% ± 5.1% and 79.2% ± 4.5%, the CV were 6.4% and 5.7% for fortification levels 120 and 260 µg mL^{-1} levels, respectively.	[231]
Imidacloprid, fenitrothion, and parathion	Chinese cabbage	Ultrasonic solvent extraction (USE)	The sample was extracted by sonication in ultrasonic water bath with acetone–petroleum ether, 5:3 (v/v)	Recoveries of the pesticides from Chinese cabbage by use of this analytical method were 80.04%–85.22%, and RSD were 4.18%–13.15%	[232]
Pyrethroids (fenpropathrin, deltamethrin, bifenthrin)	Vegetables (spinach, green soybean, and fresh kidney beans)	Ultrasonic solvent extraction (USE)	Crushed vegetables (20 g) were homogenized and fortified, petroleum ether (25 mL), acetone (15 mL), and anhydrous sodium sulfate were added, and the mixture was blended and extracted by ultrasonication for 12 min. The extract separated into two layers: the upper layer was petroleum ether and the lower layer was acetone–water. A portion (12.5 mL) of the former was concentrated to 1 mL at 45°C	The green vegetables were spiked with three pyrethroid insecticides at three different levels, 0.5, 1.0, and 5.0 mg kg^{-1} and then analyzed by HPTLC. The recovery obtained varied between 70.2% and 108.5%. The precision was 1.94%–27.94%	[233]

Carbamate residues (pirimicarb, methomyl, carbaryl, and carbofuran)	Vegetables (potato and wax gourd)	Ultrasonic solvent extraction (USE)	Crushed vegetables (20 g) were homogenized and fortified, acetone (40 mL) was added, and the mixture was blended. Next petroleum ether–dichloromethane (1:1, v/v, 40 mL) was added and blended samples were extracted by exposure to ultrasound for 5 min. The extracted sample was centrifuged at 3000 rpm for 5 min, and petroleum ether–dichloromethane extract was isolated and concentrated to 1 mL at 45°C	The green vegetables were spiked with three different levels, 1.0, 2.0, and 5.0 mg kg^{-1} and then analyzed by HPTLC. The recovery obtained varied between 70.05% and 103.7%. Accuracy was 3.35%–26.49%	[234]
6 phenylurea herbicides	Foods (apples, asparagus, carrots and wheat)	Sample cleanup SPE silica (0.5 g) column	A sample of plant material (50 g) was weighed into a blender cup. Acetone (100 mL) was added and homogenized for 5 min, next filtered by glass wool, and the blender and the filter cake were washed with acetone (2 × 25 mL). The solution was transferred to a separating funnel and water (300 mL), saturated with sodium chloride (30 mL), and dichloromethane (70 mL), and the combined mixture evaporated to near dryness at 45°C. The volume of the solution was then adjusted to exactly 10 mL with acetone–hexane (1:9); next 2 mL was transferred to a separatory funnel with hexane (2 mL) and acetonitrile (6 mL) and shaken for 5 min, and then the lower acetonitrile phase collected. The hexane was extracted again with acetonitrile (6 mL). The hexane phase was discarded and the acetonitrile phases combined and evaporated to just dryness under a gentle flow of nitrogen. The volume was adjusted to 2 mL acetone–hexane (1:9). Next 1 mL of solution was transferred to a silica (0.5 g) column previously conditioned with hexane (5 mL), washed with ether–hexane (2:8; 5 mL), pesticides eluted with acetone–ether (3:7; 5 mL), dryness, redissolved in methanol (0.25 mL)	Recovery was >80% for the ureas at 0.1 ppm	[235]

and up to 20% less than SPE with cartridges. Another modification is Speedisk, especially used for difficult samples (turbid and including suspension of matter).

SFE is mainly used for complicated systems such as soils [228,236], sediments [237], and food [238]. The advantages of SFE include high extraction efficiencies attained, good selectivity, short extraction times, simple preconcentration steps, a great reduction of volumes of toxic and environmentally hazardous solvents used, a reduction in the cost of analyses, and the feasibility of online coupling with chromatographic techniques for routine use [238]. There are also some drawbacks [215]: the choice of the extraction system is very limited. In fact, carbon dioxide with certain modifiers is virtually the only system that is routinely used. The selection and optimization of the experimental conditions are difficult and still largely empirical. The maintenance of constant experimental conditions during the procedure is demanding, and the measurement may suffer from high blank and noise levels. The procedures involving intermediate washing and drying steps may be rather lengthy.

It is expected that microwave-assisted solvent extraction (MASE), which was applied for the efficient determination of triazines in soil samples with aged residue [239], or matrix solid-phase dispersion (MSPD), which permits complete fractionation of the sample matrix components as well as elution of single compounds or several classes of compounds from the sample [240], will also find broad application in planar chromatography. Another technique, ultrasonic extraction (USE), has also proved to be a reliable extraction technique that successfully replaces classical procedures as an efficient method for determining pesticides in soils [230–233,241]. Atrazine, propham, chloropropham, diflubenzuron, α-cypermethrin, and tetramethrin were determined in soil with recoveries of 79%–103% by acetone USE [241]. USE was also used for the analysis of residues of atrazine and simazine in honey [230]. The best recovery of pesticides from spiked honey samples was achieved after extraction by USE, which was carried out in three steps for 20 min using 20 mL of benzene:water 1:1 (v/v). Recoveries obtained by USE were 92.3% ± 2.8% and 94.2% ± 3.1% for atrazine and simazine, respectively [230]. Ultrasonic solvent extraction was compared with the traditional shake-flask extraction method. Recoveries obtained by the shake-flask extraction method were 71.9% ± 5.1% and 75.7% ± 4.5% for atrazine and simazine, respectively [230]. USE and shake-flask extraction method were used for the identification of fenitrothion in apples [231]. Method of recovery was studied by analyzing five replicates of samples spiked at 120 and 260 µg mL^{-1} levels. Recoveries were in the range of 71.9%–85.1% for 120 µg mL^{-1} level and 73.4–85.4 for 260 µg mL^{-1} level. The average recoveries from the spiked samples and standard deviations (SD) were 79.5% ± 5.1% and 79.2% ± 4.5%; the coefficients of variation (CV) were 6.4% and 5.7% for fortification levels 120 and 260 µg mL^{-1}, respectively [231]. Ultrasonic solvent extraction was also used for the determination of fenitrothion, imidacloprid, and parathion in Chinese cabbage [232]. Recoveries of the pesticides from Chinese cabbage by use of this analytical method were 80.04%–85.22%, and relative standard deviations (RSDs) were 4.18%–13.15% [232]. USE and HPTLC determination of pyrethroid residues in vegetables was also described [233]. The green vegetables water spinach, green soybean, and fresh kidney beans were spiked with three pyrethroid insecticides at three different levels, 0.5, 1.0, and 5.0 mg kg^{-1} and then analyzed by HPTLC. The recovery obtained varied between 70.2% and 108.5%. The precision was 1.94%–27.94% [233].

9.4 DETECTION AND IDENTIFICATION OF PESTICIDES IN TLC

Various chemical, physical, and biochemical methods have been used in the detection and identification of pesticides. Some of them are very simple. For the detection of UV-absorbing substances simply visually by eye, TLC plates are prepared with fluorescence indicators (e.g., manganese-activated zinc-silicate). This dye absorbs light at 254 nm showing a green florescence at ~520 nm; the sample molecules inhibit light absorption on the plate with a fluorescence indicator. In the case of uncovered plates, dark spots or zones on a bright fluorescent background will indicate the positions of the components. The commonly used method is to expose the chromatographic plate to iodine vapor in a closed chamber that contains some iodine crystals. Sample spots show a dark brown color on a plate with yellow-brown background.

9.4.1 VISUALIZATION BY COLOR REACTIONS

The possibility of using reagents (especially color reagents) is one of the essential advantages of planar chromatography. Stahl and Mangold [242] reviewed the generally used common spraying "reagents." A basic source of reagents is given in the book that was edited by Jork et al. [243].

Ethylenebisdithiocarbamate fungicides and their toxic metabolite ethylenethiourea were detected in different vegetables (tomatoes and green beans treated with Perocin 75B [Zineb; Agria Co., Bulgaria]) by spraying with 2% aqueous sodium nitroferricyanide reagent after development of silica gel 60 GF$_{254}$ plates with chloroform–butanol–methanol–H$_2$O (100 + 5 + 1 + 0.5) or chloroform–ethyl acetate–metanol (3 + 2 + 1) as mobile phases [244]. The same reagent as well as Erlich's reagent (solution of 50 mg 4-dimethylaminobenzaldehyde in 50 mL methanol with 10 mL concentrated HCl) and iodine vapor were used in the detection of oxidative degradation products of ethylenebisdithiocarbamate fungicides (zineb, nabam) and their ethylenethiourea metabolites [245]. After use of the Bratton-Marshall reagent, the sensitivity of the technique for diflubenzuron was 0.1 µg, and residues in water at a concentration of 50 µg L^{-1} were determined with recoveries of 95%–97% and RSD of 2%–3%. The residues could be semiquantitatively determined at concentrations down to 125 ng L^{-1}. The zones were detected by spraying in turn with 6 M ethanolic hydrochloric acid, 1% sodium nitrite in ethanolic HCl, and 1% ethanolic N-(1-naphthyl) ethylenediamine dihydrochloride (Bratton-Marshall reagent). The layer was covered with a clean glass plate and heated to 180°C for 10 min after the first spray. The zones were scanned at 550 nm in the single-beam, single-wavelength reflectance mode [221]. A chromogenic spray reagent for the detection of monocrotophos was described [246]. The red colored spots of chloranil derivatives of standard monocrotophos and of monocrotophos extracted from viscera were observed. After alkaline hydrolysis, monocrotophos yields dimethylphosphoric acid and N-methylacetoacetamide. The methylene (–CH$_2$–) group of N-methylacetoacetamide is located at the alpha position to two carboxyl groups, which increases the reactivity of the two α-hydrogen atoms on this methylene group. This active methylene group reacts with the chromogenic reagent chloranil to give a red compound [246]. Next paper also described a spray reagent for the detection of monocrotophos, which on alkaline hydrolysis yields N-methylacetoacetamide, which in turn reacts with diazotized sulfanilamide or sulfanilic acid to give a red color [247]. The detection limit is approximately 1 µg [247]. Various visualization reagents were examined for detection of other OPs insecticides, for example, dichlorvos (DDVP). A semiquantitative determination of dichlorvos in visceral tissue (stomach, intestine, liver) was described by using a spray reagent prepared by dissolving anhydrous sodium carbonate (1.4 g) in distilled water (50 mL) and adding 2-thiobarbituric acid (2 g), and next the resulting solution diluted to 100 mL [248]. In experiments, the recovery of dichlorvos from biological materials was determined; the recovery was ~90% [248]. A spray reagent for selective detection of dichlorvos from various types of visceral tissue (stomach, intestine, liver, spleen, and kidney) by TLC was reported [249]. The limit of detection (LOD) of the reagent is approximately 10 µg [249]. A new reagent was examined for spectrophotometric quantification and TLC detection of the insecticides dichlorvos and diptrex after extraction from biological tissues, blood, and commercial formulation [250]. The reagent was prepared from strong alkali (e.g., 10% sodium hydroxide) and 0.5% aqueous sodium sulfide solution (based on "Ogston Reaction"). The color system obeys the Beer-Lambert law at λ = 401 nm in the concentration range 2–100 µg mL^{-1}. Dichloroacetaldehyde produced by alkaline hydrolysis of insecticides reacts with sodium sulfide giving yellow spots, which turn wine red after some time [250]. Selective and sensitive TLC method for the detection of dichlorvos and dimethoate with orcinol was described [251]. A yellow-brown spot for dichlorvos and yellow spots for dimethoate were observed (λ = 366 nm). The detection limit was 1 µg spot^{-1} and 15 µg spot^{-1} for dichlorvos and dimethoate, respectively [251]. Application of HPTLC for the detection of 25 OPs insecticides in human serum after acute poisoning was described [252]. The detection limits of dichlorvos, fenitrothion, malathion, methidathion, parathion, and trichlorfon in serum by the LLE method were 1.1, 0.12, 0.05, 0.6, and 0.1 μg mL^{-1}, respectively [252]. Several detection reagents for carbaryl determined by TLC were also described. In the first [253], the silica plates were developed to a distance of 10 cm with n-hexane–acetone and 8 + 2 (v/v) as mobile phase in a saturated Camag twin-trough TLC chamber. After development the plate was removed, dried in air, and heated in an oven at 100°C for approximately 5 min. After cooling, the plate was sprayed successively with 5% sodium hydroxide solution and then with a 1:1 mixture of 2% diphenylamine solution and 5% formaldehyde solution (based on "Mannich Reaction"). Blue-green spots stable for several days for carbaryl were observed [253]. The minimum LODs for carbaryl are 5 μg on TLC silica gel G plates and 1 μg on TLC silica gel 60 F_{254} plates [253]. In Ref. [254] the detection limit for carbaryl was approximately 0.1 μg $spot^{-1}$ after the development of silica TLC plate with n-hexane–acetone (4:1) as mobile phase and spraying the plate with 1% phenylhydrazine hydrochloride in an alkaline medium reagent for the detection of carbaryl [254]. Potassium hydroxide solution or p-nitro-benzenediazonium tetrafluorobromate solution was also used for detection of carbamate insecticides (e.g., carbaryl, propoxur, carbofuran), fungicides (e.g., carbendazim [Bavistin], and other pesticides [255]). The reaction between thiocarbamate herbicides and 2,6-dichlorobenzoquinone-N-chloroimine or 2,6-dibromobenzoquinone-N-chloroimine was described for the detection of these herbicides on TLC plates [256]. Plumbite reagent was used for the detection of mancozeb [257]. The reagent was prepared by dissolving lead monoxide (1 g) in aqueous sodium hydroxide (32%), and this reagent can be applied for the detection of mancozeb in soil extract. The LOD of the reagent is 0.45 μg $spot^{-1}$ [257].

9.4.2 Physical Methods of Detection

Detection of fluorescent spots under UV light at 254 or 366 nm is widely applied. AMD-HPTLC analysis of organochloride pesticides on silica gel plates was described [258]. Vizualization of organochloride pesticides was achieved by spraying the plates with 1% o-tolidine solution in ethanol followed by 15 min irradiation under UV light ($\lambda = 254$ nm). The derivatized compounds were scanned in the absorbance mode at $\lambda = 500$ nm and in the reflectance mode [258].

Detection by autoradiography was used frequently, in parallel with UV irradiation and chemical color reactions in pesticide analysis in different samples. Metabolism of flumiclorac pentyl in rats (absorption, distribution, biotransformation, and excretion) was reported with visualization of spots by UV light or by color reaction by spraying Bromocresol Purple (BCP) or 2,6-di-chlorophenolindophenol sodium salt. Radioactive spots were detected by autoradiography using x-ray films (SB-5, Kodak, Rochester, NY) [259].

Degradation of four commonly used pesticides (2,4-D, lindane, paraquat, and glyphosate) in Malaysian agricultural soils was controlled by autoradiography [260]. Detection by radioscanning was done for TLC separation of soil-bound residues of cyprodinil [261] and [^{14}C]tebupirimphos [262].

Soil TLC is a development technique in which the studied soil serves as the stationary phase. It is frequently used for the investigation of pesticide mobility through soil microstructures. Soil TLC with water or water–methanol as solvents allows the observation and measurement of the mobility of herbicides and insecticides: triazines (atrazine), phenylureas (diuron and isoproturon), biscarbamates (phenmedipham), phenylpyrazols (fipronil), and two ionic pesticides (the cationic paraquat and the anionic glyphosate) through soil microstructures (11 different sieved matrices) [263]. Autoradiography of five spots of fipronil applied at different amounts (1.25, 6.25, 46.25, 100, and 500 μmol) on a soil TLC plate developed with H_2O–methanol (3:2) as the mobile phase is shown in Figure 9.25 [263]. The effect of soil amendment using urban compost, agricultural organic amendments, and surfactants on the mobility of two sparingly soluble pesticides—diazinon and linuron—was studied by soil TLC [264]. The movement of (^{14}C) pesticide spot was detected using a Berthold TLC Tracemaster 20 linear detector [264]. The results also suggested the possibility of using organic materials and surfactants to develop physicochemical methods for preventing the pollution of soils and waters by pesticides and eliminating pesticide residues from these media [264].

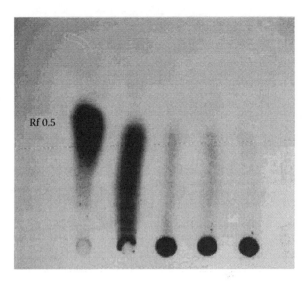

FIGURE 9.25 Occurrence of tailing when the amount of the deposit is too high for obtaining an immediate and total water solubility: autoradiography of five spots of fipronil applied at different amounts (1.25, 6.25, 46.25, 100, and 500 µmol) on a CSA plate (solvent:water–methanol, 3:2). For each spot the same amount of ^{14}C-labeled fipronil (1.25 nmol) was deposited. (From Ravanel, P. et al., *J. Chromatogr. A*, 864, 145, 1999. With permission.)

Metabolism of 7-fluoro-6-(3,4,5,6-tetrahydrophtalimido)-4-(2-propynyl)-2*H*-1,4-benzoxazin-3(4H)-one in rat was examined [265]. An unlabeled standard herbicide on TLC plates was detected by viewing under UV light (254 nm), whereas the radioactive spots were detected by x-ray films [265]. Comparative detection of fluorinated xenobiotics and their metabolites through ^{19}F NMR or ^{14}C label in plant cells was described [266]. The metabolism of the new fluorinated fungicide [*N*-ethyl-*N*-methyl-4-(trifluoromethyl)-2-(3,4-dimethoxyphenyl)benzamide] by *Acer pseudoplatanus* cells was studied concurrently through the use of ^{19}F NMR and with ^{14}C-labeled compounds [266].

9.4.3 BIOLOGICAL AND OTHER METHODS OF DETECTION

The use of biological methods in planar chromatography is justified, because they are highly specific and the detection limits are lower than those in other methods. For the detection of toxic effects in the environment, the use of biosensors, for example, for inhibition of the growth of microorganism or enzyme inhibitor, is increasingly important.

An enzymatic inhibition screening method was described using maleimide CPM (7-diethylamino-3-(4′-maleimidylphenyl)-4-methylcoumarin) as fluorogenic reagent [220]. It reacts with thiocholine released after hydrolysis of acetylcholine with acetylcholinesterase at pH 8 to form a strongly blue fluorescent background after 1–5 min, whereas the sites of enzyme inhibition are dark. OPs insecticides can be detected as dark spots with the LOD of 1–10 ng spot^{-1} (Figure 9.26).

Toxicological evaluation of harmful substances by in situ enzymatic and biological detection in HPTLC was also described [267]. The separated components were detected and quantified directly on the HPTLC chromatogram by physical and chemical methods. By coupling HPTLC with biological or biochemical inhibition tests, it was possible to detect toxicologically active substances in situ. A linear relationship was shown between the signal of the inhibition of cholinesterase and the concentration of the inhibitor using a constant enzyme concentration and a constant incubation time. Inhibition of the luminescence of Photobacterium *Vibrio fisheri* in relation to the concentration of pentachlorophenol was also examined. Measurements were done by using a densitometer ($\lambda = 553$ nm) and a videodensitometric scanner [267]. The metabolism of the cyano-oxide fungicide cymoxanil and its analogues was studied with the fungus *Botrytis cinerea* owing to their difference

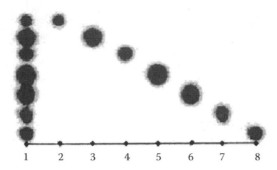

FIGURE 9.26 Chromatogram obtained from the seven OP pesticides on a silica gel TLC plate after spraying with acetyl cholinesterase, acetylthiocholine, and, finally, maleimide CPM. 1, mixture of the seven OP pesticides; 2, chlorpyriphos; 3, diazinon; 4, fenthion; 5, methidation; 6, azinophos methyl; 7, omethoate; 8, methamidophos. (From Hamada, M. and Winstersteiger, R., *J. Planar Chromatogr. Mod. TLC*, 16, 4, 2003. With permission.)

in sensitivity toward cymoxanil [268]. TLC monitoring was done on octadecyl silica plates with ion paring and it allowed the monitoring of ionizable metabolites for substrates that were demonstrated to decompose the most rapidly [268].

9.5 QUANTITATIVE DETERMINATION OF PESTICIDES: APPLICATION OF MODERN SCANNERS

Detection of separated components on a TLC plate and generating analogue curves of the chromatogram tracks for qualitative and quantitative evaluation are generally called densitometry. Densitometry can be performed in absorbance or fluorescence mode. Unlike scanning densitometry, which is based on sequential evaluation of the individual chromatogram tracks, video densitometry is based on grouping the pixels of the image according to the tracks and evaluating them on a gray scale [269]. All pixels of the track, which have the same R_F value, are averaged and can be plotted as a function of distance in the direction of development. Because monochromatic light in the range of 190–800 nm can be used and tuned to the absorption or fluorescence maximum of the individual compounds, the measurement is highly sensitive. Typical detection limits are in the low nanogram range (absorbance) or medium picogram range (fluorescence). Densitometry is usually performed before derivatization. Only substances without chromophoric groups must be chemically altered to render them detectable [269].

Application of a modern diode-array TLC scanner has several advantages, for example, [270–274]:

- The scanner can measure TLC plates simultaneously at different wavelengths without destroying the plate surface and permits parallel recording of chromatograms and in situ UV spectra in the range of 191–1033 nm; therefore, it is possible to obtain more correct identification of the compounds on a chromatogram.
- The TLC scanner DAD permits analysis of each compound at its optimum wavelength, thus offering optimum sensitivity for the detection of each component.
- The TLC scanner DAD permits to obtain a 3D chromatogram $A = f(\lambda, t)$.
- The TLC-scanner DAD gives the possibility of obtaining parallel UV spectra of a compound with comparison to the spectrum of the standard from the library of spectra.
- Software is available that allows the user access to all common parameters used in HPLC-DAD: peak purity, resolution, identification via spectral library match and so on.
- The TLC-scanner DAD is especially useful for correct identification of components, which occur in difficult, complicated mixtures, in plant extract and toxicological analysis.

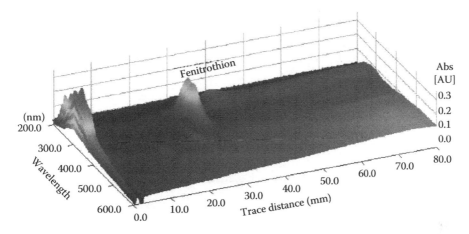

FIGURE 9.27 Three-dimensional plot obtained from an apple extract containing fenitrothion. (From Tuzimski, T., *J. Planar Chromatogr. Mod. TLC*, 18, 419, 2005. With permission.)

At present, only a limited number of articles describe fiber optical scanning in TLC, especially of pesticides. An application of fiber optical scanning densitometry for identification and quantitative analysis of fenitrothion in fresh apple juice was demonstrated [231]. Figure 9.27 shows an example of the 3D plot (scanning range × trace distance × absorbancy) obtained from the apple extract [231]. Identification was achieved by comparing the UV spectrum obtained from the extract and a fenitrothion standard. Figure 9.28 shows UV spectra obtained from fenitrothion standards at eight concentrations (100–1000 µg mL^{-1}) and the UV spectrum obtained from fenitrothion in an extract from freshly squeezed apple juice [231]. The components of two mixtures of pesticides, which were separated by 2D-TLC with adsorbent gradients of the type silica-wettable with water octadecyl silica or silica-cyanopropyl, were identified by R_F in both chromatographic systems and by comparison

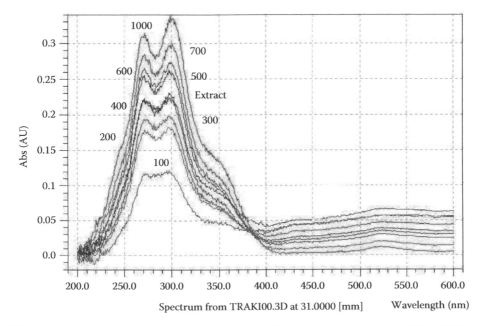

FIGURE 9.28 UV spectra obtained from fenitrothion standards at eight concentrations (100–1000 µg mL^{-1}) and from an extract of freshly squeezed apple juice containing fenitrothion. (From Tuzimski, T., *J. Planar Chromatogr. Mod. TLC*, 18, 419, 2005. With permission.)

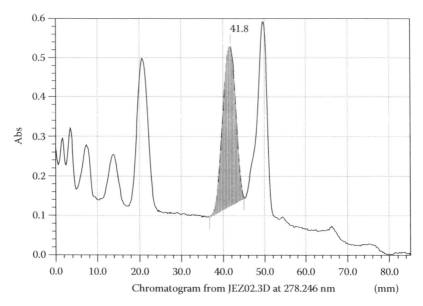

FIGURE 9.29 Chromatogram obtained of fortified water sample after SPE and TLC-DAD for optimal wavelength for clofentezine ($\lambda = 278.246$ nm). (From Tuzimski, T., *J. AOAC Int.*, 91(5), 1203, 2008. With permission.)

of UV spectra [146]. The peak-purity index is a numeral measure of the quality of coincidence between two datasets. The peak purity index is a numerical index for the quality of the coincidence between two datasets. It is given by the least-squares-fit coefficient calculated for all intensity pairs in the two datasets under consideration. The following equation is applied:

$$P = \frac{\sum_i \left(s_i - \bar{s}\right)\left(r_i - \bar{r}\right)}{\sqrt{\sum_i \left(s_i - \bar{s}\right)^2 \sum \left(r_i - \bar{r}\right)^2}}, \qquad (9.11)$$

where

s_i and r_i are the respective intensities for the same abscissa value
i is the number of data points
$\bar{s}$ and $\bar{r}$ are the average intensities of the first and second dataset

In another article an application of fiber optical scanning densitometry (TLC-DAD) and HPLC-DAD for identification and quantitative analysis of pesticides in water samples was demonstrated [225]. Dichloromethane eluates were analyzed by TLC-DAD (Figure 9.29). The identities of the bands of analytes in the water samples were confirmed by overlaying their UV absorption spectra with those of the standards of these compounds (Figure 9.30). A peak-purity index of 1 indicates that the compared spectra are identical. The least-squares-fit value of the spectrum from a fortified sample of water and a spectrum from clofentezine standard was also presented (Figure 9.31).

9.6 RESIDUE ANALYSIS OF PESTICIDES

Because of its simplicity and speed, planar chromatography if often used in research on pesticides (and their residues) belonging to various chemical classes of samples (water, soil, food, other samples). Along with the examples mentioned herein, the reader will also find practical data on the use of planar chromatography in the foregoing section.

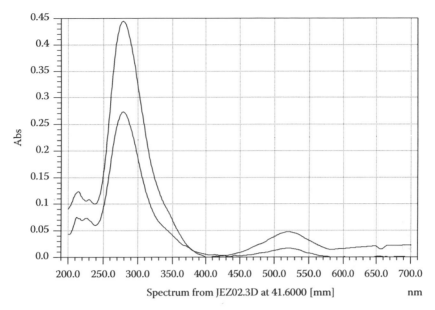

FIGURE 9.30 Comparison of UV spectrum of clofentezine standard with in situ spectrum of fortified water sample after SPE and TLC-DAD (Purity index (Pearson's *r*) $P = 0.9959$). (From Tuzimski, T., *J. AOAC Int.*, 91(3), 1203, 2008. With permission.)

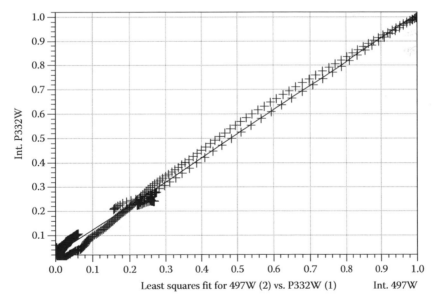

FIGURE 9.31 Least-squares fit value (obtained by cross-correlation) of spectrum from fortified sample of water and spectrum from clofentezine standard. (From Tuzimski, T., *J. AOAC Int.*, 91(3), 1203, 2008. With permission.)

9.6.1 RESIDUE ANALYSIS OF INSECTICIDES

There is a large range of synthetic insecticides, including important chemical classes commonly referred to as organochlorines (now largely obsolete), OPs compounds (or "organophosphates"), carbamates, pyrethroids (synthetic analogues of the natural pyrethrums), insect growth regulators (IGRs), and the relatively recent nicotinyl/chloronicotinyl compounds (related to naturally occurring nicotine). The organochlorines (a commonly used term referring to the persistent organochlorine pesticides, which include the cyclodienes, DDT and related compounds, lindane and the hexachlorocyclohexanes, and toxaphene) were the first widely used group of synthetic insecticides, coming into use after World War II. These chemicals were generally long-acting, controlling pests for an extended period of time. Unfortunately, their high solubility in fat and chemical stability means that they can bioaccumulate over a long time, with concentrations increasing in animals higher in the food chain. Their ability to volatilize in warm regions means they also can spread over quite long distances, with measurable concentrations being found even near the Arctic Circle and alpine areas where they have not been used. These organochlorines have been eliminated from agricultural use in most countries because of the concerns about environmental persistence, bioaccumulation, and transboundary movement. Nevertheless, they need to be considered in the context of chronic intake of pesticide residues, because they can still be found at low levels (generally decreasing with time) in a limited number of products, particularly of animal origin [275].

HPTLC with AMD gradient elution for the monitoring of insecticides (oxamyl, pirimicarb, carbaryl, malathion, phosalone, fenitrothion, tetradifon, methoxychlor) from three different groups of compounds (OPs, organochlorine, and carbamate pesticides) in soil was described [227]. Quantitative assessment was achieved by UV absorption measurement scanning the chromatograms by a "zig-zag" technique. Six benzoyl urea insecticides were analyzed by TLC-AMD [276]. Time-dependent sorption of imidacloprid in two different soils (sandy loam and silt loam) was studied [277]. Radioactive zones of interest on TLC plates were measured using a bioimaging analyzer [277].

OPs compounds are commonly used as insecticides in a variety of crops and as ectoparasiticides in animal husbandry. An enzymatic inhibition method sufficiently sensitive for rapid screening of seven OPs insecticides in water by SPE was described [220]. OP pesticides were separated on silica gel TLC plates with *n*-hexane–acetone (75 + 30, v/v) on a 10 cm distance in a saturated chamber. Spots of OP insecticides were quantified by scanning fluorescence of quenched zones at 220 nm with dual wavelength, flying-spot densitometer in the reflectance mode. Calibration plots were linear between 100 and 2000 ng for all pesticides; the correlation coefficients, r, were between 0.9994 and 0.9997 [220].

Carbamates (e.g., aldicarb and methiocarb) form an important group of insecticides. Like the OPs compounds, they inhibit acetylcholinesterase, but their effects are quicker in onset and more rapidly reversible. Chemicals that are structurally related to these carbamates have also been developed as fungicides, herbicides, and molluscicides.

The synthetic pyrethroids, which mimic the structure and action of naturally occurring pyrethrins, are very widely used as insecticides. Synthetic nicotinoids and neonicotinoids (related to the natural compound nicotine as the pyrethroids are related to pyrethrum) are gaining increasing importance. Time-dependant sorption of imidacloprid (chloro-nicotinyl insecticide) in two different soils (sandy loam and silt loam) was investigated [277]. High-performance TLC method for the determination of fenvalerate and deltamethrin in emulsifiable concentrate formulation was developed [278]. The results from the HPTLC method were comparable with those obtained by GC with flame-ionization detection [278]. Average recovery with ± SD was 97.56% ± 0.98% and 99.8% ± 0.18% for fenvalerate and deltamethrin, respectively [278].

Another class of insecticides of growing importance is the so-called IGRs, which kill insects by interfering with the normal process of juvenile development, either by disrupting hormonal processes or exoskeleton development. IGRs, from several different chemical classes, are relatively selective to specific pests, provide a reasonably long period of protection, and have not shown resistance problems.

9.6.2 Residue Analysis of Herbicides

Herbicides are pesticides used for weed control. Most modern synthetic herbicides have low mammalian toxicity, because they are designed to mainly affect specific metabolic pathways within plants. Important chemical classes of herbicides are the 1,3,5-triazines (e.g., atrazine and simazine), the ureas (e.g., diuron and isoproturon), and the sulfonylureas (e.g., chlorsulfuron and tribenuron). Determination of bensulfuron-methyl residues in tapwater was described [217]. Herbicides (atrazine, alachlor) and α-cypermethrin in water samples were determined [218]. Seven phenylurea herbicides were determined in drinking water [219]. A rapid fluorodensitometric screening method, involving thermal hydrolysis and subsequent derivatization with fluorescamine, was also developed and improved the LOD 25-fold. Separation of the reaction products of cyanazine and terbuthylazine nitrosation by TLC and 2D-TLC was reported [193]. In the next study [279] TLC methods for the analysis of 18 ^{14}C-labeled herbicides from several classes (trazine, trazinon, phenylurea, sulfonylurea, phenoxy acid, and others) were developed.

9.6.3 RESIDUE ANALYSIS OF FUNGICIDES

Next to herbicides and insecticides, fungicides are an economically very important group of pesticides. One of the major chemical classes of fungicides includes the azole compounds (e.g., propiconazole and fenbuconazole), but as for herbicides and insecticides, there are a diverse range of chemical types. The fungicide cyprodinil labeled with ^{14}C in either the phenyl or the pyrimidyl ring was incubated with four different soils under various conditions to evaluate the formation of bound residues and their subsequent plant uptake [261]. Separation of fenoxaprop-p-ethyl biodegradation products by HPTLC was described [280]. The separation was visualized by irradiation of the plates at $\lambda = 236$ nm with a UV lamp. Densitometric analysis of the fenoxaprop-p-ethyl biodegradation products was performed with Camag TLC Scanner 3 in the absorbance mode ($\lambda = 236$ nm) [280].

9.7 DETERMINATION OF RESIDUE PESTICIDES IN DIFFERENT SAMPLES

Solid-phase extraction can be used especially in water analysis, owing to the fact that it provides a high concentration ratio. It also enables satisfactory cleanup of dirty samples. Large volumes of water can be extracted with barely any effort and eluted with small volumes of organic solvent.

The pollution of water by pesticides depends on several variables, including the type and quantity of pesticides used, soil type, tillage system, and frequency of rainfalls. The properties of pesticides that determine pesticide transport from the field to surface water are solubility in water, the organic carbon/water partition coefficient, and octanol–water partition coefficient (K_{OC}, K_{OW}, respectively). Pesticides with high water solubility and low soil adsorption will move easily to the groundwater [218,275]. Examples of TLC analysis of pesticides in water and soil samples are listed in Table 9.4. Additional sample preparation methods are listed in Table 9.3 (Section 9.3). Five sulfonylurea herbicides (metsulfuron-methyl, chlorsulfuron, bensulfuron-methyl, tribenuron, chlorimuron-methyl) were separated on aluminium- or glass-backed silica gel F_{254} HPTLC plates, with two different mobile phases [217]. Bensulfuron-methyl was added to tapwater and determined after extraction by SPE on C-18 cartridges [217]. Recoveries of bensulfuron-methyl from water spiked at three levels 5, 10, and 20 μg kg^{-1} were 96%, 92%, and 96%, respectively. The detection limit was 5 ng spot^{-1} whereas LODs ranged from 2 to 8 ng spot^{-1} for herbicides. Alachlor, atrazine, and α-cypermethrin were preconcentrated from water samples by SPE and determined by TLC [218]. With the goal to determine SPE recovery from different water samples, the experiments were repeated with 250, 500, 750, and 1000 mL volumes. Recoveries of pesticides obtained by use of different solvents (hexane, dichloromethane, acetone, acetonitrile, methanol, ethanol, ethyl acetate, and 2-propanol) for SPE were reported. The seven phenylurea herbicides were separated by use of three different TLC systems and quantified densitometrically at nanogram levels [219]. For the determination of herbicides in drinking water the substances were enriched from water samples by SPE on C18 cartridges. Drinking water (1 L) was spiked with a solution of the seven phenylurea herbicides (chlorotoluron, diuron, fluometuron, isoproturon, linuron, methabenzthiazuron, and neburon) and adjusted to pH 5–6. The herbicides were extracted from water by C18 cartridges, with recoveries of 91%–102%,

TABLE 9.4
Examples of TLC Analysis of Pesticides in Water and Soil Samples

Pesticides	Stationary Phase	Mobile Phase (v/v)	Detection	Type of Sample, Detection Level (µg), Problem, and Other Techniques Applied	References
		Examples of TLC Analysis of Pesticides in Water Samples			
Sulfonylurea herbicides (metsulfuron-methyl, chlorsulfuron, bensulfuron-methyl, tribenuron, chlorimuron-ethyl) and bensulfuron-methyl residues	HPTLC—silica gel 60 F$_{254}$	Chloroform–acetone–glacial acetic acid (90 + 10 + 0.75) with presaturation of the chamber for 30 min and the plate for 15 min, toluene–ethyl acetate (50:50) with presaturation of the chamber for 15 min	TLC-scanner (λ = 201 nm)	Tapwater; SPE (C 18) Detection limits ranged from 2 to 8 ng zone^{-1}	[217]
Alachlor, atrazine, α-cypermethrin	HPTLC—RP-18F$_{254S}$	2-Propanol–water (9 + 1), 2-propanol–water (8 + 2), 2-propanol–water (7 + 3), 2-propanol–water (6 + 4)	Videoscanner (λ = 254 nm, 3CCD HV-C20 color video camera)	SPE (C18 Empore disk)	[218]
7 phenylurea herbicides (chlortoluron, diuron, fluometuron, isoproturon, linuron, methabenzthiazuron, neburon)	System I: Alugram silica gel Nano Sil Diol with fluorescent indicator UV$_{254}$. System II: NH$_2$-F$_{254S}$; System III: silica gel 60 F$_{254}$	H$_2$O–acetone–methanol (60 + 10 + 30), Chloroform–toluene (80 + 20), Benzene–triethylamine–acetone (75 + 15 + 10).	UV-lamp (λ = 254 nm or 366 nm) Densitometry (λ = 245 nm or 265 nm for methabenzthiazuron)	Drinking water; SPE (C18), quantified densitometrically at nanogram levels	[219]
7 OPs pesticides: azinophos methyl, chlorpyriphos, diazinon, fenthion, methamidophos, methidathion, omethoate	Silica gel 60 F$_{254}$	n-Hexane–acetone (75 + 30)	UV-lamp (λ = 254 nm or 366 nm). The scanning densitometry (λ = 220 nm). Bromine or iodine vapor for 2 min, inhibition by cholinesterases: the plates were sprayed 0.06% acetylcholinesterase in 0.01 M sodium phosphate (or 0.05 Tris–HCl + 0.1% Triton x-100) buffer solution at ptt-8, dried for 20 min, sprayed with 0.3 mM acetylthiocholine chloride substrate in the same buffer, and after 10 min sprayed with 0.4 mM maleimide CPM (7-Diethylamino-3-(4'-maleimidylphenyl)-4-methylcoumarin). Blue fluorescence appeared as background after 1–5 min, whereas sites of enzyme inhibition were dark	Water samples: SPE (SDB-1, C18; C18; LOD: 1–10 ng spot^{-1}	[220]

Analyte	Technique/Plate	Mobile phase	Detection	Description	Ref.
Diflubenzuron residues	HPTLC—silica gel	Methylene chloride–methanol (1:1), ethyl acetate–toluene (1:3)	Bratton-Marshall Reagent: 6 M ethanolic hydrochloric acid, 1% sodium nitrite in ethanolic HCl, and 1% ethanolic N-(1-naphthyl) ethylenediamine dihydrochloride. Densitometry ($\lambda = 550$ nm)	Determination of diflubenzuron residues in water (deionized and river) by SPE and quantitative HPTLC	[221]
265 pesticides	HPTLC—silica gel 60 F_{254}	33-Step gradient according to DIN 38407, Part 11	HPTLC/AMD Densitometry—TLC scanner	SPE (C18), LLE, screening of 265 pesticides in water by TLC with AMD. List of the limits of detection for those pesticides is reported GC-MS, HPLC.	[222]
Pesticides	HPTLC	Two gradients of different selectivity	TLC/AMD	Analysis of pesticide residues in drinking water by planar chromatography. The suitability of this method was proved for 283 pesticides and the corresponding ISO Standard has been applied. HPLC/AMD	[223]
12 pesticides and their metabolites	HPTLC—silica gel 60 F_{254}	25-Step gradient	HPTLC/AMD Densitometry at six or seven different wavelengths ($\lambda = 190, 200, 220, 240, 260, 280,$ and 300 nm)	Identification and determination of pesticides in water	[224]
Atrazine, clofentezine, chlorfenvinphos, hexaflumuron, terbuthylazine, lenacyl, neburon, bitertanol, and metamitron	HPTLC and TLC—silica gel 60 F_{254}	20%–70% ethyl acetate in n-heptane or hexane	TLC-DAD scanner ($\lambda = 191$–1033 nm)	Application of SPE (C18, C18 Polar Plus, CN, C18/ SDB-1) and HPLC-DAD and/or TLC-DAD to the determination of 21 pesticides in water samples from Łęczyńsko-Włodawskie Lake District (South-East Poland)	[225]

(continued)

TABLE 9.4 (continued)
Examples of TLC Analysis of Pesticides in Water and Soil Samples

Pesticides	Stationary Phase	Mobile Phase (v/v)	Detection	Type of Sample, Detection Level (μg), Problem, and Other Techniques Applied	References
21 pesticides	HPTLC and TLC—silica gel 60 F$_{254}$	20%–70% ethyl acetate in n-heptane or hexane	TLC-DAD scanner (λ = 191–1033 nm)	Application of SPE (C18, C18 Polar Plus, CN, C18/SDB-1), and HPLC-DAD and/or TLC-DAD to the determination of 21 pesticides in water samples from Łęczyńsko-Włodawskie Lake District (South-East Poland)	[226]

Examples of TLC Analysis of Pesticides in Soil Samples

8 insecticides (oxamyl, pirimicarb, carbaryl, malathion, phosalone, fenitrothion, tetradifon, methoxychlor) from three different groups of compounds (organophosphorus, organochlorine, and carbamate pesticides)	HPTLC—silica gel 60 F$_{254}$	Gradient (dichloromethane–acetone–n-hexane)	HPTLC/AMD Quantitative assessment by UV absorption measurement	Application of instrumental TLC and SPE (e.g., C18 Polar Plus) to the analyses of pesticide residues in grossly contaminated samples of soil GC-MS	[227]
107 pesticides	HPTLC—silica gel 60 WRF$_{254}$	Plates developed by a 35-step gradient with a solvent mixture starting with methyl-tert-butylether/acetonitrile/hexane and ending with pure hexane. The solvent composition according to DIN 38407, Part 11	HPTLC/AMD UV-Vis, Densitometry	Screening of pesticide-contaminated soil by SFE and HPTLC with AMD. The limits of detection for 20 pesticides	[228]

Imidacloprid	Silica gel 60 F$_{254}$	Acetonitrile–dichloromethane (50:50), acetonitrile–ethyl acetate–H$_2$O (70:23:7)	Radioactivity zones on TLC plates were measured using a bioimaging analyzer	Time-dependent sorption of imidacloprid in two different soils (sandy loam and silt loam)	[277]
Herbicides (2,4-D, lindane, paraquat, glyphosate)	Silica gel 60 F$_{254}$, microcrystalline cellulose TLC plate	Benzene–n-hexane–acetone (25:25:1), n-hexane–acetone (9:1), methanol–H$_2$O–0.5 N NaCl (180:60:0.3)	UV-lamp, x-ray film	Degradation of 4 commonly used pesticides in Malaysian agricultural soils (sandy loam, muck soils, anaerobic muck)	[260]
Fungicide (cyprodinil)	Silica gel 60 F$_{254}$	Toluene–methanol (90:10), chloroform–ethanol–acetic acid (90:10:1)	Radioscanning, MS-CI	Formation of soil-bound residues of cyprodinil	[261]
Herbicides and insecticide: triazines (atrazine), phenylureas (diuron and isoproturon), biscarbamates (phenmedipham), phenylpyrazols (fipronil), and two ionic pesticides (the cationic paraquat and the anionic glyphosate)	Silica gel 60 F$_{254}$, soil thin-layer plates	Light petroleum (b.p. 40°C–65°C)–ethyl acetate–formic acid–acetic acid (40:40:1:1), H$_2$O, H$_2$O–methanol (4:1), H$_2$O–methanol (3:2)	Autoradiography	Mobility of labeled pesticides through soil microstructures (11 different sieved matrices)	[263]
Diazinon and linuron	Soil thin-layer plates	H$_2$O	Berthold TLC Tracemaster 20 linear detector	Pesticide mobility	[264]

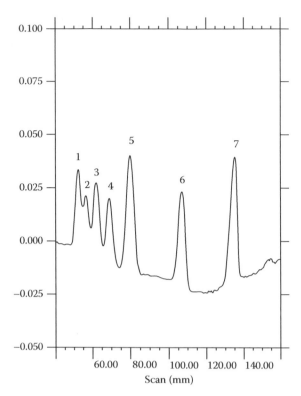

FIGURE 9.32   Separation of seven phenylureas on diol TLC plate, which was developed twice. 1, ispo-proturon; 2, fluometuron; 3, chlortoluron; 4, methabenzthiazuron; 5, diuron; 6, linuron; 7, neburon. (From Hamada, M. and Winstersteiger, R., *J. Planar Chromatogr. Mod. TLC*, 15, 11, 2002. With permission.)

except for methabenzthiazuron (87%). A new method for analyzing drinking water for these herbicides shown in Figure 9.32 used TLC on diol-modified silica gel with water–acetone–methanol (6 + 1 + 3) mobile phase, on amino-modified silica gel with chloroform—toluene (4 + 1), and on silica gel with benzene–triethylamine–acetone (15 + 3 + 2). Detection was by fluorescence quenching at 254 nm, and quantification, at nanogram levels by reflectance-mode dual-wavelength flying-spot densitometry at 245 and 265 nm. Seven OPs pesticides (azinophos methyl, chlorpyriphos, diazinon, fenthion, methamidophos, methidathion, omethoate) were determined in water at 0.1 μg L^{-1} by SPE enrichment on styrene-divinylbenzene copolymer (SDB-1) sorbent and C18 cartridges [220]. The recovery rates were between 94% and 102% for both cartridge types except for chlorpyriphos on C18; the RSD values were ±0.7%–3.7% and ±1.1%–4.9% for SDB-1 and C18 cartridges, respectively. Determination of diflubenzuron residues in water by SPE on C18 and quantitative HPTLC was also described [221]. The sensitivity of the technique for diflubenzuron was 0.1 μg, and residues in water at a concentration of 50 μg L^{-1} were determined with recoveries of 95%–97% and RSD of 2%–3%. Residues could be semiquantitatively determined at concentrations down to 125 ng L^{-1}. HPTLC with automated multiple development (HPTLC–AMD) was used to screen water samples for pesticides [222]. A universal gradient based on dichloromethane, as described in the German official method for water analysis (33-step gradient according to DIN 38407, Part 11), is applied for the screening of 283 pesticides differing widely in polarity.

Analysis of pesticide residues in drinking water by planar chromatography and AMD was also described [223]. Application of HPTLC and AMD with a 25-step gradient for the identification and determination of pesticides in water was also presented [224]. Application of SPE and HPLC-DAD and/or TLC-DAD for the determination of pesticides in water samples from Wieprz-Krzna Canal from Łęczyńsko-Włodawskie Lake District (South-East Poland) was also described [225].

Atrazine, clofentezine, chlorfenvinphos, hexaflumuron, terbuthylazine, lenacyl, neburon, bitertanol, and metamitron were enriched from canal water samples by SPE on C18/SDB-1, C18, C18 Polar Plus, and CN cartridges. The recovery rates were high for all extraction materials except for all pesticides on cyanopropyl (CN) plates, for which the values were lower. SPE was used not only for the preconcentration of analytes but also for their fractionation. The analytes were eluted with methanol and next with dichloromethane. Methanol eluates were analyzed by HPLC-DAD (Figure 9.33), and the dichloromethane eluates, with TLC-DAD. The method was validated for precision, repeatability, and accuracy. The calibration plots were linear between 0.1 and 50.0 $\mu g\ mL^{-1}$ for all pesticides, and the correlation coefficients, r, were between 0.9994 and 1.000 as determined by HPLC-DAD. Calibration plots were linear between 0.1 and 1.5 $\mu g\ spot^{-1}$ for all pesticides, and the correlation coefficients, r, were between 0.9899 and 0.9987 determined by TLC-DAD. The LOD was between 0.04 and 0.23 $\mu g\ spot^{-1}$ (TLC-DAD) and between 0.02 and 0.45 $\mu g\ mL^{-1}$ (HPLC-DAD) [225]. Application of SPE and HPLC-DAD and/or TLC-DAD to the determination of pesticides in water samples from eight lakes from Łęczyńsko-Włodawskie Lake District (South-Eastern Poland) was also described [226]. Application of HPTLC with AMD and SPE for the analysis of pesticide residues in strongly contaminated samples of soil was described [227]. The recovery level for eight insecticides was ~80% and RSD was less than ~9%. Screening of pesticide-contaminated soil by SFE and HPTLC–AMD was also described [228]. The plates were developed by a 35-step gradient with solvent mixtures starting with methyl-tert-butylether/acetonitrile/hexane and ending with pure hexane (according to DIN 38407, Part 11). The limits of detection of 20 pesticides on HPTLC plate were between 0.3 ng and 5.6 ng [228].

Agricultural pesticides are widely used as crop protection agents across the world in the production of food and feed. As a result, the residues of pesticides end up in food for human consumption. In view of the toxicological properties of pesticides, the presence of these residues in food may pose a risk to the health of consumers. Health protection and assuring the proper use of pesticides are the main reasons why governments establish legal limits for pesticide residues in raw agricultural commodities, usually as part of their national legislation for the regulations on pesticides. Food with actual residue levels in compliance with these limits is thought to be safe. Examples of TLC analysis of pesticides in food are listed in Table 9.5. A HPTLC method

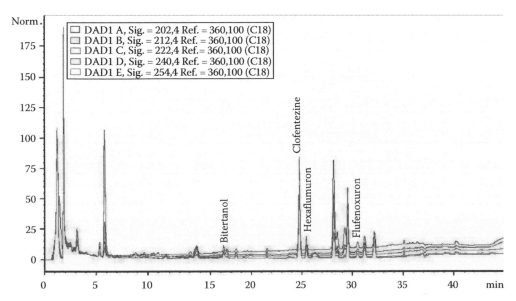

FIGURE 9.33 Chromatogram of water obtained from Wieprz-Krzna Canal (July 2007) showing four detected and quantified pesticides. (From Tuzimski, T., *J. AOAC Int.*, 91(5), 1203, 2008. With permission.)

TABLE 9.5
Examples of TLC Analysis of Pesticides in Food

Pesticides	Stationary Phase	Mobile Phase (v/v)	Detection	Type of Sample, Detection Level (μg), Problems, and Other Techniques Applied	References
Tricyclazole, thiram, and folpet	HPTLC—silica gel 60 F$_{254}$	n-Hexane–acetone (6 + 4)	Densitometry—TLC scanner	Analysis of residues of pesticides in tomatoes. LOD of tricyclazole, thiram, and folpet: 1.2×10^{-8}, 2.0×10^{-8}, 4.0×10^{-8} g, respectively	[229]
Atrazine and simazine	HPTLC—silica gel 60 F$_{254}$	n-Hexane–chloroform–acetone (60:25:15)	Videoscanning ($\lambda = 254$ nm, 3CCD HV-C20 color video camera)	Ultrasonic solvent extraction. Analysis of residues of pesticides in honey. LOQ of atrazine and simazine: 80 ng/spot, 90 ng/spot, respectively	[230]
Fenitrothion	TLC—silica gel 60 F$_{254}$	Ethyl acetate—n-heptane (20 + 80)	TLC-DAD scanner ($\lambda = 200$–600 nm)	TLC in combination with fiber optical (diode array) scanning densitometry for identification of fenitrothion in apples and fresh apple juice	[231]
Imidacloprid, fenitrothion, and parathion	HPTLC—silica gel 60 F$_{254}$	n-Hexane–acetone (7 + 3)	Densitometry—TLC scanner	Ultrasonic solvent extraction. Analysis of residues of pesticides in Chinese cabbage. LOD of imidacloprid, fenitrothion, and parathion: 5.0×10^{-9}, 2.0×10^{-8}, 1.0×10^{-8}, respectively	[232]
Pyrethroids (fenpropathrin, deltamethrin, bifenthrin)	HPTLC and TLC—silica gel 60 F$_{254}$	Toluene–petroleum ether (8 + 3), cyclohexane–chloroform (5 + 5), cyclohexane–chloroform (4.5 + 5.5)	Densitometry ($\lambda = 203$ nm) UV-lamp ($\lambda = 254$ nm or 366 nm)	Analysis of pyrethroid residues in vegetables (spinach, green soybean, and fresh kidney beans). LOD of all pyrethroids: 1.0×10^{-8} g	[233]
Carbamate residues (pirimicarb, methomyl, carbaryl, and carbofuran)	HPTLC—silica gel 60 F$_{254}$	Toluene–acetone (8 + 2), dichloromethane–acetone (8 + 2), ethyl acetate–petroleum ether (6 + 4), chloroform–petroleum ether (9 + 1)	Densitometry—TLC scanner ($\lambda = 243$ nm, 207 nm, 254 nm, or 366 nm)	Analysis of carbamate residues in vegetables (potato and wax gourd). LOD of majority of pesticides: 1.0×10^{-8} g. LOD of carbaryl: 2.0×10^{-9} g	[234]
6 phenylurea herbicides	HPTLC—silica gel 60 F$_{254}$	25-Step gradient—mobile-phase composition used for AMD with MeCN, dichloromethane, acetic acid, toluene, and hexane	HPTLC/AMD Densitometry—TLC scanner ($\lambda = 245$ nm)	The analyses of residues of 6 phenylurea herbicides in foods (apples, asparagus, carrots, and wheat). The LOD is 0.01 ppm	[235]
Ethylenebisdithiocarbamate, fungicides, and their toxic metabolite ethylenethiourea. Perocin 75B (Zineb; Agria Co., Bulgaria)	Silica gel 60 GF$_{254}$	Chloroform–butanol–methanol–H$_2$O (100 + 5 + 1 + 0.5), chloroform–ethyl acetate–methanol (3 + 2 + 1)	UV light ($\lambda = 254$ nm) or by application of 2% aqueous sodium nitroferricyanide reagent	Vegetables—tomatoes and green beans treated with Perocin 75B. HPLC—UV ($\lambda = 240$ nm)	[244]

for analysis of the residues of tricyclazole, thiram, and folpet in tomatoes was described [229]. The detection limits (LODs) of tricyclazole, thiram, and folpet were 1.2×10^{-8}, 3.0×10^{-8}, 4.0×10^{-8} g, respectively. The recoveries of these pesticides from tomatoes were between 67.66% and 98.02%, and RSD values were between 0.13% and 22.06% [229]. A method for quantitative determination of atrazine and simazine in honey samples was described [230]. The extracts from honey samples and of the blank extract were applied on HPTLC Silica Gel F_{254} plates as bands by means of Linomat IV (Camag, Muttenz, Switzerland). The chromatograms were developed in a Camag chamber to 7 cm distance by the ascending technique, under prior chamber saturation, with hexane–chloroform–acetone (60:25:15, v/v/v) as the mobile phase. Video densitometric determination of pesticides was validated for linearity and quantification limit. The LODs of atrazine and simazine were 80 and 90 ng spot^{-1}, respectively [230]. TLC in combination with fiber optical (diode array) scanning densitometry for identification of fenitrothion in apples and fresh apple juice was described [231]. For identification of fenitrothion in fresh squeezed juice from apples, a sample was spotted on a silica HPTLC plate and developed with ethyl acetate–n-heptane (20:80, v/v) as the mobile phase. Chromatograms were developed on a distance of 80 mm. The tracks were scanned in the range 200–600 nm. The contour plot of apple extract is shown in Figure 9.34. The zone of fenitrothion is between 30 and 35 mm (hR_F from [186] is 40). The instrument detection limit (IDL) for fenitrothion was also determined. The IDL for fenitrothion was 10 μg mL^{-1}. The concentration of fenitrothion in the extract from fresh squeezed apple juice 45 days after spraying apples was below the method detection limit (MDL) for fenitrothion [231]. HPTLC method was developed and validated for the analysis of residues of imidacloprid, fenitrothion, and parathion in Chinese cabbage [232]. The sample extract after USE was directly applied, as bands, to glass-backed silica gel 60F$_{254}$ HPTLC plates. The plates were developed with hexane–acetone, 7 + 3 (v/v), in an unsaturated glass twin-trough Camag chamber and analyzed by densitometry with a Camag TLC Scanner 3. The LODs of imidacloprid, fenitrothion, and parathion were 5.0×10^{-9}, 2.0×10^{-8}, and 1.0×10^{-8}, respectively [232]. The analysis of pyrethroid residues (fenpropathrin, deltamethrin, bifenthrin) in vegetables (spinach, green soybean, and fresh kidney beans) was reported [233]. The LOD value of all pyrethroids was 1.0×10^{-8} g [233]. The analysis of carbamate residues (pirimicarb, methomyl, carbaryl, and carbofuran) in vegetables (potato and wax gourd) was also described [234]. The described method uses two mobile phases

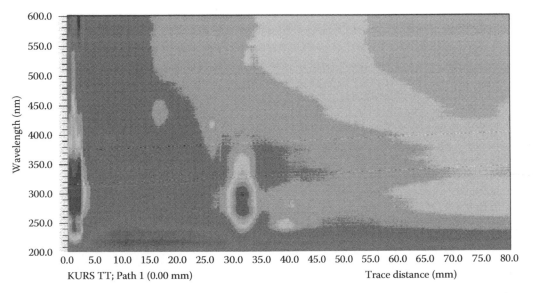

FIGURE 9.34 Contour plot of an apple juice extract containing fenitrothion. (From Tuzimski, T., *J. Planar Chromatogr. Mod. TLC*, 18, 419, 2005. With permission.)

on silica gel 60 F_{254} GLP HPTLC layers and detection by means of a TLC Scanner with UV lamp. The LODs for the majority of pesticides was 1.0×10^{-8} g, and for carbaryl, 2.0×10^{-9} g [234]. A TLC method using AMD was developed for the determination of six phenylurea herbicides in food [235]. The herbicides were extracted with acetone and purified by SPE, the extract was evaporated, the residue dissolved in acetone and spotted on silica gel plates, and chromatographed by AMD. A 25-step gradient composed of MeCN, dichloromethane, acetic acid, toluene, and hexane was used. Quantification was done by measurement of UV. The LOD was 0.01 ppm [235].

9.8 CONCLUSION

Summing up, planar chromatography is one of the principal separation techniques. It can be used for identification of known and unknown compounds, and—at least equally important—for correct identification of pesticides in environmental samples. TLC has many advantages, such as wide optimization possibilities with the chromatographic systems, special development modes, detection methods, and low-cost analysis of samples, requiring minimal sample cleanup. TLC is a chromatographic technique widely used for separation, detection, and qualitative and quantitative determination of pesticides belonging to different chemical classes.

REFERENCES

1. Geiss, F., *Fundamentals of Thin-Layer Chromatography*, Dr. Alfred Hüthig Verlag, Heidelberg, Germany, 1987.
2. Siouffi, A. M. and Abbou, M., Optimization of the mobile phase, Chapter 3, in *Planar Chromatography, A Retrospective View for the Third Millennium*, Ed. Nyiredy, Sz., Springer Scientific Publisher, Budapest, 2001.
3. Hahn-Deinstrop, E., *Applied Thin-Layer Chromatography*, Wiley-VCH, Weinheim, Germany, 2000.
4. Fried, B. and Sherma, J., *Thin-Layer Chromatography*, 4th edn., revised and expanded, Marcel Dekker Inc., New York, Basel, 1999.
5. Dzido, T. H. and Tuzimski, T., Chapter 7 in *Thin-Layer Chromatography in Phytochemistry*, Eds. Waksmundzka-Hajnos, M., Sherma, J., Kowalska, T., Taylor & Francis, Boca Raton, FL, 2008.
6. www.camag.com.
7. Jaenchen, D. E., *Handbook of Thin Layer Chromatography*, 2nd edn., Eds. Sherma, J., Fried, B., Marcel Dekker Inc., New York, 1996.
8. www.chromdes.com.
9. Dzido, T. H., Chapter 4 in *Planar Chromatography, A Retrospective View for the Third Millennium*, Ed. Nyiredy, Sz., Springer Scientific Publisher, Budapest, 2001.
10. Fenimore, D. C. and Davis, C. M., High performance thin-layer chromatography, *Anal. Chem.*, 53, 252A–266A, 1981.
11. De Brabander, H. F., Smets, F., and Pottie, G., Faster and cheaper two-dimensional HPTLC using the '4 × 4' mode, *J. Planar Chromatogr.—Mod. TLC*, 1, 369–371, 1988.
12. Dzido, T. H. and Polak, B., A coadsorption effect in a liquid–solid TLC system of the type silica – heptane + dioxane in horizontal DS chamber, *J. Planar Chromatogr.—Mod. TLC*, 6, 378–381, 1993.
13. Markowski, W. and Matysik, G., Analysis of plant extracts by multiple development thin-layer chromatography, *J. Chromatogr.*, 646, 434–438, 1993.
14. Matysik, G., Modified programmed multiple gradient development (MGD) in the analysis of complex plant extracts, *Chromatographia*, 43, 39–43, 1996.
15. Matysik, G. and Soczewiński, E., Computer-aided optimization of stepwise gradient TLC of plant extracts, *J. Planar Chromatogr.—Mod. TLC*, 9, 404–411, 1996.
16. Matysik, G., Markowski, W., Soczewiński, E., and Polak, B., Computer-aided optimization of stepwise gradient profiles in thin-layer chromatography, *Chromatographia*, 34, 303–307, 1992.
17. Matysik, G., Soczewiński, E., and Polak, B., Improvement of separation in zonal preparative thin-layer chromatography by gradient elution, *Chromatographia*, 39, 497–504, 1994.

18. Dzido, T. H. and Polak, B., in *Preparative Layer Chromatography*, Eds. Kowalska, T., Sherma, J., CRC Press, Taylor & Francis Group, Boca Raton, FL, 2006.
19. Soczewiński, E., *Chromatographic Methods. Planar Chromatography*, Vol. 1, Eds. Kaiser, R.E., Dr. Alfred Huthig, Verlag, Heidelberg, Basel, New York, 1986, pp. 79–117.
20. Kraus, L., Koch, A., and Hoffstetter-Kuhn, S., *Dünnschichtchromatographie*, Springer-Verlag, Berlin, Heidelberg, 1996.
21. Brenner, M. and Niederwieser, A., Durchlaufende dünnschicht-chromatographie, *Experientia*, 17, 237–238, 1961.
22. Geiss, F., Schlitt, H., and Klose, A., *Fresenius Z. Anal. Chem.*, 213, 331–346, 1965.
23. Geiss, F. and Schlitt, H., A new and versatile thin-layer chromatography separation system ('KS-Vario-Chamber'), *Chromatographia*, 1, 392–402, 1968.
24. Regis Technologies Inc., Morton Grove, IL.
25. Buncak, P., in GIT *Supplement Chromatographie*, 3–8, 1982.
26. Rumiński, J. K., Sandwich chamber for preparative layer chromatography, *Chem. Anal. (Warsaw)*, 33, 479–481, 1988.
27. Su, P., Wang, D., and Lan, M., Half-way development—The technique and the device, *J. Planar Chromatogr.—Mod. TLC*, 14, 203–207, 2001.
28. Lan, M., Wang, D., and Han, A new distributor for half-way development, *J. Planar Chromatogr.—Mod. TLC*, 16, 402–404, 2003.
29. Blome, J., in *High Performance Thin-Layer Chromatography*, Eds. Zlatkis, A., Kaiser, R. E., Elsevier, Institute of Chromatography, Bad Dürkheim, Germany, 1977.
30. Kaiser, R. E., in *High Performance Thin-Layer Chromatography*, Eds. Zlatkis. A., Kaiser, R. E., Elsevier, Institute of Chromatography, Bad Dürkheim, Germany, 1977.
31. Tyihák E. and Mincsovics E., Forced-flow planar liquid chromatographic techniques, *J. Planar Chromatogr.—Mod. TLC*, 1, 6–19, 1988.
32. http://www.sarstedt.com/php/main.php?newlanguage – en.
33. www.baron-lab.de.
34. Zarzycki, P. K., Simple horizontal chamber for thermostated micro-thin-layer chromatography, *J. Chromatogr. A*, 1187, 250–259, 2008.
35. Perry, J. A., Haag, K. W., and Glunz, L. J., Programmed multiple development in thin-layer chromatography, *J. Chromatogr. Sci.*, 11, 447–453, 1973.
36. Burger, K., *Fresenius Z. Anal. Chem.*, 318, 228–233, 1984.
37. Markowski, W., in *Encyclopedia of Chromatography*, Ed. Cazes, J., Marcel Dekker, Inc., New York, Basel, 2005, pp. 699–713.
38. Markowski, W., Computer-aided optimization of gradient multiple development thin-layer chromatography. III. Multi-stage development over a constant distance, *J. Chromatogr. A*, 726, 185–192, 1996.
39. Poole, C. F. and Belay, M. T., Progress in automated multiple development, *J. Planar Chromatogr.—Mod. TLC*, 4, 345–359, 1991.
40. Morlock, G. E., in *Preparative Layer Chromatography*, Eds. Kowalska, T., Sherma, J., CRC Press, Taylor & Francis Group, Boca Raton, FL, 2006.
41. Botz, L., Nyiredy, Sz., and Sticher, O., A new solid phase sample application method and device for preparative planar chromatography, *J. Planar Chromatogr.—Mod. TLC*, 3, 10–14, 1990.
42. Nyiredy, Sz. and Benkö, A., in *Proceedings of the International Symposium on Planar Separations, Planar Chromatography 2004*, Ed. Nyiredy, Sz., Research Institute for Medicinal Plants, Budakalász, Hungary, 2004, pp. 55–60.
43. Soczewiński, E. and Wawrzynowicz, T., Thin-layer chromatography as a pilot technique for the optimization of preparative column chromatography, *J. Chromatogr.*, 218, 729–732, 1981.
44. Poole, C. F. and Dias, N. C., Practitioner's guide to method development in thin-layer chromatography, *J. Chromatogr., A*, 892, 123–142, 2000.
45. Snyder, L. R., Kirkland, J. J., and Glajch J. L., Chapter 6 in *Practical HPLC Method Development*, 2nd edn., J. Wiley & Sons, New York, 1997, pp. 233–291.
46. Lide, D. R. and Frederikse, H. P. R., Eds., Fluid properties, Chapter 6 in *CRC Handbook of Chemistry and Physics*, CRC Press, Boca Raton, FL, 1995.
47. Héron, S. and Tchapla, A., *Analusis*, 21, 327–347, 1993.
48. Rohrschneider, L., Solvent characterization by gas-liquid partition coefficients of selected solutes, *Anal. Chem.*, 45, 1241–1247, 1973.
49. McReynolds, W. O., Characterization of some liquid phases, *J. Chromatogr. Sci.*, 8, 685–691, 1970.

50. Jandera, P. and Churáček, Gradient elution in liquid chromatography. I. The influence of the composition of the mobile phase on the capacity ratio (retention volume, band width, and resolution) in isocratic elution—theoretical considerations, *J. Chromatogr.*, 91, 207, 1974.

51. Karger, B. L., Gant, R., Hartkopf, A., and Weiner P. H., Hydrophobic effects in reversed-phase liquid chromatography, *J. Chromatogr.*, 128, 65–78, 1976.

52. Schoenmakers, P. J., Billiet, H. A. H., Tijssen, R., and De Galan, L., *J. Chromatogr.*, 149, 519–537, 1978.

53. Snyder, L. R., Classification of the solvent properties of common liquids, *J. Chromatogr.*, 92, 223–230, 1974.

54. Snyder, L. R., Classification of the solvent properties of common liquids, *J. Chromatogr. Sci.*, 16, 223–234, 1978.

55. Snyder, L. R., Glajch, J. L., and Kirkland, J. J., Theoretical basis for systematic optimization of the mobile phase selectivity in liquid-solid chromatography. Solvent-solute localization effects. *J. Chromatogr.*, 218, 299–326, 1981.

56. Glajch, J. L., Kirkland, J. J., Squire, K. M., and Minor, J. M., Optimization of solvent strength and selectivity for reversed-phase liquid chromatography using an interactive mixture-design statistical technique, *J. Chromatogr.*, 199, 57–79, 1980.

57. Kowalska, T. and Klama, B., On a deficiency of the concept of the eluotropic series, *J. Planar Chromatogr.*, 10, 353–357, 1997.

58. Johnson, B. P., Khaledi, M. G., and Dorsey, J. G., Solvatochromic solvent polarity measurements and retention in reversed-phase liquid chromatography, *Anal. Chem.*, 58, 2354–2365, 1986.

59. Kosower, E. M., The effect of solvent on spectra. I. A new empirical measure of solvent polarity: Z-values, *J. Am. Chem. Soc.*, 80, 3253–3260, 1958.

60. Kosower, E. M. and Mohammad, M., The solvent effect on a electron-transfer reaction of pyridinyl radicals, *J. Am. Chem. Soc.*, 90, 3271–3272, 1968.

61. Kamlet, M. J., Abboud, J. L. M., Abraham, M. H., and Taft, R. W., Linear solvation energy relationships. 23. A comprehensive collection of the solvatochromic parameters, π^*, α, and β, and some methods for simplifying in generalized solvatochromic equation, *J. Org. Chem.*, 48, 2877–2887, 1983.

62. Kamlet, M. J., Abboud, J. L. M., and Taft, R. W., The solvatochromic comparison method. 6. The π^* scale of solvent polarities, *J. Am. Chem. Soc.*, 99, 6027–6038, 1977.

63. Laurence, C., Nicolet, P., Dalati, M. T., Abboud, J. L. M., and Notario, R., The empirical treatment of solvent-solute interactions: 15 years of π^*, *J. Phys. Chem.*, 98, 5807–5816, 1994.

64. Taft, R. W. and Kamlet, M. J., The solvatochromic comparison method. 2. The α-scale of solvent hydrogen-bond donor (HBD) acidities, *J. Am. Chem. Soc.*, 98, 2886–2894, 1976.

65. Abraham, M. H., Poole, C. F., and Poole S. K., Classification of stationary phases and other material by gas chromatography, *J. Chromatogr. A*, 842, 79–114, 1999.

66. Abraham, M. H., Whiting, G. S., Shuely, W. J., and Doherty, R. M., *Can. J. Chem.*, 76, 703, 1998.

67. Abraham, M. H., Whiting, G. S., Carr, P. W., and Quyang, H., *J. Chem. Soc., Perkin Trans.*, 2, 1385, 1998.

68. Abraham, M. H., Platts, J. A., Hersey, A., Leo, A. J., and Taft, R. W., *J. Pharm. Sci.*, 88, 670, 1999.

69. Trappe, W., *J. Biochem.*, 305, 150–154, 1940.

70. Snyder, L. R., Solvent selectivity in adsorption chromatography on alumina. Non-donor solvents and solutes, *J. Chromatogr.*, 63, 15–44, 1971.

71. Gocan, S., Mobile phase in thin-layer chromatography, in *Modern Thin-Layer Chromatography*, Ed. Grinberg, E., Chromatographic Science Series, 52, Marcel Dekker, New York, 1990, pp. 427–434.

72. Neher, R., in *Thin-Layer Chromatography*, Ed. Marini-Bettolo B.G., Elsevier, Amsterdam, the Netherlands, 1964.

73. Saunders D. L., Solvent selection in adsorption liquid chromatography, *Anal. Chem.*, 46, 470, 1974.

74. Snyder, L. R., *Principles of Adsorption Chromatography*, Marcel Dekker, New York, 1968.

75. Snyder, L. R., Linear elution adsorption chromatography. VII. Gradient elution theory, *J. Chromatogr.*, 13, 415–434, 1964.

76. Snyder, L. R., Linear elution adsorption chromatography. III. Further delineation of the eluent role in separations over alumina, *J. Chromatogr.*, 8, 178–200, 1962.

77. Snyder, L. R., Linear elution adsorption chromatography. IX. Strong eluents and alumina. The basis of eluent strength. *J. Chromatogr.*, 16, 55–88, 1964.

78. Snyder, L. R., Linear elution adsorption chromatography. V. Silica as adsorbent. Adsorbent standardization, *J. Chromatogr.*, 11, 195–227, 1963.

79. Stahl, E., Ed., *Dünnschicht-Chromatographie*, Springer Verlag, Berlin 1st edn., 1962 and 2nd edn., 1967.

80. Abboutt, D. and Andrews, R. S., *An Introduction to Chromatography*, Longmans, Green, London, 1965, p. 27.

81. Soczewiński, E., Solvent composition effects in thin-layer chromatography. System of the type silica gel-electron donor solvent, *Anal. Chem.*, 41(1), 179, 1969.
82. Soczewiński, E., Quantitative retention-eluent composition relationships in partition and adsorption chromatography, Chapter 11 in *A Century of Separation Science*, Ed., Issaq, H., Marcel Dekker, New York, 2002, pp. 179–195.
83. Soczewiński, E., Mechanistic molecular model of liquid–solid chromatography: Retention–eluent composition relationships, *J. Chromatogr. A*, 965, 109, 2002.
84. Snyder, L. R., Dolan, J. W., and Gant, J. R., Gradient elution in high-performance liquid chromatography. I. Theoretical basis for reversed-phase systems, *J. Chromatogr.*, 165, 3–30, 1979.
85. Soczewiński, E. and Wachtmeister C. A., The relation between the composition of certain ternary two-phase solvent systems and R_M values, *J. Chromatogr.*, 7, 311–320, 1962.
86. Snyder, L. R., Carr, P. W., and Rutan, S. C., *J. Chromatogr. A*, 656, 537–547, 1993.
87. Qasimullach, A. A., Andrabi, S. M. A., and Qureshi, P. M., Solvent polarity as a function of R_f in thin-layer chromatography of selected nitro functions. *J. Chromatogr. Sci.*, 34, 376–378, 1996.
88. Martin, A. J. P. and Synge, R. L. M., Some applications of periodic acid to the study of the hydroxy-aminoacids of protein hydrolysates, *Biochem. J.*, 35, 294–314, 1941.
89. Martin, A. J. P., Gas-liquid partition chromatography: The separation and micro-estimation of volatile fatty acids from formic acid to dodecanoic acid, *Biochem. J.*, 50, 679, 1952.
90. Scott, R. P. and Kucera, P., Solute interactions with the mobile and stationary phases in liquid–solid chromatography, *J. Chromatogr.*, 112, 425, 1975.
91. Scott, R. P., The role of molecular interactions in chromatography, *J. Chromatogr.*, 122, 35, 1976.
92. Kowalska, T., A thermodynamic interpretation of the behavior of higher fatty alcohols and acids upon the chromatographic paper, *Microchem. J.*, 29, 375, 1984.
93. Kowalska, T., Bestimmung der aktivitätskoeffizienten im chromatographischen model "binärer lösungen", *Monatsh. Chem.*, 116, 1129, 1985.
94. Kowalska, T., *Fat Sci. Technol.*, 90, 259, 1988.
95. Ościk, J. and Chojnacka, G., Investigations on the adsorption process in thin-layer chromatography by using two-component mobile phases, *J. Chromatogr.*, 93, 167–176, 1974.
96. Ościk, J. and Chojnacka, G., The use of thin-layer chromatography in investigations of the adsorption of some aromatic hydrocarbons, *Chromatographia*, 11, 731–735, 1978.
97. Glajch, J. L. and Kirkland, J. J., Optimization of selectivity in liquid chromatography, *Anal. Chem.*, 55, 319A–336A, 1983.
98. Li, S. F. Y., Lee, H. K., and Ong, C. P., Optimization of mobile phase composition for high-performance liquid chromatographic separation by means of the overlapping resolution mapping scheme, *J. Chromatogr.*, 506, 245–252, 1990.
99. Nurok, D., Beker, R. M., Richard, M. J., Cunningham, P. D., Gorman, W. B., and Bush, C. L., *J. High Resolut. Chromatogr. Chromatogr. Commun.*, 5, 373–380, 1982.
100. Prus, W. and Kowalska, T., A novel approach to the optimization of separation quality in adsorption TLC, *J. Planar Chromatogr.—Mod. TLC*, 8, 205–215, 1995.
101. Prus, W. and Kowalska, T., A comparison of selected methods for optimization of separation selectivity in reversed-phase liquid–liquid chromatography, *J. Planar Chromatogr.—Mod. TLC*, 8, 288–291, 1995.
102. Coman, V., Măruţoiu, C., and Puiu, S., Optimization of separation of some polycyclic aromatic compounds by thin-layer chromatography, *J. Chromatogr. A*, 779, 321–328, 1997.
103. Kiridena, W. and Poole, C. F., Structure-driven retention model for method development in reversed-phase thin-layer chromatography on octadecylsiloxane-bonded layers, *J. Planar Chromatogr.—Mod. TLC*, 12, 13–25, 1999.
104. Pelander, A., Summanen, J., Yrjönen, T., Haario, H., Ojanperä, I., and Vuorela, H., Optimization of separation in TLC by use of desirability functions and mixture designs according to the 'PRISMA' method, *J. Planar Chromatogr.—Mod. TLC*, 12, 365–372, 1999.
105. Markowski, W., Soczewiński, E., and Matysik G., A microcomputer program for the calculation of RF values of solutes in stepwise gradient thin-layer chromatography, *J. Liquid Chromatogr.*, 10, 1261–1275, 1987.
106. Markowski, W. and Czapińska, K. L., Computer simulation of the separation in one- and two-dimensional thin-layer chromatography by isocratic and stepwise gradient development, *J. Liquid Chromatogr.*, 18, 1405–1427, 1995.
107. Matyska, M., Dąbek, M., and Soczewiński, E., Computer-aided optimization of liquid–solid systems in thin layer chromatography. 3. A computer program for selecting the optimal eluent composition for a given set of solutes from a database, *J. Planar Chromatogr.—Mod. TLC*, 3, 317–321, 1990.

108. Perišić-Janjić, N. U., Djaković-Sekulić, T., Jevrić L. R., and Jovanović, B. Ž., Study of quantitative structure-retention relationships for s-triazine derivatives in different RP HPTLC systems, *J. Planar Chromatogr.—Mod. TLC*, 18, 212–216, 2005.

109. Tuzimski, T. and Sztanke, K., Retention data for some carbonyl derivatives of imidazo[2,1-c][1,2,4] triazine in reversed-phase systems in TLC and HPLC and their use for determination of lipophilicity. Part 1. Lipophilicity of 8-Aryl-3-phenyl-6,7-dihydro-4H-imidazo[2,1-c][1,2,4]triazin-4-ones, *J. Planar Chromatogr.—Mod. TLC*, 18, 274–281, 2005.

110. Djaković-Sekulić, T. and Perišić-Janjić, N., Study of the characteristics and separating power of unconventional TLC supports. II. Principal-components analysis, *J. Planar Chromatogr. Mod. TLC*, 20, 7–11, 2007.

111. Flieger, J., Świeboda, R., and Tatarczak, M., Chemometric analysis of retention data from salting-out thin-layer chromatography in relation to structural parameters and biological activity of chosen sulphonamides, *J. Chromatogr. B*, 846, 334–340, 2007.

112. Tatarczak, M., Flieger, J., and Szumiło, H., Use of a graphical method to predict the retention times of selected flavonoids in HPLC from thin-layer chromatographic data *Chromatographia*, 61, 307–309, 2005.

113. Tuzimski, T., Thin-layer chromatography (TLC) as pilot technique for HPLC. Utilization of retention database (R_F) vs. eluent composition of pesticides, *Chromatographia*, 56, 379–381, 2002.

114. Dallenbach-Tölke, K., Nyiredy, Sz., Meier, B., and Sticher, O., Optimization of overpressured layer chromatography of polar naturally occurring compounds by the 'PRISMA' model, *J. Chromatogr.*, 365, 63–72, 1986.

115. Nyiredy, Sz., Dallenbach-Tölke, K., and Sticher, O., The 'PRISMA' optimization system in planar chromatography, *J. Planar Chromatogr.—Mod. TLC*, 1, 336–342, 1988.

116. Nyiredy, Sz., Dallenbach-Tölke, K., and Sticher, O., Correlation and prediction of the k' values for mobile phase optimization in HPLC, *J. Liquid Chromatogr.*, 12, 95–116, 1989.

117. Nyiredy, Sz. and Fatér, Z. S., Automatic mobile phase optimization, using the 'PRISMA' model, for the TLC separation of apolar compounds, *J. Planar Chromatogr.—Mod. TLC*, 8, 341–345, 1995.

118. Nyiredy, Sz., Solid–liquid extraction strategy on the basis of solvent characterization, *Chromatographia*, 51, S-288-S-296, 2000.

119. Nyiredy, Sz., Separation strategies of plant constituents–current status, *J. Chromatogr. B*, 812, 35–51, 2004.

120. Reich, E. and George, T., Method development in HPTLC, *J. Planar Chromatogr.—Mod. TLC*, 10, 273–280, 1997.

121. Morita, K., Koike, S., and Aishima, T., Optimization of the mobile phase by the Prisma and simplex methods for the HPTLC of synthetic red pigments, *J. Planar Chromatogr.—Mod. TLC*, 11, 94–99, 1998.

122. Pelander, A., Sivonen, K., Ojanperä, I., and Vuorela, H., Retardation behaviour of cyanobacterial hepatotoxins in the irregular part of the 'PRISMA' model for thin-layer chromatography, *J. Planar Chromatogr.—Mod. TLC*, 10, 434–440, 1997.

123. Cimpoiu, C., Hodişan, T., and Naşcu, H., Comparative study of mobile phase optimization for the separation of some 1,4-benzodiazepines, *J. Planar Chromatogr.—Mod. TLC*, 10, 195–199, 1997.

124. Tuzimski, T., A new procedure for separation of complex mixtures of pesticides by multidimensional planar chromatography, *J. Separation Sci.*, 30, 964–970, 2007.

125. Nyiredy, Sz., Fully on-line TLC/HPTLC with diode-array detection and continuous development. Part 1: Description of the method and basic possibilities, *J. Planar Chromatogr.—Mod. TLC*, 15, 454–457, 2002.

126. Perry, J. A., A new look at solvent strength selectivity and continuous development, *J. Chromatogr.*, 165, 117–140, 1979.

127. Matysik, G. and Soczewiński, E., Stepwise gradient development in thin-layer chromatography. IV. Miniaturized generators for continuous stepwise gradients, *J. Chromatogr.*, 446, 275–282, 1988.

128. Lee, Y. K. and Zlatkis, K., *Advances in Thin Layer Chromatography, Clinical and Environmental Applications*, Ed. Touchstone, J. C., Wiley-Interscience, New York, 1982.

129. Gonnord, M. F., Levi, F., and Guiochon, G., Computer assistance in the selection of the optimum combination of systems for two-dimensional chromatography, *J. Chromatogr.*, 264, 1–6, 1983.

130. Steinbrunner, J. E., Johnson, E. K., Habibi-Goudarzi, S., and Nurok, D., in *Planar Chromatography Vol. I*, Ed. Kaiser, R. E., Hüthig Verlag, Heidelberg, 1986.

131. Nurok, D., Habibi-Goudarzi, S., and Kleyle, R., Statistical approach to solvent selection as applied to two-dimensional thin-layer chromatography, *Anal. Chem.*, 59, 2424–2428, 1987.

132. Nurok, D., Strategies for optimizing the mobile phase in planar chromatography, *Chem. Rev.*, 89, 363–375, 1989.
133. Nurok, D., Kleyle, R., McCain, C. L., Risley, D. S., and Ruterbories, K. J., Statistical method for quantifying mobile phase selectivity in one- and two-dimensional overpressured layer chromatography, *Anal. Chem.*, 69, 1398–1405, 1997.
134. De Spiegeleer, B., Van den Bossche, W., De Moerlose, P., and Massart, D., A strategy for two-dimensional, high-performance thin-layer chromatography, applied to local anesthetics, *Chromatographia*, 23, 407–411, 1987.
135. Hauck, H. E. and Mack, M., Chapter 4 in *Handbook of Thin Layer Chromatography*, 2nd edn., Eds. Sherma, J., Fried, B., Marcel Dekker Inc., New York, 1996, pp. 101–128 (Figure 5, p. 115).
136. Tuzimski, T., Separation of a mixture of eighteen pesticides by two-dimensional thin-layer chromatography on a cyanopropyl-bonded polar stationary phase, *J. Planar Chromatogr.—Mod. TLC*, 17, 328–334, 2004.
137. Szabady, B. and Nyiredy, Sz., *Dünnschicht-Chromatographie in Memorian Professor Dr. Hellmut Jork*, Eds. Kaiser, R. E., Günther, W., Gunz, H., Wulff, G., InCom Sonderband, Düsseldorf, 1996, pp. 212–224.
138. Tuzimski, T. and Soczewiński, E., Chemometric characterization of the R_F values of pesticides in thin-layer chromatography on silica with mobile phases comprising a weakly polar diluent and a polar modifier. Part V, *J. Planar Chromatogr.—Mod. TLC*, 15, 164–168, 2002.
139. Tuzimski, T. and Soczewiński, E., Correlation of retention parameters of pesticides in normal- and reversed-phase systems and their utilization for the separation of a mixture of 14 triazines and urea herbicides by means of two-dimensional thin-layer chromatography, *J. Chromatogr. A*, 961, 277–283, 2002.
140. Tuzimski, T. and Soczewiński, E., Use of database of plots of pesticide retention (R_F) against mobile-phase composition. Part I. Correlation of pesticide retention data in normal- and reversed-phase systems and their use to separate a mixture of ten pesticides by 2D-TLC, *Chromatographia*, 56, 219–223, 2002.
141. Tuzimski, T. and Bartosiewicz, A., Correlation of retention parameters of pesticides in normal and RP systems and their utilization for the separation of a mixture of ten urea herbicides and fungicides by two-dimensional TLC on cyanopropyl-bonded polar stationary phase and two-adsorbent-layer Multi-K plate. *Chromatographia*, 58, 781–788, 2003.
142. Pandey, R. C., Misra, R., and Rinehart Jr., K. L., Graft thin-layer chromatography, *J. Chromatogr.*, 169, 129–139, 1979.
143. Łuczkiewicz, M., Migas, P., Kokotkiewicz, A., Walijewska, M., and Cisowski, W., Two-dimensional TLC with adsorbent gradient for separation of quinolizidine alkaloids in the herb and in-vitro cultures of several *Genista* species, *J. Planar Chromatogr.—Mod. TLC*, 17, 89–94, 2004.
144. Glensk, M., Sawicka, U., Mażol, I., and Cisowski, W., 2D TLC—Graft planar chromatography in the analysis of a mixture of phenolic acids, *J. Planar Chromatogr.—Mod. TLC*, 15, 463–465, 2002.
145. Glensk, M. and Cisowski, W., Two-dimensional TLC with an adsorbent gradient for analysis of saponins in *Silene vulgaris* Garcke, *J. Planar Chromatogr.—Mod. TLC*, 13, 9–11, 2000.
146. Tuzimski, T., Two-dimensional TLC with adsorbent gradients of the type silica-octadecyl silica and silica-cyanopropyl for separation of mixtures of pesticides, *J. Planar Chromatogr.—Mod. TLC*, 18, 349–357, 2005.
147. Tuzimski, T., Separation of multicomponent mixtures of pesticides by graft thin-layer chromatography on connected silica and octadecyl silica layers, *J. Planar Chromatogr.—Mod. TLC*, 20, 13–18, 2007.
148. Ebel, S. and Völkl, S., *Dtsch. Apoth. Ztg.*, 130, 2162–2169, 1990.
149. Markowski, W. and Soczewiński, E., Computer-aided optimization of gradient multiple-development thin-layer chromatography. I. Two-stage development, *J. Chromatogr.*, 623, 139–147, 1992.
150. Markowski, W. and Soczewiński, E., Computer-aided optimization of stepwise gradient and multiple-development thin-layer chromatography, *Chromatographia*, 36, 330–336, 1993.
151. Markowski, W., Computer-aided optimization of gradient multiple-development thin-layer chromatography. I. Multi-stage development, *J. Chromatogr.*, 635, 283–289, 1993.
152. Markowski, W., Czapińska K. L., and Błaszczak M., Determination of the constants of the Snyder-Soczewiński equation by means of gradient multiple development, *J. Liquid Chromatogr.*, 17, 999–1009, 1994.
153. Snyder, L. R. and Kirkland, J. J., *Introduction to Modern Liquid Chromatography*, 2nd edn., John Wiley & Sons, New York, 1979.
154. Gołkiewicz, W., Chapter 6 in *Handbook of Thin-Layer Chromatography*, 3rd edn. (revised and expanded), Eds. Sherma, J., Fried, B., Marcel Dekker, Inc., New York, Basel, 2003, pp. 153–173.

155. Matysik, G. and Soczewiński, E., A miniaturized for the generation of eluent composition gradients for sandwich thin-layer chromatography, *Anal. Chem. (Warsaw)*, 33, 363–369, 1988.

156. Matysik, G. and Soczewiński, E., Stepwise gradient development in thin-layer chromatography. IV. Miniaturized generators for continuous and stepwise gradients, *J. Chromatogr.*, 446, 275–282, 1988.

157. Soczewiński, E. and Matysik, G., A simple device for gradient elution in equilibrium sandwich chambers for continuous thin-layer chromatography, *J. Liquid Chromatogr.*, 8, 1225–1238, 1985.

158. Matysik, G. and Soczewiński, E., Gradient thin-layer chromatography of plant extracts, *Chromatographia*, 26, 178–180, 1988.

159. Zarzycki, P. K., Some technical problems associated with temperature-controlled thin-layer chromatography, *J. Planar Chromatogr.—Mod. TLC*, 14, 63–65, 2001.

160. Zarzycki, P. K., Simple chamber for temperature-controlled planar chromatography, *J. Chromatogr. A*, 971, 193–197, 2002.

161. Dzido, T. H., Effect of temperature on the retention of aromatic hydrocarbons with polar groups in binary reversed-phase TLC, *J. Planar Chromatogr.—Mod. TLC*, 14, 237–245, 2001.

162. Zarzycki, P. K., Nowakowska, J., Chmielewska, A., Wierzbowska, M., and Lamparczyk, H., Thermodynamic study of the retention behaviour of selected macrocycles using reversed-phase high-performance thin-layer chromatography plates and methanol-water mobile phases, *J. Chromatogr. A*, 787, 227–233, 1997.

163. Zarzycki, P. K., Wierzbowska, M., Nowakowska, J., Chmielewska, A., and Lamparczyk, H., Interactions between native cyclodextrins and *n*-alcohols studied using thermostated thin-layer chromatography, *J. Chromatogr. A*, 839, 149–156, 1999.

164. Zarzycki, P. K., Wierzbowska, M., and Lamparczyk, H., Retention and separation studies of cholesterol and bile acids using thermostated thin-layer chromatography, *J. Chromatogr. A*, 857, 255–262, 1999.

165. Hopf, P. P., Radial chromatography in industry, *Ind. Eng. Chem.*, 39, 365, 1947.

166. Caronna, G., *Chim. Ind. (Milan)*, 37, 113, 1955.

167. McDonalds, H. J., Bermes, E. W., and Shepherd, H. G., *Chromatogr. Methods*, 2, 1,1957

168. Herndon, J. F., Appert, H. E., Touchstone, J. C., and Davis, C. N., Horizontal chromatography accelerating apparatus. Description of apparatus and applications, *Anal. Chem.*, 34, 1061, 1962

169. Heftman, E., Krochta, J. M., Farkas, D. F., and Schwimmer, S., The chromatofuge, an apparatus for preparative rapid radial column chromatography, *J. Chromatogr.*, 66, 365–369, 1972.

170. Finley, J. W., Krochta, J. M., and Heftman, E., Rapid preparative separation of amino acids with the chromatofuge, *J. Chromatogr.*, 157, 435–439, 1978.

171. Hostettmann, K., Hostettmann-Kaldas, M., and Sticher, O., Rapid preparative separation of natural products by centrifugal thin-layer chromatography, *J. Chromatogr.*, 202, 154–156, 1980.

172. Botz, L., Nyiredy, S., Wehrli, E., and Sticher, O., Applicability of Empore™ TLC sheets for forced-flow planar chromatography. I. Characterization of the silica sheets, *J. Liquid Chromatogr.*, 13, 2809–2828, 1990.

173. Botz, L., Dallenbach, K., Nyiredy, S., and Sticher, O., Characterization of band broadening in forced-flow planar chromatography with circular development, *J. Planar Chromatogr.—Mod. TLC*, 3, 80–86, 1992.

174. Nyiredy, S., in Planar chromatography, Ed. Heftmann, E., *Chromatography*, Elsevier, Amsterdam, the Netherlands, Oxford, New York, Basel, 1992, pp. A109–A150.

175. Izmailov, N. A. and Schraiber, M. S., *Farmatzija*, 3, 1, 1938.

176. Ripphahn, J. and Halpaap, H., *HPTLC High Performance Thin Layer Chromatography*, Eds. Zlatkis, A., Kaiser, R. E., Elsevier, Amsterdam, the Netherlands, 1977, pp. 189–221.

177. Botz, L., Nyiredy, S., and Sticher, O., A new device for circular preparative planar chromatography, *J. Planar Chromatogr.—Mod. TLC*, 3, 401–406, 1990.

178. Nyiredy, S., *Handbook of Thin-Layer Chromatography*, 3rd edn., Eds. Sherma, J., Fried, B., Marcel Dekker, Inc., New York, Basel, 2003, pp. 307–337.

179. Kaiser, R. E., Anticircular high performance thin-layer chromatography, *J. High Resol. Chromatogr. Chromatogr Commun.*, 3, 164–168, 1978.

180. Studer, A. and Traitler, H., Sample collection, quantification, and identification in preparative anticircular planar chromatography, *J. High Resol. Chromatogr. Chromatogr. Commun.*, 9, 218–223, 1986.

181. Issaq, H. J., Triangular thin-layer chromatography, *J. Liquid Chromatogr.*, 3, 789–796, 1980.

182. Giddings, J. C., Use of multiple dimensions in analytical separations, in *Multi-dimensional Chromatography*, Ed. Cortes, H. J., Marcel Dekker, New York, 1990, pp. 251–299.

183. Nyiredy, Sz., Multidimensional planar chromatography, *LC GC Eur.*, 16, 52–59, 2003.

184. Nyiredy, Sz., Multidimensional planar chromatography, Chapter 8, in *Multidimensional Chromatography*, Eds. Mondello, L., Lewis, A. C., Bartle, K. D., Chichester, 2002, pp. 171–196.

Use of Planar Chromatography in Pesticide Residue Analysis

261

185. Tuzimski, T., Strategy for separation of complex mixtures by multidimensional planar chromatography, *J. Planar Chromatogr.—Mod. TLC*, 21, 49–54, 2008.
186. Tuzimski, T. and Soczewiński, E., Retention and selectivity of liquid-solid chromatographic systems for the analysis of pesticides (Retention database of ca. 100 pesticides), in *Problems of Science, Teaching and Therapy*. Medical University of Lublin, Poland, No 12, Lublin, October 2002, 219 pp. (limited number available on request: tomasz.tuzimski@am.lublin.pl).
187. Soczewiński, E., Tuzimski, T., and Pomorska, K., Chemometric characterization of the R_F values of pesticides in thin-layer chromatographic systems of the type silica + weakly polar diluent – polar modifier. *J. Planar Chromatogr.—Mod. TLC*, 11, 90–93, 1998.
188. Soczewiński, E. and Tuzimski, T., Chemometric characterization of the R_F values of pesticides in thin-layer chromatographic systems of the type silica + weakly polar diluent – polar modifier. Part II. *J. Planar Chromatogr.—Mod. TLC*, 12, 186–189, 1999.
189. Tuzimski, T. and Soczewiński, E., Chemometric characterization of the R_F values of pesticides in thin-layer chromatographic systems of the type silica + weakly polar diluent–polar modifier. Part III. *J. Planar Chromatogr.—Mod. TLC*, 13, 271–275, 2000.
190. Tuzimski, T., Chemometric characterization of the R_F values of pesticides in thin-layer chromatography on silica with mobile phases comprising a weakly polar diluent and a polar modifier. Part IV. *J. Planar Chromatogr.—Mod. TLC*, 15, 124–127, 2002.
191. Tuzimski, T. and Soczewiński, E., Correlation of retention data of pesticides in normal- and reversed-phase systems and utilization of the data for separation of a mixture of ten urea herbicides by two-dimensional thin-layer chromatography on cyanopropyl-bonded polar stationary phase and on two-adsorbent-layer Multi-K SC5 plate. *J. Planar Chromatogr.—Mod. TLC*, 16, 263–267, 2003.
192. Tuzimski, T. and Wojtowicz, J., Separation of a mixture of pesticides by 2D-TLC on two-adsorbent-layer Multi-K SC5 plate, *J. Liquid Chromatogr.*, 28, 277–287, 2005.
193. Zwickenpflug, W., Weiß, H., Fürst-Hunnius, N., and Richter, E., Separation of the reaction products of cyanazine and terbuthylazine nitrosation by thin-layer chromatography, *Fresenius J. Anal. Chem.*, 360, 679–682, 1998.
194. Neicheva, A. Bogdanova, and T. Konstantinova, Thin-layer chromatographic study of the synthesis of some unsaturated chlorotriazine derivatives with herbicidal activity, *J. Planar Chromatogr.—Mod. TLC*, 12, 145–149, 1999.
195. Rathore, H.S. and Varshney, C., Chromatographic behavior of dithiocarbamate fungicides on cellulose plates, *J. Planar Chromatogr.—Mod. TLC*, 20, 287–292, 2007.
196. Reuke, S. and Hauck, H.E., Thin-layer chromatography as a pilot technique for HPLC demonstrated with pesticide samples, *Fresenius J. Anal. Chem.*, 351, 739–744, 1995.
197. Tuzimski T. and Soczewiński E., Use of database of plots of pesticide retention (R_F) against mobile-phase composition. Part II. TLC as a pilot technique for transferring retention data to HPLC, and use of the data for preliminary fractionation of a mixture of pesticides by micropreparative column chromatography, *Chromatographia*, 56, 225–227, 2002.
198. Tuzimski, T., Two-stage fractionation of a mixture of pesticides by micropreparative TLC and HPLC, *J. Planar Chromatogr.—Mod. TLC*, 18, 39–43, 2005.
199. Tuzimski, T., Two-stage fractionation of a mixture of 10 pesticides by TLC and HPLC, *J. Liquid Chromatogr.*, 28, 463–476, 2005.
200. Tuzimski T. and Soczewiński E., Use of database of plots of pesticide retention (R_F) against mobile-phase compositions for fractionation of a mixture of pesticides by micropreparative thin-layer chromatography. *Chromatographia*, 59, 121–128, 2004.
201. Hansch, C., Maloncy, P. P., and Fujita, T., *Nature (London)*, 194, 178–180, 1962.
202. Biagi, L., Barbaro, A. M., Gamba, M. F., and Guerra, M. C., Partition data of penicillins determined by means of reversed-phase thin-layer chromatography, *J. Chromatogr.*, 41, 371–379, 1969.
203. Cserháti, T. and Forgács, E., Relationship between the hydrophobicity and specific hydrophobic surface area of pesticides determined by high-performance liquid chromatography compared with reversed-phase thin-layer chromatography, *J. Chromatogr. A*, 771, 105–109, 1997.
204. Biagi, G. L., Barbaro, A. M., Sapone, A., and Recanatini, M., Thin-layer chromatographic study of the lipophilicity of triazine herbicides. Influence of different organic modifiers, *J. Chromatogr.*, 625, 392–396, 1992.
205. Gozalbes, R., de Julián-Ortiz, J. V., Antón-Fos, G. M., Gálvez-Alvarez, J., and Garcia-Domenech, R., Prediction of chromatographic properties of organophosphorus insecticides by molecular connectivity, *Chromatographia*, 51, 331–337, 2000.
206. Cserháti, T. and Forgács, E., Retention of 4-cyanophenyl herbicides on water-insoluble β-cyclodextrin support, *J. Chromatogr. A*, 685, 295–302, 1994.

207. Cserháti, T., Forgács, E., Darwish, Y., Oros, G., and Illes, Z., Interaction of pesticides with a β-cyclodextrin derivative studied by reversed-phase thin-layer chromatography and principal component analysis, *J. Inclusion Phenomena and Macrocyclic Chemistry*, 42, 235–240, 2002.

208. Janicka, M., Comparison of different properties—log P, $\log_{kw}$, and φ_0—as descriptors of the hydrophobicity of some fungicides, *J. Planar Chromatogr.—Mod. TLC*, 19, 361–370, 2006.

209. Janicka, M., Ościk-Mendyk, B., and Tarasiuk, B., Planar chromatography in studies of the hydrophobic properties of some new herbicides, *J. Planar Chromatogr.—Mod. TLC*, 17, 186–191, 2004.

210. Perišić-Janjić, N. U. and Jovanović, B. Ž., Reversed-phase thin-layer chromatographic behaviour of some s-Triazine derivatives, *J. Planar Chromatogr.—Mod. TLC*, 16, 71–75, 2003.

211. Djaković-Sekulić, T., Perišić-Janjić, N. U., Sârbu, C., and Lozanov-Crenković, Z., Partial least-squares study of the effects of organic modifier and physicochemical properties on the retention of some thiazoles, *J. Planar Chromatogr.—Mod. TLC*, 20, 251–257, 2007.

212. Janicka, M., Perišić-Janjić, N. U., and Różyło, J. K., Thin-layer and overpressured-layer chromatography for evaluation of the hydrophobicity of s-triazine derivatives, *J. Planar Chromatogr.—Mod. TLC*, 17, 468–475, 2004.

213. Janicka, M., Use of thin-layer and over-pressured-layer chromatography to study the hydrophobicity of homologous s-triazines, *J. Planar Chromatogr.—Mod. TLC*, 20, 267–272, 2007.

214. Das, B., Sharif, Y., Khan, A., Das, P., and Shaheen, S. M., Organochlorine pesticide residues in catfish, *Tachysurus thalassinus* (Ruppell, 1835), from the South Patches of the Bay of Bengal, *Environ. Pollut.*, 120, 255–259, 2002.

215. Pacáková, V., Štulík, K., and Jiskra, J., High-performance separations in the determination of triazine herbicides and their residues, *J. Chromatogr. A*, 754, 17–31, 1996.

216. Font, G., Manes, J., Molto, J. C., and Picó, Y., *J. Chromatogr. A*, 642, 135–161, 1993.

217. Zhang, R., Yue, Y., Hua, R., Tang, F., Cao, H., and Ge, S., Separation of sulfonylurea herbicides and determination of bensulfuron-methyl residues in tapwater by HPTLC, *J. Planar Chromatogr.—Mod. TLC*, 16, 127–130, 2003.

218. Babić, S., Mutavdžić, D. and Kaštelan-Macan, M., SPE preconcentration and TLC determination of alachlor, atrazine and α-cypermethrin in water samples, *J. Planar Chromatogr.—Mod. TLC*, 16, 160–164, 2003.

219. Hamada, M. and Winstersteiger, R., Determination of phenylurea herbicides in drinking water, *J. Planar Chromatogr.—Mod. TLC*, 15, 11–18, 2002.

220. Hamada, M. and Winstersteiger, R., Fluorescence screening of organophosphorus pesticides in water by an enzyme inhibition procedure on TLC plates, *J. Planar Chromatogr.—Mod. TLC*, 16, 4–10, 2003.

221. Sherma, J. and Rolfe, C., Determination of diflubenzuron residues in water by solid-phase extraction and quantitative high-performance thin-layer chromatography, *J. Chromatogr.*, 643, 337–339, 1993.

222. Butz, S. and Stan, H.-J., Screening of 265 pesticides in water by thin-layer chromatography with automated multiple development, *Anal. Chem.*, 67, 620–630, 1995.

223. Morlock, G. E., Analysis of pesticide residues in drinking water by planar chromatography, *J. Chromatogr. A*, 754, 423–430, 1996.

224. De la Vigne, U., Jänchen, D. E., and Weber, W. H., Application of high-performance thin-layer chromatography and automated multiple development for the identification and determination of pesticides in water, *J. Chromatogr.*, 553, 489–496, 1991.

225. Tuzimski, T., Determination of pesticides in water samples from the Wieprz-Krzna Canal in the Łęczyńsko-Włodawskie Lake District of southeastern Poland by thin-layer chromatography with diode-array scanning and high-performance column liquid chromatography with diode array detection, *J. AOAC Int.*, 91(5), 1203–1209, 2008.

226. Tuzimski, T., Application of SPE-HPLC-DAD and SPE-TLC-DAD to the determination of pesticides in real water samples, *J. Sep. Sci.*, 31(20), 3537–3542, 2008.

227. Błądek, J., Rostkowski, A., and Miszczak, M., Application of instrumental thin-layer chromatography and solid-phase extraction to the analyses of pesticide residues in grossly contaminated samples of soil, *J. Chromatogr. A*, 754, 273–278, 1996.

228. Koeber, R. and Niessner, R., Screening of pesticide-contaminated soil by supercritical fluid extraction (SFE) and high-performance thin-layer chromatography with automated multiple development (HPTLC/AMD), *Fresenius J. Anal. Chem.*, 354, 464–469, 1996.

229. Fan, W., Yue, Y., Tang, F., and Cao, H., Use of HPTLC for simultaneous determination of three fungicides in tomatoes, *J. Planar Chromatogr.—Mod. TLC*, 20, 419–421, 2007.

230. Rezić, I., Horvat, A. J. M., Babić, S., and Kapelan-Macan, M., Determination of pesticides in honey by ultrasonic solvent extraction and thin-layer chromatography, *Ultrasonics Sonochem.*, 12, 477–481, 2005.

231. Tuzimski, T., Use of thin-layer chromatography in combination with diode-array scanning densitometry for identification of fenitrothion in apples, *J. Planar Chromatogr.—Mod. TLC*, 18, 419–422, 2005.

232. Cao, H., Yue, Y., Hua, R., Feng, T., Zhang, R., Fan, W., and Chen, H., HPTLC determination of imidacloprid, fenitrothion and parathion in Chinese cabbage, *J. Planar Chromatogr.—Mod. TLC*, 18, 151–154, 2005.

233. Ge, S., Tang, F., Yue, Y., Hua, R., and Zhang, R., HPTLC determination of pyrethroids residues in vegetables, *J. Planar Chromatogr.—Mod. TLC*, 17, 365–368, 2004.

234. Tang, F., Ge, S., Yue, Y., Hua, R., and Zhang, R., High-performance thin-layer chromatographic determination of carbamate residues in vegetables, *J. Planar Chromatogr.—Mod. TLC*, 18, 28–33, 2005.

235. Lautie, J.-P. and Stankovic, V., Automated multiple development TLC of phenylurea herbicides in plants, *J. Planar Chromatogr.*, 9, 113–115, 1996.

236. Goli, D. M., Locke, M. A., and Zablotowitz, R. M., *J. Agric. Food Chem.*, 45, 1244–1250, 1997.

237. Robertson, A. M. and Lester, J. N., *Environ. Sci. Technol.*, 28, 346–351, 1994.

238. Janda, V., Bartle, K. D., and Clifford, A. A., Supercritical fluid extraction in environmental analysis, *J. Chromatogr.*, 642, 283–299, 1993.

239. Molins, C., Hogendoorn, E. A., Heusinkveld, H. A. G., van Harten, D. C., van Zoonen, P., and Baumann, R. A., Microwave assisted solvent extraction (MASE) for the efficient determination of triazines in soil samples with aged residues, *Chromatographia*, 43, 527–532, 1996.

240. Barker, S. A., Matrix solid-phase dispersion, *J. Chromatogr. A*, 885, 115–127, 2000.

241. Babić, S., Kaštelan-Macan, M., and Petrović, M., Determination of agrochemical combination in spiked soil samples, *Water Sci. Technol.*, 37, 243–250, 1998.

242. Stahl, E. and Mangold, H. K., in *Chromatography*, Van Nostrand-Reinhold, New York, 1975, p. 164.

243. Jork, H., Funk, W., Fischer, W., and Wimmer, H., Eds., *Thin-Layer Chromatography*, VCH, Verlagsgesellschaft, Weinheim, 1990.

244. Vassileva-Alexandrova, P. and Neicheva, A., Chromatographic study of the residues of fungicidal ethylenebisdithiocarbamates and ethylenethioureas in plants after oxidative inactivation with potassium permanganate, *J. Planar Chromatogr.—Mod. TLC*, 12, 425–428, 1999.

245. Vassileva-Alexandrova, P., Neicheva, A., Ivanov, K., and Nikolova, M., Thin-layer chromatography of the degradation products of fungicidal ethylenebisdithiocarbamates on oxidation with potassium permanganate, *J. Planar Chromatogr.—Mod. TLC*, 9, 425–429, 1996.

246. Patil, V. B. and Garad, M. V., A new chromogenic spray reagent for detection and identification of monocrotophos, *J. Planar Chromatogr.—Mod. TLC*, 14, 210–212, 2001.

247. Patil, V. B. and Shingare, M. S., Thin-layer chromatographic detection of monocrotophos in biological materials, *Talanta*, 41, 2127–2130, 1994.

248. Rane, K. D., Mali, B. D., Garad, M. V., and Patil, V. B., Selective detection of dichlorvos by thin-layer chromatography, *J. Planar Chromatogr.—Mod. TLC*, 11, 74–76, 1998.

249. Patil, V. B. and Shingare, M. S., A new spray regent for selectivity detection of dichlorvos by thin-layer chromatography, *Talanta*, 41, 367–369, 1994.

250. Daundkar, B. B., Mavle, R. R., Malve, M. K., and Krishnamurthy, R., Spectrophotometric and TLC detection reagent for the insecticides dichlorvos (DDVP) and diptrex (trichlorfon), and their metabolites, in biological tissues, *J. Planar Chromatogr.—Mod. TLC*, 20, 217–219, 2007.

251. Mali, B. D., Garad, M. V., Patil, V. B., and Padalikar, S. V., Thin-layer chromatographic detection of dichlorvos and dimethoate using orcinol, *J. Chromatogr. A* 704, 540–543, 1995.

252. Fugatami, K., Narazaki, C., Kataoka, Y., Shuto, H., and Oishi, R., Application of high-performance thin-layer chromatography for the detection of organophosphorous insecticides in human serum after acute poisoning, *J. Chromatogr. B*, 704, 369–373, 1997.

253. Daundkar, B. B., Mavle R. R., Malve M. K., and Krishnamurthy, R., Detection of carbaryl insecticide in biological samples by TLC with a specific chromogenic reagent, *J. Planar Chromatogr.—Mod. TLC*, 19, 467–468, 2006.

254. Patil, V. B. and Shingare, M. S., Thin-layer chromatographic detection of carbaryl using phenylhydrazine hydrochloride, *J. Chromatogr. A*, 653, 181–183, 1993.

255. Rathore, H. S. and Begum, T., Thin-layer chromatographic behaviour of carbamate pesticides and related compounds, *J. Chromatogr.*, 643, 321–329, 1993.

256. Fodor-Csorba, K., Holly, S., and Neszmélyi, A., Bujtás, G., Characterization of reaction products formed during thin-layer chromatographic detection of thiocarbamate herbicides, *Talanta* 39, 1361–1367, 1992.

257. Rathore, H. S. and Mital, S., Spot test analysis of pesticides: TLC detection of mancozeb, *J. Planar Chromatogr.—Mod. TLC*, 10, 124–127, 1997.

258. Lodi, G., Betti, A., Menziani, E., Brandolini, V., and Tosi, B., *J. Planar Chromatogr.-Mod. TLC*, 4, 106–110, 1991.
259. Matsunaga, H., Isobe, N., Kaneko, H., Nakatsuka, I., and Yamane, S., Metabolism of pentyl 2-chloro-4-fluoro-5-(3,4,5,6-tetrahydrophthalimido)phenoxyacetate (flumiclorac pentyl, S-23031) in rats. 2. Absorption, distribution, biotransformation, and excretion, *J. Agric. Food Chem.*, 45, 501–506, 1997.
260. Cheah, U.-B., Kirkwood, R. C., and Lum, K.-Y, Degradation of four commonly used pesticides in Malaysian agricultural soils, *J. Agric. Food Chem.*, 46, 1217–1223, 1998.
261. Dec, J., Haider, K., Rangaswamy, V., Schäffer, A., Fernandes, E., and Bollag, J.-M., Formation of soil-bound residues of cyprodinil and their plant uptake, *J. Agric. Food Chem.* 45, 514–520, 1997.
262. Harlarnkar, P. P., Leimkuehler, W. M., Green, D. L., and Marlow, V. A., *J. Agric. Food Chem.*, 45, 2349–2353, 1997.
263. Ravanel, P., Liégeois, M. H., Chevallier, D., and Tissut, M., Soil thin-layer chromatography and pesticide mobility through soil microstructures. New technical approach, *J. Chromatogr. A*, 864, 145–154, 1999.
264. Sánchez-Camazano, M., Sánchez-Martín, M. J., Poveda, E., and Iglesias-Jiménez, E., *J. Chromatogr. A*, 754, 279–284, 1996.
265. Tomigahara, Y. et al., Metabolism of 7-fluoro-6-(3,4,5,6-tetrahydrophthalimido)-4-(2-propynyl)-2*H*-1,4-benzoxazin-3(4*H*)-one (S-53482) in rat. 1. Identification of a sulfonic acid type conjugate, *J. Agric. Food Chem.*, 47, 305–312, 1999.
266. Serre, A. M., Roby, C., Roscher, A., Nurit, F., Euvrard, M., and Tissut, M., Comparative detection of fluorinated xenobiotics and their metabolites through ^{19}F NMR or ^{14}C Label in plant cells, *J. Agric. Food Chem.*, 45, 242–248, 1997.
267. Tellier, F., Fritz, R., Leroux, P., Carlin-Sinclair, A., and Cherton, J.-C., Metabolism of cymoxanil and analogs in strains of the fungus *Botrytis cinerea* using high-performance liquid chromatography and ion-pair high-performance thin-layer chromatography, *J. Chromatogr. B*, 769, 35–46, 2002.
268. Weins, C. and Jork, H., Toxicological evaluation of harmful substances by in situ enzymatic and biological detection in high-performance thin-layer chromatography, *J. Chromatogr. A*, 750, 403–407, 1996.
269. Reich, E. and Schibli, A., *High-Performance Thin-Layer Chromatography for the Analysis of Medicinal Plants*, Thieme Medicinal Publisher, Inc., New York, 2006.
270. Spangenberg, B. and Klein, K.-F., New evaluation algorithm in diode-array thin-layer chromatography, in *Planar Chromatography 2001, New Milestones in TLC*, Ed. Nyiredy Sz., Research Institute for Medicinal Plants, Budakalász, Hungary, 2001, pp. 15–21.
271. Spangenberg, B. and Klein, K.-F., Fibre optical scanning with high resolution in thin-layer chromatography, *J. Chromatogr. A*, 898, 265–269, 2000.
272. Spangenberg, B., Lorenz, K., and Nasterlack, S., Fluorescence enhancement of pyrene measured by thin-layer chromatography with diode-array detection, *J. Planar Chromatogr.—Modern TLC*, 16, 331–337, 2003.
273. Spangenberg, B., Seigel, A., Kempf, J., and Weinmann, W., Forensic drug analysis by means of diode-array HPTLC using R_F and UV library search, in *Planar Chromatography 2005, Milestones in Instrumental TLC*, Ed. Nyiredy Sz., Research Institute for Medicinal Plants, Budakalász, Hungary, 2005, pp. 3–16.
274. Ahrens, B., Blankenhorn, D., and Spangenberg, B., Advanced fibre optical scanning in thin-layer chromatography for drug identification, *J. Chromatogr. B*, 772, 11–18, 2002.
275. Davies, L., O'Connor, M., and Logan, S., Chronic intake, in *Pesticide Residues in Food and Drinking Water. Human Exposure and Risk*, Eds. Hamilton, D., Crossley, S., John Wiley & Sons Ltd., Chichester, England, 2004, pp. 213–241.
276. Stan, H. J. and Wippo, U., Pesticide residue analysis in plant food products by TLC with automated multiple development, *GIT Fachz. Lab.*, 40, 855, 1996.
277. Oi, M., Time-dependent sorption of imidacloprid in two different soils, *J. Agric. Food Chem.*, 47, 327–332, 1999.
278. Sharma, K. K., High-performance thin-layer chromatographic methods for determination of fenvalerate and deltamethrin in emulsifiable concentrate formulations, *J. Planar Chromatogr.—Mod. TLC*, 15, 67–70, 2002.
279. Rasmussen, J. and Jacobsen, O. S., Thin-layer chromatographic methods for the analysis of eighteen different ^{14}C-labeled pesticides, *J. Planar Chromatogr.—Mod. TLC*, 18, 248–251, 2005.
280. Song, L., Zhao, Y., and Hua, R., Separation of fenoxaprop-p-ethyl biodegradation products by HPTLC, *J. Planar Chromatogr.—Mod. TLC*, 18, 85–88, 2005.

10 Role of Surfactants in Thin-Layer Chromatography of Pesticides

Ali Mohammad and Amjad Mumtaz Khan

CONTENTS

Surfactants constitute the most important group of detergent components. Generally, they are water-soluble surface active agents made up of a hydrophobic portion, usually a long alkyl chain, attached to a hydrophilic or water-soluble functional group.

According to the charge of the hydrophilic portion of the molecule, surfactants can be categorized into four groups:

1. Anionic
2. Nonionic
3. Cationic
4. Amphoteric

All surfactants possess the common property of lowering the surface tension when added to water in small amounts. The characteristic discontinuity in the plots of surface tension against surfactant concentration can be experimentally determined. The corresponding surfactant concentration at this discontinuity corresponds to the critical micelle concentration (CMC). At surfactant concentrations below CMC, surfactant molecules are loosely integrated into the water structure. In the region of CMC, surfactant water structure is changed in such a way that the surfactant molecules begin to build up their own structures (micelles in the interior and monolayer at the surface).

Because of the limited solubility of surfactants in water, aggregates are formed in which the hydrophobic or hydrophilic sections of the surfactants are stuck together [1]. The micelle may be represented as a globular, cylindrical, or ellipsoidal cluster [2] of individual surfactant molecules in equilibrium with its monomers [2–11]. The reverse orientation of the hydrophilic and hydrophobic portion of the surfactant in a hydrocarbon medium leads to reversed micelles [12].

Surfactants are of widespread importance in detergent industry, emulsification, lubrication, catalysis, tertiary oil recovery, and drug delivery.

10.1 ANALYTICAL ASPECTS OF SURFACTANT

As the alkyl chain length of a surfactant has a decisive influence on its physiochemical properties and consequently on various biochemical applications [13,14], each surfactant has been analyzed with respect to the uniformity of the alkyl chain by various techniques, such as thin-layer chromatography (TLC), high-performance chromatography (HPLC), or gas chromatography (GC) (after hydrolysis). Some traditional surfactants, such as Triton series, are mixtures of a variety of homologues and must correspond to a standard mixture. Surfactants consisting of only one species are characterized by a minimum purity assay, which refers to chain homologue purity. Each surfactant is further checked for appearance, solubility, identity (by FTIR or NNMR), and relevant trace impurities such as respective starting material, peroxides, UV-absorbing foreign materials, and metal traces (by ICP-AES). The stereochemical purity is checked by measuring optical rotation.

10.2 ROLE OF SURFACTANTS IN THIN-LAYER CHROMATOGRAPHY

Surfactants were first used in paper chromatography in 1963 [1]. Later, the use of surfactants was extended to other chromatographic techniques. Different versions of TLC with surfactant modifications expanded the potentialities of the method and, in some cases, provided efficient separation of mixtures, especially those containing neutral and charged organic compounds. It was demonstrated that surfactants modify both the mobile and stationary phases, imparting new properties to the phases qualitatively.

1. Aqueous micellar solutions of surfactant
   In this case, the concentration of surfactant in the mobile phase exceeds the CMC and the main carrier of separated compounds is normal. This type of chromatography was initially called pseudo-phase, and later as micellar TLC (MTLC) [7].
2. Molecular solutions of ionic surfactants
   In this case, the concentration of surfactant ions in the solution is kept lower than the CMC. The surfactant ions act as hydrophobic countermines of separated compounds. This version of chromatography is known as ion-pair TLC.

MTLC has some advantages over conventional TLC based on nonaqueous and aqueous organic mobile phases. Aqueous micellar mobile phases (MMPs) are free from certain drawbacks, such as strong smell, high volatility, flammability, and toxicity when compared with the organic solvents [2,3,15]. Other important advantages of MMPs in TLC include their low cost and biodegradability [3,15]. The high selectivity of MTLC is due to the presence of surfactant micelles in the mobile phase. The increased selectivity is based on the difference in the degree of binding of separated mixture components with micelles. Furthermore, the increased selectivity is also based on the fact that three types of equilibria are involved: solvent–sorbent, solvent–micelle, and micelle–sorbent equilibria [16].

The mobility of adsorbates depends on the concentration level of the surfactants, which actually means that each concentration corresponds to a new solvent and thus affects the selectivity. In case of ionic surfactants, the selectivity can be controlled either by changing the nature counter ion [4] or by introducing small amounts of organic solvents. Another advantage of MTLC is the possibility of obtaining lower detection limits of fluorescent and phosphorescent compounds than that of conventional TLC.

Among all types of micelles of surfactants (cationic, anionic, and nonionic) used as mobile phase, ionic surfactants were found to be most effective for MTLC [5,10,11]. Sodium dodecylsulfate (SDS) anionic surfactant was especially very useful. It is important for polar silica gel, whose surface binds cationic surfactants more strongly because of both hydrophobic and electrostatic interactions with dissociated silanol groups [3,17]. Among the cationic surfactants, cetyltrimethylammonium bromide and chloride (CTAB and CTAC) are the most frequently used surfactants. In addition, micelles of nonionic surfactants, Tween-80 [18], and Triton X-100, have also been used [19].

MMPs were used for the separation of a mixture of polycyclic aromatic hydrocarbons [4,18], pesticides [3,5], nucleotides [3], anthraquinone and 1,4-napthoquinone, vitamin K [5], aminophenols [20], amino acids [4,21], p-nitrophenol and p-nitroaniline [22,23], phenol, resorcinol, pyrogallol, pyrocatechol and hydroquinone [15,20], mixture of bromocresol green, methyl orange, methylene blue and fluorescence [20], and mixtures of different food dyes [15,24].

The dynamic modification of the surface of the adsorbent with surfactant molecules is a result of their sorption from the solution [20]. In case of TLC plates with a normal polar phase, the surface of hydrophilic silica gel is hydrophobized and gains properties of a reverse phase [10,11,25–28]. On the contrary, the surface of plates with a reversed phase (RP-8, RP-18) acquires the charge of the adsorbed. Surfactant also shows ion-exchange properties [17,21]. The characteristic of modification becomes more predictable if the sorption mechanism of different surfactants in normal and reversed phases is clear. Furthermore, the peculiarity of the adsorption of cationic, anionic, and nonionic surfactants on different stationary phases is taken into consideration [29–32].

A model describing the chromatographic behavior of dissolved compounds in MMPs was proposed by Armstrong and Nome [13]. In MTLC, with aqueous organic mobile phases, the compound is distributed not only between the mobile and stationary phases, but also in the mobile phase, between water and surfactant micelle. Therefore, the chromatographic behavior of the adsorbate depends on the combined effect of the three partition coefficients: the partition coefficient between micelle and water (K_{mw}), the partition coefficient between stationary phase and water (K_{sw}), and the partition coefficient between stationary phase and micelle (K_{sm}), as shown in Figure 10.1.

The proposed model of micellar liquid chromatography makes it possible to divide all the compounds into four groups [17,15]: (a) Compounds that are bound by micelles. The mobility of these compounds tends to increase with the increase in the concentration of the surfactant in the mobile phase. (b) Compounds that are not bound by micelles and their mobility is independent of

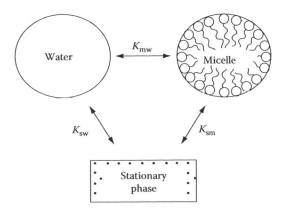

FIGURE 10.1 Three-phase model of micellar chromatography. (From Armstrong, D.W. and Nome, F., *Anal. Chem.*, 53, 1662, 1981. With permission.)

268 Handbook of Pesticides: Methods of Pesticide Residues Analysis

the concentration. (c) Compounds that are not bound by micelles and their mobility decreases with increasing concentration of the surfactant in the mobile phase ($K_{mw} < 0$). (d) High molecular compounds with an anomalously strong binding, which involves more than one surfactant micelle.

An analysis of partition coefficients K_{mw} of nitroaniline, nitrophenol, and *p,p*-dichlorodiphenyl-trichloroethane (DDE) pesticide [22,23] demonstrates that solubilization of micelles is controlled by the charge of the surfactant and nature of its counterion and solubilizate.

The length of the hydrocarbon radical of the surfactant also affects K_{mw} [23]. In MTLC, the surfactants occur in the mobile phase in both micellar and ionic forms. Because of a high concentration of the surfactant in the mobile phase, the surface of the adsorbent is saturated with surfactant ions, whose concentration in the stationary phase is nearly constant. Therefore, an increase in the total concentration of the surfactant in the eluent leads to an increase in the concentration of micelles in MMPs and decrease in the retention of adsorbates [33]. The decrease in the retention of adsorbates depends on the ratio of the parameters K_{mw} and K_{sw}.

However, a different situation is observed in ion-pair TLC, in which the mobile phase contains only ions of a surfactant. An increase in their concentration in the mobile phase increases the concentration of surfactant ions adsorbed on the stationary phase. As a result, the retention of oppositely charged adsorbates, unlike MTLC, will increase rather than decrease [33]. Another consequence of the nonconstant concentration of the surfactant at the surface of the adsorbent in ion-pair TLC is lower reproducibility of the results, when compared with MTLC. To eliminate this disadvantage, the preliminary impregnation of the adsorbent with a surfactant solution is frequently used. During elution of compounds, the mobile phase can either contain surfactant or not. Thus, three types of modifications are possible: (1) impregnation of stationary phase in the absence of surfactant in the mobile phase; (2) impregnation of stationary phase with simultaneous introduction of surfactant in the molecular form into the mobile phase; (3) introduction of ionic surfactants as ion-pair reagents only in the mobile phase that leads to dynamic modification of the stationary phase.

10.3 CHARACTERISTICS OF THE METHOD

TLC plates coated with silica gel are impregnated with ethanolic [2,21,34] or methanolic [8,17] solutions of anionic [2,21], cationic [8,21,34], or nonionic [35,36] surfactants. The impregnation of reversed phases RP-2, RP-8 with an ethanolic solution of dodecylsulfuric acid [37] and RP-18 with a solution of CTAB or paraffin [38] is also reported. The procedure for preparing adsorbent layers modified with surfactants has been described elsewhere [2,39–44]. Commercial TLC plates are impregnated in ethanolic [21,37] or methanolic [8] solutions of surfactant and dried in air.

The implications of TLC plates impregnated with surfactants are

- Change in the elution order of compounds in both normal and reversed phase modes
- Effects of the nature of the impregnating agent and its concentration
- Effects of the nature of substituents and their position in the molecules of adsorbates
- Effects of the composition of the mobile phase on the efficiency and selectivity of the separations

CTAB has the maximum effect on the sorption properties of the stationary phase and causes inversion of the elution order on the normal phase. The elution order on the stationary phase impregnated with Triton X-100 remains unchanged.

10.4 MECHANISM OF SEPARATION ON IMPREGNATED PLATES

As described earlier [2,41,42], the main separation mechanisms on impregnated stationary phase are ion exchange and distribution mechanisms. Cation exchange occurs in the case of adsorption of anionic surfactants, while anion exchange occurs in the case of adsorption of cationic surfactants.

The separation mechanism can be affected by the concentration of the impregnating surfactant and ion-exchange mechanism at higher concentrations of the surfactant [41]. Another important factor is the acidity of the eluent that can change the chemical state of the adsorbate and the mechanism of separation process [42–45].

10.5 ION-PAIR TLC

Two different modes exist. An ion-pair reagent can be introduced in the mobile phase and the stationary phase can be dynamically modified [11,46,47], while other researchers impregnate the adsorbent with an ion-pair reagent [34,48,49]. Unlike in MTLC, in ion-pair TLC, the adsorbate is distributed only between the mobile and stationary phases. An increase in selectivity arises owing to the stability and hydrophobicity of ion pairs. The efficiency of separation of adsorbed compounds is affected by the polarity of the mobile phase and the concentration of the ion-pair reagent. An increase in the concentration of water in the aqueous-methanolic mobile phase from 0%–30% leads to a decrease in the retention of diols, biphenyls, and naphthalene derivatives on Separon C_{18} and hence this improves separation [46]. A more significant improvement in the separation can be achieved by the addition of ion-pair reagent (sodium dodecylsulfate) to the mobile phase.

In reversed phase ion-pair high-performance TLC, water concentration in the mobile phase should not be higher than 80% for RP-2 and 60% for RP-8 or RP-18 [50]. The retention and separation mechanism in ion-pair TLC is complex and cannot be explained by any single model [51]. The main factors that control the separation are the concentration and hydrophobicity of the ion-pair reagent, ratio of water and organic solvent in the mobile phase, and the pH of the eluent.

10.6 THIN-LAYER CHROMATOGRAPHY

TLC is considered to be the most simple, rapid, versatile, and cost-effective method, which is applicable to the characterization and separation of a variety of multicomponent mixtures (both ionic and nonionic), except those that are volatile or reactive substances. The TLC technique has been applied for many years in the analysis of organic and inorganic substances, and in the analysis of biological, pharmaceutical, and environmental samples [52–56].

An improved version of TLC, the high-performance thin-layer chromatography (HPTLC), was introduced by Pretorius [57] in 1974, who described this technique as a high-speed TLC. This technique is based on electro-osmotic flow and has been frequently used for the separation and determination of a large number of substances [58–60]. Izmailov was the first to report about TLC, and Shraiber, in 1938, utilized the thin layer of alumina on glass plate for the separation of plant extracts [61].

The popularity of TLC as an analytical technique continues, because of its following favorable features:

1. Use of colorful reactions
2. Minimal sample cleanup
3. Wider choice of mobile and stationary phases
4. Excellent resolution power
5. High sample loading capacity
6. Handling of several samples simultaneously
7. Disposable nature of TLC plates
8. Possible use of corrosive detection reagents

Apart from TLC and HPTLC, other primary methods used for pesticide residue determinations include GC, HPLC, and immunoassay, because of their unique advantages such as high sample throughput and low operating costs. Multiple samples can be analyzed with standards on a single plate with a very low volume of solvent, with selective and sensitive detection and identification with a very wide variety of chromogenic, fluorogenic, and biological reagents coupled with spectrometric techniques as well as high resolution and accurate and precise quantification achieved on HPTLC plates. Thin-layer radio chromatography (TLRC) is used routinely for metabolism, breakdown, and other studies of pesticides in plants, animals, and in the environment. Furthermore, studies on hydrophobicity and pesticide migration through soils have also been carried out. Each year, almost 30% of TLC papers are about pesticide determinations in food, and environment TLC is widely used in a variety of pesticide studies, such as the determination of quantitative structure–activity relations (QSARs) that describe how the molecular structure, in terms of descriptors (lipophilic, electronic, and steric), affects the biological activity of a compound.

10.7 PREPARATIVE LAYER CHROMATOGRAPHY

Preparative layer chromatography (PLC) on 0.5–2 mm layers of sorbent is used to isolate and purify material in larger amounts that are normally chromatographed on 0.1–0.25 mm analytical thin layers. The antifungal compound, phenylacetic acid, produced by the antagonistic bacterium *Pseudomonas* sp. was isolated from greenhouse soil by PLC on silica gel with cyclohexane–ethyl acetate as the mobile phase. The layer fraction was scraped and eluted with diethyl ether, and the purified antifungal compound was identified ($R_F = 0.38$) by analytical silica gel TLC with cyclohexane–diethyl ether as the mobile phase [62].

10.8 TLC APPLICATIONS

A recent work showing the role of surfactants in the analysis of pesticides by TLC has been summarized in Table 10.1.

A lot of studies have been carried out on TLC of pesticides using nonsurfactant-mediated mobile-phase systems comprising a mixture of organic solvents, mixed aqueous organic solvents, and acidic and basic inorganic solvent systems. However, very little work has been reported on the use of surfactants in the mobile phase or stationary phase for the analysis of pesticides. Continuous efforts are needed to exploit the full analytical potential of surfactants in the analysis of pesticides by normal phase, reversed phase, or MTLC.

TABLE 10.1
Application of Surfactants in TLC Analysis of Pesticide

Pesticides Studied	Stationary Phase	Mobile Phase	Refs.
Diazinon, atrazine metalachlor, and acephate	Soil, soil amended with surfactants	Tetradecyl methyl ammonium bromide (cationic), lauryl sulfate (anionic), and Tween-80 (nonionic)	[1]
Diazinon, acephate, atrazine, and ethofumesate	Soil	Tetradecyl methyl ammonium bromide (cationic), sodium dodecylsulfate (anionic), and polyoxyethylene sorbitammonooleate (nonionic)	[3]
p,p′-DDT,*p,p′*-DDD,*p,p′*-DDE, and decachlorophenyl	Polyamide and alumina thin-layer sheets	Sodium dodecylsulfate (anionic) and cetyltrimethylammonium bromide (cationic)	[4]
Sulfur-, chlorine-, and phosphorous-containing pesticides	Silica, soil, and mixed layers with aqueous or sodium salt solutions	Cetyltrimethylammonium bromide (cationic) and pure solvents	[5]

REFERENCES

1. Farulla, E., Iacobelli-Turi, C., Lederer, M., and Salvetti, F., Surface active agents in paper chromatography. *J. Chromatogr.*, 1963, 12, 255–261.
2. Lepri, L., Desideri, P. G., and Heimler, D., Soap thin layer chromatography of some primary aliphatic amines. *J. Chromatogr.*, 1978, 153, 77–82.
3. Armstrong, D. and Terril, R. Q., Thin layer chromatographic separation of pesticides, decachlorobiphenyl, and nucleosides with micellar solutions. *Anal. Chem.*, 1979, 51(13), 2160–2163.
4. Armstrong, D. W. and McNeely, M., Use of micelles in the TLC separation of polynuclear aromatic compounds and amino acids. *Anal. Lett.*, 1979, 12(12), 1285–1291.
5. Armstrong, D. W., Pseudophase liquid chromatography: Applications to TLC. *J. Liq. Chrom. Rel. Technol.*, 1980, 3(6), 895–900.
6. World Health Organization, WHO gives indoor use of DDT a clean bill of health for controlling malaria, www.who.int/mediacentre/news/releases/2006/pr50/en/index.html
7. Pelizzetti, E. and Pramauro, E., Analytical applications of organized molecular assemblies. *Anal. Chim. Acta.*, 1985, 169, 1–29.
8. Van Peteghem, C. and Bijl, J., Ion-pair extraction and ion-pair adsorption thin-layer chromatography for rapid identification of ionic food dyes. *J. Chromatogr.*, 1981, 210(1), 113–120.
9. Lobe, J. (Sept 16, 2006), WHO urges DDT for malaria control strategies, www.commondreams.org/headlines06/0916–05.htm
10. Shtykov, S. N., Sumina, E. G., Parshina, E. V., and Lopukhova, S. S., *Zh. Anal. Khim.*, 1995, 50, 747.
11. Shtykov, S. N., Sumina, E. G., Smushkina, E. V., and Tyurina, N. V., *J. Planar Chromatogr.*, 2000, 13, 182.
12. Savvin, S. B., Chernova, R. K., and Shtykov, S. N., *Poverkhnostno-aktivnye veshchestva (Surfactants)*, Moscow: Nauka, 1991.
13. Armstrong, D. W. and Nome, F., Partitioning behavior of solutes eluted with micellar mobile phases in liquid chromatography. *Anal. Chem.*, 1981, 53, 1662–1666.
14. Armstrong, D. W. and Stine, J. Y., Evaluation of partition coefficients to micelles and cyclodextrins via planar chromatography. *J. Am. Chem. Soc.*, 1983, 105(10), 2962–2964.
15. Armstrong, D. W., Bui, K. H., and Barry, R. M., *J. Chem. Educ.*, 1984, 61, 457.
16. Khaledi, M. G., Bioanalytical capabilities of micellar liquid chromatography. *Trends Anal. Chem.*, 1988, 7(8), 293–300.
17. Pramauro, E. and Pelizzetti, E., *Surfactants in Analytical Chemistry.* Amsterdam, New York: Elsevier, 1996.
18. Lepri, L., Desideri, P. G., and Heimler, D., Soap thin-layer chromatography of primary aromatic amines. *J. Chromatogr.*, 1978, 1155(1), 119–127.
19. Ge, Z. and Lin, H., *Fenxi Huaxue*, 1992, 20, 1369.
20. Tabor, D. G. and Underwood, A. L., Some factors in solute partitioning between water and micelles or polymeric micelle analogues, *J. Chromatogr.*, 1989, 463(1), 73–80.
21. Sherma, J., Sleckman, B. P., and Armstrong, D. W., Chromatography of amino acids on reversed phase thin layer plates. *J. Liq. Chromatogr.*, 1983, 6(1), 95–108.
22. Armstrong, D. W. and Bui, K. H., Use of Aqueous micellar mobile phases in reverse phase TLC. *J. Liq. Chrom. Rel. Technol.*, 1982, 5(6), 1043–1050.
23. Kovacs-Hadady, K. and Szilagyi, J., Separation of minoxidil and its intermediates by overpressured layer chromatography using a stationary phase bonded with tricaprylmethylammonium chloride. *J. Chromatogr.*, 1991, 553(1–2), 459–466.
24. Sumina, E. G., Ermolaeva, E. V., Tyurina, N. V., and Shtykov, S. N., *Zavod. Lab.*, 2001, 67, 5.
25. Armstrong, D. W., Hinzet, W. L., Bui, K. H., and Singh, H. N., Enhanced fluorescence and room temperature liquid phosphorescence detection in pseudophase liquid chromatography. *Anal. Lett.*, 1981, 14(19), 1659–1667.
26. Weinberger, R., Yarmchuk, P., and Cline Love, L. J., Liquid chromatographic phosphorescence detection with micellar chromatography and postcolumn reaction modes. *Anal. Chem.*, 1982, 54, 1552–1558.
27. Campiglia, A. D., Berthod, A., and Winefordner, J. D., Solid-surface room-temperature phosphorescence detection for high-performance liquid chromatography. *J. Chromatogr.*, 1990, 508(1), 37–49.
28. Alak, A., Heilweil, E., Hinze, W. L., Oh, H., and Armstrong, D. W., Effects of different stationary phases and surfactant or cyclodextrin spray reagents on the fluorescence densitometry of polycyclic aromatic hydrocarbons and dansylated amino acids. *J. Liq. Chrom. Rel. Technol.*, 1984, 7(7), 1273–1288.

29. Berthod, A., Girard, I., and Gonnet, C., Micellar liquid chromatography, Adsorption isotherms of two ionic surfactants on five stationary phases. *Anal. Chem.*, 1986, 58, 1356–1358.

30. Berthod, A., Girard, I., and Gonnet, C., Additive effects on surfactant adsorption and ionic solute retention in micellar liquid chromatography. *Anal. Chem.*, 1986, 58, 1362–1367.

31. Berthod, A. and Roussel, A., The role of the stationary phase in micellar liquid chromatography. Adsorption and efficiency. *J. Chromatogr.*, 1988, 449(2), 349–360.

32. Borgerding, M. F. Hinze, W. L., Stafford, L. D., Fulp, G. W., and Hamlin, W. C., Investigations of stationary phase modification by the mobile phase surfactant in micellar liquid chromatography. *Anal. Chem.*, 1989, 61, 1353–1358.

33. Kord, A. S. and Khaledi, M. G., Controlling solvent strength and selectivity in micellar liquid chromatography: Role of organic modifiers and micelles. *Anal. Chem.*, 1992, 64(17), 1894–1900.

34. Szepesi, G., Vegh, Z., Gyulay, Z., and Gazdag, M., Optimization of reversed-phase ion-pair chromatography by over-pressurized thin-layer chromatography I. Over-pressurized thin-layer chromatography. *J. Chromatogr.*, 1984, 290, 127–134.

35. Garcia, A. M., Pesticide exposure and women's health. *Am. J Ind. Med.*, 2003, 44(6), 584–594.

36. Schwartz, D. A., Newsum, L. A., and Heifetz, R. M., Parental occupational and birth outcome in an agricultural community. *Scand. J. Work. Environ. Health*, 1986, 12(1), 51–54.

37. Lepri, L., Desideri, P. G., and Heimler, D., Thin-layer chromatography of amino acids and dipeptides on RP-2, RP-9 and RP-18 plates impregnated with dodecylbenzenesulphonic acid. *J. Chromatogr.*, 1981, 209(2), 312–315.

38. Wilson, I. D., Ion-pair reversed-phase thin-layer chromatography of organic acids Re-investigation of the effects of solvent pH on R_F values. *J. Chromatogr.*, 1986, 354, 99–106.

39. Lepri, L., Desideri, P. G., and Coas, V., Separation and identification of water-soluble food dyes by ion-exchange and soap thin-layer chromatography. *J. Chromatogr.*, 1978, 161, 279–286.

40. Lepri, L., Desideri, P. G., and Heimler, D., Soap thin-layer chromatography of sulphonamides and aromatic amines. *J. Chromatogr.*, 1979, 169, 271–278.

41. Kamel, F. and Hoppin, J. A., Association of pesticide exposure with neurologic dysfunction and disease. *Environ. Health Perspect.*, 2004, 112(9), 950–958.

42. Lepri, L., Desideri, P. G., and Heimler, D., Reversed-phase and soap thin-layer chromatography of amino acids. *J. Chromatogr.*, 1980, 195(1), 65–73.

43. Lepri, L., Desideri, P. G., and Heimler, D., Reversed-phase and soap thin-layer chromatography of peptides. *J. Chromatogr.*, 1980, 195(2), 187–195.

44. Lepri, L., Desideri, P. G., and Heimler, D., Reversed-phase and soap thin-layer chromatography of dipeptides. *J. Chromatogr.*, 1981, 207(3), 412–420.

45. Mohammad, A., Tiwari, S., Chahar, J. P. S., and Kumar, S., Water-in-oil microemulsion as mobile phase in thin-layer chromatographic retention studies of anions. *J. Am. Oil Chem. Soc.*, 1995, 72(12), 1533–1536.

46. Mohammad, A. and Iraqi, E., Migration behavior of aromatic amines on alumina thin layers developed with water-in-oil microemulsion. *J. Surfactants Deterg.*, 1999, 2(1), 85–90.

47. Jain, R. and Gupta, S., *J. Indian Chem. Soc.*, 1994, 71, 709.

48. Szepesi, G., Gazdag, M., Pap-Sziklay, Z., and Vegh, Z., Optimization of reversed-phase ion-pair chromatography by over-pressurized thin-layer chromatography III. Over-pressurized thin-layer chromatography using 10-camphor sulfonic acid as ion-pair reagent. *Chromatographia*, 1984, 19(1), 422–430.

49. Pesticide Data Program. USDA, www.ams.usda.gov/AMSv1.0/getfile?dDocName = STELPRDC5059863

50. Hauck, H. E. and Jost, W., *Am. Lab.*, 1983, 15(8), 72.

51. Ahuja, S., *Trace and Ultratrace Analysis by HPLC*, New York: Wiley, 1992.

52. Spackman, D. H., Stein, W. H., and Moore, S., Automatic recording apparatus for use in the chromatography of amino acids. *Anal. Chem.*, 1958, 30, 1190–1192.

53. Laskar, S., Sengupta, B., and Das, J., *J. Indian Chem. Soc.*, 1989, 66, 899.

54. Neicheva, A., Kovacheva, E., and Karageorgieu, B., Simultaneous determination of insecticides, acaricides and fungicides by thin-layer chromatography. *J. Chromatogr.*, 1990, 509(1), 263–269.

55. Marutoiu, C., Sarbu, C., Vlassa, M., Liteanu, C., and Bodoga, P., A new separation and identification method of some organophosphorus pesticides by means of temperature programming gradient thin-layer chromatography. *Analysis*, 1986, 14(2), 95–98.

56. Khan, S. and Khan, N. N., The mobility of some organophosphorus pesticides in soils as affected by some soil parameters. *Soil Sci.*, 1986, 142(4), 214–222.

57. Rathore, H. S. and Sharma, R., Sequential thin layer chromatography of carbaryl and related compounds: Thin layer chromatography. *J. Liq. Chromatogr.*, 1992, 15(10), 1703–1717.
58. Sane, R. T., Francis, M., and Khatri, A. R., High-performance thin-layer chromatographic determination of etodolac in pharmaceutical preparations. *J. Planar Chromatogr.*, 1998, 11(3), 211–213.
59. Funk, W., Cleres, L., Pitzer, H., and Donnevert, C., Organophosphorus insecticides: Quantitative HPTLC determination and characterization. *J. Planar Chromatogr.*, 1989, 2, 285–289.
60. Imrag, T. and Junker-Buchheit, A., Determination of preservatives in cosmetic products: Detection and identification of thirty selected preservatives by HPTLC. *J. Planar Chromatogr.*, 1996, 9(1), 39–47.
61. Pastene, M., Montes, M., and Vega, M., New HPTLC method for quantitative analysis of flavonoids of *Passiflora coerulea* L. *J. Planar Chromatogr.*, 1997, 10(5), 362–367.
62. Kang, J. G., Kim, S. T., and Kang, K. Y., Production of the antifungal compound phenylacetic acid by antagonistic bacterium pseudomonas sp. *Agr. Chem. Biotechnol.*, 1999, 42, 197–201.

11 Pressurized Liquid Extraction and Liquid Chromatographic Analysis of Pesticide Residues

Cristina Blasco and Yolanda Picó

CONTENTS

11.1 INTRODUCTION

Nowadays, routine pesticide-residue determination is well established and is equally carried out by either gas chromatography (GC) or liquid chromatography (LC), mostly coupled with mass spectrometry (MS) detection. Recent publications establish that LC–MS provides a wider scope and better sensitivity for all classes of pesticides excepting one, the organochlorine, for which GC–MS achieves better performance [1–5]. The reasons for this movement from GC to LC are that applied pesticides are more and more degradable, thermolabile, and polar as well as the growing interest in degradation pathways and transformation products, which are also more polar and amenable to LC analysis [6–9].

Extraction is still one of the least improved steps of the analytical procedure at the same time that it is a crucial aspect, because it predetermines the compounds that could be detected since the determination technique cannot detect a substance that is not in the extract [10]. Extraction of pesticide residues from solid samples is normally performed by shaking with an organic solvent [11,12]. However, during the last years, trends in this field have been toward less (organic) solvent consumption, faster extraction time, improved quantification (i.e., higher recoveries, better reproducibility, and drive to ever lower detection limits), and automation [13–20]. Pressurized liquid extraction (PLE) fulfills many of these criteria and has therefore been widely used in environmental persistent organochlorine pollutants (POPs) investigations since the mid-1990s [21–23]. Another important aspect related to this technique is that the United States Environmental Protection Agency (US EPA) adopted the technique in 1995 (US EPA Method 3545) [24].

This extraction uses liquids at elevated temperatures and pressures as extractants. The technique was initially named accelerated solvent extraction (ASE), because it was the term patented for a commercial device by Dionex. Within a short span of time, however, alternative names, such as pressurized fluid extraction (PFE), pressurized hot solvent extraction (PHSE), PLE, high-pressure solvent extraction (HPSE), and subcritical solvent extraction (SSE), have gradually replaced ASE, a commercial designation that bears no relationship to the basis of the technique. In revising Method 3545 in December 1995, the EPA replaced "ASE" with "PFE," which was confirmed in the November 2000 update [25]. Finally, "PLE" has been adopted by most authors and publishers. Compared with extractions at or near room temperature and at atmospheric pressure, PLE delivers enhanced performance by the increased solubility, the improved mass transfer, and the disruption of surface adsorption by the conditions applied.

Several recently published, review-type articles present information on the characteristics and applications of PLE:

- Luque-Garcia and de Castro [26] discuss the possible single and multiple couplings between PLE and other steps involved in the analytical process. The application of PLE for food (and biological) sample analysis has been summarized by Carabias-Martinez et al. [27]. Recent applications, which have been reviewed, include the determination of POPs [21–23] and organometals [28].
- Another variant of PLE is extraction at high temperatures and pressures with water. Superheated water extraction (SHWE) including many applications was recently reviewed by Smith [29], Ramos et al. [30], and Kronholm et al. [31].

Since pesticide-residue determination is a growth area for PLE application, this chapter summarizes some scientific publications dealing with extraction of these compounds from food, biological, and environmental matrices using PLE and LC. The reviewed applications mainly use LC–MS or LC–MS/MS to further determine pesticides. However, some examples of PLE combinations with LC–UV or LC–fluorescence are also included. There is a special focus on applications dealing with selective extraction procedures, which apply integrated or in-line cleanup approaches, a strategy for combining extraction and cleanup or fractionation to further decrease the time spent on sample preparation.

11.2 DESCRIPTION OF THE TECHNIQUE: ADVANTAGES AND LIMITATIONS

PLE uses conventional liquid solvents at elevated temperatures and pressures to achieve quantitative extraction from solid and semisolid samples in a short time and with a small amount of solvent [19,20,32]. Temperature rise increases solubility, diffusion rates, and mass transfer, whereas viscosity and surface tension of the solvents are less than those at room temperature [23]. Furthermore, at elevated temperature the activation energy of desorption is more readily overcome, and the kinetics of desorption and solubilization are also more favorable [33]. Pressure helps to force liquid into the pores and to keep the solvent liquid at operating temperatures [34]. Compared with extractions at/or near room temperature and at atmospheric pressure, PLE delivers enhanced performance by the increased solubility, the improved mass transfer, and the disruption of surface adsorption by the conditions applied [23,35].

Most published studies describe a similar construction of the pressurized liquid, homemade equipment. High-pressure pumps are efficient enough to pressurize the solvent and deliver it through

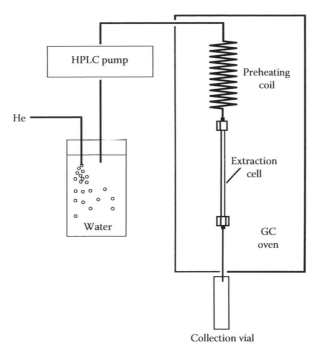

FIGURE 11.1   Sketch for a laboratory-made extraction device.

the sample [36–44]. Various heating systems (e.g., GC ovens, sand baths, or resistive heating blocks) have been applied to heat and to maintain the extraction vessel at the desired temperature (Figure 11.1).

Some others have used the commercial PLE systems (ASE system from Dionex is the most successful one) or a system based on supercritical fluid extraction (SFE) equipment, which might contain an active valve or a fixed restrictor [33,34,45,46]. Patented ASE technology is available in both automated and manual versions for all laboratory sizes and types [47]. The ASE 100 is an entry-level system capable of performing the extraction of a single sample, with cell sizes ranging from 10 to 100 mL. The ASE 200 offers fully automated extraction of 24 samples with cell sizes of 1–33 mL. The ASE 300 offers automated extraction of 12 samples, with cell sizes of 34–100 mL [48].

ASE operates by moving the extraction solvent through an extraction cell containing the sample. The sample cell is heated by direct contact with the oven. The extraction is performed by direct contact of the sample with the hot solvent. When the extraction is complete, compressed nitrogen moves all of the solvent from the cell to the vial for analysis. The filtered extract is collected away from the sample matrix, ready for analysis (Figure 11.2).

Several articles compared PLE with other techniques in food and environmental analysis. Conte et al. [49] compared PLE with traditional extraction using acetonitrile for the analysis of the herbicide flufenican in soil. Figures 11.3 and 11.4 show chromatograms of the extract (ASE and traditional) of a real sample treated in situ and of a blank extracted with ASE. The comparison of the two methods is shown in Table 11.1. The advantages of PLE are total automation of the extraction step allows a complete standardization of procedures, compared with traditional manual techniques; the direct contact between operator and solvent vapor is strongly decreased; the consumption of solvents, the subsequent storage, and disposal are limited: the volume of used solvent is a fifth of the traditional extraction volume; about a quarter of the time is required for the preparation and extraction. Blasco et al. [50] and Soler et al. [45,51] compared PLE with conventional solvent extraction using ethyl acetate for determining different pesticides and metabolites. The researchers

ASE® Schematic

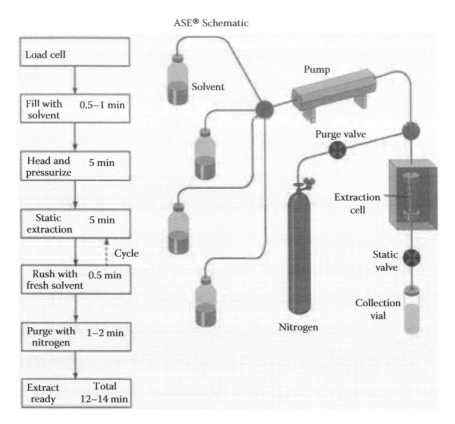

FIGURE 11.2   Scheme of the commercial ASE system from Dionex.

claimed better recoveries, low detection limits, and high speed by PLE and comparable results in the other parameters. Bichon et al. [32] have studied centrifugation with methanol, Folch, Soxhlet, and PLE in terms of extraction efficiency and time consumption. Soxhlet was tested with two solvents, dichloromethane and a mixture of dichloromethane/acetone (50:50), and compared with methanol extraction and Folch extraction with water/methanol/chloroform (1:4:8). The best recoveries were obtained with Soxhlet using dichloromethane/acetone mixture. Then, Soxhlet and PLE were compared in the same conditions (*T* fixed at 60°C to avoid phenylurea degradation). PLE provided several advantages: the extraction time was very short (30 min per sample), the recovery yield of this step was almost 100%, and the solvent volumes used were limited. Results are presented in Table 11.2. All these studies agree that the speed of the extraction process is greatly increased compared with conventional liquid–solid methods and virtually all organics can be extracted. The disadvantages of PLE are the lack of selectivity, which means that further cleanup is needed, and that the sample is too dilute for direct analysis and further concentration is required.

Any aqueous or organic solvent can conceivably be used with PLE. Water can be used effectively as solvent in PLE because the physicochemical properties of water are readily altered through changes in temperature and pressure. At room temperature, water is a too polar solvent for many pesticides, but at elevated temperatures, it becomes less polar, making it an interesting and environmental friendly alternative to organic solvents [35,52].

Subcritical water extraction (SWE) has proved an efficient alternative for the extraction of pesticide residues. However, from an analytical point of view, the most salient negative feature of its use in a continuous extraction mode is the dilution of the analytes in the extract, which calls for a preconcentration step before chromatographic analysis of the target compounds, making the automation of the analytical process difficult. The original SWE apparatus devised by Hawthorne et al. [53]

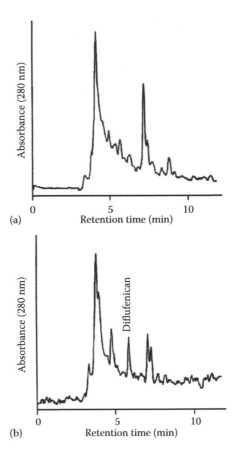

FIGURE 11.3   Chromatogram of PLE extraction of soil. Untreated (a) and sample treated with diflufenican (b). Reversed-phase liquid chromatographic conditions: mobile phase, methanol–acetonitrile–0.05 M ammonium acetate (20:70:10); flow-rate, 1.0 mL/min. (From Conte, E. et al., *J. Chromatogr. A*, 765, 121, 1997. With permission.)

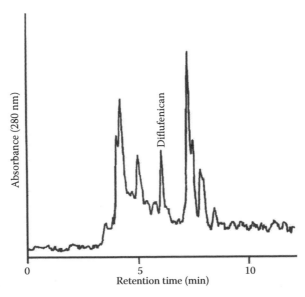

FIGURE 11.4   Chromatogram of diflufenican obtained from traditional extraction of a soil sample. Reversed-phase liquid chromatographic conditions as in Figure 11.3. (From Conte, E. et al., *J. Chromatogr. A*, 765, 121, 1997. With permission.)

TABLE 11.1
Comparison between Two Extraction Techniques

	PLE	Traditional
Solvent volume (mL)	20	100
Concentration	$15 \rightarrow 2$	$25 \rightarrow 2$
Glassware	1 vial	1 funnel, 1 bottle, 1 flask
Preparation and extraction time (min)	25	70
Average recovery (%)	96 ± 4.6	94 ± 4.3
Lower limit of detection (mg/kg)	0.01	0.01

Source: Crescenzi, C. et al., *Anal. Chem.*, 72, 3050, 2000. With permission.

TABLE 11.2
Comparative Recovery Observed with Various Extraction Techniques

Extraction	Solvent Used	Mean Recovery (%)	Time of Extraction
Folch	Water/methanol/chloroform, 1:4:8	4.5	2 h
Centrifugation	Methanol	8.6	2 h
Soxhlet	Dichloromethane	8.1	6 h
Soxhlet	Dichloromethane/acetone, 50:50	11.4	6 h
PLE	Dichloromethane/acetone, 50:50	35.3	30 min

Source: Bichon, E. et al., *J. Chromatogr. B*, 838, 96, 2006. With permission.
Note: Mean values are obtained on 10 different molecules. Each recovery yield includes the respective extraction and the same basic and nonoptimized purification.

involved analyte collection by liquid–liquid partitioning, using methylene chloride or chloroform. Later on, solvent trapping was replaced by sorbent trapping. The latter extraction technique is more efficient than the former one when affording analysis of compounds with a broad range of polarity. In different works [43,54] sorbent trapping was performed by a graphitized carbon black (GCB) cartridge set online with the extraction apparatus. After analyte re-extraction from the GCB cartridge, the final extract was analyzed by LC–MS with an electrospray interface (ESI). As with solvent trapping, the drawbacks of sorbent trapping are that the analysis time is lengthened, evaporative loss of volatile analytes can occur during solvent removal, and the sensitivity of the analysis is limited, as only a fraction of the final extract can be injected into the chromatographic apparatus. SWE coupled off-line to an LC system using a C-18 trap was shown to be an efficient device for analyzing insecticides in dust [40].

The coupling of a subcritical water extractor with a high-pressure liquid chromatograph was developed by Crescenzi et al. [39] but using a relatively complicated system involving several shut-off valves. Luque-Garcia and de Castro [38] proposed a fully automated method for the determination of acid herbicides in different types of soil. The coupling of the steps of the analytical process, namely, SWE–filtration–preconcentration–individual chromatographic separation–detection, has been developed using a flow-injection (FI) system as interface between the extractor and the chromatograph, thus allowing an easier to handle and cheaper approach than those reported previously [39], and with the possibility of including a filtration step online for the removal from the extract of in-suspension particles. Herrera et al. [37] used static/dynamic, SHWE coupled online to filtration, solid-phase extraction (SPE), and high-performance liquid chromatography (HPLC) with postcolumn derivatization for fluorescence detection of *N*-methylcarbamates in food samples. The method

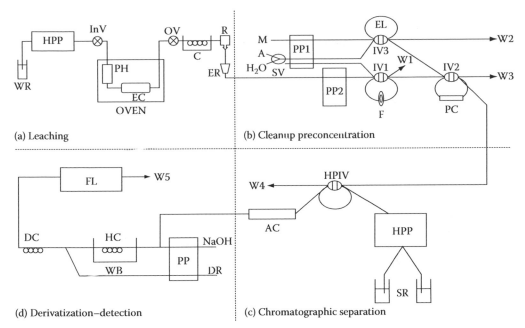

FIGURE 11.5 Experimental setup of the fully automated approach. (a) Leaching step: WR, water reservoir; HPP, high-pressure pump; InV, inlet valve; OV, outlet valve; PH, preheater; EC, extraction chamber; C, cooler; R, restrictor; ER, extract reservoir. (b) Cleanup preconcentration step: M, methanol; A, air; SV, selection valve; PP, peristaltic pump; EL, elution loop; IV, injection valve; W, waste; F, filter; PC, preconcentration column. (c) Chromatographic separation step: HPIV, high-pressure injection valve; SR, solvent reservoirs; AC, analytical column. (d) Derivatization–detection: FL, fluorimeter; DC, derivatization coil; HC, hydrolysis coil; WB, water bath; DR, derivatization reagent. (From Herrera, M.C. et al., *Anal. Chim. Acta*, 463, 189, 2002. With permission.)

was developed using the assembling shown in Figure 11.5. Automation of the steps involved in the analytical process was considered an advantage. Tajuddin and Smith [36] used online linked SHWE and superheated water separation for the analysis of triazine herbicides in compost samples. This eliminated the manual sample treatment.

In the papers published, SHWE efficiencies have also been evaluated by comparing the results with those obtained by other extraction techniques. Crescenzi et al. [43] used SHWE, Soxhlet, and sonication for extraction of acidic and nonacidic herbicides from soil. Compared with Soxhlet extraction, SWE was shown to be more efficient in extracting phenylurea herbicides. Compared with sonicated extraction, this method appeared to be more effective in removing polar acidic herbicides from the soil but less efficient in extracting the least polar acidic herbicides, 2,4-DB and MCPB. A latter study of the same authors [39] using a fully online system corroborates the previous results. Eskilsson et al. [40] compared SHWE and SFE in the extraction of carbofuran, carbosulfan, and imidacloprid from contaminated process dust remaining from seed-pellet production. The results revealed that SHWE is advantageous for polar compounds, because the solubility of the analyte in water is high enough at low temperatures. For nonpolar compounds carbon dioxide-based extraction is preferred unless the target analyte is highly thermostable.

Although SHWE is by far the most "green" extraction technique, the main disadvantages of SHWE are as follows: (1) the solutes are obtained in dilute aqueous medium and further extraction with an organic solvent is required; (2) a large number of matrix compounds are extracted as well so that further cleanup is needed; and (3) the thermal stability of the target solutes under SHWE conditions should be carefully evaluated.

11.3 SAMPLE PRETREATMENT

PLE is a more efficient form of liquid solvent extraction, so all the principles inherent to that technique apply. Sample preparation is also a key step. When the extraction is carried out with organic solvents that are water immiscible, as with Soxhlet, the ideal sample for extraction is a dry, finely divided solid. For an efficient extraction to occur, the solvent must make contact with the target analytes. The more the surface area that can be exposed in a sample the faster this will occur. Samples with large particle sizes should be ground before extraction. Efficient extraction requires a minimum particle size, generally smaller than 0.5 mm. Grinding can be accomplished with a conventional mortar and pestle or with electric grinders and mills [48].

Biological, food, and environmental samples that contain a high amount of water need to be treated before extraction to eliminate most or all of the water. This can be accomplished by several ways:

- Air-drying
- Freeze-drying (lyophilization)
- Oven-drying
- Dispersing mixing sample with sand, anhydrous sodium sulfate, or any other drying agent

Plant material can contain a varying degree of moisture and often requires treatment before extraction. Simply grinding (or mix) the plant material with the drying agent, such as Extrelut [33], Hydromatrix [46], or anhydrous sodium sulfate [51,55], using a mortar and pestle often leads to the success of the analyte isolation procedure. A ratio of plant material to drying agent of 1:1 is a good mixture, but the amount of drying agent used needs to be increased if the plant material is very wet. Frenich et al. [46] optimized a PLE process for vegetables in terms of the test portion size (portions of 3–8 g were tested). The weight of the sample/hydromatrix ratio was varied from 1/1.2 (3 g of sample) to 0.8 (8 g of sample). It was observed that ratio values of 1.1 (4 g of sample) gave the best results. Higher amounts of sample gave more aqueous extracts, whereas lower amounts showed poor precision value. Other studies [45,50], going one step further, researched different adsorbents that can serve a dual purpose in drying the sample and preventing the extraction of unwanted compounds.

An excess of water in the samples can prevent nonpolar organic solvents from reaching the target analytes. The use of more polar solvents (e.g., acetone, methanol) or solvent mixtures (e.g., hexane/acetone, methylene chloride/acetone) can also assist in the extraction of wet samples [32,56]. Sample drying before extraction is the most efficient way to handle these sample types. Oven drying and freeze-drying are other viable alternatives for sample drying prior to extraction; however, the recovery of volatile compounds may be compromised by these procedures. Bichon et al. [32] freeze-dried oysters as pretreatment for the determination of several phenylurea and triazine herbicides without any apparent loss of target compounds. Similarly, Dagnac et al. [57] air-dried soil samples before the analysis of soils contaminated with chloroacetanilides, triazines, and phenylureas without any appreciable influence on the pesticide content.

Dry food/environmental samples may not need a drying step but usually require some pretreatment. Samples such as cereals and grains should be ground and dispersed with celite before extraction. A mortar and pestle are sufficient for many samples and are excellent tools for this purpose [34].

The aggregation of sample particles, as occurs in soil sample, may prevent efficient extraction. In these cases, dispersing the sample with an inert material such as sand or diatomaceous earth will assist in the extraction process, even though water is used as the extracting solvent [56].

11.4 EXTRACTION PARAMETERS

11.4.1 Solvent

For an efficient extraction, the solvent must be able to solubilize the target analytes as much as possible without extracting other sample components. The polarity of the extraction solvent should closely match that of the target compounds. For example, Blasco et al. [50] tested ethyl acetate,

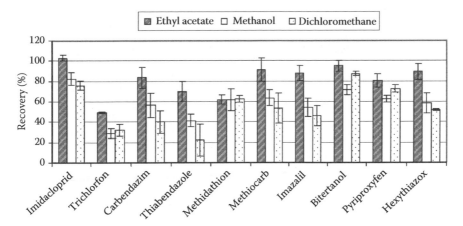

FIGURE 11.6 Effect on the extraction efficiency of different extraction solvents. Extraction conditions were temperature 75°C, pressure 1500 psi, flush 100% in one static cycle, and drying agent anhydrous sodium sulfate. Concentration ranges between 0.1 and 1 mg/kg. (From Blasco, C. et al., *J. Chromatogr. A*, 1098, 37, 2005. With permission.)

methanol, and dichloromethane to extract a set of multiclass pesticides. Results are shown in Figure 11.6. An increase in the extraction efficiency was observed from dichloromethane to ethyl acetate, except for bitertanol, trichlorfon, pyriproxyfen, and methiocarb. Dirty extracts were obtained with methanol, because it also extracts other food components such as flavonoids, carotenes, and sugars with higher efficiency than dichloromethane or ethyl acetate.

Mixing solvents of differing polarities can be used to extract a broad range of compound classes. In an elegant study for determining arylphenoxypropionic herbicides in soil using different extraction solvents, Marchese et al. [58] showed that extraction with methanol/water (80:20) is more effective than extraction with methanol/water (50:50) in the case of arylphenoxypropionic esters. Contrary to the results obtained in the previous work by Crescenzi et al. [43] water (without methanol) does not appear to be suitable for the extraction of neutral pesticides such as arylphenoxypropionic esters. This was justified by the authors as an effect of the very low solubility of these compounds in water. An evaluation was also made of extraction using organic solvents such as pure methanol and acetone, but the recoveries were largely unsatisfactory.

Frenich et al. [46] also experimented with the influence of solvent on the simultaneous PLE extraction of both semipolar and polar pesticides in vegetable samples and subsequent analysis by LC and GC. Figure 11.7 shows, as an example, the extraction efficiency of one pesticide analyzed by GC (deltamethrin) and one analyzed by LC (tebufenpyrad). The solvents selected were those normally used for pesticide-residue analysis. The extraction solvents (individually or in mixtures) tested were ethyl acetate, acetone, ethyl acetate:acetone (1:1, v/v), and ethyl acetate:acetone (3:1, v/v). For this, three aliquots of spiked cucumber samples were extracted using the following conditions: 100°C, 1500 psi, with a 5 min static period. The extraction with acetone gave the worst results based on recovery data for the target pesticides of GC and LC. The extraction with ethyl acetate showed satisfactory recoveries for the LC pesticides but not for all of the GC pesticides. The study with ethyl acetate:acetone (1:1, v/v) generally gave improved results for GC pesticides but unsatisfactory results for LC pesticides. The best results were observed for all compounds with the ethyl acetate:acetone (3:1, v/v) mixture, with recoveries of 70%–110% and precisions, expressed as relative standard deviations (RSD), that were lower than 15%. Generally, if a particular solvent has been shown to work well in a conventional procedure, it will also work well in PLE.

Compatibility with the postextraction analytical technique, the need for extract concentration (solvent volatility), and the cost of the solvent should all be considered [23]. Solvents that exhibit marginal results at ambient conditions may perform adequately under PLE conditions. Most liquid solvents, including water and buffered aqueous mixtures, can be used in PLE [35].

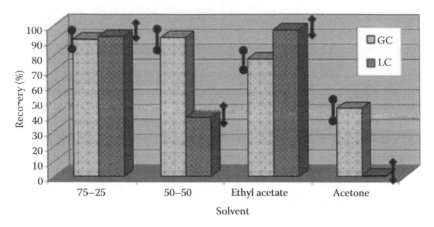

FIGURE 11.7 Influence of solvent (ethyl acetate, acetone, or mixtures of them) on the extraction efficiencies of deltamethrin (GC) and the tebufenpyrad (LC). Errors bars (RSD values) are also shown for deltamethrin (error bars with circles) and tebufenpyrad (error bars with diamonds). (From Garrido Frenich, A. et al., *Anal. Bioanal. Chem.*, 383, 1106, 2005. With permission.)

11.4.2 TEMPERATURES

Temperature is the most important parameter used in PLE extraction. As the temperature is increased, the viscosity of the solvent is reduced, thereby increasing its ability to wet the matrix and solubilize the target analytes. The added thermal energy also assists in breaking analyte matrix bonds and encourages analyte diffusion to the matrix surface [23]. On the contrary, solvents at high temperatures could degrade labile compounds. Most PLE applications operate in the 60°C–125°C range, in which, commonly a good compromise between extraction efficiency and degradation is achieved. An example of the effect of temperature is shown below for the extraction of chloroacetanilides, triazines, and phenylureas from soil [57]. Figure 11.8 contains examples of modeling curves for the herbicide recoveries in a "blank" Calcisol, as a function of the temperature, setting

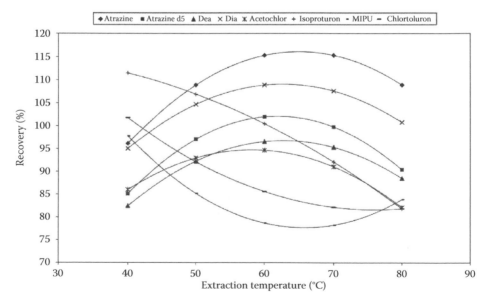

FIGURE 11.8 Models fitted by a Doehlert design for herbicide recoveries after PLE in a Luvisol sample collected between 0 and 30 cm of depth (From Dagnac, T. et al., *J. Chromatogr. A*, 1067, 225, 2005. With permission.)

the extraction time at 3 min. Recoveries increase with the temperature for triazines and acetochlor, with maxima values between 60°C and 70°C. In contrast, phenylurea recoveries decrease with the temperature increase, with a dramatic effect for isoproturon. Similarly, Bichon et al. [32] also fixed the extraction temperature at 60°C to avoid phenylurea degradation.

Because extractions are performed at even more elevated temperatures using water as solvent, thermal degradation should also be a major concern. The effect of increasing the extraction temperature on the ability of a methanol/water (80:20, v/v) solution of NaCl (0.12 M) in extracting the arylphenoxypropionic herbicides from soils was assessed by Eskilsson and Bjorklund [56]. A general enhancement of the recoveries of all the analytes was achieved by increasing the extraction temperature from 60°C to 90°C. Progressively, greater losses of Quizalofop ethyl and Haloxyfop ethyl were observed as the temperature was increased from 90°C to 120°C. Pesticides may be degraded at relatively low temperatures: Crescenzi et al. [39] found phenylurea herbicides to be degraded at 120°C. Tajuddin and Smith [36] reported decreased recoveries of chlorotriazines because of thermal degradation above 170°C.

11.4.3   PRESSURE

The effect of pressure is to maintain the solvents as liquids while above their atmospheric boiling points and to rapidly move the fluids through the systems. The pressures used in PLE are commonly well above the thresholds required to maintain the solvents in their liquid states. Changing pressure might have a little influence on recovery. However, Frenich et al. [46] reported that higher recoveries were obtained for a multipesticide extraction of spiked cucumber samples using a pressure of 1000 psi more than pressures of 1500 or 2000 psi, and the RSDs were the lowest at this pressure. With the same types of samples, Blasco et al. [50] also reported better recoveries at 1500 psi than at 2000 or 2500 psi.

11.4.4   STATIC/DYNAMIC MODES

PLE can be done using two modes, static and dynamic, or even a combination of both. In the dynamic mode the solvent is passed through the sample at a fixed flow rate. In the static one, a fixed volume is used, so the efficiency of the extraction depends on the analyte mass-transfer equilibrium between the matrix and the extractant. The commercially available extractors implement only the static mode but develop the use of static cycles to introduce fresh solvent during the extraction process, which helps to maintain a favorable extraction equilibrium. This effectively approximates dynamic extraction conditions without the need for troublesome flow restrictors to maintain pressure [23].

For example, three 3 min static cycles can be used in place of one 10 min static step. When low-temperature extractions are desired (<75°C), multiple static cycles should be used to compensate for the lack of fresh solvent [57]. On the contrary, Blasco et al. [50] checked from one to five extraction cycles for the analysis of pesticides in fruits. The extraction efficiency was constant from one to three cycles, whereas starting from the fourth cycle a remarkable decrease was noted. A justification of this behavior is that the more the cycles were used the greater the amounts of interfering substances extracted. Henriksen et al. [59] also evaluated the use of one or two extraction cycles to analyze metribuzin in soils, but no differences were observed.

11.4.5   TIME

Certain sample matrices can retain analytes within pores or other structures. Increasing the static time at elevated temperatures can allow these compounds to diffuse into the extraction solvent. Henriksen et al. [59] evaluated the extraction time for metribuzin in soils in the range of 3–10 min. A significant increase in the recovery was obtained within this range, and at 10 min the curve started to flatten out. However, a small improvement resulted from increasing the extraction time beyond 10 min. These authors conclude that the effect of static time should always be explored in conjunction with static cycles to produce a complete extraction in the most efficient way possible.

On the contrary, Blasco et al. [50] analyzed several pesticides in fruits and vegetables using different static times (from 3 to 15 min). The extraction efficiency remains constant for all the static times tested, which can be explained by the high solubility of all studied pesticides in ethyl acetate and/or the weak analyte matrix interactions. Similarly, Frenich et al. [46] indicates that results obtained with a static period of 5 min at 100°C were the best. Higher static periods or temperatures did not yield significantly better recoveries.

11.5 SELECTIVE EXTRACTION PROCEDURES USING INTEGRATED OR IN-LINE CLEANUP STRATEGIES

Interferences may be extracted along with desired analytes during an extraction process. These unwanted coextractables may interfere with analyte detection or decrease instrument performance. Traditionally, chromatographic techniques, such as gel-permeation chromatography (GPC) or SPE, are used to purify sample extracts before separation and analysis. Recent advances using PLE systems, as described in several publications, include procedures for selective removal of interferences during sample extraction, thus combining extraction and purification into a single step [60–63].

In an effort to eliminate postextraction cleanup steps, Blasco et al. [50] tested alumina, florisil, silica, and anhydrous sodium sulfate, as drying materials, for PLE in oranges and peaches. In addition, alumina was tested in the three pH ranges available (basic, neutral, and acidic). Recoveries were very similar for all the compounds, except for trichlorfon, the recovery of which decreases from 75% using acidic silica to 32% using basic alumina. A probable explanation is that trichlorfon is quickly degraded in slightly basic aqueous solutions. Figure 11.9 shows the recoveries and RSDs obtained from oranges using these sorbents, excepting basic alumina. As it can be seen, alumina and silica provided almost the same recoveries for all the compounds, except for imazalil, which is better recovered from alumina. However, RSDs obtained using alumina were lower than those obtained using silica, especially for the most polar compounds (imidaclorid, trichlorfon, carbendazim, and thiabendazole). Neutral and acidic alumina provided very similar recoveries;

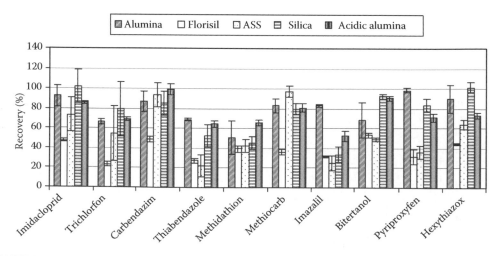

FIGURE 11.9 Effect on the extraction efficiency of different drying agents. Extraction conditions were temperature 75°C, pressure 1500 psi, flush 100% in one static cycle, and solvent, ethyl acetate. Concentration ranges between 0.1 and 1 mg/kg. (From Blasco, C. et al., *J. Chromatogr. A*, 1098, 37, 2005. With permission.)

however, slightly better recoveries were observed working with acidic alumina, particularly for trichlorfon. Florisil gave lower recoveries for all the compounds, and anhydrous sodium sulfate also gave low recoveries for thiabendazole, imazalil, bitertanol, pyriproxyfen, and hexythiazox. In addition, the last sorbent provides the dirtiest extracts with a cloudy and strong color. The optimum procedure was to disperse the sample with acidic alumina. This approach has proven successful in producing clean extracts that are ready for direct analysis.

11.6   POSTEXTRACTION TREATMENT

Sometimes, in spite of the pretreatment effort, some type of postextraction treatment is necessary to produce high-quality analytical data. Bichon et al. [32] sequentially used SPE and solvent partitioning to purify and separate phenylurea and triazine herbicides and their dealkylated degradation products in oyster. This study established the behavior of 10 phenylurea compounds on 6 different stationary phases qualified as "normal": diol, acidic and basic alumina, cyanopropyl/silica double phase, silica, and florisil. The best profile was obtained with the CN/SiOH column, which allowed for an efficient rinsing of the stationary phase, eliminating lipophilic molecules. Triazine residues, because of their higher polarity, needed a more eluotropic mobile phase (ethyl acetate/methanol; 80:20, v/v) to be eluted.

To improve the purification (elimination of polar interferences) six reversed stationary phases were compared. The best profile was obtained with a C_{18} cartridge, which permitted to eliminate most interferences with a methanol/water mixture before analyte elution.

Because of their nitrogen atoms, phenylurea and triazine residues could be retained on cation exchange columns such as OASIS MCX (Waters). This specific phase was composed of an hydrophilic lipophilic balance (HLB) polymer on which a controlled sulfonation was carried out to insert cation exchange groups. All residues of triazines and phenylureas could be retained both by the HLB polymer and the exchange ion sites. After sample loading onto the cartridges, water was used to discard polar anionic molecules. Then, one washing step involving 3 mL $H_2O/CH_3OH/NH_4OH$ (78:20:2) permitted the elimination of cationic interferences. Application of a 3 mL acetate buffer (2 M) broke ionic affinity, and analytes were finally eluted with 3 mL of $H_2O/CH_3OH/NH_4OH$ (18:80:2).

When the purification step with the three successive cartridges described above (CN/SiOH, MCX, and C_{18} consecutive cartridges) was tested, the complexity of extracts led to a saturation of cation exchange cartridge. This phenomenon was the result of a bad dissolution of dry extract in water. The most lipophilic molecules must be avoided. A liquid/liquid partition with hexane was also carried out. Therefore, to keep the analyte in aqueous phase, an addition of methanol was required. This complex protocol, summarized in Figure 11.10, led to a mean recovery of 30%, with the worse recoveries obtained for terbutylazine and 1-(3,4-dichlorophenyl)urea. Both analytes represent, respectively, the most apolar and the most polar of the compounds monitored.

Others [33,34,56] used the postextraction SPE cleanup method to reduce the time for the analysis.

As stated above, dynamic PLE usually requires a concentration step before determination, and because the extracted analytes are dissolved in a liquid—usually aqueous—phase, SPE is a very useful tool for avoiding the dilution effect. For this purpose, SPE cartridges [40–43] and columns packed with appropriate sorbents and coupled online to the extractor outlet [36–39,54] have been employed.

These methodologies have been used for the extraction of herbicides, such as phenoxy acid herbicides [39,43] and chlorophenoxy acid herbicides [38,39,43] in soil, N-methylcarbamates in food [37] and several insecticides in dust [40], obtaining in all cases recoveries higher than 81%, with preconcentration factors in the range of 20–166 [37–40,43].

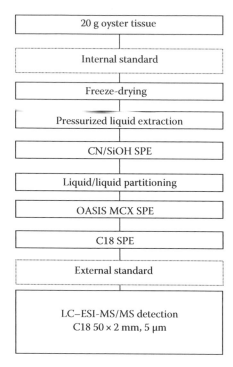

FIGURE 11.10 General analytical procedure for determining phenylurea and triazine herbicides and their dealkylated degradation products in oysters. (From Bichon, E. et al., *J. Chromatogr. B*, 838, 96, 2006. With permission.)

11.7 APPLICATIONS

The main areas of application of PLE and LC have been for solid and powdered samples, most commonly soils and environmental solids, plant materials, or animal tissues, largely because these matrices are compatible with a flow extraction system. So far there has been no report of the applications of PLE in water and only scarce ones on biological matrices probably because water or liquid biological fluid matrices are difficult to handle in closed systems. Tables 11.3 and 11.4 outlined the reported application for PLE using conventional solvents or water, respectively.

11.7.1 PLANT MATERIALS

Plant materials are a popular PLE application area. The methods reported are focused on two aspects (1) development of a multiresidue method and (2) determination of metabolites.

Different multiresidue methods were developed mainly in combination with LC–MS. A multi-residue method using LC–quadrupole ion trap–triple-stage mass spectrometry (LC–QIT–MS3) has been developed by Blasco et al. [50] for determining trace levels of pesticides in fruits. The selected pesticides can be distinguished in benzimidazoles and azoles, organophosphorus, carbamates, neonicotinoids, and acaricides. PLE has been optimized to extract these pesticide residues from oranges and peaches by studying the effect of experimental variables on PLE efficiency. Samples were extracted at high temperature and pressure (75°C and 1500 psi) using ethyl acetate as extraction solvent and acidic alumina as drying agent. The recoveries obtained by PLE ranged from 58% to 97%, and the RSDs, from 5% to 19%. The limits of quantification (LOQs) of the compounds were from 0.025 to 0.25 mg/kg.

Carabias-Martinez et al. [64] optimized the PLE extraction conditions before the LC–ESI-MS analysis in cereal samples of seven endocrine-disrupting compounds: bisphenol A (BPA), 4-*tert*-butylbenzoic acid (BBA), 4-nonylphenol (NP), 4-*tert*-butylphenol (*t*-BP), 2,4-dichlorophenol (DCP),

TABLE 11.3
The Application of PLE to Environmental and Food Analysis of Pesticide Residues

Analyte	Matrix	PLE Conditions	LC Conditions	References
10 pesticides including benzimidazoles, azoles, organophosphorus, carbamates, neonicotinoids, and acaricides	Fruits	Acidic alumina as drying agent and extraction in an ASE 200 with 22 mL cells and ethyl acetate (100% flush volume) at 75°C and 1500 psi in two cycles of 7 min static time	C_{18} column (150 × 4.5 mm, 5 μm) Gradient of methanol–water at 0.8 mL/min APCI in positive ionization mode LC–IT–MS, MS^2, and MS^3	[50]
12 pesticides (7 insecticides, 4 fungicides, and 1 herbicide)	Fruits	Acidic alumina as drying agent and extraction in an ASE 200 with 22 mL cells and ethyl acetate (100% flush volume) at 75°C and 1500 psi in two cycles of 7 min static time	C_{18} column (150 × 4.5 mm, 5 μm) Gradient of methanol–water at 0.6 mL/min APCI in positive ionization mode LC–IT–MS, MS^2, and MS^3, LC–QqQ–MS/MS, and LC–QTOF–MS/MS	[45]
Carbosulfan and seven metabolites	Oranges	Anhydrous sodium sulfate as drying agent and extraction in an ASE 200 with 22 mL cells and dichloromethane (60% flush volume) at 40°C and 2000 psi in two cycles of 5 min static time	C_{18} column (150 × 4.5 mm, 5 μm) Gradient of methanol–water at 0.8 mL/min APCI in positive ionization mode LC–IT–MS, MS^2, and MS^3	[51]
Carbosulfan and seven metabolites	Fruits and vegetables	Anhydrous sodium sulfate as drying agent and extraction in an ASE 200 with 22 mL cells and dichloromethane (60% flush volume) at 40°C and 2000 psi in two cycles of 5 min static time	C_{18} column (150 × 2.1 mm, 5 μm) Gradient of methanol–water with ammonium acetate at 0.2 mL/min APCI in positive ionization mode LC–QqTOF–MS/MS	[55]
Pentachlorofenol and other endocrine-disrupting compounds	Cereals	Hydromatrix as drying agent and extraction in an ASE 100 with 10 mL cells and methanol (200% flush volume) at 120°C and 1600 psi in one cycle of 10 min static time	C_{18} column (100 × 2.1 mm, 3.5 μm) Gradient of methanol and 0.0025 M ammonium formiate buffer adjusted to pH 3.1 with formic acid ESI in negative ionization mode LC–QqQ–MS/MS	[64]
Multiclass pesticides (31 compounds by LC)	Fruit and vegetables	Hydromatrix as drying agent and extraction in an ASE 200 with 22 mL cells and ethyl acetate:acetone (3:1) (60% flush volume) at 100°C and 1000 psi in two cycles of 5 min static time	C_{18} column (150 × 2 mm, 3 μm) Gradient of methanol and water with 2 mM ammonium formiate (buffer, pH 2.8) ESI in positive ionization mode LC–QqQ–MS/MS	[46]

(*continued*)

TABLE 11.3 (continued)
The Application of PLE to Environmental and Food Analysis of Pesticide Residues

Analyte	Matrix	PLE Conditions	LC Conditions	References
N-Methylcarbamate	Foods	Extrelut as drying agent and extraction with an ASE 200 with 33 mL cells and acetonitrile (60% flush volume) at 100°C and 2000 psi in one cycle of 5 min static time	C_{18} column (150 × 2 mm, 3 μm) Gradient of methanol and water LC with postcolumn fluorescence	[33]
450 pesticides (43 pesticides determined by LC–MS/MS)	Cereals	Celite as drying agent and extraction with an ASE 200 with 33 mL cells and acetonitrile (60% flush volume) at 80°C and 1500 psi in two cycles of 3 min static time	C_{18} column (150 × 2 mm, 3 μm) Gradient of methanol and water with 2 mM ammonium formiate (buffer, pH 2.8) ESI in positive ionization mode LC–QqQ–MS/MS	[34]
Phenylurea and triazine herbicides and their dealkylated degradation products	Oyster freeze-dried	Fontainebleau sediment and Celite as drying agent and extraction in an ASE 200 with 33 mL cells and methylene chloride and acetone (50:50, v/v) (5 min of each) at 60°C and 100 bar	C_{18} column (50 × 2 mm, 3 μm) Gradient of acetonitrile and acetic acid in water ESI in positive ionization mode LC–QqQ–MS/MS	[32]
Diflufenican	Soil	Mixed with diatomaceous earth and extraction in an ASE 200 with 11 mL cells and acetonitrile (150% flush volume) at 100°C and 2000 psi in one cycle of 4 min static time	Two C_{18} columns (300 × 3.9 mm, 4 μm) Gradient of acetonitrile–methanol–0.05 M ammonium acetate at 1 mL/min LC–UV–Vis at 280 nm	[49]
Metribuzin, deaminometribuzin, diketometribuzin, and deaminodiketometribuzin	Soil	ASE 200 with 33 mL cells and methanol–water (75:25) (60% flush volume) at 60°C and 1500 psi in one cycle of 10 min static time	C_{18} column (150 × 2 mm, 3 μm) Gradient of methanol and water with 2 mM ammonium formiate (buffer, pH 2.8) ESI in positive ionization mode LC–QqQ–MS/MS	[59,65]
Chloroacetanilides, triazines, phenylureas, and their metabolites	Soil	15 g soil was extracted with an ASE 200 with 30 mL of acetone under 101,300 kPa at temperature <80°C	C_{18} column (150 × 3 mm, 3 μm) Gradient of acetonitrile–water at flow rate of 0.4 mL/min ESI in positive ionization mode LC–MS and LC–QqQ–MS/MS	[57]

TABLE 11.3 (continued)
The Application of PLE to Environmental and Food Analysis of Pesticide Residues

Analyte	Matrix	PLE Conditions	LC Conditions	References
Arylphenoxypropionic herbicides	Soil	5 g soil + 2 g sand into the extraction cell. The analytes were extracted with 25 mL methanol–water (80:20, v/v) solution of NaCl (0.12 M). The first 2.5 mL of solvent was passed through the cell at a flow rate of 0.4 mL/min. The flow rate was then increased to 1 mL/min in 1 min	C_{18} column (250 × 4.6 mm, 5 μm) Gradient of acetonitrile–water, both solvents contained 25 mM formic acid ESI in negative ionization mode LC–QqQ–MS/MS	[56]

2,4,5-trichlorophenol (TCP), and pentachlorophenol (PCP). For the PLE procedure, methanol was selected as the extraction solvent. An experimental design approach was applied to optimize other PLE parameters. The recoveries achieved for all seven compounds were in the 81%–104% range, with RSDs of 4%–9%. An additional preconcentration step, based on SPE, after the PLE step proved to be a successful way for obtaining a more sensitive method. The detection limits achieved in corn breakfast cereals were in the 0.003–0.013 μg/g range, except for BPA, with a detection limit of 0.043 μg/g, for a sample size of 2.5 g. These values are similar to or even lower than currently legislated limits for pesticides in cereals and cereal-based foodstuffs.

Frenich et al. [46] applied PLE to the simultaneous extraction of a wide range of pesticides from food commodities. Extractions were performed by mixing 4 g of sample with 4 g of Hydromatrix and (after optimization) a mixture of ethyl acetate:acetone (3:1, v/v) as extraction solvent, a temperature of 100°C, a pressure of 1000 psi, and a static extraction time of 5 min. After extraction, the more polar compounds were analyzed by LC, and the apolar and semipolar pesticides by GC. In both cases LC and GC were coupled with MS in tandem (MS/MS) mode. The overall method (including the PLE step) was validated in GC and LC according to the criteria of the SANCO Document of the European Commission. The average extraction recoveries (at two concentration levels) for most of the analytes were in the range 70%–80%, with precision values usually lower than 15%. LOQ were low enough to determine the pesticide residues at concentrations below or equal to the maximum residue levels (MRL) specified by legislation. In order to assess its applicability to the analysis of real samples, aliquots of 15 vegetable samples were processed using a conventional extraction method with dichloromethane, and the results obtained were compared with the proposed PLE method; differences lower than 0.01 mg/kg were found. Pang et al. [34] also determined 405 pesticide residues in grain, using PLE, SPE, GC–MS, and LC–MS–MS. The method was based on appraisal of the GC–MS and LC–MS–MS characteristics of 660 pesticides, their efficiency of extraction from grain, and their purification. Samples of grain (10 g) were mixed with Celite 545 (10 g) and the mixture was placed in a 34 mL cell of an accelerated solvent extractor and extracted with acetonitrile in the static state for 3 min with two cycles at 1500 psi and at 80°C. For the 362 pesticides determined by GC–MS, half of the extracts were cleaned with an Envi-18 cartridge and then further cleaned with Envi-Carb and Sep-Pak NH2 cartridges in series. The pesticides were eluted with acetonitrile–toluene, 3:1, and the eluates were concentrated and used for analysis after being exchanged with hexane twice. For the 43 pesticides determined by LC–MS–MS the other half of the extracts were cleaned with Sep-Pak Alumina N cartridge and further cleaned with Envi-Carb and Sep-Pak NH2 cartridges. Pesticides were eluted with acetonitrile–toluene, 3:1. After evaporation to dryness the eluates were diluted with acetonitrile–water, 3:2, and used for analysis. In the linear

TABLE 11.4
The Application of SWE to Environmental and Food Analysis of Pesticide Residues

Analyte	Matrix	SWE Conditions	LC Conditions	References
N-Methylcarbamates	Food	5 g of sample in the extraction cell pressurized with 15 bar of water at 75°C for 20 min (static extraction) and at 0.5 mL/min for 30 min (dynamic extraction). This extraction system was coupled with preconcentration and determination	C_{18} column (250 × 4.6 mm, 5 μm) Methanol–water gradient at flow rate of 0.8 mL/min LC with postcolumn fluorescence	[37]
N-Methylcarbamates	Milk	3 mL milk + 12 g sand extracted in a manual device with water heated at 90°C for 5 min at 1 mL/min flow rate	C_{18} column (250 × 4.6 mm, 5 μm) Gradient of methanol water with 10 mM formic acid LC–ESI-MS	[42]
N-Methylcarbamates	Fruits and vegetables	2 g sample + 6 g sand extracted in a manual device with water heated at 50°C for 3 min at 1 mL/min flow rate	C_{18} column (250 × 4.6 mm, 5 μm) Gradient of methanol water with 10 mM formic acid LC–ESI-MS	[44]
Tricyclazole	Soil and sediment	10 g sample was extracted for 60 min at a flow rate of 1 mL/min temperatures ranging from 50°C to 150°C	C_{18} column (150 × 4.6 mm, 5 μm) Isocratic methanol–water Collection of 0.5 min fractions and analysis by ^{14}C LSC	[41]
Chlorophenoxy acid herbicides	Soil	5 g of sample in the extraction cell extracted at 1 mL/min with water at 85°C for 10 min (dynamic extraction). This extraction system was coupled with preconcentration and determination	C_{18} column (250 × 4.6 mm, 5 μm) Isocratic 1% aqueous solution of H_3PO_4–acetonitrile at a flow rate of 1.7 mL/min LC–UV at 280 nm	[38]
Triazine herbicides	Compost	0.5 g compost in an extraction cell extracted first at a flow rate of 2 mL/min at ambient temperature for 10 min and then at flow rate of 1 mL/min at 170°C for 5 min. This extraction system was coupled with preconcentration and determination	Superheated water separation Hypercarb PGC analytical column Water at a gradient of temperatures from 130°C to 220°C LC–UV at 222 nm	[36]
Nonacidic and acidic herbicides	Soil	3 g of soil mixed with 2 g of sand in the extraction cell extracted at 0.4 mL/min with water at 90°C for 6.2 min and then increased to 1 mL/min for 2 min. This extraction system was coupled with preconcentration	C_{18} column (250 × 4.6 mm, 5 μm) Gradient of methanol water with 10 mM trifluoroacetic acid LC–ESI-MS	[43]
Terbutylazine and its metabolites	Aged and incubated soils	3 g of soil mixed with 2 g of sand in the extraction cell and extracted with 10 mL of phosphate-buffered water (0.5 M, pH 7.5) at 100°C. The flow rate was 0.3 mL/min for 7 min, then increased at 1 mL/min in 2 min. This extraction system was coupled with preconcentration	C_{18} column (250 × 4.6 mm, 5 μm) Gradient of methanol water with 10 mM trifluoroacetic acid LC–ESI-MS	[54]

TABLE 11.4 (continued)
The Application of SWE to Environmental and Food Analysis of Pesticide Residues

Analyte	Matrix	SWE Conditions	LC Conditions	References
13 pesticides (polar and medium polar)	Soil	3 g of soil mixed with 2 g of sand in the extraction cell and extracted with 10 mL of phosphate-buffered water (0.5 M, pH 7.5) at 100°C. The flow rate was 0.3 mL/min for 7 min, then increased at 1 mL/min in 2 min. This extraction system was coupled with preconcentration	C_{18} column (250 × 4.6 mm, 5 μm) Gradient of methanol water with 10 mM formic acid LC–ESI–MS	[39]
Carbofuran, carbosulfan, and imidacloprid	Dust	10 g dust extracted with hot water at 250 bar and between 100°C and 150°C for 30 min	C_{18} column (250 × 4.6 mm, 5 μm) Gradient of acetonitrile–water at 1 mL/min LC–UV at 220 nm	[40]

range of each pesticide the linear correlation coefficient r was equal to or greater than 0.956% and 94% of linear correlation coefficients were greater than 0.990. At low, medium, and high fortification levels, at the LOD, twice the LOD, and 10 times LOD, respectively, recoveries ranged from 42% to 132%; for 382 pesticides, or 94.32%, recovery was from 60% to 120%. The RSD was always below 38% and was below 30% for 391 pesticides or 96.54%. The LOD was 0.0005–0.3000 mg/kg. The proposed method is suitable for determination of 405 pesticide residues in grains such as maize, wheat, oat, rice, and barley, and so on.

Soler et al. [45] optimized the extraction conditions before the LC analysis with different LC–MS instruments equipped with triple quadrupole (QqQ), quadrupole ion trap (QIT), and quadrupole-time-of-flight (QqTOF), suitable to carry out MS/MS. Twelve pesticides (acrinathrin, bupirimate, buprofezin, cyproconazole, λ-cyhalothrin, fluvalinate, hexaflumuron, kresoxim-methyl, propanil, pyrifenox, pyriproxyfen, and tebufenpyrad) and six matrices (oranges, strawberries, cherries, peaches, apricots, and pears) were taken as model. The comparison was focused on two aspects: the quantitative, covering sensitivity, precision, and accuracy, as well as the qualitative, checking the possibility to identify any metabolite present in the samples, which were not targeted in the methods. The extraction was carried out using PLE with ethyl acetate and acid alumina. Recoveries were more than 70% for all the analytes. Repeatabilities were better for the QqQ (5%–12%) than that for QIT (6%–15%) and for QqTOF (14%–19%). QqQ offered a linear dynamic range of at least three orders of magnitude, whereas those of QIT and QqTOF were two and one orders of magnitude, respectively. QqQ reached at least 20-fold higher sensitivity than those of QIT and QqTOF. However, QqQ failed to identify nontarget compounds. QIT and QqTOF were able to successfully identify the metabolite of bupirimate, ethirimol. Figure 11.11 illustrates the results obtained for a PLE extract of strawberry using QqTOF; the exact mass measurements corresponding to the compounds help to identify them. The method was applied to monitor the content in fruits, taken from agricultural cooperatives, and to calculate the estimated daily intake (EDI) to establish if there is any difference in toxicological interest.

One use of PLE gaining popularity is the determination not only of the pesticide but also of its degradation products. Soler et al. [51,55] identified and confirmed carbosulfan and seven of its main metabolites at trace levels in food by either LC–QIT–MS[3] or LC–QqTOF–MS. PLE recoveries ranged from 55% to 94%, with LOQs from 10 μg/kg (for carbosulfan, carbofuran, 3-hydroxycarbofuran, dibutylamine) to 70 μg/kg (3-keto-7-phenolcarbofuran). The method is precise, with RSDs varying between 5% and 11% for the repeatability (with-in-day) and 8% and 13% for the reproducibility (interday). This method was used to monitor the presence and fate of the target compounds

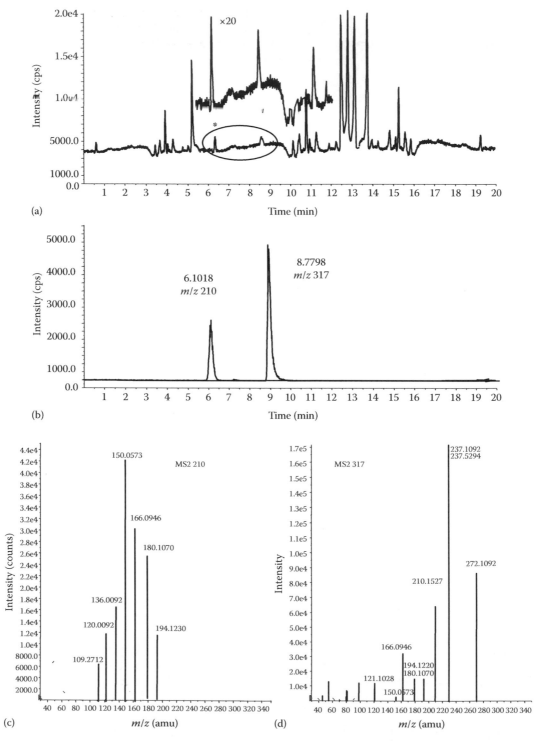

FIGURE 11.11 LC–QqTOF–MS chromatogram of a strawberry extract obtained by PLE (a) total ion chromatogram and (b) extracted ion chromatogram at tr = 6.1018 and tr = 8.7198 and accurate product ion mass spectrum of (c) precursor ion at *m/z* 210.2882 (identified as ethirimol) and (d) precursor ion at *m/z* 317.1569 (identified as bupirimate). (From Soler C. et al., *J. Chromatogr. A*, 1157, 73, 2007. With permission.)

in orange, potato, and rice crops treated with a commercial product containing carbosulfan. Field degradation studies show that carbofuran, 3-hydroxycarbofuran, and dibutylamine are the main degradation products formed in the environmental disappearance of carbosulfan.

SHWE has been scarcely used in plant material. However, two interesting examples have been reported. A simple, specific, and rapid analytical method for determining seven largely used carbamate insecticides in tomato, spinach, lettuce, zucchini, pear, and apple was presented by Bogialli et al. [44]. This method is based on the matrix solid-phase dispersion technique, with heated water as extractant, followed by LC–MS equipped with a single quadrupole and an electrospray ion source. Target compounds were extracted from the vegetal matrixes by water heated at 50°C. After acidification and filtration, 0.25 mL of any aqueous extract was injected in the LC column. MS data acquisition was performed in the selected ion-monitoring mode, selecting three ions for each target compound. Heated water appeared to be an excellent extractant, because recovery data ranged between 76% (carbaryl in spinach) and 99% (pirimicarb in spinach), with RSDs not larger than 10%. Using trimethacarb (an obsolete carbamate insecticide) as a surrogate internal standard, the accuracy of the analysis varied between 84% and 110%, with RSDs not larger than 9%. On the basis of a signal-to-noise ratio (SNR) of 10, the LOQ were estimated to range between 2 (pirimicarb) and 10 ppb (oxamyl) and were not influenced by the type of matrix. When trying to fractionate analytes by using a short chromatographic run time, marked weakening of the ion signals for oxamyl, methomyl, and aldicarb were observed. This effect was traced to polar endogenous coextractives eluted in the first part of the chromatographic run that interfered with gas-phase ion formation for carbamates. Adopting more selective chromatographic conditions eliminated this effect.

A combination of static–dynamic modes of pressurized hot water extraction (PHWE) has been used by Herrera et al. [37] for the extraction of N-methylcarbamates (oxamyl, dioxacarb, metholcarb, carbofuran, and carbaryl) from different fruits and vegetables. The selection of water as leaching agent provides a clean approach, which avoids the use of organic solvents. A flow-injection manifold coupled to the extractor has allowed the automation of the subsequent steps (namely, filtration, preconcentration, individual chromatographic separation, postcolumn derivatization, and fluorescence detection) involved in the analytical process. Good recoveries, ranging from 80% to 104%, and precision, expressed as RSD, between 3.0% and 8.4% have been achieved by the proposed method.

11.7.2 Animal Tissues and Biological Samples

There are few examples of PLE applications to animal tissues. Bichon et al. [32] developed a method for the determination of several phenylurea and triazine herbicides and their transformation products in oysters at the low microgram per kilogram level [32]. PLE of lyophilizated samples required successive SPE combined with a liquid/liquid extraction to provide relatively clean extracts for the determination in LC–MS/MS. The efficiency of the analytical method led to confirmatory CCα values ranging from 0.1 to 14 µg/kg with an RSD value ranging from 14% to 66% and a recovery yield ranging from 32% to 46% for phenylureas and from 29% to 75% for triazines. PLE combined with adjusted purification and LC–MS/MS analysis was demonstrated an efficient method for the determination of phenylureas and triazines in mollusk tissues. Performances of the developed methodology are very relevant in terms of sensitivity, specificity, accuracy, and linearity. Nevertheless, some molecules such as demethylated phenylurea, simazine, and terbuthylazine can only be semiquantified because of the demonstrated weak accuracy. In terms of sensitivity, the identification limit of the methodology described here is set at around 3 µg/kg. This method will be used to evaluate new metabolites, which can be used as relevant bioindicator molecules.

Still more restricted are the applications to biological fluids. A simple, specific, and rapid procedure for determining six largely used carbamate insecticides in bovine whole milk was presented by Bogialli et al. [42]. This method is based on SHWE followed by LC–MS equipped with a single quadrupole and an ESI. Target compounds were extracted from milk by water heated at 90°C. After acidification and filtration, 0.2 mL of the aqueous extract was injected in the LC column. MS data acquisition was performed in the selected ion-monitoring (SIM) mode, selecting three ions for each target compound. Heated water appeared to be an excellent extractant, since absolute recovery data ranged between 76% and 104% with RSD not larger than 8%. Using butocarboxim (an obsolete carbamate insecticide) as surrogate internal standard, the accuracy of the analysis at three spike levels varied between 85% and 105% with RSD not larger than 9%. On the basis of a SNR of 10, the LOQ were estimated to range between 3 ppb (propoxur) and 8 ppb (pirimicarb). The effects of temperature, volume, and flow rate of the extractant on the analyte recovery were studied. This work has again shown that the environmentally friendly and inexpensive water, besides being an effective extractant for polar and medium-polar contaminants in biological matrices, produces sufficiently clean extracts requiring little manipulation before final analysis by LC–MS. In addition, the ESI/MS detector equipped with a single quadrupole, where confirmatory ions are produced by in-source CID, provides specificity similar to that obtained by a much more expensive instrumentation, that is, MS/MS, and sensitivity sufficient for analyzing carbamate insecticides in milk.

11.7.3 Soil and Environmental Solids

The determination of pesticide residues in soil is the most popular PLE application area probably because of the compatibility of the technique with solid samples. Conte et al. [49] tried out, to reduce time and cost of analysis, a new extractor, ASE 200-Dionex, for the extraction of the herbicide diflufenican from soil. The developed method was compared to traditional extraction with a solvent. For each sample the consumption of extraction solvent was reduced to about a fifth and the time to about a quarter, compared with traditional extraction. This technique allowed extraction of about 120 samples in 5 days, working about 6 h/day.

Moreover, the best conditions of extraction were easily recognized because of the limited number of parameters that affect analyte recovery.

Henriksen et al. [59] presented a study of metribuzin degradation in soil. LC–MS–MS and ESI were used for analysis of metribuzin and the metabolites deaminometribuzin (DA), diketometribuzin (DK), and deaminodiketometribuzin (DADK). Soil samples were extracted by PLE using methanol–water (75:25) at 60°C. In general, recoveries were about 75% for metribuzin, DA, and DADK, and their detection limit in soil was 1.25 μg/kg. Lower sensitivity was observed for DK, with detection limit at 12.5 μg/kg and recovery about 50% [59]. These authors applied the same methodology to study the degradation and sorption of metribuzin and its metabolites in a Danish sandy loam top soil and subsoil from the field in question [65].

PLE technique was used by Dagnac et al. [57] for the simultaneous extraction of phenylureas, triazines, and chloroacetanilides and some of their metabolites from soils. Extractions were performed by mixing 15 g of dried soil with 30 mL of acetone under 100 atm at 50°C, during 3 min and with three PLE cycles. Before the analysis of naturally contaminated soils, each of the five representative soil matrices used as blanks (of different depths) was spiked in triplicate with standards of each parent and degradation compound at about 10, 30, and 120 μg/kg. For each experiment, isoproturon-D[6] and atrazine-D[5] were used as surrogates. Analysis of phenylureas and metabolites of triazines and phenylureas was carried out by LC–MS and LC–MS/MS in the positive mode. GC–IT–MS was used in the MS/MS mode for the parent triazines and chloroacetanilides. The average extraction recoveries were more than 85%, except for didesmethyl-isoproturon, and the quantification limits were between 0.5 and 5 μg/kg. The optimized multiresidue method was applied to soils and solids below the root zone, sampled from agricultural plots of a small French hydrogeological basin.

As an extension of extraction studies, there has also been an interest in using SHWE for determining pesticide residues in soil. This study is the link between PLE with conventional solvents and SHWE. A sensitive and specific analytical procedure for determining arylphenoxypropionic herbicides in soil samples, using ESI LC–MS, is presented by Eskilsson and Bjorklund [56]. Arylphenoxypropionic acids are a new class of herbicides used for selective removal of most grass species from any nongrass crop, commercialized as herbicide esters. Previous studies have shown that the esters undergo fast hydrolysis in the presence of vegetable tissues and soil bacteria, yielding the corresponding free acid. The feasibility of rapidly extracted propionic herbicides from soil by PLE techniques was evaluated. Four different soil samples were fortified with target compounds at levels of 5 and 20 ng/g by following a procedure able to mimic weathered soils. Herbicides were extracted by a methanol/water (80:20, v/v) solution (0.12 M) of NaCl at 90°C. After cleanup using GCB as absorbent, the extract was analyzed by LC–ESI-MS. The effect of concentration of acid in the mobile phase on the response of ESI-MS was investigated. The effects of varying the orifice plate voltage on the production of diagnostic fragment ions, and on the response of the MS detector, were also investigated. The ESI-MS response was linearly related to the amounts of analytes injected between 1 and 200 ng. The limit of detection (SNR = 3) of the method for the pesticides in soil samples was estimated to be less than 1 ng/g.

The use of subcritical water to extract tricyclazole from soils and sediments was examined by Krieger et al. [41]. The extraction efficiency and kinetics were determined as a function of temperature, sample age, sample matrix, sample size, and flow rate. The extraction temperature was the most influential experimental factor affecting extraction efficiency and kinetics, with increasing temperature (up to 150°C) yielding faster and higher efficiency extractions. Higher extraction temperatures were also important for quantitative recovery of tricyclazole from aged samples. Extraction at 50°C yielded 97% recoveries from samples aged 1 day but only 30% recoveries for samples aged 202 days, whereas extraction at 150°C yielded recoveries of 85%–100%, which were independent of incubation time and sample matrix, with the exception of one sediment that contained a large amount of organic matter. Sample extracts from SWE had generally a pale yellow color, contrasted with a dark brown color from organic solvent extractions of the same matrixes. Less sample cleanup was therefore required before analysis, with the total time for the extraction and analysis of a single sample being approximately 2 h. SWE is an effective technique for the rapid and quantitative extraction of tricyclazole from soils and sediments.

Tajuddin and Smith [36] evaluated the feasibility of analyzing rapid traces of polar and medium polar contaminants in soil by coupling online a hot phosphate-buffered water extraction apparatus to a LC/mass spectrometer system. Coupling was accomplished by using a small C-18 sorbent trap for collecting analytes and two six-port valves. The efficiency of this device was evaluated by extracting 13 selected pesticides from 200 mg of laboratory-aged soils by varying the extraction temperature, the extractant volume, and the flow rate at which the extractant passed through the extraction cell and the sorbent trap. In terms of extraction efficiency, robustness of the method, and extraction time, the best compromise was that of using 8 mL of extractant at 90°C and 0.5 mL/min flow rate. Under these conditions, recoveries of 11 out of 13 analytes ranged between 82% and 103%, whereas those of the least hydrophilic pesticides, that is, neburon and prochloraz, were 73% and 63%, respectively. By increasing the extractant volume to 60 mL, additional amounts of the two latter compounds could be recovered. Under this condition, however, the most hydrophilic analytes were in part no more retained by the C-18 sorbent trap. From a naturally 1.5 years aged soil, hot phosphate-buffered water removed larger amounts of three herbicides and hydroxyterbuthylazine (a terbuthylazine degradation product) than pure water and Soxhlet extraction. This result seems to confirm that hot phosphate buffer is also able to remove from soil those fractions of contaminants that, on aging, are sequestered into the humic acid framework.

Due to the great potential of atrazine in contaminating groundwater, its use has been banned in several countries and often replaced by terbuthylazine (CBET). Little is known on the fate of CBET in soil. Di Corcia et al. [43] developed an interesting study aimed (1) to develop a general method for analyzing CBET and its degradation products (DPs) in soil and (2) to use this method for elucidating the fate of CBET incubated in both surface and subsurface samples of an agricultural soil that had been receiving repeated CBET spills. This method involves analyte extraction from soil at 100°C by phosphate-buffered water. Analytes emerging from the extraction cell were collected by a graphitized carbon black extraction cartridge. After analyte elution with a suitable solvent mixture, the final extract was analyzed by LC–MS. From an aged soil, our method extracted altogether quantities of CBET and its DPs, respectively, 2.1 and 1.4 times larger than those by two previously reported methods. For the analytes considered, the LOQ (SNR 10) ranged between 0.22 and 5.5 nanogram per gram of soil. The laboratory CBET degradation experiment showed that (1) similar to atrazine, remarkable amounts of hydroxylated metabolites were formed; (2) when the subsoil microflora was in the presence of rather large amounts of CBET, it degraded the herbicide with a rate similar to that of the topsoil microflora.

Crescenzi et al. [39] also evaluated the feasibility of analyzing rapid traces of polar and medium polar contaminants in soil by coupling online a hot phosphate-buffered water extraction apparatus to a LC/mass spectrometer system. Coupling was accomplished by using a small C-18 sorbent trap for collecting analytes and two six-port valves. The efficiency of this device was evaluated by extracting 13 selected pesticides from 200 mg of laboratory-aged soils by varying the extraction temperature, the extractant volume, and the flow rate at which the extractant passed through the extraction cell and the sorbent trap. In terms of the extraction efficiency, robustness of the method, and extraction time, the best compromise was that of using 8 mL of extractant at 90°C and 0.5 mL/min flow rate. Under these conditions, recoveries of 11 out of 13 analytes ranged between 82% and 103%, whereas those of the least hydrophilic pesticides, that is, neburon and prochloraz, were 73% and 63%, respectively. By increasing the extractant volume to 60 mL, additional amounts of the two latter compounds could be recovered. Under this condition, however, the most hydrophilic analytes were in part no more retained by the C-18 sorbent trap. From a natural 1.5 years aged soil, hot phosphate-buffered water removed larger amounts of three herbicides and hydroxyterbuthylazine (a terbuthylazine degradation product) than pure water and Soxhlet extraction. This result seems to confirm that hot phosphate buffer is also able to remove from soil those fractions of contaminants that, on aging, are sequestered into the humic acid framework.

The hot phosphate buffer extraction apparatus was successfully coupled online to an LC/MS system through a C-18 sorbent trap and two six-port valves. From the moment the solid matrix is ready for extraction, sensitive and specific analysis of contaminants can be performed in less than 1 h. The authors have been using this device for more than 6 months, and the only trouble they encountered with it was the tendency of the small C-18 trap to be progressively plugged by a sort of clot consisting of humic material mixed with very fine soil particles. This resulted in a relatively short life of the sorbent trap, which had to be replaced after seven to eight extractions.

SWE has been coupled by Luque-Garcia and de Castro et al. [38] with filtration, preconcentration, and chromatographic analysis for the determination of acid herbicides in different types of soil. Two experimental designs were used for the optimization of the leaching step. The use of water as extractant in the continuous mode at a flow rate of 1 mL/min and 85°C was sufficient for quantitative extraction of the analytes. A static extraction time was unnecessary for reducing the extraction time to 1 h. A minicolumn containing C18-Hydra as sorbent proved an excellent material for the quantitative preconcentration of the herbicides before individual chromatographic separation. A flow-injection manifold was used as interface for coupling the four steps, thus allowing automation of the whole analytical process. Recoveries of the target analytes ranged between 94.2% and 113.1%, and repeatabilities, expressed as RSDs, were between 0.61% and 6.83%.

An online method, with a purely aqueous mobile phase, has employed linked, SHWE and superheated water separation for the analysis of triazine herbicides in spiked compost samples was proposed by Tajuddin and Smith [36]. After the SHWE, an X-Terra solid-phase trap was used to collect and focus the extracted analytes. The trapped analytes were then released by thermal desorption and passed directly to a superheated water chromatographic (SWC) separation using a PGC column. Two cleanup steps (prior to extraction and separation) were included to remove most of the interfering matrix components. The effects of the sample matrix and the extraction temperatures on the recovery of the triazines were investigated. Despite some thermal degradation of the chlorotriazines during the SWE, the online SWE–SWC method was sensitive and rapid. The coupled method could potentially reduce costs, and labor and by using only water in every stage it is compatible with the concepts of green chemistry.

Pressurized hot liquid water and steam were used by Eskilsson et al. [40] to investigate the possibilities of extracting insecticides (carbofuran, carbosulfan, and imidacloprid) from contaminated process dust remaining from seed-pellet production. The extraction temperature was the most important parameter in influencing the extraction efficiency and rate of extraction, whereas varying the pressure had no profound effect. A cleanup procedure of the water extracts using SPE was found to be necessary before final analysis by HPLC. Quantitative extraction (compared to a validated organic solvent extraction method) of imidacloprid was obtained at temperatures of 100°C–150°C within 30 min extraction time. Temperatures above 150°C were required to extract carbofuran efficiently. The most nonpolar analyte of the investigated compounds, carbosulfan, gave no detectable concentrations with PHWE. One reason might be its low solubility in water, and when attempts are made to increase its solubility by increasing the temperature, it may degrade to carbofuran. This can explain recovery values above 100% for carbofuran at higher temperatures. A comparison of the PHWE results and those obtained with SFE revealed that PHWE is advantageous for polar compounds, where the solubility of the analyte in water is high enough for lower temperatures to be used. For nonpolar compounds carbon dioxide-based extraction is preferred unless the target analyte is highly thermostable.

11.8 CONCLUSIONS AND FUTURE TRENDS

PLE has been applied with success to the extraction of pesticides from environmental, food, and biological samples for a number of years. PLE combined with LC enables rapid and accurate determination of pesticide residues. With proper sample preparation and optimization of the extraction parameters, nearly any sample currently extracted with solvents can be performed in less time and with smaller quantities of solvents using PLE.

SHWEs have been shown to be feasible with particular interest in avoiding the need for organic solvents in environmental extractions or in food samples. The method is thus environmentally friendly, cheap, and nontoxic. The equipment required is relatively simple and avoids the need for the high pressures employed in SFE. A further advance has been linkage to other chromatographic systems, and unlike carbon dioxide there is no problem with cooling and condensation. Most samples have been solid matrices, such as soils and plant materials.

Temperature is the main parameter affecting extraction efficiency and selectivity. Pesticides can be extracted quantitatively at moderate temperatures. Pressure is usually kept high enough to maintain the solvent in the liquid phase, and its effect on the extraction is usually less than that of temperature.

Accordingly, a continued growth in research activity in the area over the coming years and a concomitant increase in the volume of the literature dealing with the PLE techniques in LC pesticide-residue analysis are to be expected.

REFERENCES

1. Alder L., Greulich K., Kempe G., and Vieth B. Residue analysis of 500 high priority pesticides: Better by GC-MS or LC-MS/MS? *Mass Spectrom. Rev.*, 2006, 25, 838–865.
2. Soler C. and Pico Y. Recent trends in liquid chromatography-tandem mass spectrometry to determine pesticides and their metabolites in food. *TrAC-Trends. Anal. Chem.*, 2007, 26, 103–115.
3. Pico Y., Font G., Ruiz M.J., and Fernandez M. Control of pesticide residues by liquid chromatography-mass spectrometry to ensure food safety. *Mass Spectrom. Rev.*, 2006, 25, 917–960.
4. Pico Y., Blasco C., and Font G. Environmental and food applications of LC-tandem mass spectrometry in pesticide-residue analysis: An overview. *Mass Spectrom. Rev.*, 2004, 23, 45–85.
5. Hernandez F., Sancho J.V., and Pozo O.J. Critical review of the application of liquid chromatography/mass spectrometry to the determination of pesticide residues in biological samples. *Anal. Bioanal. Chem.*, 2005, 382, 934–946.
6. Soderberg D. Pesticides and other chemical contaminants. *J. AOAC Int.*, 2006, 89, 293–303.
7. Ahmed F.E. Analyses of pesticides and their metabolites in foods and drinks. *TrAC-Trends. Anal. Chem.*, 2001, 20, 649–661.

8. Barcelo D. and Petrovic M. Challenges and achievements of LC-MS in environmental analysis: 25 years on. *TrAC-Trends. Anal. Chem.*, 2007, 26, 2–11.
9. Kuster M., de Alda M.L., and Barcelo D. Analysis of pesticides in water by liquid chromatography-tandem mass spectrometric techniques. *Mass Spectrom. Rev.*, 2006, 25, 900–916.
10. Lambropoulou D.A. and Albanis T.A. Methods of sample preparation for determination of pesticide residues in food matrices by chromatography-mass spectrometry-based techniques: A review. *Anal. Bioanal. Chem.*, 2007, 389, 1663–1683.
11. Garcia-Reyes J.F., Ferrer C., Gomez-Ramos M.J., Molina-Diaz A., and Fernandez Albu A.R. Determination of pesticide residues in olive oil and olives. *TrAC-Trends. Anal. Chem.*, 2007, 26, 239–251.
12. Margariti M.G., Tsakalof A.K., and Tsatsakis A.M. Analytical methods of biological monitoring for exposure to pesticides: Recent update. *Ther. Drug Monit.*, 2007, 29, 150–163.
13. Esteve-Turrillas F.A., Pastor A., Yusa V., and de la Guardia M. Using semi-permeable membrane devices as passive samplers. *TrAC-Trends. Anal. Chem.*, 2007, 26, 703–712.
14. Rodriguez-Mozaz S., de Alda M.J.L., and Barcelo D. Advantages and limitations of on-line solid phase extraction coupled to liquid chromatography-mass spectrometry technologies versus biosensors for monitoring of emerging contaminants in water. *J. Chromatogr. A*, 2007, 1152, 97–115.
15. Hercegova A., Domotorova M., and Matisova E. Sample preparation methods in the analysis of pesticide residues in baby food with subsequent chromatographic determination. *J. Chromatogr. A*, 2007, 1153, 54–73.
16. Baggiani C., Anfossi L., and Giovannoli C. Solid phase extraction of food contaminants using molecular imprinted polymers. *Anal. Chim. Acta*, 2007, 591, 29–39.
17. Bogialli S. and Di Corcia A. Matrix solid-phase dispersion as a valuable tool for extracting contaminants from foodstuffs. *J. Biochem. Biophys. Methods*, 2007, 70, 163–179.
18. Lambropoulou D.A. and Albanis T.A. Liquid-phase micro-extraction techniques in pesticide residue analysis. *J. Biochem. Biophys. Methods*, 2007, 70, 195–228.
19. Raynie D.E. Modern extraction techniques. *Anal. Chem.*, 2006, 78, 3997–4003.
20. Raynie D.E. Modern extraction techniques. *Anal. Chem.*, 2004, 76, 4659–4664.
21. Bjorklund E., Nilsson T., and Bowadt S. Pressurised liquid extraction of persistent organic pollutants in environmental analysis. *TrAC-Trends. Anal. Chem.*, 2000, 19, 434–445.
22. Giergielewicz-Mozajska H., Dabrowski L., and Namiesnik J. Accelerated solvent extraction (ASE) in the analysis of environmental solid samples—some aspects of theory and practice. *Crit. Rev. Anal. Chem.*, 2001, 31, 149–165.
23. Bjorklund E., Sporring S., Wiberg K., Haglund P., and von Hillist C. New strategies for extraction and clean-up of persistent organic pollutants from food and feed samples using selective pressurized liquid extraction. *TrAC-Trends. Anal. Chem.*, 2006, 25, 318–325.
24. US Environmental Protection Agency. *EPA Method 3545, Pressurized Fluid Extraction, Test Method for Evaluating Solid Waste*, 3rd edn. Update III, EPA SW-846. US EPA, Washington DC, 1995, report.
25. US Environmental Protection Agency. *SW-846, Test Methods for Evaluating Solid Waste*, 3rd edn. Update IVB. Chapter 4, organic analyte. US EPA, Washington, DC, 2000, report.
26. Luque-Garcia J.L. and de Castro M.D.L. Coupling of pressurized liquid extraction to other steps in environmental analysis. *TrAC-Trends. Anal. Chem.*, 2004, 23, 102–108.
27. Carabias-Martinez R., Rodriguez-Gonzalo E., Revilla-Ruiz P., and Hernandez-Mendez J. Pressurized liquid extraction in the analysis of food and biological samples. *J. Chromatogr. A*, 2005, 1089, 1–17.
28. Alonso-Rodriguez E., Moreda-Pineiro J., Lopez-Mahia P., Muniategui-Lorenzo S., Fernandez-Fernandez E., Prada-Rodriguez D., Moreda-Pineiro A., Bermejo-Barrera A., and Bermejo-Barrera P. Pressurized liquid extraction of organometals and its feasibility for total metal extraction. *TrAC-Trends. Anal. Chem.*, 2006, 25, 511–519.
29. Smith R.M. Extractions with superheated water. *J. Chromatogr. A*, 2002, 975, 31–46.
30. Ramos L., Vreuls J.J., and Brinkman U.A.T. Miniaturised pressurised liquid extraction of polycyclic aromatic hydrocarbons from soil and sediment with subsequent large-volume injection-gas chromatography. *J. Chromatogr. A*, 2000, 891, 275–286.
31. Kronholm J., Hartonen K., and Riekkola M.L. Analytical extractions with water at elevated temperatures and pressures. *TrAC-Trends. Anal. Chem.*, 2007, 26, 396–412.
32. Bichon E., Dupuis M., Le Bizec B., and Andre F. LC-ESI-MS/MS determination of phenylurea and triazine herbicides and their dealkylated degradation products in oysters. *J. Chromatogr. B*, 2006, 838, 96–106.
33. Okihashi M., Obana H., and Hori S. Determination of *N*-methylcarbamate pesticides in foods using an accelerated solvent extraction with a mini-column cleanup. *Analyst*, 1998, 123, 711–714.

34. Pang G.F., Liu Y.M., Fan C.L., Zhang J.J., Cao Y.Z., Li X.M., Li Z.Y., Wu Y.P., and Guo T.T. Simultaneous determination of 405 pesticide residues in grain by accelerated solvent extraction then gas chromatography-mass spectrometry or liquid chromatography-tandem mass spectrometry. *Anal. Bioanal. Chem.*, 2006, 384, 1366–1408.
35. Navarro P., Cortazar E., Bartolome L., Deusto M., Raposo J.C., Zuloaga O., Arana G., and Etxebarria N. Comparison of solid phase extraction, saponification and gel permeation chromatography for the clean-up of microwave-assisted biological extracts in the analysis of polycyclic aromatic hydrocarbons. *J. Chromatogr. A*, 2006, 1128, 10–16.
36. Tajuddin R. and Smith R.M. On-line coupled extraction and separation using superheated water for the analysis of triazine herbicides in spiked compost samples. *J. Chromatogr. A*, 2005, 1084, 194–200.
37. Herrera M.C., Prados-Rosales R.C., Luque-Garcia J.L., and de Castro M.D.L. Static-dynamic pressurized hot water extraction coupled to on-line filtration-solid-phase extraction-high-performance liquid chromatography-post-column derivatization-fluorescence detection for the analysis of N-methylcarbamates in foods. *Anal. Chim. Acta*, 2002, 463, 189–197.
38. Luque-Garcia J.L. and de Castro M.D.L. Coupling continuous subcritical water extraction, filtration, preconcentration, chromatographic separation and UV detection for the determination of chlorophenoxy acid herbicides in soils. *J. Chromatogr. A*, 2002, 959, 25–35.
39. Crescenzi C., Di Corcia A., Nazzari M., and Samperi R. Hot phosphate-buffered water extraction coupled on line with liquid chromatography/mass spectrometry for analyzing contaminants in soil. *Anal. Chem.*, 2000, 72, 3050–3055.
40. Eskilsson C.S., Hartonen K., Mathiasson L., and Riekkola M.L. Pressurized hot water extraction of insecticides from process dust-comparison with supercritical fluid extraction. *J. Sep. Sci.*, 2004, 27, 59–64.
41. Krieger M.S., Cook W.L., and Kennard L.M. Extraction of tricyclazole from soil and sediment with subcritical water. *J. Agric. Food Chem.*, 2000, 48, 2178–2183.
42. Bogialli S., Curini R., Di Corcia A., Lagana A., Nazzari M., and Tonci M. Simple and rapid assay for analyzing residues of carbamate insecticides in bovine milk: Hot water extraction followed by liquid chromatography-mass spectrometry. *J. Chromatogr. A*, 2004, 1054, 351–357.
43. Crescenzi C., D'Ascenzo G., Di Corcia A., Nazzari M., Marchese S., and Samperi R. Multiresidue herbicide analysis in soil: Subcritical water extraction with an on-line sorbent trap. *Anal. Chem.*, 1999, 71, 2157–2163.
44. Bogialli S., Curini R., Di Corcia A., Nazzari M., and Tamburro D. A simple and rapid assay for analyzing residues of carbamate insecticides in vegetables and fruits: Hot water extraction followed by liquid chromatography-mass spectrometry. *J. Agric. Food Chem.*, 2004, 52, 665–671.
45. Soler C., James K.J., and Pico Y. Capabilities of different liquid chromatography tandem mass spectrometry systems in determining pesticide residues in food—Application to estimate their daily intake. *J. Chromatogr. A*, 2007, 1157, 73–84.
46. Frenich A.G., Salvador I.M., Martinez Vidal J., and Lopez-Lopez T. Determination of multiclass pesticides in food commodities by pressurized liquid extraction using GC-MS/MS and LC-MS/MS. *Anal. Bioanal. Chem.*, 2005, 383, 1106–1118.
47. Dionex Corporation. http://www.dionex.com/en-us/lp2570.html. 2008, report.
48. Dionex. http://www.dionex.com/en-us/index.html. 2008, report.
49. Conte E., Milani R., Morali G., and Abballe F. Comparison between accelerated solvent extraction and traditional extraction methods for the analysis of the herbicide diflufenican in soil. *J. Chromatogr. A*, 1997, 765, 121–125.
50. Blasco C., Font G., and Pico Y. Analysis of pesticides in fruits by pressurized liquid extraction and liquid chromatography-ion trap-triple stage mass spectrometry. *J. Chromatogr. A*, 2005, 1098, 37–43.
51. Soler C., Manes J., and Pico Y. Determination of carbosulfan and its metabolites in oranges by liquid chromatography ion-trap triple-stage mass spectrometry. *J. Chromatogr. A*, 2006, 1109, 228–241.
52. Bogialli S. and Di Corcia A. Matrix solid-phase dispersion as a valuable tool for extracting contaminants from foodstuffs. *J. Biochem. Biophys. Methods*, 2007, 70, 163–179.
53. Hawthorne S.B., Yang Y., and Miller D.J. Extraction of organic pollutants from environmental solids with subcritical and supercritical water. *Anal. Chem.*, 1994, 66, 2912–2920.
54. Di Corcia A., Caracciolo A.B., Crescenzi C., Giuliano G., Murtas S., and Samperi R. Subcritical water extraction followed by liquid chromatography mass spectrometry for determining terbuthylazine and its metabolites in aged and incubated soils. *Environ. Sci. Technol.*, 1999, 33, 3271–3277.
55. Soler C., Hamilton B., Furey A., James K.J., Manes J., and Pico Y. Liquid chromatography quadrupole time-of-flight mass spectrometry analysis of carbosulfan, carbofuran, 3-hydroxycarbofuran, and other metabolites in food. *Anal. Chem.*, 2007, 79, 1492–1501.

56. Eskilsson C.S. and Bjorklund E. Analytical-scale microwave-assisted extraction. *J. Chromatogr. A*, 2000, 902, 227–250.

57. Dagnac T., Bristeau S., Jeannot R., Mouvet C., and Baran N. Determination of chloroacetanilides, triazines and phenylureas and some of their metabolites in soils by pressurised liquid extraction, GC-MS/MS, LC-MS and LC-MS/MS. *J. Chromatogr. A*, 2005, 1067, 225–233.

58. Marchese S., Perret D., Gentili A., Curini R., and Marino A. Development of a method based on accelerated solvent extraction and liquid chromatography/mass spectrometry for determination of arylphenoxypropionic herbicides in soil. *Rapid Commun Mass Spectrom.*, 2001, 15, 393–400.

59. Henriksen T., Svensmark B., and Juhler R.K. Analysis of metribuzin and transformation products in soil by pressurized liquid extraction and liquid chromatographic-tandem mass spectrometry. *J. Chromatogr. A*, 2002, 957, 79–87.

60. Bandh C., Bjorklund E., Mathiasson L., Naf C., and Zebuhr Y. Comparison of accelerated solvent extraction and Soxhlet extraction for the determination of PCBs in Baltic Sea sediments. *Environ. Sci. Technol.*, 2000, 34, 4995–5000.

61. Bandoniene D., Gfrerer M., and Lankmayr E.P. Comparative study of turbulent solid-liquid extraction methods for the determination of organochlorine pesticides. *J. Biochem. Biophys. Methods*, 2004, 61, 143–153.

62. Bjorklund E., von Holst C., and Anklam E. Fast extraction, clean-up and detection methods for the rapid analysis and screening of seven indicator PCBs in food matrices. *TrAC-Trends Anal. Chem.*, 2002, 21, 39–52.

63. Numata M., Yarita T., Aoyagi Y., Tsuda Y., Yamazaki M., Takatsu A., Ishikawa K., Chiba K., and Okamaoto K. Sediment certified reference materials for the determination of polychlorinated biphenyls and organochlorine pesticides from the National Metrology Institute of Japan (NMIJ). *Anal. Bioanal. Chem.*, 2007, 387, 2313–2323.

64. Carabias-Martinez R., Rodriguez-Gonzalo E., and Revilla-Ruiz P. Determination of endocrine-disrupting compounds in cereals by pressurized liquid extraction and liquid chromatography-mass spectrometry-study of background contamination. *J. Chromatogr. A*, 2006, 1137, 207–215.

65. Henriksen T., Svensmark B., and Juhler R.K. Degradation and sorption of metribuzin and primary metabolites in a sandy soil. *J. Environ. Qual.*, 2004, 33, 619–627.

12 Analysis of Pesticides by Chemiluminescence Detection

*Laura Gámiz-Gracia, José F. Huertas-Pérez,
Jorge J. Soto-Chinchilla, and Ana M. García-Campaña*

CONTENTS

12.1   INTRODUCTION

Chemiluminescence (CL) is a detection system based on the production of electromagnetic radiation observed when a chemical reaction yields an electronically excited intermediate or product that either luminesces (direct CL) or donates its energy to another molecule that subsequently luminesces (indirect or sensitized CL). If radiation is emitted by energy transfer, then the process is normally called "chemiexcitation;" likewise, when the chemiluminogenic reaction is enzymatic and/or occurs within the living organisms, the phenomenon is known as bioluminescence (BL). For CL, minimal instrumentation is required and as no external light source is needed, the optical system is quite simple. Hence, strong background light levels are excluded, reducing the background signal, the effects of stray light, and the instability of the light source, leading to improved limits of detection (LOD). Therefore, CL is known as the "dark-field technique," making it easier to acquire the CL signal by a photomultiplier tube (PMT).

Owing to its simplicity, low cost, and high sensitivity and selectivity, CL-based detection has become quite a useful detection tool in flow injection analysis (FIA), liquid and gas chromatography (LC and GC), and capillary electrophoresis (CE), which together with its potential in immunoassays make this technique an interesting field of research in a wide variety of disciplines, including chemistry, biology, pharmaceutical, biomedical, environmental, and food analysis. In the past decade, several books, chapters, and reviews have been published related to CL and BL [1–4], their analytical applications (mainly in the liquid phase [5–10]), as well as about the use of CL as a detection mode in FIA [11,12], LC [12,13], GC [14,15], CE [12,13,16,17], and chemical and biochemical sensors [18,19].

Nevertheless, although extensive reviews have been reported about the specific application of CL reactions in different disciplines such as medical, biochemical [20–23], and food analysis [24], there has been limited number of contributions concerning the application of CL detection in the analysis of pesticide residues in environmental and toxicological analysis [25,26]. The aim of this chapter is to provide an update on the information regarding the determination of pesticides by means of CL in the last 5 years, including recent developments and applications in this field. Accordingly, this chapter is structured describing the main CL systems applied for pesticide analysis in liquid and gas phases, considering different CL reactions and detectors, as well as the application of CL in the screening analysis of pesticides.

12.2 ANALYSIS OF PESTICIDES BY CL DETECTION IN THE LIQUID PHASE

The CL phenomenon can solve certain problems concerning the monitoring of pesticide residues in the liquid phase, and can be considered as an alternative to other powerful detection modes, such as UV/Vis, fluorescence, or mass spectrometry (MS), especially in terms of sensitivity and simplicity of the required instrumentation, as stated in a review on the determination of pesticides by CL in the liquid phase [26], which updates the main applications of CL in this field. Different couplings of CL detection with high-performance liquid chromatography (HPLC), FIA, and other liquid-phase systems, as well as immunoassay applications will be discussed in this section, showing that this technique is an interesting tool in pesticide residue analysis. The most common CL reactions used in the analysis of pesticides, such as direct oxidations, peroxyoxalate (PO) reaction, tris(2,2'-bipyridine)ruthenium(II) system, and luminol reaction will be described in detail. In addition, the reported applications are also summarized in Table 12.1, according to the involved CL reaction.

12.2.1 DIRECT OXIDATIONS

Strong oxidants, such as MnO_4^- (in acidic or alkaline medium), ClO^-, $Ce(IV)$, H_2O_2, IO_4^-, Br_2, and N-bromosuccinimide have been tested under different chemical conditions to produce CL emission from different analytes. Usually, if oxidation of the molecule is known to give a fluorescent product, or if the analyte itself has a typical structure that might be fluorescent, then there is a possibility that oxidation of the analyte exhibits CL. In the following sections, the most common oxidants used in the CL analysis of pesticides will be discussed.

12.2.1.1 Potassium Permanganate

Potassium permanganate is one of the most common oxidants used in this kind of reactions and has been extensively investigated. Many substances have been determined via their direct oxidation by potassium permanganate in acidic medium. A recent review including discussion on the reaction conditions, the influence of enhancers such as polyphosphates, formaldehyde, and sulfite, and the relationship between analyte structure and CL intensity of this reaction has been published [27]. Some examples of the recently reported CL analysis of pesticides using potassium permanganate will be commented in this chapter.

TABLE 12.1

Application of CL in Liquid Phase in the Analysis of Pesticides According to the Involved Reaction

Analyte	Matrix	LOD	Recovery (%)	Refs.
	Direct Oxidation			
Carbaryl	Distilled, tap, table, river, and lake water samples	14.8 µg L^{-1}	87.0–99.0	[28]
Carbaryl and carbofuran	Lake and tap water	10 and 50 µg L^{-1}	88–106	[29]
Antu	River water, wheat, barley, and oat-grain samples	5–10 µg L^{-1}	91.4–104.9	[30]
Tsumacide	Cole leaves	0.66 µg L^{-1}	94.6–98.2	[31]
Aldicarb	Certified technical formulation and mineral water	0.069 µg L^{-1}	97.4–98.7	[32]
Asulam	Irrigation, tap, and spring water samples	500 µg L^{-1}	89.3–130.0	[33]
Propanil	Formulation and river, residual, underground, and mineral water samples	8 µg L^{-1}	97.0–103.0	[34]
Acrolein	River, irrigation, underground, and bottled mineral water, human urine, soil samples	0.1 µg L^{-1}	99.2–105.4	[35]
Karbutilate	Tap, surface, and bottled water, human urine	10 µg L^{-1}	99.4–102.3	[36]
Bromoxynil	Formulation and tap water sample	5 µg L^{-1}	104–110	[37]
Diphenamid	Residual, underground, mineral, and tap water samples, human urine	1 µg L^{-1}	98–103	[39]
Nine pyrethroids	Tomato	13–49 µg L^{-1}	80.7–107.4	[40]
Five benzoylurea insecticides	Cucumber	12–180 µg L^{-1}	78–114	[41]
3-Indolyl acetic acid	River, residual, underground, mineral, and drinking water samples	0.1 µg L^{-1}	92.6%–98.0%	[42]
Strychnine	Tap, waste, mineral, and bottled water, human urine, formulation	2 µg L^{-1}	97.3–103.8	[43]
Carbaryl	Commercial formulations, water, grain, and soil samples	28.7 µg L^{-1} (peak area), 45.6 µg L^{-1} (peak height)	93.2–101.4	[44]
Carbofuran	Cabbage	28.4 µg L^{-1}	98.5–112.0	[45]
	Peroxyoxalate Reaction			
Carbaryl, carbofuran, and propoxur	Fruits juices (apple, pineapple, and grapefruit)	3.3 and 2 µg L^{-1}	93–96	[48]
Propham and chlorpropham	Postharvest-treated potatoes	3.6 and 3.4 µg kg^{-1}	97.6–102.8	[49]
Carbaryl	Cucumber, tap, river, and ground water	9 µg L^{-1}	80.6–106.0	[50]
Total carbamate content	Natural waters	—	—	[51]

(continued)

TABLE 12.1 (continued)
Application of CL in Liquid Phase in the Analysis of Pesticides According to the Involved Reaction

Analyte	Matrix	LOD	Recovery (%)	Refs.
Tris(2,2'-Bipyridine)Ruthenium(II) system				
Carbaryl	Commercial formulation, water, soil, grain, and blood serum	12 µg L^{-1}	90–110	[54]
Carbofuran, promecarb	Soil and water	53 µg L^{-1}	97–100	[55]
		85 µg L^{-1}	99–101	
Bendiocarb, carbaryl, promecarb, and propoxur	Mineral, tap, ground, and irrigation water, apple, and pear	0.0039–0.0367 µg L^{-1} (water samples) 0.5–4.7 µg kg^{-1} (fruits)	80–102	[56]
Luminol Reaction				
Parathion	Rice	8 µg L^{-1}	96–97	[59]
Monocrotophos	Water	7 µg L^{-1}	98–102	[60]
Methamidophos	Vegetables	47 µg L^{-1}	90–109	[61]
Chlorpyrifos	Orange and pomelo	0.18 µg L^{-1}	94.4–107.4	[62]
Ziram, mancozeb, propineb	NR	2.0, 0.1, and 0.5 µg L^{-1}	NR	[63]
Carbaryl	Cucumber, tap, river, and ground water	4.9 µg L^{-1}	76.2–135.7	[64]
Carbofuran	Lettuce, tap, river, and ground water	20 µg L^{-1}	90.1–122.9	[65]
Omethoate, dichlorvos, and dipterex	Lettuce rape, spinach, leek	10 µg L^{-1}	93–113	[66]
Diclorvos, isocarbophos, and methyl parathion	Chervil leaves, cucumber peels, and leaves from trees	30, 50, and 100 µg L^{-1}	95.0–102.5	[67]
Carbofuran, carbaryl, and methiocarb	Mineral, river, and ground water samples	0.07–0.1 µg L^{-1}	87.6–107.7	[68]
Immunosensors				
Atrazine	Surface water and fruit juice	0.006 and 0.010 µg L^{-1}	87–124	[72]
Atrazine	Surface and tap water samples	0.024 and 0.017 µg L^{-1}	NR	[73]
ELISA				
DDT and related compounds	Soil and food extract, lyophilized samples of soil and fish	0.06 and 0.04 µg L^{-1}	54–136 and 70–144	[77]

NR, not reported; LOD, limit of detection.

An FIA system using potassium permanganate in sulfuric acid medium was proposed for the study of photodegradation kinetic and for the determination of carbaryl in surface water samples [28]. The method yielded an LOD of 14.8 ng mL^{-1} and was subjected to minor interferences from various organic and inorganic species likely to be found in natural waters. Similarly, an FIA method was described for the determination of carbaryl and carbofuran, taking advantage of the strong CL signal generated when these pesticides are mixed with Na_2SO_3 and $KMnO_4$ in acidic medium. The reported LOD were 10 and 50 ng mL^{-1}, for carbaryl and carbofuran, respectively, with a sample throughput of 180 h^{-1}. The proposed method was also applied to determine these pesticides in freshwaters, which produced satisfactory results [29]. The same oxidant was used in the CL determination of antu (a rodenticide), based on its reaction in sulfuric acid medium, in the presence of formaldehyde as an emission enhancer [30]. The recording of the whole CL intensity vs. time profiles was obtained using the stopped-flow technique in a continuous-flow system. This enabled the use of three quantitative parameters: rate of the light-decay reaction, maximum emission intensity, and total emission area, which are proportional to the analyte concentration. The LODs ranged from 0.005 to 0.010 µg mL^{-1}, depending on the parameter used. This procedure was applied to determine the presence of antu in different kinds of spiked samples, such as river water, wheat, barley, and oat grain samples, which demonstrated good recovery values. The acidic $KMnO_4$ FIA-CL system was also used for the determination of tsumacide, taking advantage of the fact that the alkaline degradation product of tsumacide emits CL signals on reaction with acidic $KMnO_4$ in the presence of rhodamine 6 G [31]. Under optimum conditions, the method demonstrated an LOD of 6.6×10^{-4} mg L^{-1} and was applied to determine the presence of tsumacide in cole leaves, which yielded good recovery results.

Occasionally, when direct oxidation is used for CL measurements, a previous chemical transformation of the analyte is required for obtaining derivatives with improved fluorophoric or CL properties. In this context, the photochemical reactions (irradiation with UV/Vis lamps) have been proposed for determining the CL of several pesticides. Martínez Calatayud et al. carried out various studies to determine the presence of several pesticides, namely aldicarb [32], asulam [33], propanil [34], acrolein [35], and karbutilate [36] by means of photodegradation of the pesticide, followed by an oxidation reaction, in a fully automated multicommutation-based flow-assembly, which adjusts the medium to make it suitable for photodegradation. The photodegradation is carried out in a polytetrafluoroethylene (PTFE) tubing, coiled around a low-pressure mercury lamp, which is incorporated into the flow manifold. The system includes solenoid valves that improve the analytical features of FIA, such as reproducibility, automation and reduction of sample, and reagent consumption. The developed manifold (similar in all the applications) is shown in Figure 12.1. The optimization of important variables such as flow rates, temperature of the photodegradation and CL process, selection of the photodegradation medium and time, influence and selection of photosensitizers and CL enhancers, or organized media and concentration of the oxidant, was carried before their application to the analysis of the pesticides in different samples. For instance, in the case of aldicarb, the determination is based on the iron(III) catalytic mineralization of the pesticide by UV-irradiation, followed by the reaction of the photodegraded product with potassium permanganate and quinine sulfate as sensitizer, producing a CL emission proportional to the concentration of the pesticide. For asulam, the CL reaction is obtained in a similar way, involving the irradiation of an aqueous solution of the carbamate in glycine buffer with UV light, followed by oxidation of the photoproducts with potassium permanganate in sulfuric acid medium, or, as an alternative, oxidation with alkaline ferricyanide. After testing the influence of potential interferants, the methods were applied to different samples, obtaining satisfactory results.

Furthermore, the determination of another pesticide (bromoxynil) using photolysis in basic medium with ethanol as sensitizer, followed by oxidation with potassium permanganate in a polyphosphoric acid medium, was also proposed [37]. For this purpose, a conventional FIA manifold was selected, using a modified simplex method to optimize the hydrodynamic variables involved in the system. The method allowed determination of 134 samples h^{-1}, with an LOD of 5 µg L^{-1},

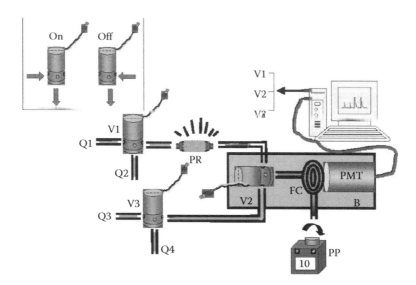

FIGURE 12.1 Optimized solenoid-valve flow-assembly for aldicarb determination by direct oxidation. Q1, aldicarb aqueous solution; Q2, medium of photodegradation (SDS 0.015% + Fe(III) 5×10^{-6} M); Q3, oxidant (KMnO$_4$ 7×10^{-4} M + H$_2$SO$_4$ 2 M); Q4, quinine sulfate 5×10^{-4} M. V1, V2, and V3, solenoid valves; PR, photoreactor; PP, peristaltic pump (flow rate =10 mL min^{-1}); PMT, photomultiplier tube; FC, spiral flow cell; B, light-tight box. (From Palomeque, M. et al., *Anal. Chim. Acta*, 512, 149, 2004. With permission.)

and it was also applied in the analysis of a water sample and formulation. This assembly was also used for the theoretical prediction of the photoinduced CL of 72 pesticides by means of molecular connectivity, a topological method that allows a unique mathematical characterization of molecular structures [38].

12.2.1.2 Hexacyanoferrate

Extending these studies and applying the FIA system, but using K$_3$[Fe(CN)$_6$] in sodium hydroxide medium as oxidant, the same research group studied the application of molecular connectivity calculations to predict the photoinduced CL behavior of a family of herbicides, grouped as amides [39]. After an exhaustive photodegradation study of the system, several compounds were theoretically studied and their CL behaviors were predicted. The method was also applied to determine the presence of diphenamid in different types of water and human urine, which demonstrated satisfactory results. Also, the potential of HPLC coupled with CL detection produced by direct oxidation was explored for the determination of nine pyrethroids (fenpropathrin, β-cyfluthrin, λ-cyhalothrin, deltamethrin, fenvalerate, permethrin, acrinathrin, τ-fluvalinate, and bifenthrin) [40] and five benzoylurea insecticides (diflubenzuron, triflumuron, hexaflumuron, lufenuron, and flufenoxuron) [41]. The CL emission took place by postcolumn irradiation of the pesticides with UV light and subsequent oxidation of the photolyzed pesticides with K$_3$[Fe(CN)$_6$] and NaOH, whose CL signal increased with the percentage of acetonitrile in the reaction medium. As the kinetic of this CL reaction was very fast, a modification in the detector was required to mix the analytes and reagents as close as possible to the detection cell. Depending on the compound, the LODs between 0.013 and 0.049 μg mL^{-1} for pyrethroids, and between 0.012 and 0.18 μg mL^{-1} for benzoylureas were obtained. These methods have been successfully applied to determine the presence of pyrethroids in tomato and benzoylureas in cucumber, previously extracted by liquid–liquid and solid–liquid extraction, which showed good results (recoveries in the range of 80.7%–107.4% and 78%–114%, respectively).

12.2.1.3 Cerium(IV)

Ce(IV) in acidic medium has also been proposed as an oxidant reagent in direct CL determinations. Recently, Martínez Calatayud et al. proposed an FIA method for the determination of the pesticide, 3-indolyl acetic acid, in water samples, whose CL emission was obtained by oxidation of the analyte with Ce(IV) in nitric acid and in the presence of β-cyclodextrine and dimethylformamide as sensitizer [42]. The method allowed the detection of 159 samples h^{-1} over the range of 0.5–15.0 mg L^{-1}, with an LOD of 0.1 μg L^{-1} and a precision of 2.7% (RSD at 2.0 mg L^{-1} of the pesticide, $n = 17$). The obtained recoveries ranged approximately 93%–98% in water samples collected from different places. The same research group also proposed the multicommutation flow assembly described earlier [32–36] for the determination of strychnine [43], including the prior photodegradation of the analyte. The method demonstrated an LOD of 2 μg L^{-1}, good precision, and it was applied in the analysis of the pesticide, 3-indolyl acetic acid, in different water samples, human urine and formulation, with recoveries in the range of 97.3%–103.8%. The same oxidant was used in an automatic FIA for the determination of carbaryl, based on the measurement of both peak height and peak area. In this case, rhodamine 6 G was used as the sensitizer [44]. The method was used for various types of matrices, such as commercial formulations, water, grain, and soil samples with recoveries higher than 93%. Previously, solid-phase extraction (SPE) was used to concentrate and separate the analyte from the matrix. The method exhibited good selectivity, as no other pesticide containing the naphthalene group was found to interfere with the determination of carbaryl. The LOD was 45.6 and 28.7 ng mL^{-1} for peak height and peak area measurements, respectively, and the RSD for 10 samples was less than 1.4% with both types of measurements. Owing to the enhancing effect of carbofuran on the CL reaction between sodium sulfite and Ce(IV) in sulfuric acid, an FIA system was proposed for determining the presence of this pesticide [45]. The proposed method is simple and shows a wide linear response range. The method provides an LOD of 2.84×10^{-8} g mL^{-1} and was applied for the analysis of carbofuran in cabbage, which demonstrated satisfactory results (recoveries ranged from 98.5% to 112.0%).

12.2.2 PEROXYOXALATE REACTION

With regard to indirect CL, one of the more efficient nonbiological systems that are frequently used is based on the so-called PO reaction, which involves the oxidation of an aryl oxalate ester with hydrogen peroxide in the presence of a fluorophore. The reaction, shown in Figure 12.2, is suggested to follow a chemically initiated electron-exchange luminescence mechanism via a high-energy intermediate, 1,2-dioxetanedione, which forms a charge complex with the fluorophore, donating one electron to the intermediate [46]. This electron is transferred back to the fluorophore, raising it to an excited state and liberating light characteristic of the fluorophore. Bis-(2,4,6,-trichlorophenyl)

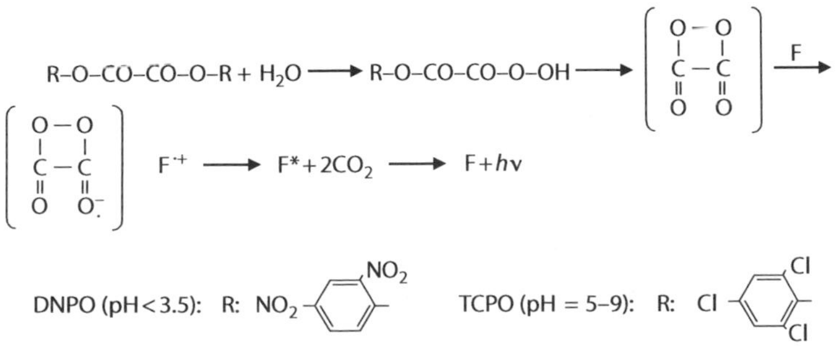

FIGURE 12.2 Possible reaction pathway for the PO–CL system. F, fluorophore; DNPO, bis-(2,4-dinitrophenyl)oxalate; TCPO, bis-(2,4,6,-trichlorophenyl)oxalate. (From Gámiz-Gracia, L. et al., *Trends Anal. Chem.*, 24, 927, 2005. With permission.)

oxalate (TCPO) and bis-(2,4-dinitrophenyl)oxalate (DNPO) are the most commonly used oxalates. The main disadvantage of this system resides in the insolubility of the above-mentioned compounds in water and their instability toward hydrolysis, which requires the use of organic solvents, such as acetonitrile, dioxane, *tert*-butanol, and ethyl acetate. This reaction can be used to determine a great number of species, such as hydrogen peroxide, compounds that are highly fluorescent (e.g., polycyclic aromatic hydrocarbons), or compounds that do not exhibit native fluorescence, but can be chemically derivatized (as some carbamates) using dansyl chloride, *ortho*-phthalaldehyde (OPA), fluorescamine, etc. [47].

Orejuela and Silva proposed a CL reaction for the analysis of carbaryl, carbofuran, and propoxur by HPLC, using TCPO and dansyl chloride as derivative reagent [48]. The hydrolysis of the carbamates is mandatory to produce an alcohol, derivatized from dansyl chloride to form the corresponding fluorescent derivative. By using cetyltrimethylammonium bromide (CTMAB) as a catalyst, the hydrolysis of carbamates and the subsequent dansylation can be simultaneously carried out in a precolumn reaction in a very short time. After their separation, the analytes can be detected using an integrated derivatization–CL detection unit based on TCPO–hydrogen peroxide system. The method was applied to determine the presence of the above-mentioned carbamates in spiked fruit juices (apple, pineapple, and grapefruit). The linear range of application was 8.0–1500, 7–1500, and 4.0–1500 µg L^{-1} with precision of 6.1%, 7.6%, and 6.3% (50 µg L^{-1} of carbamate, $n = 11$) for carbaryl, carbofuran, and propoxur, respectively. In a similar way, the presence of propham and chlorpropham was determined by HPLC, after hydrolysis of the pesticides, whose metabolites (aniline and 3-chloroaniline) were derivatized with dansyl chloride and detected by TCPO-CL reaction [49]. The method was tested in the determination of those pesticides in the postharvest-treated potatoes.

With regard to FIA analysis, two methods, one for the quantitative determination of carbaryl [50] and the other for the screening analysis of total carbamate content [51] in vegetables and different types of water, were proposed using the TCPO–hydrogen peroxide system with prior hydrolysis of the carbamates to obtain methylamine (MA), derivatized with OPA to form the fluorescent derivative. In both the cases, the manifold presented a three-channel configuration where the derivatized standard (or sample) and TCPO solutions were subsequently incorporated into a carrier of sodium dodecyl sulfate (SDS) with the aid of two manual injection valves. Thus, the TCPO solution was first injected into a valve and the standard or sample solution containing the OPA–MA derivative was injected after 5 s with another injection valve. The TCPO and the OPA–MA derivative solutions were then mixed in a reaction coil and subsequently merged with the imidazole and hydrogen peroxide streams, allowing the production of a CL emission (proportional to the concentration of MA and consequently to the concentration of the pesticide) in the detection cell placed just in front of the photomultiplier. This FIA manifold avoids the problems arising from the use of acetonitrile as solvent, as neither special tubes nor special pumps are required when using the two injection valves and micellar medium as carrier, avoiding the rapid degradation of TCPO in water. The different variables affecting the FIA system were optimized by means of a formal strategy involving the use of experimental designs [52]. In the method for the quantitative determination of carbaryl, the off-line photodecomposition reaction of carbaryl was carried out using a rod-shaped low-pressure mercury discharge lamp and a PTFE coil around it. The standard solutions of carbaryl or sample extracts were propelled using a peristaltic pump, controlling the irradiation time into the photoreactor by changing the flow rate. This method was applied in the analysis of spiked tap, river and ground water as well as spiked cucumber, with recoveries acceptable for routine analysis. In the semiquantitative method for the determination of total carbamate content, the hydrolysis of the tested carbamates (carbaryl, carbofuran, aldicarb, and promecarb) to MA was carried out in an alkaline medium with CTMAB (95°C, 5 min). This screening method (which will be commented in detail in the following section) is based on the CL measurement from the total MA–OPA concentration in the TCPO-CL system and has been established for different concentrations of total carbamates in water.

12.2.3 Tris(2,2'-Bipyridine)Ruthenium(II) System

Another CL system applied in pesticide monitoring involves the use of tris(2,2´-bipyridine) ruthenium(II) reaction, which produces an orange emission at 620 nm from the excited state $[Ru(bpy)_3^{2+}]^*$, (bpy = bipyridine), by means of the following mechanism:

   i. $Ru(bpy)_3^{2+} \rightarrow Ru(bpy)_3^{3+} + e^-$ (oxidation)
   ii. $Ru(bpy)_3^{3+} \rightarrow [Ru(bpy)_3^{2+}]^*$ (reduction with analyte)
   iii. $[Ru(bpy)_3^{2+}]^* \rightarrow Ru(bpy)_3^{2+} + h\nu$ (CL emission)

The analytical application of this reaction is based on the fact that the $Ru(bpy)_3^{3+}$ species can be reduced by a large number of potential analyte compounds or their electrochemical derivatives, via high-energy electron transfer reactions, to produce the $[Ru(bpy)_3^{2+}]^*$ excited species. A recent review comprised most of the typical applications of this CL reaction, including analysis of pesticides [53]. However, in the past, very few applications of this reaction in the analysis of pesticides have been reported.

Pérez-Ruiz et al. proposed an FIA manifold with two photochemical processes developed online for the determination of three different carbamates: carbaryl [54], carbofuran, and promecarb [55]. The methods were based on the photoconversion of the carbamate into MA, which subsequently reacts with the photogenerated $Ru(bpy)_3^{3+}$. The FIA manifold was a four-channel configuration, where the sample and $Ru(bpy)_3^{2+}$ were simultaneously injected with the aid of two rotary valves, as shown in Figure 12.3. The analytes were introduced into a water carrier merging with a phosphate buffer stream, which subsequently passes through a photoreactor and is photodegraded to MA. The $Ru(bpy)_3^{2+}$ was simultaneously injected into a water stream, which merges with the peroxydisulfate stream to achieve the photogeneration of $Ru(bpy)_3^{3+}$ along a second photoreactor. The subsequent confluence of the two combined streams in the flow cell resulted in the CL emission. The tubing between the two photoreactors and the flow cell was covered with black insulating tape to prevent fiber-optic effect resulting from the introduction of stray light into the detector. The method was applied to determine the presence of carbaryl in commercial formulations as well as spiked water, soil, grain, and blood serum, with a linear concentration range of 0.04–4.0 mg L^{-1} and a precision of 1.2% (for 0.50 mg L^{-1} of carbaryl, n = 5) [54]. The method was also applied to determine the presence of carbofuran and promecarb in spiked soil and water. The linear range of application was 0.22–11.2 and 0.41–16.6 mg L^{-1} with precisions of 0.22% and 0.26% (for 5 mg L^{-1} of carbamate) for

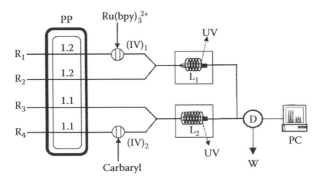

FIGURE 12.3 FIA manifold for the determination of carbaryl with the $Ru(bpy)_3^{2+}$ system. PP, peristaltic pump (with flow rates given in mL min^{-1}); R_1, water; R_2, 1.4×10^{-3} mol L^{-1} potassium peroxydisulfate and 0.05 mol L^{-1} phosphate buffer of pH 5.8; R_3, 0.15 mol L^{-1} phosphate buffer of pH 6.5; R_4, water; $(IV)_1$, $(IV)_2$, injection valves; L_1, L_2, photoreactors; D, luminometer; W, waste. (From Pérez-Ruiz, T. et al., *Anal. Chim. Acta*, 476, 141, 2003. With permission.)

carbofuran and promecarb, respectively [55]. The same research group also proposed an automated SPE-HPLC method to determine the trace concentrations of *N*-methylcarbamate (NMC) pesticides (bendiocarb, carbaryl, promecarb, and propoxur) in water and fruits, based on the postcolumn conversion of the pesticides into MA by irradiation with UV light [56]. The resultant MA was subsequently detected by CL using tris(2,2′-bipyridyl)ruthenium(III), which was online generated by photooxidation of the ruthenium(II) complex with peroxydisulfate, in a way similar to the earlier FIA methods. The intra- and interday precision values of about 0.64%–1.3% RSD (*n* = 10) and 2.2%–2.8% RSD (*n* = 15), respectively, were obtained. The LODs were within the range of 3.9–36.7 ng L^{-1} for water samples and 0.5–4.7 µg kg^{-1} for fruits.

12.2.4 LUMINOL REACTION

The best-known example (and more extensively used) in direct CL reactions is the oxidation of luminol (5-aminophthalylhydrazide) in alkaline medium, to produce the excited 3-aminophthalate anion, which emits light when relaxing to the ground state (Figure 12.4), having a quantum yield of about 0.01 in water and 0.05 in dimethyl sulfoxide (DMSO) [57]. Several oxidants such as permanganate, periodate, hexacyanoferrate(III), and hydrogen peroxide can be used. The reaction is catalyzed by metal ions (Fe(II), Cu(II), Co(II), etc.) and acts as a powerful detection system in FIA, LC, and CE, where luminol-type compounds can be used as a derivatization reagent allowing the analytes to be detected at very low levels [58], although most of its applications in the analysis of pesticides have been performed coupled with FIA.

Some pesticides and fungicides belonging to organophosphorus, carbamate, and thiocarbamate families are inhibitors of the CL or substances that are easily oxidized and acts as interferants in the luminol reaction, being indirectly determined by measuring the decrease in the CL emission. Most of the applications reported with this reaction have been implemented with FIA, exploiting the intrinsic characteristics of this technology, such as short analysis time, automation, and high precision in CL measurements. Luminol reaction is also the base for most of the CL immunoassay reactions used for the determination of pesticides, and this application will be commented in the following section.

A simple FIA-CL system, in which luminol and H$_2$O$_2$ are introduced into a mixing cell at a flow rate of 0.5 mL min^{-1}, and where the resulting solution acts as the carrier of the sample, introduced by means of an injection valve, has been applied as a highly sensitive assay for the detection of parathion in rice samples [59]. In this case, polyethylene glycol 400 surfactant was used as the enhancer of the CL signal. This enhancing effect may be attributed to the tendency of the molecules to form micelles in aqueous solution; this microenvironment not only increases the solubility of the pesticide in water and the chance to react with hydrogen peroxide and luminol in the solution, but

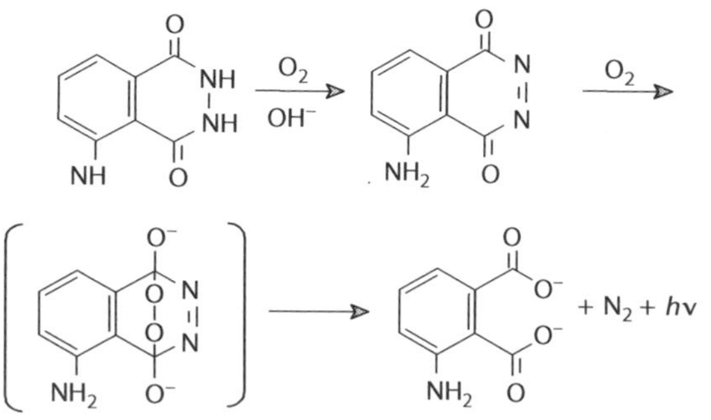

FIGURE 12.4 Proposed mechanism for the luminol CL reaction. (From Gámiz-Gracia, L. et al., *Trends Anal. Chem.*, 24, 927, 2005. With permission.)

also provides a protective environment for the excited singlet state of 3-aminophthalic acid ions. To eliminate the interference of some cations that can act as catalysts in the CL reaction, a cation-exchange column was connected to the sample line to remove these metal ions. The CL intensity was linear in the range of 0.02–0.1 mg L^{-1} of parathion, obtaining good recoveries. This method was also employed for the determination of monocrotophos, using sodium chloride instead of polyethylene glycol 400 surfactant as enhancer of the CL signal, which gave a linear range from 2×10^{-8} to 1×10^{-6} g mL^{-1} [60]. The method was successfully applied to determine the presence of monocrotophos in water samples, and a possible mechanism of the reaction was also provided.

Hydrogen peroxide has also been used as oxidant in the luminol reaction for the determination of methamidophos residue in vegetables, using SPE and FIA-CL [61]. The method is based on the enhancing effect of methamidophos on the CL reaction and provides an LOD of 0.047 μg mL^{-1} with recoveries in the range of 90%–109%. The method was successfully applied to determine the presence of methamidophos residue in some vegetable samples.

An original, rapid, and inexpensive methodology was developed by Song et al. for the determination of chlorpyrifos in spiked orange and pomelo samples, using the reaction of luminol with potassium periodate [62]. The CL intensity was inhibited in the presence of chlorpyrifos, which can be oxidized by periodate. In the proposed FIA manifold, both the CL reagents were immobilized in an online anion-exchange resin with a molar ratio of 1:2 (luminol:periodate). These reagents were quantitatively eluted with water at pH 6.5 and transferred to the six-way injection valve, and subsequently mixed with the sodium hydroxide stream. Under the optimal conditions, the decrease in the CL intensity was linear over the logarithm of concentration of chlorpyrifos from 0.48 to 484.0 ng mL^{-1}.

An FIA-CL method to determine certain dithiocarbamate fungicides, such as ziram, mancozeb, and propineb, was proposed by Kubo et al., based on the oxidation of luminol in the alkaline medium, using hexacyanoferrate(III) as catalyst/cooxidant and hexacyanoferrate(II) as depressor [63]. The behavior of the dithiocarbamates in the alkaline medium was studied by electron-spin resonance using 5,5-dimethyl-1-pyrroline N-oxide as a spin-trap agent. Hydroxyl radicals were detected in the spectra of the three fungicides, showing that these are important intermediates in the luminol CL reaction. Good sensitivity was obtained when optimized conditions were applied; however, the use of this method with real samples was not reported.

Two FIA methods for the determination of carbaryl and carbofuran using potassium permanganate as oxidant in the CL luminol reaction were proposed for the analysis of these carbamates in vegetables and different types of water [64,65]. In these papers, the different variables affecting the FIA system were optimized by means of formal strategies involving the use of experimental designs. Both the methods were satisfactorily applied in the analysis of spiked samples of tap, river, and ground water, and vegetables (cucumber or lettuce), with LODs in the lower range (μg L^{-1}), being in agreement with the current demands.

Recently, a less-common oxidant, potassium peroxodisulfate, was proposed for the simultaneous determination of three organophosphorus pesticide (OPP) residues (omethoate, dichlorvos, and dipterex) [66]. This method is based on the fact that OPPs can be decomposed into orthophosphate with potassium peroxodisulfate as oxidant under UV radiation, and that the decomposing kinetic characteristics of the pesticides with different molecular structure are significantly different. The produced orthophosphate can react with molybdate and vanadate to form the vanadomolybdophosphoric heteropoly acid, which can oxidize luminol to produce intense CL emission. The obtained data were processed chemometrically using a three-layered feed-forward artificial neural network trained by back-propagation learning algorithm. After multivariate calibration of the method, which produced an LOD of $1 \cdot 10^{-8}$ g mL^{-1}, the method was applied in the analysis of the three pesticides in different types of vegetables (lettuce, rape, spinach, and leek), obtaining recoveries ranging from 93% to 113%.

The CL luminol reaction offers promising results when it is coupled with HPLC, providing high efficiency in separation and low LODs inherent to CL, although its application in the analysis of pesticides has been limited. Though fluorescence and absorbance have been widely used as detection techniques for trace analysis of pesticides, the elimination of the excitation source in CL can reduce stray light, background emission, or light source instability. However, one of the drawbacks is that this assembly requires additional pump(s) to deliver postcolumn CL reagent. The luminol–H_2O_2 CL system coupled with HPLC has been used for the selective detection of organophosphorus insecticides (diclorvos, isocarbophos, and methyl parathion) [67]. For the assembly, a mixing coil after the mixture of CL reagents is necessary to decrease the background signal. Subsequently, this mixture is merged with the eluent from the chromatography column. The composition of mobile phase includes methanol–water at the ratio of 60:40 (v/v). A major advantage of this method over HPLC with UV detection is that the HPLC-CL method allows the analysis of vegetable samples without the need for an elaborate extraction process to remove potential interfering substances, as these substances do not provide CL signal. The method has been successfully applied in the determination of these pesticide residues in spiked samples of chervil leaves, cucumber peels, and leaves from trees.

The coupling of HPLC with CL detection using the luminol–$KMnO_4$ system for the simultaneous determination of three phenyl-N-methyl-carbamate pesticides (carbofuran, carbaryl, and methiocarb) was proposed recently [68]. The separation was reached in less than 14 min using an isocratic elution (H_2O:acetonitrile, 50:50). The CL reagents for postcolumn detection were delivered by means of a peristaltic pump and mixed in a Y-connection. Subsequently, this mixture joined the eluate from the chromatographic column just in front of the detection cell. The schematic device is shown in Figure 12.5. The optimization of variables affecting the CL reaction was carried out by means of experimental designs. Prior to HPLC-CL determination, SPE was applied to preconcentrate the sample to reach LODs below the legal maximum concentration permitted in drinking water. This method has been successfully applied in the determination of these carbamates in different water samples (tap, river, and ground water), showing satisfactory recoveries.

12.2.5 CHEMILUMINESCENCE IMMUNOASSAY

Immunoassay is a rapid, sensitive, and cost-effective technique for environmental screening, based on the ability of the immune system to produce a variety of antibodies with a high affinity for

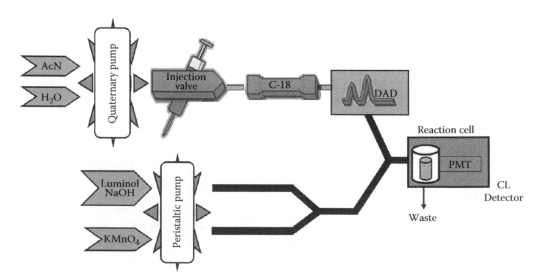

FIGURE 12.5 Proposed manifold for the analysis of phenyl-N-methyl-carbamate pesticides by HPLC-CL by means of the luminol–$KMnO_4$ system. (From Huertas-Pérez, J.F. and García-Campaña, A.M., in *31st International Symposium on High Performance Liquid Chromatography and Related Techniques (HPLC 2007)*, Ghent, Belgium, June 17–21, 2007 (Ref.: P20–28).)

foreign compounds (immunogens). In analytical chemistry, this phenomenon can be exploited by detection of this immunoreaction using labeled antibodies or antigens (i.e., compounds that can be bound by antibodies). The use of enzymes as labels in immunoassay potentially affords much more sensitive tracer detection when compared with other labels, because of the great catalytic power resulting in the generation of many product molecules from one enzyme molecule through turnover. Though the measurement of enzyme activity is strongly influenced by suboptimal variables, such as temperature, pH, salt concentration, etc., as enzyme detection is cost-effective and easy to use, enzymes have become popular labels in immunoassay, where horseradish peroxidase (HRP) is the most widely used enzyme. Thus, immunochemical techniques have gained an increasing importance in the screening and quantification of pesticides owing to their sensitivity, speed, simplicity, and low cost, allowing rapid analysis of a large number of samples. The miniaturization of immunoassays is also an increasingly useful approach, which allows not only to detect minute amounts of analyte, but also helps to build array systems for multiplex analysis, and in addition, offers experimental simplicity of handling.

The sensitivity of an immunoassay strongly depends on the affinity of specific antibodies and sensitivity of the detection method for the label. Accordingly, BL and CL have been extensively used for the ultrasensitive detection of labels in immunoassays. The methodological aspects as well as applications have been provided in some earlier books, chapters, and reviews [7,69–71]. The CL immunoassay can employ either direct labeling of antigens or antibodies with CL molecules, or labeling with enzymes detectable with CL substrates; the luminol reaction is the most widely used CL reaction, mainly owing to the enhanced CL (ECL) assays for HRP labels.

12.2.5.1 Immunosensors

Biosensors are analytical systems comprising an immobilized biological sensing element and a physical transducer. When the biological component is an antibody, the biosensor is called an immunosensor. Physical transducers can be piezoelectric, electrochemical, or optical. Theoretically, immunosensors are capable of continuous and reversible detection, but, as the antibody–antigen interactions have high-affinity constants, reversibility is difficult to achieve in practice.

Immunosensors require specific immunoreactive components to be immobilized in the solid phase. Different assay formats can be established, including the immobilized antibodies or antigen format. An alternative immunoassay format to overcome the regeneration problems typically encountered in the immobilized antibody format is based on the use of affinity proteins, such as proteins G and A. These proteins selectively bind to the fragment of crystallization (Fc) region of a wide range of immunoglobulins without interfering with the antigen-binding sites, making them very attractive as affinity-capture supports for immunobiosensing applications.

Some examples of the application of this technique in the analysis of pesticides can be found in the recent literature. Different hydrophilic polymers with long flexible chains were employed by Yakovleva et al. for modification of the silica surfaces followed by attachment of proteins A and G, and were covalently immobilized on the silicon microchips and used for developing microfluidic immunosensors based on HRP, catalyzing the CL oxidation of luminol/p-iodophenol (PIP) for the determination of atrazine [72]. The assay procedure was based on the principle of affinity-capture competitive immunoassay. In the so-called competitive immunoassay format, the antigen competes with a labeled antigen for a limited number of antibody-binding sites. At low antibody and tracer concentrations, the sensitivity of this type of assay increases, as in this condition, a variation in the amount of competing antigen has a larger influence on the change of signal that is ultimately measured. This format extends the flexibility of the assay, because a whole range of antibodies with different specificities can be bound to as well as dissociated from protein A- and G-coated surfaces. Thus, the use of the affinity-capture proteins helps in the development of generic assays, potentially applicable to any analyte. The application was developed according to the following steps: the enzyme tracer, atrazine standard or sample, and anti-atrazine antibody were mixed off-line and injected directly, or after preincubation, into the system to remove unbound immunocomplexes. Subsequently, the substrate mixture (luminol/PIP/H_2O_2) was injected and the enzyme-catalyzed CL reaction on the microchip surface was monitored via the PMT. To complete the assay cycle, the immunocomplex was removed from the affinity proteins. The schemes of the microfluidic immunosensor device and the affinity-capture competitive immunosensor format of this system are shown in Figure 12.6. All immunosensors could detect atrazine at a concentration down to

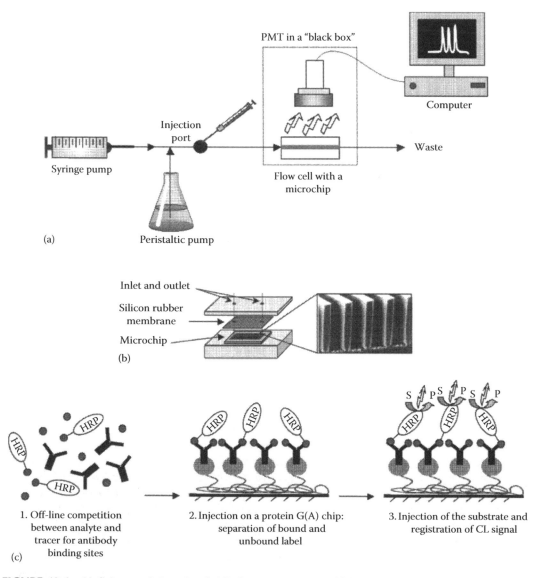

FIGURE 12.6 (a) Scheme of the microfluidic immunosensor manifold based on HRP, catalyzing the CL oxidation of luminol/p-iodophenol: a syringe pump and a peristaltic pump were used for carrier buffer and regeneration solution at flow rates of 40 and 50 mL min^{-1}, respectively. The sample, containing the enzyme tracer, atrazine standard and antibody, premixed off-line, was injected through a six-port injection valve, and the CL signal was detected by a PMT placed above the flow cell, containing the microfluidic immunosensor. (b) Detailed view of the plexi glass microchip flow cell and a magnified image of the microchip channel network. (c) Scheme of the affinity-capture competitive immunosensor format performed in the system. (From Yakovleva, J. et al., *Biosens. Bioelectron.*, 19, 21, 2003. With permission.)

μg L^{-1} level. Protein G microchips were applied in the analysis of spiked surface water and fruit juice, giving recovery values in the range from 87% to 124%, although the precision of the method needs improvement. Microfluidic immunosensors could provide multiple biological information and environmental multiplex analysis using channel arrays with different immobilized sensing elements. Using the same principle, Jain et al. developed a new CL flow immunosensor, based on a porous monolithic methacrylate and polyethylene composite disk modified with protein G and placed in a flow cell close to a PMT [73]. The performance of the disk immunosensor system was compared with a one-step continuous flow injection immunoassay (FIIA) system. The FIIA can be applied when continuous monitoring and high sample throughput are required, and its performance depends not only on the properties of the immunoreagents, but also on the characteristics of the substrate. In this case, it was found that the disk immunosensor provides lower LODs for atrazine (down to μg L^{-1} level) and the results are less influenced by the sample matrix, with a sensitivity that can be compared with that provided by microtiter plate enzyme-linked immunosorbent assay (ELISA), that is considered as more sensitive than the flow-based immunoassay techniques. Nevertheless, a main drawback of this method is that multiple assay steps are required, resulting in a lower sample throughput when compared with the column system.

The development of a portable, temperature-controlled, power-supply autonomous field immunosensor for environmental applications using 2,4,6-trinitrotoluene (TNT) as the key target has been proposed for field analysis [74]. In addition, the pesticides diuron and atrazine have been used to demonstrate the versatility of this device, whose main application is in screening analysis. Monoclonal antibodies (MAbs) are immobilized via adsorption on a gold surface with numerous pyramidal structures. The recognition reaction is enhanced in three ways: (1) via the enzymatic reaction, (2) via the pyramidal structures with the gold surface cover, and (3) via the detection of the CL of the product through a very sensitive PMT. The latter is placed directly above the pyramid tips. The immunoreagents (enzyme tracer and antibody), together with the environmental sample are located in a single-use chip, which is replaced after each measurement. This chip is the key to the versatility of the analytical system. Transport of the reagents is achieved using an automated, miniaturized flow-injection system, which is the consistent part and applicable for all analytes. The CL signal is observed to be inversely proportional to the amount of analyte present in the sample, and the LODs have been observed in the lower range of μg L^{-1}. Unfortunately, application of this method to real samples has not been reported.

Another high-sensitivity, semiquantitative method was developed for MP detection using immunochemiluminescence principle and charge-coupled device (CCD) camera [75]. The MP antibodies raised in poultry were used as a biological sensing element for the recognition of MP present in the sample. The immunoreactor column was prepared by packing the antibodies immobilized on Sepharose CL-4B through periodate oxidation method, in a glass capillary column (150 μL capacity), and placing it in the flow system. To obtain detectable light signals that could be captured by CCD camera, $K_3[Fe(CN)_6]$ was used as a signal enhancer and electron mediator along with HRP in the CL reaction. Light images generated during the CL reaction were captured by a CCD camera and further processed for image intensity, which was correlated with pesticide concentrations. Different parameters including concentrations of $K_3[Fe(CN)_6]$, luminol, urea, H_2O_2, antibody, addition sequence of reactants, and incubation time were optimized to obtain the best images. The results obtained by image analysis method showed very good correlation with competitive ELISA method, used as a reference. However, application of this method to real samples has not been reported.

12.2.5.2 ELISA

The ELISA has two major components. The first is the immunological reaction that occurs between an antigen and an antibody. This reaction is crucial and needs careful optimization. The second component is the surface where antigens or antibodies are immobilized. Biomolecules like antibodies attach to the surfaces via a variety of mechanisms controlled by the chemical properties of the surface, and also need optimization with respect to the effects of the solid phase on, e.g., reproducibility of antigen or antibody coating, nonspecific binding, sensitivity of the assay, etc. Furthermore, parameters like temperature, pH, composition of coating and assay buffer, incubation times, etc., play important roles with regard to the assay sensitivity and reliability. Although other assay containers can be used (as coated tubes or beads), the most popular one is the 96-well microtiter plate. However, ELISAs that are detected on a microplate luminometer must be performed in opaque white or black plates.

In ECL, the light yield of the HRP-catalyzed peroxidation of luminol is greatly increased by the addition of an enhancer, such as luciferase or acridan esters [69], offering the possibility of improving the sensitivity of immunoassay by at least two to three orders of magnitude when compared with the conventional colorimetric detection. The light intensity of ECL reaches a maximum, 1–2 min after the start of the reaction, thus providing a rapid detection of the analytical signal.

By exploiting this technique, Botchkareva et al. developed a heterogeneous CL flow immunoassay system for the detection of DDT and related compounds [76]. In the heterogeneous flow immunoassay, a solid support was used to immobilize either the antibody or the antigen, thus permitting the separation of free fractions from bound immunocomplexes. In this context, the authors characterized different types of immunoaffinity supports for DDT, namely beads, nylon coils, and membranes. Two basic formats were performed, using enzyme-labeled secondary and enzyme-specific MAbs, and employing the luminol/H_2O_2/HRP/PIP CL reaction. The results showed that membranes are the most-suitable support, preventing nonspecific adsorption of the different immunoreagents, and providing the lowest LOD (1 nM for p,p-DDT). The same research group optimized and characterized two conjugated-coated ELISAs–ECL systems based on MAbs of different specificity and homologous protein conjugates for the DDT and DDT-related compounds [77]. The DDT and DDT group-selective assays provided an LOD of 0.06 and 0.04 µg L^{-1}, respectively, being about four times more sensitive than the colorimetric ELISAs. Both the assays were applied in the analysis of fortified soil and methanolic extracts of foods and pesticides-free and fortified lyophilized samples of soil and fish. Using class-selective and DDT-selective CL ELISAs, recoveries between 54% and 136% and between 70% and 144%, respectively, were obtained for samples spiked with mixtures of DDT-related compounds and other organochlorine (OC) pesticides.

ELISA with CL detection was also proposed for the determination of carbofuran, carbaryl, and methiocarb in fruit juices [78]. When compared with colorimetric ELISA, the ability of the CL reagents to detect lower concentrations of HRP allowed to decrease the optimal antibody and conjugate concentrations and to reach better analytical parameters. Recovery values for both the ELISAs were around 100% and no matrix effects were observed in the analysis of fruit juices, spiked at different concentrations levels with the pesticides and diluted at a ratio of 1:20 or more.

12.3 ANALYSIS OF PESTICIDES BY CL DETECTION IN THE GAS PHASE

As in the liquid phase, CL gas-phase comprises a chemical reaction forming an excited-state product that subsequently undergoes one or more relaxation processes to attain its ground state. In gas-phase CL detection, radiative emission is usually competitive with nonradiative processes, and both the quantum yield of the reaction and the emission spectrum vary with physical conditions, such as bath gas composition, temperature, and pressure. CL detection in the gas phase has been reviewed previously, and a broad spectrum of CL reactions in the gas phase has been studied and reported [15,79,80]. Although some of them, such as nitrogen and sulfur chemiluminescence detectors (NCD and SCD, based on reaction with O_3) have been extensively used in a broad range of applications, such as petroleum characterization, food and beverage flavor analysis, and environmental monitoring [15,80], very few applications in the field of pesticide analysis have been reported [81–83], and to our knowledge, none of them in the recent years.

Thus, this section will consider only the most commonly used and commercially available gas-phase CL detector that has found an extensive application in the field of analysis of pesticides, i.e., the "flame photometric detector" (FPD) [84]. This detector has been mostly coupled to the GC, being considered as a relatively robust and cost-effective system. In FPD, the high temperature of a flame promotes chemical reactions that form key reaction intermediates and may provide additional thermal excitation of the emitting species. FPD may be used to selectively detect compounds containing sulfur, nitrogen, phosphorous, boron, antimony, arsenic, and even halogens under special reaction conditions, but commercial detectors are normally configured only for P and S detection. In a GC-FPD, the GC column extends to the sample inlet where it is mixed with oxygen (or air) and with hydrogen fuel prior to the burner head, and before entering the CL cell. Fuel-rich, hydrogen/ oxygen flames are preferred, and the combustion mixture must be optimized for each analyte. The main limitation to the sensitivity in FPD is the noise associated with the background signal, arising primarily from other flame emissions. Significant quenching by the presence of hydrocarbon solvents is another reported problem. To optimize the LOD, the PMT is positioned, and lenses are used to view a region of the flame where the signal-to-noise (S/N) ratio is greatest, while filters are employed to reduce background contributions of the flame emissions.

To solve some of the above-mentioned problems, an improved FPD called "pulse flame photometric detector" (PFPD) was developed [85,86]. The PFPD employs pulsed flame and time-resolved emission detection with gated electronics. In this design, the burner is constructed to generate a noncontinuous flame reignited at a frequency of about 1–10 Hz. This periodic interruption allows the acquisition of the time-dependent emissions from the various excited-state species present in the detector. As each species present different fluorescence lifetimes in the order of milliseconds, they can be differentiated in the time domain between the flame pulses. The added time-domain information is used to enhance the element-specific detection and increase the selectivity over hydrocarbons, substantially improving the overall performance of the FPD. The improvements include one or two orders of magnitude of sensitivity enhancement, about an order of magnitude of increased selectivity, and reduced quenching effects. A scheme of a PFPD detector is shown in Figure 12.7 [86]. To sum up, a combustible gas mixture of hydrogen and air (3) is continuously fed into the small pulsed flame chamber (6) together with the sample molecules that are eluted in the usual way

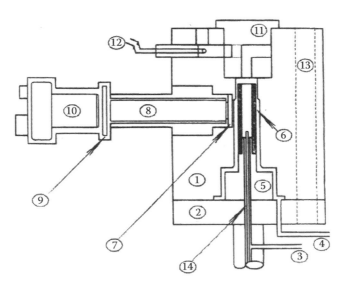

FIGURE 12.7 Schematic diagram of the PFPD design. (1) PFPD body; (2) GC-heated detector base; (3) central hydrogen-rich H₂/air mixture tube leading to the combustor; (4) outer bypass H₂/air mixture tube; (5) combustor holder; (6) quartz combustor tube; (7) sapphire window; (8) light guide; (9) colored glass filter; (10) PMT; (11) spiral igniter light shield; (12) heated wire igniter; (13) assembly guiding rod in a guiding hole; (14) column. (From Amira, A. and Jing, H., *Anal. Chem.*, 67, 3305, 1995. With permission.)

from the GC column (14). Also, the combustible gas mixture separately flows (4) to a light-shielded, continuously heated, wire igniter (12). The ignited flame is propagated back to the gas source through the pulsed flame chamber (6), and is self-terminated in a few milliseconds, as the pulsed flame cannot propagate through the small hole of the combustor holder (5) at the bottom of the pulsed flame chamber (6). The continuous gas flow creates additional ignition after a few hundred milliseconds in a pulsed periodic fashion (~3 Hz). The emitted light is transferred with a light pipe (8) through a broadband filter (9) and is detected with a PMT (10).

As the most common species detected by FPD or PFPD in the analysis of pesticides are phosphorous compounds, the first part of this section will be devoted to this application, while the second part will be focused on the sulfur compounds. In addition, Table 12.2 shows a summary of the most significant reported applications, most of them commented in the text.

12.3.1 DETERMINATION OF OPPs BY FPD OR PFPD

The mechanism of detection of phosphorus-containing compounds in an FPD is through the formation of PO, which subsequently reacts with H atoms in a fuel-rich flame to produce HPO*, emitting a light at approximately 526 nm [79].

Although some attempts have been made to couple FPD with separation techniques, such as microcolumn LC and CE for the determination of polar phosphorus-containing pesticides [87], the most important application of this detector in the field of pesticide analysis during in the past was, with a big difference, its coupling with GC for the determination of OPPs and their metabolites. In this context, most of the advances and research concerning the analysis of pesticides have been focused on the development of improved sample preparation processes. These steps are mandatory to isolate the analytes from the complex matrices, remove interfering compounds, and achieve sufficient sensitivity. In fact, sample preparation steps are often the bottleneck for combined time and efficiency in many overall analytical procedures. Thus, it is not surprising that a lot of effort has been devoted to the development of faster, safer, and more environment-friendly techniques for sample extraction and extract clean-up, prior to instrumental analysis [88–90]. Liquid–liquid extraction (LLE) and SPE [91,92] are the most common sample preparation methods for clean-up purposes. Nevertheless, these techniques are time-consuming, expensive, and especially with respect to LLE, hazardous to health owing to the high volume of potentially toxic solvents used. Thus, more environmental-friendly, economical, and miniaturized sample preparation methods are required, some of which will be discussed in this section.

As commented earlier, LLE and SPE are the most common sample preparation methods for clean-up purposes, and are mostly used together. Regarding the analysis of food samples, one of the most common matrices in the analysis of pesticides and the suitability of GC-PFPD has been demonstrated in numerous publications. Furthermore, the determination of 24 OPPs in vegetables has also been reported [93]. Pesticides were extracted with dichloromethane and analyzed without clean-up. The reliability of the results obtained by GC-PFPD was assessed by analyzing 20 samples of different fresh fruits and vegetable matrices (green beans, cucumbers, peppers, tomatoes, melons, eggplants, watermelons, and zucchini), all of them with high water content. Three pesticides were detected in those vegetable samples, and their presence was confirmed by GC with tandem mass spectrometric detection (GC-MS-MS). The LODs and limits of quantification (LOQs) ranged from 3 to 5 μg kg^{-1} and from 5 to 13 μg kg^{-1}, respectively. Recovery was between 73% and 110% with precision better than 15%. In an interesting study also concerning the analysis of foods, the OPP residues in market foods (cereals, vegetables, and fruits) in the Shaanxi area of China were investigated by analyzing the concentrations of eight OPPs by GC-FPD [94]. Extraction, clean-up, and analysis of food were performed according to the Chinese standard method. Three microliters of extract were injected into a GC equipped with the following instrument parameters: columns, glass 3.0 m × 2.5 mm packed with a mixture of 200 + 2.0% OV-17 and 4.5% DC, chromosorb WAW. The LODs (S/N = 3) were in the range of 0.002–0.006 mg kg^{-1}. The recoveries in the studied levels ranged

TABLE 12.2
Applications of GC-FPD to the Analysis of Pesticides

Analyte	Matrix	LOD	Recovery (%)	Refs.
24 OPPs	Vegetables	3–5 μg kg^{-1}	73.0–110	[93]
8 OPPs	Cereals, vegetables, fruits	2–6 μg kg^{-1}	88.0–108.0	[94]
13 OPPs	Milk	50–19 μg kg^{-1}	33.0–98.8	[95]
12 OPPs	Bovine muscle	Low μg kg^{-1}	59%–109%	[96]
45 OPPs	Vegetable oils	Low μg kg^{-1}	74.0–109.0	[97]
Acephate, methamidophos, monocrotophos	Palm oil	10 μg kg^{-1}	85.0–109.0	[98]
36 OPPs	Onion	2–10 μg kg^{-1}	61.0–94.0	[99]
37 OPPs	Peach, grapes, sweet pepper	10 μg kg^{-1}	70%–116%	[103]
20 OPPs	Apple, pear, orange juice	25 μg L^{-1}	62%–119%	[104]
Acephate	Urine	2 μg L^{-1}	102.0	[105]
O,S-DMPT	Urine	4 μg L^{-1}	108	[106]
11 OPPs	Honey, juice, pakchoi	0.003–1.0 ng g^{-1}	74.4%–105.2%	[107]
8 OPPs	Apple juice, apple, and tomato	0.003–0.09 μg kg^{-1}	55.3%–106.4%	[108]
7 OPPs	Orange juice	0.98–2.20 μg L^{-1}	76.2%–108.0%	[110]
Dichlorvos, phorate, fenitrothion, malathion, parathion, quinalphos	Lake water, apple juice, pear juice, orange juice	0.21, 0.56 μg L^{-1}	77.7%–113.6%	[111]
13 OPPs	River, well, farm water	0.001–0.02 μg L^{-1}	91%–104%	[113]
13 OPPs	River, well, farm water	0.003–0.010 μg L^{-1}	84%–125%	[114]
Dimethoate, parathion-methyl, malathion, chlorpyrifos	Vegetables	1.2–3.5 μg kg^{-1}	81.3–98.9	[115]
Dimethoate, parathion, methyl parathion, trimethyl ester, spermine	River water	0.05 μg L^{-1}	63.2–105.1	[116]
OPPs	Roots *P. grandiflorum*	1.16–4.64 μg kg^{-1}	91.9	[119]
15 OPPs	Ginkgo	1.11–4.44 μg kg^{-1}	95.2	[120]
108 OPPs	Ginseng root	25–50 μg kg^{-1}	90.0	[121]
44 OPPs	Tomato, lettuce, orange, paprika, apple, banana, broccoli, spinach, grapefruit	10–20 μg kg^{-1}	59%–97%	[122]
27 OPPs	Green tea	NR	NR	[123]

(*continued*)

TABLE 12.2 (continued)
Applications of GC-FPD to the Analysis of Pesticides

Analyte	Matrix	LOD	Recovery (%)	Refs.
23 OPPs	Beef fat	4–37 μg kg^{-1}	41.2–77%	[125]
Malathion, methidathion, parathion-ethyl, parathion-methyl, parathion-ethyl	Passiflora alata Dryander, Passiflora edulis Sims. F. flavicarpa Deg.	7–14.5 μg L^{-1}	69.3%–107.1%	[129]
48 OPPs	Tea	Low μg kg^{-1}	65%–120%	[132]
14 OPPs	Mix of apple, strawberry, plum, peach, carrot, potato, green pepper, cauliflower	0.1–1000 μg kg^{-1}	<85%	[133]
OPPs and metabolites	Olive oil	3,000–10,000 μg L^{-1}	74.0–120.0	[134]
52 nitrogen- and/or phosphorus-containing pesticides	Cabbage, lettuce, spring onion, and spinach	1–9 μg kg^{-1}	72–108	[135]
Probenazole	Soil, rice plant, paddy water	20 μg kg^{-1}	86.3–90.8	[136]
Maneb	Tomato homogenates	30 μg kg^{-1} (expressed as CS$_2$)	>70%	[137]
Thiram, ziram, maneb, zineb, mancozeb	Fruits and vegetables	5–100 μg kg^{-1}	>80%	[138]
20 OPPs	Cucumber	Low μg kg^{-1}	6.9–88.0	[141]
14 OPPs	Lycium barbarum	5–15 μg kg^{-1}	63.50–102.42	[145]

NR, not reported; LOD, limit of detection.

from 88% to 108% with an RSD < 9%. In 18 of 200 of the analyzed samples, 5 OPPs (dichlorvos, dimethoate, parathion-methyl, pirimiphos-methyl, and parathion) were found in concentrations ranging from 0.004 to 0.257 mg kg^{-1}. The conclusion of the study was that the mean levels of dimethoate in fruits and parathion in vegetables exceeded the maximum residue level (MRL) allowed in China. However, demeton, diazinon, and sumithion were not found in any sample. The results point to the need for urgent action to control the use of some excessively applied and potentially persistent OPPs, such as dimethoate and parathion. A similar study, concerning the determination of 13 OPP residues widely used as dairy cattle ectoparasiticides or in crops for animal feed, in homogenized and pasteurized Mexican milk samples was also carried out [95]. An LLE method with ethyl acetate and acetonitrile was employed, followed by measurement with GC-FPD. The LODs for 13 OPPs under study ranged between 0.0050 and 0.019 mg kg^{-1}, and recoveries ranged between 33.0% (disulfon) and 98.89%. Approximately 39.6% of the samples contained detectable levels of OPP residues. Eight samples contained residues exceeding the established MRL, and the OPPs present in these samples were dichlorvos (five samples), phorate, chlorpyrifos, and chlorfenvinphos (one sample, respectively). Once again, the study demonstrates the presence of some samples with OPP residues over the MRL values, which could be a possible risk to consumer's health, especially children. Kuivinen and Bengtsson described a rapid and simple method for the assessment of OPPs in bovine muscle using LLE and SPE [96]. After extraction with ethyl acetate, the homogenate was centrifuged and filtered through sodium sulfate. The fat was precipitated in methanol by cooling and the extract was diluted with water and passed through an SPE column (Isolute ENV+). After elution with ethyl acetate, evaporation, and redissolution, the sample was injected into a GC-FPD. The simplex method was used to optimize the GC conditions. Recoveries from bovine muscle fortified with 12 pesticides between 4 and 65 µg kg^{-1} at three different levels, ranged between 59% and 109% for 10 of them. However, the results for the two most polar pesticides (metamidophos and acephate) were not successful. The RSD were between 1% and 10% for the 10 pesticides. Furthermore, Di Muccio et al. developed a method for the determination of 45 OPP residues in vegetable oils [97]. In their method, an on-column extraction and clean-up of OPP residues in a single step by a three-cartridge system were performed. A solution of 1 g of oil in *n*-hexane was loaded into an Extrelut-NT3 cartridge (large-pore diatomaceous material). The OPP residues were extracted by eluting the cartridge with 20 mL acetonitrile, which was cleaned-up by passing through a silica and a C18 cartridge connected online to the Extrelut NT-3 cartridge. A few milligrams of lipid were carried over into the eluate, which after concentration and solvent exchange were directly amenable to determination by GC-FPD with optical filter for phosphorous compounds. In the lower concentration range (0.09–0.60 mg kg^{-1}), satisfactory results (74%–86%) were obtained for 39 OPPs; exceptions include formothion (5%), disulfoton (32%), phosalone (54%), demeton-S-Me sulfone (60%), fenthion (62%), and borderline phosphamidone (68%). In the higher concentration range (0.38–2.35 mg kg^{-1}), satisfactory results (82%–109%) were obtained for 43 OPPs, with the exceptions of formothion (48%) and disulfoton (53%). Furthermore, a fast and precise GC-PFPD method for determination of three OPPs (acephate, methamidophos, and monocrotophos) in crude palm oil was also reported [98]. The samples were extracted from the matrix by LLE with 10 mL of acetonitrile, and 2 mL of the extract was subsequently taken for the clean-up using SPE with graphite packing. The analytes were eluted from the column with 13 mL of acetonitrile followed by 2 mL of methanol. Finally, the eluates were evaporated to dryness and reconstituted with 1 mL of acetone. The method achieved an LOD of 0.01 µg g^{-1}, calculated from weighted least squares data. Recoveries for the three OPPs at three different levels of fortification ranged from 85% to 109% (RSD = 15%). Ueno et al. developed an efficient and reliable multiresidue method for 36 OPPs in onion and Welsh onion using GC-PFPD system [99]. The samples were previously extracted with acetonitrile, and the acetonitrile layer was separated by salting-out. The extract was cleaned-up with gel permeation chromatography, and then with a tandem silica-gel/pressure-sensitive adhesive (PSA) minicolumn. Finally, the combined eluate was concentrated to near dryness and dissolved in 2 mL of acetone with the addition of 0.2 mL of an internal standard solution (4 µg mL^{-1} of triphenyl phosphate). To eliminate possible interferences from the matrix, the acetone eluate was diluted eightfold. The average rate of recovery of OPPs was 93% for onion and 94% for welsh onions, except for methamidophos and acephate (61%–68%), with the RSD usually <10% ($n = 5$), and LODs between 0.002 and 0.01 mg kg^{-1}.

Fast GC-MS has been recognized as a potential tool for pesticide analysis, which can provide high sample throughput and laboratory efficiency [100–102]. Recently, several fast GC systems with direct resistive heating as the underlying principle have become commercially available, in which a capillary column is inserted into a resistively heated metal tube (resistive heating-gas chromatography, RH-GC) or enclosed in a resistively heated toroid-formed assembly (low thermal mass chromatography, LTM-GC). The RH-GC typically uses a short column (5 m) encased within a steel tube, which is connected to a power supply and resistively heated. The steel tube has high thermal conductivity and a relatively low thermal mass allowing rapid ramping of temperatures, up to 1200°C min^{-1}, and rapid cooling, for fast GC cycle times. Besides reducing the analysis time (more than 10 times when compared with conventional GC), RH-GC improves the detectability of analytes, providing narrower peaks and better retention-time repeatability. A rapid and robust RH-GC-FPD method for the routine screening of 37 OPPs was developed by Patel et al. [103]. The use of carboFrit (porous carbon plug) insert in the GC liner protects the column, improving the robustness of the method and avoiding the necessity for a clean-up step after a simple extraction with ethyl acetate. Linearity over the range of 0.001–0.5 µg mL^{-1} (0.004–0.2 mg kg^{-1} equivalent) was obtained. The method was validated in peach, grapes, and sweet pepper samples, and used for the screening of OPPs in these samples. The reporting limits were 0.01 mg kg^{-1} for all 37 OPPs in peach and grapes, and for 36 of them in sweet pepper. Mean recoveries ranged from 70% to 116% for samples spiked at 0.01 and 0.1 mg kg^{-1} of pesticides, with associated RSD ≤ 20%. The results were in good agreement with those obtained by MS confirmation. Using this method, a total of 20 samples can be screened in around 3 h. The same authors coupled RH-GC-FPD with programmable temperature vaporization (PTV) for sensitive and rapid determination of 20 OPPs [104]. The PTV injection of volumes ≥10 µL was employed to improve the LODs in GC analysis of pesticides, providing decreased analyte discrimination and improved transfer of analytes that readily degrade under hot splitless conditions. Thus, PTV injection allows the introduction of large volumes onto the RH-GC system, improving the LODs, and avoiding the need for an off-line concentration of sample extracts. Using this system, the 20 selected pesticides were separated in less than 6 min. Two different extraction methods, using ethyl acetate and acetonitrile, were tested for the validation of the method in apple, pear, and orange juice. The method based on ethyl acetate provided a linear range between 0.0025 and 0.1 µg mL^{-1}. Average recoveries between 80% and 110% with RSDs less than 10% were obtained for apple, pear, and orange juice spiked with 0.01 mg kg^{-1} of pesticides, with the exceptions of methamidophos, acephate, and omethoate in orange juice, which showed mean recoveries between 62% and 73%. For these pesticides, acetonitrile, a more polar solvent, produced higher recoveries between 92% and 104%, probably owing to the more favorable partition coefficient for these pesticides between orange juice and solvent. The PTV-RH-GC-FPD method showed good agreement with the results obtained with GC-MS for samples containing incurred residues of chlorpyrifos, phosmet, and/or azinphos-methyl.

Urine has also been a sample of interest, as exposure to pesticides can be estimated by measuring the concentration of the pesticide or its metabolites in this fluid. Thus, the OPP, acephate, in human urine was analyzed by a sensitive GC-PFPD method developed by LePage et al. [105]. Urine was diluted with water and acetone, adjusted to a neutral pH, and partitioned twice in acetone–methylene chloride (1 + 1, v/v), with NaCl added to aid separation. The solvent-reduced organic-phase extracts were clarified by activated charcoal SPE and then adjusted to a final volume with the addition of a D-xylose analyte protectant solution to reduce the matrix enhancement effects. The LOD and LOQ were established at 2 and 10 µg L^{-1}, respectively. The average recovery from urine fortified with 10–500 µg L^{-1} of acephate was 102% ± 12% ($n = 32$). Furthermore, a rugged and sensitive GC-PFPD method to monitor urinary concentration of O,S-dimethyl hydrogen phosphorothioate (O,S-DMPT), a specific biomarker of exposure to insecticide methamidophos, was also proposed [106].

The treatment of the urine samples consisted of C18 SPE clean-up and lyophilization at a low temperature to prevent losses of possibly highly volatile and unstable O,S-DMPT metabolite. The dried residue was derivatized using N-methyl-N-($tert$-butyldimethylsilyl)-trifluoroacetamide and 1% $tert$-butyldimethylchlorosilane (MTBSTFA + 1% TBDMCS) in acetonitrile. Samples and standards were injected into the splitless programmable injector port. The injection volume for each sample and standard was 2 μL. The PFPD flame was supported by hydrogen at a flow rate of 20 mL min^{-1} and air at a rate of 44 mL min^{-1}. The LOD for this method was 0.004 mg L^{-1} with an LOQ of 0.02 mg L^{-1} in urine. The mean recovery value for O,S-DMPT from 17 urine samples fortified at different concentrations was 108%, with an RSD of 12%.

Solid-phase microextraction (SPME) is a solvent-free isolation method used to extract organic compounds from aqueous samples. The analytes are extracted by absorption over the fiber that is directly exposed to the sample or to the headspace. Analytes are extracted until the partition equilibrium has been reached. After this step, the fiber is introduced into the GC injector where the analytes are thermally desorbed, and subsequently separated and quantified. Using this technique, Yu et al. prepared the sol–gel derived bisbenzo crown ether/hydroxyl-terminated silicone-oil SPME coating using allyloxy bisbenzo 16-crown-5 trimethoxysilane as precursor [107]. This new coating was used for the extraction coupled with GC-FPD to determine 11 OPPs in honey, orange juice, and pakchoi (e.g., Chinese vegetable). The extraction efficiencies of the new coating were studied and optimized by adjusting the ionic strength (saturated solution with NaCl for all the samples), the extraction temperature and dilution ratios of samples (32°C, 55°C, and 50°C for water, 1:5 diluted honey, and 1:50 diluted juice), and the extraction time, which was 60 min. The linear range was 0.5–200, 1.0–500, and 1.0–200 ng g^{-1} for honey, juice, and pakchoi, respectively. The LODs varied from 0.003 to 1 ng g^{-1} for OPPs in food samples. Recoveries obtained for samples spiked at 20 μg L^{-1} were in the range of 74.4%–105.2% with an RSD between 2.1% and 15%. Following the same research, three kinds of vinyl crown ether polar fibers were prepared using sol–gel process and used for a method based on SPME and GC-FPD to analyze eight OPPs in food samples [108]. Compared with commercial fibers (85 μm polyacrylate and 65 μm polydimethylsiloxane-divinylbenzene), the new coatings showed higher extraction efficiency and sensitivity for OPPs; specifically, the benzo-15-crown-5 coating was the most effective for the target analytes and was selected for the determination of OPPs in water, apple juice, apple, and tomato. Optimum extraction was achieved with saturated solutions of NaCl, an extraction time of 45 min, and a temperature of 70°C. Desorption time and temperature were 5 min and 270°C, respectively. Dilution ratios were 1:30 for juice, 1:50 for apple, and 1:70 for tomato. The LODs were of range 0.0015–0.081 ng g^{-1} for water, 0.003–0.075 ng g^{-1} for apple juice, 0.032–0.09 ng g^{-1} for apple, and 0.0042–0.076 ng g^{-1} for tomato samples. Recovery studies carried out at 5, 10, and 20 ng g^{-1} showed results higher than 70.5% for the target analytes, except for diazinon and parathion, which were 55.3% and 60.6%, respectively, in tomato samples spiked at 5 ng g^{-1}.

Single drop microextraction (SDME) is a solvent-minimized sample pretreatment procedure, based on the distribution of analytes between a microdrop of extraction solvent at the tip of a microsyringe needle and the aqueous phase [109]. The microdrop is exposed to an aqueous sample for a prescribed time and the analyte is extracted into the drop. After the extraction, the microdrop is retracted back into the microsyringe and injected into the analytical equipment. The SDME, in conjunction with GC-FPD, has been applied for the determination of OPPs in water and fruit juices. Accordingly, Zhao et al. described a simple and fast method for the analysis of OPPs (ethoprophos, diazinon, parathion methyl, fenitrothion, malathion, isocarbophos, and quinaphos) in orange juice [110]. The optimization of the variables affecting the extraction (such as organic solvent, drop volume, agitation rate, extraction time, and salt concentration) was carried out in fortified water, using chlorpyrifos as internal standard. The optimized parameters were 1.6 μL toluene microdrop, 5 mL water sample, 400 rpm stirring rate, 15 min extraction time, and salting out with 5% (w/v) NaCl. Extraction of OPPs in juice entails a previous dilution of 1:25 (v/v) with water, obtaining recoveries above 80%. Linear range of 10–500 μg L^{-1} for all the target analytes and LOD below 2.5 μg L^{-1} were obtained for orange juice. The repeatability and the reproducibility of the method showed an acceptable precision with an RSD below 20%. A similar method was proposed for the determination of dichlorvos, phorate, fenitrothion, malathion, parathion, and quinalphos in fortified lake water and fruit juices (apple, pear, and orange juice) [111]. In this case, the optimized values were 1.5 μL of toluene microdrop in 2 mL of aqueous sample, 600 rpm of stirring rate, sample pH between 5 and 6, extraction temperature of 20°C, and without salting out because (with the exception of dichlorvos) the extraction efficiency for OPPs decreases with the addition of NaCl. Tributyl phosphate was used as internal standard. The authors compared this extraction method with cycle-flow SDME [112], which provided smaller enrichment factor and less sensitive analysis. Linear response in water samples ranged between 1.0 and 50 ng mL^{-1} for dichlorvos and between 0.5 and 50 ng mL^{-1} for the rest of the pesticides, and LODs (S/N = 3) were between 0.21 and 0.56 ng mL^{-1} with RSDs ranging 1.7%–10%. Recoveries between 90.7% and 106.5% (RSD < 8.5%) were obtained for lake water and between 77% and 113.6% (RSD < 13.4%) for juices. In addition, a precise and reproducible method for the determination of 13 OPPs in farm, river, and well water based on SDME followed by GC-FPD, was also proposed [113]. A modified 1 μL microsyringe was used to improve the drop volume and injection as well as to increase the drop stability. The enrichment factor obtained in this method ranged from 540 to 830, with the linear range of 0.01–100 μg L^{-1}, and LODs from 0.001 to 0.005 μg L^{-1} for most of the analytes, with the exception of azinphose methyl and Co-ral, which were 0.015 and 0.020 μg L^{-1}, respectively.

Dispersive liquid–liquid microextraction (DLLME) has been used for the determination of 13 OPPs in water samples [114]. In this LLE method, the appropriate mixture of extraction solvent (12 μL of chlorobenzene) and disperser solvent (1 mL of acetone) is rapidly injected into the aqueous sample (5 mL) by a syringe. Thereby, a cloudy solution is formed, because of the formation of fine droplets of chlorobenzene, which is dispersed among the sample solution. The OPPs in water solution are extracted into the fine droplets of chlorobenzene, which are sedimented in the bottom of a conical test tube by centrifugation. A portion of the sediment is injected into the GC-FPD for separation and determination of the OPPs. Under the optimal conditions, the enrichment factors ranged between 789% and 1070%, the linear range was 10–100,000 pg mL^{-1}, and the LODs were between 3 and 20 pg mL^{-1}. The method was applied in the analysis of OPPs in spiked river, well, and farm water, obtaining extraction recoveries between 84% and 125%.

In a recent work, Xi and Dong developed a simple, sensitive, and rapid method for separating and enriching four OPPs from vegetables by "solvent sublation" [115]. In this adsorptive bubble separation technique, the hydrophobic compounds in water are adsorbed on the bubble surfaces of an ascending gas stream and then collected in an immiscible liquid layer (usually an organic solvent lighter than water) placed on top of the water column. Solvent sublation presents, among other advantages, a simultaneous separation and enrichment of the analyte of interest. The determination of OPPs was carried out by GC-FPD, after extraction and solvent sublation processes. All the sample preparation variables (as effects of organic solvent, nitrogen flow rate, pH of the solution, sublation time) were optimized. The LODs ranged from 1.2 to 3.5 μg kg^{-1}, and the recoveries of spiked vegetable samples ranged from 81.3% to 98.9%, with good RSD values.

Xu et al. developed a membrane extraction-GC method for the determination of five OPPs and related compounds in river water samples [116]. In this method, surface-modified acetic cellulose membranes were used to extract analytes in water samples, and the extracted analytes were back-extracted into a small amount of methanol (5 mL), which subsequently was analyzed by GC-PFPD. Seven types of surface-modified acetic cellulose membranes were investigated. Finally, glutaraldehyde membrane was chosen, as it provided the best recoveries, being easily used in the operation field. The total analysis time per sample was about 16 min. The LOD for each analyte was 0.05 μg L^{-1} and the recoveries for the target analytes in spiked water samples were >63%.

Pressurized liquid extraction (PLE, also called accelerated solvent extraction) is based on the use of aqueous or organic solvents at a high pressure and/or temperature without reaching the critical point. This technique presents the major advantages of automatability, reduced time and solvent requirements, and great flexibility of solvent mixtures [117]. These features make it especially suitable for environmental and food analysis [117,118]. PLE along with SPE has been proposed for a rapid sample clean-up in the determination of OPPs in the roots of *Platycodon grandiflorum* [119] and *Ginkgo* leaves [120]. In this method, the OPPs were concentrated by using an SPE cartridge and quantitatively analyzed and confirmed by GC-FPD. The average recovery was 91.9% (RSD = 4.3%) for roots of *P. grandiflorum*, and 95.2% (RSD = 4.6%) for *Ginkgo* leaves. The methods showed acceptable accuracy and precision while minimizing environmental concerns, time, and labor, providing LOD in the low range of mg kg^{-1}.

12.3.1.1 Multiresidue Analysis by Simultaneous Use of FPD and PFPD with Other Detectors

The FPD and PFPD detectors in GC have been used simultaneously with other selective detector, such as electron capture detection (ECD), halogen-specific detector (XSD) or MS, allowing the development of multiresidue analytical methods, in which several families of pesticides are analyzed at the same time, taking advantage of their intrinsic characteristics and the selectivity of each detector. These hyphenated methods are used mainly for screening purposes, where a rapid yes/no response is needed, and a confirmation by MS is sometimes required.

In a recent method, the determination of 108 OPPs in dried ground ginseng root has been proposed [121]. In this method, pesticides were extracted from the sample using acetonitrile/water saturated with salts, followed by solid-phase dispersive clean-up, and analyzed by capillary GC with electron ionization MS in selective ion monitoring mode (GC-MS/SIM) and GC-FPD in phosphorus mode. Confirmation was achieved using GC-MS, whereas the use of a megabore column in GC-FPD was used for quantification of some of the nonpolar OPPs without the use of matrix matched standards or standard addition. The quantification was achieved from 0.050 to 5.0 µg g^{-1} ($R^2 > 0.99$) for a majority of the pesticides using both detectors. The LODs for most of the pesticides were 0.025–0.05 µg g^{-1} using GC-FPD. Recovery studies were performed by fortifying the dried ground ginseng root samples with concentrations of 0.025, 0.1, and 1.0 µg g^{-1}, resulting in recoveries of >90% for most of the pesticides analyzed by GC-FPD. Lower (<70%) and higher (>120%) recoveries were most likely owing to pesticide lability or volatility, matrix interference, or inefficient desorption from the solid-phase sorbents. In the method described by Okihashi et al. [122] up to 180 pesticide residues belonging the OPPs, OC and pyrethroid pesticides were determined in different food matrices. FPD detector was used for OPPs and negative chemical ionization (NCI) mode MS for OC and pyrethroids. Sample treatment consisted of extraction with acetonitrile, followed by salting-out step with anhydrous $MgSO_4$ and NaCl. Subsequently, a centrifugation step was carried out for removing sediment and water simultaneously, followed by a clean-up step using graphitized carbon black and primary secondary amine (GCB/PSA) SPE cartridges. Recovery studies were carried out in tomato, lettuce, orange, and paprika spiked with 0.05 µg g^{-1} of pesticides and in apple, banana, broccoli, spinach, and grapefruit spiked with 0.1 µg g^{-1}. It was found that OPPs had lower RSDs than other pesticides, and consequently, it was speculated that GC-FPD was more accurate than GC-MS, although some OPPs were measured with GC-MS because of interference in broccoli samples. Recoveries between 70% and 110% with RSD below 25% were obtained for all the target analytes. Five tested pesticides showed low recoveries and/or high RSDs in tested crops, therefore, the method was considered as a screening procedure for these compounds.

As commented earlier, fast GC-MS has been recognized as a potential tool for pesticide multiresidue analysis. Recently, a fast method based on LTM-GC-MS/PFPD has been proposed for the analysis of pesticides in complex samples [123]. LTM-GC systems provide fast temperature programming rates combined with rapid cooldown and short equilibration times for shortest possible analytical cycle times. Using LTM-GC system, maximal heating rate of 30°C s^{-1} and cooldown time of less than 1 min (e.g., from 280°C to 40°C) could be achieved. The LTM-GC-MS consisted of one injector and two column modules (dual LTM-GC-MS), and PFPD was coupled with dual LTM-GC-MS for simultaneous detection in dual-column separation. This system (shown in

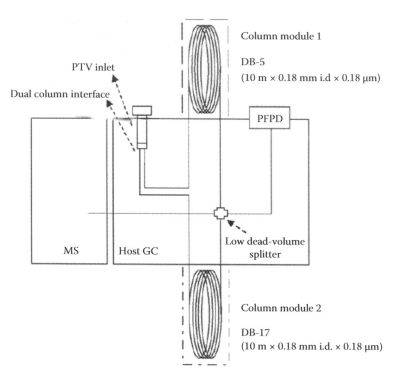

FIGURE 12.8 Schematic diagram for Dual LTM GC-MS/PFPD. (From Sasamoto, K. et al., *Talanta*, 72, 1637, 2007. With permission.)

Figure 12.8) was evaluated for the separation of 82 pesticide mixtures including 27 OPPs, and applied in the analysis of pg mL^{-1} levels of pesticides in a brewed green tea sample with dual stir bar sorptive extraction method (dual SBSE) [124].

A method using simultaneous PFPD and micro-ECD was developed and validated for the analysis of 23 OPPs and 17 OC pesticides in animal fat [125]. The GC-PFPD + μECD system used a single injector and column, but the flow was split in the ratio of 1:1 after the chromatographic separation to the two detectors, using a press-fit deactivated glass splitter. The PFPD was used in the phosphorous mode to detect OPPs, although N-containing OPPs could be sensitively detected with μECD as well. The method involves a gel permeation chromatography step after extraction to separate fats from the pesticide. Recoveries ranging from 60% to 70% with 10%–20% RSD, were obtained for most of the compounds (except for the polar OPPs methamidophos, acephate, and omethoate) when the method was applied to beef fat. The lowest concentration levels used in the calibration curve were between 4 and 37 ng g^{-1}. Many other pesticides could be detected with this method, which is useful for screening applications with a subsequent confirmation of the analyte identity.

Supercritical fluid extraction (SFE) has been proposed for the analysis of pollutants and pesticides in different samples [88,89,126]. In general, SFE methods developed for analysis of pesticide residues are faster, simpler, less expensive, and environmentally safer than conventional solvent-based methods, but one important disadvantage is that SFE of pesticides from these types of samples presents an elevated matrix dependence, and the variables related to the preparation of the SFE sample are, in general, more critical than those affecting the extraction process. It has been stated that the real factors that determine the effectiveness of the method are the type and amount of material added to the sample and the presence of water, salts, or modifiers in the SFE sample [127,128]. Zuin et al. [129] described a fast method for the determination of OPPs and OCs in medicinal plant *Passiflora* by means of SFE followed by GC-ECD-FPD. When the method was applied to real samples, results were confirmed using an MS detector applying the same chromatographic conditions as for the GC-ECD-FPD. The SFE conditions were 100 bar and 40°C (pure CO_2, ρ = 0.62 g mL^{-1}), 5 min equilibration time, 10 min dynamic extraction time at 1 mL min^{-1} and restrictor temperature of 45°C, octadecylsilica (ODS) trap and elution with 1 mL n-hexane at 2 mL min^{-1}. This sample treatment permits the direct analysis without prior cleaning procedure. Obtained LOD for OPPs determined by FPD varied from 7 ng mL^{-1} for fenthion to 14.5 ng mL^{-1} for methidathion. Mean recoveries of 69.8%–107.1% were obtained, with precision of 1.4%–14.7% (RSD). The same detectors (in different GC equipments) and extraction technique have been used for the analysis of pesticides in "gazpacho" (a table-ready food composite containing crude vegetables, white bread, vegetable oil, water, and other minor components) [130]. Thus, SFE was used for the extraction of 17 organohalogen pesticides and OPPs using anhydrous magnesium sulfate as drying agent. However, this work is focused on the study of the effects of different parameters of the SFE (such as fat content in gazpacho composites, magnesium sulfate/gazpacho ratio, supercritical fluid volume, pressure, temperature, and static modifier additions) on SFE recoveries from spiked gazpacho samples. Analyses were performed by GC-FPD, GC-ECD, and GC-MS detectors. Quantitative analysis of dichlorvos, methamidophos, acephate, diazinon, chlorpyrifos-methyl, chlorpyrifos, triazophos, and pyrazophos were performed by GC-FPD. No LODs were reported with regard to recovery studies on different gazpacho samples under different SFE conditions. Automated SFE using carbon dioxide (CO_2) and clean-up by SPE with GCB have also been used in a method developed for the analysis of OPP residues in grain (wheat, maize, and rice) and dried foodstuffs (cornflakes and kidney beans) by GC-FPD and GC-MS [131]. The method allows residues to be determined down to 0.05 mg kg^{-1}.

GC-ECD-FPD was also proposed in another multiresidue method for determining 84 pesticides in tea [132], using a GC system equipped with dual-column and dual-tower autosampler. However, a prior extraction step with organic solvents and clean-up and enrichment with SPE were required. Recoveries ranged 65%–120% (RSD 0.34%–16%) for spiked tea samples at a concentration level of 0.02–3.0 mg kg^{-1}.

A novel approach to sorptive extraction called solvent in silicone tube extraction (SiSTEx) was used in conjunction with GC with simultaneous detection by PFPD and XSD for the determination of 14 OPPs and 22 OC pesticides [133]. SiSTEx is a form of open tubular sorptive extraction, in which a piece of silicone tubing (4 cm long, 1.47 mm ID, 1.96 mm OD in this method) is attached to the cap of a 20 mL glass vial that contains the aqueous sample. The tubing is plugged at the end dangling in the sample solution, and desorption solvent (40 µL of acetonitrile) is added by a syringe into the inner tube volume through a septum in the cap. A stir bar is used to mix the sample for a certain time (60 min), allowing the partition of the analytes into the tubing where they diffuse across the silicone, and partition into the acetonitrile. The final acetonitrile containing the concentrated analytes is then analyzed. The flow was split at the end of the analytical column toward the two detectors (33% of the eluent went to the PFPD). The method using SiSTEx was applied in the analysis of pesticides in fruit and vegetables, making possible the detection of 26 of the 36 pesticides at a concentration level of 10 ng g^{-1} with average RSD of 11%, and 44-fold lower LODs in matrix extracts.

GC-FPD has also been used jointly with the nitrogen phosphorous detector (NPD) for determination of OPP residues and their metabolites in virgin olive oil samples [134]. Forty-eight samples of virgin olive oil were collected directly from olive mills during three harvest periods (1999–2002). Analytes were extracted by liquid–liquid partitioning with solvents of different polarity. The target compounds were determined by GC-FPD or GC-NPD detection. In the case of positive samples, the findings were confirmed using columns of different polarity or by GC-MS. During the validation process, the sensitivity and linearity of the response of both FPD and NPD detectors to the analytes were examined by the injection of standard solutions. The LODs (S/N = 3) were in the range of 0.003–0.01 mg mL^{-1} for the target analytes, and the LOQ values ranged from 0.002 to 0.007 mg kg^{-1}. The recoveries were acceptable (74%–120%). Furthermore, a dual-column GC with NPD and FPD has been proposed for a multiresidue method for the determination of 52 nitrogen- and/ or phosphorous-containing pesticides in different vegetables [135]. Samples were extracted with acetonitrile and the separated acetonitrile layer was cleaned-up by a salting-out step, and subsequently purified by gel permeation chromatography that divided the pesticide eluate into two fractions; the pesticide fractions were purified by a two-step minicolumn clean-up in which the second pesticide fraction was loaded on a silica-gel minicolumn. After a Florisil minicolumn was inserted on the silica-gel minicolumn, the first pesticide fraction was loaded on the tandem minicolumn, which was eluted with acetone-petroleum ether (3 + 7). The final combined eluate was subjected to analysis by GC-NPD-FPD. Recoveries of the 52 pesticides from fortified cabbage, lettuce, spring onion, and spinach ranged from 72% to 108% (RSD of 2%–17%), except for methamidophos and chlorothalonil. The LODs of the pesticides ranged from 0.001 to 0.009 mg kg^{-1}.

12.3.2 Determination of Sulfur Pesticides by FPD

Although less frequently used than for OPPs, FPD has also been used as a specific detector for sulfur-containing pesticides. In the case of sulfur compounds, the CL reaction is the result of a full combustion of the compound in the flame, resulting in the formation of sulfur atoms that recombine to form S_2 in an electronically excited state, which relaxes to the ground state by emission of light with maximum intensities at 284 and 294 nm [79]. The following are the examples of this detection mode.

Yi and Lu described a simple and efficient method for determination of the fungicide probenazole in soil, rice plant, and paddy water by GC-FPD [136]. Depending on the sample to be analyzed, different sample treatments were optimized. Thus, for soil samples, the surface soil (0–15 cm) was air-dried, crushed with a hammer, and passed through a 40 mesh screen. Fifty grams of soil was soaked overnight in a 250 mL cone flask with 150 mL of acetone, filtrated and evaporated with rotary evaporator until 2 mL, approximately. The 2 mL concentrated extract was passed through a silica gel column (3 g), and eluted with 100 mL of acetone: n-hexane (10:90, v/v). The eluate was evaporated to dryness and recomposed in 2 mL with acetone: n-hexane (5:95, v/v). For rice plant, sample (50 g) was sieved to 2 mm, extracted with acetone, and evaporated in rotary evaporator until a final volume of 10 mL. The sample was then extracted with sodium chloride (50 mL) and dichloromethane (110 mL), and subsequently concentrated to 2 mL for analysis. Finally, for water samples, 100 mL of paddy water sample was extracted and cleaned-up by liquid–liquid partition with acetone (50 mL), sodium chloride (20 g), and dichloromethane (110 mL). The sample was concentrated to 5 mL for further column clean-up. The following clean-up steps were the same as for soil samples. The LOD of probenazole (S/N ratio of 3) was 0.02 mg kg^{-1}, with the minimum detectable limit of 5×10^{-10} g. The recoveries were in the range of 88.0%–87.0%, 86.3%–90.9%, and 87.0%–90.8% for soil, rice and water, respectively.

The determination of maneb, a dithiocarbamate pesticide, by means of GC-FPD in the sulfur mode has also been reported [137]. This work examined the effect of storage at 5°C and thermal treatments, cooking at 100°C for 15 min and sterilization at 121°C for 15 min, on maneb stability in tomato homogenates. Maneb residues were determined according to the official EN 12396–2 GC method based on the measurement of carbon disulfide released upon heating of the sample (50 g) with hydrochloric acid and stannous (II) chloride. Thus, carbon disulfide collected in the headspace of a 250 mL gastight flask was determined by headspace GC with a FPD in the sulfur mode. The results revealed that no significant losses of maneb were observed during cold storage for up to 6 weeks. Conversely, thermal treatment resulted in substantial degradation of maneb with extensive conversion to its toxic metabolite ethylenethiourea (ETU). After cooking, only 26% ± 1% of initial maneb residues remained in the samples, while the conversion to ETU was 28% ± 1% (mol mol^{-1}). Sterilization eliminated the residues of the parent compound giving rise to conversion to ETU up to 32% ± 1% (mol mol^{-1}).

In a recent work, several N,N-dimethyldithiocarbamate (thiram, ziram) and ethylenebis (dithiobamate) (maneb, zineb, mancozeb) have been analyzed in fruits and vegetables by GC-FPD, with prior microwave-assisted extraction (MAE [117]) [138]. Residues were extracted from the plant matrixes and hydrolyzed to CS_2 in a single step in the presence of 1.5% $SnCl_2$ in 5 N hydrochloric acid using a microwave oven operating in the closed-vessel mode. The evolved CS_2, trapped in a layer of isooctane overlaying the reaction mixture was analyzed by GC-FPD. The LODs and LOQs were in the range of 0.005–0.1 mg kg^{-1}, and the recoveries were >80% (RSD < 20%).

12.4 SCREENING ANALYSIS OF PESTICIDES BY CL DETECTION

Public concern over pesticide residues has risen notably during the last decade and their accurate determination in food and environmental samples is gaining great importance. Nevertheless, in the case of routine quality control laboratories, this determination could be unnecessary as the only information usually required is whether the concentration of a particular compound is over or under a given MRL. In such a case, laboratories could be interested in a binary "yes/no" result in relation to the presence of the compound above its MRL, avoiding the time-consuming procedure involved in a quantitative analysis. These are the cases when fast qualitative methods become of special relevance [139]. In general, screening methods have some of the following characteristics [140]: (a) they tend to have a qualitative rather than quantitative emphasis; (b) they normally involve little or no sample treatment; (c) they are rapid and the response is used for immediate decision making; and (d) the response obtained sometimes requires confirmation by a conventional alternative. Despite the fact that qualitative methods do not provide the same amount of information as quantitative methods, some advantages make them suitable for the screening of groups of compounds with a regulated limit, where the key point is stating whether or not a sample complies with legislation instead of giving an accurate concentration result. Some of these advantages are especially useful in routine laboratories, such as the avoidance of continuous recalibrations, a higher sample throughput, the achievement of a quicker binary response to the analytical problem, the reduction of the maintenance operations to achieve a satisfactory instrumental response, and the easier establishment of quality control programs based on control chart as long as fewer number of concentrations are under study [141]. In this section, two different CL methods are developed, one in the gas phase and other in the liquid phase, as examples of the applicability of CL in screening analysis.

12.4.1 EXAMPLE OF SCREENING ANALYSIS IN THE GAS PHASE

A qualitative method for the screening of OPPs in cucumber samples by GC-PFPD after extraction with ethyl acetate and sodium sulfate has been proposed [141]. Confirmation of compounds was performed by GC-MS. The dual output channel of the PFPD allows the simultaneous collection of two signals depending on the filters used by the experimenter. In this case, phosphorous and sulfur chromatograms were obtained for each particular analysis notably improving the confirmation capability of the method owing to the fact that most of the OPPs also contain sulfur. The screening GC-PFPD test was based on the possibilities of committing type I (false positive, usually 0.05) and II (false-negative, usually 0.05) errors at the same time when stating that an analysis performed with an unknown sample is negative or nonnegative. To do so, an interval was developed around the average peak area obtained from 10 injection replicates of standard solutions at the MRL concentrations made-up with clean extracts of cucumber obtained in an identical way as the real samples. The interval size was observed to be directly dependent on the precision of the measurements, as well as on the values of type I and II errors that were selected as admissible. These limits in the intervals (screening limits) also depend on the precision of the overall experiment by means of the introduction of a standard deviation in its formulation. The expression for its calculation is: $\bar{A} \pm \Delta(\alpha,\beta,\nu)s_{MRL}$, where $\bar{A}$ is the average peak area obtained from 10 injection replicates of standard solutions at the MRL concentrations, $\Delta(\alpha,\beta,\nu)$ is the noncentral parameter of a noncentral t-distribution with ν degrees of freedom, α and β are the accepted probabilities of committing type I and II errors, respectively, and s_{MRL} is the standard deviation of the replicated peak area values obtained for every pesticide under study. The use of the Δ statistics, instead of the more usual Student's t-test, assures the probability of accepting a sample as being in compliance with a certain limit when the concentration of the analyte is above the specification limit (false negative).

Once the screening limits (upper and lower) were established, the instrumental responses corresponding to the unknown samples were directly compared with them. The comparison offers three possibilities, as follows:

1. The instrumental response of the unknown sample is lower than the lower screening limit.
2. The instrumental response of the unknown sample is between the values of (or equal to) the lower and upper screening limits.
3. The instrumental response of the unknown sample is higher than the higher screening limit.

Case (1) indicates the end of the analysis. In such a case, the unknown sample would be labeled as negative. Contrarily, cases (2) and (3) label the samples as nonnegatives and direct them to confirmation by GC-MS. Figure 12.9 shows the flowchart that summarizes the whole process of extraction, screening, and further confirmation. This work also proposed a full validation method that takes advantage of the information obtained from the replicates used for the establishment of the screening limits. The method allows a rapid and accurate identification of the studied OPPs until the ng mL^{-1} range for forbidden OPPs and above the MRL concentration of the rest of the samples, and could be applied to other vegetables.

12.4.2 Example of Screening Analysis in the Liquid Phase

The usefulness of an FIA-CL method using the PO reaction has been shown in the screening of total NMC content in ground water [51]. After alkaline hydrolysis (5 min, 95°C) and OPA derivatization reaction of NMCs (1 min ultrasounds, 20°C), the sample was measured in the FIA manifold, which is described in earlier studies [50,51,142]. The screening method is based on the establishment of

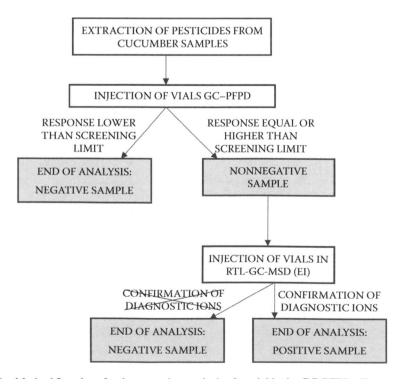

FIGURE 12.9 Method flowchart for the screening analysis of pesticides by GC-PFPD. (From Aybar Muñoz, J. et al., *Talanta*, 60, 433, 2003. With permission.)

a screening uncertainty interval for different concentration levels, each one characterized by two figures of merits: sensitivity and specificity. In some aspect, this is similar to the PFPD-MS method described in the earlier section, but some modifications have been included. Thus, the different steps of the screening methods are summarized as follows:

Step 1. A specified limit (SL) was established, expressed as the concentration of analyte in the sample. This limit is often set by regulatory bodies, but also by the quality management system in manufacturing companies or by the final user of the analysis object [143]. In this case, the target is to determine if the total content of four NMCs (carbaryl, carbofuran, aldicarb, and promecarb) in ground water is above or below a certain concentration level with a certain probability value. Different concentration thresholds were assayed to evaluate the screening method at different SLs. In this way, depending on the sample to be tested, a different SL could be selected, according to the requirements of the final application.

Step 2. Each SL is converted into an instrumental signal, defined as "specification limit signal" (r_{SL}). Thus, each r_{SL} corresponds to the mean CL intensity obtained for a given SL, at a certain concentration level. With the purpose of calculating the different r_{SL}, different aliquots of water spiked with several proportions of the pesticides were prepared by triplicate at each selected SL, hydrolyzed, derivatized, and measured by triplicate. From these sets of independent samples, the mean of the instrumental response (r_{SL}) and an associated standard deviation, was calculated, establishing a "screening uncertainty interval" for each SL.

Step 3. The screening uncertainty interval ($u_{screening}$) is calculated from a coverage interval defined as (Equation 12.1):

$$u_{screening} = r_{SL} \pm \Delta(\alpha, \beta, \nu) \sqrt{\frac{MS_{SL}}{m}} \qquad (12.1)$$

where MS_{SL} is the mean square for a given value of SL, considering that different mixtures have been measured at a same SL concentration level, and is calculated as

$$MS_{SL} = \frac{\sum_i \sum_j (x_{ij} - \bar{x}_i)^2}{N - r} \qquad (12.2)$$

where

m is the number of replicates that will be performed on an unknown sample (usually $m = 1$)
x_{ij} denotes the different measurements, where j represents each replicate performed for the spiked mixture i
$\bar{x}_i$ is the mean value for mixture i
N is the number of independent determinations performed to calculate r_{SL}
r is the number of different spiked solutions prepared at each SL

Step 4. Once the screening limits were established for each NMCs concentration, the instrumental signal (r_i) corresponding to the samples spiked with different NMCs concentrations were directly compared with them. As a result, it could be decided whether their concentration level complies with the limit or not. The comparison offers three possibilities as follows:

1. The instrumental signal of the unknown sample is below the lower screening limit (negative values).
2. The instrumental signal of the unknown sample is between the values of (or equal to) the lower and upper screening limits (nonnegatives values).
3. The instrumental signal of the unknown sample is above the higher screening limit (positive values).

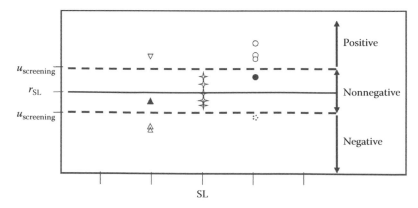

In case (1), the unknown sample would be labeled as negative (i.e., it contains a total carbamate concentration lower than SL); in case (3), the sample would be considered as positive (it contains a total carbamate content higher than SL); on the other hand, in case (2), the sample would be considered as nonnegative, as it cannot be assured that the sample has a total carbamate content lower than SL, however, it can neither be considered as negative (so, it is a possible positive sample). These three different situations of the classification rule are shown in Figure 12.10.

Considering the r_i obtained for each SL, where the reliability of the screening was determined by estimating the number of true and false positives, true and false negatives, and nonnegative values given by the method. With this purpose, the following parameters were defined [144]:

$$\text{Sensitivity} = \frac{tp}{tp + fn + nn} \times 100$$

$$\text{Specificity} = \frac{tn}{tn + fp} \times 100$$

where
"tp" is the number of true positives
"fn" is the number of false negatives
"nn" is the number of nonnegatives
"tn" is the number of true negatives
"fp" is the number of false positives

Sensitivity is thus, the proportion of true positives provided by the method with respect to the total number of real positives, including the values within the defined interval (nonnegatives). The statistical meaning of sensitivity can be explaining in terms of "power" of the method, defined as $(1 - \beta)$. In a similar way, specificity is statistically defined as the proportion of true negatives with respect to the total number of real negatives, and is related to the level of significance, α.

The method provides good specificity, as the number of false negative was equal to 0, and the sensitivity was also very good, as no false positives were found and only nonnegative values were obtained. However, a different SL could be selected depending on the particular application or the matrix of interest.

FIGURE 12.10 Scheme of the different simulated situations of analysis. SL, specified limit; r_{SL}, specification limit signal; ✦, responses obtained for the SL; △, true negatives; ○, true positives; ∴, false negatives; ▽, false positives; bold, nonnegatives. (From Soto-Chinchilla, J.J. et al., *Anal. Chim. Acta*, 541, 113, 2005. With permission.)

REFERENCES

1. T.A. Nieman, Chemiluminescence, in: F.A. Settle (ed.), *Handbook of Instrumental Techniques for Analytical Chemistry*, Marcel Dekker, New York, 1997, pp. 541–559.
2. R.A. LaRossa (ed.), *Bioluminescence Methods and Protocols in Methods in Molecular Biology*, The Humana Press, Totowa, NJ, 1998.
3. A.M. García-Campaña, W.R.G. Baeyens, L. Cuadros-Rodríguez, F. Alés-Barrero, J.M. Bosque-Sendra, and L. Gámiz-Gracia, Potential of chemiluminescence and bioluminescence in organic analysis, *Curr. Org. Chem.* 6 (2002) 1.
4. A.M. García-Campaña and W.R.G. Baeyens (eds.), *Chemiluminescence in Analytical Chemistry*, Marcel Dekker, New York, 2001.
5. M.G. Sanders, K.N. Andrew, and P.J. Worsfold, Trends in chemiluminescence detection for liquid separations, *Anal. Commun.* 34 (1997) 13 H.
6. L.P. Palilis, A.C. Calokerinos, W.R.G. Baeyens, Y. Zhao, and K. Imai, Chemiluminescence detection in flowing streams, *Biomed. Chromatogr.* 11 (1997) 85.
7. W.R.G. Baeyens, S.G. Schulman, A.C. Calokerinos, Y. Zhao, A.M. García-Campaña, K. Nakashima, and D. De Keukeleire, Chemiluminescence-based detection: Principles and analytical applications in flowing streams and in immunoassays, *J. Pharm. Biomed. Anal.* 17 (1998) 941.
8. A.M. García-Campaña, W.R.G. Baeyens, X. Zhang, E. Smet, G. Van Der Weken, K. Nakashima, and A.C. Calokerinos, Detection in the liquid phase applying chemiluminescence, *Biomed. Chromatogr.* 14 (2000) 166.
9. A.M. García-Campaña and W.R.G. Baeyens, Principles and recent analytical applications of chemiluminescence, *Analusis* 28 (2000) 686.
10. A.M. García-Campaña and F.J. Lara Vargas, Trends in the analytical applications of chemiluminescence in the liquid phase, *Anal. Bioanal. Chem.* 387 (2007) 165.
11. P. Fletcher, K.N. Andrew, A.C. Calokerinos, S. Forbes, and P.J. Worsforld, Analytical applications of flow injection with chemiluminescence detection—A review, *Luminescence* 16 (2001) 1.
12. N.D. Danielson, Analytical applications: Flow injection, liquid chromatography, and capillary electrophoresis, in: J.A. Bard (ed.), *Electrogenerated Chemiluminescence*, Marcel Dekker, Inc., New York, 2004, pp. 397–444.
13. K. Tsukagoshi, Chemiluminescence detection for capillary electrophoresis and liquid chromatography. A review, *Sci. Eng. Rev.—Doshisha University* 45 (2005) 168.
14. A. Townshend (ed.), *Encyclopedia of Analytical Science*, Vol. 1, Academic Press, London, 1995, p. 621.
15. X. Yan, Detection by ozone-induced chemiluminescence in chromatography, *J. Chromatogr. A* 842 (1999) 267.
16. X.J. Huang and Z.L. Fang, Chemiluminescence detection in capillary electrophoresis, *Anal. Chim. Acta* 414 (2000) 1.
17. A.M. García-Campaña, L Gámiz-Gracia, W.R.G. Baeyens, and F. Alés-Barrero, Derivatization of biomolecules for chemiluminescent detection in capillary electrophoresis, *J. Chromatogr. B* 793 (2003) 49.
18. X.R. Zhang, W.R.G. Baeyens, A.M. García-Campaña, and J. Ouyang, Recent developments in chemiluminescence sensors, *Trends Anal. Chem.* 18 (1999) 384.
19. H.Y. Aboul-Enein, R.I. Stefan, J.F. Van Staden, X.R. Zhang, A.M. García-Campaña, and W.R.G. Baeyens, Recent developments and applications of chemiluminescence sensors, *Crit. Rev. Anal. Chem.* 30 (2000) 271.
20. C. Dodeigne, L. Thunus, and R. Lejeune, Chemiluminescence as diagnostic tool. A review, *Talanta* 51 (2000) 415.
21. A. Roda, P. Pasini, M. Guardigli, M. Baraldini, M. Musiani, and M. Mirasoli, Bio- and chemiluminescence in bioanalysis, *Fresen. J. Anal. Chem.* 366 (2000) 752.
22. A. Roda, M. Guardigli, P. Pasini, and M. Mirasoli, Bioluminescence and chemiluminescence in drug screening, *Anal. Bioanal. Chem.* 377 (2003) 826.
23. L.J. Kricka, Clinical applications of chemiluminescence, *Anal. Chim. Acta* 500 (2003) 279.
24. S. Girotti, E. Ferri, S. Ghini, A. Roda, P. Pasini, G. Carrea, R. Bovara, S. Lodi, G. Lasi, J. Navarro, and P. Raush, Luminescent techniques applied to food analysis, *Quím. Anal.* 16 (1997) S111.
25. A.M. Jiménez and M.J. Navas, Chemiluminescent methods in agrochemical analysis, *Crit. Rev. Anal. Chem.* 27 (1997) 291.
26. L. Gámiz-Gracia, A.M. García-Campaña, J.J. Soto-Chinchilla, J.F. Huertas-Pérez, and A. González-Casado, Analysis of pesticides by chemiluminescence detection in the liquid phase, *Trends Anal. Chem.* 24 (2005) 927.

27. J.L. Adcock, P.S. Francis, and N.W. Barnett, Acidic potassium permanganate as a chemiluminescence reagent—A review, *Anal. Chim. Acta* 601 (2007) 36.
28. G.Z. Tsogas, D.L. Giokas, P.G. Nikolakopoulos, A.G. Vlessidis, and N.P. Evmiridis, Determination of the pesticide carbaryl and its photodegradation kinetics in natural waters by flow injection-direct chemiluminescence detection, *Anal. Chim. Acta* 573–574 (2006) 354.
29. A. Waseem, M. Yaqoob, and A. Nabi, Flow-injection determination of carbaryl and carbofuran based on KMnO$_4$-Na$_2$SO$_3$ chemiluminescence detection, *Luminescence* 22 (2007) 349.
30. J.A. Murillo Pulgarín, L.F. García Bermejo, and J.A. Rubio Aranda, Development of time-resolved chemiluminescence for the determination of Antu in river water, wheat, barley, and oat grain samples, *J. Agric. Food Chem.* 53 (2005) 6609.
31. H. Liu, Y. Hao, J. Ren, P. He, and Y. Fang, Determination of tsumacide residues in vegetable samples using a flow-injection chemiluminescence method, *Luminescence* 22 (2007) 302.
32. M. Palomeque, J.A. García Bautista, M. Catalá Icardo, J.V. García Mateo, and J. Martínez Calatayud, Photochemical-chemiluminometric determination of aldicarb in a fully automated multicommutation based flow-assembly, *Anal. Chim. Acta* 512 (2004) 149.
33. A. Chivulescu, M. Catalá Icardo, J.V. García Mateo, and J. Martínez Calatayud, New flow-multicommutation method for the photo-chemiluminometric determination of the carbamate pesticide asulam, *Anal. Chim. Acta* 519 (2004) 113.
34. J.R. Albert-García, M. Catalá Icardo, and J. Martínez Calatayud, Analytical strategy photodegradation/chemiluminescence/continuous-flow multicommutation methodology for the determination of the herbicide propanil, *Talanta* 69 (2006) 608.
35. T. Gamazo Climent, J.R. Albert-García, and J. Martínez Calatayud, Photo-induced chemiluminometric determination of acrolein in a multicommutation flow assembly, *Anal. Lett.* 40 (2007) 629.
36. C.M.P.G. Amorim, J.R. Albert-Garcia, M.C.B.S. Montenegro, A.N. Araujo, and J. Martínez Calatayud, Photo-induced chemiluminometric determination of Karbutilate in a continuous-flow multicommutation assembly, *J. Pharm. Biomed. Anal.* 43 (2007) 421.
37. Z. Pawlicova, J.R. Albert-García, I. Sahuquillo, J.V. García Mateo, M. Catalá Icardo, and J. Martínez Calatayud, Chemiluminescent determination of the pesticide bromoxynil by on-line photodegradation in a flow-injection system, *Anal. Sci.* 22 (2006) 29.
38. I. Sahuquillo Ricart, G.M. Antón-Fos, M.J. Duart, J.V. García Mateo, L. Lahuerta Zamora, and J. Martínez Calatayud, Theoretical prediction of the photoinduced chemiluminescence of pesticides, *Talanta* 72 (2007) 378.
39. A. Czescik, D. López Malo, M.J. Duart, L. Lahuerta Zamora, G.M. Antón Fos, and J. Martínez Calatayud, Photo-induced chemiluminescence-based determination of diphenamid by using a multicommuted flow system, *Talanta* 73 (2007) 718.
40. M. Martínez Galera, M.D. Gil García, and R. Santiago Valverde, Determination of nine pyrethroid insecticides by high-performance liquid chromatography with post-column photoderivatization and detection based on acetonitrile chemiluminescence, *J. Chromatogr. A* 1113 (2006) 191.
41. M.D. Gil García, M. Martínez Galera, and R. Santiago Valverde, New method for the photo-chemiluminometric determination of benzoylurea insecticides based on acetonitrile chemiluminescence, *Anal. Bioanal. Chem.* 387 (2007) 1973.
42. A.I. Pimentel Neves, J.R. Albert-García, and J. Martínez Calatayud, Chemiluminometric determination of the pesticide 3-indolyl acetic acid by a flow injection analysis assembly, *Talanta* 71 (2007) 318.
43. A.S. Alves Ferreira, J.R. Albert-García, and J. Martínez Calatayud, Chemiluminometric photo-induced determination of Strychnine in a multicommutation flow assembly, *Talanta* 72 (2007) 1223.
44. J.A. Murillo Pulgarín, A. Alañón Molina, and P. Fernández López, Automatic chemiluminescence-based determination of carbaryl in various types of matrices, *Talanta* 68 (2006) 586.
45. Z. Xie, X. Ouyang, L. Guo, X. Lin, and G. Chen, Determination of carbofuran by flow-injection with chemiluminescent detection, *Luminescence* 20 (2005) 226.
46. R. Bos, N.W. Barnett, G.A. Dyson, K.F. Lim, R.A. Rusell, and S.P. Watson, Studies on the mechanism of the peroxyoxalate chemiluminescence reaction. Part 1. Confirmation of 1,2-dioxetanedione as an intermediate using 13C nuclear magnetic resonance spectroscopy, *Anal. Chim. Acta* 502 (2004) 141.
47. M. Stigbrand, T. Jonsson, E. Pontén, K. Irgum, and R. Bos, Mechanism and applications of peroxyoxalate chemiluminescence, in: A.M. García-Campaña and W.R.G. Baeyens (eds.), *Chemiluminescence in Analytical Chemistry*, Marcel Dekker, New York, 2001, Chapter 7.
48. E. Orejuela and M. Silva, Monitoring some phenoxyl-type *N*-methylcarbamate pesticide residues in fruit juices using high-performance liquid chromatography with peroxyoxalate-chemiluminescence detection, *J. Chromatogr. A* 1007 (2003) 197.

49. E. Orejuela and M. Silva, Determination of propham and chlorpropham in postharvest-treated potatoes by liquid chromatography with peroxyoxalate chemiluminescence detection, *Anal. Lett.* 37 (2004) 2531.
50. J.J. Soto-Chinchilla, A.M. García-Campaña, L. Gámiz-Gracia, L. Cuadros-Rodríguez, and J.L. Martínez Vidal, Determination of a *N*-methylcarbamate pesticide in environmental samples based on the application of photodecomposition and peroxyoxalate chemiluminescent detection, *Anal. Chim. Acta* 524 (2004) 235.
51. J.J. Soto-Chinchilla, L. Gámiz-Gracia, A.M. García-Campaña, and L. Cuadros-Rodríguez, A new strategy for the chemiluminescent screening analysis of total *N*-methylcarbamate content in water, *Anal. Chim. Acta* 541 (2005) 113.
52. L. Gámiz-Gracia, L. Cuadros-Rodríguez, E. Almansa-López, J.J. Soto-Chinchilla, and A.M. García-Campaña, Use of highly efficient Draper-Lin small composite designs in the formal optimisation of both operational and chemical crucial variables affecting a FIA-chemiluminescence detection system, *Talanta* 60 (2003) 523.
53. B.A. Gorman, P.S. Francis, and N.W. Barnett, Tris(2,2′-bipyridyl)ruthenium(II) chemiluminescence, *Analyst* 131 (2006) 616.
54. T. Pérez-Ruiz, C. Martínez-Lozano, V. Tomas, and J. Martín, Chemiluminescence determination of carbofuran and promecarb by flow injection analysis using two photochemical reactions, *Analyst* 127 (2002) 1526.
55. T. Pérez-Ruiz, C. Martínez-Lozano, V. Tomas, and J. Martín, Flow injection chemiluminescence determination of carbaryl using photolytic decomposition and photogenerated tris (2,2′-bipyridyl) ruthenium(III), *Anal. Chim. Acta* 476 (2003) 141.
56. T. Pérez-Ruiz, C. Martinez-Lozano, and M.D. García, Determination of *N*-methylcarbamate pesticides in environmental samples by an automated solid-phase extraction and liquid chromatographic method based on post-column photolysis and chemiluminescence detection, *J. Chromatogr. A* 1164 (2007) 174.
57. Y. Rakicioğlu, J.M. Schulman, and S.G. Schulman, Applications of chemiluminescence in organic analysis, in: A.M. García-Campaña and W.R.G. Baeyens (eds.), *Chemiluminescence in Analytical Chemistry*, Marcel Dekker, New York, 2001, Chapter 5.
58. M. Yamaguchi, H. Yoshida, and H. Nohta, Luminol-type chemiluminescence derivatization reagents for liquid chromatography and capillary electrophoresis, *J. Chromatogr. A* 950 (2002) 1.
59. X. Liu, J. Du, and J. Lu, Determination of parathion residues in rice samples using a flow injection chemiluminescence method, *Luminescence* 18 (2003) 245.
60. J.X. Du, X.Y. Liu, and J. Lu, Determination of monocrotophos pesticide by flow injection chemiluminescence method using luminol-hydrogen peroxide system, *Anal. Lett.* 36 (2003) 1029.
61. X.Z. Li, T.T. Guan, C. Zhou, J.Q. Yin, and Y.H. Zhang, Solid phase extraction chemiluminescence determination of methamidophos on vegetables, *Chem. Res. Chinese Univ.* 22 (2006) 21.
62. Z. Song, S. Hou, and N. Zhang, A new green analytical procedure for monitoring sub-nanogram amounts of chlorpyrifos on fruits using flow injection chemiluminescence with immobilized reagents, *J. Agric. Food Chem.* 50 (2002) 4468.
63. H. Kubo, Y. Tsuda, Y. Yoshimura, H. Homma, and H. Nakazawa, Chemiluminescence of dithiocarbamate fungicides based on the luminol reaction, *Anal. Chim. Acta* 494 (2003) 49.
64. J.F. Huertas-Pérez, A.M. García-Campaña, L. Gámiz-Gracia, A. González-Casado, and M. del Olmo Iruela, Sensitive determination of carbaryl in vegetal food and natural waters by flow-injection analysis based on the luminol chemiluminescence reaction, *Anal. Chim. Acta* 524 (2004) 161.
65. J.F. Huertas-Pérez, L. Gámiz-Gracia, A. González-Casado, and A.M. García-Campaña, Chemiluminescence determination of carbofuran at trace levels in lettuce and waters by flow-injection analysis, *Talanta* 65 (2005) 980.
66. B. Li, Y. He, and C. Xu, Simultaneous determination of three organophosphorus pesticides residues in vegetables using continuous-flow chemiluminescence with artificial neural network calibration, *Talanta* 72 (2007) 223.
67. G. Huang, J. Ouyang, W.R.G. Baeyens, Y. Yang, and C. Tao, High-performance liquid chromatographic assay of dichlorvos, isocarbophos and methyl parathion from plant leaves using chemiluminescence detection, *Anal. Chim. Acta* 474 (2002) 21.
68. J.F. Huertas-Pérez and A.M. García-Campaña, HPLC-chemiluminescence method for the analysis of carbamate pesticides in water samples, in: *31st International Symposium on High Performance Liquid Chromatography and Related Techniques (HPLC 2007)*, Ghent, Belgium, June 17–21, 2007 (Ref.: P20–28).

69. G. Zomer and M. Jacquemijns, Application of novel acridan esters as chemiluminogenic signal reagents in immunoassay, in: A.M. García-Campaña and W.R.G. Baeyens (eds.), *Chemiluminescence in Analytical Chemistry*, Marcel Dekker, Inc., New York, 2001, Chapter 18.

70. G. Gübitz, M.G. Schmid, H. Silviaeh, and H.Y. Aboul-Enein, Chemiluminescence flow-injection immunoassays, *Crit. Rev. Anal. Chem.* 31 (2001) 167.

71. A. Roda, P. Pasini, M. Mirasoli, E. Michelini, and M. Guardigli, Biotechnological applications of bioluminescence and chemiluminescence, *Trends Biotechnol.* 22 (2004) 295.

72. J. Yakovleva, R. Davidsson, M. Bengtsson, T. Laurell, and J. Emméus, Microfluidic enzyme immunosensors with immobilised protein A and G using chemiluminescence detection, *Biosens. Bioelectron.* 19 (2003) 21.

73. S.R. Jain, E. Borowska, R. Davidsson, M. Tudorache, E. Pontén, and J. Emnéus, A chemiluminescence flow immunosensor based on a porous monolithic methacrylate and polyethylene composite disc modified with Protein G, *Biosens. Bioelectron.* 19 (2004) 795.

74. I.M. Ciumasu, P.M. Krämer, C.M. Weber, G. Kolb, D. Tiemann, S. Windisch, I. Frese, and A.A. Kettrup, A new, versatile field immunosensor for environmental pollutants, *Biosens. Bioelectron.* 21 (2005) 354.

75. R.S. Chouhan, K. Vivek Babu, M.A. Kumar, N.S. Neeta, M.S. Thakur, B.E. Amitha Rani, A. Pasha, N.G.K. Karanth, and N.G. Karanth, Detection of methyl parathion using immuno-chemiluminescence based image analysis using charge coupled device, *Biosens. Bioelectron.* 21 (2006) 1264.

76. A.E. Botchkareva, F. Fini, S. Eremin, J.V. Mercader, A. Montoya, and S. Girotti, Development of a heterogeneous chemiluminescent flow immunoassay for DDT and related compounds, *Anal. Chim. Acta* 453 (2002) 43.

77. A.E. Botchkareva, S.A. Eremin, A. Montoya, J.J. Manclús, B. Mickova, P. Rauch, F. Fini, and S. Girotti, Development of chemiluminescent ELISAs to DDT and its metabolites in food and environmental samples, *J. Immunol. Meth.* 283 (2003) 45.

78. B. Mickova, P. Rauch, A. Montoya, E. Ferri, F. Fini, and S. Girotti, The determination of *N*-methylcarbamate pesticides using enzyme immunoassays with chemiluminescent detection, *Czech J. Food Sci.* 22 (2004) 280.

79. J.E. Boulter and J.W. Birks, Gas-phase chemiluminescence detection, in: A.M. García-Campaña and W.R.G. Baeyens (eds.), *Chemiluminescence in Analytical Chemistry*, Marcel Dekker, Inc., New York, 2001, Chapter 13.

80. X. Yan, Unique selective detectors for gas chromatography: Nitrogen and sulfur chemiluminescence detectors, *J. Sep. Sci.* 29 (2006) 1931.

81. L.O. Courthaudon and E.M. Fujinari, Nitrogen-specific gas chromatography detection based on chemiluminescence, *LC.GC* 9 (1991) 732.

82. H.C.K. Chang and L.T. Taylor, Use of sulfur chemiluminescence detection after supercritical fluid chromatography, *J. Chromatogr.* 517 (1990) 491.

83. H.C.K. Chang and L.T. Taylor, Sulfur-selective chemiluminescence detection after packed-capillary-column high-performance liquid chromatography, *Anal. Chem.* 63 (1991) 486.

84. S.S. Brody and J.E. Chaney, Flame photometric detector. Application of a specific detector for phosphorus and for sulfur compounds sensitive to subnanogram quantities, *J. Gas Chromatogr.* 4 (1966) 42.

85. S. Cheskis, E. Atar, and A. Amirav, Pulsed-flame photometer: A novel gas chromatography detector, *Anal. Chem.* 65 (1993) 539.

86. A. Amira and H. Jing, Pulsed flame photometer detector for gas chromatography, *Anal. Chem.* 67 (1995) 3305.

87. E.W.J. Hooijschuur, C.E. Kientz, J. Dijksman, and U.A.T. Brinkman, Potential of microcolumn liquid chromatography and capillary electrophoresis with flame photometric detection for determination of polar phosphorus-containing pesticides, *Chromatographia* 54 (2001) 295.

88. A. Hercegová, M. Dömötörová, and E. Matisová, Sample preparation methods in the analysis of pesticide residues in baby food with subsequent chromatographic determination, *J. Chromatogr. A* 1153 (2007) 54.

89. N. Fidalgo-Used, E. Blanco-González, and A. Sanz-Medel, Sample handling strategies for the determination of persistent trace organic contaminants from biota samples, *Anal. Chim. Acta* 590 (2007) 1.

90. I. Mukherjee and P.K. Sharma, Advances in extraction techniques of pesticide residues in food and environment, *Res. J. Chem. Environ.* 10 (2006) 90.

91. F.J. Schenck, S.J. Lehotay, and V. Vega, Comparison of solid-phase extraction sorbents for cleanup in pesticide residue analysis of fresh fruits and vegetables, *J. Sep. Sci.* 25 (2002) 883.

92. M. Sakata, in: O. Suzuki and K. Watanabe (eds.), *Drugs and Poisons in Humans*, Springer GmbH, Berlin, 2005, pp. 535–544.

93. I. Martínez Salvador, A. Garrido Frenich, F.J. González Egea, and J.L. Martínez Vidal, Determination of organophosphorus pesticides in vegetables by GC with pulsed flame-photometric detection, and confirmation by MS, *Chromatographia* 64 (2006) 667.

94. Y. Bai, L. Zhou, and J. Wang, Organophosphorus pesticide residues in market foods in Shaanxi area, China, *Food Chem.* 98 (2006) 240.

95. H. Salas, M. González, M. Noa, N. Perez, G. Diaz, R. Gutierrez, H. Zazueta, and I. Osuna, Organophosphorus pesticide residues in Mexican commercial pasteurized milk, *J. Agric. Food Chem.* 51 (2003) 4468.

96. J. Kuivinen and S. Bengtsson, Solid-phase extraction and cleanup of organophosphorus pesticide residues in bovine muscle with gas chromatographic detection, *J. Chromatogr. Sci.* 40 (2002) 392.

97. A. Di Muccio, A.M. Cicero, A. Ausili, and S. Di Muccio, Determination of organophosphorus pesticide residues in vegetable oils by single-step multicartridge extraction and cleanup and by gas chromatography with flame photometric detector, *Meth. Biotechnol.* 19 (2006) 263.

98. Y. Chee Beng, K. Ainee, D. Selvarajan, R.O. Mohd, Y. Muhd, and M.N. Mohd Razali, Determination of acephate, methamidophos and monocrotophos in crude palm oil, *Eur. J. Lipid Sci. Technol.* 108 (2006) 960.

99. E. Ueno, H. Oshima, I. Saito, H. Matsumoto, and H. Nakazawa, Determination of organophosphorus pesticide residues in onion and welsh onion by gas chromatography with pulsed flame photometric detector, *J. Pestic. Sci.* 28 (2003) 422.

100. K. Mastovská, J. Hajslová, M. Modula, J. Krivánková, and V. Kocourek, Fast temperature programming in routine analysis of multiple pesticide residues in food matrices, *J. Chromatogr. A* 907 (2001) 235.

101. K. Mastovska and S.J. Lehotay, Practical approaches to fast gas chromatography-mass spectrometry, *J. Chromatogr. A* 1000 (2003) 153.

102. M. Kirchner, E. Matisova, S. Hrouzkova, and J.D. Zeeuw, Possibilities and limitations of quadrupole mass spectrometric detector in fast gas chromatography, *J. Chromatogr. A* 1090 (2005) 126.

103. K. Patel, R.J. Fussell, R. Macarthur, D.M. Goodall, and B.J. Keely, Method validation of resistive heating-gas chromatography with flame photometric detection for the rapid screening of organophosphorus pesticides in fruit and vegetables, *J. Chromatogr. A* 1046 (2004) 225.

104. K. Patel, R.J. Fussell, D.M. Goodall, and B.J. Keely, Application of programmable temperature vaporisation injection with resistive heating-gas chromatography flame photometric detection for the determination of organophosphorus pesticides, *J. Sep. Sci.* 29 (2006) 90.

105. J. LePage, V. Hebert, E. Tomaszewska, J. Rothlein, and L. McCauley, Determination of acephate in human urine, *J. AOAC Int.* 88 (2005) 1788.

106. E. Tomaszewka and V.R. Hebert, Analysis of *O,S*-dimethyl hydrogen phosphorothioate in urine, a specific biomarker for methamidophos, *J. Agric. Food Chem.* 51 (2003) 6103.

107. J. Yu, C. Wu, and J. Xing, Development of new solid-phase microextraction fibers by sol–gel technology for the determination of organophosphorus pesticide multiresidues in food, *J. Chromatogr. A* 1036 (2004) 101.

108. L. Cai, S. Gong, M. Chen, and C. Wu, Vinyl crown ether as a novel radical crosslinked sol–gel SPME fiber for determination of organophosphorus pesticides in food samples, *Anal. Chim. Acta* 559 (2006) 89.

109. E. Psillakis and N. Kalogerakis, Developments in single-drop microextraction, *Trends Anal. Chem.* 21 (2002) 54.

110. E. Zhao, L. Han, S. Jiang, Q. Wang, and Z. Zhou, Application of a single-drop microextraction for the analysis of organophosphorus pesticides in juice, *J. Chromatogr. A* 1114 (2006) 269.

111. Q. Xiao, B. Hu, C. Yu, L. Xia, and Z. Jiang, Optimization of a single-drop microextraction procedure for the determination of organophosphorus pesticides in water and fruit juice with gas chromatography-flame photometric detection, *Talanta* 69 (2006) 848.

112. L.B. Xia and B. Hu, Single-drop microextraction combined with low-temperature electrothermal vaporization ICPMS for the determination of trace Be, Co, Pd, and Cd in biological samples, *Anal. Chem.* 76 (2004) 2910.

113. F. Ahmadi, Y. Assadi, S.M.R. Milani Hosseini, and M. Rezaee, Determination of organophosphorus pesticides in water samples by single drop microextraction and gas chromatography-flame photometric detector, *J. Chromatogr. A* 1101 (2006) 307.

114. S. Berijani, Y. Assadi, A. Mansoor, M.R. Milani Hosseini, and E. Aghaee, Dispersive liquid–liquid microextraction combined with gas chromatography-flame photometric detection: Very simple, rapid and sensitive method for the determination of organophosphorus pesticides in water, *J. Chromatogr. A* 1123 (2006) 1.

115. Y. Xi and H. Dong, Application of solvent sublation for the determination of organophosphorous pesticides in vegetables by gas chromatography with a flame photometric detector, *Anal. Sci.* 23 (2007) 295.

116. P. Xu, D. Yuan, S. Zhong, and Q. Lin, Determination of organophosphorus pesticides and related compounds in water samples by membrane extraction and gas chromatography, *Environ. Monit. Assess.* 87 (2003) 155.

117. S. Morales Muñoz, J.J. Luque-García, and M.D. Luque de Castro, Approaches for accelerating sample preparation in environmental analysis, *Crit. Rev. Environ. Sci. Technol.* 33 (2003) 391.

118. J.A. Mendiola, M. Herrero, A. Cifuentes, and E. Ibañez, Use of compressed fluids for sample preparation: Food applications, *J. Chromatogr. A* 1152 (2007) 234.

119. X. Yi, Q. Hua, and Y. Lu, Determination of organophosphorus pesticide residues in the roots of *Platycodon grandiflorum* by solid-phase extraction and gas chromatography with flame photometric detection, *J. AOAC Int.* 89 (2006) 225.

120. X. Yi and Y. Lu, Multiresidue determination of organophosphorus pesticides in ginkgo leaves by accelerated solvent extraction and gas chromatography with flame photometric detection, *J. AOAC Int.* 88 (2005) 729.

121. J. Wong, M. Hennessy, D. Hayward, A. Krynitsky, I. Cassias, and F. Schenck, Analysis of organophosphorus pesticides in dried ground ginseng root by capillary gas chromatography-mass spectrometry and -flame photometric detection, *J. Agric. Food Chem.* 55 (2007) 1117.

122. M. Okihashi, Y. Kitagawa, K. Akutsu, H. Obana, and Y. Tanaka, Rapid method for the determination of 180 pesticide residues in foods by gas chromatography/mass spectrometry and flame photometric detection, *J. Pestic. Sci.* 30 (2005) 368.

123. K. Sasamoto, N. Ochiai, and H. Kanda, Dual low thermal mass gas chromatography–mass spectrometry for fast dual-column separation of pesticides in complex sample, *Talanta* 72 (2007) 1637.

124. E. Baltussen, P. Sandra, F. David, and C. Cramers, Stir bar sorptive extraction (SBSE), a novel extraction technique for aqueous samples: Theory and principles, *J. Microcol. Sep.* 11 (1999) 737.

125. J. Zrostlíková, S.J. Lehotay, and J. Hajslová, Simultaneous analysis of organophosphorus and organochlorine pesticides in animal fat by gas chromatography with pulsed flame photometric and microelectron capture detectors, *J. Sep. Sci.* 25 (2002) 527.

126. G. Anitescu and L.L. Tavlarides, Supercritical extraction of contaminants from soils and sediments, *J. Supercrit. Fluids* 38 (2006) 167.

127. A. Valverde, A.R. Fernández-Alba, M. Contreras, and A. Agüera, Supercritical fluid extraction of pesticides from vegetables using anhydrous magnesium sulfate for sample preparation, *J. Agric. Food Chem.* 44 (1996) 1780.

128. S.J. Lehotay, Supercritical fluid extraction of pesticides in foods, *J. Chromatogr. A* 785 (1997) 289.

129. V.G. Zuin, J.H. Yariwake, and C. Bicchi, Fast supercritical fluid extraction and high-resolution gas chromatography with electron-capture and flame photometric detection for multiresidue screening of organochlorine and organophosphorus pesticides in Brazil's medicinal plants, *J. Chromatogr. A* 985 (2003) 159.

130. A. Aguilera, M. Brotons, M. Rodríguez, and A. Valverde, Supercritical fluid extraction of pesticides from a table-ready food composite of plant origin (Gazpacho), *J. Agric. Food Chem.* 51 (2003) 5616.

131. K.N.T. Norman and S.H.W. Panton, Application of supercritical fluid extraction for the analysis of organophosphorus pesticide residues in grain and dried foodstuffs, *Meth. Biotechnol.* 19 (2006) 311.

132. H.P. Li, G.C. Li, and J.F. Jen, Fast multi-residue screening for 84 pesticides in tea by gas chromatography with dual-tower auto-sampler, dual-column and dual detectors, *J. Chinese Chem. Soc.* 51 (2004) 531.

133. M. Jánská, S.J. Lehotay, K. Mastovská, J. Hajslová, T. Alon, and A. Amirav, A simple and inexpensive "solvent in silicone tube extraction" approach and its evaluation in the gas chromatographic analysis of pesticides in fruits and vegetables, *J. Sep. Sci.* 29 (2006) 66.

134. E. Botitsi, P. Kormali, S. Kontou, A. Mourkojanni, E. Stavrakaki, and D. Tsipi, Monitoring of pesticide residues in olive oil samples: Results and remarks between 1999 and 2002, *Int. J. Environ. Anal. Chem.* 84 (2004) 231.

135. E. Ueno, H. Oshima, I. Saito, and H. Matsumoto, Determination of nitrogen- and phosphorus-containing pesticide residues in vegetables by gas chromatography with nitrogen-phosphorus and flame photometric detection after gel permeation chromatography and a two-step minicolumn cleanup, *J. AOAC Int.* 86 (2003) 1241.

136. X. Yi and Y. Lu, Residues and dynamics of probenazole in rice field ecosystem, *Chemosphere* 65 (2006) 639.

137. S. Kontou, D. Tsipi, and C. Tzia, Stability of the dithiocarbamate pesticide maneb in tomato homogenates during cold storage and thermal processing, *Food Addit. Contam.* 21 (2004) 1083.

138. E. Papadopoulou-Mourkidou, E.N. Papadakis, and Z. Vryzas, Application of microwave-assisted extraction for the analysis of dithiocarbamates in food matrices, *Meth. Biotechnol.* 19 (2006) 319.

139. M. Valcárcel, S. Cárdenas, and M. Gallego, Qualitative analysis revisited, *Crit. Rev. Anal. Chem.* 30 (2000) 345.

140. R. Muñoz-Olivas, Screening analysis an overview of methods applied to environmental, clinical and food analyses, *Trends Anal. Chem.* 23 (2004) 203.

141. J. Aybar Muñoz, E. Fernández González, L.E. García-Ayuso, A. González Casado, and L. Cuadros-Rodríguez, A new approach to qualitative analysis of organophosphorus pesticide residues in cucumber using a double gas chromatographic system: GC-pulsed-flame photometry and retention time locking GC-mass spectrometry, *Talanta* 60 (2003) 433.

142. L. Gámiz-Gracia, L. Cuadros-Rodríguez, E. Almansa-López, J. Soto-Chinchilla, and A.M. García-Campaña, Use of highly efficient Draper-Lin small composite designs in the formal optimisation of both operational and chemical crucial variables affecting a FIA-chemiluminescence detection system, *Talanta* 60 (2003) 523.

143. A. Pulido, I. Ruisánchez, R. Boqué, and F.X. Rius, Estimating the uncertainty of binary test results to assess their compliance with regulatory limits, *Anal. Chim. Acta* 455 (2002) 267.

144. A. Pulido, I. Ruisánchez, R. Boqué, and F.X. Rius, Uncertainty of results in routine qualitative analysis, *Trends Anal. Chem.* 22 (2003) 647.

145. L. Li, F. Liu, C. Qian, S. Jiang, Z. Zhou, and C. Pan, Determination of organophosphorus pesticides in *Lycium barbarum* by gas chromatography with flame photometric detection, *J. AOAC Int.* 90 (2007) 271.

13 Simple and Affordable Methods: Spectrophotometric, Thin-Layer Chromatographic, and Volumetric Determination of Pesticide Residues

Hamir Singh Rathore and Shafiullah

CONTENTS

345

13.1 INTRODUCTION

One must do something no matter how little for those who need help. Something that brings no reward other than the joy of being allowed to do it.

Albert Schweizer

It would be a tough challenge for India and similar developing countries to produce enough food for the growing population (almost twenty million a year), protecting plant, animal, and human health, and at the same time conserving the environment. Green revolution technologies have doubled the yield of rice and wheat. Green revolution has been made possible only with the help of agrochemicals, particularly pesticides [1]. Still a considerable quantity of food products is destroyed because of pests.

The process of chemical crop protection is a profit-induced poisoning of the environment. The magnitude of the threat is considerable to humans and the environment through either deliberate or ignorant misuse of pesticides (e.g., using parathion to treat head lice). This threat is considerably greater in developing countries where there is little awareness of the danger of pesticide use and inadequate user protection.

However, even in the highly industrialized countries, users are at considerable risk through the intensive use of pesticides, despite more knowledge of the dangers and widespread user protection. Pesticide residues problems cannot therefore be regarded as specific only to developing countries.

In most of the laboratories situated in the developing countries, expensive, sophisticated, and ultrasensitive instruments have either not been installed so far due to shortage of funds or they are lying idle due to lack of maintenance facilities.

Therefore, it was thought worthwhile to describe some simple, affordable, and easily operative methods, such as spectrophotometry, thin-layer chromatography, and volumetry to determine the active ingredients of pesticide formulations as well as pesticide residues in environmental samples.

These methods are very useful and valuable to those people in the developing countries who are directly affected and suffer due to pesticides.

13.2 SIMULTANEOUS SPECTROPHOTOMETRIC DETERMINATION OF ATRAZINE AND DICAMBA IN WATER BY PARTIAL LEAST SQUARES REGRESSION

13.2.1 THEORY AND SIGNIFICANCE

Nowadays, herbicides are used in a wide variety of crops to control pests. According to the US Environmental Protection Agency, atrazine (2-chloro-4-ethylamino-6-isopropylamino-1,3,5-triazine) (ATR) (Figure 13.1) has a low acute toxicity and is used to control broadleaf weeds and some grassy weeds in corn, sorghum, sugarcane, wheat, macadamia nuts, pasture, conifers, and woody ornamentals, among others. Triazine is a colorless crystalline material. It is stable in neutral, weakly acidic, and weakly alkaline media. It is hydrolyzed to the herbicidally inactive hydroxy derivative in strong acids and alkalis and at higher temperature in neutral media. Its solubility is 28 g L^{-1} in water, 183 g kg^{-1} in dimethyl sulfoxide, 52 g kg^{-1} in chloroform, 28 g kg^{-1} in ethyl acetate, 18 g kg^{-1} in diethyl ether, and 0.36 g kg^{-1} in n-pentane at 20°C. It is a selective systemic herbicide, absorbed principally through the roots and also through the foliage, with translocation acropetally in the xylem and accumulation in the apical meristems and leaves. It inhibits photosynthesis and interferes with other enzymic processes. It is toxic to mammals (acute oral LD$_{50}$ for rats 3080 mg kg^{-1}), birds, and fish.

Dicamba (3,6-dichloro-2-methoxybenzoic acid) (DIC) (Figure 13.1) is a postemergence herbicide used to control weeds, docks, bracken, and brush, to mention some examples. It is available as colorless crystals, which melt at 114°C–116°C. It is resistant to oxidation and hydrolyses under normal conditions. It is stable in acids and alkalis. Its solubility is 6.5 g L^{-1} in water, 922 g L^{-1} in ethanol, 916 g L^{-1} in cyclohexanone, 810 g L^{-1} in acetone, 260 g L^{-1} in dichloromethane, 130 g L^{-1} in toluene, and 78 g L^{-1} in xylene at 25°C. It is a selective systemic herbicide absorbed by the leaves and roots, with ready translocation throughout the plant via both the symplastic and apoplastic systems. It acts as an auxin-like growth regulator. It is toxic to mammals (acute oral LD$_{50}$ for rats 1707 mg kg^{-1}), birds, and fish.

Both herbicides can be used in combination; therefore, some formulations with the two active ingredients are available. Partial least squares regression type-1 (PLS-1) is a powerful analytical tool [2], which is generally used for the resolution of multicomponent systems with chemical or spectral interference drawbacks. Hence, a method for simultaneous determination of atrazine and dicamba in water is described.

13.2.2 REAGENTS

Prepare stock solution of ATR and DIC (80 mg mL^{-1}) in methanol/water (1:1, v/v). Store the solutions at 4°C, and protect against light. These solutions are stable for at least 1 month. Prepare the working solutions daily by appropriate dilution. Prepare a buffer solution of KH$_2$PO$_4$/NaKHPO$_4$ (0.1 M), and adjust its pH 7 with 0.1 M NaOH.

13.2.3 PROCEDURE

Carry out the individual calibration for the two herbicides under study. Transfer the appropriate volumes of the working solution (0.3–4.9 mg mL^{-1} of ATR and 0.8–8.8 mg mL^{-1} DIC) of each

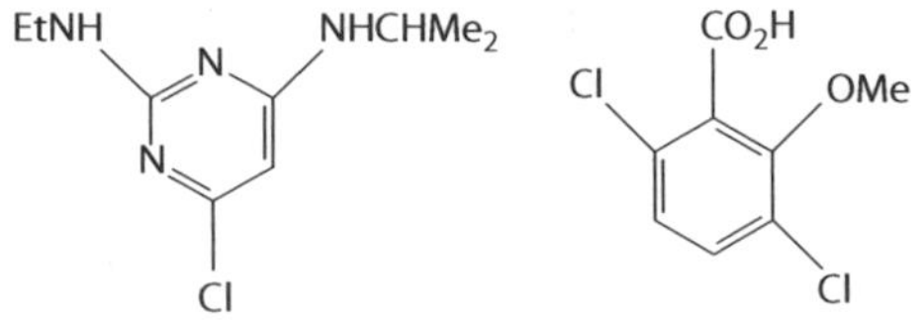

FIGURE 13.1 Atrazine (ATR) and dicamba (DIC).

herbicide to 10 mL volumetric flask, add 1 mL of buffer solution, add methanol to complete the volume to 2 mL, and make up the solution to the mark with distilled water. Record the absorption spectra in the range of 200–300 nm against a blank. Prepare a calibration set of samples in the same way for the resolution of binary mixtures by PLS-1.

Mix the adequate volumes of the working solutions of ATR and DIC with 1 mL of the buffer solution and methanol to complete 1 mL; finally dilute each sample to 10 mL with ultrapure water. Record the absorption spectra in the range of 200–300 nm against a blank. Their composition is described in Table 13.2. Prepare a synthetic set of samples to evaluate the ability of prediction on the calibration model.

For the analysis of water, spike 250 mL of each sample with different concentrations of compounds of interest. Filter each sample with a nylon membrane (0.2 µm of pore size) and pass through an solid phase extraction (SPE) cartridge, with a flow rate of 20 mL min^{-1}. Dry the contents of the cartridge with nitrogen and elute the analyte with 4 mL of methanol. Remove the eluent with N$_2$. Dissolve the residue in 1 mL of methanol and transfer quantitatively to a 10 mL standard flask. Finally, prepare the sample by using the procedure described above.

13.2.4 Results and Discussion

There is a partial overlapping of both spectra of ATR and DIC (Figure 13.2). Therefore, the simultaneous determination of these herbicides requires (a) the use of a separation technique before their determination, or (b) the application of a chemometric technique for the resolution of the binary system. The second option is chosen owing to its simplicity, rapidity, and low cost.

According to Beer's law, ATR exhibits a linear working range of 0.3–4.9 µg mL^{-1} at 222.2 nm and DIC of 0.8–8.8 µg mL^{-1} at 203.4 nm. Calibration functions and correlation coefficients (R) are given in Table 13.1.

The limit of detecting ($L_D = 3$ s m^{-1}) and quantification ($L_C = 10$ s m^{-1}) were calculated by means of the standard deviation of the analytical signals (s), at 222.2 nm for ATR or 203.4 nm for DIC, in a set of 10 blank samples; "m" represents the slope of each calibration curve. For the estimation of RSD, sets of 10 samples of ATR (2.9 µg L^{-1}) and DIC (3.5 µg L^{-1}) were also prepared.

Working conditions for PLS-1 were (a) mean center as preprocessing strategy for spectral data, (b) cross internal validation leaving out one sample by iteration, (c) a maximum of nine factors for

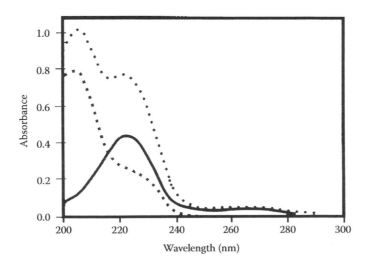

FIGURE 13.2 Absorption spectra of ATR 2.2 mg mL^{-1} (—), DIC 5.3 µg mL^{-1} (■), and their mixture in 2.6 and 5.9 mg mL^{-1}, respectively (●), under the experimental conditions proposed. (From Amador-Hernandez, J. et al., *J. Chil. Chem. Soc.*, 50(20), 461, 2005. With permission.)

TABLE 13.1
Spectrophotometric Data

Equation	ATR $A = 0.0369°C + 0.1391°C$	DIC $A = 0.1393°C + 0.2172°C$
R	0.9990	0.9998
L_D (μg mL^{-1})	0.11	0.25
L_C (μg mL^{-1})	0.35	0.85
RSD (f)	1.6	2.8

Source: Amador-Hernandez, J. et al., *J. Chil. Chem. Soc.*, 50(20), 461, 2005. With permission.

the construction of a calibration model, and (d) a spectral range of 200–240 nm (201 independent variables). The composition of the calibration set of samples used is given in Table 13.2.

PRESS (prediction error sum of squares) as a function of the number of factors was estimated to identify the factors required in the construction of the calibration model; the criterions of the *F*-test and the first local minimum, as well as cumulative variance, were taken into account during the optimization. Two factors were selected as optimum for ATR and three for DIC; the later required an additional factor probably because of the presence of the absorption band of the blank in the same spectral region as this herbicide. The statistical parameters of *R* (the correlation coefficient between theoretical and estimated concentration values), SEC (standard error of calibration), RMSD (the average error index in the analysis), and REP (the error average percentage in the set) were calculated to evaluate the prediction capability of the calibration model by PLS-1; the summary of the results is given in Table 13.3.

TABLE 13.2
Calibration Set of Samples Used (μg mL^{-1})

Sample	ATR (μg mL^{-1})	DIC (μg mL^{-1})	Sample	ATR (μg mL^{-1})	DIC (μg mL^{-1})
1	0.5	1.0	19	0.5	5.7
2	1.4	1.0	20	1.4	5.7
3	2.3	1.0	21	0.5	7.7
4	3.2	1.0	22	0.3	—
5	4.1	1.0	23	0.5	—
6	5.0	1.0	24	0.8	—
7	0.5	2.2	25	1.5	—
8	1.4	2.2	26	2.2	—
9	2.3	2.2	27	2.9	—
10	3.2	2.2	28	3.5	—
11	4.1	2.2	29	4.2	—
12	0.5	3.3	30	4.9	—
13	1.4	3.3	31	—	1.8
14	2.3	3.3	32	—	3.5
15	3.2	3.3	33	—	4.4
16	0.5	4.6	35	—	7.0
17	1.4	4.6	36	—	8.8
18	2.3	4.6	—	—	—

Source: Amador-Hernandez, J. et al., *J. Chil. Chem. Soc.*, 50(20), 461, 2005. With permission.

TABLE 13.3
Statistical Parameters

Parameter Factors	ATR	DIC
Factors	2	3
Cumulative variance (%)	99.989	99.996
R	0.999	0.998
SEC	0.068	0.151
RMSD	0.067	0.149
RED (%)	4.520	5.640

Source: Amador-Hernandez, J. et al., *J. Chil. Chem. Soc.*, 50(20), 461, 2005. With permission.

TABLE 13.4
Synthetic Mixtures Used

Sample	ATR (μg mL^{-1})	DIC (μg mL^{-1})	Sample	ATR (μg mL^{-1})	DIC (μg mL^{-1})
1	0.6	3.5	9	4.2	6.4
2	1.5	2.5	10	0.7	4.2
3	4.0	3.6	11	1.6	4.8
4	4.8	8.8	12	2.6	5.7
5	3.0	2.9	13	1.2	4.6
6	1.8	1.5	14	0.8	1.8
7	3.3	1.2	15	4.6	1.1
8	3.7	2.3	—	—	—

Source: Amador-Hernandez, J. et al., *J. Chil. Chem. Soc.*, 50(20), 461, 2005. With permission.

Then, the optimized calibration models proposed for ATR and DIC by PLS-1 were used to estimate the concentration of both compounds in the synthetic mixtures (Table 13.4).

The mean recovery percentages, SEP (standard error of prediction), and REP (%) obtained for this set of samples (Table 13.5) are with good results in all cases.

Comparison between SEC and SEP allows to identify an over- or subfitting calibration model, with more or less factors than strictly necessary. In this case, the magnitudes of SEC and SEP are similar, which confirm the adequate selection of factors in both cases. The REP (%) for calibration samples was higher than that for synthetic mixtures; the first set includes samples without one of the compounds, which highlight the differences between theoretical and estimated concentrations. Even so, both values are of the same order for the two herbicides.

Samples of 500 mL of tap, well, and seawater were stored in borosilicate containers at 4°C during less than 7 days before the analysis. Aliquots of 250 mL were spiked with different amounts of the compounds of interest and analyzed by the proposed method. Some tap water samples were directly analyzed by spectrophotometry (without SPE), to test the prediction capability of the PLS-1 models with real samples. The rest of the set of tap, well, and seawater samples were treated by SPE before the detection step. Comparison between expected and calculated concentrations is satisfactory in all cases.

TABLE 13.5
Statistical Parameters

Parameter	ATR	DIC
$X \pm S^a$	99.3	99.9
SEP	0.077	0.177
REP (%)	0.33	4.88

Source: Amador-Hernandez, J. et al., *J. Chil. Chem. Soc.*, 50(20), 461, 2005. With permission.

[a] Mean recovery percentage ± standard deviation.

13.3 SIMULTANEOUS HIGH-PERFORMANCE LIQUID CHROMATOGRAPHIC DETERMINATION OF AZADIRACHTIN

13.3.1 THEORY AND SIGNIFICANCE

The neem tree, *Azadirachta indica*, is a tropical plant that is well known for its pesticidal properties. Many studies have demonstrated that its seed contains abundant limonoids and simple terpenoids that are responsible for its biological activity. Among the limonoids, azadirachtin A (commonly referred to as azadirachtin) is considered to be the most important active principle due to its various effects on insects and has gained considerable attention as potential nontoxic, biodegradable, and natural pesticide (Figure 13.3). Furthermore, azadirachtin also has great application in herbal medicine/healthcare products, especially for major skin diseases and for antimalarial, antituberculosis, antiworms, anticlotting, blood detoxifier, antiviral, antiperiodontitic, antibacterial, and antifungal products. For example, studies on cytotoxicity of azadirachtin in human glioblastoma cell lines indicate that azadirachtin can affect reproductive integrity and cell division.

However, the utility of azadirachtin as a biopesticide or herbal medicine is greatly limited by its low water solubility and instability as a result of its propensity under mild acidic, basic, and photolytic conditions. In the past decades, many efforts have been made for the synthesis of new azadirachtin derivatives and to identify their structure-bioactivity relationship to overcome the difficulties mentioned earlier. It is toxic to mammals (acute oral LD_{50} for rats 2820 mg kg^{-1}), birds, and fish.

A simple method [3] of estimation of azadirachtin in neem-based pesticides is described here.

13.3.2 REAGENTS

Dissolve 2 mg of active ingredient (A.I.) of neem-based pesticides in 50 mL of 90% methanol in water. Shake well and allow layers to separate in the case of emulsifiable concentrate (EC) samples.

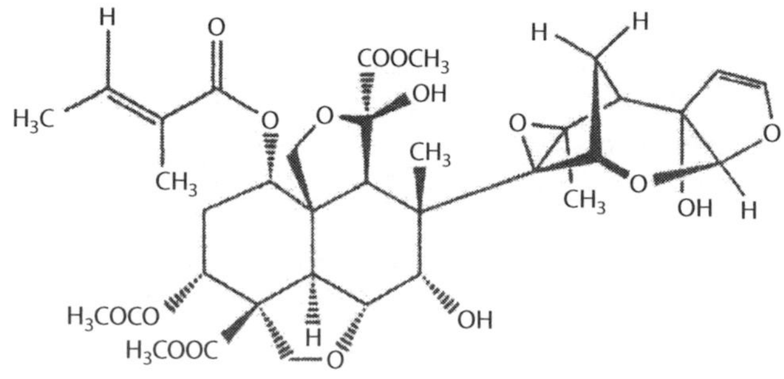

FIGURE 13.3 Azadirachtin.

13.3.3 PROCEDURE

Pack about 500 mg of RPC_{18} powder (30–40 µm particle size) in all glass injection syringes to give an adsorbent height of about 2 cm. Prewet this column with a few mL of 90% methanol in water. Quantitatively transfer 2 mL of aforesaid analyte into the column. Elute the A.I. from the column by adding at a time a small volume (about 1 to 2 mL) with 90% methanol in water. Collect the eluate into a 10 mL standard flask until the volume is up to the mark. Inject 20 µl of it into HPLC operated as per the conditions suggested below. Column: RPC_{18} (Particle size 5 µ) packed in an SS column of 25 cm length/4.6 mm i.d.; mobile phase: acetonitrile/water (25:65); flow rate: 10 mL min⁻¹; detector: 215 nm/sensitivity 0.005 AUFS; and retention time: 10 min (approx.).

13.3.4 RESULTS AND DISCUSSION

Five micron particle size may be substituted by a ten micron column. The former may separate Aza-A from Aza-B, which is not required for routine quality control. The latter may not separate these two isomers and gives a total of both isomers, which will be sufficient. After elution of A.I., HPLC should be allowed to run till all peaks from previous injection appear. This may take as much as about 60 min, which may differ from sample to sample. A guard column should be employed. It is preferable to have a separate column only for this purpose. Sometimes poor quality of water may interfere. Use only triple distilled water from an all-glass distillation assembly. Freshly distilled water may be tried if this problem still persists. Periodically regenerate the HPLC column as per standard procedure. For cleanup readymade Sep-Pak cartridges may also be used. Standards of Aza-A and Aza-B are available in Indian Agricultural Research Institute (IARI), New Delhi. The estimated value should not be less than the declared nominal value.

13.4 SPECTROPHOTOMETRIC DETERMINATION OF CARBARYL INSECTICIDE IN FORMULATIONS

13.4.1 THEORY AND SIGNIFICANCE

New groups of pesticides such as the carbamate were invented for plant protection at doses of 1.0–1.5 kg ha⁻¹ with a reversible mode of action, less persistence, systemic action, and adequate potency against crop pests. Carbamate pesticides were considered to be less toxic than organophosphates.

A carbamate pesticide is a white crystalline solid that is readily soluble in polar organic solvent. Its solubility is 120 mg L⁻¹ in water at 30°C, 400–480 g kg⁻¹ in dimethylformamide, 400–450 g kg⁻¹ in dimethyl sulfoxide, 200–300 g kg⁻¹ in acetone, 200–250 g kg⁻¹ in cyclohexanone, 100 g kg⁻¹ in isopropanol, and 100 g kg⁻¹ in xylene at 25°C.

1-Naphthyl methyl carbamate is being used to control pests in apples, brassicas, lettuce, pees, and so on. It is also used as growth regulator and earthworm killer in turf. Carbaryl is toxic to mammals, fish, and bees. It causes inhibition of blood cholinesterase, and its maximum acceptable concentration in work-place atmosphere over an 8 h working period is 5 mg m⁻³. Its acute oral LD_{50} and acute dermal LD_{50} for rats are 400–850 mg kg⁻¹ and >4000 mg kg⁻¹, respectively.

Hardon, Brunink, and Vander Pol [4] determined carbaryl in treated apples by coupling it with diazotized sulfanilamide to give a red dyestuff, which was measured at 520 µm. The rapid and specific method described below [5] for analyzing "Col" containing 40%–50% (w/w) of active ingredient depends on extraction of the insecticide into chloroform and formation of the red color by coupling with diazotized sulfanilic acid in strong alkali. The maximum color is developed in 10 min and is stable for a period of further 10 min, permitting differential absorptiometry to be applied (Figure 13.4).

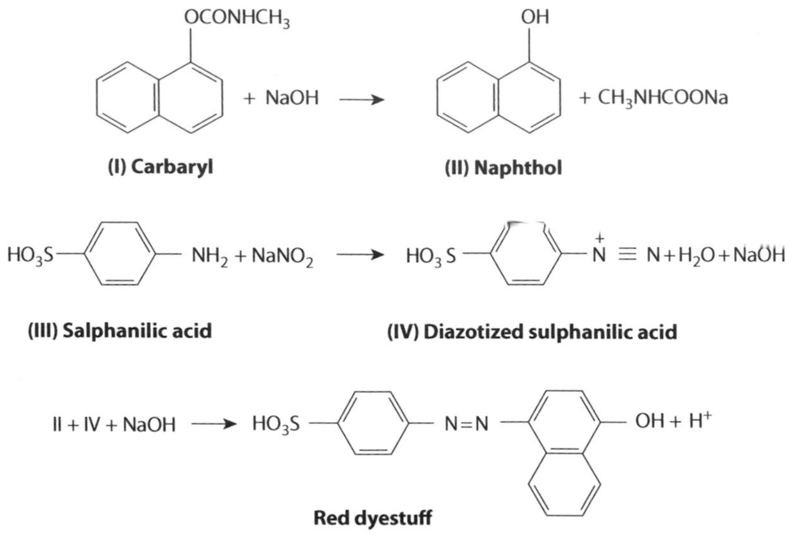

FIGURE 13.4 From carbaryl to a red dyestuff.

13.4.2 REAGENTS

Prepare a fresh solution (0.3% w/v) of sodium nitrite each day, sulfanilic acid solution (0.2% w/v) by heating 0.2 g of sulfanilic acid, make up the cooled solution to 100 mL with water, NaOH (4 N) as usual, and standard carbaryl solution by dissolving 0.050 g of pure carbaryl in methanol, and dilute to 500 mL with methanol. Then 1 mL of solution ≡ 0.1 mg of carbaryl.

13.4.3 PROCEDURE

Weigh accurately about 0.5 g of well-mixed sample into a 250 mL separating funnel. Add 50 mL of water and extract successively with two 100 mL and one 50 mL portion of chloroform by shaking vigorously for 1 min. Collect the combined extract in a 250 mL standard flask, fill up to the mark with chloroform, and mix. Filter the extract through a Whatman No. 40 filter paper. Dilute 10 mL of the filtrate to 200 mL with chloroform, and call it solution A.

Transfer by pipette 10 mL of solution A to a 100 mL boiling flask, and evaporate chloroform under reduced pressure, using moderate suction, on a water bath maintained at about 60°C. Dissolve the residue in 10 mL of methanol and transfer the solution to a 100 mL standard flask. Wash the flask twice with 5 mL portions of methanol and pass the washings into the standard flask.

At the same time, dilute 2.0, 3.0, 4.0, and 5.0 mL of standard solution, equivalent to 0.2, 0.3, 0.4, and 0.5 mg of carbaryl, respectively, to 20 mL with methanol in four 100 mL standard flasks. Add to each flask 20 mL of water and 5 mL each of sodium nitrite and sulfanilic acid solutions, mix, then set the flasks aside for 10 min. Add, by a fast-running pipette, 10 mL of 4 N NaOH, dilute the contents to the mark with water, and mix. Measure the optical densities of solutions, after 10 min of mixing at 520 μm in a 1 cm optical cell against the 0.2 mg standard of carbaryl as reference. Draw the calibration graph relating optical densities of standards to concentrations. Compute the carbaryl content by interpolation. Call this quantity y mg.

$$\text{Carbaryl content, percent, w/w} = \frac{50 \times y}{\text{weight of sample}}$$

TABLE 13.6
Recovery of Carbaryl by the Proposed Method

1-Naphthol Added % (w/w)	Carbaryl Added % (w/w)	Carbaryl Found % (w/w)
0.0	50.0	50.0, 50.5, 50.8, 49.6
2.5	50.0	50.9, 51.1
5.0	50.0	49.2, 49.7
7.5	50.0	50.8, 49.8
10.0	50.0	51.2, 50.4
12.5	50.0	51.4, 51.9
15.0	50.0	52.7, 53.7

13.4.4 RESULTS AND DISCUSSION

Recoveries of carbaryl were determined in 16 samples of laboratory-prepared "Col," each containing 50% (w/w) of carbaryl and dispersing agents, together with varying amounts of 1-naphthol. The results obtained (Table 13.6) show that 1-naphthol present up to 10% (=20% of carbaryl) had no appreciable effect on the determination. The average recovery, based on a 12 carbaryl "Col" containing 0%–10% of 1-naphthol, was found to be 100.7% with a standard deviation of ±1.3%.

In the carbaryl formulation, 1-naphthol is the major contaminate, either present as an impurity or formed by the decomposition of carbaryl during storage. Under the procedure used, 1-naphthol gives a yellow color at the diazotization stage, and this is intensified when the solution is rendered alkaline. This color has an absorption maximum at 400–420 μm, and the optical density, measured at 520 μm, is about one-tenth of that obtained from an equivalent amount of carbaryl. Thus, under normal conditions separation of carbaryl from 1-naphthol is not required. If a substantial quantity of 1-naphthol is suspected, indicated in the final solution by an orange red instead of a red color, a portion of solution A should be washed twice with dilute NaOH and then with water to remove 1-naphthol, before the chloroform is evaporated, and then the color is developed.

The method under study is satisfactory for analyzing carbaryl in a "Col" technical sample, formulations, and so on.

13.5 SPECTROPHOTOMETRIC DETERMINATION OF CYPERMETHRIN INSECTICIDE IN VEGETABLES

13.5.1 THEORY AND SIGNIFICANCE

Cypermethrin (Ambush, Atrobam, Biothrin) or (RS)-cyano-3-phenoxy (IRS, 3RS), (2,2-dichloro-vinyl)-2,2-dimethylcyclopropanecarboxylate is a digestive and contact insecticide effective against a wide range of pests, particularly leaf- and fruit-eating Lepidoptera and Coleoptera in cotton, fruits, vegetables, vines, tobacco, and other crops. Therefore it is widely used to control insect pests of vegetables. Cypermethrin is toxic to mammals (acute oral LD_{50} for rats 251 mg kg^{-1}), birds, fish, and bees. The pyrethroid insecticides containing nitrile group, viz. cypermethrin, have been identified as highly effective contact insecticides.

Its solubility is 0.01 mg L^{-1} in water, >450 g l^{-1} in acetone, chloroform, cyclohexanone, and xylene, and 103 g L^{-1} in hexane at 20°C. It is relatively stable in neutral and weakly acidic media, with optimum stability at pH 4. It is hydrolyzed in alkaline media. It is relatively stable in light and thermally stable up to 220°C.

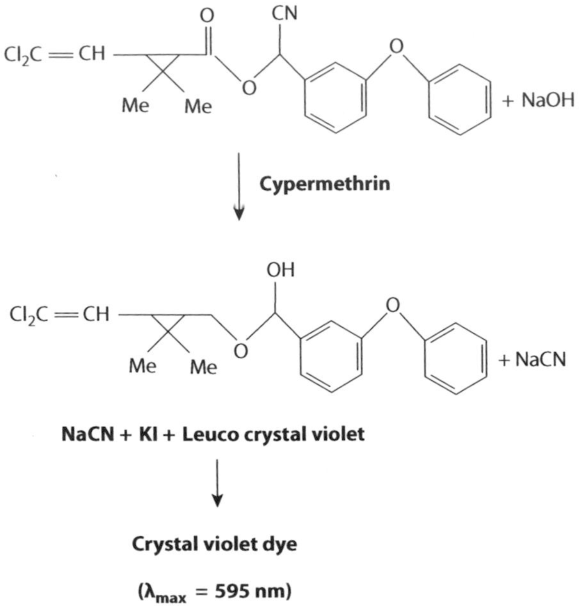

FIGURE 13.5 Production scheme of crystal violet dye starting from cypermethrin. (From Janghel, E.K. et al., *J. Braz. Chem. Soc.*, 18(3), 1, 2007. With permission.)

Chromogenic reagents such as phosphomolybdic acid, palladium chloride, silver nitrate, and copper (II) acetate have been reported to be selective for pyrethroids containing the nitrile group. Several instrumental methods are known to determine cypermethrin residues.

This method [6] is based on the fact that cypermethrin is hydrolyzed to give cyanide ion, which reacts with potassium iodide and LCV to produce a crystal violet dye of λ_{max} 595 nm in acidic medium (Figure 13.5). The reagent is selective for cypermethrin and it obeys Beer's law in the range of 0.12–0.68 ppm of cypermethrin.

13.5.2 REAGENTS

Prepare a stock solution of cypermethrin (1 mg mL^{-1}, Syngenta Crop Protection Private Limited, India). Prepare standard solutions of different concentrations by diluting stock solution with deionized water. Dissolve LCV (250 mg, Eastman Kodak Co.) (LCV) in water (200 mL) containing (3 mL of 85%) phosphonic acid and make up the final volume to the mark in a 1 L standard flask with water. The chemical name and structure of LCV are 4,4',4''-methylidynetris (*N,N'*-dimethyl aniline) and (CH[C$_6$H$_4$N (CH$_3$)$_2$]$_3$), respectively.

13.5.3 PROCEDURE

Preparation of calibration curve

Take an analyte containing 3.0 to 17 μg of cypermethrin in a 25 mL graduated cylinder and add 1.0 mL of 20% NaOH to it. Keep the solution at room temperature for complete hydrolysis. Neutralize the reaction mixture, make slightly acidic with 4 M phosphoric acid, and treat with 1 mL of 0.1% potassium iodide to liberate iodine. Then add 1 mL of LCV, shake thoroughly, and keep for 15 min for full color development of the crystal violet dye. Make up the colored solution up to the mark with water. Measure the absorbance at 595 nm against a reagent blank.

Determination of cypermethrin

Collect samples of vegetables, fruits, or foliage from agricultural fields where cypermethrin was sprayed as an insecticide. Macerate the sample (25 mg) with two 20 mL portions of ethanol/demineralized water (1:1), filter through a thin cotton cloth, and centrifuge the filtrate at 1850 g for 10 min.

In the case of vegetables and fruits transfer the filtrate quantitatively into a 50 mL volumetric flask and fill up to the mark with 50% ethanol. Take the aliquots of supernatant in a 25 mL graduated cylinder, add 1.0 mL of 20% NaOH, and keep it at room temperature for 10 min for complete hydrolysis. Neutralize the reaction mixture and make it acidic with 4 M phosphoric acid, treat with 1 mL of 0.1% potassium iodide and 1 mL of LCV solution, mix thoroughly, and keep for 15 min for full color development. Finally, make up the solution to the mark with water and measure the absorbance at 595 nm against a reagent blank.

In the case of foliages, pass the filtrate through a silica gel column (10×1 cm) fitted with 5 mg silica gel, which was found to be sufficient for removal of chlorophyll and other interfering materials present in the extracted sample. Wash the column with 10 mL of 50% ethanol, collect the washings in a 25 mL volumetric flask, and analyze aliquots spectrophotometrically as above.

In the case of surface water, collect the sample from the main stream of a river. Filter the sample through a Whatman No. 40 filter paper. Analyze aliquots of the water sample using the above-mentioned procedure.

13.5.4 RESULTS AND DISCUSSION

The crystal violet dye formed in the proposed method shows maximum absorbance at 595 nm. All spectral measurements were carried out against demineralized water as the reagent blank, which showed negligible absorption at this wavelength.

The colored product obeys Beer's law in the range 3.0–17 µg of cypermethrin per 25 mL of final solution at 595 nm. The molar absorptivity and Sandell's sensitivity [7] were found to be 3.3×10^5 L mol^{-1} cm^{-1} and 0.054 µg cm^{-2}, respectively.

It is a rapid, simple, and sensitive method. The chromogenic reagent used is sensitive as well as selective for pyrethroid insecticides containing the nitrile group. The lower limit of detection is 0.003 µg per 25 mL.

The recovery for the dosed sample of water, tomatoes, apples, cauliflowers, and cotton foliages has been found to be 8.3%–99% and of blood, urine, and cysteine to be 92%–98%.

13.6 SPECTROPHOTOMETRIC DETERMINATION OF DDT RESIDUES IN FOOD GRAINS

13.6.1 THEORY AND SIGNIFICANCE

DDT, 1,1,1-trichloro-2,2-bis-(p-chlorophenyl)ethane, is practically insoluble in water, readily soluble in aromatic and chlorinated solvents, and moderately soluble in polar organic solvent and petroleum oils. Its solubility is 1000 g L^{-1} in cyclohexanone and dioxane, 850 g L^{-1} in dichloroethane, 770 g L^{-1} in benzene, 720 g L^{-1} in trichloroethylene, 600 g L^{-1} in xylene, 500 g L^{-1} in acetone, 310 g L^{-1} chloroform, 270 g L^{-1} in diethylether, 270 g L^{-1} in ethanol, and 40 g L^{-1} in methanol at 270°C. It possesses a high degree of persistence when applied to a solid surface. It is used to control many species of insects in a very wide range of crops; flying and crawling insects in households, animal houses, and stored products; flies, mosquitoes, cockroaches, bugs, and other insects in public health; and also used as a wood preservative and as an animal ectoparasiticide. It penetrates the internal tissue with which it reacts to liberate a molecule of HCl, and this proves fatal to insects.

DDT, the first pesticide, was first used in the United States by the military for delousing and as a preventive medicine against typhus fever. Later in 1944, DDT was successfully sprayed in

Naples (Italy) over 3,000,000 people for the control of locust-carried typhus. Its discoverer, the Swiss chemist, Paul Miller, won the Noble Peace Prize in 1948.

In spite of its tremendous service to humanity, DDT was banned in the United States and then all over the world because of concerns about its long-term health effects and because housefly that was conquered earlier with DDT has now developed resistance to it. It is toxic to mammals (acute oral LD_{50} is 113 mg kg^{-1} for rats), fish, and aquatic life.

DDT stays in animal bodies, soil, vegetation, and water for a long time, so there is growing interest in the analysis of its residues in flora and fauna. A spectrophotometric method is based on the nitration of DDT with fuming nitric acid/concentrated sulfuric acid (1:1, v/v). The nitrated DDT reacts with alcoholic potash containing urea to produce a blue color. The amount of DDT is determined colorimetrically/spectrophotometrically at 610 nm [8,9] using a calibration curve prepared from a pure p,p' isomer of DDT. The tentative reaction mechanism is given in Figure 13.6.

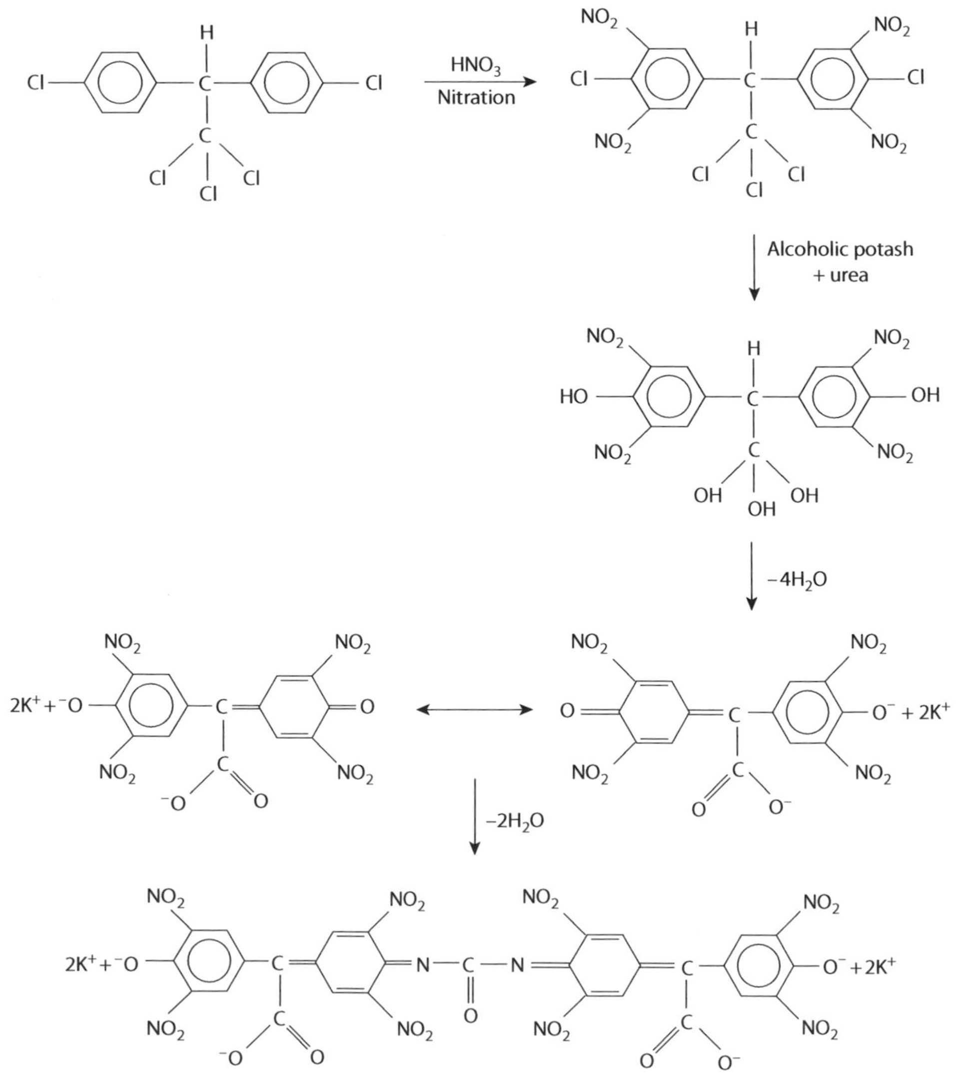

Blue color product

FIGURE 13.6 From DDT to a blue color product. (From Yuen, S.H., *Analyst*, 90, 569, 1965. With permission.)

13.6.2 REAGENT

Prepare a solution of alcoholic potassium hydroxide (5% w/v) by refluxing 5 g of it in 100 mL of absolute alcohol until potassium hydroxide dissolves completely, and then add 2 g of urea into the solution. Filter the solution if necessary. Shake neutral alumina (100 g) with 6 mL of distilled water for 2 h, and keep it overnight before use. Saturate *N,N'*-dimethyl formamide (chromatographic grade) with hexane before use. Prepare the nitrating mixture by mixing equal volumes of fuming nitric acid and concentrated sulfuric acid.

13.6.3 PROCEDURE

Extraction
Powder 100 g of the grain sample in a hand or an electric grinder. Extract with 150 mL hexane in a Soxhlet apparatus for 4 h. Remove most of the hexane so that the final volume of the extract is about 5 mL.

Cleanup
Partition with dimethylformamide: Transfer the extract into a 125 mL separating funnel. Use 20 mL more of hexane in small quantities to wash the extraction flask and transfer the same to the separating funnel. Extract the total hexane extract with 10 mL of *N,N'*-dimethylformamide solution. Set aside the mixture for 2–3 min to permit the phases to separate and drain the clear dimethyl formamide into another 125 mL separating funnel, taking care to retain the emulsions in the first funnel. Reextract the hexane solution with more 10 mL portion of dimethylformamide, and discard the hexane. Collect the total dimethylformamide extract and wash with 10 mL hexane to remove any trace of fat. Extract the hexane used for this washing with another 10 mL of dimethylformamide. Add this 10 mL to the original 30 mL extract of dimethylformamide. Transfer the total dimethylformamide extract to a 250 mL separating funnel containing 200 mL of the sodium sulfate solution. Shake the contents of the funnel for 2 min vigorously and allow the same to stand for 20 min to permit the hexane dissolved in the dimethylformamide to separate. Gather the hexane droplets by gentle swirling and drain out the aqueous phase. Dry the stem of the funnel with a filter paper and retain the final hexane layer for the next step.

Chromatography over alumina: Mount a chromatographic column over a Kuderna-Danish evaporator and pour a slurry of 10 g activated neutral alumina with 6% hexane into it. Allow the slurry to settle keeping the alumina covered with hexane. Add more hexane, and cover the surface of the alumina with a 5 cm layer of anhydrous sodium sulfate. Allow the hexane level to fall to the level of the upper sodium sulfate surface.

Add the hexane extract from the top to the column, wash the separating funnel with three 2 mL portions of hexane, and add to the column. Elute the column with 100 mL of hexane. Evaporate the eluate down to about 1 mL and the extract is ready for the next step.

Nitration of DDT: Transfer the cleaned up hexane extract or an aliquot of it depending on the level of the DDT in the sample to a 30 mL ground glass-stoppered test tube. Evaporate hexane carefully and completely with a current of dry air by adding one drop of ethylene glycol to prevent the loss of DDT. Add 2 mL of the nitrating acid, stopper the test tube after lubricating it with a drop of concentrated sulfuric acid, and heat on a boiling water bath for 1 h. Cool and slowly add 10 mL cold water. Cool again and transfer to a 50 mL separating funnel using a small volume of water to wash the test tube. Pipette 5 mL of benzene into the separating funnel and shake it vigorously for 1 min. Allow the layers to separate, and drain off the lower aqueous layer. Wash the benzene layer with 10 mL portions of 2% aqueous NaOH until the aqueous layer is colorless. Dry the stem of the funnel with a filter paper and transfer the benzene solution into a glass-stoppered test tube containing about 0.5 g of sodium sulfate.

Pipette 2 mL of the benzene extract containing the nitrated DDT into a dry ground glass-stoppered test tube. Add 4 mL of alcoholic potash containing urea, and measure the absorbance of blue color at 610 nm in a spectrophotometer using reagent blank as reference solution.

Calculations: Prepare a standard curve by following the procedure as detailed using hexane solutions of *p,p'*-DDT in concentrations of 0 to 50 μg. By using the standard curve, determine the amount of residues in ppm.

13.6.4 RESULTS AND DISCUSSION

DDT residues at the 1 ppm level can be determined by the above-mentioned procedure. A level of 0.1 ppm of DDT can be determined by coupling the enrichment process that is, the suitable reduction in the volume of the final solution before the addition of a coloring reagent.

13.7 VOLUMETRIC DETERMINATION OF DDT IN FORMULATIONS

13.7.1 THEORY AND SIGNIFICANCE

DDT is a solid in the form of white needles. Its melting point (m.p.) is 108.5°C–109°C and vapor pressure (v.p.) is 1.5×10^{-7} mm/20°C. DDT undergoes dehydrochlorination in alkaline solutions. It changes to the noninsecticidal 1,1'-(2,2-dichloroethyllidine)-bis (-4-chlrobenzen) (DDE). This reaction is catalyzed by ferric chloride, aluminum chloride, and UV light. It is an effective insecticide, particularly for mosquitoes, flies, and crop pests. It is responsible for the, almost, eradication of malaria-carrying mosquitoes and thus making the world free from malaria. Unfortunately, it is toxic to humans and plays havoc with several species of useful birds and fish. Its use has been recently banned in the United States and some other countries.

The estimation of DDT is based on the fact that one molecule of DDT gives out a molecule of KCl when treated with alcoholic potash, so the consumption of potash is proportional to DDT [9].

$$(C_6H_4Cl)_2CHCCl_3 + KOH \rightarrow (C_6H_4Cl)_2 C = CCl_2 + KCl + H_2O$$

The chloride of the KCl solution so obtained is treated with excess of standard silver nitrate solution, and the residual silver nitrate is determined by titration with standard thiocyanate solution. Now silver chloride is more soluble than silver thiocyanate and would react with the thiocyanate as follows:

$$AgCl (s) + SCN^- = AgSCN (s) + Cl^-$$

It is therefore necessary to remove the silver chloride by filtration.

13.7.2 REAGENT

Prepare 0.028 N silver nitrate solution by dissolving 4.79 g of silver nitrate in water and making up the solution to the mark in a 1 L standard flask by adding water. Prepare 1 N potash by dissolving 56 g of KOH in a small volume of water and making up the solution to the mark in 1 L standard flask by adding purified alcohol. Dissolve 40 g of iron alum in 100 mL of water to obtain a saturated solution of iron alum for use as an indicator. Dissolve ammonium thiocyanate (2.15 g) in water and make up volume in a 1 L standard flask by adding water. Standardize it with silver nitrate solution.

13.7.3 PROCEDURE

Sample preparation
Weigh a sample of insecticide containing 25–200 mg DDT (0.025–0.20 g DDT powder) in a conical flask. Add 50 mL alcoholic potash (chloride-free solution). Fit up a reflux condenser and reflux alcohol for 15 min on a water bath. Add 100 mL water, cool, and add 3 mL HNO_3.

Titration

Mix well the above solution, add 25 mL of standard silver nitrate solution into it, and shake well so as to precipitate all chloride. Add more silver nitrate solution if necessary and ensure an excess of silver nitrate. Add 2 mL of saturated iron alum and 20 mL of nitric acid (free from nitrous acid). Shake well and titrate excess of silver nitrate with standard NH_4CNS solution until a faint pink color develops. Carry out a blank titration also in the same way and determine the end point by taking alternately NH_4CNS solution and $AgNO_3$ solution in a conical flask and the burette.

Calculate results considering

$$1 \text{ mg Cl} = 10 \text{ mg pure DDT}$$

Calculations

$$\% \text{ DDT in the given sample} = \frac{v}{w}$$

where
  v is the volume (mL) of 0.028 N silver nitrate consumed by KCl
  w is the weight (g) of given sample

13.8 VOLUMETRIC DETERMINATION OF DECAMETHRIN INSECTICIDE

13.8.1 THEORY AND SIGNIFICANCE

Pyrethrum has been used for the control of harmful insects in many countries. It has a low order of toxicity to warm-blooded animals and produces no harmful residues on food crops. Its acute oral LD_{50} and acute dermal LD_{50} for rats are 1500 mg kg^{-1} and >1800 mg kg^{-1}, respectively. Although pyrethrum was formerly used as finely ground flowers, today it is used as dusts and sprays. The structure of the active principle in pyrethrum was established in the 1950s. Staudinger and Ruzicka [10] found two active compounds, pyrethrin I and pyrethrin II, which are esters of a keto alcohol (pyrethrolone) and two acids (chrysanthemum monocarboxylic acid and C. dicarboxylic acid). Two more compounds were identified, and the four possible compounds have been named as pyrethrins I and II and cinerins I and II (Figure 13.7). These compounds are all viscous liquids, soluble in a variety of solvents but not in water. The relative toxicities of pyrethrin I, pyrethrin II, cinerin I, and cinerin II are 100, 23, 71, and 18, respectively. Cinerins are more stable than pyrethrins. Now it has been found that the insecticidal property of pyrethrum is due to five esters (two pyrethrins, two cinerins, and jasmolin II) that are present mostly in the achenes of the flower (0.7%–3.0%).

Pyrethrins when applied to the insect nerve produce arrhythmic spontaneous discharge at 0.01 to 0.1 ppm and a reversible blocking of conduction at 1 ppm. Paralyzed insects exhibit characteristic vacuolization of the nerve tissue. A threefold increase in dosage is required to produce the same percentage of mortality as that of knockdown, since there is a substantial recovery after rapid knockdown when pyrethroids are applied. Pyrethrins are practically insoluble in water and readily soluble in organic solvents, for example, alcohols, chlorinated hydrocarbons, nitromethane, and kerosene. Their oxidation and loss of insecticidal activity occur in light and air. They rapidly hydrolyze in the presence of alkalis, again with a loss of insecticidal activity. Synergists have a stabilizing effect on pyrethrins.

The unstable nature of pyrethrins to light and oxygen coupled with the establishment of the structure of pyrethrins stimulated research on synthetic pyrethroids. From 1964, a group of compounds such as K-othrin, Kodethrin, Properthrin, and Enprothrins, were synthesized, but they could not excel the natural pyrethrins as all of them were unstable to light and temperature. The first synthetic pyrethroid, permethrin, possessing high activity against insects, low mammalian toxicity, and greatly increased stability, was described in 1973.

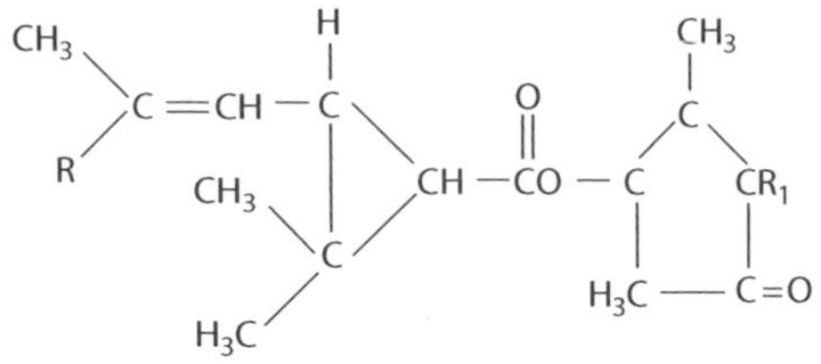

| R= | | R₁= |

Pyrethrin-I	$-CH_3$	$-CH_2CH=CHCH=CH-$
Pyrethrin-II	$-COOCH_3$	$-CH_2CH=CHCH-CH_2-$
Cinerin-I	$-CH_3$	$-CH_2CH=CHCH_3$
Cinerin-II	$-COOCH_3$	$-CH_2CH=CHCH_3$
Jasmolin-II	$-COOCH_3$	$-CH_2CH=CHCH_2CH_3$

FIGURE 13.7 Different pyrethrins.

With the structural requirements for the photostable pyrethroids established, many new compounds that retained the chemical and stereochemical features of parent materials and their biological activity were synthesized during 1973–1977. In 1974, Okno and others [10] described highly active compounds such as Cypermethrin, Fenvalerate, Decamethrin, and so on. These compounds are viscous lipophilic liquids with a high boiling point and volatility. They are practically insoluble in water, highly photostable, biodegradable, have low mammalian toxicity, and do not leave residues in the biological systems. They are of low volatility, low polarity, and nonsystemic. Their translocation is minimum in the environment, and consequently they cause less contamination. As the synthesis of pyrethroids is more complex, they are costly on weight basis. However, they are found to be effective even at 1/10th to 1/50th of the commonly used rates of conventional pesticides, so the synthetic pyrethroids are likely to be competitive.

The empirical formula of decamethrin is $C_{22}H_{19}Br_2NO_3$, which has a molecular weight of 505.2. When it is treated with sodium metal in isopropanol, it splits sodium bromide, which can be titrated with standard silver nitrate solution using ferric alum as indicator [11,14].

$$C_{22}H_{19}Br_2NO_3 + Na \rightarrow C_{22}H_{21}NO_3 + 2NaBr$$

$$NaBr + AgNO_3 \rightarrow AgBr + NaNO_3$$

13.8.2 REAGENTS

Dissolve 5 g of phenolphthalein in 500 mL of ethanol and add 500 mL of water with constant stirring, and filter, if precipitate forms. Weigh out accurately 8.494 g of AR silver nitrate, dissolve it in water, and make up to 500 mL in a standard flask. This gives a 0.1 N solution of silver nitrate. To prepare approximately 0.1 N ammonium thiocyanate solution, weigh out about 8.4 g AR ammonium thiocyanate and dissolve it in 1 L of water in a standard flask. Shake well and standardize it by titrating with 0.1 M silver nitrate solution. Dissolve 0.1 g eosin in 100 mL of 70% ethanol or by dissolving 0.1 g of the sodium salt in 100 mL of water.

13.8.3 PROCEDURE

Accurately weigh a sample containing about 0.15 g of decamethrin in a 250 mL conical flask. Add 5 mL of benzene and 30 mL of isopropanol into it. Shake well and add 2.5 g of sodium metal and reflux slowly for about 2 h. Cool the contents and destroy excess of sodium metal by adding 50% aqueous isopropanol drop by drop with constant stirring. Boil the reaction mixture again for 10 min. Cool again, add 20 mL of water, and shake well. Now add 4 mL of formaldehyde and keep for 10 min. Add 2–4 drops of phenolphthalein and acidify with nitric acid/acetic acid. Add 1 mL of eosin and titrate with 0.1 N $AgNO_3$ till the precipitate assumes a magenta color. Alternatively add 25 mL of 0.1 N $AgNO_3$, 2 mL of nitrobenzene, and 5 mL of ferric. Shake well and titrate the excess of silver nitrate with 0.1 N ammonium thiocyanate to a brick red end point.

$$\% \text{ Decamethrin in the given sample} = \frac{v \times 2.525}{w}$$

where
   v is the volume of 0.1 N silver nitrate consumed by NaBr
   w is the weight (g) of the sample taken

13.8.4 RESULTS AND DISCUSSION

The most suitable adsorption indicator is eosin [14], which can be used in dilute solutions and even in the presence of 0.1 M nitric acid, but in general, acetic acid solutions are preferred. With eosin indicator, the silver bromide flocculates approximately 1% before the equivalent point, and the local development of a red color becomes more and more pronounced with the addition of silver nitrate solution. At the end point the precipitate assumes a magenta color.

13.9 SPECTROPHOTOMETRIC DETERMINATION OF FENVALERATE INSECTICIDE

13.9.1 THEORY AND SIGNIFICANCE

Fenvalerate (Figure 13.8), cyano(3-phenoxyphenyl)methyl 4-chloro-alpha-(1-methylethyl)benzeneacetate is a clear viscous liquid, yellow in color, with a mild chemical odor, and more stable in acidic solution than in alkaline solution. It is soluble in acetone, alcohol, ether, xylene, and kerosene. It is almost insoluble in water. It is toxic to mammals (acute oral LD_{50}, for rats 451 mg kg^{-1}), birds, fish, and bees. It is used for control of insects on cotton and a range of other crops.

Fenvalerate absorbs ultraviolet radiations at 278 nm, so it can be determined by UV spectrophotometry [9,12].

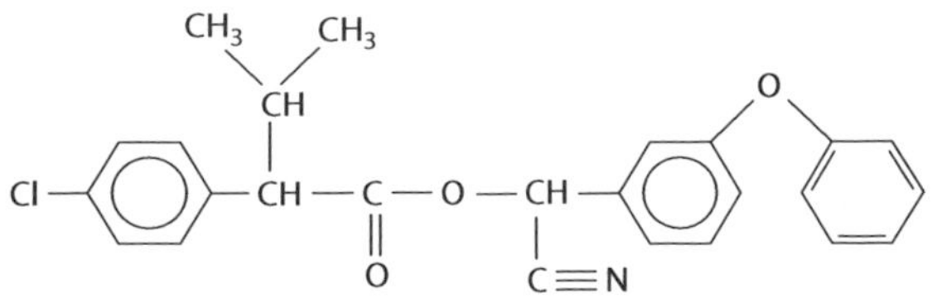

FIGURE 13.8 Fenvalerate.

13.9.2 REAGENTS

Dissolve fenvalerate (100 mg, commercial grade) in 100 mL of chloroform. Dissolve 100 mg of fenvalerate formulation in 100 mL of chloroform.

13.9.3 PROCEDURE

Take a known volume of the standard solution of fenvalerate (0.5–2.5 mL) in a 25 mL standard flask, dilute, and fill up to the mark with chloroform. Mix the solution thoroughly, and record its absorbance at 278 nm using chloroform as a blank. Make a calibration curve by plotting the absorbance versus concentration of fenvalerate (mg mL^{-1}). Similarly take a known volume of the solution of an unknown sample in a 25 mL standard flask, dilute, fill up to the mark with chloroform, and measure its absorbance at 278 nm using chloroform as a blank. Determine the concentration of the unknown with the help of a calibration curve.

$$\% \text{ Fenvalerate in the given samples} = \frac{\text{mg of fenvalerate from curve}}{\text{mg of fenvalerate sample taken}} \times 100$$

13.10 COLORIMETRIC DETERMINATION OF GLYPHOSATE RESIDUES IN WATER

13.10.1 THEORY AND SIGNIFICANCE

Glyphosate, N-(phosphanomethyl)glycine (Figure 13.9a), is a new broad-spectrum herbicide being increasingly used to control annual and perennial weeds on noncrop sites, such as rights of way, fence rows, and commercial turfs. In agriculture, glyphosate also has a wide application and use as a preemergent herbicide in no-till corn and sorghum, and has been production and in pasture renovation. Glyphosate is soluble in water (12 g L^{-1} at 25°C). It is insoluble in common organic solvents. Its alkali metal and amine salts are readily soluble in water. Glyphosate and its metabolite, amino-methylphosphonic acid (Figure 13.9b), have been found to degrade rapidly in soil with little or no residual effect.

Several analytical methods have been reported for the detection and determination of glyphosate and its metabolite. Differential pulse polarography, the amino acid analyzer-colorimetry method, thin-layer chromatography (TLC), and gas chromatography (GC) have been used for the determination of the N-butyl-N-trifluoroacetyl derivative of glyphosate. Although these methods appear technically sound and represent significant accomplishments, generally, they are lengthy and expensive and require specialized equipment.

Hence, a simple and rapid chemical method [13,14] is described for determining glyphosate in pure and natural waters. In this method, the organic phosphate in glyphosate is oxidized with hydrogen peroxide to the orthophosphate, which is then measured colorimetrically as phosphomolybdate heteropoly blue complex at 830 nm.

Orthophosphate and molybdate ions condense in acidic solution to give molybdophosphoric acid (phosphomolybdic acid), which upon selective reduction (with hydrazinium sulfate) produces a blue color, due to molybdenum blue of uncertain composition. The intensity of blue color is

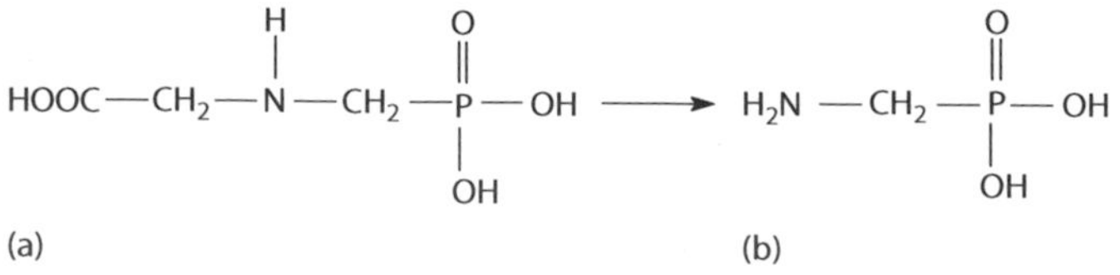

(a)   (b)

FIGURE 13.9   (a) Glyphosate and (b) aminomethylphosphonic acid.

proportional to the amount of phosphate initially incorporated in the heteropoly acid. If the acidity at the time of reduction is 0.5 M in sulfuric acid and hydrazinium sulfate is the reductant, the resulting blue complex exhibits maximum absorption at 820–830 nm. A few colored compounds of molybdenum are given below:

$$\left[PO_4 \cdot 12MoO_3\right]^{-3}$$

Yellow phosphomolybdate ion
MoO_3 molybdenum blue

$$\left[P_2Mo_{18}O_{62}\right]^{-6} \leftrightarrow \left[PMo_{12}O_{40}\right]^{-3}$$

Blue heteropoly phosphomolybdate

Complex ($\lambda_{max} = 830$ nm)

13.10.2 Reagents

Prepare molybdate solution by dissolving 12.5 g of AR sodium molybdate ($Na_2MoO_4 \cdot 2H_2O$) in 5 M-sulfuric acid and dilute to 500 mL with 5 M sulfuric acid. Prepare standard phosphate solution by dissolving 0.2197 g of AR potassium dihydrogen phosphate in deionized water and dilute to 1 L in a graduated flask; then 12 mL = 0.05 mg P. Dilute as appropriate. Prepare glyphosate solution by dissolving an appropriate quantity of glyphosate (94% purity, Monsanto chemical Co., St. Louis, MO) in deionized water. Prepare hydrazinium sulfate solution by dissolving 1.5 g of AR hydrazinium sulfate in deionized water and dilute to 1 L.

13.10.3 Procedure

Take 50 mL aliquots of the three water samples in volumetric flasks and spike with 50–100 µg (1–20 ppm) of glyphosate. Adjust to pH of about 5.0 for consistency with either 0.1 N NaOH or 0.1 N HCl. Add 1.0 mL of 30% hydrogen peroxide to each sample (caution should be exercised in the handling of this oxidant: rubber gloves are recommended to avoid skin burns); when larger volumes of water are analyzed, the samples are concentrated to about 50 mL before adding hydrogen peroxide.

Boil the peroxide-treated water samples at a moderate rate to dryness. Cool the samples and then add 20–30 mL of 0.25 M HCl solution to each flask to dissolve the residue. Oxidize the glyphosate to orthophosphate by this procedure, which requires about 40 min. Neutralize the contents with dilute 4 M NaOH solution.

The sample solution should contain not more than 0.1 mg of phosphorus as the orthophosphate in 25 mL and should be neutral. Transfer 25 mL solution to a 50 mL Pyrex graduated flask. Add 5.0 mL of the molybdate solution, followed by 2.0 mL of the hydrazinium sulfate solution, dilute to the mark with distilled water, and mix well. Immerse the flask in a boiling water bath for 10 min. Remove and cool rapidly. Shake the flask, adjust the volume, and measure the absorbance at 830 nm against either deionized water or reagent blank.

Construct the calibration curve using the standard phosphate solution, in the usual manner.

13.10.4 Results and Discussion

The average recovery of glyphosate from distilled and natural water samples was greater than 95% (Table 13.7). The analysis time for one glyphosate determination was 1.5 h for distilled and runoff waters and 2.0 h for the river water.

TABLE 13.7
Recovery of Glyphosate[a] as Orthophosphate from Water

Glyphosate Added (µg)	Distilled Water Found (µg)	%	River Water Found (µg)	%	Corrected (µg)/%	Runoff Water Found (µg)	Corrected (µg)	%
0.0[b]	0.0	—	0.0		—	0.0	—	—
0.0[c]	0.0	—	47.6	—	—	45.0	—	—
50	49.5	99.0	95.6	48.0	96.0	93.6	46.8	93.6
100	91.4	91.4	147.1	99.5	99.5	153.2	108.2	108.2
200	198.5	99.9	252.1	204.5	102.3	247.0	202.0	101.0
400	401.8	100.5	441.0	203.4	98.4	N.R.	—	—
1000	N.R.	—	970.0	922.4	92.2	N.R.	—	—
Average		97.2 ± 2.4			97.7 ± 3.4			100.9 ± 5.9

Source: Glass, R.L. *Anal. Chem.*, 53, 921, 1981. With permission.
Note: N.R.—not run.
[a] Average values of at least four determinations (Final averages include standard deviations).
[b] The oxidation step was omitted.
[c] Background determinations.

Interference from unknown substances in the natural water samples presented some analytical difficulties. In the final solutions of the fortified river water, the stability of the heteropoly phosphomolybdate blue was markedly improved when aliquots of these samples were first reduced with zinc under acidic conditions. This treatment required an additional 0.5 h. Recovery from river and runoff water samples was corrected for background phosphorus, presumably, due to some inorganic or organic phosphates that formed the orthophosphate upon oxidation. In blank determinations of the fortified natural water samples, an equivalent of nearly 45 µg of glyphosate was measured and used for background corrections. No orthophosphates were formed in blank determination when the hydrogen peroxide was omitted in the oxidation step (Table 13.7).

13.11 POTENTIOMETRIC DETERMINATION OF ISOPROTURON HERBICIDE

13.11.1 THEORY AND SIGNIFICANCE

Isoproturon (Figure 13.10), (I), 3-(4-isopropyl phenyl)-1,1-dimethylurea, exists as colorless crystals. It melts at 155°C–156°C. It is very stable to light, acids, and alkalis. It is hydrolytically cleaved by strong alkalis on heating. It is sparingly soluble in water (27 mg L^{-1}) at 20°C and readily soluble in alcohols, ketones, esters, aromatic hydrocarbons, and chlorinated hydrocarbons. It is soluble in methanol (56 g L^{-1}), dichloromethane (63 g L^{-1}), benzene (6 g L^{-1}), and hexane (0.1 g L^{-1}) at 20°C.

It is a selective systemic herbicide, absorbed by the roots and leaves, with translocation. It inhibits photosynthesis. It is used for pre- and postemergence control of annual grasses and many annual broad-leaved weeds in spring and winter wheat (except durum wheat), spring and winter barley, winter rye, and triticale. Isoproturon is toxic to mammals (acute oral LD$_{50}$ for rats >2000 mg kg^{-1}), birds, and fish. It is not toxic to bees.

It can be determined potentiometrically [9,12] using standard perchloric acid in dioxane.

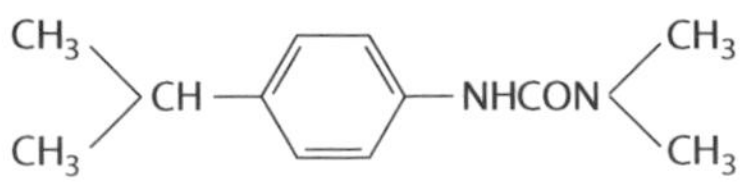

FIGURE 13.10 Isoproturon.

13.11.2 REAGENTS

Prepare 0.1 N perchloric acid in dioxane.

13.11.3 PROCEDURE

Dip the electrode system in acetic anhydride overnight to obtain equilibrium. Weigh 0.25 g of the sample in a beaker and add 50 mL of acetic anhydride and cover it with a lid containing three holes. Fix the electrodes, thermometer, and burette in these holes. Cool the contents to 5°C by placing the system in an ice bath and switch on the magnetic stirrer. Titrate the contents with 0.1 N perchloric acid in dioxane and record the change in potential (m volt) after each addition. Standardize the method by using a standard solution of isoproturon. Find the end point and calculate the concentration as follows:

$$\% \text{ Isoproturon in the given sample} = \frac{v \times 20.6}{w}$$

where
  v is the volume (mL) of 0.1 N perchloric acid consumed by isoproturon
  w is the weight (g) of the given samples

13.12 VOLUMETRIC DETERMINATION OF MALATHION INSECTICIDE IN FORMULATIONS

13.12.1 THEORY AND SIGNIFICANCE

Malathion is widely used in the control of stored grain pests, insect vectors of malaria and encephalitis, household pests, animal pests, and plant pests. Malathion, diethyl (dimethoxy phosphiothioylthio) succinate, is a clear amber liquid of boiling point 156°C–157°C at 0.7 mmHg. It is relatively stable in neutral aqueous media and decomposed by acids and alkalis. It is soluble in water (145 mg L^{-1} at 25°C) and miscible with most organic solvents, such as alcohols, esters, ketones, ethers, and aromatic hydrocarbons. It is slightly soluble in petroleum ether and some mineral oils. Malathion is toxic to mammals, (acute oral LD_{50} for rats 28,000 mg kg^{-1}), birds, fish, and bees. Thus, there is a growing interest in the analysis of malathion residues. Volumetric methods [9,15] for malathion are based upon the hydrolysis of malathion to give O,O-dimethyl phosphorodithiote (DPD) and then the formation of a complex/salt of DPD with Ag(I) or Cu(II) or Bi(III). The first category of volumetry is based on the formation of a white stable precipitate of DPD with Ag(I) and the detection of end point with the help of adsorption indicator, dichlorofluorescein. The second category of volumetry is based on the determination of the amount of Cu(II) or Bi(III) left after complexation with DPD. The third category is based on the oxidation of DPD to O,O-dimethyl phosphoric acid with chloramine T and detection of end point iodimetrically (Figure 13.11).

Fajans [15] introduced a useful type of indicator for precipitation reactions as a result of his studies on the nature of adsorption. The action of these indicators is due to the fact that at the equivalence points the indicator is absorbed by the precipitate, and during the process of adsorption a change occurs in the indicator, which leads to a substance of different color. They have therefore been termed adsorption indicators.

13.12.2 REAGENTS

Dissolve dichlorofluorescein (0.1 g) in 100 mL of ethanol (60%–70%) or dissolve sodium dichlorofluoresceinate (0.1 g) in 100 mL of distilled water. Dissolve malathion (0.90–1.00 g) in 50 mL of distilled water and fill up the volume to the mark in a 50 mL standard volumetric flask. Prepare the solution of phenol by dissolving it (30 g) in 100 mL of methanol.

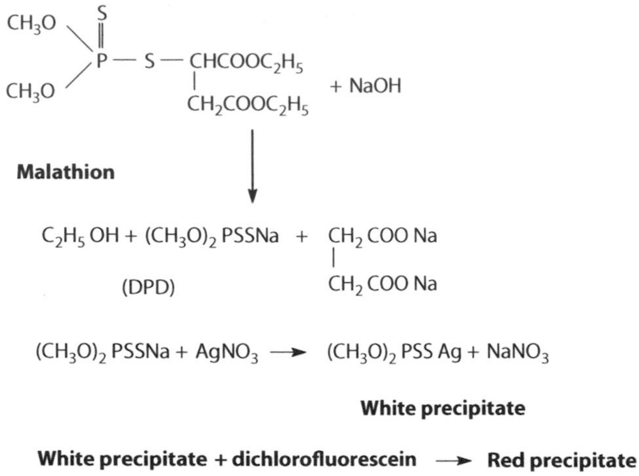

FIGURE 13.11   Malathion hydrolysis and precipitation of DPD with $AgNO_3$.

13.12.3 PROCEDURE

Prepare a chromatography column (300 × 22 m i.d.) by filling 60–100 mesh Florisil (10 cm) and anhydrous sodium sulfate (1 cm). Moisten the column with 40 mL of petroleum ether. Add a known volume of malathion liquid formulation (0.90–1.00 g of active ingredient) to the column and then elute malathion with 100 mL of ethyl ether/petroleum ether (1:1, v/v) at the rate of 5 mL min^{-1}. Evaporate the eluate, dissolve the residue in methanol, and fill the solution up to the mark in a 50 mL standard flask with methanol. Take 10 mL of aliquot in a 250 mL conical flask and add to it 2 mL of 3 N NaOH and 2 mL of phenol solution (10%) in methanol. Mix up the solution and allow to stand for 20 min. Adjust the pH to 6.5–8.0 by adding dilute HNO_3. Add 15 drops of indicator solution (0.1% dichlorofluorescein) and dilute to 100 mL with distilled water. Titrate the solution with 0.1 N silver nitrate to the point at which the precipitate formed coagulates and red color appears on the surface. Calculate the concentration of malathion as follows:

$$\% \text{ Malathion in the given sample} = \frac{(v \times N \times 165)}{\text{gram of sample}}$$

where
  v is the volume (mL) of $AgNO_3$
  N is the normality of $AgNO_3$

13.12.4 RESULTS AND DISCUSSION

If the solution is acid, neutralization may be effected with chloride-free calcium carbonate, sodium tetraborate, or sodium hydrogen carbonate; the AR substances are suitable. Mineral acid may also be removed by neutralizing most of the acid with ammonia solution and then adding an excess of AR ammonium acetate. Titration of the neutral solution, prepared with calcium carbonate, by the adsorption indicator method is rendered easier by the addition of 5 mL of 2% dextrin solution; this offsets the coagulating effect of the calcium ion [14]. If solution is basic, it may be neutralized with chloride free nitric acid, using phenolphthalein as indicator.

13.13 SPECTROPHOTOMETRIC DETERMINATION OF MALATHION RESIDUES IN SOIL

13.13.1 THEORY AND SIGNIFICANCE

Malathion (S-1,2-bis ethoxycarbonyle ethyl O,O-dimethyl phosphodithioate) is a broad-spectrum insecticide used for the control of sucking and chewing insects, including aphids, houseflies, mosquitoes, scale insects, and spider mites on fruits, ornamentals, beans, vegetables, and stored products. It is considered as one of the safe pesticides. When malathion is treated with copper sulfate in acidic medium the following yellow complex [9,16] is obtained, which absorbs at 420 nm and follows Beer-Lambert's law (Figure 13.12).

13.13.2 REAGENTS

Prepare stock solution of malathion by shaking 20 g of malathion (5% malathion dust, Singhal pesticides Pvt. Ltd., Agra) with 50 mL of ethanol in a 250 mL conical flask for 15 min. Filter the supernatant liquid through a Whatman No. 42 filter paper. Treat the undissolved portion with 25 mL of fresh ethanol and repeat the treatment with 25 mL of fresh ethanol. Filter and store the supernatant liquid in the same flask. Finally fill up a volume of 100 mL with ethanol in a standard flask. Take 2 mL of the stock solution, dilute by adding 10 mL of distilled water, and then extract malathion with 10 mL of chloroform. Remove chloroform from the extract under reduced pressure and dissolve the residue in 2 mL of ethanol. Dilute the ethanolic solution (2 mL) with distilled water and make up the total volume to 100 mL in a standard flask. Mix the contents thoroughly and store it at 5°C. Prepare 0.06 M Cu(II) sulfate pentahydrate solution in distilled water.

13.13.3 PROCEDURE

Take different volumes of standard malathion solution (1–3 mL of 1070 mg L^{-1} in acetone) in a 60 mL separating funnel and add 2 mL of 3 M NaOH followed by 10 mL of distilled water. Acidify the contents with 4 M HNO$_3$ and ascertain the acidification with a litmus paper. Add 10 mL of carbon tetrachloride and 2 mL of 0.06 M CuSO$_4$ to it. Shake it for 1 min and take out the yellow layer. Measure

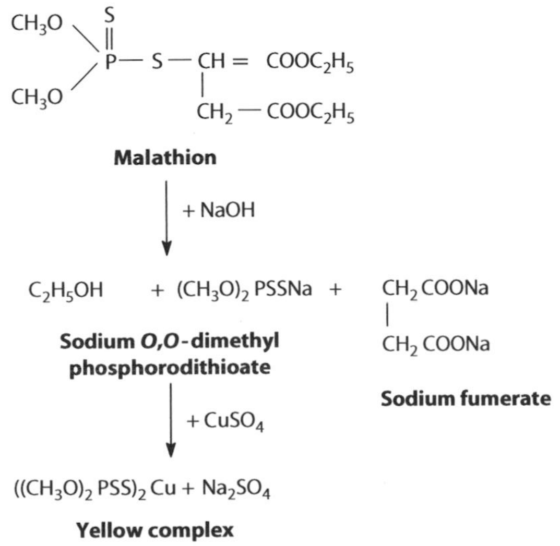

FIGURE 13.12 Formation of yellow complex with Malathion and copper sulfate.

the absorbance of yellow solution at 420 nm against a blank containing the above reagents except malathion. Make a calibration curve by plotting absorbance versus concentration of malathion.

Take a sample of soil contaminated with malathion, dry in air, grind, and mix with 0.25 g of activated charcoal and 0.50 g of Florisil. Pack the admixture in a glass column (50 × 1.5 cm i.d.) containing a 3 cm layer of sodium sulfate over a cotton plug. Elute the column with 200 mL of ethanol for a minimum period of 4 h.

Collect the eluent, transfer it into a 500 mL distillation flask, and distill off the solvent at 80°C. Dissolve the dried residues in 10 mL of ethanol and transfer into a 125 mL separating funnel. Wash the distillation flask with the same volume of ethanol. Transfer the washings into the separating funnel and add 2 mL of 3 M NaOH into it. Mix thoroughly and acidify the contents with 4 M HNO_3 using litmus paper. Add 10 mL of carbon tetrachloride and 2 mL of 0.06 M copper sulfate solution, shake well, and take out the yellow layer. Measure the absorbance as above and calculate the concentration of malathion by using the calibration curve.

13.13.4 RESULTS AND DISCUSSION

The aforesaid single-step extraction and cleanup technique removes most of the interfering coextractives. Malathion can be analyzed in the eluent, spectrophotometrically.

The spectrophotometric method for organophosphates given by Orloski [17] was used in the present study. It is a convenient and sensitive method for analyzing a large number of samples. At constant temperature Beer's law is followed between 1.0 and 3.0 mg of malathion, and below 1.0 mg the absorptivity decreases.

13.14 DETERMINATION OF METHYL PARATHION RESIDUES IN FOOD GRAINS AND VEGETABLES

13.14.1 THEORY AND SIGNIFICANCE

Methyl parathion, O,O-dimethyl O-(4-nitrophenyl) phosphorothionate is also known as dimethyl phosphorothionate. It is pale yellow liquid when pure. It has a b.p. of 157°C–162°C at 0.6 mmHg, m.p. 6°C, and v.p. 3.78×10^{-5} mmHg at 20°C. It is soluble in water (20–25 ppm) and petroleum hydrocarbons.

It is used for control of many insects of economic importance, being especially effective for boll weevil control. It is also effective against green leafhopper, stem borers, armyworm, cutworm, rice caseworms, leaf folders, and rice bugs. It is toxic to mammals (acute oral LD_{50} for rats 14–24 mg kg^{-1}), birds, fish, and bees.

The solution of methyl parathion is reduced with zinc and acetic acid–hydrochloric acid mixture to yield amino group. The amino group is titrated with standard nitrite solution [9,18,19]. At the end point a piece of potassium iodide starch paper turns intense blue–black due to the reaction of excess nitrite solution (Figure 13.13).

13.14.2 REAGENTS

Prepare a mixture of acetic acid and hydrochloric acid by mixing glacial acetic acid and concentrated hydrochloric acid (9:1). Prepare a standard solution of sodium nitrite (0.1 M) by dissolving 6.909 g of sodium nitrite in water. Fill up the solution to the mark in a 1 L standard flask with distilled water.

13.14.3 PROCEDURE

Take an accurately weighed sample containing 0.8 to 1.0 g of methyl parathion in a 250 mL separator funnel containing 100 mL of benzene and 1.0 to 2.0 g of anhydrous sodium sulfate. Add 25 mL

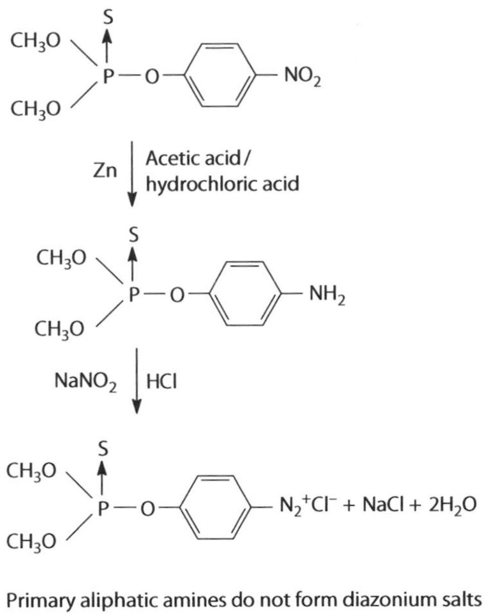

FIGURE 13.13   Reduction of methyl parathion with zinc, acetic acid, and hydrochloric acid.

of 10% chilled sodium bicarbonate solution and very carefully as well as slowly extract the free *p*-nitrophenol. Drain out the clear yellow colored aqueous layer. Repeat this process until the total extraction of the *p*-nitrophenol, that is, absence of yellow product.

Transfer the benzene layer to a 500 mL round-bottom flask quantitatively. Remove the solvent by distillation, and add 35 mL of acetic acid–hydrochloric acid mixture and 3.0 g of zinc dust. Cover the flask with a funnel and heat the contents on a steambath for 30 min. Add 25 mL of concentrated hydrochloric acid and continue the heating until the zinc is completely dissolved. Cool and transfer the contents of the round-bottom flask into a conical flask quantitatively by washing with distilled water. Add 5 g of potassium bromide, cool by adding ice, and stir with the help of a magnetic stirrer. Titrate the mixture with standard sodium nitrite solution (0.1 N) as rapidly as the external indicator detection permits. Dip a glass rod into the solution to be tested and then touch it to a piece of potassium iodide starch paper. If the paper turns intense blue–black within 15–20 s the end point has come. Calculate the concentration of methyl parathion:

$$\% \text{ Methyl parathion in the given sample} = \frac{v \times N \times 26.30}{w}$$

where
v is the volume (mL) in of 0.1 N sodium nitrite consumed by the amine
w is the weight (g) of the given sample

13.14.4   RESULTS AND DISCUSSION

Methyl parathion is separated from the impurity, *p*-nitrophenol, by treating the benzene solution with cold (10°C) 10% sodium bicarbonate solution. The benzene solution of methyl parathion is reduced to give an amino group with zinc in the presence of acetic acid and hydrochloric acid. The amino group compound so obtained is titrated with standard sodium nitrite solution.

13.14.5 SPECTROPHOTOMETRIC METHOD

The color reaction [14] based on coupling of diazotized *o*-toludine, diazotized sulfanilic acid, and diazotized sulfanilimide with carbaryl or α-naphthol has been used for spectrophotometric determination of latter compounds. This color reaction can also be used for the determination of methyl parathion as follows: Take appropriate volume (10–100 μg) of colorless diazotized product of methyl parathion (Figure 13.13), add 2 mL of 1% α-naphthol, place the solution for 5–10 min with occasional shaking, and then add 0.5 mL of 0.1 N KOH. Make up the total volume to 5 mL with distilled water, mix thoroughly, and measure the absorbance at 520 nm against a blank. Use a calibration curve for calculating the concentration of methyl parathion in the given sample. The color is stable for 30 min.

13.15 SPECTROPHOTOMETRIC DETERMINATION OF PARAQUAT DICHLORIDE PESTICIDE

13.15.1 THEORY AND SIGNIFICANCE

Bipyridylium herbicide, Paraquat dichloride, 1,1'-dimethyl-4,4'-bipyridylium dichloride is a colorless, crystalline, and hydroscopic solid. It decomposes at about 300°C. It is used for broad-spectrum control of broad-leaved weeds and grasses in fruit orchards (including citrus), plantation crops (bananas, coffee, cocoa palms, etc.), vines, olives, tea, etc. It is also used for general weed control and control of aquatic weeds. It is soluble in water (700 g L^{-1}) at 20°C, sparingly soluble in lower alcohols, and practically insoluble in most other organic solvents. It is stable in neutral and acidic media but readily hydrolyzed in alkaline media. It is photochemically decomposed by UV irradiation in an aqueous solution. Its aqueous solution and soluble concentrate are available in the market. It is irritating to the skin and eyes. Paraquat dichloride is toxic to mammals (acute oral LD$_{50}$ for rats 150 mg kg^{-1}), birds, and fish. It is not toxic to bees. Paraquat dichloride (Figure 13.14a) reacts [9,19] with sodium dithionite (Figure 13.14b) in alkaline medium to yield a blue color, which can be measured quantitatively at 600 nm by using a spectrophotometer (Figure 13.14).

13.15.2 REAGENTS

Prepare 1% solution of sodium dithionite in 0.1 N NaOH (make fresh solution to be used in 1 h). Dry Paraquat dichloride at 110°C for 5 h. Dissolve 0.1728 g of it in water and fill up the volume to 500 mL with water. One milliliter of this solution contains 0.035 g Paraquat dichloride.

13.15.3 PROCEDURE

Weigh a sample containing 0.8 g of Paraquat dichloride and transfer it to a 250 mL volumetric flask. Dissolve the sample in water and make total volume up to the mark. Name this solution as A. Dilute 10 mL of this solution to 250 mL using a standard flask and name this solution as B. Pipette out 10 mL of solution B and transfer it separately in five different 100 mL standard flasks. Add 6.0, 8.0, 10.0, 12.0, and 14.0 mL of standard solution of Paraquat dichloride in these flasks. Dilute the contents of each flask to about 80 mL. Then add 10 mL of sodium dithionate solution to all the five

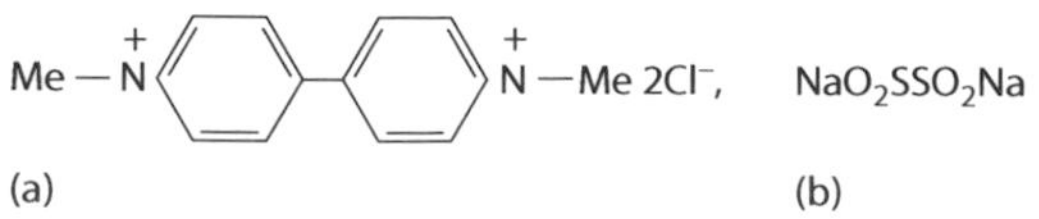

$$\text{Me} - \overset{+}{\text{N}} \underset{}{\bigcirc\!\!\!-\!\!\!\bigcirc} \overset{+}{\text{N}} - \text{Me } 2\text{Cl}^-, \qquad \text{NaO}_2\text{SSO}_2\text{Na}$$

(a)   (b)

FIGURE 13.14   (a) Paraquat dichloride and (b) sodium dithionite.

flasks separately and finally fill up the solution to the mark with water. Mix thoroughly and measure the optical density immediately at 600 nm. Compare the optical density with permanent reference standard.

13.15.4 RESULTS AND DISCUSSION

The optical density should be measured within 2 min to avoid the error due to fading of color. Make a calibration curve with the standard solution of Paraquat dichloride.

13.15.5 CALCULATION

Calculate the concentration of paraquat dichloride by using the calibration curve.

13.16 VOLUMETRIC DETERMINATION OF THIRAM FUNGICIDE

13.16.1 THEORY AND SIGNIFICANCE

Thiram is tetramethyl thiuram disulfide. It is a white crystalline powder, m.p. 155°C–156°C (technical grade 146°C). It has a negligible vapor pressure at room temperature and is slightly soluble in water (30 mg L^{-1}); it is soluble in ethanol (>10 mg L^{-1}), acetone (80 mg L^{-1}), and chloroform (230 mg L^{-1}). It is stable under ordinary conditions but it is decomposed by acids. It is noncorrosive except to iron and copper. It is irritating to skin, eyes, and the respiratory tract.

It is used as fungicide with protective action and as wildlife repellent. It is commonly used for the control of *Botrytis* on lettuce, soft fruits, grapes, vegetables; ornamental scab and *Gloeosporium* rot on apples and pears; rusts on ornaments; stem diseases of cucurbits; *Monilia* spp., shot-hole, and witches broom in stone fruits; peach leaf curl on peaches, and so on. Thiram is toxic to mammals (acute oral LD_{50} for rats 864 mg kg^{-1}) and fish. It is not toxic to bees. Thiram yields a salt of dimethyl amine on digestion with dilute mineral acid [9,20]. After boiling the volatile sulfur compounds, the amine is liberated by steam distillation under alkaline condition and estimated by titration with a standard acid.

$$(CH_3)_2N - \overset{\overset{\displaystyle S}{\|}}{C} - S - S - \overset{\overset{\displaystyle S}{\|}}{C} - N(CH_3)_2 + 2H_2SO_4 \longrightarrow 2CH_3NH_3HSO_4 + 2CS_2$$

$$CH_3NH_3HSO_4 + 2NaOH \longrightarrow CH_3NH_2 + Na_2SO_4 + 2H_2O$$

$$CH_3NH_2 + HCl \longrightarrow CH_3NH_3Cl$$

13.16.2 REAGENTS

Dissolve 20 g of boric acid in 1 L of water containing 12 mL of 0.05% methyl red and 0.7 mL of 0.35% methylene blue solutions. Prepare 0.2 N hydrochloric acid in water and standardize it. Prepare 0.05% methyl red indicator.

13.16.3 PROCEDURE

Weigh accurately sufficient quantity of the prepared sample containing 0.5 to 0.8 g of the active ingredient into a 150 mL round-bottomed flask and add to it 50 mL of 5 M H_2SO_4. Fix a water-cooled refluxed condenser into the flask and heat the contents slowly to boiling. Add some glass beads into the flask to avoid spurting of the reaction mixture. Allow to reflux for 1 h, cool, and transfer the contents quantitatively to the distillation flask of the steam distillation assembly. Conduct steam distillation until 200 mL of distillate is collected. Stop the distillation and reject the distillate, which contains carbon disulfide. Add a few drops of methyl red indicator to the contents of the distillation flask followed by 7.5 M NaOH solution until alkaline, add 10 mL in excess, and steam distill out the liberated amine. Collect about 250 mL of the distillate in the receiver containing 50 mL of boric acid solution and titrate with 0.2 N HCl to violet the end point. Carry out a blank test on the reagents.

A simple apparatus consists of a 1 L flat-bottom flask provided with a long glass safety tube dipping well below the surface of water and passing into a rubber-stoppered, round-bottomed flask through a bent tube. The flask is also provided with a separator funnel and a steam outlet tube, which is connected to a water-cooled upright condenser through a splash head. The distillate from the condenser is collected in a 500 mL conical flask by a piece of rubber tubing, which dips just below the liquid surface in the conical flask placed in a cold bath.

13.16.4 CALCULATION

$$\% \text{ Thiram in the given sample} = \frac{v \times 1.2044}{w}$$

where

v is the volume (mL) of 0.2 N HCl consumed by the amine
w is the weight of the given sample

13.17 SIMULTANEOUS DETERMINATION OF TRICYCLAZOLE, THIRAM, AND FOLPET IN TOMATO BY HPTLC

13.17.1 THEORY AND SIGNIFICANCE

Tricyclazole, 5-methyl-1,2,4-triazolo [3,4-b] benzothiazole (Figure 13.15a), is a crystalline solid, which melts at 187°C–180°C. Its solubility is 1.6 g L^{-1} in water, 10.4 g L^{-1} in acetone, 25 g L^{-1} in methanol, and 2.1 g L^{-1} in xylene at 25°C. It is a systemic fungicide, absorbed rapidly by the roots, with translocation through the plant. It is used to control rice blast (*Pyricularia oryzae*) in transplanted and direct seeded rice and can be applied as a flat drench, transplant root soak, foliar application, or seed treatment.

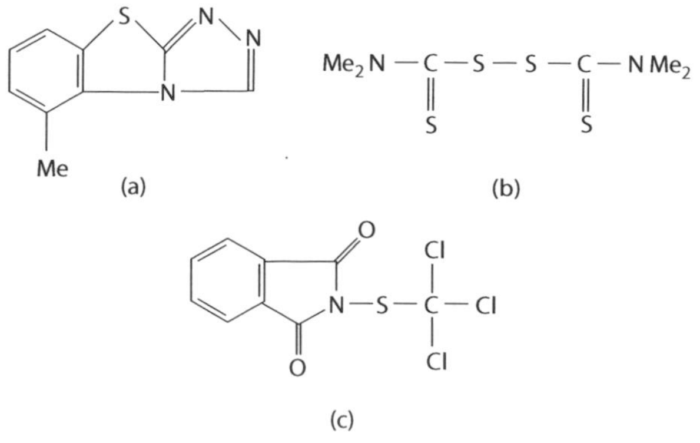

FIGURE 13.15   (a) Tricyclazole, (b) Thiram, and (c) Folpet.

Thiram, tetramethylthiuram disulfide (Figure 13.15b), exists in the form of colorless crystals of m.p. 155°C–156°C. It is used to control root rot and *Botrytis* in asparagus and snow mould and dollar spot on turf. It is used in seed treatment for control of damping off and emergence diseases in vegetables, legumes, and so on. It is also used as a repellent for birds, rabbits, rodent, and deer.

Folpet, *N*-(trichloromethylthio) phthalimide (Figure 13.15c), exists as colorless crystals (technical 90%–95%, yellow powder) of m.p. 177°C (with decomposition). It is stable in dry state, slowly hydrolyzed by moisture at room temperature, and rapidly hydrolyzed in concentrated alkalis at elevated temperature. It is foliar fungicide with protective action. Folpet is used to control downy mildews, powdery mildews, leaf spot diseases, scab, *Gloeoporium* rots, *Botrytis*, *Alternaria*, *Pythium*, and *Rhizoctonia* spp. in pome fruit, stone fruit, soft fruit, citrus fruit, vines, olives, hops, potatoes, lettuce, cucurbits, onions, leeks, celery, tomatoes, and ornamentals. It is toxic to mammals (acute oral LD_{50} for rats >10,000 mg kg^{-1}), birds, and fish. It is not toxic to bees. Tomato is one of the popular vegetables in the world. The monitoring of multiresidues of pesticides in vegetables is very important, because it involves public health, environmental monitoring, and foreign trade aspects. Reversed-phase high-performance liquid chromatography (RP-HPLC) with UV detection, liquid chromatography–electrospray ionization–tandem mass spectrometry, and GC with different detectors has usually been used for the analysis of pesticide residues in tomato. Simple, quick, and inexpensive TLC is used for screening of pesticide residues. Now it has been replaced by highly reproducible, efficient, high-performance thin-layer chromatography (HPTLC). Obviously, HPTLC is an improved form of TLC and now it is used for the detection as well as quantification of pesticide residues. In this experiment a HPTLC method [21] for multiresidue determination of the three fungicides is described. Fungicides are extracted and concentrated by using the cleaning ability of thin layer and no additional purification procedure.

13.17.2 Reagents

Prepare stock calibration solutions (2 g L^{-1} of folpet, 1 g L^{-1} of thiram, and 0.6 g L^{-1} of tricyclazole) in methanol and store in glass-stoppered bottles at 4°C. Dilute 1 mL of each of the stock solutions in 10 mL. Prepare the admixtures of the three fungicides by mixing them in a given ratio. Use fresh solutions daily.

13.17.3 Procedure

Sample preparation: Spike 10 g of tomato sample with known concentration of the fungicide/fungicides solution into a 150 mL conical flask with stopper. Mix the fortified samples after equilibrium with 50 mL of acetone/dichloromethane (1:1, v/v) and 3 g anhydrous sodium sulfate, and then extract them by mechanical vibration at room temperature for 30 min. Filter the extract through a glass funnel containing 5 g anhydrous sodium sulfate into a 150 mL round-bottomed flask. Repeat the extraction with fresh 50 mL of the solution, filter it, collect the filtrate in the same flask, and evaporate the solvent at 34°C–40°C to dryness. Dissolve the residue in 3 mL methanol, and concentrate the solution to 2 mL by using nitrogen evaporator at 35°C.

Chromatography: Apply the analyte by means of a Linomat applicator (Camag Muttenz, Switzerland) equipped with a 100 µL syringe as 6.0 mm bands to the 10 × 20 cm glass-backed silica gel 60 F 254 HPTLC plates (E. Merck Germany) previously prewashed with methanol and activated at 110°C for 30 min. Develop the plates in hexane/acetone (6:4, v/v) to 70 mm by linear ascending technique in an unsaturated glass twin-trough Camag chamber. Remove the mobile phase in a stream of air and measure the absorbance at 235 nm densitometrically using a deuterium lamp. Record the peak heights for all the tracks.

13.17.4 Results and Discussion

Hexane/acetone (6:4, v/v) is the optimum mobile phase for the development of chromatogram. It gives dense and compact spots and well-resolved peaks corresponding to R_f values 0.26, 0.65, and 0.77 of tricyclazole, thiram, and folpet, respectively.

TABLE 13.8
Regression Equations and Detection Limits of Fungicides

Fungicide	R_f	Regression Equation	Correlation Coefficient (%)	Detection Limit (ng)
Tricyclazole	0.26	$Y = 0.3209x + 22.31$	0.9909	12
Thiram	0.65	$Y = 0.3334x + 10.12$	0.9931	30
Folpet	0.77	$Y = 0.2462x + 22.32$	0.9980	40

Source: Fan, W. et al., *J. Planar Chromatog.*, 20(6), 419, 2007. With permission.

Construct the calibration curve for each fungicide by plotting the peak height (*y*-axis) against the concentration of the fungicide (*x*-axis). The linear regression equations for the selected fungicides are given in Table 13.8. The limit of detection is also given in this table.

Recoveries of tricyclazole, thiram, and folpet from spiked tomato at fortification levels 0.12–3.0 mg kg^{-1}, 0.2–5.0 mg kg^{-1}, and 0.4–10.0 ng kg^{-1}, respectively, are found to be 67.66%–73.92%, 181.98%–92.62%, and 85.99%–98.02%, respectively, and relative standard deviations (RSD) of which are 4.63%–22.06%, 0.31%–7.23%, and 0.13%–4.33%, respectively. The precision and accuracy of the method under study are generally fit for the analysis of fungicide residues in tomato. The method is specific for the determination of fungicides in tomato. No interfering substance on the chromatoplate is found in samples of tomato.

13.18 VOLUMETRIC DETERMINATION OF ZIRAM IN PESTICIDE FORMULATIONS

13.18.1 THEORY AND SIGNIFICANCE

Ziram is zinc dimethyl dithiocarbamate (Figure 13.16a). It is a white powder, m.p. 240°C, has negligible vapor pressure, is slightly soluble in water, moderately soluble in acetone, and soluble in dilute alkali and carbon disulfide. It is toxic to mammals (acute oral LD$_{50}$ for rats 1400 mg kg^{-1}) and fish. It is not toxic to bees. It is stable under ordinary conditions, but it is decomposed by acids, and it is noncorrosive except to iron and copper.

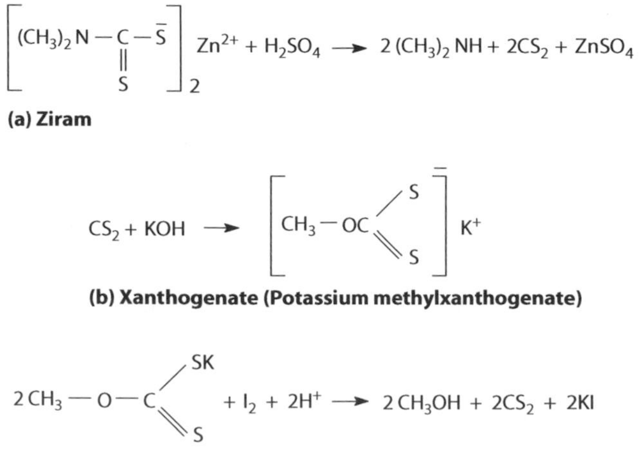

(a) Ziram

(b) Xanthogenate (Potassium methylxanthogenate)

FIGURE 13.16   (a) Ziram and (b) xanthogenate (potassium methylxanthogenate).

When ziram is digested with sulfuric acid and the liberated carbon disulfide is allowed to react with alcoholic potassium hydroxide, potassium methylxanthogenate (Figure 13.16b) is formed. The latter so produced is titrated with iodine [22]. The following reaction mechanism may be proposed:

13.18.2 REAGENTS

Prepare 10% lead acetate solution (w/v) in double-distilled water and 2 N potassium hydroxide solution by dissolving 112 g KOH pellets in 500 mL anhydrous methanol. Filter through cotton and add additional 500 mL anhydrous methanol. Dissolve 20 g of iodate-free potassium iodide in 30–40 mL of distilled water in a glass-stoppered 1 L graduated flask. Weigh about 12.7 g of AR or resublimed iodine on a watch glass on a rough balance (never on an analytical balance because of the iodine vapors) and transfer it by means of a small funnel into the concentrated potassium iodide solution. Insert the glass stopper into the flask, and shake in the cold until all the iodine has dissolved. Allow the solution to acquire room temperature, and fill up to the mark with distilled water. The iodine solution is best preserved in small glass-stoppered bottles. These bottles should be filled completely and kept in a cool and dark place.

13.18.3 PROCEDURE

Weigh accurately a sufficient quantity of the material containing 0.3 g of ziram and transfer it into a 200 mL reaction flask and connect it to two absorbers, one of them containing lead acetate solution to precipitate sulfide and the others containing solution of potassium hydroxide in methanol. The temperature of the absorbent containing the potassium hydroxide solution should be maintained at 25°C ± 2°C throughout the process of determination. Apply suction to the system and adjust the bubbling rate of 10–12 bubbles per second in the two absorbers. Add 50 mL of sulfuric acid (20%) through an inlet tube and boil the contents. Continue boiling and suction and quantitatively transfer the contents of the potassium hydroxide absorber into a 500 mL flask and wash the flask with distilled water, taking care not to use more than 100 mL of the same.

Cool the flask and neutralize accurately with 30% acetic acid solution using phenolphthalein as the indicator. Add starch indicator solution and titrate against 0.1 N iodine solution until the color changes, that is, the blue color appears.

13.18.4 CALCULATION

$$\% \text{ Ziram in the given sample} = \frac{v \times 1.529}{w}$$

where
v is the volume (mL) of 0.1 N iodine solution
w is the weight (g) of the given sample

13.18.5 RESULTS AND DISCUSSION

The procedure above is much more sensitive because of the use of a solution of starch as indicator. Starch reacts with iodine in the presence of iodide to form an intensely blue-colored complex, which is visible at a very low concentration of iodine. The sensitivity of the color reaction is such that a blue color is visible when the iodine concentration is 2×10^{-5} M and the iodide concentration is greater than 4×10^{-4} M at 20°C. The color sensitivity decreases with increasing temperature of the solution; thus at 50°C it is about 10 times less sensitive than at 25°C. The sensitivity decreases on the addition of solvents, such as ethanol. No color is obtained in solutions containing 50% ethanol or more. It cannot be used in a strongly acid medium because of hydrolysis of the starch solution. The procedure under study is also applicable to determine other dithiocarbamate fungicides such as zineb.

13.19 SIMPLE AND SELECTIVE SPECTROPHOTOMETRIC DETERMINATION OF ZIRAM, ZINEB, AND FERBAM IN COMMERCIAL SAMPLES AND FOODSTUFFS USING PHENYLFLUORONE

13.19.1 THEORY AND SIGNIFICANCE

Ziram, zineb, and ferbam (Figure 13.17) are well-know dithiocarbamate fungicides, which are widely used against a variety of plant pathogenic fungi.

Ziram is extensively used on vegetables and also applied to some fruit crops. It is the most stable of the metallic dithiocarbamates. It is nonphytotoxic except for zinc-sensitive plants. It does not build up in the soil and is rapidly decomposed by weathering.

Zineb is used on a variety of fruits and vegetables, especially on potato seed pieces and tomatoes. Nabam has been replaced to considerable extent by this compound. Zineb also results from combining nabam with zinc sulfate in the spray tank. Zineb, zinc ethylenebis(dithiocarbamate) (polymeric), is a pale yellow powder, which decomposes on heating. It is soluble in pyridine and carbon disulfide, slightly soluble in water (10 mg L^{-1}), and insoluble in common organic solvents. It is toxic to mammals (acute oral LD$_{50}$ for rats is 75,200 ng kg^{-1}) and fish. It is not toxic to bees. Ferbam, iron tris(dimethyldithiocarbamate), is a black powder, which decomposes above 180°C. It is stable in closed containers. It tends to decompose on exposure to moisture and heat and prolonged storage. It is soluble in organic solvents, such as chloroform, pyridine, acetonitrile, and acetone. Its solubility in water is 130 mg L^{-1}. It is toxic to mammals (acute oral LD$_{50}$ for rats is >17,000 mg kg^{-1}) moderately toxic to fish, and not toxic to bees.

The principal uses of ferbam are in the control of apple scab and cedar apple rust and tobacco blue mold. It is also applied as a protective fungicide to other crops against many fungus diseases.

The following spectrophotometric method [23] is relatively simple, rapid, sensitive, and selective versus those reported in the literature (volumetry, polarography, HPLC, and spectrophotometry). In this method ziram and zineb are converted into a zinc-phenylfluorone complex and ferbam into an

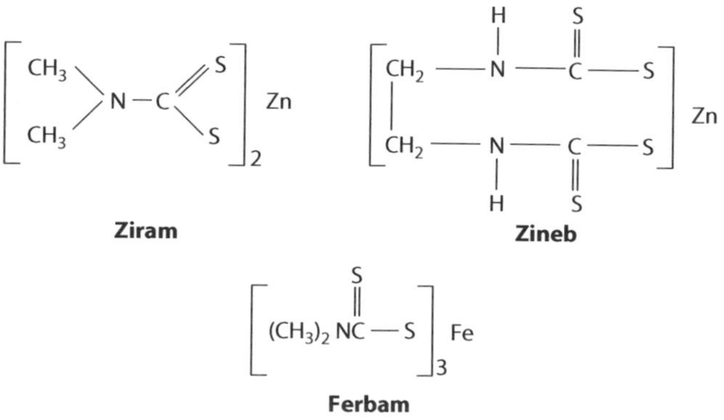

FIGURE 13.17   Ziram, zineb, and ferbam.

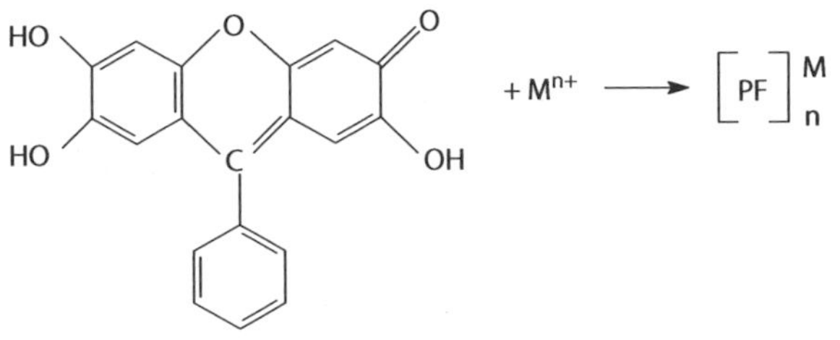

FIGURE 13.18   Metal-phenylfluorone (PF) complex.

iron-phenylfluorone complex. These metal-phenylfluorone [2,3,7-trihydroxy-9-phenyl-6-fluorone (PF)] complexes are suitable for spectrophotometric determination (Figure 13.18). The ratio of PF and dithiocarbamate has been found to be 1:1. The reaction mechanism is given below:

13.19.2   Reagents

Prepare the stock solutions of pure (100%) ziram and zineb (1 g L^{-1}) by dissolving 100 mg of this reference material in 100 mL of 0.1 M NaOH. Prepare a stock solution of pure (100%) ferbam (1 g L^{-1}) by dissolving 100 mg of this reference material in 100 mL of acetonitrile and further diluting the resultant solution with acetonitrile. Dissolve 0.5 g L^{-1} of pure PF in ethanol containing 0.5–1.0 mL of concentrated hydrochloric acid. Dissolve (5 g L^{-1}) of cetylpyridinium bromide (CPB) in distilled water. Prepare pyridine solution (25%, v/v) in distilled water. Prepare NaOH solution (1×10^{-3} M) by dissolving 0.04 g L^{-1} of it in distilled water. Prepare boric acid buffer solution of pH 8.0 as usual.

13.19.3   Procedure

Calibration curve: Take the known volume of the sample solutions containing 2–30 μg of ziram, 2–27 μg of zineb, and 30–100 μg of ferbam in different 10 mL standard flasks. Add the following reagents into these solutions: (a) 2.0 mL of buffer solution (pH 8.0) in ziram and zineb while 2.0 mL of 0.1 M NaOH in ferbam, (b) 1.5 mL of PF in ziram and zineb and 2.0 mL of PF in ferbam, (c) 2.0 mL each of CPB and pyridine in ziram and zineb, and (d) volume of all the three (ziram, zineb, and ferbam) solutions was separately made up to 10 mL with distilled water. Mix the reaction mixture thoroughly, allow to stand for 5 min, and measure the absorbance at 570 nm in the case of ziram and zineb and at 600 nm in the case of ferbam against a reagent blank prepared under similar conditions. Prepare a calibration curve by plotting absorbance versus concentration of the analyte.

Determination of ziram, zineb, and ferbam in crops: The method was applied for the determination of ziram and zineb in cabbage and potatoes and ferbam in fortified wheat grains and apples.

Take the known volumes (corresponding to 0.40–4.00 mg kg^{-1}, Table 13.9) of ziram (0.1 g L^{-1} in acetonitrile), zineb (0.1 g L^{-1} in dimethyl sulfoxide), and ferbam (0.1 g L^{-1} in acetonitrile) solutions and crush with 20 g of crop sample with pestle and mortar and stir mechanically with 100 mL chloroform for 1 h in the case of ziram and ferbam and with 50 mL of dimethyl sulfoxide in the case of zineb. Filter the mixture and wash the residue in the funnel with chloroform or dimethyl sulfoxide (3×10 mL). Heat the extract to reduce to 20 mL in a water bath at 70°C–90°C. Remove the solvent totally by blowing hot air. Dissolve the residues of ziram and zineb in 0.1 M NaOH and those of ferbam in acetonitrile and determine the contents by the procedure given under the calibration curve. Analyze the untreated sample and make the necessary corrections if required.

TABLE 13.9
Determination of Ziram, Zineb, and Ferbam in Crops

| Dithiocarbamate | Crop | Amount of Dithiocarbamate (mg kg⁻¹) | | |
		Taken	Found	RSD %
Ziram	Cabbage	0.40	0.396	2.5
		0.80	0.793	2.1
		1.20	1.195	2.0
	Potatoes	0.40	0.389	2.4
		0.80	0.785	2.3
		1.20	0.193	2.1
Zineb	Cabbage	0.40	0.387	2.4
		0.80	0.798	2.4
		1.20	1.197	2.2
	Potatoes	0.40	0.391	1.9
		0.80	0.790	1.8
		1.20	1.197	1.6
Ferbam	Wheat	1.00	0.996	2.3
		2.00	1.990	2.3
		4.00	4.000	1.9
	Apples	1.00	0.991	2.9
		2.00	1.985	2.5
		4.00	3.996	2.4

Source: Malik, A.K. et al., *J. Environ. Monit.*, 2, 367, 2000. With permission.

Note: Each result is the average of five sets of experiments.

13.19.4 RESULTS AND DISCUSSION

The results obtained show that this method gives the same color with ziram as well as zineb, so it fails to distinguish between ziram and zineb. However, the following test can be used to distinguish between ziram and zineb. When a sample is treated with 2 mL of 2% sodium molybdate solution in the presence of 4 M H_2SO_4, the formation of a yellow color in cold shows the presence of ziram. The yellow product can be extracted in isobutyl methyl ketone. When the residue is treated again with 2 mL of 2% sodium molybdate solution and heated to boiling for about 5 min, the formation of a blue color indicates the presence of zineb.

The method is also applicable for determining ziram, zineb, and ferbam in commercial samples. The formulation is dissolved in acetonitrile for ferbam and in 0.1 M NaOH for ziram and zineb and is determined by using the procedure given under the calibration curve. The results obtained (Table 13.9) are in good agreement with those obtained by spectrophotometry for the yellow complex of copper(II) dithiocarbamate.

ACKNOWLEDGMENT

Hamir Singh Rathore is thankful to the All India Council of Technical Education, New Delhi, India for the award of Emeritus Fellowship.

REFERENCES

1. Begum, T. 1993. Studies on planar chromatographic methods for use in pesticides analysis. PhD dissertation, Aligarh Muslim University, Aligarh, India.
2. Amador-Hernandez, J., Velazquez-Manzanares, M., Gutierez-Ortiz, M.D.R., Hernandez-Carlos, B., Peral-Torres, M., and Lopez-de-Alba, P.L. 2005. Simultaneous spectrophotometric determination of atrazine and dicamba in water by partial least squares regression. *J Chil Chem Soc* 50(20): 461–464.
3. Liu, Y., Chen, G.S., Chen, Y., and Lin, J. 2005. Inclusion complexes of azadirachtin with native and methylated cyclodextrins: Solubilization and binding ability. *Bioorg Med Chem* 13(12): 4037–4042.
4. Hardon, H.J., Brunink, H., and Van der Pol, E.W. 1960. The determination of 1-naphthyl methylcarbamate (seven) residues in apples. *Analyst* 85: 187–189.
5. Yuen, S.H. 1965. Spectrophotometric determination of carbaryl in insecticide formulations. *Analyst* 90: 569–571.
6. Janghel, E.K., Rai, J.K., Rai, M.K., and Gupta, V.K. 2007. A new sensitive spectrophotometric determination of cypermethrin insecticide in environmental and biological samples. *J Braz Chem Soc* 18(3): 1–5.
7. Sandell, E.B. 1959. *Colorimetric Determination of Traces of Metals*. Interscience Publishers Inc., New York.
8. Saxena, S.K. 1989. Studies on analysis of some organic pollutants in water. PhD dissertation, Aligarh Muslim University, Aligarh, India.
9. Chopra, S.L. and Kanwar, J.S. (eds.) 1991. Pesticides. In *Analytical Agriculture Chemistry*, 4th edition, Kalyani Publisher, New Delhi, pp. 382–383.
10. Ramulu, U.S.S. 1995. Chemistry of insecticides and fungicides. Mohan Primilani for Oxford & Publishing Co. Pvt. Ltd., New Delhi.
11. Chopra, S.L. and Kanwar, J.S. (eds.) 1991. Pesticides. In *Analytical Agriculture Chemistry*, 4th edition, Kalyani Publisher, New Delhi, pp. 417–418.
12. Rathore, H.S. and Shafiullah. 2005. Unpublished work.
13. Glass, R.L. 1981. Colourimetric determination of glyphosate in water after oxidation to orthophosphate. *Anal Chem* 53: 921–923.
14. Bassett, J., Denney, R.C., Jeffery, G.H., and Mendham, J. 1978. *Vogel's Textbook of Quantitative Inorganic Analysis*. 4th edtion, Longman, London, p. 756.
15. Gupta, S. 1987. Separation, detection and estimation of some form chemicals. PhD dissertation, Aligarh Muslim University, Aligarh, India.
16. Singh, K.K. 1994. Physico-chemical studies on Adsorption and persistence of some organophosphate and carbamate pesticides in soils. PhD dissertation, Aligarh Muslim University, Aligarh, India.
17. Hill, A.C. 2006. Nature of the yellow copper complex produced in certain analytical methods for the determination of malathion. *J Sci Food Agric* 20(1): 4–7.
18. Rangaswamy, J.R. and Majumdar, S.K. 1974. Colorimetric method for estimation of carbaryl and its residues on grains. *J Assoc Off Anal Chem* 57(3): 592–594.
19. Sharma, R. 1995. Development of simple and inexpensive methods for the analysis of some pesticides. PhD dissertation, Aligarh Muslim University, Aligarh, India.
20. Rathore, H.S. and Shafiullah. 2006. Unpublished work.
21. Fan, W., Yue, Y., Tang, F., and Cao, H. 2007. Use of HPTLC for simultaneous determination of three fungicides in tomatoes. *J Planar Chromatog* 20(6): 419–421.
22. Varshney, G. 2007. Detection, determination and synthesis of some dithiocarbamate fungicides. PhD dissertation, Aligarh Muslim University, Aligarh, India.
24. Malik, A.K., Kapoor, J., and Rao, A.L.J. 2000. Simple and sensitive spectrophotometric determination of ziram, zineb and ferbam in commercial samples and foodstuffs using phenylfluorone. *J Environ Monit* 2: 367–371.

14 Recent Trends in Sample Preparation for Pesticide Analysis

Chanbasha Basheer, Suresh Valiyaveettil, and Hian Kee Lee

CONTENTS

14.1   INTRODUCTION

Pesticides are important classes of persistent organic pollutants (POPs) that are commonly found in the environment. The endocrine-disrupting effects of POPs are believed to have a major impact on wildlife. Exposure to POPs in the environment and/or through the food chain may affect the immune system, induce abnormal thyroid function, decrease fertility in humans, and may also cause imposex in birds and marine species, altering the sex ratios of the populations [1]. Most of the pesticides are resistant to physical, chemical, and biological degradation, transported over long distances, and accumulate in both the aquatic and terrestrial food webs [2,3]. The presence of these pollutants in remote areas is a result of two phenomena: (a) long-range transportation through air and water from areas where they are still in use (though it is clear that although they are banned, they are still being used illegally), and (b) traces from previous usage [4]. Pesticides are generally divided into five types, based on their functional groups and toxicity. These are the organophosphorus pesticides, organochlorine pesticides, carbamates, pyrethroids, and herbicides.

The Stockholm Convention on POPs was implemented in the year 2001 to reduce and eliminate POPs, and a great deal of emphasis was placed on evaluating the risks posed by the toxicity of these chemicals. This convention was signed by 151 governments [5,6]. The POPs are mainly transported by volatilization from contaminated soils and water, which is another important source for their presence in the atmosphere [7]. The major removal mechanism of the pesticide compounds from the atmosphere is through either wet or dry deposition. Wet deposition of POPs in both aquatic and terrestrial ecosystems is particularly important in places such as tropical countries that receive abundant rainfall.

Analysis of pesticides at trace levels in the environmental and food samples poses special challenges for the analytical chemists, as the pesticides are present at low concentrations. Generally, they belong to different categories (classes) across a wide range of polarity with high bioaccumulative properties. A wide spectrum of volatility and thermal stability is also involved, necessitating the use of instrumental systems.

Analysis of real samples provides key information about the samples, which is necessary to make decisions about human health and environmental toxicity. Rapid selective precise analysis and accurate results will always be the important objectives for an analytical chemist. Extraction techniques play a major role in analytical chemistry. However, decades-old extraction techniques are still being used in many laboratories. Most of these techniques require the use of large volumes of solvent, and are time-consuming and labor-intensive. In the past decade, novel and simple microscale or miniaturized analytical methods have become increasingly popular.

For the extraction of pesticides in liquid samples, solventless and solvent-minimized techniques have attracted great attention. For solid samples, Soxhlet extraction and accelerated solvent extraction (ASE), supercritical fluid extraction (SFE), and microwave-assisted extraction (MAE) have been the primary techniques employed. This chapter is focused on the recent works on the sample preparation of liquid, solid samples, and novel instrumental approaches on pesticide analysis, which have been published recently.

14.2 LIQUID SAMPLES

Selection of a suitable method of analysis depends on the nature of the analytes and the sample matrix. In case of liquid samples, i.e., for water and wastewater analysis, the analytes can be isolated in various ways, such as extracting the analytes into a liquid phase as in liquid–liquid extraction (LLE). More recently, miniaturized techniques such as single-drop microextraction (SDME) and hollow-fiber membrane liquid-phase microextraction (HFM-LPME) have emerged. Another approach is by trapping the analytes on a solid sorbent, such as in solid-phase extraction (SPE), the more-recent solid-phase microextraction (SPME), stir-bar sorptive extraction (SBSE), and polymer-coated hollow-fiber microextraction (PC-HFME).

14.2.1 LIQUID–LIQUID EXTRACTION

LLE is one of the classical methodologies employed using a separatory funnel and has been officially used in United States Environmental Protective Agency (USEPA) and European commission standard methods. LLE is a convenient and easy-to-use extraction technique for liquid matrices. Immiscible liquid solvents, such as n-hexane, benzene, and ethyl acetate, have been used for the extraction of nonpolar pesticides and dichloromethane, chloroform–methanol, diethyl ether, or other polar solvents are generally used for polar pesticide extraction. This technique is mainly used for the extraction of nonvolatile and semivolatile pesticides. In this method, the water sample is shaken with an appropriate volume of a suitable organic solvent so that migration of organic compounds from the aqueous phase to the organic phase occurs. Sodium chloride or other suitable salt is often added to the mixture to avoid foam formation during extraction to enhance the efficiency of the process. Finally, the analyte containing organic solvent is preconcentrated, usually under reduced pressure. The LLEs can be performed in various ways, such as discontinuous liquid extraction, continuous extraction, online LLE, and countercurrent extraction.

Discontinuous liquid extraction is a multistep and most conventional extraction method. Continuous extraction is performed when distribution constant value is low and the volume of the samples is large. Online LLE methodology is a sensitive, rapid method, which reduces the volume of organic solvent and sample, but introduces more complex instrumentations. Countercurrent extraction is mainly used for complex samples with similar distribution. However, LLE has several drawbacks, such as usage of large volumes of toxic organic solvent, potential loss of analytes, and a considerable time required for extraction. To overcome these drawbacks, miniaturized solvent extraction approaches have been developed in recent years, such as those mentioned earlier.

14.2.2 SOLVENT-MINIMIZED EXTRACTION APPROACHES

A recent trend in analytical chemistry is the development of novel miniaturized extraction procedures for various organic compounds including pesticides. SDME [8], headspace SDME [9], continuous-flow microextraction (CFME) [10], HFM-LPME [11,12], and HFM protected liquid–liquid–liquid extraction (LLLE) [13] are some of these extraction procedures that have been applied for pesticide analysis.

These miniaturized techniques eliminate the disadvantages of conventional extraction methods. They are inexpensive and offer considerable freedom in selecting appropriate solvents (although the choice is not limitless) for the extraction of different analytes. As very little solvent is used, exposure to toxic organic solvent is minimal. The important feature of these procedures is that almost all the extracted organic solvents can be injected into the analytical instrument to improve the detection limits.

14.2.2.1 Single-Drop Microextraction

SDME is a very simple extraction technique in which a single drop acts as the extracting phase. Normally, 0.5–2.5 μL of solvent is used. SDME has received much attention owing to its simplicity. The extracting phase must have a sufficiently high surface tension to form a drop that is exposed to the sample solution [14]. After extraction, the single drop is directly injected into the instrument. SDME can be used in both static and dynamic modes. The aqueous sample solution is agitated with a magnetic stirrer and a fixed volume of the solvent is withdrawn into the syringe. The needle is then inserted through the septum of a sample vial in such a way that its tip is immersed in the sample. The plunger is depressed to expose an organic drop to the stirred solution for a specific period of time. Finally, the single drop is retracted into the microsyringe and transferred into a sample vial for instrumental analysis. In the case of dynamic mode, all these steps are carried out automatically, using a syringe pump. After the first report on organochlorine extraction using SDME by Zhao and Lee [15], there have been many reports on SDME for various target analytes, including pesticides [16]. However, SDME has some limitations. It is not suitable for complex sample matrices and the stability of the solvent is always a major concern. To overcome this, the use of hollow-fiber supported LPME was proposed, which was first reported by Zhu et al. [17] in which the organic solvent is protected by using a short hollow fiber.

14.2.2.2 Hollow-Fiber Membrane Liquid-Phase Microextraction

An alternative concept for LPME based on the use of single, cost-effective, disposable, and porous, hollow fiber made up of polypropylene was introduced recently [9–12,18–21] (see Figure 14.1).

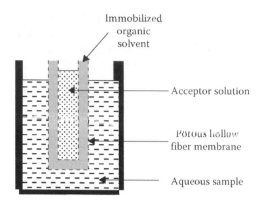

FIGURE 14.1 Basic extraction setup in HFM-LPME.

In the hollow-fiber membrane liquid-phase microextraction (HFM-LPME) device, the extractant solvent is contained within the lumen (channel) of a porous hollow fiber, such that it is not in direct contact with the sample solution. As a result, samples may be stirred or vibrated vigorously without any loss of the solvent during extraction. HFM-LPME is a more robust and reliable alternative of LPME, as the solvent is "protected." In addition, the equipment used is very simple and inexpensive. Polypropylene HFM is normally used for HFM-LPME, because it is highly compatible with a broad range of organic solvents. In addition, with a pore size of approximately 0.2 μm, polypropylene immobilizes a range of organic solvents used in HFM-LPME.

The acceptor solution may be the same organic solvent as that immobilized in the pores, resulting in extraction of the analyte (A) in a two-phase system in which the analyte is collected in an organic phase:

$$A_{sample} \longleftrightarrow A_{acceptor\ organic\ phase}$$

Two-phase LPME may be applied to most analytes with a substantially higher solubility in a water-immiscible organic solvent than that in an aqueous medium. The acceptor solution in this mode is directly compatible with gas chromatography (GC) analysis, whereas evaporation of solvent and reconstitution in an aqueous medium is required for high-performance liquid chromatography (HPLC) or capillary electrophoresis (CE) analysis.

The following method is recommended for the extraction of organochlorine pesticides. A microsyringe (10 μL) equipped with a cone-tipped needle (0.47 mm OD) (Hamilton, Reno, Nevada) is used for the extraction. It is first filled with organic solvent, and the syringe needle is tightly fitted to one end of the open 1.3 cm length of the polypropylene HFM. The latter is impregnated with a 5 mL aliquot of toluene for 3 s to open or dilate the membrane pores. It is then immersed in the 5 mL sample solution and stirred. The syringe plunger is pushed down so that the HFM is filled with solvent originally in the syringe barrel. Extraction takes place between the sample and the solvent-containing porous fiber. After extraction, the solvent in the HFM is withdrawn into the syringe, and the HFM is discarded. The syringe plunger is then pushed down until 1 μL of the extract remains in the syringe. The extract is then injected into a gas chromatography–mass spectrometry (GC–MS) system. This extraction method is highly sensitive with detection limits of subpart per billion ranges. Application of this HFM-LPME for pesticide analysis has been recently reviewed by Lambropoulou and Albanis [22].

14.2.2.3 Solvent-Bar Microextraction

Jiang and Lee have proposed a new methodology [23] for the determination of organic pollutants in aqueous samples, which could achieve high enrichment factor, with good linearity, repeatability, and high recoveries. SBME extractions are carried out as follows: an HFM is carefully cut manually into segments of 2.0 cm length. Each segment is ultrasonically cleaned in acetone and dried in air before use. One end of the membrane is flame-sealed. An 8 μL aliquot of *n*-octanol is withdrawn into the microsyringe with cone-needle tip and introduced into the HFM. The other end is also sealed. The effective length of the solvent bar is ~1.5 cm and the organic solvent volume inside the channel is ~3.0 μL. The "solvent bar" is then placed in the aqueous solution for extraction. The volume of the aqueous solution is typically 10 mL in a 12 mL vial. The solution is stirred (700 rpm) for 10 min during the extraction. Subsequently, the solvent bar is removed from the sample solution, one end of the hollow fiber is trimmed off, and a 1 μL aliquot of the analyte-enriched solvent is withdrawn into the syringe with a flat-cut needle tip. This is directly injected into the GC–MS system for analysis.

14.2.2.4   Liquid–Liquid–Liquid Microextraction

In LLLME, the final extracting (acceptor) solution may be another aqueous phase providing a three-phase system, in which the analytes (A) are extracted from an aqueous sample through an intermediary thin film of organic solvent impregnated in the pores of the fiber wall, and into the final solution that is generally set at a different pH from that of the sample solution:

$$A_{sample} \longleftrightarrow A_{organic\ phase} \longleftrightarrow A_{acceptor\ aqueous\ phase}$$

The two-phase (aqueous-to-organic) system is more suitable for GC, whereas the three-phase LPME system is suitable for HPLC and CE analysis, as the final extract is aqueous. Generally, both the methods, being based on diffusion, are promoted by high partition coefficients. The three-phase system is known as LLLME. This is similar to conventional LLE in which there is an additional, back-extraction step. One application of this technique for pesticide analysis is described as follows.

Phenoxy herbicides are very polar compounds that are highly soluble in water. Therefore, two-phase HFM-LPME and LLE usually afford low extraction recoveries. Wu et al. [13] reported a single-step three-phase LLLME technique for phenoxy herbicide extraction. The extraction procedure is as follows: a 10 mL aliquot of sample solution is first acetified with HCl, and a 7 μL aliquot of the acceptor phase (NaOH) is withdrawn into a syringe. The syringe needle is then inserted into the HFM, and the acceptor solution is introduced into it. The fiber is then immersed in the organic solvent for 10 s for impregnation, to dilate the pores of the HFM. The extraction is performed by holding the HFM in the solution. After extraction, 5 μL of the acceptor solution is withdrawn from the fiber, and injected into the HPLC system.

14.2.3   Sorbent-Based Extraction Techniques

There has been a considerable interest in developing new selective and sensitive extraction methods based on selective sorptive extraction procedures [24]. Here, the analytes are extracted using suitable sorbents coated on a supporting device. The selectivity of the sorbent is an important parameter to be taken into account. Some sorbent-based extraction methods include SPE, SPME, and SBSE that are commercially available. More recently, PC-HFME and micro-solid-phase extraction (μ-SPE) have been developed for pesticide analysis. The extraction procedures and their applications in pesticide analysis in water samples are discussed in the following section.

14.2.3.1   Solid-Phase Extraction

SPE is a classical and conventional extraction method that has been used for a wide range of pesticide analysis. SPE functions on the adsorption principle, in which analytes are trapped on

TABLE 14.1
Commonly Used Commercial Sorbents for SPE

Solid-Support	Phase	Modifications to Surface Range (m²/g)	Surface-Area Range (m²/g)	Particle-Size Range (μ)	Pore-Size Range (Å)	Retention Mechanism
Silica	C18	Octadecyl (polymeric)	450–550	50–60	65–75	RP
Silica	C18	Octadecyl (monomeric)	280–320	50–60	120–140	RP
Silica	C8	Octyl	450–550	50–60	60–75	RP
Silica	PH	Phenyl	450–550	50–60	60–75	RP
Silica	CN	Cyanobutyl	450–550	50–60	60–75	RP + NP
Silica	NH₂	Aminopropyl	450–550	50–60	60–75	RP + IEX
Silica	SCX	Phenylsulfonic acid	450–550	50–60	60–75	IEX
Silica	SAX	Me₂(propyl) ammonium Cl⁻	450–550	50–60	60–75	IEX
Silica	Silica	Acidic, neutral	250–600	50–60	60–75	NP
Alumina	Alumina	Acidic, neutral, basic	100–150	50–300	100–120	NP
Florisil	Florisil	None	300–600	50–200	60–80	NP
Polymer	SDB	None	500–1000	75–150	50–300	RP

RP, reverse-phase sampling conditions; NP, normal-phase sampling conditions; IEX, ion-exchange sampling conditions; SCX, strong cationic exchanger; SAX, strong anionic exchanger.

the sorbent, and subsequently eluted with a minimal volume of solvent. This methodology has some disadvantages, such as irreversible adsorption, possible reaction between analytes and sorbent materials, and low recoveries [25]. Various kinds of sorbent devices are available commercially, such as syringe-barrels, cartridges, and disks. A syringe-barrel column is the most popular SPE configuration. The sorbent bed is held in place by porous polyethylene frits and the syringe-barrel is typically manufactured from highly pure materials. Cartridges have no significant reservoir capacity and are fitted with both male and female luer lock fittings. To meet the various needs of contemporary applications, there is an ever-increasing demand for novel and more effective sorbents, and some of the commercially available sorbents for SPE are given in Table 14.1 [26,27].

Ballesteros and Parrado proposed a novel continuous SPE procedure [28] to reduce analysis time, increase the sensitivity, minimize the solvent volume for extraction, and avoid risk of contamination. This system is coupled to a preconcentration step and used to extract organophosphorus pesticides from aqueous samples. In SPE, the choice of sorbent can be optimized based on the target pesticides, but C_{18}-bonded silica and styrene/divinyl benzene copolymer phases are the most frequently used. SPE has been widely applied to water [29,30] and food samples [31,32]. Recently, SPE-based sample preparation methods have been reported with novel adsorbents, such as immunoadsorbents and molecularly imprinted polymers [25–27].

14.2.3.2 Solid-Phase Microextraction

Pawliszyn and his coworkers developed this microscale technique in the late 1980s [33,34]. They introduced it as a solvent-free sample preparation technique that could serve as an alternative to the traditional extraction procedures, such as LLE, Purge & Trap (P&T), static headspace, and SPE. SPME preserves all the advantages of SPE, while eliminating the main disadvantages of low analyte recovery, plugging, and solvent use. This technique utilizes a short and thin solid rod of fused silica (typically 1 cm long, 0.1 μm outer diameter), coated with sorbent, normally a polymer. The coated fused silica (SPME fiber) is attached to a metal rod. The entire assembly (fiber holder) may be described as a modified syringe. In the standby position, the fiber is withdrawn into a protective

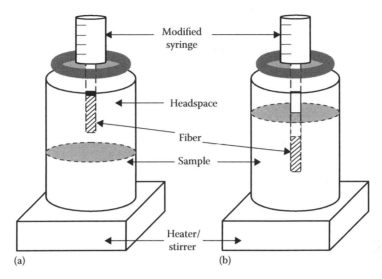

FIGURE 14.2 The two modes of SPME: (a) headspace SPME and (b) direct immersion-SPME.

stainless-steel sheath. For sampling, a liquid or solid sample is placed in a vial and capped with a septum-lined cap. The sheath, which also acts as a needle, is pushed through the septum and the plunger is lowered, introducing the fiber into the vial, where it is immersed directly into the liquid sample or is held in the headspace. Analytes in the sample are adsorbed on the fiber. After a predetermined time, the fiber is withdrawn into the protective sheath, which is then removed from the sampling vial. Immediately after that, the sheath is inserted through the septum of a GC injector, the plunger is pushed down, and the fiber is forced into the injector where the analytes are thermally desorbed and separated on the GC column. The duration of desorption step is usually 1–2 min. After desorption, the fiber is withdrawn into its protective sheath and the sheath is removed from the GC injector.

There are two approaches to SPME sampling of volatile organics: direct and headspace, as shown in Figure 14.2 [34,35]. In direct sampling, the fiber is placed into the sample matrix, and in headspace sampling, the fiber is placed in the headspace of the sample. SPME has been interfaced to HPLC, CE, and Fourier transform infrared spectroscopy (FTIR), in addition to GC [36–38], for the determination of various kinds of analytes [39]. Several sorbent polymers are commercially available for SPME, such as polydimethylsiloxane (PDMS) (which is normally used for pesticides and volatile halogenated compounds), polyacrylate (PA), or a mixture of PA with carbowax (CW), and/or polydivinylbenzene (DVB). It has been established that the fiber can be reused 100 times or more.

The advantages of SPME techniques are as follows:

- It is an equilibrium technique and is therefore selective for volatile and semivolatile compounds.
- Time required for the analyte to reach an equilibrium between the coated fiber and sample is relatively short.
- Solvent-less extraction and injection eliminates the problems with regard to solvent use and disposal.
- By sampling from headspace, SPME can extract analytes from very complex matrices.
- All the analytes collected on the solid phase can be injected into the GC for further analysis.
- The method is simple, fast, less expensive, and easily automated, although at a significant capital cost.

The disadvantages of SPME are as follows:

- Often only a small fraction of the sample analytes is extracted by the coated fiber.
- Quantification in SPME requires strict and careful calibration.
- Carryover may result owing to incomplete desorption.
- The fiber is easily fragile, partially, and hence, it cannot be directly immersed in "dirty" matrices.

With regard to the last disadvantage, Basheer et al. developed the hollow-fiber membrane-protected-SPME [40] or simply, HFM-protected-SPME, to enhance the detection limits for the analysis of herbicides using challenging matrices. The experimental procedure for HFM-SPME is very similar to that of the conventional SPME, except that a porous polypropylene HFM is used to protect the SPME fibers during the extraction of pesticides from matrices, such as milk and waste-water samples without pretreatment. The internal diameter of the HFM (600 µm) is large enough to accommodate the SPME fiber. The SPME fiber assembly is inserted into a 7 cm long (one end is flame-sealed) HFM, so that the latter completely covered the stainless-steel tubing and the poly-meric fiber. A long-neck 10 mL vial is filled with 5 mL of the sample with known pH. The HFM-protected SPME fiber is exposed to the sample solution to attain the extraction equilibrium. During extraction, the polymeric fiber is immersed in the sample solution, as in conventional SPME (about half of the HFM-protected stainless-steel tubing was also immersed). Each sample is stirred vigor-ously (1200 rpm) during the sorption step using a stir bar. After extraction, the HFM is discarded. The fibers can be reused for up to 50 analyses.

14.2.3.3 Stir-Bar Sorptive Extraction

SBSE was developed by Baltussen et al. [41] to improve the enrichment factor and surpass the more limited extraction capacity of SPME. A glass stirrer bar (10 or 20 mm) is coated with a (0.5 and 1.0 mm thickness, respectively, are available) thick layer of PDMS to give a large surface area of sor-bent phase, leading to a higher phase ratio with higher sample capacity. Transfer of the analyte from the bar is achieved either by a dedicated thermal desorption unit or elution with a solvent.

A number of reviews on SBSE were published recently. For example, Sandra et al. [42] gave examples of food analysis; Demyttenaere et al. [43] compared SBSE with SPME for the analysis of alcoholic beverages and concluded that SBSE was more sensitive; Blasco et al. [44] investigated the use of SBSE for the analysis of pesticides in oranges. However, PDMS was the only sorbent available, and they concluded that SBSE, in general, was not suitable for polar pesticides.

14.2.3.4 Polymer-Coated Hollow-Fiber Microextraction

This is a novel methodology for extraction of organochlorine pesticides from aqueous sam-ples [24]. In PC-HFME, an amphiphilic polymer is coated on the walls of an HFM; owing to the high porosity, a high-active surface area is expected. A polymer-coated HFM is placed in a 4 mL sample vial for extracting analytes. The sample solution is stirred (at 1000 rpm) using a magnetic stirrer. The stirring aids the immersion and tumbling movement of the fiber in the aqueous sample solution during extraction. After the extraction, the fiber is removed with a pair of tweezers. The extracted analytes on the fiber are desorbed using hexane (the coated polymer is insoluble in hexane) in a crimper vial, via ultrasonication. The fiber is then removed and discarded, and the hexane is evaporated to dryness with a gentle stream of nitrogen gas. The extract is reconstituted to 20 µL with the same solvent. Finally, 2 µL of the reconstituted extract is injected into the GC–MS. By using the PC-HFME, lower detection limits can be achieved when compared with LPME and SPME [24].

14.3 SOLID SAMPLES

14.3.1 Extraction of Organics from Solid Matrices

The extraction and recovery of an analyte from a solid matrix can be regarded as a five-stage process [45]:

1. The desorption of the compound from the active sites of the matrix
2. Diffusion into the matrix itself
3. Solubilization of the compound in the extractant
4. Diffusion of the compound in the extractant
5. Collection of the extracted compounds

In environmental applications, the first step is usually the rate-limiting step, as analyte–matrix interactions are very difficult to overcome and predict. Consequently, the optimization strategy will strongly depend on the nature of the matrix to be extracted. The currently available extraction methods for pesticide analysis are as follows:

1. Soxhlet extraction and Soxtec
2. SFE
3. Pressurized fluid extraction (PFE)
4. MAE

14.3.1.1 Soxhlet and Soxtec Extraction

Soxhlet extraction is commonly used as the benchmark method for validating and evaluating other extraction techniques. A Soxtec apparatus not only reduces the extraction time to 2–3 h, when compared with 60–48 h in Soxhlet, but also decreases the solvent use from 250–500 mL per extraction to 40–50 mL. Two to six samples can be extracted simultaneously with a single Soxtec apparatus [46]. In general, solvent consumption for both these methods is significantly high.

14.3.1.2 Supercritical and Pressurized Liquid Extraction

SFE is also a very popular technique for environmental analysis. It is an appropriate technique for the analysis of the less volatile compounds. However, it has limitations with respect to the range of analytes that can be extracted simultaneously. Nevertheless, for a particular semivolatile analyte or a narrow selection of analytes, this technique is preferable over solvent extraction. Kurt Zosel developed a natural product extraction with supercritical carbon dioxide (CO_2), for the first time in the 1960s. Supercritical fluids have a viscosity close to gases with high permeability, and their properties can be tuned based on the target analytes by varying the temperature, leading to improved extraction efficiency [47]. Analytical SFE was traditionally focused on CO_2 as the extracting solvent. As CO_2 is a nonpolar solvent, small amount of alcohols or other polar organic solvents are added in an attempt to increase the fluid polarity. SFE has been applied to extract pesticides from fruits, vegetables, and baby foods [48–52]. Based on the target pesticides and sample matrix, the extraction procedure can be modified, for example, organophosphorus pesticides in wheat sample may be extracted as follows [53]: 7–10 g of sample is placed inside the extraction cell and a fused-silica capillary tube (30 cm × 100 μm ID) is attached to the outlet of the extractor as a restrictor (to maintain the pressure within the system). Tubes with PTFE caps are used as collection vessels. The restrictor is passed through the cap, and immersed in the collection solvent (about 20 mL of hexane). CO_2 is then delivered at flow-rates of 0.7–1.5 mL/min by a syringe pump to the extraction cell containing the sample. Temperature between 40°C and 100°C and pressure between 72.5 and 482.6 bar may be applied.

14.3.1.3   Pressurized Fluid Extraction

A new technique known as PFE appeared around a decade ago. This technique is commercially termed as accelerated solvent extraction (ASE, which is a Dionex trademark). More generally, it has been called pressurized liquid extraction (PLE), pressurized solvent extraction (PSE), or enhanced solvent extraction (ESE). In PFE, the extractant is maintained in its liquid state. To achieve elevated temperatures, pressure is applied inside the extraction cell. In this way, temperatures around 100°C–200°C may be attained with common organic solvents. In fact, at such high temperatures and pressures, the solvent may be considered as being in a subcritical state, with interesting mass-transfer properties.

PFE has the capability of performing fast, efficient extractions owing to the use of elevated temperatures, as the decrease in solvent viscosity helps to disrupt the analyte–matrix interactions and increases diffusion coefficients. In addition, high temperature favors solubilization of the compounds. Furthermore, the pressure favors the penetration of the solvent into the matrix, which again aids in extraction. Consequently, this technique is of growing interest, and numerous commercial systems are under use. PFE has been recognized as an official method by the USEPA, and the method has enabled efficient screening of selected semivolatile pesticides and other compounds in soils [54,55].

Owing to extraction under elevated pressure and temperature, PFE represents an exceptionally effective extraction technique with the advantages of shorter extraction times and lower consumption of solvents when compared with Soxhlet extraction. It allows the universal use of solvents or solvent mixtures with different polarities and individually variable pressures of 5–200 atm (0.3–20 MPa) to maintain the extraction solvent in a liquid state, and temperatures ranging from room temperature to 200°C to increase the extraction rate [56]. A number of applications have been reported in the literature, including the extraction of pesticide residues in grain [57], muscle of chicken, pork, and lamb [58], vegetable matrices [59], adipose and organ tissue [60], animal feed [61], and fish samples [62].

14.3.1.4   Microwave-Assisted Extraction

Initially, microwave heating was used for sample digestion and extraction owing to its improved efficiencies, reduced extraction time, low solvent consumption, and high level of automation, when compared with the conventional extraction techniques [63,64]. The major advantage is that microwave energy is absorbed by the extractant, which in turn transfers it to the sample in the form of heat. Based on the sample matrix and target pesticides, extraction temperature can be optimized to achieve higher extraction efficiencies [65]. Owing to the dipole rotation and ionic-conductance effects of microwaves on matter, heating with microwaves is instantaneous and occurs homogeneously, leading to very fast extractions [63,66,67]. In most applications, the extraction solvent is selected as the medium to absorb microwave energies. Alternatively (for thermolabile compounds), the microwaves may be absorbed only by the matrix, resulting in heating of the sample and release of the solutes into the unheated solvent.

Microwave energy may be applied to samples in two ways: in closed vessels (under controlled pressure and temperature) or in open vessels (at atmospheric pressure) [63,68], termed as pressurized MAE and focused MAE, respectively. In open vessels, the temperature is limited by the boiling point of the solvent, at atmospheric pressure. However, in closed vessels, the temperature may be elevated by simply applying the appropriate pressure. The MAE has been useful to extract pesticides from vegetables [69], sesame seeds [70], tree leaves [71], marine biological tissues [72], and Chinese teas [73]. Table 14.2 shows the comparison of the advantages and disadvantages of Soxhlet and Soxtec, SFE, PLE, and MAE.

TABLE 14.2
Advantages and Disadvantages of Various Techniques

Technique	Advantages	Disadvantages
Soxhlet	Matrix independent	Slow (up to 24–48 h)
	Inexpensive equipment	Large amount of solvent (500 mL)
	Unattended operation	Mandatory evaporation of extract
	Rugged, benchmark method	
	Filtration not required	
Soxtec	Matrix independent	Relatively slow (2 h)
	Inexpensive equipment	
	Less solvent (50 mL)	
	Evaporation integrated	
	Filtration not required	
SFE	Fast (30–75 min)	Matrix dependent
	Minimal solvent use (5–10 mL)	Small sample size (2–10 g)
	CO_2 is environmental-friendly	Expensive equipment
	Controlled selectivity	Limited applicability
	Filtration not required	
	Evaporation not needed	
PFE	Fast (12–18 min)	Expensive equipment
	Small amount of solvent (30 mL)	Cleanup necessary
	Large amount of sample (100 g)	
	Automated	
	Easy-to-use	
	Filtration not required	
MAE	Fast (10–30 min)	Polar solvent needed
	High sample throughput	Cleanup mandatory
	Small amount of solvent (30 mL)	Filtration required
	Large amount of sample (20 g)	Expensive equipment
		Degradation possible

14.4 RECENT ANALYTICAL TRENDS FOR PESTICIDE ANALYSIS

Some interesting novel applications for pesticide analysis from various sample matrices are listed in Table 14.3.

14.5 CONCLUSIONS

Pesticide detection methods are becoming more specific and sensitive; yet, to achieve the desired specificity and sensitivity, there is still a need for careful sample preparation. Trace-level determination of the target compounds in complex samples, such as milk, blood, or other biological matrices, is particularly important as it can account for a significant amount of variability of the extraction method. Conventional extractions, such as LLE and Soxhlet extraction, are not selective enough to meet the needs of environmental monitoring, food safety, and food regulatory requirements. However, selective, simple, and miniaturized sample preparation methods that are environmentally benign and that can be applied to routine pesticide analysis are more desirable. Thus, it is expected that a great deal of effort will be expended along this line in the future.

TABLE 14.3
Recent Trends in the Pesticide Analysis using Various Matrices

Pesticides Studied	Sample Matrix	Sample Preparation Method	Comments	LODs	LOQs	Ref.
2,4-D, atrazine, chlorpyrifos, malathion, permethrin, and propoxur	Air and soil	Aerosol time-of-flight mass spectrometer (ATOFMS)	Direct determination of single aerosol particles	1–100 ppm	—	[74]
Chloronicotinyl pesticides	Vegetables	Liquid chromatography/time-of-flight mass spectrometry (LC/TOF-MS)	Solvent extraction	0.002–0.01 mg/kg	—	[75]
Chlorinated pesticides	Black-crowned night heron (*Nycticorax nycticorax*) egg samples	Solvent extraction	HPLC cleanup and GC–MS	0.03–1.09 ng/g (wet wt)	—	[76]
~130 Pesticides	Vegetables	Solvent extraction	Triple quadrupole (QqQ) mass spectrometry (MS) as detection system in GC	0.01–3.21 µg/kg	—	[77]
Dinitroaniline pesticides	Tobacco smoke	Solvent extraction (ultrasonication)	Electron monochromator-mass spectrometry (EM-MS)	—	—	[78]
Organophosphorus	Drinking water	Liposome-based nano-biosensors	Optical fluorescent	10^{-10} M	—	[79]
Organochlorine and organophosphorus	Animal liver	Solvent extraction, gel permeation chromatography (GPC), and a matrix solid-phase dispersion (MSPD)	Electron impact ionization tandem mass spectrometry (GC–(EI-)MS/MS) using a triple quadrupole (QqQ) analyzer	0.01–8.7 µg/kg	25 and 50 µg/kg	[80]
Organochlorines (OCs)	Human semen	Solvent extraction	—	0.01 ppb	1 ppt	[81]
Organochlorine and organophosphorus	Food samples	Matrix solid-phase dispersion (MSPD) technique	Automated online gel permeation chromatography–gas chromatograph–mass spectrometer (GPC–GC–MS)	3–79 (µg/kg)	27–262 (µg/kg)	[82]

Pesticide	Sample	Method	Remarks			Reference
Pyrethroids	Urine samples	Enzyme-linked immunosorbent assay (ELISA)	Does not require sample preconcentration or cleanup	ng/mL	—	[83]
Herbicide (diuron)	Water samples	Time-resolved fluorescence immunoassay (TR-FIA)	Solvent extraction followed by solid-phase dispersive cleanup	20 ng/L	—	[84]
Organophosphorus	Ginseng root	Gas chromatography with electron ionization mass spectrometry in selective ion monitoring mode (GC–MS/SIM) and flame photometric detection (GC–FPD) in phosphorus	Solvent extraction followed by low-temperature precipitation	0.005–0.50 µg/g	—	[85]
Insecticides and triazine herbicides	Raw and processed olives	GC-electron capture detection (ECD)	Immunosorbents (ISs)-based analysis	0.01 mg/kg	0.05 mg/kg	[86]
Chloroanisole group pesticides	White wine samples	(antibody-antigen) solid-phase extraction (SPE) followed by ELISA analysis	New instrument	About 200–400 ng/L	—	[87]
Organophosphorus and organosulfurous pesticides	Standard solution	Microcountercurrent flame photometric detector (µ-cc-FPD) was adapted and optimized for ultrafast GC	Accelerated solvent extraction (ASE)	12 pg	—	[88]

REFERENCES

1. National Research Council, *Hormonally Active Agents in the Environment*, National Academic Press, Washington, DC, 1999.
2. HW Vallack, DJ Bakker, I Brandt, E Brorström-Lundén, A Brouwer, KR Bull, and C Gough et al. Controlling persistent organic pollutants—What next? *Environ Toxicol Pharmacol* 63: 143–207, 1998.
3. M Porta, N Malats, M Jariod, JO Grimault, J Rifa, A Carrato, and A Guarner et al. Serum concentrations of organochlorine compounds and K-ras mutations in exocrine pancreatic cancer. *Lancet* 354: 2125–2130, 1999.
4. HA Alegria, TF Bidleman, and TJ Shaw. Organochlorine pesticides in ambient air of Belize, Central America. *Environ Sci Technol* 34: 1953–1958, 2000.
5. UNEP. List of signatories and ratifications to the Stockholm Convention on POPs. 2003. http://www.pops.int/documents/signature/signstatus.htm (accessed May 5, 2009).
6. Ministry of the Environment, Singapore: http://www.env.gov.sg/faq/faq2.htm#q8 (accessed November 12, 2008).
7. M Venier and RA Hites. Chiral organochlorine pesticides in the atmosphere. *Atmos. Environ.*, 41(4): 768–775, 2007.
8. EC Zhao, LJ Han, SR Jiang, Q Wang, and Z Zhou. Application of a single-drop microextraction for the analysis of organophosphorus pesticides in juice. *J Chromatogr A* 1114(2): 269–273, 2006.
9. G Shen and HK Lee. Headspace liquid-phase microextraction of chlorobenzenes in soil with gas chromatography-electron capture detection. *Anal Chem* 75: 98–103, 2003.
10. Y He and HK Lee. Continuous flow microextraction combined with high-performance liquid chromatography for the analysis of pesticides in natural waters. *J Chromatogr A* 1122(1–2): 7–12, 2006.
11. C Basheer, B Rajasekhar, and HK Lee. Determination of organic micropollutants in rainwater using hollow fiber membrane/liquid-phase microextraction combined with gas chromatography–mass spectrometry. *J Chromatogr A* 1016: 11–20, 2003.
12. C Basheer, HK Lee, and JP Obbard. Determination of organochlorine pesticides in seawater using liquid-phase hollow fiber membrane microextraction and gas chromatography–mass spectrometry. *J Chromatogr A* 968: 191–199, 2002.
13. J Wu, KH Ee, and HK Lee. Automated dynamic liquid–liquid–liquid microextraction followed by high-performance liquid chromatography-ultraviolet detection for the determination of phenoxy acid herbicides in environmental waters. *J Chromatogr A* 1082: 121–127, 2005.
14. B Buszewski and T Ligor. Single-drop extraction versus solid-phase microextraction. *LC–GC Eur* 15: 92–97, 2002.
15. L Zhao and HK Lee. Determination of phenols in water using liquid phase microextraction with back extraction combined with high-performance liquid chromatography. *J Chromatogr A* 931: 95–105, 2001.
16. L Xu, C Basheer, and HK Lee, Developments in single-drop microextraction. *J Chromatogr A* 1152: 184–192, 2007.
17. L Zhu, KH Ee, L Zhao, and HK Lee. Analysis of phenoxy herbicides in bovine milk by means of liquid–liquid–liquid microextraction with a hollow-fiber membrane. *J Chromatogr A* 963: 335–343, 2002.
18. S Pedersen-Bjergaard and KE Rasmussen. Liquid–liquid–liquid microextraction for sample preparation of biological fluids prior to capillary electrophoresis. *Anal Chem* 71: 2650–2656, 1999.
19. S Pedersen-Bjergaard and KE Rasmussen. Liquid-phase microextraction and capillary electrophoresis of acidic drugs. *Electrophoresis* 21: 579–585, 2000.
20. TG Halvorsen, S Pedersen-Bjergaard, and KE Rasmussen. Reduction of extraction times in liquid-phase microextraction. *J Chromatogr B* 760: 219–226, 2001.
21. L Zhu and HK Lee. Liquid–liquid–liquid microextraction of nitrophenols with a hollow fiber membrane prior to capillary liquid chromatography. *J Chromatogr A* 924: 407–414, 2001.
22. DA Lambropoulou and TA Albanis. Liquid-phase micro-extraction techniques in pesticide residue analysis. *J Biochem Biophys Methods* 70: 195–228, 2007.
23. X Jiang and HK Lee. Solvent bar microextraction. *Anal Chem* 76: 5591–5596, 2004.
24. C Basheer, V Suresh, R Renu, and HK Lee. Development and application of polymer-coated hollow fiber membrane microextraction to the determination of organochlorine pesticides in water. *J Chromatogr A* 1033: 213–22, 2004.
25. IJ Liska. Fifty years of solid-phase extraction in water analysis—Historical development and overview. *J Chromatogr A* 885: 3–16, 2000.
26. MC Hennion. Solid-phase extraction: Method development, sorbents, and coupling with liquid chromatography. *J Chromatogr A* 856: 3–54, 1999.

27. CF Poole. New trends in solid-phase extraction. *Trends Anal Chem* 22: 362–373, 2003.
28. E Ballesteros and MJ Parrado. Continuous solid-phase extraction and gas chromatographic determination of organophosphorus pesticides in natural and drinking waters. *J Chromatogr A* 1029: 267–273, 2004.
29. R Carabias-Martinez, E Rodriguez-Gonzalo, and E Herrero-Hernandez. Behaviour of triazine herbicides and their hydroxylated and dealkylated metabolites on a propazine-imprinted polymer: Comparative study in organic and aqueous media. *Anal Chim Acta* 559: 186–194, 2006.
30. R Carabias-Martinez, E Rodriguez-Gonzalo, and E Herrero-Hernandez. Determination of triazines and dealkylated and hydroxylated metabolites in river water using a propazine-imprinted polymer. *J Chromatogr A* 1085: 199–206, 2005.
31. A Juan-Garcia, G Font, and Y Pico. Quantitative analysis of six pesticides in fruits by capillary electrophoresis-electrospray-mass spectrometry. *Electrophoresis* 26: 1550–1561, 2005.
32. C Arthur and J Pawliszyn. Solid phase microextraction with thermal desorption using fused silica optical fibers. *Anal Chem* 62: 2145–2148, 1990.
33. R Berlardi and J Pawliszyn. The application of chemically modified fused silica fibers in the extraction of organics from water matrix samples and their rapid transfer to capillary columns. *Water Pollut Res J Can* 24: 179–183, 1989.
34. Z Zhang and J Pawliszyn. Headspace solid-phase microextraction. *Anal Chem* 65: 1843–1852, 1993.
35. B Page and G Lacroix. Application of solid-phase microextraction to the headspace gas chromatographic analysis of halogenated volatiles in selected foods. *J Chromatogr A* 648: 199–211, 1993.
36. J Chen and J Pawliszyn. Solid-phase microextraction coupled to high-performance liquid-chromatography. *Anal Chem* 67: 2350–2533, 1995.
37. J Pawliszyn. *Solid Phase Microextraction, Theory and Practice*, John Wiley and Sons, New York, 1997.
38. MA Allam and SF Hamed. Application of FTIR spectroscopy in the assessment of live oil adulteration. *J Applied Sci Res* 3: 102–108, 2007.
39. Supelco. *SPME Application Guide*, Supelco, Bellefonte, PA, USA 2001.
40. C Basheer and HK Lee. Hollow fiber membrane-protected solid-phase microextraction of triazine herbicides in bovine milk and sewage sludge samples. *J Chromatogr A* 1047: 189–194, 2004.
41. E Baltussen, P Sandra, F David, and C Cramers. Stir bar sorptive extraction SBSE, a novel extraction technique for aqueous samples: Theory and principles. *J Microcolumn Sep* 11: 737–747, 1999.
42. P Sandra, F David, and B Tienpont. Stir-bar sorptive extraction of trace organic compounds from aqueous matrices. *LC–GC Eur* 16: 410–417, 2003.
43. JCR Demyttenaere, JIS Martínez, R Verhé, P Sandra, and ND Kimpe. Analysis of volatiles of malt whisky by solid-phase microextraction and stir bar sorptive extraction. *J Chromatogr A* 985: 221–232, 2003.
44. C Blasco, G Font, and Y Pico. Comparison of microextraction procedures to determine pesticides in oranges by liquid chromatography–mass spectrometry. *J Chromatogr A* 970: 201–212, 2002.
45. J Pawliszyn. Kinetic model of supercritical fluid extraction. *J Chromatogr Sci* 31: 31–37, 1993.
46. EPA Method 3540C, *Soxhlet Extraction, Test Methods for Evaluating Solid Waste*, EPA, Washington, DC 1996.
47. RM Smith. Supercritical fluids in separation science—The dreams, the reality and the future. *J Chromatogr A* 856: 83–115, 1999.
48. BL Halvorsen, C Thomsen, T Greibrokk, and E Lundanes. Determination of fenpyroximate in apples by supercritical fluid extraction and packed capillary liquid chromatography with UV detection. *J Chromatogr A* 880: 121–128, 2000.
49. A Valverde-Garcia, AR Fernandez-Alba, A Aguera, and M Contreras. Extraction of methamidophos residues from vegetables with supercritical fluid carbon dioxide. *J AOAC Int* 78: 867–873, 1995.
50. JC Chuang, MA Pollard, M Misita, and JM Van Emon. Evaluation of analytical methods for determining pesticides in baby food. *Anal Chim Acta* 399: 135–142, 1999.
51. SJ Lehotay and KI Eller. Development of a method of analysis for 46 pesticides in fruits and vegetables by supercritical fluid extraction and gas chromatography/ion trap mass spectrometry. *J AOAC Int* 78: 821–830, 1995.
52. SJ Lehotay, N Aharonson, E Pfeil, and MA Ibrahim. Development of a sample preparation technique for supercritical fluid extraction for multiresidue analysis of pesticides in produce. *J AOAC Int* 78: 831–838, 1995.
53. DH Kim, GS Heo, and DW Lee. Determination of organophosphorus pesticides in wheat flour by supercritical fluid extraction and gas chromatography with nitrogen–phosphorus detection. *J Chromatogr A* 824: 63–70, 1998.
54. EPA Method 3545A, *Pressurized Fluid Extraction, Test Methods for Evaluating Solid Waste*, EPA, Washington, DC 1998.

55. JA Fisher, MJ Scarlett, and AD Stott. Accelerated solvent extraction: An evaluation for screening of soils for selected U.S. EPA semivolatile organic priority pollutants. *Environ Sci Technol* 31: 1120–1127, 1997.
56. FE Ahmed. Analyses of pesticides and their metabolites in foods and drinks. *Trends Anal Chem* 20: 649–661, 2001.
57. GF Pang, YM Liu, CL Fan, JJ Zhang, YZ Cao, XM Li, ZY Li, YP Wu, and TT Guo. Simultaneous determination of 405 pesticide residues in grain by accelerated solvent extraction then gas chromatography-mass spectrometry or liquid chromatography-tandem mass spectrometry *Anal Bioanal Chem* 384: 1366 1400, 2006.
58. AG Frenich, JLM Vidal, ADC Sicilia, MJG Rodriguez, and PP Bolanos. Multiresidue analysis of organochlorine and organophosphorus pesticides in muscle of chicken, pork and lamb by gas chromatography-triple quadrupole mass spectrometry. *Anal Chim Acta* 558: 42–52, 2006.
59. C Marcic, G Lespes, and M Potin-Gautier. Pressurized solvent extraction for organotin speciation in vegetable matrices. *Anal Bioanal Chem* 382: 1574–1583, 2005.
60. K Saito, A Sjodin, CD Sandau, MD Davis, H Nakazawa, Y Matsuki, and DG Patterson. Development of an accelerated solvent extraction and gel permeation chromatography analytical method for measuring persistent organohalogen compounds in adipose and organ tissue analysis. *Chemosphere* 57: 373–381, 2004.
61. S Chen, M Gfrerer, E Lankmayr, X Quan, and F Yang. Optimization of accelerated solvent extraction for the determination of chlorinated pesticides from animal feed. *Chromatographia* 58: 631–636, 2003.
62. W Zhuang, B McKague, D Reeve, and J Carey. A comparative evaluation of accelerated solvent extraction and Polytron extraction for quantification of lipids and extractable organochlorine in fish. *Chemosphere* 54: 467–480, 2003.
63. CS Eskilsson and E Björklund, Analytical-scale microwave-assisted extraction. *J Chromatogr A* 902: 227–250, 2000.
64. PL Buldini and RL Sharma. Recent applications of sample preparation techniques in food analysis. *J Chromatogr A* 985: 47–70, 2002.
65. V Camel. Microwave-assisted solvent extraction of environmental samples. *Trends Anal Chem* 19: 229–248, 2000.
66. JRJ Pare, JMR Belanger, and SS Stafford. Microwave-assisted process MAP: A new tool for the analytical laboratory. *Trends Anal Chem* 13: 176–184, 1994.
67. HM Pylypiw Jr, TL Arsenault, CM Thetford, and MJI Mattina. Suitability of microwave-assisted extraction for multiresidue pesticide analysis of produce. *J Agric Food Chem* 45: 3522–3528, 1997.
68. M Letellier and H Budzinski. Microwave assisted extraction of organic compound. *Analysis* 27: 259–271, 1999.
69. SB Singh, GD Foster, and SU Khan. Determination of thiophanate methyl and carbendazim residues in vegetable samples using microwave-assisted extraction. *J Chromatogr A* 1148: 152–157, 2007.
70. EN Papadakis, Z Vryzas, and E Papadopoulou-Mourkidou. Rapid method for the determination of 16 organochlorine pesticides in sesame seeds by microwave-assisted extraction and analysis of extracts by gas chromatography–mass spectrometry. *J Chromatogr A* 1127: 6–11, 2006.
71. M Barriada-Pereira, MJ González-Castro, S Muniategui-Lorenzo, P López-Mahía, D Prada-Rodríguez, and E Fernández-Fernández. Determination of 21 organochlorine pesticides in tree leaves using solid-phase extraction clean-up cartridges. *J Chromatogr A* 1061: 133–139, 2004.
72. S Bayen, HK Lee, and JP Obbard. Determination of polybrominated diphenyl ethers in marine biological tissues using microwave-assisted extraction. *J Chromatogr A* 1035: 291–294, 2004.
73. L Cai, J Xing, L Dong, and C Wu. Application of polyphenylmethylsiloxane coated fiber for solid-phase microextraction combined with microwave-assisted extraction for the determination of organochlorine pesticides in Chinese teas. *J Chromatogr A* 1015: 11–21, 2003.
74. JR Whiteaker and KA Prather. Detection of pesticide residues on individual particles. *Anal Chem* 75: 49–56, 2003.
75. I Ferrer, EA Thurman, and AR Fernández-Alba. Quantitation and accurate mass analysis of pesticides in vegetables by LC/TOF-MS. *Anal Chem* 77: 2818–2825, 2005.
76. S Chu, C Hong, BA Rattner, and PC McGowan. Methodological refinements in the determination of 146 polychlorinated biphenyls, including non-ortho- and mono-ortho-substituted PCBs, and 26 organochlorine pesticides as demonstrated in Heron eggs. *Anal Chem* 75: 1058–1066, 2003.
77. AG Frenich, MJ González-Rodríguez, FJ Arrebola, and JLM Vidal. Potentiality of gas chromatography-triple quadrupole mass spectrometry in Vanguard and Rearguard methods of pesticide residues in vegetables. *Anal Chem* 77: 4640–4648, 2005.

78. V Vamvakaki and NA Chaniotakis. Pesticide detection with a liposome-based nano-biosensor. *Biosens Bioelectron* 22: 2848–2853, 2007.
79. AG Frenich, PP Bolaños, and JLM Vidal. Multi residue analysis of pesticides in animal liver by gas chromatography using triple quadrupole tandem mass spectrometry. *J Chromatogr A* 1153: 194–202, 2007.
80. N Pant, R Kumar, N Mathur, SP Srivastava, DK Saxena, and VR Gujrati. Chlorinated pesticide concentration in semen of fertile and infertile men and correlation with sperm quality. *Environ Toxicol Pharmacol* 23: 135–139, 2007.
81. LB Liu, Y Hashi, YP Qin, HX Zhou, and JM Lin. Development of automated online gel permeation chromatography–gas chromatograph mass spectrometry for measuring multiresidual pesticides in agricultural products. *J Chromatogr B* 845: 61–68, 2007.
82. HJ Kim, KC Ahn, SJ Ma, SJ Gee, and BD Hammock. Development of sensitive immunoassays for the detection of the glucuronide conjugate of 3-phenoxybenzyl alcohol, a putative human urinary biomarker for pyrethroid exposure. *J Agric Food Chem* 55: 3750–3757, 2007.
83. MA Bacigalupo and G Meroni. Quantitative determination of diuron in ground and surface water by time-resolved fluoroimmunoassay: Seasonal variations of diuron, carbofuran, and paraquat in an agricultural area. *J Agric Food Chem* 55: 3823–3828, 2007.
84. JW Wong, MK Hennessy, DG Hayward, AG Krynitsky, I Cassias, and FJ Schenck. Analysis of organophosphorus pesticides in dried ground Ginseng root by capillary gas chromatography-mass spectrometry and -flame photometric detection. *J Agric Food Chem* 55: 1117–1128, 2007.
85. EJ Avramides and S Gkatsos. A multiresidue method for the determination of insecticides and triazine herbicides in fresh and processed olives. *J Agric Food Chem* 55: 561–565, 2007.
86. N Sanvicens, EJ Moore, GG Guilbault, and MP Marco. Determination of haloanisols in white wine by immunosorbent solid-phase extraction followed by enzyme-linked immunosorbent assay. *J Agric Food Chem* 54: 9176–9183, 2006.
87. S Kendler, SM Reidy, GR Lambertus, and RD Sacks. Ultrafast gas chromatographic separation of organophosphorus and organosulfur compounds utilizing a microcountercurrent flame photometric detector. *Anal Chem* 78: 6765–6773, 2006.
88. NS Tulve, PA Jones, MG Nishioka, RC Fortmann, CW Croghan, JY Zhou, A Fraser, C Cave, and W Friedman. Pesticide measurements from the first national environmental health survey of child care centers using a multi-residue GC/MS analysis method. *Environ Sci Technol* 40: 6269–6274, 2006.

Part III

Pesticides and the Environment

15 Medicinal Plants, Pesticide Residues, and Analysis

Fatma U. Afifi, Rima M. Hajjo, and
Abdelkader H. Battah

CONTENTS

15.1 MEDICINAL PLANTS

The medicinal use of herbs is as old as mankind itself. The poisonous and beneficial healing properties of plants were discovered by man, through trial and error, in search for food. Scientists from early civilizations (Egyptian, Chinese, Indian, Aztec, Greek, and Moslem) enriched our current knowledge in herbal medicine. Their manuscripts are a living proof showing the use of plants for medicinal purposes. With time, plants became the backbone of the earliest generation of pharmacologically active compounds. Today, despite modern technological advances, a large number of potentially active drugs still originate from plants [1–6].

The pharmaceutical importance of medicinal plants lies in their production of organic compounds possessing pharmacological properties. Currently, in the Western World, botanical products are widely used as food ingredients, supplements, over the counter (OTC) drug products, and phytomedicines. As a result of globalization, plant products are found on the shelves of drugstores and health-food stores located continents away from the plants' native habitat. Herbal preparations and especially herbal teas, marketed extensively with emphasis on their medicinal properties, are gaining popularity worldwide. This is mainly owing to their continuous use throughout history, and the belief that they are free from harmful side effects [7–9]. Products originating from plants are available in various forms: dried crude plants, crude extracts, or standardized extracts. In these preparations, plants are used in the processed or nonprocessed form. Herbalists, found mainly in developing countries, are the primary users of dried local flora collected from natural reservoirs or cultivated fields.

Medicinal plants play an important role in both commercial and healthy-lifestyle aspects. The lucrative business with the medicinal and culinary plants accounts for billions of U.S. dollars in the United States and overseas. Toward the end of the twentieth century, World Health Organization (WHO) estimated that an impressive 80% of the world's population would rely mainly on natural medicines, with plant-originated medicines as the main line of this trend (in developed countries) or tradition (in developing countries). In developed countries, many have turned to herbalism as a form of healthcare after identifying its effectiveness and relative inexpensiveness [10–13]. It has been suggested that herbal medicine is estimated to be three to four times more commonly practiced than conventional medicine [14].

In addition to the use of plants for curative purposes, they are widely used in the preparation of herbal beverages and as culinary herbs. Hence, the use of medicinal plants is a persistent aspect of the modern-day healthcare and lifestyle. Subsequently, one expects that the WHO guidelines on good agricultural and collection practices (GACP) for medicinal plants are followed when dealing with medicinal plants obtained by cultivation [15]. Moreover, it is also assumed that for this subgroup of medicinal plants, the application of fertilizers and pesticides will be supervised and monitored by national agencies.

The safety and quality of raw medicinal plant materials and their manufactured products depend on factors that may be classified as intrinsic (genetic) or extrinsic (environment, collection methods, cultivation, harvest, postharvest processing, transport, and storage practices) [15,16]. Herbs can get contaminated easily during any stage of growth and processing. This would subsequently result in the deterioration of the final products and their lack of necessary safety and quality. As medicinal plants and herbal medicines are usually used for prolonged periods of time, their contamination with toxic pollutants introduces the possibility of chronic health hazards [17]. Moreover, contaminants can interfere and change the chemical composition of the plants.

The WHO has been concerned with the need for quality assurance of herbal products, including testing for inadvertent contamination. Contamination sources of medicinal plants in the unprocessed form are diverse: adulteration with toxic botanicals, toxic metals, microorganisms and microbial toxins, radioactivity, fumigation agents, and pesticides. Finely powdered plant materials and final dosage forms can be additionally adulterated with synthetic and animal drug substances. Therefore, each contaminant must be investigated individually. However, this chapter emphasizes on contamination of medicinal plants with pesticides. Medicinal plants are liable to contain pesticide residues that may accumulate from agricultural practices, such as spraying, storage, transportation, or soil treatment during cultivation. Persistent pesticide residues in the soil pose as another potential source of contamination in the surrounding environment. These sources of contamination are also applicable for wild medicinal plants [9,18–20].

15.2 PESTICIDES

15.2.1 DEFINITIONS

A pesticide can be simply defined as a specific mixture of active and inert ingredients used to control pests. However, there are more precise definitions, such as those provided by the European Pharmacopoeia (EP) or the Federal Environmental Pesticide Control Act (FEPCA). The EP defines a pesticide to be "any substance, or mixture of substances, intended for preventing, destroying or controlling any pest, unwanted species of plants or animals causing harm during or otherwise interfering with the production, processing, storage, transport or marketing of vegetable drugs." The item includes substances intended for use as growth regulators, defoliants or desiccants and any substance applied to crops either before or after harvest to protect the commodity from deterioration during storage and transport [21]. The FEPCA defines a pesticide to be "any substance or mixture of substances intended for preventing, destroying, repelling or mitigating any pest/insect, rodent, nematode, fungus, weed, other forms of terrestrial or aquatic plant or animal life or viruses or bacteria or other microorganisms, except viruses, bacteria or other microorganisms on or in living man or other animals, which the Administrator declares to be a pest—and any substance or mixture of substances intended for use as a plant regulator, defoliant or desiccant" [22].

The selection, use, and handling of pesticides are a complex issue. Several hundreds of chemicals are classified and registered as pesticides in multiple formulations, with variations in their interaction with organisms (target and nontarget) and the environment. There are more than 700 pesticides currently in use, and more than 3000 basically different formulations of pesticides. Ausubel reported that since 1945, the overall pesticide use has raised by 3300% [23–26].

The primary concern of many farmers in developing countries—especially those working on a small scale—is to grow and sell crops to survive and exist. Such farmers are usually poor and unaware of the dangerous implications of pesticide misuse [27]. Moreover, their scientific background is insufficient to appropriately appreciate the mechanism of action of herbicides and weedicides. This understanding is of utmost importance when using such chemicals, as they target the biochemistry shared between the parasitic weeds and beneficial crops.

15.2.2 PESTICIDES AND THEIR BENEFITS

It is well accepted that the use of any biologically active compound can be accompanied by various degrees of toxic reactions or adverse effects. Therefore, even the safe and biologically degradable pesticides may harm the end-consumers: human and animals. Although the hazards of pesticides are evident when used irrationally and excessively, their positive and essential contribution to health and economy is worth mentioning.

Plants are the main source of food and are liable to being attacked by a wide range of pests. Pesticides eliminate or, at least, minimize the occurrence of certain arthropods and other vector-borne diseases. This, in turn, increases the production of plant-based food and fibers. However, if pesticides use is controlled and monitored, then protection from pests through each developmental stage, transport, and storage can be achieved. Needless to say, international and national regulations protecting the users and consumers should be strictly followed.

15.2.3 CLASSIFICATION OF PESTICIDES

Pesticides can be classified according to their function, chemical origin, intended use, mode of action, and toxicity. According to the former classification, the well-known classes include herbicides, rodenticides, fungicides, and insecticides. Additional subgroups based on specific target pests are larvicides, miticides, acaricides, nematicides, or silvicides. Furthermore, attractants and repellents, although mostly devoid of "-cidal activity," are also considered as pesticides as they are used for pest control [28,29]. The WHO recommends classifying pesticides based on hazard into five classes on the basis of LD_{50} values for rat as [30]:

- Extremely hazardous (Ia)
- Highly hazardous (Ib)
- Moderately hazardous (II)
- Slightly hazardous (III)
- Unlikely to present hazard in normal use (III+)

However, a classification based on the chemical structure of pesticides is the most useful for analytical purposes. Such a classification incorporates [25]:

- Organochlorines (OC or chlorinated hydrocarbons) and related pesticides: e.g., clofenotane (DDT), aldrin, chlordane, hexachlorocyclohexane (HCH), and hexachlorobenzene (HCB).
- Chlorinated phenoxyalkano acid herbicides: e.g., 2,4-dichlorophenoxy-acetic acid (2,4-D) and 2,4,5-trichlorophenoxy-acetic acid.
- Organophosphorous pesticides: e.g., malathion, parathion, demeton, and chlorpyrifos.
- Carbamate insecticides: e.g., carbaryl.
- Dithiocarbamate fungicides: e.g., ferbam, maneb, thiram, and zineb.
- Inorganic pesticides: e.g., aluminum phosphide, calcium arsenate, and lead arsenate.
- Pesticides of plant origin: e.g., tobacco leaf and nicotine, neem extract, pyrethrum flower, pyrethrum extract, pyrethroids, derris root, and rotenoids.
- Miscellaneous: pesticides classified as miscellaneous vary in their chemical structures, toxicities, physical, and chemical properties. Nevertheless, some members of this class might share common mechanisms of action. Bromopropylate, ethylene oxide, methyl bromide, and ethylene dibromide are some of the well-known miscellaneous pesticides.

It is expected that pesticides belonging to the same group will exhibit similar pharmacologic effects based on the similarities of their structures. In this aspect, it is worth mentioning that the toxicological effects of the individual pesticides within each group differ widely, and hence, each pesticide, regardless of their structural similarities, should be assessed and evaluated independently [25].

15.2.4 PESTICIDE RESIDUES

Pimentel reported that from the millions of tons of pesticides utilized worldwide, only 1% reaches the target pest. The remaining 99% of these more or less toxic chemicals, on the other hand, are released indiscriminately into the environment [31]. Such data indicate the alarming possibility, whereby any living organism can be contaminated with different pesticides without being their direct target. A residue will result when a crop, edible animal (commodity), or medium of the environment (air, water, soil, wildlife, etc.) is treated with a chemical or exposed unintentionally to it by drift, irrigation water, feed, or any other mode. The pesticide residues can be detected in the commodity or medium at the time of exposure and for some period afterward [32]. The detection of these chemicals depend on several factors, such as concentration, solubility, volatility, biodegradation and stability of the chemicals, type of pesticide formulation, time span between exposure and detection process, growth stage of the plant material, the environment in which the pesticide resides, and finally, the methods and precision applied in pesticide detection. The magnitude of the residue at any point of time will depend on the treatment, exposure level, and the rate at which the residue dissipates from the commodity. While most of the pesticides have very short residual actions, only the chlorinated hydrocarbons and other related pesticides (e.g., aldrin, chlordane, DDT, dieldrin, HCH), in addition to a few organophosphorous pesticides (e.g., carbophenothion) have a long residual action [9,32]. Therefore, WHO suggests that if the length of the exposure to pesticides is unknown, then the medicinal plant material should be tested for the presence of organically bound chlorine and phosphorous [9]. However, the analysis and identification of such substances may not always be easy. It is widely accepted that some plants convert pesticides into insoluble residues unextractable using conventional laboratory procedures [33]. Furthermore, some plant species (squash and carrots) can concentrate chlorinated compounds found in the environment during their growth cycle [34]. Moreover, it should be acknowledged that different plants and herbs exhibit different capabilities in retaining pesticide residues. An earlier study showed that chlorothalonil is suitable for dill, garlic, chives, and mint, but causes excessive residues in rosemary [35].

15.2.5 Sources of Contamination of Medicinal Plants with Pesticides

Needless to say, pesticide residues are expected to be found in almost all plants, including grains, vegetables, fruits, and medicinal plants. Even the so-called organically raised crops are not necessarily pesticide-free, and the use of inappropriate methods of detection can easily yield false negative results [34]. The vast majority of countries follow the directives for the controlled and integrated use of agrochemicals. Moreover, many countries have their own restrictions for the use of pesticides. Nevertheless, cultivated and naturally grown herbs found on the herbalist-shelves can still be contaminated with pesticide residues. Some of the reported reasons for contamination are [19]

- Origin of herbs from countries where pesticide-application regulations are not rigorously applied or even neglected.
- Accidental contamination of herbs with pesticides either by migration from the neighboring cultivars or owing to their application for the elimination of other predator pests.
- Control of socioeconomic damages, which dictates the use of pesticides when dealing with the cultivation of medicinal plants on a large scale.
- Persistent residual pesticides lingering in the environment.

Published reports on plant contamination with pesticide residues found in the literature are primarily on crop plants, vegetables, and fruits. The reported data about medicinal plants, however, is limited [36–49]. It is of high priority that governments should reinforce the necessity of having control laboratories capable of pesticide screening in accordance with WHO and Food and Agriculture Organization (FAO) guidelines [13]. This is of high significance, as the general belief in most societies is that "plants and phytomedicines are safe." With this notion, plant infusions are popular among families with newborns, infants, nursing mothers, and elderly members. As teas are widely accepted drinks worldwide, recent publications have suggested that while screening for pesticides in the commonly used medicinal plants, infusions should be the focus of investigation instead of the plants used for their preparation. The transfer of the pesticides from the solid matrix to the infusion is not total, and depends on the plant constituents and several other physicochemical factors [50–52]. It is toxicologically interesting to determine the actual amount of pesticide that is transferred from the plant material to the infusion during its preparation. Subsequently, the actual human intake of pesticides can be measured, thus, making it possible to evaluate the chronic toxicological effect of herbal teas in habitual consumers [53].

15.3 ANALYSIS OF PESTICIDES

15.3.1 General Aspects

A general analytical method cannot be applied unconditionally to all medicinal plants and their products. Medicinal plants, or even pharmaceuticals, can sometimes contain totally dissimilar constituents. Consequently, customized methods must be developed or at least revised for each plant. When selecting a method for the analysis of a mixture of pesticides, several factors should be taken into consideration for the validation of the chosen protocol. The availability of equipments and the cost-effectiveness are the primary criteria that influence the selection of an analytical method.

The chosen methodology should generate reliable and reproducible results. The following WHO guidelines should be considered to maximize the efficiency of the chosen method [9]:

- The samples should be tested as quickly as possible after collection. This is to minimize the occurrence of any physical or chemical changes.
- To prevent pesticides degradation, plants and their extracts should be protected from exposure to light.
- The type of container or wrapping material used should not interfere with the sample or affect the analytical results.
- To prevent solvent interference, only pesticide-free grade solvents must be used.
- The simplest and quickest extraction and cleanup procedure should be used for the separation of matrix substances.
- The process of concentrating solutions should be undertaken with great care, especially when evaporating the remaining solvent traces, to avoid loss of pesticide residues.
- When preparing standard solutions of pesticides, solubility of the pesticides and stability of the solutions must be taken into consideration.

15.3.2 CHALLENGES IN THE ANALYSIS OF PESTICIDE RESIDUES

Since the advent of synthetic organic pesticides, a variety of techniques and solvents have been used to extract and analyze pesticide residues. The expectations from pesticide analysis in the twenty-first century can be summarized as follows [54]:

- Expansion of the range of pesticides monitored by a single analysis
- Detection of analytes at lower levels and with greater precision
- Increasing the confidence in the validity of the residue data
- Reducing the analysis turnaround time
- Reducing the usage of hazardous chemicals
- Improving the cost-effectiveness

15.3.3 PESTICIDE-RESIDUE ANALYSIS—METHODS

There are two general analytical methods for the determination of pesticide residues in food, plants, and environmental samples. These are [32]

Single residue method (SRM): SRM is used for the quantitative determination of a single pesticide in the samples. The SRM can be implemented, provided the sample is known or suspected to contain a specific pesticide.

Multiresidue method (MRM): MRM is capable of detecting and quantifying more than one pesticide in more than one sample. With this method, simultaneous qualitative and quantitative determination of several pesticides and their conversion products can be carried out. The former serves a screening purpose, whereby rapid determination of the presence of a pesticide along with its tolerance level can be made. The latter, as the name suggests, quantifies pesticide residues in a given sample. The MRMs are commonly used by governmental agencies for surveillance and monitoring. Compared with SRM, MRM has the advantage of being a time- and workload-saving method.

15.3.4 METHOD VALIDATION

Method validation includes procedures demonstrating that a particular method used for quantitative measurements of analytes in a given matrix is reliable and reproducible for the intended use [55].

Full validation is recommended when implementing an analytical method for the first time, while partial validation can be used when an analytical method is shared among the laboratories [56]. The inclusion of particular parameters in the validation depends on the application, test samples, and domestic or international guidelines or regulations, as applicable [57].

15.3.5 LIMIT OF DETECTION AND LIMIT OF QUANTIFICATION

Limit of detection (LOD) is the lowest concentration of an analyte that can be determined to be different, with a high degree of confidence, from the blank or background. It can be also referred to as the concentration of an analyte in a sample that gives rise to a peak with a signal-to-noise ratio (S/N) of 3. Analysis of samples at different concentrations can provide the necessary information to calculate the detection limits [32,58,59].

Limit of quantification (LOQ) refers to the level above which the quantitative results can be reported with a specified level of confidence. The LOQ is generally several times higher than LOD, reflecting the fact that most operators are not confident in reporting a residue whose signal is only twice that of the background [32].

15.4 SAMPLE PREPARATION

15.4.1 EXTRACTION AND CLEANUP OF PESTICIDE RESIDUES FROM PLANTS

When performing pesticide analysis in plants, the extraction and cleanup methods should be selected to suit the nature of the suspected pesticide(s) and matrix substances. It is essential to consider the method of analysis to be used prior to initiating sample extraction.

"Extraction" aims to separate as much of the pesticide as possible, while minimizing the isolation of complex matrix compounds and other contaminants that might interfere with the analysis. The quality of extraction will, in turn, influence the adequacy of the result analysis. Hence, the isolation and cleanup processes are of utmost importance for the success of the residue analysis. Therefore, it is essential to have good knowledge of the physicochemical properties (molecular size, solubility, and volatility) of the analyte. This is especially true when dealing with analytes of low concentration. Early investigators used nonpolar solvents to extract organic pesticides. Subsequently, mixtures of polar and nonpolar solvents were utilized. As both the types of solvents failed to extract pesticide residues satisfactorily, attempts intensified to identify the solvents agreeing with the plant matrix and solubility of the pesticide residues [32,60,61]. In this regard, a literature review indicates that a large number of single solvents and solvent mixtures were utilized to extract specific pesticide residues [62,63].

Two commonly used extraction methods are [57,62,64]

- Luke's method (Method I)—known as Pesticide Analytical Manual (PAM) 302 or Analysis of the Association of Official Analytical Chemists (AOAC) method 985.22—is suitable for the extraction and cleanup of nonfatty samples (containing <10% fat or oil).
- Mill's method (Method II)—known as PAM 304 or AOAC acetonitrile partitioning method—is applied to fatty/oily samples.

"Cleanup" refers to the step(s) carried out to purify extracts, allowing more definitive identification of the residues at lower quantitative limits, and minimizing the adverse effects on instrumentation used in the analysis. Well-accepted cleanup procedures, based on solvent partition and column chromatography, suitable for fatty and nonfatty plant materials, are reported in the PAM of the Food and Drug Administration (FDA) and the official methods of the AOAC. Methods from the latter organization are, in fact, subjected to rigorous levels of validation, where a collaborative study of the method by several different laboratories is conducted [7,32].

"The Luke's method" is a multiclass, multiresidue, rapid, and reproducible method. In this method (Method I), acetone is used as an extracting solvent. Acetone is preferred as it is completely miscible with water, thus allowing a good penetration into the aqueous part of the plant. In the last evaporation step, prior to gas-chromatography (GC) analysis, solvent transfer is an important and delicate step, to ensure that no traces of dichloromethane (DCM) would enter into the GC column, and consequently, into the detector. While using the Luke's method can result in loss of some pesticide residues, its speed, reproducibility, reliability, and broad utilities has ensured its dominance in the majority of FDA enforcement analyses in the 1980s and 1990s [32,42,44]. However, a drawback of this method is the application of DCM. The environmental impact of chlorinated solvents makes it less desirable from an environmentalist's stand. In Germany, extraction with DCM has been gradually replaced by extraction with ethyl acetate–cyclohexane (1:1, v/c) [65].

Acetonitrile partitioning in Method II is known as the "Mill's method." Oily materials and pesticides are extracted exhaustively with petroleum ether (PE) in a Soxhlet apparatus and subsequently partitioned with acetonitrile and PE. Anhydrous sodium sulfate, added to the ground plant materials in the extraction thimble, removes moisture and helps to disintegrate the sample. Pesticide residues are isolated from fat by partition of PE and acetonitrile. Most of the fat will be retained in PE, while the residues partition into acetonitrile, in proportion to their partition coefficient in the system. In the subsequent steps, residues in acetonitrile are partitioned back into PE when added water reduces their solubility in acetonitrile. Again, solvent transfer in the last evaporation step after the extract is cleaned on Florisil column is important in ensuring that no traces of acetonitrile enter into the column [57,62].

15.4.2 FLORISIL COLUMN CLEANUP

A variety of methods have been used for the cleanup of medicinal plant matrices [48,66]. Conventional adsorption column chromatography is a widely used procedure for the removal of impurities. To trap analytes, the technique uses several sorbents, including alumina, silica gel, and Florisil. While compounds unstable on Florisil or alumina may be recovered intact from silica gel, polar residues can be cleaned on a charcoal column.

Florisil, a single-step cleanup procedure, is one of the preferred cleanup methods in pesticide-residue analysis. Florisil is a magnesia-bonded silica gel sharing many of its characteristics. It is extremely polar in nature and is ideal for the isolation of polar compounds from nonpolar matrices. It is a highly selective adsorbent often used to separate pesticides of a relatively narrow polarity range from extraneous interfering compounds of different polarity [65]. Elution with approximately 6% diethyl ether in PE recovers the less polar analytes, primarily OC pesticides. On the other hand, elution using approximately 15% diethyl ether in PE moderately recovers polar OCs and organophosphates (OPs). Alternatively, usage of a 50% elution fraction will recover more polar OPs [24,62,65,67]. As is the case with any type of column-packing, Florisil column-packing should be performed meticulously. Air bubble formation should be avoided and the uniformity of column-packing should be maintained to prevent the channeling of the used solvent through the voids. Additionally, the column should not be left to dry. If these column-packing thumb-rules were not followed, residue separation and cleanup can be of inadequate efficiency.

15.4.3 NOVEL EXTRACTION AND CLEANUP METHODS

In addition to the classical multiresidue extraction methods (Luke's and Mill's method), several others have been reported in the recent literature. These include the solid-phase extraction (SPE), solid-phase microextraction (SPME), matrix solid-phase dispersion (MSPD), stir-bar sorptive extraction (SBSE), microwave-assisted solvent extraction (MASE), microwave-assisted extraction (MAE), and supercritical fluid extraction (SFE or SCFE). These new extraction methods have several advantages. They can be classified as "environment-friendly" as they depend primarily on water as an extraction solvent, thus minimizing, or even eliminating, the use of organic solvents. Moreover, they offer very high recoveries, have a shorter analysis time, cope well with critical samples of small size, and the additional cleanup is not obligatory. Such novel techniques are widely used for the determination of different classes of pesticides in various fruits, vegetables, milk and other food, water, and soil samples. The SPE columns can provide a rapid cleanup for extracts containing pesticide residues in complex matrices, by retaining the latter and facilitating the elution of the former. Subsequently, analysis of the pesticide residues is achieved by gas chromatography–mass spectrometry (GC–MS), liquid chromatography–mass spectrometry (LC–MS), liquid chromatography–tandem mass spectrometry (LC–MS–MS), gas chromatography–tandem mass spectrometry (GC–MS–MS) [68–78].

However, published reports in the literature are very limited when it comes to applying these novel methods in the extraction of pesticide residues from medicinal plants. The SBSE is reported to have been successfully applied as a reproducible method for the analysis of pesticides in *Passiflora alata* herbal teas, wine, and other beverages [53,79]. Tang et al. developed two MRMs for the determination of 15 pesticides in the medicinal herb *Isatis indigotica* Fort. and its formulations [80]. Their analytical procedure is based on ultrasonic assisted extraction and liquid–liquid extraction, followed by capillary GC determination of the pesticide residues. The MSPD-microextraction (MSPD-ME) has been reported as a useful extraction method for the screening of OC and OP pesticide residues in some medicinal herbs using GC [73]. On the other hand, Campillo et al. analyzed 10 different pesticides belonging to OCs, OPs, as well as pyrethrine pesticides using SPME and GC-atomic emission detection (GC-AED). They concluded that SPME is a low-cost alternative to common extraction methods, as it integrates sampling, extraction, preconcentration, and cleanup into a single step. Furthermore, avoidance of the use of organic solvents is another advantage [50]. The SPE and GC–MS determination was also reported for the determination of OCs in three different medicinal plants of Brazil [81]. Zuin et al. extolled the advantages of SFE while applying it for the extraction of OC and OP pesticide multiresidues from *P. alata* and *P. edulis*. These include high concentration capability, cleanliness and safety, quantification properties, expeditiousness, simplicity, and selectivity [82]. A second advanced method of extraction, namely headspace-solid-phase microextraction (HS-SPME), was successfully applied in the GC–ECD analysis of OC and OP pesticide residues extracted from *P. alata*, *P. edulis*, and *P. incarnata* [83]. Currently, new cleanup procedures using minicolumns are paving the path for automation of the process, whereby the procedure will be faster, more cost-effective, and less labor-intensive. In conclusion, it is expected that the trend in analytical technologies (miniaturization, automation, hyphenation, simplification and reduction of processing time, and consumption of toxic solvents) will pave the way for the development of newer techniques.

15.5 SAMPLE DETERMINATION

Despite the continuous progresses in analytical chemistry, sample preparation prior to final determination still remains an essential task. Sample determination is concerned with the detection and quantification of pesticide residues. The methodology used can sometimes be complicated, as the selection of an appropriate method may depend on several factors such as the nature of the sample or the equipment available [84].

"Detection" is obtaining a qualitative and quantitative response with respect to the pesticide present. Analytical instruments, such as high-performance liquid chromatographs (HPLC) and GC are frequently used to separate individual pesticides. To improve both detectability and selectivity, it has become common practice to use both HPLC and GC in connection with mass spectrometry (MS) [7,24,69,70,85].

The GC has been the predominant tool in pesticides multiresidue methodology for over 30 years. It has been widely used for the detection of pesticide residues exhibiting high stability and low polarity. As GC involves an interaction between vapor and liquid phases, its application is restricted to analytes that can be vaporized without degradation. While derivatization has been carried out for nonvolatile analytes, HPLC remains a better alternative for nonvolatile and heat-labile pesticides. Still, WHO recommends the use of conventional column chromatography (stationary-phase aluminum oxide R and chloroform as eluent) for the determination of certain pesticides (desmetryn, prometrin, simazine) in several commonly used medicinal plants (such as chamomile, thyme, balm, caraway, mint, and fennel) [9]. Owing to its simplicity, selectivity, and availability, WHO also recommends the use of thin-layer chromatography (TLC), as it is a cheap method for the screening, qualitative identification, and quantitative determination of different pesticides [9]. In the 1980s, high-performance thin-layer chromatography (HPTLC) was introduced as an important chromatographic technique along with the GC and HPLC. Nevertheless, the practical applications of HPTLC in the analysis of pesticide residues are extremely rare.

A number of other nonchromatographic techniques, such as capillary electrophoresis (CE), immunoassay, biosensors, and spectroscopy have also been used to determine pesticide residues [86–89].

Gas chromatography: Separation in GC is achieved by differences in the distribution of analytes between mobile and stationary phases, causing them to move through the column at different rates and elute at different times. During their passage through the column, analytes injected into the same solution separate from one another according to their different vapor pressures and selective interactions with the liquid phase. When analytes elute from the column and enter a detector, the detector responds to the presence of a specific element or functional group within the molecule [85].

Columns: Separation among the analytes in GC takes place within the column. Columns are available in several different physical configurations, and each offers advantages and disadvantages with regard to pesticide-residue determinations [62]. The two basic types of GC columns currently used are

- Packed columns, in which the liquid phase is immobilized as a film of particles of fine-mesh solid support and packed into 2–4 mm internal diameter (i.d.) columns.
- Open tubular capillary columns, in which the liquid stationary phase is immobilized as a film onto the interior walls of a capillary tube. Capillary columns are further distinguished by their internal diameter: wide bore (0.53 mm i.d.), traditional (0.25–0.32 mm i.d.), and narrow bore (≤0.25 mm i.d.).

Until the 1980s, packed columns were used exclusively. However, since the 1990s, the capillary columns with greatly enhanced column efficiencies have been preferred. In all GC columns, the identity of the stationary phase is the primary factor dictating the types of separations achievable [32,85,90]. Solid stationary phases are required to have a large surface area, low sorptive activity toward pesticides, and remain chemically inert. Alternatively, the liquid stationary phases are composed primarily of silicone-based oils with high temperature stability and different polarities. The carrier gas (mobile phase) is also integral to GC operation. However, since only inert gases can be used as carriers, the list of options is limited. Nitrogen is the most commonly used carrier gas owing to its availability and cost. Operating parameters that affect column efficiency, including column temperature, carrier-gas identity, and flow rate, provide additional variables that can be adjusted to achieve separations required for the analyses [32,85,90]. Confirmation of the presence of pesticide residues should be carried out by the use of two chromatographic columns of different polarities [32,85,90].

Detectors: The advantage of using GC is that the gases emerging from the column can be monitored with great sensitivity using a variety of universal and specific detectors. The detectors used for the pesticide-residue analysis are of the selective type. This is because even following a highly efficient cleanup, a food, plant, or soil extract, may still contain hundreds of potential interferences, many at a much higher concentration than the analyte(s) of interest. While numerous types of detectors are available, two of them are widely used. These are

1. *Electron-capture detector (ECD)*: ECD is widely applied for the analysis of polyhalogenated compounds, to which it is particularly sensitive. The detector responds to the loss of electrical signal when the electrons produced by radioactive beta-emitters are captured by the organic analytes, as they elute from the GC column. The relatively high standing current, produced by the beta particles and their descendent thermal electrons produced following ionization of nitrogen or another carrier gas by the beta particles, is reduced when the electron capture takes place. In general, polyhalogenated compounds, compounds with electron withdrawing groups, those with conjugated carbonyl systems, and those with sulfur, respond well to ECD. With regard to medicinal plants, the detector's major limitation is the wide variety of naturally occurring compounds possessing at least one of the above-mentioned structural features, thus, presenting an array of potential interferences and necessitating the use of efficient cleanup methods [32,85,90]

2. *Nitrogen–Phosphorous detector (NPD)*: NPD is an expensive detector, with good detection limits for the trace analysis of OPs, carbamates, urea, triazine, and other classes of pesticides [90,91]. The GC column effluent impinges onto the surface of an electrically heated and polarized alkali source in the presence of air/hydrogen plasma. Subsequently, ionization occurs and the flow of ions between the plasma phase and an ion collector is amplified and recorded. The detector response to analytes results when compounds containing nitrogen or phosphorous elute from the column [62].

Although the two previously mentioned detectors are mainly used for the pesticide-residue detection by GC, the use of AED has also been reported. The highly selective nonmass-spectrometry-based AED can be considered as an alternative detector, as pesticides, in general, contain heteroatoms [50].

HPLC: HPLC is the advanced form of the liquid chromatography (LC), used in pesticide determination since mid-1970s. A major advantage of HPLC is that it is applicable to virtually any organic analyte, regardless of its volatility and thermal stability. On the other hand, the lack of a readily adaptable array of selective detectors comparable with those available for GC is the main disadvantage. In general, the versatility is much greater in LC than in GC because both stationary and mobile phases affect the separation, and a wide range of both the phases can be used in the former. Depending on the characteristics of the stationary phase, different modes of operation can be achieved with HPLC. In adsorption HPLC (normal- or straight-phase mode), the stationary phase is more polar than the organic mobile phase, while in partition HPLC, normal- or reversed-phase mode operations can be applied. In the latter mode, the mobile phase is more polar than the stationary phase. Several types of column-packing (porous, nonporous, spherical/irregular in shape, based on naturally occurring minerals or synthetically manufactured ones) are available. Among them, silica gel and C_{18} silica gel are two of the most commonly used types for normal- and reversed-phase chromatographic applications [24,85,90,92].

Among the big array of detectors, the UV–visible absorbance, and fluorescence detectors are the most widely used detection methods for pesticides. This variety of columns and detectors indicates that the selectivity of the method can be readily adjusted. As with the GC, HPLC can be attached to the MS. The powerful features of HPLC–MS, which include efficient separation, identification, and quantification of the analytes, make this technique very attractive for pesticide-residue analysis [24,85,90,92].

15.6 PESTICIDE-RESIDUE DETERMINATION IN A MEDICINAL PLANT: *TRIGONELLA FOENUM-GRAECUM*

The possible contamination of *T. foenum-graecum* with pesticide residues was evaluated using the Luke's method and Florisil column cleanup for the simultaneous determination of multiresidue pesticides. The work described in this section was performed at the authors' laboratory.

15.6.1 MATERIALS AND METHODS

All the solvents used were of pesticide residue (PR) grade (Scharlau, Barcelona). Pesticide standards were purchased from Ehrenstorfer (Augsburg, Germany). Standards, their purity levels, and the groups to which they belong are listed in Table 15.1. With the exception of *trans*-chlordane (purchased as a 10 μg/mL standard solution), the stock solutions of the individual pesticides (1000 μg/mL) were prepared and stored at −20°C. Individual dilutions were prepared as needed and stored at 4°C. Seven mixed standard solutions of the CHs, pyrethroids, and miscellaneous pesticides, and one mixed solution of the OP pesticides and primicarb (a carbamate) were prepared with concentrations of range 0.01–0.5 μg/mL (Table 15.1). Concentrations were selected to suite the sensitivity of the detectors used. An internal standard solution of endrin (0.03 μg/mL) and bromophos (0.2 μg/mL) was also prepared and added to all the mixed pesticide standard solutions and extracts in the final

TABLE 15.1
Groups of the Pesticides and the Percentage of Purity of Their Standards

Pesticide[a]	Group	Purity (%)	Conc[b] (μg/mL)	Pesticide	Group	Purity (%)	Conc (μg/mL)
Folpet	Misc.[c]	99	0.5	β-Endosulfan	OC	97.5	0.1
HCB	OC[d]	99.7	0.02	*p,p*-DDT	OC	99.3	0.05
α-HCH	OC	97.5	0.01	Bromopropylate[e]	OC	99.7	0.1
Quintozene	OC	99.8	0.05	Fenpropathrin	PY[f]	92	0.1
γ-HCH	OC	99.4	0.05	Tetramethrin	PY	100	0.5
β-HCH	OC	98.5	0.02	Tetradifon	OC	99.1	0.05
Vinclozolin	Misc.	97	0.05	Permethrin	PY	97.5	0.5
Chlorothalonil	OC	98.5	0.2	Cypermethrin	PY	94.8	0.3
Dicofol	OC	97.5	0.1	Deltamethrin	PY	99.6	0.4
Penconazole	Misc.	99.5	0.05	Methacrifos	OP[g]	95.5	0.2
trans-Chlordane	OC	100	0.03	Formothion	OP	100	0.5
Procymidone	Misc.	99	0.1	Primicarb	C[h]	98.3	0.2
o,p-DDE	OC	97.5	0.03	Chlorpyrifos	OP	99.5	0.2
p,p-DDE	OC	99.3	0.03	Bromophos	OP	99.9	0.2
Dieldrin	OC	96	0.03	Phosalone	OP	98	0.5
Endrin	OC	98	0.03	Pyrazophos	OP	99.7	0.3
o,p-DDT	OC	97	0.2				

Source: Hajou, R.M.K. et al., *Pharm. Biol.*, 43(6), 554, 2005. With permission.
[a] The pesticide names used here are as quoted in *The Pesticide Manual* [106].
[b] Concentration of the prepared pesticide's stocks.
[c] Miscellaneous.
[d] Organochlorines.
[e] Bromopropylate is a Br-DDT analogue, in some references it is classified as a miscellaneous pesticide [13].
[f] Pyrethroid.
[g] Organophosphate.
[h] Carbamate.

step prior to the GC analysis. All the pesticide standard solutions and dilutions were prepared in acetone–hexane (10:90, % v/v). Before use, Florisil 60–100 mesh (Aldrich, USA) was activated in an oven (Mod.N7/H, Nr.66341—Naber, West Germany) at 675°C for 6 h. Activated Florisil was kept in 500-mL glass flasks with glass stoppers and stored at 130°C in Memmert's oven (Schawbach, West Germany). Anhydrous sodium sulfate analytical reagent (AR) grade (Merck, Germany) was heated in Memmert's oven (Schawbach, West Germany) at 130°C for 5 h, and then stored in 500 mL glass jars with glass stoppers in a desiccator (Pragati, India). Sodium chloride AR grade (Nottingham, U.K.) and Whatman filter paper (Cat. No. 1002 110—Medicell International Ltd., U.K.) were used.

The equipments used included a high-speed blender with a stainless steel jar (Moulinex, France), a Rotavapor (R110—Büchi, Switzerland), a cooler circulator (Julabo, Germany), chromatographic tubes with Teflon stopcocks and course fritted glass plate (22 mm i.d. × 300 mm—Quickfit, England), and microliter syringes (Hamilton Bonadus AG, Switzerland). All glassware was subjected to thorough rinsing using soap and deionized water, then washed with acetone, and oven-dried overnight (100–130°C).

15.6.1.1 Plant Material

Five dry *T. foenum-graecum* samples (T1–T5) were purchased from different herbal shops in Jordan. As the study was concerned with the possible contamination of this medicinal plant found in the Jordanian market, no enquiries were made about the origin of the purchased samples. One-kilogram of the sample, from different sources, was immediately treated after delivery to the laboratory [93].

The plant samples were authenticated by comparing them with the herbarium specimens.

15.6.1.2 Chromatographic Instrumentation

Determination of CHs, pyrethroids, and miscellaneous pesticides: An HP-5890 series II GC equipped with an HP-608 capillary column (30 m, 0.53 mm i.d., 0.5 μm of film thickness) with the stationary phase comprising 50% phenyl methylpolysiloxane and ^{63}Ni ECD was used. The GC instrument was operated under the following conditions: injector in the split mode (split ratio, 1:17), injector temperature of 250°C, detector temperature of 300°C, argon–methane (5:95, % v/v) as the carrier gas at a flow rate of 1 mL/min or as the make-up gas at a flow rate of 24 mL/min. The column temperature was initially held at 80°C for 1 min, then programmed at 30°C/min to 180°C, followed by 5°C/min to 200°C, and 10°C/min to 280°C and held for 14 min.

Determination of OP pesticides: An HP-5890 series II GC, equipped with an HP-1 capillary column (25 m, 0.2 mm i.d., 0.5 μm of film thickness) with the stationary phase comprising 100% dimethyl polysiloxane and NPD was used. The instrument was operated under the following conditions: injector in the split mode (split ratio, 1:10), injector temperature of 225°C, detector temperature of 280°C, helium as the carrier gas with flow rate of 1 mL/min, and detector-gas flow rates were 3–3.5 mL/min for hydrogen and 100 mL/min for air. Column temperature was initially held at 90°C for 2 min, then programmed at 20°C/min to 150°C, followed by 6°C/min to 270°C and held for 15 min.

Confirmation of identity: An HP-5890 Series II GC, equipped with an HP-5 capillary column (30 m, 0.25 mm i.d., 0.25 μm of film thickness) with the stationary phase comprising 5% diphenyl and 95% dimethyl polysiloxane, and ^{63}Ni ECD was used. The instrument was operated under the following conditions: injector in the split mode (split ratio, 1:17), injector temperature of 280°C, detector temperature of 300°C, carrier gas was helium with a flow rate of 2 mL/min, make-up gas was argon–methane (5:95, % v/v) with a flow rate of 30 mL/min, column temperature was initially held at 80°C for 2 min, then followed by 30°C/min to 175°C, and at 10°C/min to 225°C and held for 2 min, then at 20°C/min to 280°C and held for 10 min.

Extraction and partitioning: About 10 g of powdered plant samples were extracted with water/acetone, followed by liquid–liquid partitioning in PE/DCM as described in Method I.

Florisil column cleanup: In this study, the original AOAC Florisil column cleanup method was modified by eluting the column with 250 mL of DCM–PE (20:80, % v/v), and 150 mL of DCM. The combined eluates were evaporated and, before reaching dryness, solvent transfer using few milliliters of *n*-hexane was performed. The evaporation was continued until only a thin film of solvent remained in the flask. The internal standard solution was used to adjust the final volume to 5 mL. This sample was applied to GC analysis.

Determination of pesticides retention times (t_R) *and relative retention times* (*RRTs*): For the determination of t_R for each individual pesticide, 1 µL of the 1.0 µg/mL of pesticide solution was injected into the GC column. Standard mixtures of the pesticides were prepared in the concentrations listed in Table 15.1. One microliter of each standard mixture was also injected into the GC column. The pesticides were identified by comparing their t_Rs and RRTs (Table 15.2).

Limits of detection (*LOD*): LOD of the used instruments, equipped with ECD and NPD, were determined for each pesticide by successive dilution of the standard mixed pesticide solution, followed by repetitive injections into the GC column. Serial dilution experiments provided the necessary information to calculate the detection limits [58,59]

Recovery tests: The recovery test was evaluated with all the 33 pesticides used in this study. This was performed by spiking the plant samples with a concentrated mixed pesticide solution in concentrations ranging from 0.01 to 0.5 µg/mL. The spiked plant samples were then extracted according to the proposed method [62]. Samples used in recovery tests were chosen to be pesticide-free, and tested using the same methodology. To evaluate the recoveries of the residue analytical procedure without being affected by interferences from the plant matrices, spiking of water, instead of plant samples, was also carried out. All the tests were carried out in triplicate. Recovery studies were performed at one concentration level, differing from one pesticide to another [94].

Reagent blank analysis: Reagent blank analysis was performed using only the reagents (without plant sample) to determine whether there were any detector responses that could be mistaken for the pesticide residues. These blank analyses were performed once every week.

Internal standards: Endrin and bromophos were selected as the internal standards as they were rarely detected in the previous work carried out in Jordan [95].

Control samples: Chromatograms of plant extracts, found to be free from contamination with pesticide residues, were used for comparison purposes with chromatograms of plant samples that were contaminated. These pesticide-free plant samples (control samples) were used for fortification purposes in the recovery studies.

Residue analysis: For residue analysis, the purchased samples (T1–T5) were ground mechanically and sieved through No. 60 mesh sieve. About 10 g of the samples were extracted according to the AOAC method and cleaned on a Florisil column.

15.6.2 Results and Discussion

In this study, some modifications were made on the original method: first, the quantity of the medicinal plant employed for the analysis was reduced to about ½ or even ¼ the amount suggested by Luke's method. Consequently, the consumption of all used reagents was reduced, without decline in the performance. Second, rotavapor was used instead of the Kurdena–Danish concentrator.

The LODs for GC–ECD and GC–NPD were 0.0008–0.05 ppm for the OCs and the miscellaneous pesticides, 0.02–0.1 ppm for pyrethroids, and 0.006–0.5 ppm for OP pesticides (Table 15.2). The mean recoveries of the studied pesticides from spiked water samples ranged from 83% to 120% for the AOAC method 985.22 (Table 15.3), and from 72% to 120% for the AOAC method combined

TABLE 15.2
Types of Detectors, Pesticides' Retention Times, Relative Retention Times, and Limits of Detection

Nos.	Pesticide	Detector	$t_R{}^a$	RRT[b]	LOD[c] (pg)
1[d]	Folpet	ECD	11.552	0.490	0.05
2	HCB	ECD	13.552	0.586	0.00086
3	α-HCH	ECD	14.453	0.612	0.0008
4	Quintozene	ECD	15.427	0.654	0.001
5	γ-HCH	ECD	16.029	0.679	0.0025
6	β-HCH	ECD	16.283	0.690	0.0022
7	Vinclozolin	ECD	17.309	0.734	0.003
8	Chlorothalonil	ECD	17.367	0.736	0.01
9	Dicofol	ECD	19.885	0.843	0.003
10	Penconazole	ECD	20.491	0.868	0.007
11	trans-Chlordane	ECD	20.896	0.886	0.003
12	Procymidone	ECD	21.128	0.895	0.003
13	o,p-DDE	ECD	21.434	0.908	0.004
14	p,p-DDE	ECD	22.090	0.936	0.0056
15	Dieldrin	ECD	22.401	0.949	0.006
16	Endrin	ECD	23.597	1.000	0.006
17	o,p-DDT	ECD	23.845	1.011	0.01
18	β-Endosulfan	ECD	24.160	1.024	0.01
19	p,p-DDT	ECD	24.835	1.053	0.01
20	Bromopropylate	ECD	26.331	1.116	0.01
21	Fenpropathrin	ECD	26.447	1.121	0.06
22	Tetramethrin	ECD	27.102	1.149	0.02
23	Tetradifon	ECD	29.029	1.230	0.005
24	Permethrin	ECD	31.708	1.344	0.06
			34.785	1.474	
25	Cypermethrin	ECD	35.094	1.487	0.07
			35.551	1.507	
26	Deltamethrin	ECD	47.622	2.018	0.1
27[e]	Methacrifos	NPD	10.132	0.542	0.0075
28	Formothion	NPD	15.387	0.822	0.019
29	Primicarb	NPD	15.640	0.836	0.01
30	Chlorpyrifos	NPD	18.080	0.966	0.024
31	Bromophos	NPD	18.710	1.000	0.5
32	Phosalone	NPD	26.320	1.407	0.1
		ECD	29.343	1.244	0.14
33	Pyrazophos	NPD	27.655	1.478	0.006

Source: Hajou, R.M.K. et al., Pharm. Biol., 43(6), 554, 2005. With permission.

a t_R: retention time.
b RRT: relative retention time = t_R(pesticide)/t_R(internal standard).
c LOD: limit of detection.
d Pesticides 1–26 are numbered according to sequence of elution from HP-608 GC column on ECD.
e Pesticides 27–33 are numbered according to sequence of elution from HP-5 GC column on NPD.

TABLE 15.3
The Spiked Level of Each Pesticide, Mean Recovery, RSD, Relative Errors, and Total Errors for Method I

Nos.	Pesticide	Added Absolute Amount (µg)	Mean Recovery[a] % ± SD	RSD (%)	Relative Error	Total Error
1	Folpet	2.5	92 ± 1.2	1.3	8	10.6
2	HCB	0.1	114 ± 6.5	5.7	4	15.4
3	α-HCH	0.05	105 ± 6.5	6.2	5	17.4
4	Quintozene	0.25	98 ± 5.3	5.4	2	12.8
5	γ-HCH	0.25	112 ± 6.1	5.4	12	22.8
6	β-HCH	0.1	84 ± 1.6	1.9	16	19.8
7	Vinclozolin	0.25	96 ± 2.4	2.5	4	9.0
8	Chlorothalonil	1.0	87 ± 5.0	5.7	13	24.4
9	Dicofol	0.5	117 ± 4.2	3.4	17	23.8
10	Penconazole	0.25	93 ± 5.9	6.3	7	19.6
11	trans-Chlordane	0.15	92 ± 3.4	3.7	8	15.4
12	Procymidone	0.5	99 ± 1.2	1.2	1	3.4
13	o,p-DDE	0.15	99 ± 6.5	6.6	1	14.2
14	p,p-DDE	0.15	100 ± 6.5	6.5	0	13.0
15	Deldrin	0.15	95 ± 3.6	3.8	5	12.6
16	Endrin	0.15	97 ± 1.0	1.0	3	5.0
17	o,p-DDT	1.0	120 ± 9.0	7.5	20	35.0
18	β-Endosulfan	0.5	96 ± 7.0	7.3	4	18.6
19	p,p-DDT	0.25	91 ± 5.0	5.5	9	20.0
20	Bromopropylate	0.5	92 ± 5.6	6.1	8	20.2
21	Fenpropathrin	0.5	95 ± 4.2	4.4	5	13.8
22	Tetramethrin	2.5	99 ± 7.0	7.1	1	15.2
23	Tetradifon	0.25	90 ± 2.9	3.2	10	16.4
24	Permethrin	2.5	98 ± 3.0	9.1	2	20.2
25	Cypermethrin	1.5	99 ± 3.0	3.1	1	7.2
26	Deltamethrin	2.0	83 ± 2.5	3.0	17	23.0
27	Methacrifos	1.0	118 ± 2.8	2.4	18	22.8
28	Formothion	2.5	120 ± 7.1	5.9	20	31.8
29	Primicarb	1.0	104 ± 3.6	3.5	4	11.0
30	Chlorpyrifos	1.0	98 ± 2.8	2.9	2	7.8
31	Bromophos	1.0	106 ± 3.4	3.4	6	12.8
32	Phosalone	1.0	114 ± 5.0 (NPD) 77±7.0 (ECD)	4.4	14	22.8
33	Pyrazophos	1.5	104 ± 4.2	4.0	4	12.0

Source: Hajou, R.M.K. et al., *Pharm. Biol.*, 43(6), 554, 2005. With permission.

[a] Recovery % is the mean value for three runs.

with Florisil cleanup, except for deltamethrin (26%), formothion (14%), and pyrazophos (21%) (Table 15.4). For spiked *T. foenum-graecum* samples extracted according to the AOAC method 985.22 and cleaned on a Florisil column, the mean recoveries were in the range 72%–116% (Table 15.5). Tetramethrin, formothion, and pyrazophos showed low recoveries, while *p,p*-DDT exhibited a recovery of >150%.

Two parameters were calculated to determine the accuracy and precision of the used method:

TABLE 15.4
The Spiked Level of Each Pesticide, Mean Recovery, RSD, Relative Errors and Total Errors for Method I and Florisil Cleanup

Nos.	Pesticide	Added Absolute Amount (µg)	Mean Recovery[a] % ± SD	RSD (%)	Relative Error	Total Error
1	Folpet	2.5	86 ± 2	3.2	14	20.4
2	HCB	0.1	107 ± 9.0	8.4	7	23.8
3	α-HCH	0.05	93 ± 1.4	1.5	7	10.0
4	Quintozene	0.25	95 ± 4.3	4.5	5	14.0
5	γ-HCH	0.25	100 ± 4.0	4.0	0	8.0
6	β-HCH	0.1	83 ± 1.4	1.7	17	20.4
7	Vinclozoline	0.25	94 ± 3.8	4.0	6	14.0
8	Chlorothalonil	1.0	80 ± 3.4	4.3	20	28.6
9	Dicofol	0.5	118 ± 5.0	4.2	18	26.4
10	Penconazole	0.25	90 ± 3.6	4.0	10	18.0
11	trans-Chlordane	0.15	91 ± 4.1	4.5	9	18.0
12	Procymidone	0.5	101 ± 6.0	5.9	1	12.8
13	o,p-DDE	0.15	92 ± 4.0	4.3	8	16.6
14	p,p-DDE	0.15	93 ± 1.4	1.5	7	10.0
15	Deldrin	0.15	92 ± 3.2	3.5	8	15.0
16	Endrin	0.15	96 ± 9.0	9.4	4	22.8
17	o,p-DDT	1.0	113 ± 4.0	3.5	13	20.0
18	β-Endosulfan	0.5	95 ± 2.5	2.6	5	10.2
19	p,p-DDT	0.25	120 ± 7.0	5.8	20	31.6
20	Bromopropylate	0.5	89 ± 4.2	4.7	11	20.4
21	Fenpropathrin	0.5	91 ± 9.2	10.1	9	29.2
22	Tetramethrin	2.5	26 ± 7.5	28.8	74	131.6
23	Tetradifon	0.25	88 ± 4.9	5.6	12	23.2
24	Permethrin	2.5	92 ± 3.0	3.3	8	14.6
25	Cypermethrin	1.5	92 ± 2.0	2.2	8	12.4
26	Deltamethrin	2.0	85 ± 2.5	2.9	15	20.8
27	Methacrifos	1.0	93 ± 2.3	2.4	7	11.8
28	Formothion	2.5	14 ± 1.6	11.4	86	108.8
29	Primicarb	1.0	Not recovered			
30	Chlorpyrifos	1.0	103 ± 6.4	6.2	3	15.4
31	Bromophos	1.0	111 ± 3.6	3.2	11	17.4
32	Phosalone	1.0	76 ± 2.0 (NPD)	2.6	24	29.2
			72 ± 1.2 (ECD)	1.7	28	31.4
33	Pyrazophos	1.5	21 ± 2.0	9.5	79	98.0

Source: Hajou, R.M.K. et al., *Food Chem.*, 88, 469, 2004. With permission.

[a] Recovery % is the mean value for three runs.

1. Relative standard deviation (RSD), a measure of method's precision:

$$RSD = SD/\% \, recovery \times 100 \, (SD : standard \, deviation).$$

2. Relative error (RE), a measure of method's accuracy:

$$RE = \frac{100 - \% \, recovery \times 100}{100}.$$

TABLE 15.5
The Spiked Level of Each Pesticide, Mean Recovery, RSD, Relative Errors, and Total Errors in Fortified *Trigonella foenum-graecum* Samples

Nos.	Pesticide	Added Absolute Amount (μg)	Mean Recovery[a] % ± SD	RSD (%)	Relative Error	Total Error
1	Folpet	2.5	74 ± 1.5	2.0	26	30.0
2	HCB	0.1	103 ± 5.0	4.9	3	12.8
3	α-HCH	0.05	101 ± 6.5	6.4	1	13.8
4	Quintozene	0.25	94 ± 3.8	4.0	6	14.0
5	γ-HCH	0.25	98 ± 4.0	4.1	2	10.2
6	β-HCH	0.1	82 ± 2.0	2.4	18	22.8
7	Vinclozolin	0.25	93 ± 3.2	3.4	7	13.8
8	Chlorothalonil	1.0	81 ± 3.1	3.8	19	26.6
9	Dicofol	0.5	116 ± 2.5	2.2	16	20.4
10	Penconazole	0.25	91 ± 4.8	5.3	9	19.6
11	*trans*-Chlordane	0.15	91 ± 5.2	5.7	9	20.4
12	Procymidone	0.5	92 ± 3.5	3.8	8	15.6
13	*o,p*-DDE	0.15	93 ± 4.5	4.8	7	16.6
14	*p,p*-DDE	0.15	93 ± 3.5	3.8	7	14.6
15	Dieldrin	0.15	93 ± 4.1	4.4	7	15.8
16	Endrin	0.15	101 ± 1.0	1.0	1	3.0
17	*o,p*-DDT	1.0	114 ± 0.5	0.4	14	14.8
18	β-Endosulfan	0.5	89 ± 3.5	3.9	11	18.8
19	*p,p*-DDT	0.25	>150			
20	Bromopropylate	0.5	90 ± 6.0	6.6	10	23.2
21	Fenpropathrin	0.5	94 ± 5.9	6.3	6	18.6
22	Tetramethrin	2.5	29 ± 1.0	3.4	71	77.8
23	Tetradifon	0.25	90 ± 6.5	7.2	10	24.4
24	Permethrin	2.5	101 ± 3.0	3.0	1	7.0
25	Cypermethrin	1.5	100 ± 5.0	5.0	0	10.0
26	Deltamethrin	2.0	101 ± 3.5	3.5	1	8.0
27	Methacrifos	1.0	93 ± 1.1	1.2	7	9.4
28	Formothion	2.5	103 ± 1.3	1.3	3	5.6
29	Primicarb	1.0	Not recovered			
30	Chlorpyrifos	1.0	103 ± 2.5	2.4	3	7.8
31	Bromophos	1.0	107 ± 3.1	2.9	7	12.8
32	Phosalone	1.0	76 ± 3.5 (NPD)	4.6	24	33.2
			72 ± 6.5 (ECD)	9.0	28	46.0
33	Pyrazophos	1.5	22 ± 1.1	5.0	78	88.0

[a] Recovery % is the mean value for three replicate runs.

REs of 20% or less are considered as satisfactory. If the best method available gives <80% recovery, it may still be used, provided the percent recovery is reproducible [24]. Sometimes, it is useful to calculate the method's total error, where both RSD and RE are included:

$$\text{Total error} = \text{RE} + 2\text{RSD}.$$

Total errors tend to run high in trace analyses. A total error of <50% is considered good, 50%–100% acceptable, and occasionally methods with >100% total error can still be usable, if no better method exists [24,96]. The values of RSD, relative errors, and total errors are listed in Tables 15.3 through 15.5.

Representative GC–ECD chromatograms of a reagent blank using AOAC official method 985.22 prior and post Florisil cleanup are shown in Figures 15.1 and 15.2, respectively. Figure 15.3 represents a contaminated *T. foenum-graecum* sample. Unfamiliar peaks observed in chromatograms of the reagent blank extracts are assumed to originate from the contaminants in sodium

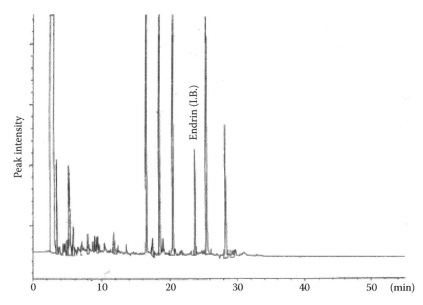

FIGURE 15.1 Representative GC–ECD chromatogram of a reagent blank using AOAC official method 985.22 prior Florisil cleanup. (From Hajou, R.M.K. et al., *Pharm. Biol.*, 43(6), 554, 2005. With permission.)

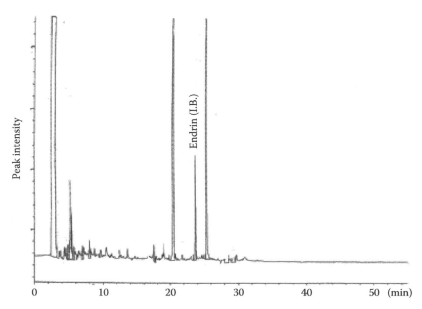

FIGURE 15.2 Representative GC–ECD chromatogram of a reagent blank using AOAC official method 985.22 post Florisil cleanup.

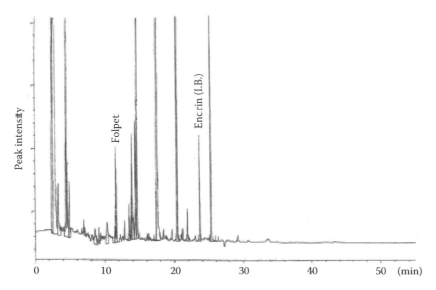

FIGURE 15.3 Representative GC–ECD chromatogram of a contaminated *T. foenum-graecum* sample.

sulfate or sodium chloride, as these were of AR grade, while all the other chemicals used were of PR grade. Literature review revealed that sodium sulfate is one of the likely sources of background peaks in blank extracts [62]. Furthermore, chromatograms of control samples also showed unidentified peaks. They may correspond, most probably, to the different constituents of the medicinal plant materials that can be identified on the detector. These peaks are different from those originating from the contaminants already present in the reagent blanks. Control sample analysis was also carried out in accordance with the recovery tests to make sure that the fortified plant samples were originally pesticide-free.

In this study, the minimization of complicating background peaks verified the feasibility of the proposed Florisil cleanup method with such plant samples and enabled the analysis of pesticide residues at higher limits of quantification. The used elution systems were able to extract most of the studied pesticides. Only the highly polar pesticides showed low recoveries. Theoretically, to elute them completely from the Florisil column, the polarity of the elution solvent should be increased. Such an action, however, would be on the expense of eluting other polar coextractives that might complicate the GC chromatogram. Polar residues are usually cleaned on a charcoal column [62,67].

Pesticide residues present in the real samples were identified tentatively by comparing the RRTs of the suspected peaks with those of the injected standards, and were then quantified using the following equation:

$$C_s (\text{mg/kg plant}) = (A_s/A_{is} \times C_{st} \times 5\,\text{mL} \times F \times R)/(A_{st}/A_{ist} \times \text{weight})$$

where
C_s is the concentration of pesticide residues in sample in mg/kg dry plant material
C_{st} is the concentration of the pesticide in the mixed pesticide standard solution
A_s is the average peak area obtained for the pesticide found in sample
A_{is} is the average peak area obtained for the internal standard injected with the sample
A_{st} is the peak area obtained for the pesticide in the mixed pesticide standard solution

TABLE 15.6
Pesticide Residues in *Trigonella foenum-graecum*
Samples (T1–T5) and Their Concentrations

Sample	Pesticides Found	Concentration (mg/kg)
F1	α-HCH	0.001
F2	Penconazole	0.101
	β-endosulfan	0.023
F3	Permethrin	0.742
	Cypermethrin	0.166
F4	β-HCH	0.021
	Chlorothalonil	0.061
F5	Folpet	0.041

A_{ist} is the peak area obtained for the internal standard found in the mixed pesticide standard solution

R is the recovery factor calculated from 100% recovery

5 is the final volume (V_{final}) of the analyzed sample in milliliter

F is the extraction factor

The identity of the detected pesticides was confirmed using GC–ECD equipped with another column of a different polarity, namely HP-5. The analyzed *T. foenum-graecum* samples were contaminated with OCs, pyrethroid, and miscellaneous pesticide residues (Table 15.6). The pesticides found in these samples were: folpet, α-HCH, β-HCH, chlorothalonil, penconazole, permethrin, and cypermethrin. However, no OP pesticide residues were detected in the analyzed samples, which could be owing to (1) the generally shorter half-life of these pesticides and (2) the applied methods of analysis, which may not be suitable for extracting such pesticides, which are relatively more polar than the OCs and pyrethroids.

15.6.3 Conclusions

Most of the plant samples studied were contaminated at least with one pesticide residue. The number of pesticides found reflects the diversity of the compounds used. The pesticide residues detected in the analyzed medicinal plant samples were those that can be extracted with the proposed methods. The inability to detect pesticide residues in some plant samples does not necessarily mean that these samples were pesticide-free. Some pesticides may exist, but could not be extracted by the applied methods. It is also possible that some pesticide residues may be present in concentrations below the detection limits achieved in this study and, consequently, could not be detected. In general, the concentrations of most pesticide residues found in this study were below the tolerance levels. For the rest of the pesticides, the concentrations were well below the maximum residue limits (MRLs). However, these low concentrations do not indicate that such plants do not pose any harm to human beings, unless necessary studies are conducted at long-term intervals to assess the chronic toxicities of such pollutants.

No OP pesticides were detected in the current study. Furthermore, it is reported in literature that the absence of OP pesticide residues in plant or water samples is frequently encountered [97].

15.7   STRUCTURES OF THE PESTICIDES TESTED IN *TRIGONELLA FOENUM-GRAECUM*

The following is a schematic presentation of the tested pesticides:

Folpet   Hexachlorobenzene   Hexachlorocyclohexane

Quintozene   Vinclozolin   Chlorothalonil

Dicofol   Penconazole

Chlordane

Procymidone

p,p-DDT

p,p-DDE

Dieldrin

Endrin

Endosulfan

Bromopropylate

Tetradifon

Pyrethroids

Tetramethrin

Permethrin

Cypermethrin

Deltamethrin

Fenpropathrin

Organophosphorous pesticides

Methacrifos

Formothion

Chlorpyrifos

Bromophos

Phosalone

Pyrozophos

Carbamates

Primicarb

15.8 TOXICOLOGICAL PROFILE OF THE SCREENED PESTICIDES

The toxicological evaluation of pesticide residues in medicinal plant materials should be based on the likely intake of the material by consumers. In general, the intake of residues from medicinal plant materials should not be >1% of the total intake from all the sources, including food and drinking water. Certain plant materials may contain extremely high levels of pesticide residues, but the levels that remain after the extraction of the active constituents or preparing teas are much lower. It is therefore important to determine the actual quantity of the residues consumed in the final dosage form [9,37,45,46,50]

The highest dose (mg pesticides/kg bodyweight per day) that produces no observable toxic effects in the most sensitive species is called the nonobservable effect level (NOEL). This is derived from chronic toxicity tests and is used to set the acceptable daily intake (ADI) for humans.

$$ADI = NOEL \times [Safety\ factor(1/100\ to\ 1/2000)]$$

Some countries have established national requirements for residue limits in medicinal plant materials. The limits for some pesticides are indicated in the EP [21]. However, when such requirements do not exist, or if the limits are not listed in the EP, then the following formula may be used:

$$MRL = (ADI \times E \times 60)/(MDI \times 100)$$

where
   MRL is the maximum residue limit of pesticide (mg/kg plant material)
   ADI is the maximum acceptable daily intake of pesticide as published by FAO–WHO (mg/kg of body mass)
   E is the extraction factor that determines the transition rate of the pesticide from the plant material to the dosage form
   MDI is the mean daily intake of medicinal plant material

The 60 in the numerator represents the mean adult body weight, while the denominator incorporates a consumption factor of 100, reflecting the fact that no more than 1% of the total pesticide residue consumed should be derived from the medicinal plant material [9,21].

The amount of residue detected in the plant will depend on the time of analysis (analyzed immediately after spraying or after a period of time has elapsed) as well as on the spraying levels. This could explain our findings of low and high levels of the same pesticide in different analyzed samples.

Hexachlorobenzene (HCB) is very persistent in the environment owing to its chemical stability [62]. Although its former use as an active ingredient in some insecticides and fungicides is banned, HCB is still found in the food chain [98]. However, HCB was not detected in *T. foenum-graecum*. Yet, low concentrations of HCB were detected in samples of *Mentha piperita, Origanum syriacum,* and *Pimpinella anisum* purchased in Jordan [42–44]. While HCB is rarely used in Jordan as a pure pesticide, it is still present as a contaminant of the commonly used fungicide, quintozene. It is also produced as a by-product from the production of chlorinated solvents and is subsequently transferred between the locations and countries owing to its relatively high volatility. Hence, any release of this chemical into air or water will inevitably lead to accumulation of residues in the food chain [98].

α- *and* β-*hexachlorohexane (α-HCH and β-HCH)* were detected in two of the *T. foenum-graecum* samples in concentrations 0.001 and 0.021 mg/kg plant (Table 15.6). Though these two compounds are not used as pesticides, they are found in the technical grade Lindane (γ-HCH), which is converted to the α- and β-isomers under environmental conditions. However, Lindane is no longer manufactured in the United States. The Environmental Protection Agency (EPA) canceled most of the agricultural and dairy uses of Lindane as it has been associated with a carcinogenic risk. However, this compound and its isomers are still being detected in the Jordanian environment even though it has been banned from use in the Kingdom since the early 1980s. This might indicate the illegal use of this pesticide inside Jordan [99].

trans-Chlordane is an OC insecticide, which is highly persistent in soils with a half-life of about 4 years. Several studies have shown chlordane residues in excess of 10% of the initially applied amount, 10 years or more after application, but no data are available about its breakdown in vegetation. It should be mentioned that the use of chlordane was banned in 1988, owing to concerns about its carcinogenic risk [100].

Dieldrin is an OC insecticide and a by-product of the pesticide, aldrin. It is a persistent, bioaccumulative, and toxic pollutant targeted by the EPA. Its residues are still found in our environment from the past uses [101].

Endrin is a stereoisomer of dieldrin. It is highly toxic to man and domestic animals. However, compared with dieldrin, the degree of persistence of eldrin is lower as it isomerizes to the nontoxic ketone on exposure to light [102].

Dichlorodiphenyl trichloroethane (*DDT*) and its breakdown products are highly persistent in the environment with reported half-lives of 2–15 years. They have similar physical and chemical properties. However, they do not appear to be taken up or stored by plants to a great extent [103]. Although DDT was banned from agricultural use in Jordan in 1980, the compound and its breakdown products are still found in the Jordanian environment and, in low concentration, in some medicinal plants [7,99,104].

It is reported that *p,p*-DDT and *o,p*-DDT were detected in *O. syriacum*, while dichlorophenyl dichloroethylene (DDE), which is one of the breakdown products of DDT, was found in *M. piperita* samples. Both the plants are readily available in the Jordanian market [43,44]. The presence of DDT residues without any breakdown products might indicate that the contamination of the plant material with such pesticide residues is recent. Earlier studies showed that DDT was not translocated into alfalfa or soybean plants, and only trace amounts were observed in carrots and radishes. However, no similar studies were available concerning medicinal plants. The morphological features of *O. syriacum* leaves may aid in preserving and storing such residues. The concentration of *p,p*-DDT was greater than that of the *o,p*-isomer in both the *O. syriacum* samples. This could be explained by the following facts: (1) the low concentration of the *o,p*-DDT isomer and its breakdown product DDE in the technical grade material and (2) the higher degradation rates of the *o,p*-isomers than the *p,p*-isomers [44,105,106].

β-*Endosulfan* is categorized as a restricted use pesticide (RUP). It is an OC insecticide and acaricide. It is used on a wide variety of food crops, including tea, coffee, fruits, vegetables, and grains. Technical endosulfan is made up of a mixture of two isomers, the α- and β-isomers. In *T. foenum-graecum*, only the β-isomer was detected. It is reported in literature that endosulfan is moderately persistent in the soil environment with an average half-life of 50 days. The two isomers have different degradation times in soil. The half-life of the α-isomer is 35 days and the β-isomer is 150 days [107]. The findings of this study are consistent with these half-lives.

Folpet is a protective leaf-fungicide, used on fruits and vegetables, and for seed and plant-bed treatment [106]. In the late-1980s, the EPA cancelled the use of this chemical [108]. In this study, folpet was detected in samples obtained as dry materials from herbal shops. The date of collection and the origin (local or imported) of these samples were unknown. Hence, residues in these samples could have originated from the intentional use of this pesticide on these medicinal plants, or on nearby other agricultural crops.

Quintozene is a fungicide used for seed dressing or soil treatment to control a wide range of fungi species in crops, such as potatoes, wheat, onions, lettuce, tomatoes, garlic, and others. The EPA currently describes this chemical as a potential health hazard, given its widespread use on a multitude of crops, which may lead to cumulative exposure. Quintozene is of varying persistence in the environment; various half-lives in the soil have been reported, ranging from less than 3 weeks to over a year [109]. Plants can take up this compound from both soil and water, and it may be translocated throughout the plant. Supplement makers using herbal products are advised to monitor for quintozene and the related pesticide HCB, because the levels of these two have been found to be as high as 1–20 mg/kg in some medicinal plants [110].

Vinclozolin is a nonsystemic fungicide with low to moderate persistence in the soil, and is partially broken down by soil microorganisms [106]. A survey in the European Union (EU) member states and Norway reported that vinclozolin was the seventh most frequently encountered residue [111]. The EU appears satisfied with this compound, but the EPA remains committed to phasing out most of its use, because it is a proven endocrine disrupter, causing antiandrogenic effects [112]. However, the U.S. regulators emphasize that the potential impacts on endocrine disruption are yet to be weighed. They are also concerned with the additional impacts, as vinclozolin may share a common mechanism of toxicity with other fungicides, such as procymidone and iprodione. As medicinal plants, especially in Jordan, are grown in the same field with other agricultural crops, they may be exposed unintentionally to pesticides intended for use on other plants.

Chlorothalonil is a broad-spectrum OC fungicide used to control fungi that threaten vegetables, trees, small fruits, ornamentals, and other agricultural crops. It is classified by the EPA as a general use pesticide (GUP). This chemical is moderately persistent, with a half-life of 3 weeks in soil, while its residues may remain on the aboveground crops at harvest and then dissipate over time. Chlorothalonil is fairly persistent on plants, depending on the rate of application [113].

Dicofol is an OC miticide used on a wide variety of fruit, vegetable, ornamental, and field crops. It is moderately persistent in the soil with a half-life of 60 days, but residues on treated plants have been observed to remain unchanged for up to 2 years [107].

Penconazole is a systemic conazole fungicide first evaluated in 1992. This pesticide was observed to be stable for at least 16 months in apples and grapes under frozen conditions [106].

Procymidone is a dicarboximide fungicide. It is a potential antiandrogenic pesticide that shares a common mechanism of action with some other pesticides. This should be considered when assessing the toxicological impacts of such pesticides, especially when they are present in the same sample at the same time [25].

Bromopropylate is a miticide and is only recommended to growing plants. The concentration of its residues is proportional to the amount of active ingredient applied. Multiple applications have an additive effect and result in higher residues. Dissipation of residues is mainly owing to weathering and growth dilution, and its half-life period on most fruits is 3 weeks. However, on leafy crops, such as hops and tea, the rate of dissipation is much greater [114].

Fenpropathrin is a pyrethroid insecticide–miticide that is stable at an environmental pH between 6–8, and temperature of 25°C. It degrades under soil aerobic conditions with a half-life of 33–34 days [115]. However, there is no published evidence of carcinogenic or mutagenic effects.

Tetramethrin is mostly used for indoor pest control. Rapid degradation occurs when a thin film of tetramethrin is exposed to sunlight. No data are available on the exact levels of tetramethrin in the environment. However, with the proper current domestic pattern of use, the environmental exposure is expected to be very low. Furthermore, degradation of tetramethrin to less toxic products is rapid [116].

Tetradifon is widely used as an acaricide. There are no indications, at present, that it causes an environmental-pollution problem. Based on the data available, it is observed that tetradifon does not present any short-term threats to the environment. It is persistent but does not bioaccumulate significantly in fish. As no long-term data are available, its hazards cannot be adequately evaluated [117]. Furthermore, the Codex Alimentarius has not yet set a standard MRL for this chemical.

Permethrin is a broad-spectrum synthetic pyrethriod insecticide, used against a variety of pests on different agricultural crops. It is used in greenhouses, homes, gardens, and for termite control. Consequently, one could be exposed to its residues from a multitude of sources. Permethrin is reported to show no teratogenic or mutagenic activity, but the evidence regarding the carcinogenicity of this chemical is still inconclusive [118]. Permethrin was detected in one of the studied *T. foenum-graecum* samples from the Jordanian market (0.742 mg/kg plant), below the concentrations of the MRL stipulated by the EP for this pesticide (1.0 mg/kg). It is interesting to detect such residues in dry medicinal plant materials purchased from herbal shops, as permethrin is biodegradable and of a low to moderate persistence in the soil environment, with reported half-lives of 30–38 days. Treated apples, grapes, and cereal grains were observed to contain less than 1.0 mg/kg of permethrin at harvest time [119]. Subsequently, it was important to unveil the source of permethrin residues in our samples studied. After a prolonged survey and discussion with the herbalists, it was discovered that some pyrethroids are sprayed on the stored medicinal plants during transport or even at the herbal shops. This could explain the high levels of permethrin residues found in these samples.

Cypermethrin is widely used in Jordan to control many pests on a wide range of agricultural crops, and its residues are frequently encountered in the Pesticide Residue Lab/Ministry of Agriculture. Cypermethrin was found in *T. foenum-graecum*, in concentrations below the MRL stipulated by the EP for this pesticide, which is 1.0 mg/kg plant. This pesticide has a moderate persistence in soil with half-lives between 4 days and 8 weeks [120]. Earlier studies showed that when cypermethrin was applied on wheat, its quantification immediately after spraying was 4 ppm. This reading declined to 0.2 ppm 27 days later, and no cypermethrin was detected in the grain. Similar loss patterns have been observed on treated lettuce and celery crops, which might also be applicable to medicinal plants [102,121]. Interestingly, cypermethrin was separated into three peaks in the present study under the applied GC conditions, whereas, it usually consists of four peaks [94].

Pyrazophos is an OP systemic pesticide widely used to control powdery mildew in fruits. In Jordan, it was found to be an excellent fungicide on squash and cucumber in greenhouses [122].

Deltamethrin is an insecticide with quick initial action and short residual effects. Within the plant tissue, it is rapidly translocated to control sucking insects and certain Dipteria [106].

Methacrifos is an insecticide and acaricide, used primarily for the control of arthropod pests in stored products [106].

Formothion is a contact and systemic insecticide and acaricide, and is widely used for the protection of a variety field crops, fruit trees, and vegetables. It is metabolized in plants to different substances, such as dimethoate or omethoate [106].

Chlorpyrifos is a broad range, nonsystemic insecticide, used for the control of various crop pests in soil and on foliage. Owing to its high volatility, it also protects the untreated surfaces [106].

The concentration of a certain chemical in a plant sample is not the only factor affecting the toxicology and safety of this compound. Many factors may interfere with such issue: (1) the number of pesticides found in each sample, (2) some pesticides may possess similar mechanisms of toxicity, such as vinclozolin and procymidone, hence, their presence in the same samples may increase the risk of toxicity [123], and (3) one is exposed to pesticide residues from a multitude of dietary sources, such as fruits, vegetables, milk, meats, eggs, and many other natural or processed food items. If each item is contaminated by a variety of pesticides (as is the case with medicinal plants), the safety of ingesting such food must be questioned.

It should be realized; that there might be other pesticide residues contaminating the studied samples which were not extracted and consequently detected, because the used methodologies were, simply unsuitable. This, in addition to the relatively short half-lives of the OP pesticides, could be a possible reason for the absence of OP pesticide residues in the analyzed samples.

In spite of some similarity in the chemical structure and pharmacological effects between pesticides in the same group, the individual pesticides within each group differ widely in toxicity and in their storage capacity. Therefore, each pesticide must be evaluated separately [25].

Earlier studies proved that washing plants would reduce the concentration of pesticide residues and consequently, their toxicity. Particularly, the concentration of these residues is further reduced in teas prepared from contaminated medicinal plants [37,124].

REFERENCES

1. Abu-Irmaileh B. E. and Afifi F. U. Herbal medicine in Jordan with special emphasis on commonly used herbs. *J. Ethnopharmacol.* 89, 193–197 (2003).
2. Fowler M. W. Review: Plants, medicines and man. *J. Sci. Food Agric.* 86, 1797–1804 (2006).
3. Goldmann P. Herbal medicines today and the roots of modern pharmacology. *Ann. Intern. Med.* 135, 594–600 (2001).
4. Kinghorn A. D. Pharmacognosy in the 21st century. *Pharmacy Pharmacol.* 53, 135–138 (2001).
5. Phillipson J. D. Phytochemistry and medicinal plants *Phytochemistry* 56, 237–243 (2001).
6. Rates S. M. K. Plants as source of drugs. *Toxicon* 39, 603–613 (2001).
7. Hajou R. M.-K. Determination of some pesticide residues in selected medicinal plants commonly used in Jordan. MSc thesis, University of Jordan, Amman, Jordan (2003).
8. Schulz V. Herbal medicinal plants, evidence of efficacy and safety. 46th Annual Congress of Society for Medicinal Plant Research, August 31–September 4, Vienna, (1998).
9. WHO, *Quality Control Methods for Medicinal Plant Materials.* World Health Organization Publications, Geneva, Switzerland, pp. 1–4; 47–60 (1998).
10. Kayne S. B. *Complementary Therapies for Pharmacists.* Pharmaceutical Press, London, pp. 139–144 (2002).
11. WHO, *Legal Status of Traditional Medicine and Complementary Alternative Medicine, A Worldwide Review.* World Health Organization Publications, Geneva, Switzerland, (2001).
12. WHO, *Research Guidelines for Evaluating the Safety and Efficacy of Herbal Medicines.* World Health Organization, Regional office for Western Pacific, Manila, Philippines, (1993).
13. WHO, *World Health Organization: Regulatory Situation of Herbal Medicines. A Worldwide Review.* World Health Organization Publications, Geneva, Switzerland (1998).
14. Geddes & Grosset. *Herbal Remedies and Homeopathy.* Children's Leisure Products Limited, New Lanark, Scotland, pp. 5–11 (2001).
15. WHO, *WHO Guidelines on Good Agricultural and Collection Practices (GACP) for Medicinal Plants,* World Health Organization Publications, Geneva, Switzerland (2003).
16. Zhou S., Koh H.-L., Gao Y., Gong Z., and Lee E. J. D. Herbal bioactivation: The good, the bad and the ugly. *Life Sci.* 74, 935–968 (2004).
17. Dwivedi S. K. and Dey S. Medicinal herbs: A potential source of toxic metal exposure for man and animals in India. *Arch. Environ. Health* 57(3), 229–231 (2002).
18. Chan K. Some aspects of toxic contaminants in herbal medicines. *Chemosphere* 52, 1361–1371 (2003).
19. De Smet P. A. G. M. Toxicological outlook on the quality assurance of herbal remedies. In: *Adverse Effects of Herbal Drugs.* De Smet P. A. G. M., Keller K., Haensel R., and Chandler R. F., eds. Springer Verlag, Berlin, Vol. 1 pp. 1–72 (1992).
20. Newall C. A., Anderson L. A., and Phillipson D. J. *Herbal Medicines, A Guide for Health-Care Professionals.* The Pharmaceutical Press, London, pp. 31–32 (1996).
21. EP, *European Pharmacopoeia*, 4th ed., Council of Europe, Strasbourg (2002).
22. Hayes W. J. Jr. *Toxicology of Pesticides*, Williams & Wilkins, Baltimore, MD, p. 9 (1975).
23. Ausubel K. *Seeds of Change-The Living Treasure*, 1st ed., Harper San Francisco, New York, pp. 103–174, (1994).
24. Fong G. W., Moye A. H., Seiber J. N., and Toth J. P. *Pesticide Residues in Foods.* John Wiley & Sons, New York, pp. 10, 39, 45, (1999).
25. Hayes W. J. and Laws E. R. *Handbook of Pesticide Toxicology.* Academic Press, New York, Vols. 1 and 2, (1991).
26. U.S. Fish and Wildlife Service, *Feature Series: Pesticides and Wildlife*, Lakewood, CO, Vol. 1 No. 3 (2002).
27. Shaw L. Regulation and monitoring of pesticides in Indonesia. *Pesticide News* 44, 10 (1999).
28. Marer P. J. *The Safe and Effective Use of Pesticides.* University of California Statewide Integrated Pest Management Project, Division of Agriculture and Natural Resources Publication 3324, Oakland, CA, pp. 73–126 (1988).
29. Richardson M. L. *Chemistry, Agriculture and the Environment.* The Royal Society of Chemistry, Cambridge, U.K. (1991).
30. WHO/FAO, Pesticide residues in food. Report of the 1983 Joint Meeting of the FAO Panel of Experts on Pesticide Residues in Food and the Environment and the WHO Expert Group on Pesticide Residues. Rome: Food and Agriculture Organization of the United Nations (FAO Plant Production and Protection Paper 56) (1984).

31. Pimentel D. Amounts of pesticides reaching the target pest: Environmental impacts and ethics. *Agric. Environ. Ethics* 8, 17–29 (1995).

32. Seiber J. N. The analytical approach. In: *Pesticide Residues in Food.* Fong W. G., Moye H. A., Seiber J. N. and Toth J. P. eds. John Wiley & Sons Inc., New York, Chaps. 1 and 2, pp. 1–16, 17–61 (1999).

33. Sandermann H. Jr. Bound and unextractable pesticidal plant residues: Chemical characterization and consumer exposure. *J. Environmental Sci.* 19(2), 205–209 (2007).

34. Raloff J. The pesticide shuffle. *Sci. News* 149, 174–175 (1996).

35. RIRDC, *Determining whether Pesticide Application Rates for Culinary Herbs Meet MRL Requirements.* Rural Industries Research and Development Corporation, Australia, (2000).

36. Abou-Arab A. K. and Abou Donia M. A. Pesticide residues in some Egyptian spices and medicinal plants as affected by processing. *Food Chem.* 72(4), 439–445 (2001).

37. Ali S. L. Bestimmung der Pestizieden Rueckstaende und anderer bedenklicher Verunreinigungen sowie toxische Metallspuren in Arzneipflanzen. I. Mitt.: Pestizied Rueckstaende in Arzneidrogen. *Pharm. Ind.* 45, 1156–1156 (1983).

38. Ahmed M. T., Loutfy N., and Yousef Y. Contamination of medicinal herbs with organophosphorous insecticides. *Bull. Environ. Contam. Toxicol.* 66, 421–426. (2001).

39. Benecke R., Ennet D., and Frauenberger H. Pesticide residues in drugs from wild-growing medicinal plants. *Pharmazie* 43, 348–351 (1987).

40. Dogheim S. M., El-Ashraf M. M., Alla S. A., Khorshid M. A., and Fahmy S. M. (2004), Pesticides and heavy metals levels in Egyptian leafy vegetables and some aromatic medicinal plants. *Food Addit. Contam.* 21(4), 323–330.

41. Gupta A., Parihar N. S., and Bhatnagar A. Lindane, chlorpyriphos and quinalphos residues in mustard seed and oil. *Bull. Environ. Contam. Toxicol.* 67, 122–125 (2001).

42. Hajou R. M. K., Afifi F. U., and Battah A. H. Comparative determination of multi-pesticide residues in *Pimpinella anisum* using two different AOAC methods. *Food Chem.* 88, 469–478 (2004).

43. Hajou R. M. K., Afifi F. U., and Battah A. H. Determination of multipesticide residues in. *Mentha piperita. Pharm. Biol.* 43(6), 554–562 (2005).

44. Hajou R. M. K., Afifi F. U., and Battah A. H. Multiresidue pesticide analysis of the medicinal plant. *Origanum syriacum. Food Addit. Contam.* 24(3), 274–279 (2007).

45. Pluta J. Studies on contamination of vegetable drugs with halogen derivative pesticides. I. Changes of concentrations of halogen derivatives in herbal raw materials within the period of 1980–1984. *Pharmazie* 43, 121–123 (1988).

46. Pluta J. Studies on contamination of vegetable drugs with halogen derivative pesticides. II. Concentrations of halogen derivative pesticides in vegetable blends and herbal granulated products produced in Poland in 1980–1984. *Pharmazie* 43, 348–351 (1988).

47. Sendren D., Miere D., Mocanu A., and Tamas M. Organochlorinated pesticide residues in dry plants used for medicinal tea. *Clujul Med.* 68(3), 373–379 (1995).

48. Schilcher H. Determination of residues in medicinal plants and drug preparations. *Anal. Chem.* 321(4), 342–351 (1985).

49. Schilcher H., Peters H., and Wank H. Pestiziede und Schwermetalle in Arzneipflanzen und Arzneipflanzen-Zubereitungen. *Pharm. Ind.* 49(2), 203–211 (1987).

50. Campillo N., Penalver R., and Hernandez-Cordoba M. Pesticide analysis in herbal infusions by solid-phase microextraction and gas chromatography with atomic emission detection. *Talanta* 71, 1417–1423 (2007).

51. Ozbey A. and Uygun U. Behaviour of some organophosphorous pesticide residues in thyme and stinging nettle tea during infusion process. *Int. J. Food Sci. Tech.* 42, 380–383 (2007).

52. Ozbey A. and Uygun U. Behaviour of some organophosphorous pesticide residues in peppermint tea during infusion process. *Food Chem.* 104, 237–241 (2007).

53. Bicchi C., Cordero C., Iori C., Rubiolo P., Sandra P., Yariwake J. H., and Zuin V. G. SBSE-GC-ECD/FPD in the analysis of pesticide residues in *Passiflora alata* Dryander herbal teas. *J. Agric. Food Chem.* 51, 27–33 (2003).

54. Leandro C. C., Hancock P., Fussell R. J., and Keely B. J. Ultra-performance liquid chromatography for the determination pesticide residues in foods by tandem quadrupole mass spectrometry with polarity switching. *J. Chromatogr. A* 1144, 161–169 (2007).

55. Prichard E., MacKay G. M., and Pionts J. eds. *Trace Analysis: A Structured Approach to Obtaining Reliable Results.* The Royal Society of Chemistry, Cambridge, U.K., pp. 1–119 (1996).

56. FDA, *Guidance for Industry: Bioanalytical Method Validation.* U.S. Department of Health and Human Service Center for Drug Evaluation and Research, Rockville, IN, pp. 2–21 (2001).

57. AOAC, Official Methods of Analysis, 17th ed., Association of Official Analytical Chemists, Gaithersburg, MD, Vol. 1, Ch. 1, pp. 19–22; Ch. 10, pp. 2–10 (2000).

58. Boyd-Boland A. A. and Pawliszyn J. B. Solid-phase microextraction of nitrogen-containing herbicides. *J. Chromatogr. A* 704, 163–172 (1995).

59. Lehotay S. J. and Valverde-García A. Evaluation of different solid-phase traps for automated collection and cleanup in the analysis of multiple pesticides in fruits and vegetables after supercritical fluid extraction. *J. Chromatogr. A* 765, 69–84 (1997).

60. Steinwandter H. Universal extraction and cleanup methods. In: *Analytical Methods for Pesticides and Plant Growth Regulators*. Sherma J. ed., Academic Press, San Diego, CA, Ch. 2, pp. 35–73 (1989).

61. Wheeler W. B., Edelstein R. L., and Thompson N. P. Extraction of pesticide residues from plants In: *Pesticide Chemistry, Human Welfare and the Environment*, Vol. 4, *Pesticide Residues and Formulation Chemistry*. Miyamoto J., Kearney P. C., Greenhalgh R., and Drescher N. eds. Pergamon Press, New York, pp. 49–54 (1983).

62. FDA, *Pesticide Analytical Manual of the FDA*. U.S. Department of Health and Human Services, Washington, DC, Vol. I, pp. 1, 13, 14 (1994).

63. EPA, *Manual of Methods for Pesticides in Human and Environmental Samples*. Environmental Protection Agency, Washington, DC (1980).

64. Luke M. A., Froberg J. E., and Matsumoto H. T. Extraction and cleanup of organochlorine organophosphate, organonitrogen and hydrocarbon pesticides in produce for determination by gas-liquid chromatography. *J. AOAC Int.* 58, 1020–1026 (1975).

65. Hoff G. R. and Zoonen P. Trace analysis of pesticides by gas chromatography. *J. Chromatogr. A* 843 (1–2), 312–322 (1999).

66. Ling Y. C., Teng H. C., and Cartwright C. Supercritical fluid extraction and cleanup of organochlorine pesticides in Chinese Herbal Medicine. *J. Chromatogr. A* 835(1–2), 145–157 (1999).

67. Manirakiza P., Coaci A., and Schepens P. Single cleanup and GC-MS quantitation of organochlorine pesticide residues in spice powder. *Chromatographia* 52(11–12), 787–790 (2000).

68. Sharif Z., Man Y. B. C., Hamid N. S. A., and Keat C. C. Determination of organochlorine and pyrethroid pesticides in fruit and vegetables using solid phase extraction clean-up cartridges. *J. Chromatogr. A* 1127, 254–261 (2006).

69. Albero B., Sanchez-Brunete C., and Tadeo J. Multiresidue determination of pesticides in juice by solid-phase extraction and gas chromatography-mass spectrometry. *Talanta* 66, 917–924 (2005).

70. Di Muccio A., Fidente P., Barbini D. A., Dommarco R., Seccia S., and Morrica P. Application of solid-phase extraction and liquid chromatography-mass spectrometry to the determination of neonicotinoid pesticide residues in fruit and vegetable. *J. Chromatogr. A* 1108, 1–6 (2006).

71. Ling T., Xiaodong M., and Chongjiu L. Application of gas chromatography-tandem mass spectrometry (GC-MS-MS) with pulsed splitless injection for the determination of multiclass pesticides in vegetables. *Anal. Let.* 39, 985–996 (2006).

72. Chu X.-G., Hu X.-Z., and Yao H.-Y. Determination of 266 pesticide residues in apple juice by matrix solid-phase dispersion and gas chromatography-mass selective detection. *J. Chromatogr. A* 1063, 201–210 (2005).

73. Tang F., Yue D., Hua R. M., and Cao H. Q. Matrix solid-phase dispersion microextraction and determination of pesticide residues in medicinal herbs by gas chromatography with a nitrogen-phosphorous detector. *J. AOAC Int.* 89(2), 498–502 (2006).

74. Papadakis E. N., Vryzas Z., and Papadapoulou-Mourkidou E. Rapid method for the determination of 16 organochlorine pesticides in sesame seeds by microwave-assisted extraction and analysis of extracts by gas chromatography-mass spectrometry. *J. Chromatogr. A* 1127, 6–11 (2006).

75. Halvorsen B. L., Thomsen C., Greibrokk T., and Lundanes E. Determination of fenpyroximate in apples by supercritical fluid extraction and packed capillary liquid chromatography with UV detection. *J. Chromatogr. A* 880(1–2), 121–128, (2000).

76. Sagratini G., Manes J., Giardina D., Damiani P., and Pico Y. Analysis of carbamate and phenylurea pesticide residues in fruit juices by solid-phase microextraction and liquid chromatography-mass spectrometry. *J. Chromatogr. A* 1147, 135–143 (2007).

77. Fytianos K., Raikos N., Theodoridis G., Velinova Z., and Tsoukali H. Solid phase microextraction applied to the analysis of organophosphorous insecticides in fruits. *Chemosphere* 65, 2090–2095 (2006).

78. Granby K., Andersen J. H., and Christensen H. B. Analysis of pesticides in fruit, vegetables and cereals using methanolic extraction and detection by liquid chromatography-tandem mass spectrometry. *Anal. Chim. Acta* 520, 165–176 (2004).

79. Sandra P., Tienpoint B., Vercammen J., Tredoux A., Sandra T., and David F. Stir bar sorptive extraction applied to the determination of dicarboximide fungicides in wine. *J. Chromatogr. A* 928, 117–126 (2001).

80. Tang F., Yue Y., Hua R., Ge S., and Tang J. Development of methods for the determination of 15 pesticides in medicinal herbs *Isatis indigotica* Fort. by capillary gas chromatography with electron capture or flame photometric detection. *J. AOAC Int.* 88(3), 720–728 (2005).

81. Rodrigues M. V. N., Magalhaes P. M., Reyes F. G. R., and Rath S. GC-MS determination of organochlorine pesticides in medicinal plants harvested in Brazil. *Toxicol. Lett.* 164 (Suppl. 1), S242 (2006).

82. Zuin V. G., Yariwake J. H., and Bicchi C. Fast supercritical fluid extraction and high-resolution gas chromatography with electron-capture and flame photometric detection for multiresidue screening of organochlorine and organophosphorous pesticides in Brazil's medicinal plants. *J. Chromatogr. A* 985, 159–166 (2003).

83. Zuin V. G., Lopes A. L., Yariwake J. H., and Augusti F. Application of a novel sol-gel polydimethylsiloxane-poly(vinyl alcohol) solid-phase microextraction fiber for gas chromatographic determination of pesticide residues in herbal infusions. *J. Chromatogr. A* 1056, 21–26 (2004).

84. Cortes J. C. M. and Berrada H. Urea pesticide residues in food. In: *Handbook of Food Analysis.* Nollet L. M. L. ed., Marcel Dekker, New York, Vol. 2, pp. 1065–1099 (2004).

85. Chamberlain J. *The Analysis of Drugs in Biological Fluids.* CRC Press, Boca Raton, FL, pp. 119–130, 139–161 (1995).

86. Rodriguez R., Pico Y., Font G., and Manes J. Analysis of thiabendazole and procymidone in fruits and vegetables by capillary electrophoresis-electrospray mass spectrometry. *J. Chromatogr. A* 949 (1–2), 359–366 (2002).

87. Lawrence J. F., Menard C., Hennion M.-C., Pichon V., Le Goffic F., and Durand N. Use of immunoaffinity chromatography as a simplified clean-up technique for the liquid chromatographic determination of phenylurea herbicides in plant material. *J. Chromatogr. A* 732, 277–281 (1996).

88. Stevenson D. Immuno-affinity solid-phase extraction. *J. Chromatogr. B* 745(1), 39–48 (2000).

89. Zeng K., Yang T., Zhong P., Zhou S., Qu L., He J., and Jiang Z. Development of an indirect competitive immunoassay for parathion in vegetables. *Food Chem.* 102, 1076–1082 (2007).

90. Miller J. M. *Chromatography Concepts and Contrasts* John Wiley & Sons Inc. New Jersey, pp. 67–92, 141–182, 183–276 (2005).

91. Kolb B. and Bischoff J. A new design of a thermionic nitrogen and phosphorous detector for GC. *J. Chromatogr. Sci.* 12, 625–629 (1974).

92. Tribaldo E. B. Carbamate pesticide residues in food. In: *Handbook of Food Analysis.* Nollet L. M. L. ed., Marcel Dekker, New York, Vol. 2, pp. 1177–1209 (2004).

93. Their H. and Zeumer H. *Manual of Pesticide Residue Analysis.* VCH publishers, Germany, Vol. 1, pp. 17–20 (1987).

94. Obana H., Akutsu K., Okihashi M., and Hori S. Multiresidue analysis of pesticides in vegetables and fruits using two-layered column with graphitized carbon and water absorbent polymer. *Analyst* 126, 1529–1534 (2001).

95. Al-Okla K. Determination of pesticide residue levels in the processed tomato and the impact of processing on residues in Jordan. MSc thesis, University of Jordan, Amman, Jordan (1998).

96. McFarren E. F., Liskka R. J., and Parker J. H. Criterion for judging acceptability of analytical methods. *Anal. Chem.* 42, 358–365 (1970).

97. Al-Nasir F., Jiries A. G., Batarseh M. I., and Beese F. Pesticides and trace metals residue in grape and home made wine in Jordan. *Environ. Monit. Assess.* 66(3), 253–263 (2001).

98. WHO, *Pesticide Residue Series 3: Hexachlorobenzene,* World Health Organization Publications, Geneva, Switzerland (1973).

99. Abu-Hilal D. A. Organochlorine pesticide and PCB pollution levels in the Gulf of Aqaba. MSc thesis, University of Jordan, Amman, Jordan (1994).

100. U.S. ATSDR, *Toxicological Profile for Chlordane (ATSDR/TP-89/06),* U.S. Agency for Toxic Substances and Disease Registry, Atlanta, GA (1989).

101. EPA, *Persistent Bioaccumulative and Toxic (PBT) Chemical Program: Aldrin/Dieldrin,* Environmental Protection Agency, Washington, DC (2002).

102. Khan A., Rao R. A., and Rathore H. S. Organochlorine pesticide residues in food. In: *Handbook of Food Analysis.* Nollet L. M. L. ed., Marcel Dekker, New York, Vol. 2, pp. 1101–1175 (2004).

103. WHO, *Environmental Health Criteria 83: DDT and its Derivatives Environmental Effects.* World Health Organization Publications, Geneva, Switzerland (1989).

104. EPA, *Environmental Fate and Effect Division, Pesticide Environmental Fate On Line Summary: DDT*, Environmental Protection Agency, Washington, DC (1989).

105. Chowdhury A. B. M., Nasir U., Jepson P. C., Howse P. E., and Ford M. G. Leaf surfaces and the bioavailability of pesticide residues. *Post Manage. Sci.* 57(5), 403–412 (2001).

106. Worthing C. R. *The Pesticide Manual. A World Compendium*. British Crop Protection Council, Lavenham Press, U.K., pp. 179–180, 208, 232–235, 434–435, 539 (1991).

107. Wauchope R. D., Buttler T. M., Hornsby A. G., Augustijn Beckers P. W. M., and Burt J. P. Pesticide properties database for environmental decision making. *Rev. Environ. Contam. Toxicol.* 123, 1–157 (1992).

108. EWG, *Forbidden Fruit; Illegal Pesticides in the U.S. Food Supply*. Eldrich S., Wiles R., and Campbell C. eds., Environmental Working Group, Washington, DC (1995).

109. EPA, Pesticide Abstracts: 75: 98, 78: 2944, 79: 1635, 80: 246, 81. Environmental Protection Agency, Washington, DC, 1968–1981 (1983).

110. U.S. National Library of Medicine, *Hazardous Substances Databank*, U.S. National Library of Medicine, Bethesda, MD (1995).

111. EC, *Monitoring for Pesticide Residues in European Union and Norway-Report*, European Commission, Belgium (1997).

112. Gray L. E., Ostby J., Monosson E., and Kelce W. R. Environmental antiandrogens: Low doses of the fungicide Vinclozolin alter sexual differentiation of the male rat. *Toxicol. Ind. Health* 15, 65–79 (1999).

113. Vettorazzi G. *International Regulatory Aspects for Pesticide Chemicals*. CRC Press, Boca Raton, FL (1979).

114. WHO, World Health Organization: *Pesticide Residue Series 3: Bromopropylate*. World Health Organization Publications, Geneva, Switzerland (1973).

115. Cornell University, Fenpropathrin (Danitol) chemical fact sheet 12/89, New York (1989).

116. IPCS, *Environmental Health Criteria 98, Tetramethrin*. International Program on Chemical Safety, Geneva (1990).

117. IPCS, *Environmental Health Criteria 11, Tetradiofon*. International Program on Chemical Safety, Geneva (1987).

118. Hallenbeck W. H. and Cunningham-Burns K. M. *Pesticides and Human Health*, Springer Verlag, New York (1985).

119. Kidd H. and James D. R. *The Agrochemical Handbook*. The Royal Society of Chemistry Information Services, Cambridge, U.K. (1991).

120. EPA, *Pesticide Fact Sheet Number 199: Cypermethrin, Office of Pesticides and Toxic Substances*. Environmental Protection Agency, Washington, DC (1989).

121. Wescott N. D. and Reichle R. A. Persistence of deltamethrin and cypermethrin on wheat and sweet clover. *Environ. Sci. Part B* 22(1), 91–101 (1987).

122. Mustafa T. M., Al-Rifai J. H., and Al-Shuraiqi Y. T. Residues of pyrozophos in cucumber in the central highlands of Jordan. *Dirasat* 21B(92), 7–14 (1994).

123. Gray L. E. Xenoendocrine disrupters: Laboratory studies on male reproductive effects. *Toxicol. Let.* 102–103, 331–335 (1998).

124. Soliman K. M. Changes in concentration of pesticide residues in potatoes during washing and home preparation. *Food and Chem. Toxicol.* 39 (8), 887–891 (2001).

16 Sample Preparation and Quantification of Pesticide Residues in Water

N. C. Basantia, S. K. Saxena, and Leo M. L. Nollet

CONTENTS

16.1 INTRODUCTION

The term "pesticides" indicates industrial chemicals used to kill insects (insecticides), destroy unwanted plants (herbicides), prevent mold and mildew (fungicides), kill spiders (acaricides), rodents (rodenticides), algae (algicides), and plant growth regulators. By far, the majority of pesticides are insecticides. These agrochemicals are used to protect crops from insects, pests, and weeds, which results in a substantial increase in the yields of foodstuffs. Therefore, the use of pesticides on raw agricultural commodities has grown rapidly over the years. It has been estimated that the use of pesticides in some developing countries is increasing at a rate of more than 10% every year. The improper use or excessive application of these pesticides leads to many problems owing to the higher level of residues in food.

Nowadays, monitoring of pesticide residues in food and water is a priority objective for food-processing professionals to get an extensive evaluation of food quality and avoid possible risks to human health. A large number of multiresidue extraction methods have been developed over the years. The most frequently used methods employ solvent extraction followed by liquid–liquid partitioning chromatographic cleanup and gas chromatography (GC) with selective detection. The tentative identification is by comparison of the retention times of standard reference material (SRM) with sample extracts, and conformation using thin-layer chromatography (TLC), by analysis using different polarity columns, and most recently through mass spectrometry (MS).

In the first part of this chapter, the basic steps in sampling and analysis of pesticide residues and the basic principles of analytical instruments to analyze and detect the pesticide residues, not only in water but also in food, are discussed.

In the second part, four analysis methods for detecting pesticide residues in water are in brief and the recent literature of the detection methods is reviewed.

16.2 MAJOR GROUP OF PESTICIDE RESIDUES AND THEIR PHYSICOCHEMICAL PROPERTIES

Pesticides can be classified into the following groups:

- Organochlorine pesticides (e.g., DDT, lindane, aldrin, chlordane, endosulfan, etc.)
- Organophosphorus pesticides (e.g., dichlorovos, monochrotophos, ethion, malathion, chloropyrifos)
- Carbamate pesticides (e.g., carbaryl, carbofuran, aldicarb, carbanolate, propoxur)
- Synthetic pyrethroids (e.g., cypermethrin, deltamethrin, fenvalerate, fluvalinate, flufenprox)
- Urea pesticides (e.g., diafenthiurun, fhicofuron, sulcouron, etc.)
- Chlorinated acidic pesticides

In Tables 16.1 and 16.2, the characteristics of the pesticides are enumerated. The numbers in the first column of Table 16.1 are used in other tables throughout the chapter.

16.3 BASIC STEPS IN PESTICIDE ANALYSIS IN FOOD AND WATER

16.3.1 SAMPLING

The main objective of the sampling procedure is to obtain a representative sample. Therefore, all the steps are important in the sampling procedure [3].

16.3.1.1 Collection of Primary Samples

Primary samples must consist of sufficient material for the laboratory-required samples. The position from which a primary sample is taken in the lot should preferably be chosen randomly;

TABLE 16.1
Classification of Pesticides according to Target Pest

Numbers	Chemical Class	Representative Pesticides of the Group	LD$_{50}$ (mg/kg)[a]	Water Solubility (mg/L at 25°C)
		Insecticides		
1	Carbamates	Aldicarb	0.9	4.93 g/L (pH 7, 20°C)
2		Aminocarb	<51	Slightly soluble
3		Bendiocarb	40–156	280 (pH 7, 20°C)
4		Carbaryl	500–850	120 (20°C)
5		Carbofuran	8	320 (20°C)
6		Methiocarb	135	27 (20°C)
7		Oxamyl	5.4	280 g/L
8		Pirimicarb	147	3,000 (pH 7.4, 20°C)
9		Propoxur	128	1.9 g/L (20°C)
10	Organochlorines	Aldrin	67	0.01–0.2
11		Chlordane	457–590	0.1
12		Dieldrin	40–87	0.186 (20°C)
13		DDT	113–118	0.0077 (20°C)
14		Dicofol	587–595	0.8
15		Endosulfan	70	0.33 (22°C)
16		Endosulfan sulfate	—	0.117
17		Endrin	—	0.23
18		Heptachlor	147–220	0.056
19		Lindane (γ-HCH)	88–270	7.3
20		Methoxychlor	6,000	0.1
21	Organophosphorus compounds	Azinphos-ethyl	12	4.5 (20°C)
22		Azinphos-methyl	9	28 (20°C)
23		Chlorpyrifos	135–163	1.4
24		Diazinon	300–400	60 (20°C)
25		Dichlorvos	50	8 g/L
26		Dimethoate	290–325	23.8 g/L (pH 7, 20°C)
27		Fenthion	250	4.2 (20°C)
28		Malathion	1,375–2,800	145
29		Parathion ethyl	2	11 (20°C)
30		Parathion methyl	6	55 (20°C)
31		Phorate	1.6–3.7	50
32		Pirimiphos-methyl	2,050	8.6 (pH 7.3, 30°C)
33	Synthetic pyrethroids	Allethrin	585–1,100	Practically insoluble
34		Bifenthrin	54.5	0.1
35		Bioallethrin	709–1,042	4.6
36		Cyflurthrin	500	2.2 (pH 7, 20°C)
37		Cyhalothrin	114–166	0.004 µg/L (20°C)
38		Cypermethrin	250–4,150	0.004 (pH 7)
39		Deltamethrin	135	<0.2 µg/L
		Acaricides		
40	Dinitro compounds	Dinocap	980–1,190	<0.1
41	Organochlorines	Chlorobenzilate	2,784–3,880	10 (20°C)
42		Tetradifon	>14,700	0.08 (20°C)
43	Organotin compounds	Cyhexatin	540	<1
44	Others	Fenazaquin	134–138	0.22 (20°C)

(continued)

TABLE 16.1 (continued)
Classification of Pesticides according to Target Pest

Numbers	Chemical Class	Representative Pesticides of the Group	LD_{50} (mg/kg)[a]	Water Solubility (mg/L at 25°C)
45		Rotenone	132–1,500	15 (100°C)
46		Tebufenpyrad	595–997	2.8
Fungicides				
47	Antibiotics	Blasticidin-S	55.9–56.8	>30 g/L (20°C)
48	Azole compounds	Bitertanol	>5,000	2.9 (20°C)
49		Flusilazole	674–1,100	54 (pH 7.2, 20°C)
50		Flutriafol	1,140–1,480	130 (pH 7, 20°C)
51		Plochloraz	1,600	34.4
52	Benzimidazoles	Benomyl	>10,000	4 (pH 3–10)
53		Carbendazim	>15,000	29 (pH 4)
54		Thiabendazole	3,100	10 g/L (pH 2)
55		Thiophanate-methyl	6,640–7,500	Practically insoluble
56	Chlorine-substituted aromatics	Dichlone	1,300	0.1
57		Dicloran	4,040	6.3 (20°C)
58		Quintozene	>5,000	0.1 (20°C)
59	Dithiocarbamates	Mancozeb	>5,000	6–20
60		Maneb	6,750	Practically insoluble
61		Metam-sodium	1,700–1,800	722 g/L (20°C)
62		Propineb	>5,000	10 (20°C)
63		Thiram	1,800	18
64		Zineb	5,200	10
65		Ziram	320	0.03 (20°C)
66	Inorganics	Copper hydroxide	1,000	2.9 (pH 7)
67		Copper sulfate	800–1,200	230.5 g/L
68	Morpholine	Dodemorph	2,465–3,944	<100
69		Tridemorph	480	11.7 (pH 7, 20°C)
70	Organochlorine	Chlorothalonil	>10,000	0.9
71		Hexachlorobenzene (HCB)	10,000	0.006
72	Organomercurials	Phenylmercury acetate	24	4.37 g/L (15°C)
73	Organotin compounds	Fentin acetate	140–298	9 (pH 5, 20°C)
74	Phthalimides	Captafol	5,000–6,200	1.4 (20°C)
75		Captan	9,000	3.3
76		Folpet	>10,000	1
77	Piperazines	Triforine	>16,000	9 (20°C)
78	Pyrimidines	Ethirimol	6,340	150 (pH 7.3, 20°C)
Rodenticides				
79	Hydroxycoumarins	Brodifacoum	0.27	<10 (pH 7, 20°C)
80		Difenacoum	1.8–2.45	<10 (pH 7)
81		Flocoumafen	0.25	1.1
82		Warfarin	186	17 (20°C)
83	Indadione anticoagulants	Diphacinone	2.3	0.3
84		Pindone	280	18
85	Others	Ergocalciferol	56	50
86		Zinc phosphide	45.7	Practically insoluble
Molluscicides				
87	Others	Metaldehyde	630	260 (30°C)
88		Niclosamide	5,000	1.6 (pH 6.4, 20°C)
89		Triphenmorphone	—	—

TABLE 16.1 (continued)
Classification of Pesticides according to Target Pest

Numbers	Chemical Class	Representative Pesticides of the Group	LD_{50} (mg/kg)[a]	Water Solubility (mg/L at 25°C)
		Nematicides		
90	Organophosphorus compounds	Fenamiphos	6	700 (20°C)
91		Fosthiazate	—	9.85 g/L (20°C)
92	Others	1,2 Dibromoethane	146–420	4.3 g/L
93		DCIP	503	1.7 g/L
		Herbicides		
94	Aryloxyalkanoic acids	2,4-D	639–764	311 (pH 1)
95		2,4-DB	370–700	46
96		MCPA	900–1,160	734
97		MCPB	4,700	44
98		Mecoprop (MCCP)	930–1,166	734
99	Benzonitriles	Bromoxynil	190	130
100		Dichlobenil	4,460	18
101		Ioxynil	110	50
102	Carbanilates and carbamates	Asulam	>4,000	5 g/L
103		Carbetamide	11,000	3.5 g/L (20°C)
104		Chlorpropham	5,000–7,500	89
105		Propham	5,000	250 (20°C)
106		Triallate	1,100	4
107	Chlorinated aliphatic acids	Dalapon sodium	7,570–9,330	900 g/L
108		TCA-sodium	>2,000	1.2 kg/L
109	Chlorocetanilides	Alachlor	930–1,200	242
110		Butachlor	2,000	20
111		Metazachlor	2,150	430 (20°C)
112		Metolachlor	2,780	488
113		Propachlor	550–1,700	613
114		Thenylchlor	>5,000	11
115	Pheylureas or substituted ureas	Chlorotoluron	>10,000	74
116		Diuron	3,400	42
117		Fenuron	6,400	3.85 g/L
118		Isoproturon	1,826	65 (22°C)
119		Linuron	1,500–4,000	81
120		Siduron	>7,500	18
121	Phosphonoamino acids	Glufosinate	1,620–2,000	1,370 g/L (22°C)
122		Glyphosate	5,600	12 g/L
123	Pyridozinones and pyridinones	Amitrole	1,100–24,600[a]	280 g/L (23°C)
124		Fluridone	10,000	12 (pH 7)
125		Pyrazon (chloridazon)	2,140–3,830	340 (20°C)
126	Thiocarbamates	Butylate	>3,500	36 (20°C)
127		Cycloate	2,000–3,190	75 (20°C)
128		EPTC	>2.000	375
129		Molinate	369–450	88 (20°C)
130		Pebulate	1,120	60 (20°C)
131	Triazines	Atrazine	1,869–3,080	33 (20°C)
132		Desmetryne	1,390	580 (20°C)
133		Metribuzin	2,000	1.05 g/L (20°C)

(*continued*)

TABLE 16.1 (continued)
Classification of Pesticides according to Target Pest

Numbers	Chemical Class	Representative Pesticides of the Group	LD_{50} (mg/kg)[a]	Water Solubility (mg/L at 25°C)
134		Propazine	>7,000	5.0 (20°C)
135		Simazine	>5,000	6.2 (20°C)
136		Terbuthylazine	1,590–>2,000	8.5 (20°C)
137		Terbutryn	2,500	22 (20°C)
	Plant Growth Regulators, Desiccants, Defoliants			
138	Azoles	Paclobutrazol	1,300–2,000	26 (20°C)
139		Uniconazole	1,790–2,020	8.41
140	Hydrazides	Maleic hydrazide	>5,000	4.51 kg/L
141	Organophosphates	S, S, S-Tributyl phosphorotrithioate (DEF 6')	250	2.3 (20°C)
142	Phenol derivatives	Pentachlorophenol (PCP)	210	80 (30°C)
143	Quaternary ammonium compounds (bipyridiliums)	Chlormequat	807–966	>1 kg/L
144		Diquat	231	700 g/L (20°C)
145		Paraquat	157	700 g/L (20°C)
146	Synthetic auxins	2-(1-Naphthyl) acetamide (NAD)	1,690	39 (40°C)
147		2-(1-Naphthyl) acetic acid	1,000–5,900	420

Source: Tribaldo, E.B., Analysis of pesticides in water, in *Handbook of Water Analysis*, Nollet, L.M.L. (ed.), 2nd edn, CRC Press, Boca Raton, FL, 2007.

[a] LD50 for rats (oral).

however, if this is physically impractical, then it should be from random position in the accessible parts of the lot. The number of units required for a primary sample should be determined based on the minimum size and the number of laboratory samples required. For plant, egg, and dairy products, where more than one primary sample is taken from a lot, each sample should contribute an approximately similar proportion to the bulk sample.

16.3.1.2 Preparation of Bulk Samples

The primary samples should be combined and mixed well, if practicable, to form the bulk sample. Alternatively, the unit should be allocated randomly to replicate the laboratory samples at the time of taking the primary samples, when: (1) mixing to form the bulk sample is inappropriate or impractical; (2) units are damaged by the process of mixing or subdivision of the bulk sample; or (3) large units cannot be mixed to produce a more uniform residue distribution.

16.3.1.3 Preparation of the Laboratory Sample

The sample sent to or received by the laboratory must be a representative quantity of the material removed from the bulk sample. However, if the bulk sample is larger than that required for a laboratory sample, then it should be divided to provide a representative portion, where required replicate laboratory samples should be withdrawn at this stage or may be prepared.

The laboratory sample must be placed in a clean inert container that provides secure protection from contamination, damage, and leakage. The container should be sealed and securely labeled. Spoilage must be avoided, e.g., fresh samples should be kept cool, while frozen samples must remain frozen.

TABLE 16.2
Physiochemical Properties of Pesticides

Numbers	Common Name	IUPAC Name	Chemical Formula	Major Uses
109	Alachlor	(2-Chloro-N (2,6-diethyl phenyl)-N-methoxy-methyl) acetamide	$C_{14}H_{20}ClNO_2$	Control of annual grasses and broad-leaved weeds in maize, cotton, brassicas, oil seed, sugarcane, and soybean
10/12	Aldrin and dieldrin	(1R, 4S, 4aS, 5S, 8R, 8aR) 1,2,3,4,10,10-Hexachloro 1,4,4a,5,8,8a-hexahydro 1,4,5,8-dimethanonaphthalene (HHDN)	$C_{12}H_8Cl_6$	Highly effective insecticide for soil-dwelling pests and wooden structures against termites and wood borers
11	Chlordane	1,2,4,5,6,7,8-Octachloro-2,3,3a,4,7,7a-hexahydro 4,7-methano 1-H indene	$C_{10}H_6Cl_8$	It is a versatile broad-spectrum insecticide used mainly for nonagricultural purposes
13	DDT and its derivatives	Dichlorodiphenyl trichloro ethane	$C_{14}H_9Cl_{15}$	It is a nonsystemic contact insecticide. It is used for control of yellow fever, sleeping sickness, malaria, and insect-transmitted diseases
18	Heptachlor	1,4,5,6,7,8,8-Heptachlor 3a,4,7,7a-tetrahydro-4,7 methano 1H-indene	$C_{10}H_5Cl_7$	It is used for soil treatment, seed treatment (maize, small grains), control of ants, cat worms, termites, and many other insects
18	Heptachlor epoxide	2,3,4,5,6,7,7-Heptachlor 1a,1b,5,5a,6,6a-hexahydro 2,5-methano-2H-indene (1,2b) oxirane	$C_{10}H_5Cl_7O$	It is not available commercially, but is an oxidation product of heptachlor
18	Hexachlor benzene (BHCS)	Hexachlor benzene	C_6Cl_6	It is a selective fungicide used to control dwarf bunt of wheat as well as soil- and seed-borne diseases.
19	Lindane	γ-Hexachloro cyclohexane	$C_6H_6Cl_6$	It is used as an insecticide on fruit and vegetable crops, for seed treatment, and is also used as therapeutic pesticide
	Monocro-tophos	Phosphoric acid dimethyl [1-methyl-3-(methylamino) 3-oxo-1-propenyl] ester	$C_7H_{14}NO_5P$	Used as acaricide
9	Dichlorvos	2,2-Dichloro vinyl dimethyl phosphate	$C_4H_7Cl_2O_4P$	Used as insecticide
23	Chlorpyrifos	O-O-diethyl O-(3,5,6-trichloro-2-pyridyl) phosphorothioate	$C_9H_{11}Cl_3NOPS$	Used as insecticides and acaricide
28	Malathion	O-O-dimehyl 5-(1,2-dicarbethoxyethyl) phosphoro dithioate	$C_{10}H_{19}O_6PS_2$	Used as acaricide and insecticides
31	Phorate	O-O-diethyl 5-[(ethythio)methyl] phosphoro dithioate	$C_7H_{17}O_2PS_3$	Used as insecticides
1	Aldicarb	2-Methyl-2(methylthio) propionaldehyde O-methyl carbamoyloxime	$C_7H_{14}N_2O_2S$	Control nematodes, insects, and mites in soil and wide variety of crops, citrus fruits, potatoes, soybean, and tobacco

(continued)

TABLE 16.2 (continued)
Physiochemical Properties of Pesticides

Numbers	Common Name	IUPAC Name	Chemical Formula	Major Uses
5	Carbofuran	2,3-Dihydro-2,2-dimethyl beozofuran-7-yl-methyl carbamate	$C_{12}H_{15}NO_3$	It is a systemic acaricide, insecticide, and nematocide. It is used mainly on cereals, citrus fruit, grapes, potatoes, rice, sugarcane, and vegetables
38	Permethrin	3-Phenoxy benzyl (1RS)-cis, trans-3(2,2-dichlovinyl)-2,2 dimethyl cyclopropane carboxylate	$C_{21}H_{20}Cl_2O_3$	It is a contact insecticide effective against a broad range of pests in agriculture forestry and public health
131	Atrazine	6-Chloro-N-ethyl N'-isopropyl-1,3,5-triazine-2,4-diamine	$C_{18}H_{14}ClN_5$	Used as selective pre- and postemergence herbicide for the control of weeds in asparagus, serghum, sugarcane, and pineapple
135	Simazine	6-Chloro-N, N'-diethyl-1,3,5-triazine-2,4-diylamine	$C_7H_{12}ClN_5$	It is a pre-emergence herbicide used to control broad-leaved and grass weeds in berries, citrus fruits, maize, sugarcane, tea, and orchards
94	2,4D	2,4-Dichloro phenoxy acetic acid	$C_8H_6O_3Cl_2$	It is a systemic herbicide widely used throughout the world to control broad-leaved weeds in cereal cropland on lawn and pastures
96	MCPA	4-Chloro-2-methyl phenoxy acetic acid	$C_9H_9ClO_3$	It is a systemic hormone-type selective herbicide. It is used to control of annual and perennial weeds in cereals and grasslands
118	Isoproturon	3-(4-Isopropylphenyl)-1,1-dimethyl urea	$C_{12}H_{18}N_2O$	It is a systemic herbicide and in control of annual grasses and broad-leaved weeds

Sources: World Health Organization, *Guidelines for Drinking Water Quality*, 3rd edn, WHO, Geneva, 2004; O'Neil, M. J., *The Merck Index: An Encyclopedia of Chemicals, Drugs, and Biologicals*, 14th edn, Wiley, New York, 2006.

16.3.1.4 Preparation of the Analytical Sample

The part of the commodity to be analyzed, i.e., the analytical sample, should be separated as soon as possible. One must keep in mind that some parts are not analyzed. For example, the stones of the stone fruit are not analyzed, but the residue level is calculated assuming that they are included but contain no residue.

The analytical sample should be comminuted and mixed well to enable representative analytical portions. The size of the analytical portion is the function of the analytical method and efficiency of mixing.

The methods for comminuting and mixing should not affect the residue present in the analytical sample. Furthermore, the analytical sample should be processed under appropriate special conditions, e.g., at subzero temperature to minimize adverse effects.

16.3.2 EXTRACTION

Pesticide residues are extracted from the food products using different extraction procedures, such as (1) solvent extraction or (2) solid-phase extraction (SPE) [3].

16.3.2.1 Solvent Extraction

Pesticide residues are extracted from the food products using various solvent and solvent combinations, depending on the type of food products and physicochemical properties of the analyte. Water-miscible solvents are used to extract pesticide residues from high-moisture products. Acetone, acetonitrile, and methanol are used to extract nonionic residues from fruits and vegetables. Variation in polarity may affect the degree to which each solvent can extract any particular residue. Furthermore, various other reasons may also influence the choice of solvent. For example, though acetone and acetonitrile have similar extraction capability, acetone is preferred to acetonitrile, because unlike acetonitrile, acetone is less toxic, has lower boiling point, does not affect the detectors adversely, and does not form a two-phase system when analyzing fruits. Liquid partitioning of residues from the initial extractant to a nonaqueous solvent is a step common to most multiresidue methods (MRMs). Certain commodities also present greater challenges to the extraction process, and the methods may include special steps. For example, dry products are extracted by a combination of organic solvent and water to make up for the absence of water in the commodity itself, and several studies support the use of water/acetonitrile for this purpose.

16.3.2.2 Solid-Phase Extraction

The SPE works on the principle of liquid chromatography, achieved by using strong but reversible interactions between the analytes and the surface of the stationary phase. The SPE has several advantages over liquid partition, such as

- Rapid sample preparation
- Higher recoveries without the formation of emulsion
- Saving of solvent and hence reduction in both material cost and cost of disposal
- High precision of analytical results owing to the use of disposal cartridges

Four steps are necessary for carrying out SPE. These steps, namely, (1) conditioning of the sorbent; (2) application of the sample; (3) washing, and (4) elution of adsorbed analyte, should be optimized to obtain maximum recovery [4].

16.3.3 Cleanup

Cleanup steps are designed to purify extracts to permit a more definitive identification of residues at lower limits of quantification, and to minimize the adverse effects on detection. However, almost all cleanup steps adsorb, destroy, or otherwise remove at least some residues from the extract. Furthermore, residues can often be detected, but not reliably quantified in an uncleaned extract.

Many cleanup steps involve chromatography of the extract solution on a column or cartridge. Choices of the column/cartridge material and eluting solvents are based on the chemicals that need to be recovered. Increasing the polarity of the eluent permits recovery of more polar residues, but decreases the degree of cleanup as more coextractives are also eluted.

16.3.4 Identification and Quantification

Chromatographic analyses of the samples are carried out by introducing the sample into the chromatographic system and selecting a type of mobile phase, stationery phase, and the type of detector, depending on the properties of the analyte. Analytes are quantified using either external or internal standards. Confirmation techniques include analysis on a second column with a dissimilar stationery phase, using GC–MS, high-pressure liquid chromatography–MS (HPLC–MS), or HPLC–UV at two different wavelengths.

The calculation of the results depends on the type of calibration, i.e., external or internal calibration, and the calibration model used, i.e., linear or nonlinear [5].

For external standard and linear calibration, the concentration of each analyte in the sample is determined by comparing the detector response that may be the peak area or height with the response for that analyte in the initial calibration.

16.3.5 Validation of the Method

Method validation is necessary to ensure that an analytical methodology is accurate, reproducible, and rugged over the specified range in which an analyte will be analyzed. It provides an assurance of reliability during normal use and is referred to as the process of providing documented evidence that the method does what it is intended to do. Method validation includes eight steps: (1) precision, (2) accuracy, (3) limit of detection (LOD), (4) limit of quantification (LOQ), (5) specificity, (6) linearity and range, (7) ruggedness, and (8) robustness [6].

1. Precision: The measure of the degree of repeatability of an analytical method under normal operation and is usually expressed as RSD%. Precision should be performed at three different levels: (a) repeatability, (b) intermediate precision, and (c) reproducibility.
2. Accuracy: The measure of exactness of an analytical method or the closeness of agreement between the true value and the value found.
3. Specificity: The ability to accurately and specifically measure the analyte of interest in the presence of other components expected to be present in the sample matrix. It is a measure of the degree of interference from components, such as other active ingredients, recipient impurities, and degradation products, ensuring that the peak response is owing to the single component only.
4. LOD: This is defined as the lowest concentration of an analyte in a sample that can be detected. It is a limit test that specifies whether an analyte is above or below a certain value. It is expressed as a concentration at a specified signal-to-noise ratio (S/N ratio), usually 3.
5. LOQ: This is defined as the lowest concentration of an analyte in a sample that can be determined with acceptable precision and accuracy under the stated operational conditions of the method. The S/N ratio of 10:1 is used to determine the LOQ and is considered as a good rule of thumb.

6. Linearity and range: The ability of the method to elicit test results that are directly proportional to the analyte concentration within a given range. Range is the interval between the upper and lower levels of the analyte to be detected.

7. Ruggedness: The degree of reproducibility of the results obtained under varying conditions and expressed as RSD%. These conditions include different laboratories, analysts, instruments, reagents, number of days, etc.

8. Robustness: The capacity of a method to remain unaffected by small deliberate variations in the method parameters. The robustness of a method is evaluated by varying the method parameters, such as pH, ionic strength, temperature, etc.

16.3.6   STANDARD REFERENCE MATERIAL OR CERTIFIED REFERENCE MATERIAL

Two classes of materials are recognized by the International Organization for Standardization/ Committee for Reference Materials (ISO/REMCO) [7]: certified reference material (CRMs) and reference materials (RMs).

A CRM is certified by a Metrological Institution. However, the RMs are of lesser quality than CRMs, which must possess established traceability. The U.S. National Institute of Standards and Technology (NIST) [8] produces SRMs that are equivalent to ISO/REMCO CRMs. The RMs are also available commercially and are usually accompanied by a certificate indicating traceability to the standards of a national metrological institution.

Pure standards of analytes and internal standards should have a known purity, and each standard must be uniquely identified. Furthermore, they should be stored at low temperature.

16.3.7   QUALITY CONTROL OF PESTICIDE RESIDUE ANALYSIS

Quality control is necessary to achieve accurate and precise results [9].

1. *Sampling, transport, processing, and storage of sample*
   Laboratory samples should be taken as discussed earlier. The samples must be transported in clean containers, with robust packaging. Very fragile or perishable products should be frozen to avoid spoilage and then transported in dry ice or some similar substance. On receipt, each laboratory sample must be allocated a reference code. Furthermore, sample preparation, sample processing, and subsampling must be carried out before a visible deterioration occurs.

2. *Maintenance of SRM*
   SRMs or calibration solutions should be prepared, tested for purity, stored, and used as per the guidelines described in Section 16.3.6.

3. *Extraction and concentration*
   Temperature, pH, and other parameters must be controlled if these parameters affect the extraction efficiency, analyte stability, or solvent losses. Great care must be taken when extracts are evaporated to dryness. The evaporation temperature should be as low as possible. When extracts are diluted to a fixed volume, accurately calibrated vessels of not less than 1 mL capacity should be used and further evaporation should be avoided.

4. *Contamination and interferences*
   Samples must be separated from each other and from other sources of potential contamination during transit and storage at the laboratory for pest control. Containers, solvents, reagents, filter aids, etc., should be checked for possible contamination. Furthermore, extracts should be kept out of contact with seals. Analysis of reagent blanks should be carried out to identify sources of interference in the equipment or materials used.

5. *Analytical calibration, representative analytes, matrix effects, and chromatographic integration*
   Accurate calibration is dependent on correct identification of the analyte. Responses used to quantify residues must be within the dynamic range of the detector. Extracts containing

high levels of residues may be diluted to bring them within the calibrated range. Residues below the lowest calibrated level (LCL) should be considered uncalibrated and reported as less than LCL.

Possible matrix effects should be assessed at method-validation step. Furthermore, blanks may be used for calibration purposes. Matrix effect is best calibrated by standard addition. Standard addition is the addition of a known quantity of an analyte to one of the two duplicate analytical samples immediately prior to extraction. If the pesticide is determined as a degradation product or derivative, then the calibration solution should be the pure standard of that degradation product or derivative.

6. *Analytical methods and analytical performance*
   The method must be tested to assess for sensitivity, mean recovery, and precision. The analytical method should be capable of providing mean recovery within the range. As far as possible, the recovery of all components defined by the MRL should be determined routinely.

   When single recovery result is low (<60%), the batch sample should be reanalyzed. However, if the recovery is low but consistent (i.e., demonstrating good precision) and the basis for this is well established, then a lower recovery may be acceptable.

7. *Proficiency testing and analysis of reference materials*
   The laboratory must participate regularly in relevant proficiency tests. When the accuracy achieved or any of the tests are questionable or unacceptable, the problems should be investigated and rectified. In-house RMs may be analyzed regularly to help in providing evidence of analytical performance.

8. *Confirmation of results*
   Negative results must be interpreted with caution. Confirmation of positive results must be supported by the appropriate calibration and recovery determination. Additional confirmation requirements for all positive results, especially those close to LOQs, must be decided on a case-by-case basis.

   If detectors of limited specificity are employed, then a second chromatographic column of different polarity provides only limited confirmatory evidence. If required, it may be confirmed using a more specific technique. The MS is one of these specific techniques.

9. *Reporting of results*
   Results should be normally expressed in mg/kg or μg/kg. Residues below LOQ should be reported as less than LOQ. In general, the residue data must not be adjusted for recovery. If they are adjusted for recovery, then this must be stated. When two or more test portions are analyzed, the arithmetic mean of the most accurate results obtained from each portion should be reported.

It is essential to maintain uniformity in reporting the results. In general, results <0.1 mg/kg should be rounded to one significant figure, those ≥0.1 and <10 mg/kg should be rounded to two significant figures, and those >10 mg/kg may be rounded to three significant figures or to a whole number.

16.3.8 GOOD LABORATORY PRACTICES FOR PESTICIDE RESIDUE ANALYSIS

The reliability of the analytical results, particularly in pesticide residue analysis, depends not only on the availability of reliable analytical methods, but also on the experience of the analyst and maintenance of a good practice in the analysis of pesticides. The guidelines for good laboratory practices (GLP) consist of three interrelated parts [10]: (1) the analyst, (2) the laboratory, and (3) the analysis.

16.3.8.1 The Analyst

As in the pesticide residue analysis, the analyte concentrations are in the range of μg/kg to mg/kg, the analysis is challenging, and attention toward every detail is essential.

- The analyst should have an appropriate professional qualification and be experienced and competent in residue analysis.
- The staff must be fully trained and experienced in the correct use of the apparatus, and in appropriate laboratory skills.
- They must have an understanding of the principles of pesticide residue analysis and know the requirement of an analytical quality assurance (AQA) system.
- They must understand the purpose of each stage in the method, the importance of following the methods exactly as described, and noting any possible deviation.
- They must also be trained in the evaluation and interpretation of the data.

16.3.8.2 The Laboratory

- The laboratory and its facilities must be designed to allow tests to be allocated for maximum safety and minimum chance of contamination.
- Under ideal conditions, separate rooms should be designated for sample receipt and storage, sample preparation, extraction and cleanup, and detection.
- The area used for extraction and cleanup must meet solvent laboratory specification, and all fume extraction facilities must be of high quality.
- Only small volumes of solvents should be held in the working area, and the bulk of the solvent should be stored separately away from the main working area. All the records should be kept up-to-date.
- All the laboratories require pesticide reference standards of known and acceptably high purity. Analytical standards should be available for all parent compounds for which the laboratory monitors the samples, and the metabolites are included in MRLs.
- All analytical standards, stock solutions, and reagents should be properly labeled including preparation date, analyst's identification, solvent used, and storage conditions. The compounds whose integrity could be influenced by degradative processes must be clearly labeled with an expiry date and stored under appropriate conditions. Reference standards must be kept under suitable conditions.

16.3.8.3 The Analysis

- All glassware, reagents, organic solvents, and water should be checked for possible interfering contaminants before use by analysis of reagent blanks.
- Soaps-containing germicides, insect sprays, perfumes and cosmetics, etc., can give rise to interference problems that can be particularly significant when an electron capture detector (ECD) is used. Therefore, they should be banned by the staff in the laboratory. Lubricants, sealants, plastics, natural and synthetic rubbers protective gloves, oil from ordinary compressed airlines, and manufacturing impurities in thimbles, filter papers, glass wool, and cotton wools may also give rise to contamination.
- Chemical reagents and general laboratory solvents may contain, adsorb, or absorb compounds that interfere in the analysis. Therefore, it is necessary to purify reagents and generally, it is necessary to use redistilled solvents. Redistilled water is preferred to deionized water.
- Contamination of glassware, syringes, and gas chromatographic columns can arise from contact with previous samples or extracts. All the glassware should be cleaned with a detergent solution and rinsed thoroughly with distilled water. Glassware to be used for trace analysis must be kept separately and must not be used for any other purposes.
- Pesticide reference standards should always be stored at a suitable temperature in a room separate from the main residue laboratory. Concentrated analytical standard solutions and extracts should not be kept in the same storage area.

- Apparatus-containing PVC should be regarded as a possible source of contamination and should not be allowed in the residue laboratory. Other materials containing plasticizers should also be regarded as source of contamination, but PTFE and silicon rubbers are usually acceptable.
- Analytical instruments should be kept in a separate room.

Sample processing and subsampling should be carried out using procedures as described in Section 16.3.1, to get a representative analytical portion and should not have any effect on the concentration of residue present.

- Samples that cannot be analyzed immediately should be stored at 1°C–5°C away from direct sunlight, and analyzed within a few days. Samples received deep-frozen must be kept frozen. If samples are required to be stored for a longer period, then the storage temperature should be −20°C, which prevents enzymatic degradation.
- When samples are to be frozen, the analytical test portions must be taken prior to freezing, to minimize the possible effect of water separation as ice crystals during storage. Care must be taken to ensure that the entire test portion is used in the analysis. Furthermore, the containers must not leak. Neither the containers used for storage nor their caps or stoppers should allow migration of the analytes into the storage compartment.

The following steps are important in the analysis:

- Standard operation procedures (SOPs) should be used for all operations. These should contain full working instructions as well as information on applicability, expected performance, internal quality-control requirements, and calculation of results. Any deviation from an SOP must be recorded and authorized by the analyst in charge.
- Always a validated method should be used for the analysis. Individual (single residue) methods should be fully validated with all analytes and sample materials specified for the purpose, or using sample matrices representative of those to be tested by the laboratory. Preference should be given to methods having multiresidue and multimatrix applicability. The use of representative analytes or matrices is important in validating the method.
- Group-specific methods (GSM) should be validated initially with one or more representative commodities and a minimum of two representative analytes selected from the group.
- Performance of the instrument and the method should be verified from time to time. Only the confirmed data should be reported. When no residues are detected, the values should be reported as less than the LCL or the established detection limits.

16.4 BASIC PRINCIPLES OF ANALYTICAL INSTRUMENTS

16.4.1 HIGH-PRESSURE LIQUID CHROMATOGRAPHY

High-pressure liquid chromatography (HPLC), or simply liquid chromatography, is a separation technique based on a solid stationary phase and liquid mobile phase [11]. Separations are achieved by partition, absorption, or ion-exchange processes. The HPLC are advantageous than GC for the analysis of organic compounds, because (1) the compounds to be analyzed are dissolved in a suitable solvent and most separations take place at room temperature; (2) most of the compounds that are nonvolatile and thermally unstable can be separated without decomposition or the necessity for making volatile derivatives.

After separation, different detectors are available for use:

- Spectrophotometric detectors (fixed wavelength, variable wavelength, diode-array detectors)
- Differential refractometer detectors
- Fluorimetric detectors
- Potentiometric, voltammetric, or polarographic electrochemical detectors

16.4.2 Gas Chromatography

Gas chromatography is a technique for the separation of thermally stable and volatile organic and inorganic compounds. Gas liquid chromatography (GLC) accomplishes separation by partitioning the components of a mixture between a moving (mobile) gas phase and stationary liquid phase held on a solid support depending on their polarity.

The kind of detection depends on the specific nature of the compounds [12].

The detectors used very frequently are

- Electron capture detector
- Nitrogen/phosphorus detector

In GC, two basic types of columns are generally used: packed columns and capillary columns [12].

Packed columns are constructed from tubing of stainless steel, nickel, or glass. The inner diameters may range from 1.6 to 9.5 mm, with the usual length of 3 m.

Capillary columns have an internal diameter of 1 mm or less. These are usually constructed with silica.

16.4.3 GC–MS

The GC–MS instrument has the following components and functions:

- Separation
    - GC
- Ionization source
    - Electron impact ionization (EI)
    - Chemical ionization (CI)
- Mass analyzer
    - Quadrupole
    - Ion trap
- Detector
    - Electron multiplier
    - Photo multiplier
- Data system

1. *GC*: Samples are introduced into an MS through a gas chromatograph. After introducing the sample into the GC column, the compounds are separated using a suitable temperature programming. The compound, after being eluted from the column, is transferred to the ion source through the heated transfer line. This transfer line should be as short as possible with independent and uniform heating facilities. The ion source, mass filter, and detector should be under vacuum. The vacuum system makes it possible for the ions to move from the ion source to the detector without colliding with other ions and molecules.
2. *Ionization source*: Sample molecules are introduced into the ion source through the sample inlet. Before the MS can analyze a sample, the sample molecules must be ionized. In the ion source, the molecules go through ionization and fragmentation process.
3. *Electron impact ionization*: In electron impact ionization, a molecule emerging from the GC column is ionized by interaction with a stream of relatively high-energy electrons.
4. *CI*: In CI, the sample molecules are introduced into an excess of reagent gas, typically methane or ammonia.
5. *Mass analyzers*: The ions that are formed either by EI or CI technique must be separated by their mass to change ratio. Four types of mass filters (analyzers) exist:

a. Radio frequency (both quadruple filter and ion trap)
b. Time-of-flight (TOF)
c. Fourier transform: ion cyclotron resonance (ICR or FTMS)
d. Magnetic stirrer

16.5 SELECTED MRMS FOR WATER

16.5.1 QUANTIFICATION OF ORGANOCHLORINE PESTICIDE RESIDUES IN WATER BY GC

Method 8081 determines the concentrations of various organochlorine pesticides in extracts from solid and liquid matrices [13]. The method uses fused-silica, open-tubular, capillary columns, and electrolytic conductivity detector (ECD or ELCD) as detection modes.

Liquid samples are extracted with hexane–acetone (1:1) or methylene chloride–acetone (1:1) using Soxhlet, pressurized fluid extraction (PFE), or ultrasonic extraction.

Cleanup may be with alumina, Florisil, silica gel, gel permeation chromatography, or sulfur.

16.5.2 QUANTIFICATION OF NITROGEN- AND PHOSPHORUS-CONTAINING PESTICIDES IN WATER

The AOAC official method 991.07 is a gas chromatographic method for nitrogen- and phosphorus-containing pesticides in finished drinking water [14].

The sample is extracted with CH_2Cl_2. This extract is separated and dried with anhydrous Na_2SO_4. Subsequently, it is solvent-exchanged with methyl-*tert*-butyl ether and concentrated to 5 mL. Pesticides are separated by capillary GC using nitrogen–phosphorus detector.

16.5.3 QUANTIFICATION OF N-METHYL CARBAMOYLOXIMES AND N-METHYL CARBAMATES IN WATER BY DIRECT AQUEOUS INJECTION HPLC WITH POSTCOLUMN DERIVATIZATION

The EPA method 531.2 is an HPLC method to determine *N*-methyl carbamoyloximes and *N*-methyl carbamates in finished drinking waters [15]. A water sample is filtered, and the analytes are chromatographically separated by injecting up to 1000 μL of it into an RP-HPLC system (C_{18} column). After elution from the column, the analytes are hydrolyzed in a postcolumn reaction with 0.075 M NaOH at 80°C–100°C to form methyl amine. The methyl amine reacts with *o*-phthalaldehyde (OPA) and 2-mercaptoethanol (or *N,N*-dimethyl-2-mercaptoethylamine), forming a highly fluorescent iso-indole, detected by a fluorescence detector. The analytes are quantitated using the external standard technique.

16.5.4 CHLORINATED ACIDIC PESTICIDE RESIDUES IN FINISHED DRINKING WATER

The AOAC official Method 992.32 is a method for chlorinated acidic pesticide residues in finished drinking water [16]. It is a gas chromatographic method using an ECD.

Laboratory samples are treated with sodium hydroxide to hydrolyze the esters of the analytes, washed with solvent to remove extraneous organic material, and acidified. The chlorinated acids are ether-extracted and converted to methyl esters using diazomethane. Excess derivatizing reagent is removed by solvent wash and activated magnesium silicate cleanup column. The esters are determined by capillary column GC using ECD.

16.5.5 LITERATURE REVIEW OF PESTICIDE RESIDUE ANALYSIS METHODS IN WATER

In Tables 16.3 through 16.9, the extraction, cleanup, detection, and confirmation methods of pesticide residues in water are enumerated [184,185].

TABLE 16.3
Analytical Methods for the Pretreatment of Water Samples and Chromatographic Determination of Pesticides

Pesticides[a]	Type of Water	Extraction Technique[b]	Cleanup Technique[c]	Derivatization Reagent[d]	Determination Technique[e]	Chromatographic Column/Stationary Phase	Analytical Figures of Merit[f]	References
Gas Chromatography								
4, 5, 9, phenols	River, pond, waste	Column continuous SPE (XAD-2, C_{18}) Eluted with ethyl acetate	—	—	GC/FID Online	HP-1 (15 m)	DL: 0.7–1 µg/L Rec: 94%–103.5% RSD: 1.9%–3.9%	[17]
4, 6, 9, 53, others	River	LLE CH_2Cl_2	SPE Florisil column	Trifluoroacetyl anhydride or diazomethane	CG/MS (SIM)	Ultra-2 (25 m)	DL: 0.014–0.18 µg/L Rec: 83%–127% RSD: 2.6%–22.6%	[18]
10, 11, 12, 13, 15, 17, 18, 71, others	River, sea	LLE CH_2Cl_2, hexane, light petroleum	SPE Silica gel column	—	GC/ECD	Methyl-5% phenyl silicone (30 m)	Range: 5.5–20.6 ng/L Rec: 71%–101.2%	[19]
12, 13, 15, 19, others	River, marine, lake, pond, well	LLE CH_2Cl_2		—	GC/ECD GC/NPD	Nonpolar and semipolar	DL: 0.02–0.3 µg/L Rec: 70%–120% RSD: 5%–20%	[20]
4, 119, others	Freeze-dried	LLE (hexane–CH_2Cl_2) SFE	SPE Florisil cartridges	—	GC/NPD	DB-225 (15 m)	—	[21]
13, 15, 18, 23, 24, 26, 39, 104, others	Tap, mineral, river, sea. sewage	LLE Ethyl acetate	—	—	GC/AED	HP-5 (30 m)	DL: 20–100 pg RSD: 4.3%–8.2%	[22]
2, 3, 4, 5, 6, 9	Aqueous solution	Continuous LLE Ethyl acetate	—	Acetic anhydride	GC/FID Online	HP-17 (10 m)	DL: 0.2–0.4 mg/L RSD: 1.9%–3.9%	[23]
16, 20, 111, 115, 118, 131, others	River, lake	Continuous LLE Toluene	—	—	GC/FID GC/MS Online	HP-5; HP-5MS (20 m)	DL: 1.6–11.9 ng/L RSD: 4.2%–25.6%	[24]

(continued)

TABLE 16.3 (continued)
Analytical Methods for the Pretreatment of Water Samples and Chromatographic Determination of Pesticides

Pesticides[a]	Type of Water	Extraction Technique[b]	Cleanup Technique[c]	Derivatization Reagent[d]	Determination Technique[e]	Chromatographic Column/Stationary Phase	Analytical Figures of Merit[f]	References
131	Aquifer, tap	SPE (C$_{18}$)-cartridge Eluted with ethyl acetate	—	—	GC/MS	DB-1 (30 m)	DL: 0.14–0.38 ±pt Rec: 67%–84% RSD: 15%	[25]
24, 27, 28, 29, 131, 135, 137, others	River	SPE (C$_{18}$)-disk Eluted with CH$_2$Cl$_2$	—	—	LC–GC/NPD Online TOTAD	Methyl-5% phenyl silicone (30 m)	DL: 0.04–1.54 ng/L	[26]
12, 28, 30, 75, others	Wetland	SPE (C$_{18}$)-cartridge Eluted with CH$_2$Cl$_2$–acetonitrile	—	—	GC/ECD GC/ MS–MS tandem	HP-1 (60 m) DB5-MS (30 m)	DL: 2–26 ng/L Rec: 70%–133% RSD: 5.3%–17.4%	[27]
24, 26, 27, 28, 29, 30, others	River, pond, well, tap	SPE (C$_{18}$)-column Continuous Eluted with ethyl acetate	—	—	GC/NPD GC/FID	HP-5 (30 m)	DL: 50–130 ng/L Rec: 93.8%–114.5% RSD: 2.9%–4.3%	[28]
1, 4, 5, 6, 8, 9, others	Ultrapure	SPE (C$_{18}$)-cartridge SPME	—	—	GC–MS	DB-5 (30 m)	DL < 0.1 µg/L Rec: 64%–85% RSD: 10%–2 %	[29]
13, 15, 19	Well	SPE (C$_{18}$)-disks Eluted with methanol	—	—	GC/ECD	HP-608 (30 m)	Rec: 72.4%–104.9%	[30]
23, 24, 25, 71, 135, others	River	SPE (PLRP-S)-cartridge Automatic (Prospekt System)	—	—	GC/MS PTV Online	NB-5 (30 m)	Rec: 67.3%– 60.6% RSD: 2.0%–13.5%	[31]
10, 13, 15, 19, 26, 27, 28, 29, 30, 71, 119, 131, 135, others	Drinking, sea, river, waste	SPE (Oasis-60)-cartridge Eluted with CH$_2$Cl$_2$-acetonitrile Automatic	—	—	GC/MS	HP-5 MS (30 m)	DL: ng/L Rec: 70%–122%	[32]
10, 12, 13, 15, 20, 28, 29, 71, 131, others	River, sea, tap, irrigation, waste	SPE (PLRP-S)-cartridge Automatic (Prospekt System)	—	—	GC/MS PTV Online	HP-5 MS (28 m)	DL: 1–36 ng L Rec: 29%–105% RSD: 1%–35%	[33]

Compounds	Water type	Extraction method		Analysis method	Column	Detection/recovery/RSD	Ref.
10, 15, 16, 19, others	Tap, well, river, swamp, pond	SPE (C$_{18}$)-column Continuous	—	GC/ECD GC/MS	Methyl-5% phenyl silicone (30 m)	DL: 0.01–0.1 µg/L Rec: 92% RSD: 3.5%–5.9%	[34]
57, 74, 75, 76, others	Ground, sea, river, lake	SPME (PDMS, PA, CW-DMS, PDMS-DVB fibers)	—	GC/MS GC/ECD	DB-1 DB-5 MS (30 m)	DL: 1–60 ng/L Rec: 70%–124.4% RSD: 3%–14%	[35]
133, others	River	SPME (PA, PDMS fibers)	—	GC/ECD GC/NPD	HP-5 (30 m)	DL: 0.002–3 µg/mL RSD: 1.9%–27.7%	[36]
10, 12, 13, 15, 17, 18, 19, others	Ground	SPME (CW-DMS, CAR-PDMS, DVB-CAR-PDMS fibers)	—	GC/ECD	SPB-5 (30 m)	DL: 0.4–11.6 ng/L RSD < 10%	[37]
10, 11, 12, 13, 15, 17, 18, 19, 24, 25, 26, 28, 29, 30, 37, 38, 71, 131, 134, 135, 136, others	Ground, drinking	SPME (PDMS, CW-DMS, PDMS-DVB fibers)	—	GC/ECD GC/TSD	MDN-5 (30 m)	DL: 1–50 ng/L Rec: 77.4%–131.7% RSD: 5.8%–56.2%	[38]
28, 29, 131, others	Ground	SPME (PDMS-DVB fibers)	—	GC/MS (SIM)	DB-5 (30 m)	DL: 2–8 µg/L	[39]
24, 27, 28, 29, 30, others	River, sea, lake, tap	SPME (PDMS, CW-DVB, PDMS-DVB, PA fibers)	—	GC/NPD GC/MS	DB-5 DB-5 MS (30 m)	DL: 2–90 ng/L Rec: 80%–114% RSD: 3%–15%	[40]
10, 11, 13, 15, 16, 17, 19, 24, 25, 28, 29, 37, 38, 39, 71, 131, 134, 135, 136, others	Ground	SPME (PDMS-DVB fibers)	—	GC/MS GC/MS–MS tandem	CPSil-8 CB (30 m)	DL: 0.001–5.2 µg/L RSD: 4%–35.3%	[41]
109, 112, 118, 131, 136, others	Rain	SPME (PDMS, PA, CW-DVB fibers)	—	GC/MS–MS tandem	DB-5 MS (30 m)	DL: 0.01–0.05 µg/L RSD: 8.5–14%	[42]
10, 12, 13, 15, 16, 17, 18, others	Field	MA-HS-SPME (PDMS, PA, CW-DVB, PDMS-DVB fibers)/ LLE	—	GC/ECD	DB-608 (30 m)	DL: 2–70 ng/L Rec: 39%–118% RSD: 1.9%–16.5%	[43]
4, 24, 25, 28, 129, 135, others	Mineral, river	HS-SPME (PDMS, PDMS-DVB, CW-DVB fibers) automatic	—	GC/MS (PTV)	Rtx-5MS (30 m)	DL: 0.01–10 µg/L Rec: 45.2%–159.7% RSD: 2.5%–18.2%	[44]

(continued)

TABLE 16.3 (continued)
Analytical Methods for the Pretreatment of Water Samples and Chromatographic Determination of Pesticides

Pesticides[a]	Type of Water	Extraction Technique[b]	Cleanup Technique[c]	Derivatization Reagent[d]	Determination Technique[e]	Chromatographic Column/Stationary Phase	Analytical Figures of Merit[f]	References
13, 17, 71, others	Lake	HS-SPME (PDMS-DVB fibers)	—	—	GC/ECD	AT.SE-54 (15 m)	DL: 0.83–13 ng/L Rec: 71.5%–115.5% RSD ≤ 11.8%	[45]
10, 12, 13, 15, 17, 19, 71, others	Tap, reservoir, sea	LPME (polypropylene fibers)	—	—	GC/MS	DB-5 (30 m)	DL: 13–59 ng/L Rec: 77.3%–99.9% RSD < 14%	[46]
131, 134, 135, others	Aqueous solution	LPME (polypropylene fibers)	—	—	GC/MS	DB-5 (30 m)	DL: 7–63 ng/L Rec: 91.8%–105.7% RSD < 3.5%	[47]
10, 12, 13, 16, 17, 19, others	Sea	HFM, SPME, LPME (PH-PPP)	—	—	GC/MS	DB-5 (30 m)	DL: 1–8 ng/L Rec: 76.1%–109.1% RSD: 1.9%–5.9%	[48]
24, 27, 28, 29, 30, others	Tap, river, lake	LPME (hollow fiber membrane)	—	—	GC/MS	DB-5 MS (30 m)	RSD: 8.6%–9% DL: 10–73 ng/L Rec: 57%–102%	[49]
10, 12, 13, others	Aqueous solution	LPME (polypropylene hollow fiber membrane)	—	—	GC/MS (IT)	DB-5 (30 m)	DL: 0.1 µg/L	[50]
21, 28, 90, 109, others	Drinking, surface, ultrapure	SDME	—	—	GC/NPD	MDN-55 (30 m)	DL: 0.2–5 µg/L RSD: 5%–13%	[51]
5, 25, others	Drinking, river	LPME (hollow fiber membrane)	—	—	GC/NPD	DB-5 (30 m)	DL: 1–72 ng/L Rec: 80%–104% RSD: 4.5%–10.7%	[52]
13, 18, 71, others	River	SPMDs elution by dialysis or sonication	GPC	—	GC/ECD	DB-17, DB-5 (60 m)	DL: 0.02–0.12 pg Rec: 89%–103% RSD: 2.9%–81%	[53]

Analytes	Water source	Extraction	SPE	Derivatization	Detection	Column	Results	Ref
94, 96, 97, 98, others	River, waste, milli-Q	SPME (PDMS, CW-DMS, CAR-PDMS, PDMS-DVB fibers)	—	MTBSTFA On-fiber silylation	GC/MS	BP-5 (30 m)	QL: 4–30 ng/L Rec: 86%–115% RSD: 4%–12%	[54]

Liquid Chromatography

Analytes	Water source	Extraction	SPE	Derivatization	Detection	Column	Results	Ref
9	River, sea, rain, drinking	LLE (CH$_2$Cl$_2$)	—	—	LC/MS (APCI and ESI)	Phenomenex ODS (150 mm)	DL: 1–10 µg/L	[55]
29, 131, others	Freeze-dried	LLE (n-hexane–CH$_2$Cl$_2$) SFE (CO$_2$)	SPE Florisil cartridges	—	LC/DAD	LiChrospher 100 RP-18 (125 mm)	—	[21]
Metsulfuron-methyl, ethametsulfuron	Sea, tap, bottled mineral	Continuous LLE (CH$_2$Cl$_2$)	—	—	LC/DAD	RP-18 (150 mm)	DL: 0.05–0.1 µg/L Rec: 83%–95% RSD: 7%–9.2%	[56]
94, 96, 98, 101, 115, 116, 118, 119, others	Tap	SPE (C$_{18}$)-cartridge Automatic Eluted with acetonitrile	—	—	LC/DAD	Spherisorb-ODS-2, RP-18 (250 mm)	DL: 0.025–0.1 µg/L Rec: 75%–112% RSD: 1%–10%	[57]
109, 115, 116, 118, 119, 131, 137, others	Ground, river	SPE (C$_{18}$, Oasis HLB)-cartridge Eluted with ethyl acetate	—	—	LC/DAD	Spherisorb-ODS-2 RP-18 (250 mm)	DL: 4–25 µg/L RSD: 3.8%–14.8%	[58]
Permethrin, tau-fluvalinate	Drinking	SPE (C$_{18}$)-cartridge Eluted with n-hexane	—	—	LC/DAD	Chiralcel OJ (250 mm) Chiral column	DL: 0.12–0.14 µg/L Rec: 103%–113% RSD: 4%–10%	[59]
4, 5, 7, 8, 115, 116, 118, 119, 131, 134, 135, others	Drinking, surface	SPE (C$_{18}$; Oasis HLB, Bond Elut)-cartridge Eluted with methanol-acetonitrile	—	—	LC/MS (ESI)	Inertsil-ODS-3 RP-18 (100 mm)	DL: 0.1–0.50 µg/L Rec: 67.7%–105.2% RSD: <12.6%	[60]
1, 4, 5, 6, 7, 9, 116, 119, 120, others	Lake, well, cistern, pond, drinking	SPE (C$_{18}$)-column Eluted with acetonitrile-water	—	—	LC/MS (ESI)	Zorbax RP-18 (150 mm)	DL: 41–210 pg/mL Rec: 75%–124% RSD: 11%–16%	[61]
Acetamiprid, imidacloprid, thiacloprid, thiamethoxam	Drinking	SPE (Lichrolut EN)-cartridge Eluted with ethyl acetate-methanol	—	—	LC/MS (ESI)	Lichrospher 100 RP-18 (150 mm)	DL: 0.01 µg/L Rec: 95%–104% RSD: 0.5%–2.8%	[62]

(continued)

TABLE 16.3 (continued)
Analytical Methods for the Pretreatment of Water Samples and Chromatographic Determination of Pesticides

Pesticides[a]	Type of Water	Extraction Technique[b]	Cleanup Technique[c]	Derivatization Reagent[d]	Determination Technique[e]	Chromatographic Column/Stationary Phase	Analytical Figures of Merit[f]	References
144, 145, others	Surface	SPE (graphitized carbon back) cartridge Automatic (OSP-2A)	—	—	LC/UV Online	Superspher RP-8 (250 mm)	DL: 1–2 ng Rec: 94%–99% RSD: 1%–13%	[63]
Chlorsulfuron, prosulfuron, nicosulfuron, others	Tap	SPE (MIP)-cartridge Eluted with methanol-HAc	—	—	LC/UV	Inertsil ODS3 (250 mm)	Rec: 8.5%–14% RSD: 3%–38%	[64]
1, 4, 5, 6, 7, 9, others	River, drinking	SPE (PLR-S, C$_{18}$)-cartridge; automatic (OSP-2A, Prospekt)	—	OPA	LC/FL Online	Superspher RP-8 (250 mm)	DL: 30–50 ng/L Rec: 60%–108.4% RSD: 2%–10%	[65]
4, 29, 119, 131, 135, others	Freeze-dried	SPE (C$_{18}$)-cartridge Automatic (Prospekt)	—	—	LC/UV (enzymic detector) Online	RP-18 (150 mm)	DL: 0.01–0.05 µg/L	[66]
4, 5, 6, 8, 9, 21, 30, 104, 105, 115, 116, 117, 118, 119, 131, 135, 136, 137, others	River, drinking	SPE (PLR-S, PRP-1, C$_8$)-column-continuous	—	—	LC/MS Thermospray Online	Lichrospher 60 RP-8 (125 mm)	DL: 1–100 ng/L Rec > 60% RSD: 1%–15%	[67]
5, 116, 118, 119, 131, 135, 136, 137, others	Aqueous solution	SPE (Lichrolut EN)-column-continuous	—	—	LC/DAD LC/UV TAD	Superspher 60 RP-18 (125 mm)	DL: 100 ng/L	[68]
143, 144, 145, others	Drinking	SPE (C$_8$, C$_{18}$, PLRP-S)-cartridges Continuous	—	—	LC/MS-MS (ESI)	XTerra MS, RP-8 (100)	DL: 0.01–0.04 µg/L Rec: 94%–114.3% RSD: 8.5%–14.5%	[69]
76, 116, others	Sea	SPE (Lichrolut EN)-cartridges Continuous	—	—	LC/MS (APCI) (SIM)	Kromasil 100 RP-18 (250 mm)	LD: 0.005–0.3 µg/L Rec: 85%–99% RSD: 1%–8%	[70]
1, 3, 4, 5, 6, 7, 9, 105, others	Drinking, tap	SPE (C$_{18}$)-cartridges Continuous	—	OPA	LC/FL LC/DAD	Varian RP18 (150 mm)	DL: 0.1–30 µg/L RSD: 2.1%–8.1%	[71]

Analytes	Sample	Extraction		Method	Column	Results	Ref.
5, 6, 8, 23, 24, 26, 27, 28, 116, 129, 135, 136, 137, others	Ground, surface	SPE (C_{18})-cartridges Continuous	—	LC/MS–MS (ESI)	Polar ABZ+ (100 mm)	DL: 0.5–60 ng/L Rec: 43%–118% RSD: 2%–21%	[72]
23, 24, 116, 118, 119, 131, 134, 135, others	River, ground, drinking	SPE (PLR-S)-cartridge Automatic (Profexs, Prospekt)	—	LC/DAD LC/MS (APCI)	Lichrospher 100 RP-18 (125 mm)	DL: 1.7–250 ng/L Rec: 64%–113% RSD: 0.4%–27.4%	[73]
7, 97, others	River, tap	SPE (Vim-DVB, Oasis HPB, Lichrolut EN)-column Automatic	—	LC/UV	Kromasil 100, RP-18 (250 mm)	DL: 0.1–0.2 µg/L Rec: 57%–86% RSD: 1%–13%	[74]
24, 26, 94, 96, 115, 116, 118, 129, 131, 135, 136, others	River, sand filtration, ground, drinking, activated carbon filtration, ozonization	SPE (hysphere resin GP, PLRP-s)-cartridges Automatic (Prospekt-2)	—	LC/MS–MS (ESI)	Purospher START RP-18 (125 mm)	DL: 0.004–2.8 ng/L Rec: 9%–111% RSD: 2%–12.1%	[75]
115, 118, 119, others	Lake	SPME (PDMS-DVB, CW-TPR fibers)	—	LC/UV coupled with SPME	RP-18 (150 mm)	DL: 0.5–5.1 ng/mL RSD: 1%–5.9%	[76]
131, 135, others	Lake	SPME (PDMS, PDMS-DVB, CW-TPR fibers)	—	LC/UV coupled with SPME	RP-18 (150 mm)	DL: 1.2–3.4 ng/mL Rec: 83%–112.9% RSD: 2.4%–8.8%	[77]
Fenitrothion and others	River	SPME (PDMS-DVB fibers)	—	LC/DAD DCDA coupled with SPME	Spherisorb ODS2 (100 mm)	DL: 1.2–11.8 ng/mL Rec: 92.5%–108.2% RSD < 12.5%	[78]
63, nabam, azamethiphos	Tap	SPME (PDMS fibers)	—	LC/UV coupled with SPME	RP-18 (250 mm)	DL: 1–10 ng/mL Rec: 95.5%–99.5% RSD: 2.4%–3.5%	[79]
98, 120, others	River	In-tube SPME (porous DVB capillary)	—	LC/UV coupled with SPME	Wakosil-Agri-9 (250 mm)	DL: 0.9–4.1 ng/mL Rec: 64.4%–97.7% RSD: 1.1%–7.3%	[80]

(continued)

TABLE 16.3 (continued)
Analytical Methods for the Pretreatment of Water Samples and Chromatographic Determination of Pesticides

Pesticides[a]	Type of Water	Extraction Technique[b]	Cleanup Technique[c]	Derivatization Reagent[d]	Determination Technique[e]	Chromatographic Column/Stationary Phase	Analytical Figures of Merit	References
2,4-DCBA, 2,4-DCPA, fenoprop, others	Drain, pond	D-LLLME Automatic	—	—	LC/UV	Phenomenex, RP-18 (250 mm)	DL: 0.1–0.4 r g/mL; Rec: 85%–1C7%; RSD: 3.9%–˜.5%	[81]
4, 5, 6, 9, other	Distilled, tap, reservoir tank, irrigation, sea, waste	—	—	SDS and Brij-35, surfactants	LC/UV (MLC)	Kromasil 100, RP-18 (250 mm)	DL: 30–80 µg/L; Rec: 95%–1C4%; RSD: <6%	[82]
45	River	—	—	—	LC/UV LC/MS (TOF)	Kromasil RP-18 (packed capillary column)	DL: 10 pg/mL; Rec: 95%; RSD: 5%–8.9%	[83]
24, 26, 28, 29, 30, others	Purified waste, ground	Micellar extraction (Genapol X-800, POLE)	—	—	LC/DAD	Nova-Pak RP-18 (150 mm)	DL: 0.03–23.45 µg/L; Rec: 27%– 05%; RSD: 0.5%–5.5%	[84]
Methamidophos, acephate	Ultrapure, tap, surface runoff	SPE (Strata X) Automatic (ASPEC XIi)	—	Ammonium molybdate and thiamine/NH₃	LC/FL	Ultraspher RP-18 (250 mm)	DL: 4–12 µg/L; Rec: 96%–101%; RSD: <2.1%	[85]
1, 4, 5, 6, 7, 116, 118, 119, 131, 135, 136, others	River	SPE (Oasis HLB) cartridges	—	—	LC/MS (ESI)	Atlantis RP-18 (150 mm)	DL: 10–5C ng/L; Rec: 60%–110%; RSD: 4.6%–15%	[86]
118	River, Milli-Q	SPE (sol-gel immunosorbent)	—	—	LC/MS-MS (ESI)	Phenophenex MAX-RP (150 mm)	DL: 5–15 ng/L; Rec: 90%–92%; RSD: 3.3%–4.8%	[87]
135, 136, 137, other	Surface	—	—	—	LC/MS (TOF) (ESI)	X-Terra RP-18 (250 mm)	RSD: 0.3%–2.4%	[88]
38, 109, 131	Aqueous solution	SPE-MAE	—	—	TLC/UV HPTL	RP-18 plates	Rec: 93.2%–105.3%; RSD: 3.1%–6.3%	[89]

Other Chromatographic Methods

Number of pesticides	Water	Sample preparation	Maleimide-CPM	Detection	Column	DL/Rec/RSD	Ref.
22, 23, 24, 27, others	Natural	SPE (C$_{18}$, SDB-1) cartridge	—	TLC/scanning fluorescence	Silica gel 60 F254	DL: 1–10 ng; Rec: 94%–102%; RSD: 0.7%–4.9%	[90]
4, 5, 6, 9	Natural	SPE (C$_{18}$) column	PNBDF	HPTLC/densitometric scanning	Silica-gel plates	Rec: 82.5%–112%; RSD: 7.5%	[91]
1, 2, 3, 4, 5, 6, 8, 9, 13, 15, 24, 75, 94, 95, 104, 105, 117, 126, 127, 128, 130, 131, 136, 137, others	Drinking	SPE (C$_{18}$) cartridge LLE	—	AMD-HPTLC/UV Online	HP-plate, 60 F 254 silica gel	DL: 5–250 ng; RSD: 1%–2%	[92]
1, 3, 4, 6, 109, 116	Aqueous solution	—	—	SFC/MS	Capillary	RSD: 6.4%	[93]
4	Aqueous solution	SPE online	—	SFC/DAD	—	DL: 5 ppb	[94]
1, 5, 104, 105, 115, 117, 119, 136, others	River	SPE (C$_{18}$, PLRP-S, LiChrolut EN) online	—	SFC/DAD	Hypersil silica (250 mm) packed	DL: 0.4–2.5 µg/L; Rec: 79.4–108.3%; RSD: 5.6%–14%	[95]

Source: Tribaldo, E.B., Analysis of pesticides in water, in *Handbook of Water Analysis*, Nollet, L.M.L. (ed.), 2nd edn. CRC Press, Boca Raton, FL, 2007.

a Number of pesticides (see Table 20.1).

b LLE, liquid–liquid extraction; MAE, microwave-assisted extraction; SPE, solid-phase extraction; SPME, solid-phase microextraction; LPME, liquid-phase microextraction; SDME, solvent-drop microextraction; D-LLLME, dynamic liquid–liquid–liquid microextraction; SFE, supercritical fluid-extraction; MIP, molecularly imprinted polymers sorbent; SPMD, device for semipermeable membrane extraction; PDMS, polydimethylsiloxane-coated fiber; PA, polyacrylate-coated fiber; CW-DMS, Carbowax-divinylbenzene fiber; PDMS-DVB, polydimethylsiloxane divinylbenzene fiber; CAR-PDMS, carboxen-polydimethylsiloxane coated fiber; DVB-CAR-PDMS, divinylbenzene carboxen-polydimethylsiloxane-coated fiber; CW-TPR, carbowax-template resin; HS-SPME, headspace solid-phase microextraction; MA-HS-SPME, microwave-assisted headspace-solid-phase microextraction; HFM, porous hollow fiber membrane; PH-PPP, polydydroxylated polyparapheylene.

c SPE, solid-phase extraction; GPC, gel permeation chromatography.

d OPA, o-phthaldehyde; MTBSTFA, *N*-methyl-*N*-(tert-butyldimethylsilyl)-trifluoroacetamide reagent; SDS, sodium dodecyl sulfate; PNBDF, p-nitrobenzenediazonium fluorobate.

e GC, gas chromatography; LC, liquid chromatography; TLC, thin-layer chromatography; HPTLC, high-performance thin-layer chromatography; SFC, supercritical fluid chromatography; MLC, micellar liquid chromatography; AED, atomic emission detector; ECD, electron capture detector; NPD, nitrogen–phosphorus detector; FID, flame ionization detector; TSD, thermionic-specific detector; UV, ultraviolet detector; DAD, diode-array detector; FL, fluorescence detector; MS, mass spectrometry; ESI, electrospray ionization; SIM, single-ion monitoring; DCAD, direct current amperometric detector; IT, ion trap; IR, infrared; NMR, nuclear magnetic resonance; APCI, atmospheric pressure; CI, chemical ionization interface; PTV, programmed-temperature vaporizing; TAD, thermally assisted desorption; TOF, time-of-flight; AMD, automated multiple development.

f DL, detection limit; QL, quantification limit; Rec, recovery; RSD, relative standard deviation; Range, linear range.

TABLE 16.4
Nonchromatographic Methods for the Pretreatment of Water Samples and Determination of Pesticides

Pesticides[a]	Type of Water	Extraction Technique[b]	Reagent[c]	Column[d]	Determination Technique[e]	Analytical Figures of Merit	References
			Capillary Electrophoresis				
28, others	Tap	SPE (C$_8$)-disks Eluted with ethyl ether, ethyl acetate	Borate buffer CM-β-CD	Fused-silica column (65 cm)	CE/UV (EKC)	DL: 0.1 mg/mL Rec: 107% RSD: 9.8%	[96]
4, 117, 131, 135, 137, others	River	SPE (C$_{18}$) Column-continuous	Na$_2$HPO$_4$ SDS	Fused-silica column (47 cm)	CE/UV (MEKC)	DL: 0.01–0.03 µg/mL Rec: 90%–114%	[97]
25, 28, others	Tap, runoff	SPE (Strata)-cartridges-automatic (ASPEC Xlii)	Phosphate buffer SDS	Fused-silica column (80 cm)	CE/UV (MEKC)	DL: 7–150 µg/L Rec: 50%–101% RSD: 0.8%–3%	[98]
Diclosulam, florasulam, metosulam, others	Mineral, stagnant	SPE (C$_{18}$)-cartridges-automatic (Vac-Master Manifold)	Formic acid–ammonium carbonate	Fused-silica column (50 cm)	CE/DAD (SWMR)	DL: 131–342 ng/L Rec: 55%–110% RSD: 2.1%–8.8%	[99]
2, 4, 5, 6, 9	Well, river, pond	SPE (LiChrolut EN) cartridges	Borate–phosphate–SDS buffer	Fused-silica column (50 cm)	CE/DAD (MEKC)	DL: 22–85 ng/L Rec: 82.2%–108.2% RSD: 2.6%–7.4%	[100]
1, 2, 5, 7, 9, others	Aqueous solutions	—	OPA-on column TD Borate–CTAB buffer	Fused-silica column (90 cm)	CE/FL (MEKC)	DL < 0.05 mg/L	[101]
131, 134, 135, others	Mineral, tap, CRM	SPE (PS-DVB 3M) disks	Borate–SDS–methanol buffer	Fused-silica column (40 cm) REPSM	CE/DAD (MEKC)	DL: 3.3–8.5 µg/L Rec: 95%–110% RSD: 4%–6.4%	[102]
1, 5, 12, 13, 94, 96, 115, 118, others	Drinking	SPE (C$_{18}$) cartridges	Borate–SDS	Fused-silica column (40 cm)	CE/UV (MEKC)	DL: 41–460 ng/L Rec: 48%–98% RSD: 4.1%–28%	[103]

Compounds	Sample	Sample preparation	Buffer/medium	Column	Technique (mode)	Analytical data	Ref.
1, 2, 4, 5, 7, 8, 9, other	Aqueous solutions	—	GSDS–ammonium acetate buffer	Fused-silica column (50 cm) (RMMs)	CE/UV CE/MS (ESI) (MEKC)	DL: 0.04–2 µg/L	[104]
4, 9, 116, 119, 131, 134, 135, others	Drinking	SPE (C_{18}) cartridges	Borate–SDS	Fused-silica column (50 cm) (RMMs)	CE/DAD (MEKC)	DL: 2–46 µg/L	[105]
131, 135, 144, 145	Well	SPE (C_{18}) cartridges	Na_2HPO_4–SDS	Fused-silica column (40 cm)	CE/DAD (MEKC)	DL: 0.6–1.9 µg/L; Rec: 80%–95%; RSD: 6%–10%	[106]
94, 115, 116, 119, 131, 135, 136, others	Well, tap, river	SPE (LiChrolut EN, C_8, C_{18}) cartridges	Borate–SDS	Fused-silica column (80 cm)	CE/UV (MEKC)	DL: 0.2–0.5 mg/L; Rec: 12.4%–91%; RSD: 1.4%–3.9%	[107]
115, 116, 117, 118, 119, 131, 135, others	River	SPE (C_{18}) cartridges	Borate–phosphate–SDS buffer	Fused-silica column (100 cm)	CE/UV CE/voltammetry (MEKC)	DL: 0.2–6.5×10^{-6} M; Rec: 85%–102%; RSD: 5.6%	[108]
1, 2, 3, 4, 5, 6, 7, 9, other	Tap	—	Phosphate–ammonium acetate buffer–acetonitrile	Column with C_{18} (25 cm)	CE/DAD (CEC)	DL: 2×10^{-5} M; RSD: 1.5%–6.5%	[109]
4, 9, others	Tap, lake	—	Phosphate–ammonium acetate buffer–acetonitrile	Column with C_{18}, C_8 (25 cm)	CE/DAD (CEC)	DL: 7×10^{-9} to 5×10^{-8} M	[110]
121, 122, other	River	—	Phosphate–ammonium acetate buffer	Fused-silica column (65 cm)	CE ICP-MS	DL: 0.11–0.19 mg/L; Rec: 94.5%–105.8%; RSD: 1.1%–5.3%	[111]
Immunochemical Methods							
131, 134, 135 (hydroxy-)	Milli-Q, tap, rain	—	Phosphate buffer Tween-20	—	ELISA UV: 450 nm	DL < 0.01 µg/L; Rec: 102%; RSD < 3%	[112]

(continued)

TABLE 16.4 (continued)
Nonchromatographic Methods for the Pretreatment of Water Samples and Determination of Pesticides

Pesticides[a]	Type of Water	Extraction Technique[b]	Reagent[c]	Column[d]	Determination Technique[e]	Analytical Figures of Merit[f]	References
Fenitrothion	Tap, river, bottled, purified	—	Phosphate buffer Tween-20	—	ELISA UV: 490 nm	DL: 0.3 µg/L; I_{50}: 6 µg/L; Rec: 99.5%–122.5%; RSD: 5.8%–16.6%	[113]
13, others	Tap, river	—	Phosphate buffer EDTA	—	ELISA Noncompetitive	DL: 8 µg/L; Rec: 79.2%–116.2%; RSD: 2%–7%	[114]
28	Ground, surface	—	Phosphate buffer KCl, NaCl, Tween-20	—	ELISA DAD	DL: 0.11 µg/L; I_{50}: 1.58 µg/L; Rec: 98%	[115]
36, 37, 38, 39, others	Industrial	—	Acetate buffer Dimethyl sulfoxide Tween-20	—	ELISA UV-vis: 450/650 nm	I_{50}: 6–205 µg/mL; Rec: 91.5%–109.4%; RSD: 1.6%–3.6%	[116]
131, 134, 135, 136, others	Ground, lake, river, tap, waste	—	Phosphate buffer Tween-20	Protein G column	EFIA UV: 405 nm	DL: 0.1 µg/L	[117]
15, 26, 38, 39, others	Milli-Q, surface, ground, waste	—	Photo-bacteria (Vibrio fischeri NRRL B-111 77) (Pseudomonas putida)	—	Electrochemical Biosensor Bioassay	QL < 0.1 µg/L; I_{50}: 0.89–6.1 mg/L	[118]
5, paraoxon	Tap	—	AChE innibition Phosphate buffer	—	Biosensor (EnFET)	DL: 2.21–2.75 µg/L	[119]
113, 116, 131, others	Aqueous solution	—	Tyrosinase inhibition	—	Conductometric Biosensor	DL: 1 µg/L; RSD: 5%	[120]

Analytes	Water	Sample Prep	Reagents	Method	Results	Ref.
1, 4, 5, other	Aqueous solution	—	AChE imibition	Amperometric Coulometric Biosensor (TCNQ)	DL: 0.2–1.5 µg/L; RSD: 6.5%–18.6%	[121]
118	River, Milli-Q, certified	—	—	Optical biosensor coupled to FIA system (TIRF)	DL: 0.01–0.17 µg/L; I_{50}: 1.03–1.80 µg/L; RSD: 0.6%–7.2%	[122]
144, 145, others	Deionized	—	—	Optical biosensor (FBDOCI)	DL: 0.04–0.10 µg/L; Rec: 79%–128%; RSD <10%	[123]
131	Aqueous solution	—	TSTU	Piezoelectric biosensor	DL: 0.025 µg/L	[124]
23, others	River	—	—	Piezoelectric biosensor (BZE-DADOO)(QCN)	DL: 0.02–48.5 µg/L; Rec: 60.8%–89.6%; RSD: 8%–14%	[125]
Spectrophotometric Methods						
53, 54, others	Ground	SPE (C_{18}) cartridges	—	Spectrofluorimetry (PDS)	Range: 0.1–100 µg/L; Rec: 90%–119%	[126]
53, 54, others	Aqueous solution	—	—	Spectrofluorimetry (EEM-PARAFAC)	DL: 0.15–20 µg/L; Rec: 93.3%–111.1%; RSD: 2%–3%	[127]
4	Ground, river, tap	—	TCPO, OPA, MA	Chemiluminescence FIA	DL: 9 µg/L; Rec: 89%–106%; RSD: 2.8%	[128]
4	Surface	SPE (C_{18}) cartridges	Ce (IV), rhodamine 6G	Chemiluminescence FIA	DL: 28.7–45.6 µg/L; Rec: 93.2%–96.1%; RSD: 1.4%	[129]
1	Mineral bottle	—	KMnO$_4$, quinine sulfate	Chemiluminescence FIA-multicommutation	DL: 0.069 µg/L; Rec: 98.1%–98.7%; RSD: 3.7%	[130]
102	Irrigation, tap, spring	—	KMnO$_4$/H$_2$SO$_4$ K$_3$Fe(CN)$_6$/ NaOH glycine	Chemiluminescence FIA-multicommutation	DL: 40–500 µg/L; RSD: 4.1%–11.2%	[131]

(continued)

TABLE 16.4 (continued)
Nonchromatographic Methods for the Pretreatment of Water Samples and Determination of Pesticides

Pesticides[a]	Type of Water	Extraction Technique[b]	Reagent[c]	Column[d]	Determination Technique[e]	Analytical Figures of Merit[f]	References
			Electrochemical Methods				
8	River, tap, spring, sea	SPE (MIP)-columns Eluted with methanol–water–HAc	—	—	Voltammetry (DPV) (HMDE)	DL: 4.1 µg/L Rec: 76%–102% RSD: 6.5%–7.4%	[132]
4, 5, 9, others	Lake, pond, tap	LLE (Cl_2CH_2)-evaporation, redissolved with methanol–water	$NaClO_4/NaOH/$ $HClO_4$	—	Voltammetry (DPV) (GCE) RBF-ANN	DL: 0.42–0.73 mg/L Range: 1–30 mg/L Rec: 93%–122% RSD: 5.6%	[133]
131	River	—	Britton-Robinson buffer	—	Voltammetry (SWV) (HMDE) LC	DL: 2 µg/L Rec: 92%–116% RSD < 6.1%	[134]
131	River	—	Britton-Robinson buffer	—	Voltammetry (SWV) (HMDE) FIA	DL: 2×10^{-8} M Rec: 96.2%–99.3% RSD: 5.2%–3.1%	[135]
145	Natural	—	$K_3Fe(CN)_6$, KCl	—	Voltammetry (SWV) Gold microelectrode	DL: 4.51 µg/L Rec: 89.5%–95% RSD: 1.5%–1.8%	[136]
Buprofezin	Tap, treated waste	—	Britton-Robinson buffer	—	Voltammetry (ASV) (HMDE)	DL: 2.2 µg/L Rec: 96.6%–96.2% RSD: 3.4%	[137]

| 94 | River, tap, well | LLE (Cl$_2$CH$_2$) | Britton-Robinson buffer | — | Voltammetry (ASV) (HMDE) | DL: 50 µg/L Rec: 80%–95% RSD: 0.2%–1.6% | [138] |

Source: Tribaldo, E.B., Analysis of pesticides in water, in *Handbook of Water Analysis*, Nollet, L.M.L. (ed.), 2nd edn, CRC Press, Boca Raton, FL, 2007.

a Number of pesticides (see Table 20.1).

b LLE, liquid–liquid extraction; SPE, solid-phase extraction; MIP, molecularly imprinted polymers sorbent.

c CM-β-CD, carboxymethylated β-cyclodextrin; SDS, sodium dodecyl sulfate; AChE, acetylcholinesterase enzyme; FIA, flow-injection analysis; TD, thermal decomposition; TSTU, O-(N-succinimidyl)-N,N,N',N'-tetramethyluronium tetrafluoroborate; OPA, o-phthaldehyde; MA, methylamine; TCPO, *bis*(2,4,6-trichlorophenyl)oxalate.

d REPSM, reverse electrode polarity stacking mode; RMMs, reverse migrating micelles.

e CE, capillary electrophoresis; EKC, electrokinetic chromatography; MEKC, micellar electrokinetic chromatography; CEC, capillary electrochromatography; DPV, differential pulse voltammetry; SWMR: stacking with matrix removal; ELISA, enzyme-linked immunosorbent assay; EFIA, enzyme flow immunoassay; DDP, differential pulse polarography; UV, ultraviolet detection; DAD, diode-array detector; FL, fluorescence detector; MS, mass spectrometry; ICP-MS, inductively coupled plasma-mass spectrometric detection; ESI, electrospray ionization; EnFET, enzyme field-effect transistor; TCNQ, 7,7,8,8-tetracyanoquinodimethane-modified biosensor; FIA, flow injection analysis system; TIRF, total internal reflection fluorescence (transducer); FBDOCI, four-band disposable optical capillary immunosensor; BZE-DADOO, inhibitor benzoylecgonine-1,8-diamino-3,4-dioxaoctane; QCN, quartz crystal nanobalance; PDS, piecewise direct standardization; EEM-PARAFAC, three-dimensional excitation-emission matrix fluorescence and parallel factor analysis; HMDE, hanging mercury drop electrode; GCE, glassy carbon electrode; RBF-ANN, radial basis function-artificial neural networks; SWV, square wave voltammetry; LC, liquid chromatography; ASV, adsorptive stripping voltammetry.

f DL, detection limit; QL, quantification limit; Rec, recovery; RSD, relative standard deviation; Range, linear range; I_{50}, analyte concentration that reduces the assay signal to 50% of the maximum value.

TABLE 16.5
Selected Liquid–Liquid Extraction Procedures for Extracting Fungicides (F) and Herbicides (H) from Water Samples

Compound	Sample	Solvent	Quantification Technique	Recovery (%)	References
MRM (F, H)	1 L SW	3 × 60 mL CH_2Cl_2	LC–UV, MS	88–90	[140]
Sulfonylureas (H)	0.5 L SW (pH 3)	100 mL CH_2Cl_2	GC–ECD	80–92	[141]
EBDCs (F)	0.5 L SW	3 × 50 mL chloroform–hexane (3:1)	LC–MS	77–81	[142]
Tetrachloro-terephthalate and 2 metabolites (H)	0.05 L DW acidified	16 mL diethyl ether:petroleum ether (50:50)	GC–MS	92–104	[143]
12 triazines (H)	0.7 mL PW	0.45 mL tert-buthylmethylether	GC–NPD	46–112	[144]
7 chlorophenoxy acid (H)	0.8 mL RW[n]	0.8 mL n-hexane	GC–MS	—	[145]

Source: Bogialli, S. and Di Corcia, A., Fungicide and herbicide residues in water, in *Handbook of Water Analysis*, Nollet, L.M.L. (ed.), 2nd edn, CRC Press, Boca Raton, FL, 2007.
MRM, multiresidue method; SW, surface water; LC, liquid chromatography; UV, ultraviolet detector; MS, mass spectrometric detector; GC, gas chromatography; ECD, electron capture detector; EBCDs, ethylene (bis)dithiocarbamates; DW, drinking water; PW, pure water; NPD, nitrogen–phosphorous detector; RW, river water.

TABLE 16.6
Selected SPE Procedures for Extracting Fungicides (F) and Herbicides (H) from Water Samples

Compound	Sample	Mode	Sorbent	Eluent Phase	References
5 sulfonylureas (H)	1 L tap water, rainwater, and RW	Offline	MIP	2 mL CH_2Cl_2/CH_3OH (90/10, v/v)	[146]
5 triazines (H)	0.5 L tap water, 0.1 L GW	Offline	PS-DVB disk	12 × 1 mL CH_3CN	[147]
16 sulfonylureas (H)	10 mL RW and SW	Offline	Polyclonal antibodies	10 mL of CH_3CN/H_2O (30/70, v/v)	[148]
52 compounds in MRM (H)	0.5 L SW, 2 L GW, 4 L DW	Offline	0.5 g GCB	2 mL CH_3OH, then 7 mL CH_2Cl_2/CH_3OH (80/20) for base-neutral herbicides; 7 mL CH_2Cl_2/CH_3OH (80/20) 25 mM formic acid for acid herbicides	[150]
Paraquat, diquat, and difenzoquat (H)	0.25 L tap water, RW	Offline	(690 mg) C-18 silica, 190 mg porous graphitic carbon	2 mL 8% CH_3OH in 6.0 M HCl for C-18 silica. 2 mL TFA/CH_3CN (20/80) for porous graphitic carbon	[153]
22 herbicides and 11 fungicides	0.5 L RW	Offline	PS-DVB, 265 mg	3 mL acetone, 3 mL hexane, and 3 mL ethyl acetate	[156]
10 sulfonyl- and phenylureas (H)	0.5 L RW	Offline	60 mg of copolymer of polydivinylbenzene-co-N-vinylpyrrolidone	3 mL CH_3CN	[159]
Atrazine and isoproturon (H)	0.5 L DW	Offline	500 mg C-18 silica	2 × 4 mL CH_3OH	[164]
10 phenylureas (H)	1 L RW	Offline	C-18 silica disk	Ethyl acetate followed by ethyl acetate–methylene chloride	[165]
15 phenylureas and triazines (H)	0.5 L RW	Offline	Strong anion exchange	2 × 4 mL CH_3OH	[166]
6 phenylcarbamates (H)	0.01–0.05 mL SW, DW	Online	C-18 silica	$CH_3OH/CH_3CN/H_2O$ gradient	[167]
16 carbamates (F, H), ureas and thioureas (H)	25 μL DW, SW, GW, cistern, and well water	Online	C-18 silica	CH_3CN/H_2O, 0.1% formic acid, gradient elution	[61]
Carbendazim (F), triazines, carbamates, phenylureas, and their metabolites (H)	1.33 mL SW, GW	Online	C-18 silica	CH_3CN/H_2O 0.01% formic acid	[168]
5 sulfonylureas (H)	0.12 L RW, DW	Online	C-18 silica disk	0.5 M $Na_2CO_3–NaHCO_3$ buffer	[169]

(continued)

TABLE 16.6 (continued)
Selected SPE Procedures for Extracting Fungicides (F) and Herbicides (H) from Water Samples

Compound	Sample	Mode	Sorbent	Eluent Phase	References
7 alkyl-thio-s-triazines (H)	0.5 L RW	Offline	Poly-tetrafluoroethylene-membrane	0.10 M sulfuric acid	[170]
Chloroanilines, sulfamides, phthalimides, and oxazolidines (F)	3 mL MW, SW, GW, RW	SPME, offline	PA, PDMS, CW–DVB, PDMS–DVB	Thermal desorption	[35]
4-Chloro-3-methylphenol and dichlofluanid (F)	3 mL MW, SW, RW	SPME, offline	PA	Thermal desorption	[171]
8 phenoxy acids and dicamba (H)	11 RW, waste water, well water	SPME, offline	PDMS, PA, CAR–PDMS, PDMS–DVB, CW–DVB	Thermal desorption	[174]

Source: Bogialli, S. and Di Corcia, A., Fungicide and herbicide residues in water, in *Handbook of Water Analysis*, Nollet, L.M.L. (ed.), 2nd edn, CRC Press, Boca Raton, FL, 2007.
RW, river water; MIP, molecularly imprinted polymers; GW, ground water; PS-DVB, polystyrene–divinylbenzene copolymers; SW, surface water; MRM, multiresidue method; DW, drinking water; GCB, graphitized carbon black; TFA, trifluoroacetic acid; MW, marine water; PA, polyacrylate; PDMS, polydimethylsiloxane; CW, carbowax; CAR, carbonex.

TABLE 16.7
Selected Capillary Column Gas Chromatographic Methods for Determining Fungicides (F) and Herbicides (H) in Water Samples

Compound	Derivatizing Agent	Column Characteristics	Injection Device	Detector	LOD (µg/L)	References
7 phenoxy acids (H)	Dimethyl sulfate	DB-XLB, 30 m × 0.25 mm i.d., 0.25 µm film thickness	On-column	MS	0.010–0.060	[145]
Phenoxy acids (H)	N-methyl-N-(tert-butyldimethylsilyl)trifluoroacetamide	VF 5MS, 30 m × 0.25 mm i.d., 0.25 µm film thickness	Split/splitless	MS, ion trap	0.0003–0.004	[158]
7 fungicides (F)	Direct analysis	DB-1, 30 m × 0.32 mm i.d., 0.25 µm film thickness	Split/splitless	ECD, MS	0.001–0.040	[35]
6 ureas (H)	Direct in aniline and aminotriazine degradation form	14% cyanopropylphenyl + 86% BP10, 30 m × 0.25 mm i.d., 0.25 µm film thickness, dimethylpolysiloxane	—	NPD	0.04–0.1	[172]
Chlorothalonil (F)	Direct analysis	DB-5, 30 m × 0.32 mm i.d., 0.5 µm film thickness	Split/splitless	ECD	2.9–9.2	[173]
4 phenoxy acids (H)	Pentafluorobenzyl bromide and benzyl bromide	DB-5 30 m × 0.31 mm i.d., 1 µm film thickness	Split/splitless	MS	0.2–1	[175]
6 fungicides (F)	Direct analysis	DB-5 MS, 30 m × 0.32 mm i.d., 0.25 µm film thickness	Split/splitless	ECD	0.004–0.025	[176]
MRM	Direct analysis	XTI-5, 30 m × 0.25 mm i.d., 0.25 µm film thickness	Split/splitless	MS	—	[177]
16 phenoxy acids (H)	Diazometane	DB-1, 30 m × 0.32 mm i.d., 0.25 µm film thickness	Split/splitless	MS	0.005–0.02	[178]
8 phenoxy acids (H)	Pentafluorobenzyl bromide or BF$_3$:methanol (50:50, m:m)	PTE-5, 30 m × 0.32 mm i.d., 0.25 µm film thickness	Split/splitless	ECD, MS	0.5–1000	[179]

Source: Bogialli, S. and Di Corcia, A., Fungicide and herbicide residues in water, in *Handbook of Water Analysis*, Nollet, L.M.L. (ed.), 2nd edn, CRC Press, Boca Raton, FL, 2007.
LOD, limit of detection; MS, mass spectrometry; ECD, electron capture detector; NPD, nitrogen–phosphorous detector.

TABLE 16.8
Selected Liquid Chromatographic Methods for Determining Fungicides (F) and Herbicides (H) in Water Samples

Compound	Sample	Column	Mobile Phase	Detector	LC D (μg/L)	References
MRM (F, H)	DW, RW, GW	C-18 silica (5 μm) 25 cm × 4.6 mm	CH₃OH/H₂O 10 μM TFA gradient elution	MS	0.0004–0.0009	[139]
52 compounds in MRM (H)	DW, SW, GW	C-18 silica (5 μm) 25 cm × 4.6 mm	CH₃OH/H₂O 1 mM CH₃COONH₄ gradient elution	MS	0.003–0.01	[150]
MRM (F, H)	SW	C-18 silica (5 μm) 25 cm × 4.6 mm	CH₃CN/H₂O, H₂O/CH₃OH 0.1% acetic acid both gradient elution	DAD, MS	0.0–0.10	[152]
Sulfonylureas, imidazolinones, sulfonamides (H)	SW	C-8 silica, (5-μm) 25 cm × 4.6 mm	20/80 CH₃CN/H₂O 0.15% acetic acid	DAD, MS	0.0–0.03	[154]
8 phenylureas and sulfonylureas (H)	Tap water and RW	N-isopropylacrylamide (5 μm), 150 mm × 4.6 mm	10 mM ammonium acetate, isocratic elution	UV-240 nm	0.3–1.5	[157]
12 phenylureas (H)	RW	C-18 (3 μm) 15 cm × 2.1 mm	H₂O/CH₃OH gradient elution	Ion trap	0.008–0.036	[161]
3 ciclohexanedione oxime and 2 metabolites (H)	RW	C-8 silica (5 μm) 15 cm × 2.1 mm	H₂O formic acid (0.1%)/CH₃CN/ gradient elution	DAD, MS	0.04, 0.08	[162]
7 chloroacetanilide metabolites (H)	SW, GW	C-18 silica (5 μm) 25 cm × 3.0 mm	0.3% acetic acid in 24/36/40 H₂O/CH₃OH/CH₃CN, isocratic elution	DAD, MS	0.05–0.2	[163]
3 dithiocarbamates (Ziram, maneb, zineb) (F)	RW	C-18 silica (5 μm) 25 cm × 4.6 mm	2–3 mM sodium acetate aqueous solution and methanol (70/30)	DAD/UV-260–287 nm	3–9	[180]
Triazines, phenylureas (H)	GW, SW	C-18 (5 μm), 2.1 × 250 mm	H₂O/CH₃OH 0.01% formic acid, gradient	TOFⁿ, Q-TOFⁿ, and MS	0.005–0.021	[182]

$Source:$ Bogialli, S. and Di Corcia, A., Fungicide and herbicide residues in water, in *Handbook of Water Analysis*, Nollet, L.M.L. (ed.), 2nd edn, CRC Press, Boca Raton, FL, 2007.

LOD, limit of detection; MRM, multiresidue method; DW, drinking water; RW, river water; GW, groundwater; TFA, trifluoroacetic acid; MS, mass spectrometer detector; SW, surface water; DAD, diode array detector; UV, ultraviolet detector; TOF, time-of-flight detector; Q-TOF, quadrupole-time-of-flight detector.

TABLE 16.9
Selected Mass Spectrometric Methods for Analyzing Herbicides (H) and Fungicides (F) in Water

Compounds	Separation Technique and Interface	Mode of Ionization	Detector	Acquisition Mode	References
Chlorophenoxy acid (H)	GC	EI	Ion trap	Full scan, PI	[149]
Arylphenoxypropionic herbicides (H)	LC–ESI	CID	Single Q	SIM, PI/NI	[151]
MRM (F)	GC, LC–APCI	EI, CID	Single Q, ion trap	Full scan, PI	[155]
Phenoxy acid (H)	GC	EI	Single Q	Full scan, PI	[158]
Phenyl and sulfonylureas (H)	LC-ESI	CID	Single Q	Full scan, PI/NI	[159]
Phenyl and sulfonylureas (H)	LC-ESI	CID	Single Q	SIM, PI	[160]
Chloroacetanilides (H)	LC–ESI	CID	Single Q	SIM, NI	[163]
MRM (H, F)	LC–ESI	CID	Triple Q	SRM, PI/NI	[168]
MRM (F)	GC	EI	Single Q	SIM, PI	[35]
Triazines (H)	LC–ESI	CID	Q-TOF	Full scan, PI	[181]
Acidic herbicides	LC–APCI	CID	Triple Q	SRM, PI/NI	[183]

Source: Bogialli, S. and Di Corcia, A., Fungicide and herbicide residues in water, in *Handbook of Water Analysis*, Nollet, L.M.L. (ed.), 2nd edn, CRC Press, Boca Raton, FL, 2007.

GC, gas chromatography; EI, electronic impact; PI, positive ionization mode; LC, liquid chromatography; ESI, electrospray; CID, collision-induced dissociation; Q, quadrupole; SIM, selected ion monitoring; NI, negative ionization mode; APCI, atmospheric pressure chemical ionization; SRM, selected reaction monitoring; Q-TOF, quadrupole-time-of-flight.

REFERENCES

1. WHO, *Guidelines for Drinking Water Quality*, 3rd edition. World Health Organization, Geneva, 2004.
2. O'Neil, M. J., *The Merck Index: An Encyclopedia of Chemicals, Drugs, and Biologicals*, 14th edition. Wiley, New York, 2006.
3. Miller, G., *Manuals of Food Quality Control. 13. Pesticide Residue Analysis in the Food Control Laboratory*. FAO Food and Nutrition Paper No. 14/13, FAO, Rome, 1992.
4. http://www.sigmaaldrich.com/Graphics/Supelco/objects/4600/4538.pdf
5. EPA Method 8000B, Determinative Chromatographic Separations, Revision-2: 1–46, 1996. http://www.epa.gov/sw-846/pdfs/8000B.pdf
6. Feldsine, P., Abeyta, C., and Andrews, W. H., Collaborative study for method validation. *J. AOAC Int.*, 85(5), 1187–1200, 2002.
7. http://isotc.iso.org/livelink/livelink?func=LL&objId=347489&objection=browselsort=name
8. http://ts.nist.gov/measurementservices/referencematerials/index.cfm
9. Quality Control Procedures for Pesticide Residues Analysis, Document No. SANCO/10232/2006, 2006. http://ec.europa.eu/food/plant/resources/qualcontrol_en.pdf
10. Guidelines on Good Laboratory Practice in Pesticides Residue Analysis. CAC/GL-40-1993, Revision 1, Rome, 2003.
11. Pomeraz, Y. and Meloan, C. E., *Food Analysis: Theory and Practice*, 3rd edition. Chapman & Hall, New York, 1994.
12. Willard, H. H., Meritt, L. L., Dean, J. A., and Settle, F. A., *Instrumental Methods of Analysis*. Wadsworth Publishing Company, Belmont, CA, 1988.

13. AOAC Official Method 8081A, Organochlorine Pesticides by Gas Chromatography, 1996. http://www.aoac.org/PSD_Methods/11.pdf

14. AOAC Official Method 991.07, Nitrogen- and Phosphorus-Containing Pesticides in Finished Drinking Water, 1991. http://www.food126.com/standard/edit/UploadFile4/2007811185031305.pdf

15. EPA Method 531.2, Measurement of N-Methylcarbamoyloximes and N-Methylcarbamates in Water by Direct Aqueous Injection HPLC with Postcolumn Derivatization. http://www.epa.gov/safewater/methods/pdfs/met531_2.pdf

16. AOAC Official Method 992.32, Chlorinated Acidic Pesticide Residues in Finished Drinking Water, 2005. http://www.aoac.org/PSD_Methods/3.pdf

17. Ballesteros, E., Gallego, M., and Valcárcel, M., Online preconcentration and gas chromatographic determination of N-methylcarbamates and their degradation products in aqueous samples, *Environ. Sci. Technol.*, 30, 2071, 1996.

18. Okumura, T., Imamura, K., and Nishikawa, Y., Determination of carbamate pesticides in environmental samples as their trifluoroacetyl or methyl derivatives by using gas chromatography-mass spectrometry, *Analyst*, 120, 2675, 1995.

19. Fatoki, O. S. and Awofolu, R. O., Methods for selective determination of persistent organochlorine pesticide residues in water and sediments by capillary gas chromatography and electron-capture detection, *J. Chromatogr. A*, 983, 225, 2003.

20. Kinshimba, M. A. et al., The status of pesticide pollution in Tanzania, *Talanta*, 64, 48, 2004.

21. Alzaga, R. et al., Comparison of supercritical-fluid extraction and liquid-liquid extraction for isolation of selected pesticides stored in freeze-dried water samples, *Chromatographia*, 38, 502, 1994.

22. Viñas, P. et al., Determination of pesticides in waters by capillary gas chromatography with atomic emission detection, *J. Chromatogr. A*, 978, 249, 2002.

23. Ballesteros, E., Gallego, M., and Valcárcel, M., Automatic determination of N-methylcarbamate pesticides by using a liquid-liquid extractor derivatization module coupled on-line to a gas chromatograph equipped with a flame ionization detector, *J. Chromatogr.*, 633, 169, 1993.

24. Lüthje, K., Hyötyläinen, T., and Riekkola, M. L., On-line coupling of microporous membrane liquid-liquid extraction and gas chromatography in the analysis of organic pollutants in water, *Anal. Bioanal. Chem.*, 378, 1991, 2004.

25. Cai, Z. et al., Determination of atrazine in water of low- and sub-parts-per-trillion levels by using solid-phase extraction and gas chromatography/high-resolution mass spectrometry, *Anal. Chem.*, 65, 21, 1993.

26. Pérez, M. et al., Pesticide residue analysis by off-line SPE and on-line reversed-phase LC-GC using the through-oven-transfer adsorption/desorption interface, *Anal. Chem.*, 72, 846, 2000.

27. Martínez Vidal, J. L. et al., Pesticide trace analysis using solid-phase extraction and gas chromatography with electron-capture and tandem mass spectrometric detection in water samples, *J. Chromatogr. A*, 867, 235, 2000.

28. Ballesteros, E. and Parrado, M. J., Continuous solid-phase extraction and gas chromatographic determination of organophosphorus pesticides in natural and drinking waters, *J. Chromatogr. A*, 1029, 267, 2004.

29. Carabias-Martínez, R. et al., Behaviour of carbamate pesticides in gas chromatography and their determination with solid-phase extraction and solid-phase microextraction as preconcentration steps, *J. Sep. Sci.*, 28, 2130, 2005.

30. Shukla, G. et al., Organochlorine pesticide contamination of ground water in the city of Hyderabad, *Environ. Int.*, 32, 244, 2006.

31. Sasano, R. et al., On-line coupling of solid-phase extraction to gas chromatography with fast solvent vaporization and concentration in an open injector liner. Analysis of pesticides in aqueous samples, *J. Chromatogr. A*, 896, 41, 2000.

32. Lacorte, S. et al., Broad spectrum analysis of 109 priority compounds listed in the 76/464/CEE Council Directive using solid-phase extraction and GC/EI/MS, *Anal. Chem.*, 72, 1430, 2000.

33. Brossa, L. et al., Determination of endocrine-disrupting compounds in water samples by on-line solid-phase extraction-programmed-temperature vaporisation-gas chromatography-mass spectrometry, *J. Chromatogr. A*, 998, 41, 2003.

34. Columé, A. et al., Evaluation of an automated solid-phase extraction system for the enrichment of organochlorine pesticides from waters, *Talanta*, 54, 943, 2001.

35. Lambropoulou, D. A., Konstantinou, I. K., and Albanis, T. A., Determination of fungicides in natural waters using solid-phase microextraction and gas chromatography coupled with electron-capture and mass spectrometric detection, *J. Chromatogr. A*, 893, 143, 2000.

36. Sampedro, M. C. et al., Solid-phase microextraction for the determination of systemic and non-volatile pesticides in river water using gas chromatography with nitrogen-phosphorus and electron-capture detection, *J. Chromatogr. A*, 893, 347, 2000.
37. Pérez-Trujillo, J. P. et al., Comparison of different coatings in solid-phase microextraction for the determination of organochlorine pesticides in ground water, *J. Chromatogr. A*, 963, 95, 2002.
38. Conçalves, C. and Alpendurada, M. F., Multiresidue method for the simultaneous determination of four groups of pesticides in ground and drinking waters, using solid-phase microextraction-gas chromatography with electron-capture and thermionic specific detection, *J. Chromatogr. A*, 968, 177, 2002.
39. Tomkins, B. A. and Ilgner, R. H., Determination of atrazine and four organophosphorus pesticides in ground water using solid phase microextraction (SPME) followed by gas chromatography with selected-ion monitoring, *J. Chromatogr. A*, 972, 183, 2002.
40. Lambropoulou, D. A., Sakkas, V. A., and Albanis, T. A., Validation of an SPME method, using PDMS, PA, PDMS-DBV, and CW-DVB SPME fiber coatings, for analysis of organophosphorus insecticides in natural waters, *Anal. Bioanal. Chem.*, 374, 932, 2002.
41. Conçalves, C. and Alpendurada, M. F., Solid-phase micro-extraction-gas chromatography-(tandem) mass spectrometry as a tool for pesticide residue analysis in water samples at high sensitivity and selectivity with confirmation capabilities, *J. Chromatogr. A*, 1026, 239, 2004.
42. Sauret-Szczepanski, N., Mirabel, P., and Wortham, H., Development of an SPME-GC-MS/MS method for the determination of pesticides in rainwater: Laboratory and field experiments, *Environ. Pollut.*, 139, 133, 2006.
43. Li, H. P., Li, G. C., and Jen, J. F., Determination of organochlorine pesticides in water using microwave assisted headspace solid-phase microextraction and gas chromatography, *J. Chromatogr. A*, 1012, 129, 2003.
44. Sakamoto, M. and Tsutsumi, T., Applicability of headspace solid-phase microextraction to the determination of multi-class pesticides in waters, *J. Chromatogr. A*, 1028, 63, 2004.
45. Dong, C., Zeng, Z., and Yang, M., Determination of organochlorine pesticides and their derivations in water after HS-SPME using polymethylphenylvinylsiloxane-coated fiber by GC-ECD, *Water Res.*, 39, 4204, 2005.
46. Basheer, C., Lee, H. K., and Obbard, J. P., Determination of organochlorine pesticides in seawater using liquid-phase hollow fibre membrane microextraction and gas chromatography-mass spectrometry, *J. Chromatogr. A*, 968, 191, 2002.
47. Shen, G. and Lee, H. L., Hollow fiber-protected liquid-phase microextraction of triazine herbicides, *Anal. Chem.*, 74, 648, 2002.
48. Basheer, C. et al., Development and application of polymer-coated hollow fiber membrane microextraction to the determination of organochlorine pesticides in water, *J. Chromatogr. A*, 1033, 213, 2004.
49. Lambropoulou, D. A. et al., Single-drop microextraction for the analysis of organophosphorus insecticides in water, *Anal. Chim. Acta*, 516, 205, 2004.
50. Yan, C. H. and Wu, H. F., A liquid-phase microextraction method, combining a dual gauge microsyringe with a hollow fiber membrane, for the determination of organochlorine pesticides in aqueous solution by gas chromatography/ion trap mass spectrometry, *Rapid Commun. Mass Spectrom.*, 18, 3015, 2004.
51. López-Blanco, C. et al., Determination of carbamates and organophosphorus pesticides by SDME-GC in natural water, *Anal. Bioanal. Chem.*, 383, 557, 2005.
52. Lambropoulou, D. A. and Albanis, T. A., Application of hollow fiber liquid phase microextraction for the determination of insecticides in water, *J. Chromatogr. A*, 1072, 55, 2005.
53. Setková, L. et al., Fast isolation of hydrophobic organic environmental contaminants from exposed semipermeable membrane devices (SPMDs) prior to GC analysis, *J. Chromatogr. A*, 1092, 170, 2005.
54. Rodríguez, I. et al., On-fiber silylation following solid phase microextraction for the determination of acidic herbicides in water samples by gas chromatography, *Anal. Chim. Acta*, 537, 259, 2005.
55. Sun, L. and Lee, H. K., Stability studies of propoxur herbicide in environmental water samples by liquid chromatography-atmospheric pressure chemical ionization ion-trap mass spectrometry, *J. Chromatogr. A*, 1014, 153, 2003.
56. Chao, J. et al., Determination of sulfonylurea herbicides by continuous-flow liquid membrane extraction on-line coupled with high-performance liquid chromatography, *J. Chromatogr. A*, 955, 183, 2002.
57. Nouri, B. et al., High-performance liquid chromatography with diode-array detection for the determination of pesticides in water using automated solid-phase extraction, *Analyst*, 120, 1133, 1995.
58. Carabias Martínez, R. et al., Evaluation of surface and ground-water pollution due to herbicides in agricultural areas of Zamora and Salamanca (Spain), *J. Chromatogr. A*, 869, 471, 2000.

59. Yang, G. S. et al., Separation and simultaneous determination of enantiomers of tau-fluvalinate and permethrin in drinking water, *Chromatographia*, 60, 523, 2004.
60. Nogueira, J. M. F., Sandra, T., and Sandra, P., Multiresidue screening of neutral pesticides in water samples by high performance liquid chromatography-electrospray mass spectrometry, *Anal. Chim. Acta*, 505, 209, 2004.
61. Wang, N. and Budde, W. L., Determination of carbamate, urea, thiourea pesticides and herbicides in water, *Anal. Chem.*, 73, 997, 2001.
62. Seccia, S. et al., Multiresidue determination of nicotinoid insecticide residues in drinking water by liquid chromatography with electrospray ionization mass spectrometry, *Anal. Chim. Acta*, 553, 21, 2005.
63. Ibáñez, M., Picó, Y., and Mañes, J., On-line determination of bipyridylium herbicides in water by HPLC, *Chromatographia*, 45, 402, 1997.
64. Bastide, J. et al., The use of molecularly imprinted polymers for extraction of sulfonylurea herbicides, *Anal. Chim. Acta*, 542, 97, 2005.
65. Hiemstra, M. and de Kok, A., Determination of *N*-methylcarbamate pesticides in environmental water samples using automated on-line trace enrichment with exchangeable cartridges and high-performance liquid chromatography, *J. Chromatogr. A*, 667, 155, 1994.
66. Marty, J. L. et al., Validation of an enzymatic biosensor with various liquid-chromatographic techniques for determining organophosphorus pesticides and carbaryl in freeze-dried waters, *Anal. Chim. Acta*, 311, 265, 1995.
67. Sennert, S. et al., Multiresidue analysis of polar pesticides in surface and drinking water by on-line enrichment and thermospray LC-MS, *Fresen. J. Anal. Chem.*, 351, 642, 1995.
68. Renner, T., Baumgarten, D., and Unger, K. K., Analysis of organic pollutants in water at trace levels using fully automated solid-phase extraction coupled to high-performance liquid chromatography, *Chromatographia*, 45, 199, 1997.
69. Castro, R., Moyano, E., and Galceran, M. T., Determination of quaternary ammonium pesticides by liquid chromatography-electrospray tandem mass spectrometry, *J. Chromatogr. A*, 914, 111, 2001.
70. Gimeno, R. A. et al., Monitoring of antifouling agents in water samples by on-line solid-phase extraction-liquid chromatography-atmospheric pressure chemical ionization mass spectrometry, *J. Chromatogr. A*, 915, 139, 2001.
71. García de Llasera, M. P. and Bernal-González, M., Presence of carbamate pesticides in environmental waters from the northwest of México: Determination by liquid chromatography, *Water Res.*, 35, 1933, 2001.
72. Hernández, F. et al., Rapid direct determination of pesticides and metabolites in environmental water samples at sub-μg/l level by on-line solid-phase extraction-liquid chromatography-electrospray tandem mass spectrometry, *J. Chromatogr. A*, 939, 1, 2001.
73. López-Roldán, P., López de Alda, M. J., and Barceló, D., Simultaneous determination of selected endocrine disrupters (pesticides, phenols and phthalates) in water by in-field solid-phase extraction (SPE) using the prototype PROFEXS followed by on-line SPE (PROSPEKT) and analysis by liquid chromatography-atmospheric pressure chemical ionisation-mass spectrometry, *Anal. Bioanal. Chem.*, 378, 599, 2004.
74. Fontanals, N. et al., Solid-phase extraction of polar compounds with a hydrophilic copolymeric sorbent, *J. Chromatogr. A*, 1030, 63, 2004.
75. Kampioti, A. A. et al., Fully automated multianalyte determination of different classes of pesticides, at picogram per litre levels in water, by on-line solid-phase extraction-liquid chromatography-electrospray-tandem mass spectrometry, *Anal. Bioanal. Chem.*, 382, 1815, 2005.
76. Lin, H. H., Sung, Y. H., and Huang, S. D., Solid-phase microextraction coupled with high-performance liquid chromatography for the determination of phenylurea herbicides in aqueous samples, *J. Chromatogr. A*, 1012, 57, 2003.
77. Huang, S. D., Huang, H. I., and Sung, Y. H., Analysis of triazine in water samples by solid-phase microextraction coupled with high-performance liquid chromatography, *Talanta*, 64, 887, 2004.
78. Sánchez-Ortega, A. et al., Solid-phase microextraction coupled with high performance liquid chromatography using on-line diode-array and electrochemical detection for the determination of fenitrothion and its main metabolites in environmental water samples, *J. Chromatogr. A*, 1094, 70, 2005.
79. Aulakh, J. S., Malik, A. K., and Mahajan, R. K., Solid phase microextraction-high pressure liquid chromatographic determination of Nabam, Thiram and Azamethiphos in water samples with UV detection: preliminary data, *Talanta*, 66, 266, 2005.
80. Hirayama, Y., Ohmichi, M., and Tatsumoto, H., Simple and rapid determination of golf course by in-tube solid-phase microextraction coupled with liquid chromatography, *J. Health Sci.*, 51, 526, 2005.

81. Wu, J., Ee, K. H., and Lee, H. K., Automated dynamic liquid–liquid–liquid microextraction followed by high-performance liquid chromatography-ultraviolet detection for the determination of phenoxy acid herbicides in environmental waters, *J. Chromatogr. A*, 1082, 121, 2005.

82. Gil-Agustí, M. et al., Chromatographic determination of carbaryl and other carbamates in formulations and water using Brij-35, *Anal. Lett.*, 35, 1721, 2002.

83. Holm, A. et al., Determination of rotenone in river water utilizing packed capillary column switching liquid chromatography with UV and time-of-flight mass spectrometric detection, *J. Chromatogr. A*, 983, 43, 2003.

84. Padrón Sanz, C. et al., Micellar extraction of organophosphorus pesticides and their determination by liquid chromatography, *Anal. Chim. Acta*, 524, 265, 2004.

85. Pérez-Ruiz, T. et al., High-performance liquid chromatographic assay of phosphate and organophosphorus pesticides using a post-column photochemical reaction and fluorimetric detection, *Anal. Chim. Acta*, 540, 383, 2005.

86. Belmonte Vega, A., Garrido Frenich, A., and Martinez Vidal, J. L., Monitoring of pesticides in agricultural water and soil samples from Andalusia by liquid chromatography coupled to mass spectrometry, *Anal. Chim. Acta*, 538, 117, 2005.

87. Zhang, X. et al., Development and application of a sol-gel immunosorbent-based method for the determination of isoproturon in surface water, *J. Chromatogr. A*, 1102, 84, 2006.

88. Ibañez, M. et al., Use of quadrupole time-of-flight mass spectrometry in environmental analysis: Elucidation of transformation products of triazine herbicides in water after UV exposure, *Anal. Chem.*, 76, 1328, 2004.

89. Mutavdzic, D. et al., SPE-microwave-assisted extraction coupled system for the extraction of pesticides from water samples, *J. Sep. Sci.*, 28, 1485, 2005.

90. Hamada, M. and Wintersteiger, R., Fluorescence screening of organophosphorus pesticides in water by an enzyme inhibition procedure on TLC plates, *J. Planar Chromatogr.*, 16, 4, 2003.

91. McGinnis, S. C. and Sherma, J., Determination of carbamate insecticides in water by C_{18} solid-phase extraction and quantitative HPTLC, *J. Liq. Chromatogr.*, 17, 151, 1994.

92. Butz, S. and Stan, H. J., Screening of 265 pesticides in water by thin-layer chromatography with automated multiple development, *Anal. Chem.*, 67, 620, 1995.

93. Murugaverl, B., Voorhees, K. J., and Deluca, S. J., Utilization of a benchtop mass-spectrometer with capillary supercritical fluid chromatography, *J. Chromatogr.*, 633, 195, 1993.

94. Medvedovici, A., David, V., and Sandra, P., Fast analysis of carbaryl by on-line SPE-SFC-DAD, *Rev. Roum. Chim.*, 45, 827, 2000.

95. Toribio, L. et al., Packed-column supercritical fluid chromatography coupled with solid-phase extraction for the determination of organic microcontaminants in water, *J. Chromatogr. A*, 823, 163, 1998.

96. García-Ruíz, C. et al., Enantiomeric separation of organophosphorus pesticides by capillary electrophoresis. Application to the determination of malathion in water samples after preconcentration by off-line solid-phase extraction, *Anal. Chim. Acta*, 543, 77, 2005.

97. Hinsmann, P. et al., Determination of pesticides in waters by automatic on-line solid-phase extraction-capillary electrophoresis, *J. Chromatogr. A*, 866, 137, 2000.

98. Pérez-Ruiz, T. et al., Determination of organophosphorus pesticides in water, vegetables and grain by automated SPE and MEKC, *Chromatographia*, 61, 493, 2005.

99. Hernández-Borges, J. et al., Determination of herbicides in mineral and stagnant waters at ng/l levels using capillary electrophoresis and UV detection combined with solid-phase extraction and sample stacking, *J. Chromatogr. A*, 1070, 171, 2005.

100. Molina, M., Pérez-Bendito, D., and Silva, M., Multi-residue analysis of *N*-methylcarbamate pesticides and their hydrolytic metabolites in environmental waters by use of solid-phase extraction and micellar electrokinetic chromatography, *Electrophoresis*, 20, 3439, 1999.

101. Wu, Y. S., Lee, H. K., and Li, S. F. Y., A fluorescence detection scheme for capillary electrophoresis of *N*-methylcarbamates with on-column thermal decomposition and derivatization, *Anal. Chem.*, 72, 1441, 2000.

102. Turiel, E. et al., On-line concentration in micellar electrokinetic chromatography for triazine determination in water samples: Evaluation of three different stacking modes, *Analyst*, 125, 1725, 2000.

103. Fung, Y. S. and Mak, J. L. L., Determination of pesticides in drinking water by micellar electrokinetic capillary chromatography, *Electrophoresis*, 22, 2260, 2001.

104. Molina, M. et al., Use of partial filling technique and reverse migrating micelles in the study of *N*-methylcarbamate pesticides by micellar electrokinetic chromatography-electrospray ionization mass spectrometry, *J. Chromatogr. A*, 927, 191, 2001.

105. Da Silva, C. L., de Lima, E. C., and Tavares, M. F. M., Investigation of preconcentration strategies for the trace analysis of multi-residue pesticides in real samples by capillary electrophoresis, *J. Chromatogr. A*, 1014, 109, 2003.
106. Acedo-Valenzuela, M. I. et al., Determination of neutral and cationic herbicides in water by micellar electrokinetic capillary chromatography, *Anal. Chim. Acta*, 519, 65, 2004.
107. Shakulashvili, N. et al., Simultaneous determination of various classes of pesticides using micellar electrokinetic capillary chromatography in combination with solid-phase extraction, *Chromatographia*, 60, 145, 2004.
108. Chicharro, M. et al., Multiresidue analysis of phenylurea herbicides in environmental waters by capillary electrophoresis using electrochemical detection, *Anal. Bioanal. Chem.*, 382, 519, 2005.
109. Tegeler, T. and El Rassi, E., On-column trace enrichment by sequential frontal and elution electrochromatography. 1. Application to carbamate insecticides, *Anal. Chem.*, 73, 3365, 2001.
110. Tegeler, T., and El Rassi, Z., On-column trace enrichment by sequential frontal and elution electrochromatography. II. Enhancement of sensitivity by segmented capillaries with z-cell configuration— Application to detection of dilute samples of moderately polar and nonpolar pesticides, *J. Chromatogr. A*, 945, 267, 2002.
111. Wuilloud, R. G. et al., The potential of inductively coupled plasma-mass spectrometric detection for capillary electrophoretic analysis of pesticides, *Electrophoresis*, 26, 1598, 2005.
112. Bruun, L. et al., New monoclonal antibody for the sensitive detection of hydroxy-*s*-triazines in water by enzyme-linked immunosorbent assay, *Anal. Chim. Acta*, 423, 205, 2000.
113. Watanabe, E. et al., Enzyme-linked immunosorbent assay based on a polyclonal antibody for the detection of the insecticide fenitrothion. Evolution of antiserum and application to the analysis of water samples, *J. Agric. Food Chem.*, 50, 53, 2002.
114. Anfossi, L. et al., Development of a non-competitive immunoassay for monitoring DDT, its metabolites and analogues in water samples, *Anal. Chim. Acta*, 506, 87, 2004.
115. Brun, E. M. et al., Evaluation of a novel malathion immunoassay for groundwater and surface water analysis, *Environ. Sci. Technol.*, 39, 2786, 2005.
116. Mak, S. K. et al., Development of a class selective immunoassay for the type II pyrethroid insecticides, *Anal. Chim. Acta*, 534, 109, 2005.
117. Bjarnason, B. et al., Enzyme flow immunoassay using a Protein G column for the screening of triazine herbicides in surface and waste water, *Anal. Chim. Acta*, 426, 197, 2001.
118. Farré, M. et al., Pesticide toxicity assessment using an electrochemical biosensor with *Pseudomonas putida* and a bioluminescence inhibition assay with *Vidrio fischeri*, *Anal. Bioanal. Chem.*, 373, 696, 2002.
119. Flores, F. et al., Development of an EnFET for the detection of organophosphorus and carbamate insecticides, *Anal. Bioanal. Chem.*, 376, 476, 2003.
120. Anh, T. M. et al., Conductometric tyrosinase biosensor for the detection of diuron, atrazine, and its main metabolites, *Talanta*, 63, 365, 2004.
121. Nunes, G. S., Jeanty, G., and Marty, J. L., Enzyme immobilization procedures on screen-printed electrodes used for the detection of anticholinesterase pesticides. Comparative study, *Anal. Chim. Acta*, 523, 107, 2004.
122. Mallat, E. et al., Part per trillion level determination of isoproturon in certified and estuarine water samples with a direct optical immunosensor, *Anal. Chim. Acta*, 426, 209, 2001.
123. Mastichiadis, C. et al., Simultaneous determination of pesticides using a four-band disposable optical capillary immunosensor, *Anal. Chem.*, 74, 6064, 2002.
124. Pribyl, J. et al., Development of piezoelectric immunosensors for competitive and direct determination of atrazine, *Sensor. Actuat. B.*, 91, 333, 2003.
125. Halámek, J. et al., Sensitive detection of organophosphates in river water by means of a piezoelectric biosensor, *Anal. Bioanal. Chem.*, 382, 1904, 2005.
126. Garrido Frenich, A. et al., Standardization of SPE signals in multicomponent analysis of three benzimidozolic pesticides by spectrofluorimetry, *Anal. Chim. Acta*, 477, 211, 2003.
127. Rodríguez-Cuesta, M. J. et al., Determination of carbendazim, fuberidazole and thiabendazole by three-dimensional excitation-emission matrix fluorescence and parallel factor analysis, *Anal. Chim. Acta*, 491, 47, 2003.
128. Soto-Chinchilla, J. J. et al., Determination of a *N*-methylcarbamate pesticide in environmental samples based on the application of photodecomposition and peroxyoxalate chemiluminescent detection, *Anal. Chim. Acta*, 524, 235, 2004.

129. Murillo Pulgarín, J. A., Alañón Molina, A., and Fernández López, P., Automatic chemiluminescence-based determination of carbaryl in various types of matrices, *Talanta*, 68, 586, 2006.
130. Palomeque, M. et al., Photochemical-chemiluminometric determination of aldicarb in a fully automated multicommutation based flow-assembly, *Anal. Chim. Acta*, 512, 149, 2004.
131. Chivulescu, A. et al., New flow-multicommutation method for the photo-chemiluminometric determination of the carbamate pesticide asulam, *Anal. Chim. Acta*, 519, 113, 2004.
132. Mena, M. L. et al., Molecularly imprinted polymers for on-line preconcentration by solid-phase extraction of pirimicarb in water samples, *Anal. Chim. Acta*, 451, 297, 2002.
133. Ni, Y., Qiu, P., and Kokot, S., Simultaneous voltammetric determination of four carbamate pesticides with the use of chemometrics, *Anal. Chim. Acta*, 537, 321, 2005.
134. Dos Santos, L. B. O., Abate, G., and Masini, J. C., Determination of atrazine using square wave voltammetry with the hanging mercury drop electrode (HMDE), *Talanta*, 62, 667, 2004.
135. Dos Santos, L. B. O., Silva, M. S. P., and Masini, J. C., Developing a sequential injection-square wave voltammetry (SI-SWV) method for determination of atrazine using a hanging mercury drop electrode, *Anal. Chim. Acta*, 528, 21, 2005.
136. De Souza, D. and Machado, S. A. S., Electrochemical detection of the herbicide paraquat in natural water and citric fruit juices using microelectrodes, *Anal. Chim. Acta*, 546, 85, 2005.
137. Ibrahim, M. S., Al-Magboul, K. M., and Kamal, M. M., Voltammetric determination of the insecticide buprofezin in soil and water, *Anal. Chim. Acta*, 432, 21, 2001.
138. Maleki, N., Safavi, A., and Shahbaazi, H. R., Electrochemical determination of 2, 4-D at a mercury electrode, *Anal. Chim. Acta*, 530, 69, 2005.
139. Crescenzi, C. et al., Development of a multiresidue method for analyzing pesticide traces in water based on solid-phase extraction and electrospray liquid chromatography mass spectrometry, *Environ. Sci. Technol.*, 31, 479, 1997.
140. Jeannot, R. and Sauvard, E., High-performance liquid chromatography coupled with mass spectrometry applied to analyses of pesticides in water. Results obtained in HPLC/MS/APCI in positive mode, *Analysis*, 27, 271 1999.
141. Thompson, D. G. and MacDonald, L. M., Trace-level quantitation of sulfonylurea herbicides in natural water, *J. Assoc. Off. Anal. Chem.*, 75, 1084, 1992.
142. Hanada, Y. et al., LC-MS studies on characterization and determination of N-N'-ethylenbisdithiocarbamate fungicides in environmental water sample, *Anal. Sci.*, 18, 441.
143. Carpenter, R. A., Hollowell, R. H., and Hill, K. M., Determination of the metabolites of the herbicide dimethyl tetrachloroterephthalate in drinking water by high-performance liquid chromatography with gas chromatography/mass spectrometry confirmation, *Anal. Chem.*, 69, 3314, 1997.
144. Teske, J., Efer, J., and Engewald, W., Large volume PTV injection: Comparison of direct water injection and in-vial extraction for GC analysis of triazines, *Chromatographia*, 47, 35, 1998.
145. Catalina, M. I. et al, Determination of chlorophenoxy acid herbicides in water by in situ esterification followed by in-vial liquid–liquid extraction combined with large-volume on-column injection and gas chromatography-mass spectrometry, *J. Chromatogr. A*, 877, 153, 2000.
146. Zhu, Q. et al., Selective trace analysis of sulfonylurea herbicides in water and soil samples based on solid-phase extraction using a molecularly imprinted polymer, *Environ. Sci. Technol.*, 36, 5411, 2003.
147. Turiel, E. et al., Molecular recognition in a propazine-imprinted polymer and its application to the determination of triazines in environmental samples, *Anal. Chem.*, 73, 5133, 2001.
148. Degelmann, P. et al., Determination of sulfonylurea herbicides in water and food samples using sol-gel glass-based immunoaffinity extraction and liquid chromatography-ultraviolet/diode array detection or liquid chromatography-tandem mass spectrometry, *J. Agric. Food Chem.*, 54, 2003, 2006.
149. Ding, W.-H., Liu, C.-H., and Yeh, S.-P., Analysis of chlorophenoxy acid herbicides in water by large-volume on-line derivatization and gas chromatography-mass spectrometry, *J. Chromatogr. A*, 896, 111, 2000.
150. Curini, R. et al., Solid-phase extraction followed by high-performance liquid chromatography-ion spray interface-mass spectrometry for monitoring of herbicides in environmental water, *J. Chromatogr. A*, 874, 187, 2000.
151. D'Ascenzo, G. et al., Determination of arylphenoxypropionic herbicides in water by liquid chromatography-electrospray mass spectrometry, *J. Chromatogr. A*, 813, 285, 1998.
152. Jeannot, R. et al. Application of liquid chromatography with mass spectrometry combined with photodiode array detection and tandem mass spectrometry for monitoring pesticides in surface waters, *J. Chromatogr. A*, 879, 51, 2000.

153. Carneiro, M. C., Puignou, L., and Galceran, M. T., Comparison of silica and porous graphitic carbon as solid-phase extraction materials for the analysis of cationic herbicides in water by liquid chromatography and capillary electrophoresis, *Anal. Chim. Acta*, 408, 263, 2000.

154. Rodriguez, M. and Orescan, D. B., Confirmation and quantitation of selected sulfonylurea, imidazolinone, and sulfonamide herbicides in surface water using electrospray LC/MS, *Anal. Chem.*, 70, 2710, 1998.

155. Fernandez-Alba, A. R. et al., Comparison of various sample handling and analytical procedures for the monitoring of pesticides and metabolites in ground waters, *J. Chromatogr. A*, 823, 35, 1998.

156. Tanabe, A. et al., Seasonal and spatial studies on pesticide residues in surface waters of the Shinano river in Japan, *J. Agric. Food Chem.*, 49, 3847, 2001.

157. Ayano, E. et al., Analysis of herbicides in water using temperature-responsive chromatography and an aqueous mobile phase, *J. Chromatogr. A*, 1069, 281, 2005.

158. Rodriguez Pereiro, I. et al., Optimisation of a gas chromatographic-mass spectrometric method for the determination of phenoxy acid herbicides in water samples as silyl derivatives, *Anal. Chim. Acta*, 524, 249, 2004.

159. Carabias-Martinez, R. et al., Simultaneous determination of phenyl- and sulfonylurea herbicides in water by solid-phase extraction and liquid chromatography with UV diode array or mass spectrometric detection, *Anal. Chim. Acta*, 517, 71, 2004.

160. Ayano, E. et al., Determination and quantitation of sulfonylurea and urea herbicides in water samples using liquid chromatography with electrospray ionization mass spectrometric detection, *Anal. Chim. Acta*, 507, 215, 2004.

161. Draper, W. M., Electrospray liquid chromatography quadrupole ion trap mass spectrometry determination of phenyl urea herbicides in water, *J. Agric. Food Chem.*, 49, 2746, 2001.

162. Marek, L. J., Koskinen, W. C. and Bresnahan, G. A., LC/MS analysis of cyclohexanedione oxime herbicides in water, *J. Agric. Food Chem.*, 48, 2797, 2000.

163. Hostetler, K. A. and Thurman, E. M., Determination of chloroacetanilide herbicide metabolites in water using high-performance liquid chromatography-diode array detection and high-performance liquid chromatography/mass spectrometry, *Sci. Total Environ.*, 248, 147, 2000.

164. Rodriguez-Mozaz, S., Lopez de Alda, M. J., and Barceló, D., Fast and simultaneous monitoring of organic pollutants in a drinking water treatment plant by a multi-analytebiosensor followed by LC–MS validation, *Talanta*, 69, 377, 2006.

165. Ruberu, S. R., Draper, W. M., and Perera, S. K., Multiresidue HPLC methods for phenyl urea herbicides in water, *J. Agric. Food Chem.*, 48, 4109, 2000.

166. Ferrer, I., Barceló, D., and Thurman, E. M., Double-disk solid-phase extraction: Simultaneous cleanup and trace enrichment of herbicides and metabolites from environmental samples, *Anal. Chem.*, 71, 1009, 1999.

167. Hidalgo, C. et al., Automated determination of phenylcarbamate herbicides in environmental waters by on-line trace enrichment and reversed-phase liquid chromatography-diode array detection, *J. Chromatogr. A*, 823, 121, 1998.

168. Hernández, F. et al., Rapid direct determination of pesticides and metabolites in environmental water samples at sub-μg/l level by on-line solid-phase extraction-liquid chromatography—Electrospray tandem mass spectrometry, *J. Chromatogr. A*, 939, 1, 2001.

169. Liu, J. et al., Trace analysis of sulfonylurea herbicides in water by on-line continuous flow liquid membrane extraction-C18 pre-column liquid chromatography with ultraviolet absorbance detection, *J. Chromatogr. A*, 995, 21, 2003.

170. Megersa, N. and Jönsson, J. A., Trace enrichment and sample preparation of alkylthio-*s*-triazine herbicides in environmental waters using a supported liquid membrane technique in combination with high-performance liquid chromatography, *Analyst*, 123, 225, 1998.

171. Peñalver, A. et al., Solid-phase microextraction of the antifouling Irgarol 1051 and the fungicides dichlofluanid and 4-chloro-3-methylphenol in water samples, *J. Chromatogr. A*, 839, 253, 1999.

172. Berrada, H., Font, G., and Moltó, J. C., Indirect analysis of urea herbicides from environmental water using solid-phase microextraction, *J. Chromatogr. A*, 890, 303, 2000.

173. Chen, S.-F., Su, Y.-S., and Jen, J.-F., Determination of aqueous chlorothalonil with solid-phase microextraction and gas chromatography, *J. Chromatogr. A*, 896, 105, 2000.

174. Rodriguez, I. et al., On-fibre silylation following solid-phase microextraction for the determination of acidic herbicides in water samples by gas chromatography, *Anal. Chim. Acta*, 537, 259, 2005.

175. Nilsson, T. et al., Derivatisation/solid-phase microextraction followed by gas chromatography-mass spectrometry for the analysis of phenoxy acid herbicides in aqueous samples, *J. Chromatogr. A*, 826, 211, 1998.

176. Pan, H. and Ho, W., Determination of fungicides in water using liquid phase microextraction and gas chromatography with electron capture detection, *Anal. Chim. Acta*, 527, 61, 2004.
177. Kruawal, K. et al., Chemical water quality in Thailand and its impacts on the drinking water production in Thailand, *Sci. Total Environ.*, 340, 57, 2005.
178. Li, N. and Lee, H. K., Sample preparation based on dynamic ion-exchange solid-phase extraction for GC/MS analysis of acidic herbicides in environmental waters, *Anal. Chem.*, 72, 3077, 2000.
179. Boucharat, C., Desauziers, V., and Le Cloirec, P., Experimental design for the study of two derivatization procedures for simultaneous GC analysis of acidic herbicides and water chlorination by-products, *Talanta*, 47, 311, 1998.
180. Weissmahr, K. W., Houghton, C. L., and Sedlak, D. L., Analysis of the dithiocarbamate fungicides ziram, maneb, and zineb and the flotation agent ethylxanthogenate by ion-pair reversed-phase HPLC, *Anal. Chem.*, 70, 4800, 1998.
181. Ibañez, M. et al., Use of quadrupole time-of-flight mass spectrometry in environmental analysis: Elucidation of transformation products of triazine herbicides in water after UV exposure, *Anal. Chem.*, 76, 1328, 2004.
182. Hernandez, F. et al., Comparison of different mass spectrometric techniques combined with liquid chromatography for confirmation of pesticides in environmental water based on the use of identification points, *Anal. Chem.*, 76, 4349, 2004.
183. Dijkman, E. et al., Study of matrix effects on the direct trace analysis of acidic pesticides in water using various liquid chromatographic modes coupled to tandem mass spectrometric detection, *J. Chromatogr. A*, 926, 113, 2001.
184. Tribaldo, E. B., Analysis of pesticides in water. In: *Handbook of Water Analysis*, L. M. L. Nollet (Ed.), 2nd edition. CRC Press, Boca Raton, FL, 2007.
185. Bogialli, S. and Di Corcia, A., Fungicide and herbicide residues in water. In: *Handbook of Water Analysis*, L. M. L. Nollet (Ed.), 2nd edition. CRC Press, Boca Raton, FL, 2007.

17 Analysis of Pesticide Residues in Milk, Eggs, and Meat

Claudio De Pasquale

CONTENTS

17.1 INTRODUCTION

Currently, the world population is growing at an annual rate of 1.2%, i.e., 77 million people per year. Countries like India, China, Pakistan, Nigeria, Bangladesh, and Indonesia account for half of this annual increment. The world population increased from 2.5 billion in 1950 to 6.1 billion in the year 2000. It is expected to reach between 7.7 and 10.6 billion, i.e., 9.1 billion in the year 2050, based on the estimation.

The population growth rate will be low during the 50 years from 2000 to 2050. The United Nations estimates that there will be little variation in the population of the most developed regions (1.2 billion) in the next 50 years, owing to the low fertility levels. On the other hand, population in the least developed regions is expected to increase from 4.9 billion in the year 2000 to 8.2 billion in the year 2050; these estimates lead to the belief that some degree of decline in fertility will occur [1]. Considering the decline of population growth in the developed regions of the world, probably 95% of the global population increase will take place in the developing countries.

An adult person needs an average energy of 2900 kcal/day for an efficient daily routine. In the developed countries, the average consumption of food provides about 3500 kcal/day, whereas in poor countries people may not even obtain 2000 kcal/day, and thus suffer from malnourishment. This has to be connected with the different diets all over the world, where people get their daily supplies of calories from different sources. In Europe and North America, this supply is largely obtained from livestock products, whereas in many other regions, the calories supply is primarily obtained from cereal grains. Moreover, 80% of the poor people living in the rural areas of the developing countries take their livelihood directly from agriculture with diets that are deficient in micronutrients (minerals, vitamins, etc.) and amino acids [2]. It has been estimated that millions of people, including 6 million children under the age of 5 years, die each year as a result of hunger.

At present, the critical challenge is, therefore, to produce more food and ensure food security regionally to alleviate poverty and malnourishment and, at the same time, to improve human health and welfare. Considering this goal, the fact that the growing population and greater requirements for food and fiber create increased pressures on agricultural production and the need for effective pest and pesticide management have to be weighed up. The globalization of markets and trading has also increased the demands for environmental management, as invasive and exotic species threaten native habitats and indigenous organisms, and furthermore alter the production of food and fiber [3,4].

Although new biological, chemical, and management technologies are continuously being developed to provide more sustainable production alternatives, it is expected that the use of pesticides will continue to be an essential tool in the integrated pest management. Pesticides are biologically active compounds designed to interfere with metabolic processes [5,6].

Pesticides are defined by the United States Federal Insecticide, Fungicide, and Rodenticide Act (FIFRA) as a substance or mixture intended to prevent, destroy, repel, or mitigate any pest including insects, rodents, and weeds [7].

They include not only insecticides but also herbicides, fungicides, disinfectants, and growth regulators. Pesticides have been used in simple forms since early times, but the modern use of synthetic pesticides began in the second half of the twentieth century.

It has to be recognized that many public health benefits have been realized by the use of synthetic pesticides [8]. For instance, the supply of food has become safer and more plentiful, and the occurrence of vector-borne disease has been dramatically reduced. Despite the obvious benefits of pesticides, their potential impact on the environment and public health is substantial. The most recent U.S. EPA public sales and usage report estimates that over 2.4 billion kg of the pesticide-active ingredients were applied worldwide in 1997 [9].

17.2   PESTICIDES CLASSIFICATION

Pesticides are generally categorized according to their persistence in the environment.

Organochlorine (OC) pesticides are considered as persistent pesticides. These pesticides have long environmental half-lives and tend to bioaccumulate in humans and other animals, and thus biomagnify up to 70,000 times in the food chain [10,11]. Since migratory birds and other animals are at the top of the food chain, they carry these persistent compounds all the way through and consequently transfer them to humans, the topmost in the food chain [12]. Another way in which these persistent compounds are transported is through a series of evaporation, deposit (condensation), and rain-off steps; this is the so-called "grasshopper effect" [13]. Through these means, these persistent chemicals are moved thousands of miles from their place of application. OC pesticides were used extensively in the United States as insecticides in the mid-twentieth century. OC pesticides include the cyclodienes, hexachlorocyclohexane isomers, and DDT and its analogues (e.g., DDE, methoxyclor, and dicofol). Nine of the OC pesticides as well as polychlorinated dibenzo-p-dioxins, furans, and biphenyls are the subject of the Stockholm Convention on Persistent Organic Pollutants (POPs), which was held in May 2001; this treaty calls for an immediate ban of the production, import, export, and use of most of these POPs following the disposal guidelines. The nine pesticides are aldrin, chlordane, DDT, dieldrin, endrin, heptachlor, hexachlorobenzene, mirex, and toxaphene.

Even if the proposed agreement is ultimately ratified by 50 countries and thus enters into force, the persistent OC compounds will continue to be monitored in the ecosystems, including the humans. This is because of their toxicity (known animal toxicity, known and suspected human toxicity), and the possibilities of human exposure, primarily via the food chain.

Nonpersistent pesticides are also called contemporary pesticides or current-use pesticides. The development and production of these pesticides escalated after the most persistent pesticides were banned beginning since the mid-1970s. By nature, these pesticides do not persist appreciably in the environment; most decompose within several weeks because of their exposure to sunlight and water. In addition, these pesticides do not tend to bioaccumulate. These contemporary pesticides are structurally diverse and have varied mechanisms of action. Organophosphates (OPs), carbamates, synthetic pyrethroids, phenoxyacid herbicides, triazine herbicides, chloroacetanilide herbicides are among the classes included within this group.

OP pesticides are made up of a phosphate (or thio- or dithiophosphate) moiety and an organic moiety. In most cases, the phosphate moiety is *O,O*-dialkyl substituted. These pesticides are potent cholinesterase inhibitors. They can bind covalently with the serine residue in the active site of acetyl cholinesterase, thus preventing its natural function in the catabolism of neurotransmitters. This action does not only occur in target insects, but also in wildlife and humans.

Carbamate insecticides have the same mechanism of toxicity action as the OP insecticides, except that their effects are less severe. The most popular of these pesticides for residential uses are carbaryl (Sevin) and propoxur (Baygon). Many carbamates such as aldicarb and methomyl are also used in agricultural applications.

Pyrethrins are naturally occurring chemicals produced by chrysanthemums, which exhibit a pesticidal effect on insects. Natural pyrethrins include many isomeric forms usually classified as pyrethrin I and II isomer sets. Synthetic pyrethroids are man-made chemicals that are produced to mimic the effective action of natural pyrethrins. Their chemical structures typically consist of a chrysanthemic acid analogue esterified most often with a ringed structure. Pyrethroids are nonsystemic pesticides having contact and stomach actions. Some pyrethroids also have a slight repellent effect. In most formulations, piperonyl butoxide is added as a synergist. In the last several years, the use of synthetic pyrethroids has escalated, as the use of the more toxic OP and carbamate insecticides has been curtailed. Many products such as Raid brand pesticides that are commonly found in retail stores for home use contain pyrethroids such as permethrin and deltamethrin, to eliminate household pests such as ants and spiders.

Triazines are pre- and postemergence herbicides used to control broad-leafed weeds and some annual grasses. These herbicides inhibit the photosynthetic electron transport in certain plants. Atrazine is the most studied triazine herbicide. It is also one of the most heavily applied pesticides.

Phenoxyacid herbicides are postemergence growth inhibitors used to eliminate unwanted foliage or weeds. The most common phenoxyacid herbicides are 2,4-dichlorophenoxyacetic acid (2,4-D) and 2,4,5-trichlorophenoxyacetic acid (2,4,5-T). These two herbicides were combined in equal proportions to make Agent Orange, the herbicide applied in the jungles of Vietnam, Laos, and Cambodia and in the agricultural regions of Vietnam in the late 1960s and early 1970s, during the Vietnam War. Owing to their high toxicities and persistence, 2,3,7,8-tetrachlorodibenzo-*p*-dioxin along with other chlorinated dioxins and furans, 2,4,5-T have been banned for most applications. Although 2,4-D also contains small amounts of persistent chlorinated dioxins and furans, it is still the most abundantly applied residential pesticide. It is commonly found in home and garden stores in its ester or salt form in combination with other herbicides such as dicamba or mecoprop for application on lawns.

Chloroacetanilides are pre-emergence systemic herbicides that work by preventing protein synthesis and root elongation in plants [14]. The herbicides are *N,N*-disubstituted anilines. The individual chloroacetanilides usually differ by their alkyl substituents on the aniline ring. Metolachlor and alachlor are two of the most abundantly applied herbicides [15].

Fungicide active principles constitute another category of pesticides. Although they are widely used, they are not the most studied class in foodstuff. Hexachlorobenzene and pentachlorophenol (PCP) are OC pesticides but are also categorized as fungicides. Other fungicides are captan, folpet, dichloran, chlorothalonil, metalaxyl, and vinclozolin. Most of them are previously evaluated by their metabolite products.

It is important to understand the environmental fate of pesticides, their metabolites assessing, and their potential exposure and associated risks to human health and environment. The application of pesticides to targeted areas inevitably results in the transport of a portion of these chemicals and their degradation products surrounding nontarget areas. The detection of pesticides and their degradation products in soil [16,17], water [18,19], air [20,21], and food [22,23] is of greater concern, and new instrumental techniques are continuously being sought for better detection and monitoring.

17.3 FOOD AND PESTICIDES CONTAMINATION

The worldwide consumption of pesticides, about 2 million tonnes per year, is split into 24% in the United States, 45% in Europe, and 25% in the rest of the world. The worldwide consumption of pesticides includes 47.5% of herbicides, 29.5% of insecticides, 17.5% of fungicides, and others account only for 5.5% [24].

The analytical determination of pesticide residues presents problems with disputes ranging from moderately to very difficult ones. In terms of biological activity, pesticides show an ample range of purposes, and the word "pesticide" includes a great number of chemical compounds characterized by their physical and chemical properties. Pesticides are toxic by definition and in this way, they pose risk to human health by their transport through food or environmental exposure. Because of the above-mentioned reasons, the limits of the analytical procedures in the contaminated matrices such as food, animal tissue, vegetable, fruit, agricultural product, water, soil, and ground water are very low from sub μg/kg to mg/kg, according to the regulatory controls and the human health food trade decisions. Most pesticides can be determined by gas or liquid chromatographic techniques and mass spectrometric detection.

The greatest problem of any analytical procedure is directly related to the matrix extraction method that has to be done before the analytical instrumental determination. In this respect, some analytes require single residue methods (SRMs) with a very high cost for a single determination. The most common approach to pesticides monitoring is the multiple residual method (MRM). Considering the worldwide scientific research development, focused on the minimization of critical factors affecting the production of acceptable results, the knowledge limits, of the above-mentioned analytical advances, appear to decline gradually.

In food science, the determination of pesticide levels can be divided into two different analytical approaches: the first is the determination of the absolute concentration limit, and the second is exceeding a specific concentration limit with yes/no answers. The two approaches appear to be considerably useful to the safety of human health, the first determining the human exposure and the second investigating about the pesticides use, i.e., after good agriculture practice (GAP) registration. Pesticide residue analyses play a very important role in estimating the human and the environment exposure to various allowed compounds in the diverse agriculture practices within and outside the European countries, in which restrictive laws help us to control the environmental pollution phenomena. Since the analyte concentrations are generally very low and sample matrix are too complex, we have to consider background interferences to perform measurements at very low levels. Therefore, samples cannot be analyzed without careful sample preparation steps, and the extraction methods have to provide a fraction in which the target analytes have been concentrated, while the interfering matrix components have been minimized.

The U.S. Food and Drug Administration (FDA) defines fatty food as having ≥2% fat composition and nonfatty food as having ≤2% fat [25]. Despite this categorization, the fact that there is a great difference in the analyses of high fat samples like meat lard and milk with 3% fat has to be considered; taking this into account, a correct analytical approach should divide the sample terminology category into nonfatty (<2% fat), low fatty (2%–20% fat), and high fatty (>20%) food, considering the fat content calculated on a wet weight basis.

17.4 ANALYTICAL STANDARD PROCEDURES

For any analytical procedure related to the pesticide residues, the specific purpose of the results has to be clear. Generally, screening analyses are requested by a customer to resolve two main problems that are reciprocally connected: (1) to the monitoring of GAP by means of examination of the international Codex or national legislation specific request, and (2) to the residue data analyses that could be used to study the human exposure. The correspondence of foodstuff to the Codex registration mechanism encourages the activities of fair international trade in food considering the promotion of health and the economic interest of the consumers. In this way, the analytical investigation has to ensure that the maximum residue limits (MRLs) are not exceeded.

Generally, pesticides occur in food at trace concentration levels. Trace levels have to be intended as part per million (ppm), i.e., 1 μg for 1 g of food, or less. As previously described, the coextraction of different interfering compounds together with target pesticides from food matrices such as milk, egg, or meat has an effect on the analytical response.

The first step of an analytical procedure is to define the operating modes of sample preparation, extraction, cleanup, or isolation and determination that include separation and detection actions. These single acts interact within any single defined procedure to know the concentration of pesticides in food matrices. They appear to be more restrictive and definite when they are applied to the low- or high-fat matrices, and they are related to the physical and chemical characteristics of pesticides and food matrices. It has to be considered that most of the organic compounds applied during the agriculture practice tend to accumulate in fat. However, new technologies improve the timing needed to conclude the analytical timetable, improving the low detection limit, and the laboratory productivity.

The first step in the analyses of food is to provide a representative composite sample, from which one or more subsamples are to be taken for analyses; all samples have to be handled in the same manner to prevent possible contaminations and avoid the loss of volatile pesticides and inaccurate results in the analysis of subsamples. All these operations, to obtain a homogeneous composite sample, are labor- and time-consuming, but they have to be considered as the first step of the experimental design. These steps consist of the physical separation of food samples.

Most analytical extraction methods are designed predominantly to extract pesticides from 200 to 250 g or less of food samples employing solvents such as hexane, dichloromethane, ethyl acetate, or acetone [26]. The solvent is blended with the foodstuff and can be further homogenized using an ultrasound generator bath to increase the solid–liquid surface contact and decrease the gradient extraction phenomena. In this extraction step, salts like sodium chloride or sodium sulfate can be added to absorb water. The extraction timing is associated with the pesticide and matrix physical and chemical characteristics as the choice of organic solvent. The most common problems are generally associated with an incomplete pesticide recovery, a third phase or a solvent layer formation that mystifies the partitioning process. Different experimental solutions could be applied to solve these problems: the first one involves selecting a more efficient solvent, and the second one in adding salt to the sample/solvent combination to breakdown the emulsion phase.

For these problems, diverse organic solvents were adopted and could be changed considering both the matrices and the pesticide properties. Within these considerations, analytical parameters such as sample and solvent volumes, time of contact, agitation, or other analytical actions to determine pesticides and their degradation products have to be exploited.

Fipronil, a phenylpyrazole insecticide, transferred from feed to milk [27] was found under its sulfone form by an extraction with ethyl acetate and a fat removal performed by a liquid/liquid extraction with (1:6, v/v) acetonitrile–isooctane solution. Endosulfan [28], an OC insecticide of the cyclodiene group, was detected in goat milk by an extraction procedure with (1:1, v/v) hexane–acetone solution. A multiresidue investigation for OCPs in bovine milk [29] was performed by using hexane as the organic extracting phase. The same multiresidual approach was reported in bovine milk for herbicide and fungicide residues by using solid phase extraction (SPE) method [30].

A number of different approaches and extraction methods such as solid phase microextraction (SPME) or pressure solvent extraction are named in Table 17.1.

Moreover, in eggs and meat pesticide residues, matrix solid phase dispersion (MSPD) [31–33], accelerated solvent extraction (ASE) [34,35], which is the name of the pressurized liquid extraction (PLE) system, supercritical fluid extraction (SFE) [36], hexane, acetonitrile, ethyl acetate, or acetone solvent organic phases in single or mixed solution [37–40] were used to perform pesticide isolations before the purification step (Tables 17.2 and 17.3).

TABLE 17.1
Residue Analysis of Pesticides in Milk

Method	Analytes	Extraction	Analytical System	Capillary Columns	I.S. Type	Recovery (%)	LOD ng/mL	RSD (%)
Le Faouder et al. [27]	Fipronil	Ethyl acetate/liquid–liquid extraction (acetonitrile/isooctane)/Florisil SPE	GC-MS/MS	OV1	Not available	Not available	0.01	Not available
Nag et al. [28]	Endosulfan	Hexane extraction/column chromatography	GC-ECD	CP Sil CB	Not available	Not available	Not available	Not available
Sharma et al. [29]	MRLs (OCPs)	De Faubert Maunder et al. [53] solvent method and cleanup technique of Veirov and Aharonson [54]	GC-ECD	BP5	Not available	79–85	0.04–0.7	Not available
Bogialli et al. [30]	MRLs (herbicides and fungicides)	SPE extraction	LC-MS	C-18 (packed)	sec-Butylazine	82–120	0.008–1.4	≤11
Pagliuca et al. [55]	MRLs (OPPs)	Acetone:acetonitrile (1:4)/dichloromethane extraction/cleanup with SPE C18	GLC-NPD (column in column)	ZB5 and ZB50	Parathion-ethyl	59–117	1	7–28
Basheer and Lee [56]	Triazine	HFM-protected SPME	GC-MS	DB-5	Fluoranthene d-10	57–107	0.003–0.013	4.30–12.37
Battu et al. [57]	MRLs (OCPs, OPPs and synthetic pyrethroids)	Dichloromethane extraction in a glass column	GC equipped with ECD/NPD	OV17 + OV210/OV101 (packed on Chromosorb/gas Chrom Q)	Not available	>90	0.01	1.14–2.29
Salas et al. [58]	MRLs (OPPs)	Ethyl acetate/extraction with acetonitrile saturated with hexane	GC-FID	DB-1	Not available	60.04–98.89	5–19	5–20

Reference	Analyte	Extraction method	Detection	Column		Recovery (%)		
Zhu et al. [59]	Phenoxy herbicides	Liquid–liquid–liquid microextraction with a hollow fiber membrane	HPLC	Inertsil-ODS-2	Not available	70.8–77.0	0.5	4.56–7.02
John et al. [60]	MRLs (OCPs)	Extraction with a mixture of acetone, acetonitrile, and hexane/Florisil cleanup	GC-ECD	OV17 + OV210 (packed on Chromosorb)	Not available	95–98	Not available	Not available
Di Muccio et al. [61]	Pyrethroids	Mixing with acetonitrile and ethanol/absorbing into a macroporous diatomaceous material/cleanup by SEC	GC-ECD with twin systems	DB-1 DB1701	Not available	60–119	Not available	2.5–14.4
Waliszewski et al. [62]	MRLs (OCPs)	Extraction with petroleum ether	GC-ECD	DB-5	Not available	91.0–99.1	0.01–0.03	<10
Martinez et al. [63]	MRLs (OCPs)	Chromatographic column eluting hexane or a mixture of hexane and methylene chloride	GC-ECD	QF (packed on Chromosorb)	Not available	0–100	5	Not available
Losada et al. [64]	MRLs (OCPs)	Extraction by Suzuki et al./cleanup of hexane extracts with an official method	GC-ECD	DC-200 (packed on Chromosorb)	Not available	84.5–98.2	1	Not available

TABLE 17.2
Residue Analysis of Pesticides in Meat

Method	Analytes	Extraction	Analytical System	Columns	I.S. Type	Recovery (%)	LOD (ng/g)	RSD (%)
Garrido Frenich et al. [33]	MRLs (OCPs and OPPs)	Two methods of extraction: Liquid–solid extraction with high-speed homogenizer using ethyl acetate/GPC MSPD with C18 sorbent/Florisil cleanup and ethyl acetate elution	GC-(EI)-MS/MS using a triple quadrupole analyzer	DB-5	Caffeine	70–115	Not available	≤20
Garrido Frenich et al. [35]	MRLs (OCPs and OPPs)	Ethyl acetate extraction using polytron for 2 min of homogenized meat samples	GC-MS/MS using a triple quadrupole analyzer	DB-5	Not available	70–110	Not available	Not available
Garrido Frenich et al. [65]	MRLs (OCPs and OPPs)	Extraction of homogenized meat samples mixed with sodium sulfate and ethyl acetate in plytron/soxhlet and ASE/cleanup by GPC	GC-MS/MS using a triple quadrupole analyzer	DB-5	Caffeine	70.0–92.9	<2.0	<10
Juhler René [36]	MRLs (OPPs)	SFE/collection on a Florisil trap/elution with heptane and then with acetone	GC-NPD	DB1701	Not available	78–95	10–30	Not available
Juhler René [40]	MRLs (OPPs)	Ethyl acetate extraction/cleanup by SPE on C18 minicolumns	GC-NPD	DB1701	Not available	Not available	1–20	Not available

TABLE 17.3
Residue Analysis of Pesticides in Eggs

Method	Analytes	Extraction	Analytical System	Columns	I.S. Type	Recovery (%)	LOD (ng/g)	RSD (%)
Bolaños et al. [66]	MRLs (OCPs, OPPs and PCBs)	MSPD method using C18 as sorbent/ethyl acetate and acetonitrile saturated in n-hexane, with simultaneous cleanup with Florisil in-line	GC-QqQ-MS/MS	VF-5 ms	Caffeine and 3′-fluoro-2,4,4′-trichlorobiphenyl	70–110 and 70–106	≤2.25	<20
Valsamaki et al. [32]	MRLs (OCPs and PCBs)	MSPD method using Florisil as the sorbent material and dichloromethane:hexane (1:1) as the eluting system	GC-ECD	DB-5	PCB 27	82–110	<0.7	<8
Herzke et al. [34]	MRLs (OCPs and PCBs)	Acetone:n-hexane (1:9)/ASE/GPC	GC-MS in negative chemical ionization (NICI)	DB-5	[^{13}C]-isotope labeled α- and γ-HCH, p,p′-DDE, HCB, and PCB 28, 52, 118, 153, 180	80–100	Not available	Not available
Lehotay et al. [37]	MRLs	MeCN extraction/SPE with C18 cleanup	GC MS-MS	DB-5	Terbufos	53–100	10	8–40
Schenck [38]	MRLs (OCPs and OPPs)	Acetonitrile extraction/SPE cleanup	GC-FPD (for OPPs) and GC-ECD (for OCPs)	DB-5 (FPD) DB-225 (ECD)	Not available	86–108 OCPs 61–149 OPPs	<0.5	Not available
Naoto et al. [39]	MRLs (OCPs)	n-Hexane:acetonitrile (2:1) extraction	GC-ECD	SPB1	Not available	85–102	2	Not available

Intensive and time-consuming cleanup, such as gel permeation chromatography (GPC), is usually needed to remove the coextracted fat material prior to the instrumental analytical step. This represents the cleanup step usually achieved by a combination of partitioning and purification, and the latter accomplished by preparative chromatography. The degree of cleanup will depend on the extracting efficiency of the selected solvent, which should dissolve the target pesticides, leaving most of the interference compounds. Therefore, it has to be specified that different approaches have to be considered during investigation of high-fat matrix such as animal fat (>20%) or low-fat matrices including milk, meat from poultry, pork, and eggs (2%–20%). In fact, in the first case, there is no option for the use of a nonpolar solvent to dissolve fat and extract the pesticide residues, and there is no reason to investigate a high recovery of rather polar pesticides in high-fat matrices, because only lipophilic analytes are likely or known to occur. In the second case, lipophilic and hydrophilic pesticides can be considered, and the analytical methods and the less time-consuming postextraction clean-up can be designed to have a wide polarity range of target compounds. In this case, one of the most useful organic solvents is acetonitrile (MeCN), because it gives high recoveries of a wide polarity range of pesticides and yet it does not significantly dissolve high nonpolar fats or high polar proteins, salt, and sugar that commonly occur in food [37,41]. Special and targeted modifications to the cleanup step may significantly improve the efficiency of pesticides detection.

The preparative chromatography used for purification is of two types: adsorptive and size exclusion.

The first type is based on the interaction between a chemical dissolved in a solvent and the adsorbent surface of particles of the chromatographic material, which is normally placed in a large glass column. The extracted sample, deposited on the top of the column, is eluted with various types of organic solvents to achieve the separation of pesticides that can be obtained only when they are eluted in fractions, which are different from those of the coextractive compounds.

Materials such as Florisil, alumina, silica gel, and carbon are generally used for various purification purposes considering their intrinsic characteristics. Florisil is particularly suitable for the cleanup of fatty food since it removes most interference compounds when eluted with nonpolar solvents. Alumina can be identified as a good Florisil substitute; it does not present batch variability as Florisil, but similar to Florisil, is not mainly adoptable for OP pesticides, since it results in pesticide decomposition processes. Silica gel is very suitable for polar pesticide isolations in animal tissues; on the other hand, it does not perform a good cleanup in the specific case of plant tissue. Carbon adsorbs principally nonpolar and high-molecular-weight pesticides, it works well with chlorophyll but not with wax and its cleanup action is affected by pretreatments.

The second type or size exclusion method (gel permeation) uses the molecular size to separate compounds by using the same material of the adsorption chromatography column.

The separation by molecular size is a result of the small holes obtained in the design of the columns that do not occur on the adsorption columns. The small holes of the particles placed in the columns retard the elution of the smaller molecules.

The greater analytical performances of the GPC are associated with the recovery of pesticide, because there is no loss. Clearly, the analytical apparatus is more expensive and a medium pressure pump is required.

As previously described, the cleanup is the most time-consuming step in the analytical procedures.

Today, fairly new technologies like SPE or SPME are suitable to perform extraction and cleanup in a single step.

These technologies are attractive because they minimize the expensive use of analytical grade organic solvents that are environmentally sensitive wastes. Like in many other analytical techniques, the true versatility of SPE and SPME can only be realized by the understanding and proper control of the physiochemical interactions among the analytical procedures, the matrices, and pesticides' chemical and physical natures.

The SPE technique is based on the use of cartridge that is also known as an accumulator or concentrator column and it is a simple, rapid, and inexpensive method, which does not require the preparation or maintenance of expensive apparatus.

The technique can speed up the cleanup as well as extraction. The majority of SPE extractions are "retentive," since a packed sorbent retains the target analytes, while interference material passes through the column to waste. For low fatty matrices such as milk, eggs, or meat, the techniques appear particularly useful and as very powerful as a fast cleanup step in the case of polar pesticides. The sorbents are classified by their retention mechanisms or primary interaction with the target analytes.

The most common extraction mechanisms are the reverse phase, ion exchange, or normal phase. The "reverse phase" is more suitable to extract hydrophobic or even polar organic analytes from an aqueous sample/matrix [42], graphitized carbon black (GCB) named as carbograph is used in an MRLs method from milk sample [30]. GCB is a nonspecific sorbent generally hydrophobic in nature [43]. It has been shown that a GCB cartridge is much more efficient than a C18 cartridge [44] and is extensively used in the analysis of pesticide residues [45].

The ion exchange mechanisms are used to extract charged analytes from low-ion strength aqueous or organic samples. Charged sorbents are used to retain analytes of the opposite charge. For example, positively charged analytes containing amines are retained on negatively charged "cation exchangers" such as sulfonic or carboxylic acid. On the other hand, negatively charged analytes containing sulfonic acid or carboxylic acid groups are retained on a positively charged "anion exchanger." In ion exchange mechanisms, only species of proper charge are retained by the adsorbent phase. The mechanism works on specific strong coulombic interactions between the two phases: the sorbent and the analytes. Owing to these peculiar characteristics, cation exchange cartridges are commonly used for the extraction of basic compounds from biological samples.

The third kind of absorbent phase works through normal phase retention mechanisms, and it is particularly useful for the extraction of polar analytes from nonpolar organic solvents. The stationary phase retention is based on hydrogen bonding and dipole–dipole interactions between the two phases sorbent and analytes. The interaction can be optimized by playing on the polarity of the conditioning solvent or the solvent used to treat the sample/matrix. Generally, it is possible to use one of the three different sorbent phases to extract a wide range of target analytes depending on the solvent and the sample matrix; appreciable differences occur in the selectivity, recovery, and purification. Considering the analyses of pesticides in meat, milk, or eggs, this technique appears particularly suitable for the cleanup purposes instead of more expensive, time-consuming, and less environment-friendly techniques mentioned earlier.

A fairly new technique more recently developed, SPME, has been widely adopted as a reliable and rapid alternative giving results similar from both a qualitative and a quantitative point of view to those obtained by conventional solvent-extraction methods. SPME is a relatively new technique, devised by Pawliszyn and coworkers [46] which, by combining both extraction and concentration in a single step, provides for procedures which are fast, simple to use, easy to couple with chromatographic analysis, and can achieve good sensitivity. Isolation of the analytes is achieved using a fused silica fiber coated with an appropriate material. SPME appears to offer real analytical advantages in pesticides environmental analysis, and has been recently applied to studies of various organic substances, including petroleum derivatives and pesticides, in water, in soil, and other more complex environmental matrices such as foodstuff [47,48]. SPME extraction with the transport of analytes from the matrix into the coating fiber begins as soon as the coated fiber is placed in contact with the sample. The technique is an equilibrium process, in which the amount of the extracted analytes by the coating depends on fiber coating/sample matrix distribution constant, the volume of the coating fiber, the sample volume, and the initial concentration of the given analyte. The sample matrix has to be considered as a single and homogeneous phase. There is a direct proportional relationship between sample concentration and the amount of analyte extracted. This is the basis for a quantitative response. The extraction is "by definition" nonexhaustive; therefore, constant distribution conditions between phases and careful timing of extraction are necessary. Three types of extractions can be performed using SPME: direct extraction by immersion, head space extraction, and membrane protection approach. The most useful method for food analyses is the head space mode (HS-SPME). HS-SPME is an equilibrium process involving headspace and the polymeric fiber stationary phase [49]. For a particular pesticide compound, the equilibrium between the gas phase and adsorption on the fiber has its own partition coefficient [50]. The overall recovery process is influenced by a number of experimental parameters, each one requiring optimization to obtain the maximum sensitivity, while maintaining a linear quantitative response. The choice of the sampling mode and the kind of fiber coating has a very significant role in the impact of the extraction kinetic. The extraction temperature has a significant effect on the kinetics of the process by determining the vapor pressure of the analytes. Since the efficiency of the extraction process is dependent on the distribution constant of the analytes between the fiber and sample, differences are appreciable in the analytical response obtained from HS-SPME using different SPME stationary phases.

Different polymeric stationary phases such as polydimethylsiloxane (PDMS) with the highest efficiencies for nonpolar analytes, polyacrylate (PA), or carbowax/divinylbenzene for polar compounds (CW-DVB) can be used in determining the sensitivity of the method. SPME allows rapid extraction and transfer to the analytical instrument. As a result, the analytical process can be accelerated and human errors associated with the analytical extraction and purification treatment steps can be minimized.

After the above isolation activities of the pesticide from the sample matrices, the determination of the target compounds is usually associated with the gas chromatography or liquid chromatography separation.

Historically, gas chromatography appears to be the most useful technique for the determination of the quantitative amount of pesticides in the food sample matrices. High-performance separations occur in thin film capillary column with an internal diameter less than 0.1 mm and a length longer than 10 m, sometimes reaching 100 m for specific applications. Different stationary phases are available for different purposes and they can be used to obtain the best separation and improve the detection limit. The sample placed in the beginning of the column is flash evaporated in the inlet port and is transported by a gas carrier H_2 or He through the column. The partition between the stationary phase and the target analyte determinates the retention time of the different compounds. The analytes go to the detector that produces the instrumental response. Different types of detectors work on different theoretical principles, and in such a case, they detect only certain classes of chemicals. Electron capture detector (ECD), thermionic detector such as nitrogen and phosphorous detector (NPD), flame photometric detector (FPD) or flame ionization detector (FID) and mass selective detector (MSD) are used for different analytical detection reasons.

The high-performance liquid chromatography is the second most frequently used technique to determine very polar and low-volatile pesticides after their extraction from different food matrices. The separation occurs in packed columns. A small volume of sample is placed at the top of the column and the solvent, the mobile phase, is pumped through the column with high pressure. The distribution of target analytes between the stationary and the mobile phase determines the retention time of pesticides. Clearly, the pesticide's fate is directly correlated to its affinity with stationary phase, a higher affinity determines a higher retention time. Different kinds of detectors are used to determine pesticides amount: UV absorption, fluorescence, conductivity, electrochemical and mass spectrometer detector. Many pesticides adsorb the UV light at different wavelengths; therefore, UV detector appears as one of the most useful detectors to determine the hazard compound at a very low level. The UV detector is very powerful allowing the selection of the most suitable frequency for a target analyte, and giving the possibility to exclude the determination of other unwanted compounds.

17.5   CONCLUSIONS

Chromatographic analytical techniques are commonly used for determining pesticides and their metabolites in food samples. Gas chromatography is used for determining nonpolar pesticides, whereas liquid chromatography is more suitable for determining polar and nonvolatile pesticides [51,52].

These techniques are necessary for an appropriate identification of various analytes alone or simultaneously. However, they are not free from disadvantages. In addition, we have to remember expensive instrumentation, their maintenance together with the presence of highly skilled personnel. Moreover, the analysis of each particular sample is preceded by time-consuming sample preparation, usually taking from several hours to several days. The preparation of samples and the time and expense involved in classical analytical methods (i.e., sampling, sample preparation, and laboratory instrumental analysis) limit the overall number of samples that can be analyzed. There is a real need for developing fast, easy-to-use, robust, sensitive, cost-effective, and field-analytical techniques. Immunoassays (IAs) meet these requirements and many pesticides can be analyzed and monitored at regulatory levels without any sample preparation method.

ACKNOWLEDGMENTS

The author thanks Prof. Giuseppe Alonzo for scientific collaboration, Prof. ssa Silvia Marzetti for the valuable discussions on the manuscript, and Dr. ssa Manuela Giulivi for technical help. This work was supported by the Italian Ministry of Education, University and Scientific Research.

REFERENCES

1. UN (2005). World Population Prospects. The 2004 Revision. Highlights. Population Division, Department of Economic and Social Affairs. United Nations, NY; Cohen, J.E.; Human population grows up. *Sci. Am.* (Special issue) (2005), 26–33.
2. FAO (2002). The State of Food Insecurity in the World 2002. Economic and Social Department, Food and Agriculture Organization of the United Nations, Rome.
3. Judge C. A., Neal J. C., and Derr J. F.; Response of Japanese stiltgrass (*Microstegium vimineum*) to application timing, rate, and frequency of postemergence herbicides. *Weed Technol.*, 19 (2005), 912–917.
4. Wu Q. J., Zhao J. Z., and Taylor A. G., Shelton A. M.; Evaluation of insecticides and application methods against *Contarinia nasturtii* (Diptera: Cecidomyiidae), a new invasive insect pest in the United States. *J. Econ. Entomol.*, 99 (2006), 117–122.
5. Matsumura F.; *Toxicology of Insecticides*, 2nd edn., Plenum Press, New York, 1985.
6. Manahan S. E.; *Toxicological Chemistry*, 2nd edn., Lewis Publishers, Ann Arbor, MI, 1992.
7. Laws E. R. and Hayes W. J.; *Handbook of Pesticide Toxicology*, Academic Press, San Diego, CA, 1991.
8. Committee on Pesticides in the Diets of Infants and Children; *Pesticides in the Diets of Infants and Children*, National Academy Press, Washington, DC, 1993.
9. Pesticides Industry Sales and Usage: 1996 and 1997 Market Estimates; *Pesticides Industry 1994 and 1995 Sales and Usage*, U.S. Environmental Protection Agency, Washington, DC, 1999.
10. Borga K., Gabrielsen G. W., and Skaare J. U.; Biomagnification of organochlorines along a Barents Sea food chain. *Environ. Pollut.*, 113 (2001), 187–198.
11. Albanis T. A., Hela D., Papakostas G., and Goutner V.; Concentration and bioaccumulation of organochlorine pesticide residues in herons and their prey in wetlands of Thermaikos Gulf, Macedonia, Greece. *Sci. Total Environ.*, 182 (1996), 11–19.
12. Bro-Rasmussen F.; Contamination by persistent chemicals in food chain and human health. *Sci. Total Environ.*, 188 (Suppl. 1) (1996), 45–60.
13. Wania F. and Mackay D.; Global fractionation and cold condensation of low volatility organochlorine compounds in polar regions. *Ambio. Stockholm [AMBIO]*, 22(1) (1993), 10–18.
14. Tomlin C. D. S.; *The Pesticide Manual*, British Crop Protection Council, Farnham, U.K., 1997.
15. *EPA's Pesticide Programs*. U.S. Environmental Protection Agency, Washington, DC, 1991.

16. Kumar M., Gupta S. K., Garg S. K., and Kumar A.; Biodegradation of hexachlorocyclohexane-isomers in contaminated soils. *Soil Boil. Biochem.*, 38 (2006), 2318–2327.
17. Loague K. and Soutter L. A., Desperately seeking a cause for hotspots in regional-scale groundwater plumes resulting from non-point source pesticide applications. *Vadose Zone J.*, 5 (2006) 204–221.
18. Hoffman R. S., Capel P. D., and Larson S. J.; Comparison of pesticides in eight U.S. urban streams. *Environ. Toxicol. Chem.*, 19 (2000) 2249–2258.
19. Harman-Fetcho J. A., Hapeman C. J., McConnell L. L., Potter T. L., Rice C. P., Sadeghi A. M., Smith R. D., Bialek K., Sefton K. A., and Schaffer B. A.; Pesticide occurrence in selected South Florida canals and Biscayne Bay during high agricultural activity. *J. Agric. Food Chem.*, 53 (2005), 6040–6048.
20. Buehler S. S., Basu I., and Hites R. A.; Causes of variability in pesticide and PCB concentrations in air near the Great Lakes. *Environ. Sci. Technol.*, 38 (2004), 414–422.
21. Goel A., McConnell L. L., and Torrents A.; Wet deposition of current use pesticides at a rural location on the Delmarva Peninsula: Impact of rainfall patterns and agricultural activity. *J. Agric. Food Chem.*, 53 (2005), 7915–7924.
22. Ferrer I. and Thurman E. M.; Liquid chromatography/time-of-flight/mass spectrometry (LC/TOF/MS) for the analysis of emerging contaminants; TrAC-Trends. *Anal. Chem.*, 22 (2003) 750–756.
23. Richardson S. D.; Water analysis: Emerging contaminants and current issues. *Anal. Chem.*, 79 (2007), 4295–4324.
24. Gupta P. K.; Pesticide exposure—Indian scene. *Toxicology*, 198 (2004), 83–90.
25. U.S. Food and Drug Administration; *Pesticide Analytical Manual Vol. I, Multiresidual Methods*, 3rd edn., U.S. Department of Health and Human Services, Washington, DC, 1994.
26. van der Hoff G. R., van Beuezekom, A. C., Brinkman U. A., Baumann R. A., and van Zoonen, P.; Determination of organochlorine compounds in fatty matrices Application of rapid off-line normal-phase liquid chromatographic clean-up. *J. Chromatogr. A*, 754 (1996), 487–492.
27. Le Faouder J., Bichon E., Brunschwig P., Landelle R., Andrea F., and Le Bizec B.; Transfer assessment of fipronil residues from feed to cow milk. *Talanta*, 73 (2007), 710–717.
28. Nag S. K., Mahanta S. K., Raikwar M. K., and Bhadoria B. K.; Residues in milk and production performance of goats following the intake of a pesticide (endosulfan). *Small Ruminant Res.*, 67 (2007), 235–242.
29. Sharma H. R., Kaushik A., and Kaushik C. P.; Pesticide residues in bovine milk from a predominantly agricultural state of Haryana, India. *Environ. Monit. Assess.*, 129 (2007), 349–357.
30. Bogialli S., Curini R., di Corcia A., Laganà A., Stabile A., and Sturchio E.; Development of a multi-residue method for analyzing herbicide and fungicide residue in bovine milk on solid-phase extraction and liquid chromatography-tandem mass spectrometry. *J. Chromatogr. A*, 1102 (2006), 1–10.
31. Plaza Bolaõs P., Garrido Frenich A., and Martinez Vidal J. L.; Application of gas chromatography-triple quadrupole mass spectrometry in the quantification-confirmation of pesticides and polychlorinated biphenyls in eggs at trace levels. *J. Chromatogr. A*, 1167 (2007), 9–17.
32. Valsamaki V. I., Boti V. I., Sakkas V. A., and Triantafyllos A.; Albanis, Determination of organochlorine pesticides and polychlorinated biphenyls in chicken eggs by matrix solid phase dispersion. *Anal. Chim. Acta*, 573–574 (2006), 195–201.
33. Garrido Frenich A., Plaza Bolaños P., and Martínez Vidal J. L.; Multiresidue analysis of pesticides in animal liver by gas chromatography using triple quadrupole tandem mass spectrometry. *J. Chromatogr. A*, 1153 (2007), 194–202.
34. Herzke D., Kallenborn R., and Nygård T.; Organochlorines in egg samples from Norwegian birds of prey: Congener-, isomer- and enantiomer-specific considerations. *Sci. Total Environ.*, 291 (2002), 59–71.
35. Garrido Frenich A., Romero-González R., Martínez-Vidal J. L., and Plaza-Bolaños P., Cuadros-Rodríguez L., Herrera Abdo M. A.; Characterization of recovery profiles using gas chromatography-triple quadrupole mass spectrometry for the determination of pesticide residues in meat samples. *J. Chromatogr. A*, 1133 (2006), 315–321.
36. Juhler René K.; Supercritical fluid extraction of pesticides from meat: A systematic approach for optimisation. *Analyst*, 123 (1998), 1551–1556.
37. Lehotay S. J., Lightfield A. R., Harman-Fetcho J. A., and Donoghue D. J.; Analysis of pesticide residues in eggs by direct sample introduction/gas chromatography/tandem mass spectrometry. *J. Agric. Food Chem.*, 49 (2001), 4589–4596.
38. Schenck F. J.; Determination of organochlorine and organophosphorus pesticide residues in eggs using a solid phase extraction cleanup. *J. Agric. Food Chem.*, 48 (2000), 6412–6415.
39. Naoto F., Asako O., Miyuki N., Yukari M., and Kunio O.; Simple and rapid extraction method of total egg lipids for determining organochlorine pesticides in the egg. *J. Chromatogr. A*, 830 (1999), 473–476.
40. Juhler René K.; Optimized method for the determination of Organophosphorus pesticides in meat and fatty matrices. *J. Chromatogr. A*, 786 (1997), 145–153.

41. Bennet D. A., Chung A. C., and Lee S. M.; Multiresidue method for analysis of pesticides in liquid whole milk. *J. AOAC*, 80 (1997), 1065–1077.
42. De Pasquale C., Ainsley J., Charlton A., and Alonzo G.; Use of SPME extraction to determine Organophosphorus pesticides adsorption phenomena in water and soil matrices. *Int. J. Environ. Anal. Chem.*, 85(15) (December 20, 2005), 1101–1115.
43. Lagana A., Marino A., Fago G., and Martinez B. P.; Multiresidue analysis of phenylurea herbicides in crops by HPLC with diode array detector. *Analysis*, 22 (1994), 63–69.
44. Di Corcia A., Samperi R., Marcomini A., and Samperi R.; Graphitized carbon black extraction cartridges for monitoring polar pesticides in water. *Anal. Chem.*, 65 (1993), 907–912.
45. Tekel J. and Hatrik S. J.; Pesticide residue analyses in plant material by chromatographic methods: Clean-up procedures and selective detectors. *J. Chromatogr. A*, 754 (1996), 397–410.
46. Arthur C. and Pawliszyn J., Solid phase microextraction with thermal desorption using fused silica optical fibers. *Anal. Chem.*, 62 (1990), 2145–2148.
47. Gòrecki T., Yu X., and Pawliszyn J.; Theory of analyte extraction by selected porous polymer SPME fibers. *Analyst*, 124 (1999), 643–649.
48. Madgic S., Boyd-Boland A., Jinno K., and Pawliszyn J.; Analysis of organophosphorus insecticides from environmental samples using solid-phase microextraction. *J. Chromatogr. A*, 736 (1996), 219–228.
49. Pawliszyn J.; *Solid Phase Microextraction Theory and Practice (SPME)*, Wiley-VCH, New York, 1997.
50. Zhang Z. and Pawliszyn J.; Headspace solid-phase microextraction. *Anal. Chem.*, 65 (1993), 1843–1852.
51. Ahmed F. E.; Analyses of pesticides and their metabolites in foods and drinks. *Trends Anal. Chem.*, 20(11) (2001), 649–661.
52. Santos F. J. and Galceran M. T.; The application of gas chromatography to environmental analysis. *Trends Anal. Chem.*, 21(9–10) (2002), 672–685.
53. De Faubert Maunder M. J., Egan H., Godly E. W., Hammond E. W., Roburn J., and Thompson J.; Clean up of animals fats and dairy products for the analysis of chlorinated pesticide residues. *Analyst*, 89 (1964), 169–174.
54. Veirov D. and Aharonson N.; Simplified fat extraction with sulphuric acid as clean-up procedure for residue determination of chlorinated hydrocarbons in butter. *J. AOAC*, 61 (1978), 253–260.
55. Pagliuca G., Gazzotti T., Zironi E., and Sticca P.; Residue analysis of organophosphorus pesticides in animal matrices by dual column capillary gas chromatography with nitrogen–phosphorus detection. *J. Chromatogr. A*, 1071 (2005), 67–70.
56. Basheer C. and Lee H. K.; Hollow fiber membrane-protected solid-phase microextraction of triazine herbicides in bovine milk and sewage sludge samples. *J. Chromatogr. A*, 1047 (2004), 189–194.
57. Battu R. S., Singh B., and Kang B. K.; Contamination of liquid milk and butter with pesticide residues in the Ludhiana district of Punjab state, India. *Ecotoxicol. Environ. Saf.*, 59 (2004), 324–331.
58. Salas J. H., González M. M., Noa M., Pérez N. A., Díaz G., Gutiérrez R., Zazueta H., and Osuna I.; Organophosphorus pesticide residues in Mexican commercial pasteurized milk. *J. Agric. Food Chem.*, 51(15) (2003), 4468–4471.
59. Zhu L., Ee K. H., Zhao L., and Lee H. K.; Analysis of phenoxy herbicides in bovine milk by means of liquid–liquid–liquid microextraction with a hollow-fiber membrane. *J. Chromatogr. A*, 963 (2002), 335–343.
60. John P. J., Bakore N., and Bhatnagar P.; Assessment of organochlorine pesticide residue levels in dairy milk and buffalo milk from Jaipur City, Rajasthan, India. *Environ. Int.*, 26 (2001), 231–236.
61. Di Muccio A., Pelosi P., Attard Barbini D., Generali T., Ausili A., and Versori F.; Selective extraction of pyrethroid pesticide residues from milk by solid-matrix dispersion. *J. Chromatogr. A*, 765 (1997), 51–60.
62. Waliszewski S. M., Paradìo V. T., Waliszewski K. N., Chantiri J. N., Aguirre A. A., Infanzòn R. M., and Riviera J.; Organochlorine pesticide residues in cow's milk and butter in Mexico. *Sci. Total Environ.*, 208 (1997), 127–132.
63. Martinez M. P., Angulo R., Pozo R., and Jodral M.; Organochlorine pesticides in pasteurized milk and associated health risks. *Food Chem. Toxicol.*, 35 (1997), 621–624.
64. Losada A., Fernández N., Diez M. J., Teran M. T., Guarcia J. J., and Sierra M.; Organochlorine pesticide residue in bovine milk from León (Spain). *Sci. Total Environ.*, 181 (1996), 133–135.
65. Garrido Frenich A., Martìnez Vidal J. L., Cruz Sicilia A. D., González Rodríguez M. J., and Bolaños P. P.; Multiresidue analysis of organochlorine and organophosphorus pesticides in muscle of chicken, pork and lamb by gas chromatography–triple quadrupole mass spectrometry. *Anal. Chim. Acta*, 558 (2006), 42–52.
66. Bolaños P. P., Garrido Frenich A., and Martínez Vidal J. L.; Application of gas chromatography-triple quadrupole mass spectrometry in the quantification-confirmation of pesticides and polychlorinated biphenyls in eggs at trace levels. *J. Chromatogr. A*, 1167 (2007), 9–17.

18 Determination of Pesticide Residues in Fruits and Vegetables by Using GC–MS and LC–MS

Li-Bin Liu, Yan Liu, and Jin-Ming Lin

CONTENTS

18.1 INTRODUCTION

Most farmers use agricultural chemicals on their farms. Many of these chemicals are used to control pests and are known as pesticides. Included under the heading of pesticides are herbicides, insecticides, rodenticides, fungicides, and others. Although all of these affect humans, more people are poisoned by insecticides and herbicides.

18.1.1 TYPES OF PESTICIDES

18.1.1.1 Chemical Pesticides

Some examples of chemically related pesticides follow:

Organophosphorus pesticides (OPPs): These pesticides affect the nervous system by disrupting the enzyme that regulates acetylcholine, a neurotransmitter. Most organophosphates are insecticides. They were developed during the early nineteenth century, but their effects on insects, which are similar to their effects on humans, were discovered in 1932. Some are very poisonous. However, they are usually not persistent in the environment.

Carbamate pesticides: They affect the nervous system by disrupting an enzyme that regulates acetylcholine, a neurotransmitter. The enzyme effects are usually reversible. There are several subgroups within the carbamates.

Organochlorine insecticides: They were Commonly used in the past, but many have been removed from the market due to their health and environmental effects and their persistence.

Pyrethroid pesticides: They were developed as a synthetic version of the naturally occurring pesticide pyrethrin, which is found in chrysanthemums. They have been modified to increase their stability in the environment. Some synthetic pyrethroids are toxic to the nervous system.

18.1.1.2 Biopesticides

Biopesticides are certain types of pesticides derived from such natural materials as animals, plants, bacteria, and certain minerals. For example, canola oil and baking soda have pesticidal applications and are considered biopesticides. At the end of 2001, there were approximately 195 registered biopesticide active ingredients and 780 products. Biopesticides fall into three major classes:

1. Microbial pesticides consist of a microorganism as the active ingredient. Microbial pesticides can control many different kinds of pests, although each separate active ingredient is relatively specific for its target pests. For example, there are fungi that control certain weeds and other fungi that kill specific insects.

The most widely used microbial pesticides are subspecies and strains of *Bacillus thuringiensis*, or Bt. Each strain of this bacterium produces a different mix of proteins and specifically kills one or a few related species of insect larvae. Although some Bts control moth larvae found on plants, other Bts are specific to larvae of flies and mosquitoes. The target insect species are determined by whether the particular Bt produces a protein that can bind to a larval gut receptor, thereby causing the insect larvae to starve.

2. Plant-incorporated-protectants (PIPs) are pesticidal substances that plants produce from genetic material that has been added to the plant. For example, scientists can take the gene for the Bt pesticidal protein and introduce the gene into the plant's own genetic material. Then the plant, instead of the Bt bacterium, manufactures the substance that destroys the pest. The protein and its genetic material, but not the plant itself, are regulated by EPA.

3. Biochemical pesticides are naturally occurring substances that control pests by nontoxic mechanisms. Conventional pesticides, by contrast, are generally synthetic materials that directly kill or inactivate the pest. Biochemical pesticides include substances such as insect sex pheromones that interfere with mating as well as various scented plant extracts that attract insect pests to traps. Because it is sometimes difficult to determine whether a substance meets the criteria for classification as a biochemical pesticide, EPA has established a special committee to make such decisions.

18.1.1.3 Symptoms of Pesticide Poisoning

Pesticides can be considered according to their chemical basis. Most of the more toxic pesticides fall into the following chemical groups: organophosphates, carbamates, and bipyridyls.

18.1.1.3.1 Organophosphates and Carbamates

Organophosphates and carbamates are insecticides that affect humans by inhibiting the production of the enzyme cholinesterase, which is important in the correct functioning of the nervous system. If a sufficient amount of cholinesterase is not produced, muscle reactions will become erratic and the victim may display symptoms soon after exposure. However, delayed reaction times up to 12 h may occur from exposure to parathion, guthion, or thimet. An exception to this would be cases resulting from accumulations received in small doses repeated frequently.

Symptoms of poisoning from organophosphates and carbamates may vary in the order of appearance depending on how exposure occurred. Mild cases of poisoning from organophosphates and carbamates will include some or all of the following: headache, fatigue, dizziness, loss of appetite with nausea and stomach cramps, blurred vision (tearing and shrinking of the size of the pupils), sweating, slobbering, vomiting, diarrhea, slowed heartbeat, and muscle rippling. Moderate cases of poisoning will progress from dilated pupils and secretions from the eyes, nose, mouth, lungs, and skin to unconsciousness and seizures.

18.1.1.3.2 Bipyridyls

A number of poisonings occur each year from chemicals such as paraquat and diquat. These are bipyridyls and affect the skin, nails, mucous membranes, gastrointestinal tract, and the respiratory system.

Symptoms of poisoning from bipyridyls vary according to how the chemical entered the body. Exposure of the skin usually causes irritation drying and cracking. If exposure is repeated, fingernails may start to show irregular growth and may also turn black around the cuticles or entirely black. If droplets are inhaled, irritation of the nose and throat usually occurs. Repeated or prolonged exposures may cause nosebleeds.

If ingested (swallowed) paraquat or other bipyridyls usually cause severe lung tissue damage. Immediately following ingestion, the victim will experience pain in the mouth, throat, chest, and abdominal area. These symptoms may be followed by vomiting, diarrhea, and muscle aches. Symptoms may subside and in a few days evidence of kidney and liver damage such as jaundice and urinary disorders may appear. Three to fourteen days after ingestion, coughing, difficult breathing, and fluid buildup in the lungs may occur. Total recovery will not be accomplished.

18.2 ANALYTICAL METHODS OF PESTICIDE RESIDUES

Nowadays, with the increasing focus on food safety, it has become an important issue to determine the residual pesticides accurately in the food. However, the complicated matrix of agricultural products may affect the accuracy of the analysis. All kinds of pretreatment methods are developed and provided for the need of complex samples analysis. As far as fruits and vegetables are concerned, the pretreatment includes two steps, extraction and cleanup. The common extraction methods are homogenizing,

Soxhlet extraction, microwave-assisted extraction (MAE), supercritical fluid extraction (SFE), and accelerated solvent extraction (ASE). Cleanup methods include solid-phase extraction (SPE), matrix solid-phase dispersion (MSPD) technique, and gel permeation chromatography (GPC).

Mass spectrometry (MS) could be used for qualitative and quantitative analysis by characteristic fragments and overcomes the disadvantages of conventional determination methods. The combination of chromatography and MS becomes one of the powerful tools for the screening, confirmation, and quantification of complicated organic pollutants. Gas chromatography mass spectrometry (GC–MS) is mainly used for volatile compounds, and liquid chromatography mass spectrometry (LC–MS) is useful for nonvolatile, thermal unstable, and polar compounds without derivation. Herein, the determination of residual pesticides in fruits and vegetables by using GC–MS and LC–MS was introduced.

18.2.1 Determination of Pesticide Residues by GC–MS

In past 30 years, most routine pesticide residue analysis has been done by GC in combination with electron capture detection (ECD), nitrogen–phosphorous detection (NPD), and flame photometric detection (FPD). There were some difficulties in confirmation of results due to their complex matrix, and the use of a further gas chromatograph equipped with a different type of column or detector has been necessary. With the developing of the MS technique, simultaneous determination and confirmation of pesticide residues could be obtained with GC–MS in one analytical run, which improves the analytical accuracy and shortens the analytical time. Because of a uniform database of electron ionization (EI), the methods based on GC–MS to analyze the pesticide residues were usually commended by many authority agencies and listed as standard analytical methods. However, the relatively low sensitivity obtained for some pesticide/commodity combinations in full-scan mode, selected ion monitoring (SIM), chemical ionization (CI), or tandem mass spectrometry (MS–MS) could improve sensitivity and selectivity. In the last several years, other extraction techniques coupled with GC–MS or MS detector have been developed rapidly. In Table 18.1 are shown the extraction techniques coupled with GC–MS or MS detection developed in the last few years.

18.2.2 Determination of Pesticide Residues by LC–MS

Compared with GC, methods based on LC to analyze pesticide residues were applied more rarely in the past, because traditional UV, diode array, and fluorescence detectors are often less selective and sensitive. However, in the last few years, electrospray ionization (ESI) or atmospheric pressure chemical ionization (APCI) in combination with MS (quadrupole, ion trap, or time-flight mass analyzers) that have several orders of magnitude of sensitivity than classical detectors and LC–MS have become widely accepted as the preferred techniques for the identification and quantification of polar and thermally labile pesticides, such as carbamates, phenylureas, OPPs, quaternary ammonium compounds, triazines, and chlorinated phenoxy acids. Overviews about applications of LC–MS in pesticide residue analysis were reported [23–26]. Pico et al. [27] reviewed the pesticide residue determination in fruit and vegetables by LC–MS in 2000. Recent developments of LC–MS are shown in Table 18.2.

18.3 EXPERIMENT I: RAPID ANALYSIS OF MULTIRESIDUAL PESTICIDES IN AGRICULTURAL PRODUCTS USING GC–MS

About 97 pesticides were studied in this work, including organophosphorus, organochlorine, organonitrogen, carbamate, and thiocarbamate substances, which were spiked into seven kinds of agricultural products at the 100 ng mL^{-1} level. The samples were cleaned using the slightly modified MSPD technique and were analyzed using GC–MS. The combination of the MSPD technique and the GC–MS realized the rapid determination of the 97 pesticides with acceptable recoveries.

TABLE 18.1
Analysis of Pesticides in Fruits and Vegetables by GC–MS

Analytes	Matrix	Extraction Method	Determination Methods	Detection Limit	Refs.
Pyrethroid	Strawberry	Extract with 50% aqueous acetonitrile solution under microwave then extract using 100 μm PDMS fiber	GC–EI-MS	14 μg mL^{-1}	[1]
110 pesticides	Spinach and orange	Extract with ethyl acetate, then pass through cleanup column consisting of two layers of water-absorbent polymer (upper) and graphitized carbon (lower)	GC–EI-MS, GC–NCI–MS	—	[2]
55 pesticides	Vegetable samples	Extract with a mixture of ethyl acetate and sodium sulfate	GC–PCI/EI-MS-MS	0.07–4.21 μg kg^{-1}	[3]
Dimethoate, pyrimethanil, vinclozolin, furalaxyl, oxadixyl	Lettuce	Extract with a mixture of ethyl acetate, Na$_2$SO$_4$, and NaHCO$_3$	LV-DMI–GC–TOF-MS	0.0025 μg mL^{-1}	[4]
21 multiclass pesticides	Tomato	Extract with ethyl acetate/cyclohexane (1:1) followed by cleanup with GPC columns	GC–EI-MS	—	[5]
31 multiclass pesticides	Green bean, cucumber, pepper, tomato, eggplant, watermelon, melon, marrow	Extract with dichloromethane	GC–EI-MS-MS	0.01–2.6 ng g^{-1}	[6]
81 multiclass pesticides	Vegetable	Extract with dichloromethane	GC–CI/EI-MS-MS	0.01–1.92 ng g^{-1}	[7]
97 multiclass pesticides	Potato, cabbage, carrot, apple, orange, cucumber	Extract with acetonitrile, NaCl, and anhydrous MgSO$_4$, then MgSO$_4$ and PSA sorbent was added into acetonitrile layer	GPC–GC–EI-MS	3–79 μg kg^{-1}	[8]
65 multiclass pesticides	Avocado	Extract with the ethyl acetate–cyclohexane mixture pressurized liquid extraction clean with GPC	Low-pressure-GC–EI-MS–MS	0.01–2.50 μg kg^{-1}	[9]
20 multiclass pesticides	Apple and peach	Extract with ethyl acetate/cyclohexane (1:1) followed by cleanup with GPC	GC × GC–TOF-MS	0.2–30 ng mL^{-1}	[10]
78 pesticides	Tomato and onion	Homogenized with ethyl acetate, then added anhydrous Na$_2$SO$_4$ and centrifuged, then NH$_2$ adsorbent and anhydrous MgSO$_4$ were added into the extract	Low-pressure GC–EI-MS–MS	0.1–0.4 μg kg^{-1}	[11]
20 pesticides	Peach	Homogenized with ethyl acetate, then added anhydrous Na$_2$SO$_4$, and centrifuged, then made up with cyclohexane and cleanup by GPC	GC–TOF-MS	0.5–25 μg kg^{-1}	[12]

(continued)

TABLE 18.1 (continued)
Analysis of Pesticides in Fruits and Vegetables by GC–MS

Analytes	Matrix	Extraction Method	Determination Methods	Detection Limit	Refs.
Diazinon, fenitrothion, fenthion, parathion ethyl, bromophos methyl, bromophos ethyl, and ethion	Strawberry and cherry	HS-extract using 100 μm PDMS fiber	GC–EI-MS	<13 μg kg^{-1}	[13]
Diazinon, parathion-methyl, fenitrothion, malathion, fenthion, chlorpyrifos-ethyl, bromophos-methyl, methidathion, azinphos-methyl, permethrin (cis) and (trans)	Pear, orange, apple, and grape	Analyte was eluted with ethyl acetate from C8-bonded silica	GC–EI-MS	4–90 μg kg^{-1}	[14]
102 pesticides	Leek	Extract with acetone and dichloromethane (4:3, v/v) and cleanup by GPC and SPE tube	GC–EI-MS	<0.01 mg kg^{-1}	[15]
90 pesticides	Apple, green bean, orange	Extract with acetone aqueous and clean on the cross-linked polystyrene divinyl benzene column	GC–EI-MS	0.01–0.02 mg kg^{-1}	[16]
30 pesticides	Green bean, cucumber, pepper, tomato, eggplant, watermelon, melon, pea, and marrow	Extract with dichloromethane	GC–CI/EI-MS-MS	30 ng kg^{-1} to 6 μg kg^{-1}	[17]
300 pesticides	Lettuce, pear, grape	Extract with methanol and SBSE stir bar coated with PDMS was added	Thermal desorption -PTV-GC–EI-MS	—	[18]
130 pesticides	Cucumber	Extract with acetonitrile	GC–EI-triple quadrupole MS	0.05–0 μg kg^{-1}	[19]
72 pesticides	Tomato, cucumber pepper	Extract with dichloromethane	Low-pressure GC–EI-MS-MS	0.1–1 μg kg^{-1}	[20]
18 pesticides	Apple	Acetonitrile extraction followed by salting out and purification on SPE–NH$_2$ columns and solvent exchange to toluene; acetonitrile extraction and dispersive-SPE cleaning with PSA sorbent	GC–EI-MS	<0.005 mg kg^{-1}	[21]
13 pesticides	Grape and pineapple	Extract with ethyl acetate	Auto-DMI–GC–TOF-MS	1–10 ng g^{-1}	[22]

TABLE 18.2
Analysis of Pesticides in Fruits and Vegetables by LC–MS

Analytes	Matrix	Extraction Method	Determination Methods	Detection Limit	Refs.
Oxamyl, methomyl, pirimicarb, aldicarb, Propoxur, carbofuran, carbaryl, trimethacarb	Tomato, spinach, lettuce, zucchini, pear, and apple	Extract with hot water	LC–ESI-MS	2–7 ng mL^{-1}	[28]
Methoxyfenozide	Artichoke, cucumber, and squash	Extract with methanol/aqueous 0.1N HCl, 9:1 (v/v), then separate with hexane	LC–MS–MS	5 ng mL^{-1}	[29]
Imidacloprid, trichlorfon, carbendazim, thiabendazole, methidathion, methiocarb, imazalil, bitertanol, pyriproxyfen, hexythiazox	Orange and peach	Pressurized liquid extraction with ethyl acetate	LC–IT-MS–MS	0.025–0.25 mg kg^{-1}	[30]
Trichlorfon	Kaki	Extract with acetonitrile	LC–ESI-MS–MS	0.006–0.013 mg kg^{-1}	[31]
23 pesticides	Cucumber, tomato pepper	Hollow fiber supported liquid membrane; extraction with 1% trioctylphosphine dihexyl in ether oxide	LC–ESI-MS	0.06–2.7 mg kg^{-1}	[32]
24 pesticides	Apple, lemon, and tomato	Extract with acetone and partition with cyclohexane/ethyl acetate (50:50, v/v)	LC–ESI-MS–MS	—	[33]
Acetamiprid, imidacloprid, thiacloprid, thiamethoxam	Peach, pear, courgette, celery, and apricot	Extract with acetone aqueous solution, then pass though Extrelut-NT20 cartridges	LC–ESI-MS	0.02–0.1 mg kg^{-1}	[34]
12 pesticides	Oranges, strawberries, cherries, peaches, apricots, and pears	Pressurized liquid extraction with ethyl acetate and acid alumina	LC–triple quadrupole/ion-trap/quadrupole TOF-MS	0.01–0.0001 (QqQ); 0.4–0.005 (QIT); 0.4–0.01 mg kg^{-1} (QqTOF)	[35]

(continued)

TABLE 18.2 (continued)
Analysis of Pesticides in Fruits and Vegetables by LC–MS

Analytes	Matrix	Extraction Method	Determination Methods	Detection Limit	Refs.
Bupirimate, hexaflumuron, tebufenpyrad, buprofezin, pyriproxyfen, fluvalinate	Orange	Extract with ethyl acetate	LC–triple quadrupole/ion-trap/quadrupole-MS	0.005–0.2 mg kg^{-1}	[36]
171 pesticides	Lettuce, orange, apple, cabbage, and grape	Extract with acetone/dichloromethane/light petroleum	LC–MS–MS	0.01 mg kg^{-1}	[37]
Azadirachtin abamectin	Orange	Extract with acetonitrile	LC–ESI–MS–MS	0.007 mg kg^{-1}	[38]
22 carbamates	Grape, onion	MSPD with Bondesil C8	LC–APPI–MS–MS	0.5–5 ng mL^{-1}	[39]
13 carbamates	Onion, tomato, orange, and grapes	MSPD with Bondesil C8	LC–APCI-MS–MS	0.001–0.01 mg kg^{-1}	[40]
Nitenpyram, thiamethoxam, imidacloprid, acetamiprid, thiacloprid,	Bell pepper, cucumber, eggplant, grape, grapefruit, Japanese radish leaf/root, peach, pear, potato, and tomato	Extract with methanol and pass through graphitized carbon cartridge	LC–APCI-MS–MS	0.01–0.02 mg kg^{-1}	[41]
Acephate, methamidophos, monocrotophos, omethoate, oxydemeton-methyl vamidothion	Cabbage grapes	Extract with ethyl acetate and a solvent switch to 0.1% acetic acid/water	LC–APCI-MS–MS	0.001–0.004 mg kg^{-1}	[42]
Diflubenzuron, triflumuron, lufenuron, flufenoxuron, teflubenzuron, chlorfluazuron, hexaflumuron	Lemon, pear tomato	Extract with acetone and partition into ethyl acetate/cyclohexane	LC–ESI-MS–MS	—	[43]
Tetramethrin, allethrin, fenpropathrin, lambda-cyhalothrin, cypermethrin, deltamethrin, fenvalerate, bioresmethrin, permethrin bifenthrin	Cauliflower lettuce, cabbage, carrot, green pepper, and legume	Hexane/acetone (90:10, v/v) elute from florisil SPE cartridge	LC–ESI/ion trap-MS	0.01–0.04 mg kg^{-1}	[44]
Carbendazim, thiabendazole, imazalil hexythiazox, methiocarb, imidacloprid	Orange	Extract with ethyl acetate	LC–APCI-MS–MS	0.001–0.3 mg kg^{-1}	[45]
74 pesticides	Green salad, tomatoes, and strawberry	Extract with ethyl acetate	LC–ESI-MS–MS	0.01 mg kg^{-1}	[46]

Analytes	Matrix	Sample preparation	Analytical method	Detection limit	Reference
52 pesticides and metabolites	Tomato, lemon, raisin, and avocado	Extract with MeOH:H$_2$O (80:20, v/v) 0.1% HCOOH, and then a cleanup step using OASIS HLB SPE cartridges	LC–ESI–triple quadrupole-MS	0.01 mg kg^{-1}	[47]
38 pesticides	Grape, kiwi, strawberry, spinach, lemon, peach, and nectarine	Extract with ethyl acetate	LC–API-MS–MS	0.01–0.1 mg kg^{-1}	[48]
15 pesticides	Pepper, broccoli, tomato orange, lemon, apple, and melon	Extract with ethyl acetate	LC–TOF-MS	0.0005–0.03 mg kg^{-1}	[49]
20 pesticides	Cucumber, tomato, pepper, green bean, eggplant, zucchini, and melon watermelon	Extract with ethyl acetate	LC–ESI-MS–MS	0.4–21 µg kg^{-1}	[50]
14 carbamate and organophosphorus	Tomato, apple, carrot, cabbage	Extract with acetonitrile; and then prepare by dispersive-SPE with PSA	LC–ESI-MS	0.05–2 mg kg^{-1}	[51]
Aldicarb, aldicarb sulfoxide/sulfone, carbaryl, fenobucarb, methiocarb, methomy, oxamyl, pirimicarb	Spinach, tomato, potato, apple, cucumber, and mandarin	Extract with ethyl acetate	LC–ESI-MS–MS	0.005 µg g^{-1}	[52]
52 pesticides	Orange, potato	Extract with acetonitrile; and then prepare by dispersive-SPE with PSA	LC–ESI-MS–MS with polarity switching	—	[53]
Buprofezin hexythiazox	Orange, strawberry, and pear	Extract with acetone	LC–QTOF-MS	—	[54]

18.3.1 Recording of Observations

The GC–MS-QP2010 used in this study was obtained from Shimadzu Company (Kyoto, Japan). An RTX-5ms column was used for the separation of the pesticides.

The samples were extensively crushed to achieve good sample homogeneity. Once homogenized, the samples were stored at −25°C until GC–MS analysis was carried out.

To prepare the samples, 10 g of a previously homogenized food material was transferred into a suitable glass vessel. Ten milliliters acetonitrile was then added to each sample using an adjustable volume solvent dispenser. The glass vessels were capped before mixing on a Vortex mixer for 1 min at optimum speed. Once the initial sample mixing was completed, 1 g NaCl and 4 g anhydrous $MgSO_4$ were added and immediately mixed on a Vortex mixer for 1 min. To separate the phases, the samples were centrifuged for 10 min at 3000 rpm. Using an adjustable repeating pipette, 1.0 mL aliquot of the upper acetonitrile layer was transferred into a 1.5 mL flip-top microcentrifuge vial containing 150 mg anhydrous $MgSO_4$ and 50 mg primary secondary amine (PSA) sorbent. The vial was tightly capped and shaken on a Vortex mixer for 1 min before the extracts were centrifuged for 5 min to separate the solids from the solution. The solution was then transferred into an autosampler vial for GC–MS analysis. For spiked samples, standard pesticides were spiked into the samples before adding acetonitrile for extraction; the other steps were the same as described above.

18.3.2 Computation of Results

18.3.2.1 The Chromatogram of Mixed Standard Pesticides

The chromatogram of the 97 standard pesticides (500 ng mL^{-1}) is shown in Figure 18.1.

18.3.2.2 Detection Limit

The estimated limit of detection (LOD) and limit of quantitation (LOQ) calculated as the concentration that produced a signal equal to 3 times and 10 times the background noise level, respectively, for the three analyzed matrices—potato, apple, and rice—were obtained, which are listed in Table 18.3.

18.3.2.3 Recovery Test

Recovery of the 97 pesticides (100 ng mL^{-1}) spiked into potato, cabbage, carrot, apple, orange, cucumber, and rice was investigated using GC–MS. To calculate the recovery, the spiked sample from each agricultural product was prepared thrice, and the unspiked samples were also investigated. Some of the average recovery results and the relative standard deviations (RSD) are shown

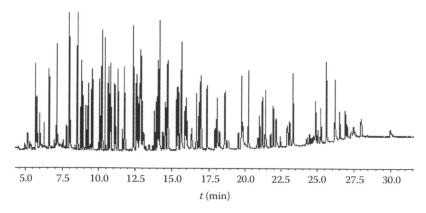

FIGURE 18.1 Chromatogram of 97 standard pesticides (500 ng mL^{-1}). (From Libin, L. et al., *Chinese J. Anal. Chem.*, 34(6), 783, 2006. With permission.)

TABLE 18.3
Estimated LOD and LOQ Calculated as the Concentration that Produced a Signal Equal to 3 Times and 10 Times the Background Noise Level, Respectively

	LOD (ng g^{-1})			LOQ (ng g^{-1})		
Pesticides	Potato	Apple	Rice	Potato	Apple	Rice
Eptam	15	56	90	50	186	297
Butylate	65	47	52	216	156	173
α-Benzene hexachloride	58	14	23	193	47	77
δ-Benzene hexachloride	52	30	71	173	100	236
Carbaryl	27	54	47	90	180	156
Diethofencarb	26	26	30	86	86	100
Fenthion	8	6	7	27	20	23
Chlorpyrifos	25	32	33	83	107	110
Tebufenpyrad	12	9	22	40	30	73
Pyridaben	12	11	10	40	37	33

TABLE 18.4
Results of the Recovery Test

	Potato		Cabbage		Carrot	
Pesticides	Mean (%)	RSD (%)	Mean (%)	RSD (%)	Mean (%)	RSD (%)
Eptam	69	0.4	69	1.6	70	2.7
Butylate	68	1.9	69	2.4	69	0.6
α-Benzene hexachloride	76	2.4	79	1.0	79	1.7
δ-Benzene hexachloride	66	6.4	82	2.6	75	4.6
Carbaryl	78	4.8	124	12.3	78	11.7
Diethofencarb	86	3.5	84	2.9	89	5.3
Fenthion	76	1.4	81	1.6	83	1.8
Chlorpyrifos	70	1.2	71	3.7	73	4.9
Tebufenpyrad	89	2.6	91	2.2	92	4.8
Pyridaben	89	0.4	96	2.6	103	3.8

Mean: average recovery; RSD: relative standard deviation.

in Table 18.4. More pesticides in vegetables and fruits showed acceptable recoveries than those in rice because of the more complicated matrix of rice. On the whole, the MSPD technique reduced the interferences of the matrix to a large degree and could be applied to a more complicated matrix such as rice. Furthermore, the low RSD values (usually <10%) indicated the high reproducibility of the system.

18.4 EXPERIMENT II: DEVELOPMENT OF AUTOMATED ONLINE GPC–GC–MS FOR MEASURING MULTIRESIDUAL PESTICIDES IN AGRICULTURAL PRODUCTS

In addition to preparation strategies, factors such as time, cost, and even the ease of operation are also important in the practical determination of residual pesticides. For those who routinely need to analyze or repeatedly analyze large numbers of samples (e.g., the Food Safety Authority), these issues are especially significant. Considering the above points, one approach is to develop a general method that can be applied to a diverse range of pesticide chemistry by automating the sample clean-up and determination steps without compromising accuracy and precision.

GPC is a recently developed and popular postextraction cleanup method and is highly effective in removing high-molecular-weight interferences, such as lipids, proteins, and pigments, before analysis by GC, GC MS, high performance liquid chromatograph (HPLC), and LC–MS. The use of GPC greatly reduces instrument downtime, extends column life, and results in increased analytical precision and accuracy. In addition, GPC has indicated the potential for automated analysis with LC or GC.

An automated online GPC–GC–MS for measuring residual pesticides in agricultural products is proposed. This newly developed system can determine 97 pesticides in 90 min using just one-fortieth of the solvent used in conventional GPC applications. After extraction, samples are injected into the automated online GPC–GC–MS system. The combination of the MSPD technique and online GPC–GC–MS system enables to accomplish a high throughput of pesticide determinations at lower cost, without adversely affecting the recovery results.

18.4.1 RECORDING OF OBSERVATIONS

The GPC–GC–MS system studied herein consists of an LC-10Avp series and a GC–MS-QP2010 equipped with a large-volume injection device (PTV-2010, Shimadzu, Kyoto, Japan). Acetone/cyclohexane (3/7, v/v) are used as the mobile phase of GPC, and the flow rate is set to 0.1 mL min⁻¹.

Sample preparation is the same as that described in Section 18.3.

A schematic flow diagram of the GPC–GC–MS system is shown in Figure 18.2. In this experiment, the injection volume is 10 µL, and the volume of the sample loop is set to 200 µL.

18.4.2 COMPUTATION OF RESULTS

In optimizing the transfer of solvent between the GPC and GC–MS systems, the GPC mobile-phase flow rate is reduced compared with that in conventional GPC. In this application, a flow rate of 0.1 mL min⁻¹ is used with a 2 mm i.d. GPC column; this resulted in a 40-fold reduction in solvent consumption compared with that in conventional GPC column applications. To investigate the fraction time of the pesticides, two marker molecules were selected. Fluvalinate (MW = 502.9) is used as the upper molecular weight marker and chinomethionate (MW = 234.3) is used as the lower molecular weight marker, corresponding to a retention time of between 3.023 and 4.925 min (the GPC

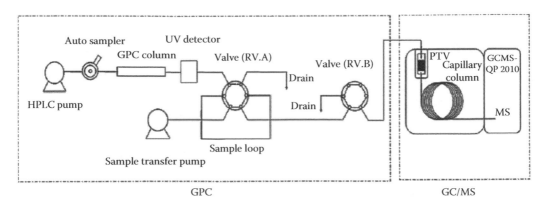

FIGURE 18.2 Schematic flow diagram of the GPC–GC–MS system. (From Liu, L.B. et al., *J. Chromatogr., B*, 845, 61, 2007. With permission.)

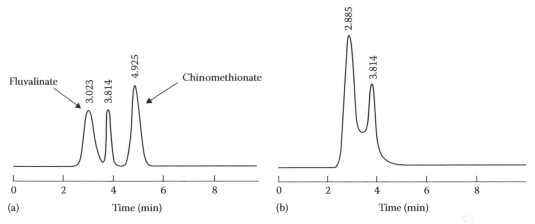

FIGURE 18.3 GPC chromatograms obtained at a UV wavelength of 210 nm. Acetone/cyclohexane (3/7, v/v) were used as the mobile phase of GPC, and the flow rate was set to 0.1 mL min^{-1}. The GPC column was kept at 40°C in the column oven. (a) GPC chromatogram of two marker molecules at 1000 ng mL^{-1} (fluvalinate (MW = 502.9) as the upper molecular weight marker and chinomethionate (MW = 234.3) as the lower molecular marker). (b) GPC chromatogram of rice sample spiked with a mixture of 97 standard pesticides at 0.1 mg kg^{-1}. (From Liu, M. et al., *J. Chromatogr. A*, 1097, 183, 2005. With permission.)

chromatograms of the two marker molecules and the rice sample spiked with 97 pesticides standard are shown in Figure 18.3). In this study, GPC eluent from 2.9 to 4.9 min is fractionated by the sample loop.

Ninety-seven target pesticides are spiked at a concentration level of 0.1 mg kg^{-1} into several kinds of agricultural products, including potato, cabbage, carrot, apple, orange, cucumber, and rice. Total ion chromatograms (TIC) of the spiked potato sample (adding standard pesticides 0.1 mg kg^{-1}) and the unspiked potato sample are shown in Figure 18.4.

Table 18.5 provides estimated LOD and LOQ, calculated as the concentration that produced a signal that was 3 times and 10 times the background noise level, respectively, for three analyzed matrices: potato, apple, and rice.

Recovery of the pesticides (0.1 mg kg^{-1}) spiked into potato, cabbage, carrot, apple, orange, cucumber, and rice was investigated via GPC–GC–MS. To calculate the recovery, the spiked sample from each agricultural product was prepared three times and the unspiked samples were also investigated. The same recovery test was investigated using a conventional GC–MS system, with the results also shown in Table 18.6. Comparing the average recovery results, it appears that more pesticides have acceptable recovery in GPC–GC–MS than that in GC–MS. For example, in GPC–GC–MS, 83 of 97 pesticides showed acceptable recovery for orange, but in GC–MS, only 54 pesticides showed acceptable recovery. The inherent characteristics of GPC have proved highly advantageous in sample pretreatment to minimize matrix interferences associated with limited solvent extraction protocols. In this regard, GPC–GC–MS has superior recovery performance to GC–MS.

18.5 EXPERIMENT III: SIMULTANEOUS DETERMINATION OF CARBAMATE AND OPPS IN FRUITS AND VEGETABLES BY LC–MS

Instead of LC/UV due to lower sensitivity, HPLC with fluorescence detection by postcolumn derivatization is now the most widely used method for the analysis of carbamate pesticides in foods. Most OPPs are easily analyzed by GC. Therefore, carbamates and OPPs are usually analyzed by LC and GC, respectively.

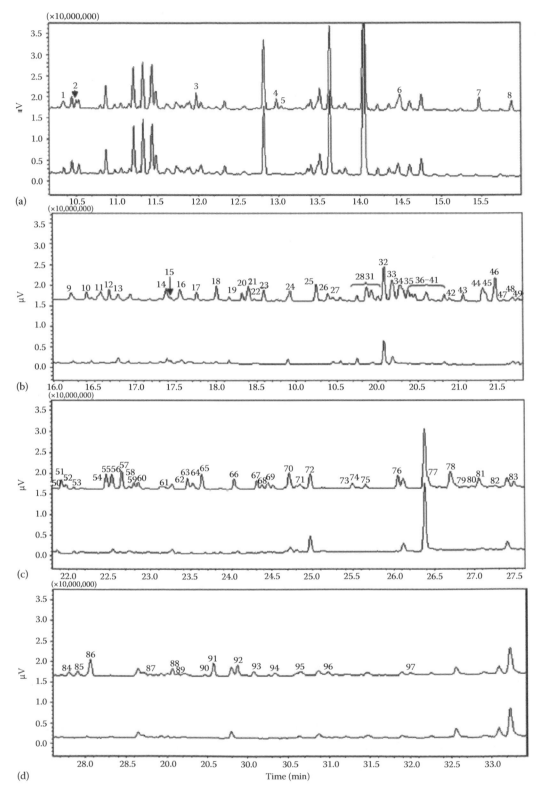

FIGURE 18.4 TIC of potato spiked with 97 pesticides at 0.1 mg kg⁻¹ (upper chromatogram) and unspiked potato (lower chromatogram). Multiple chromatograms (a, b, c, d) are used to show all the pesticides.

TABLE 18.5
Estimated LOD and LOQ Calculated as the Concentrations that Produced a Signal Equal to 3 Times and 10 Times the Background Noise Level, Respectively, in GPC–GC–MS

Pesticide	LOD (µg kg^{-1})			LOQ (µg kg^{-1})		
	Potato	Apple	Rice	Potato	Apple	Rice
EPTC	9	18	13	31	59	44
Butylate	38	79	30	127	262	101
α-BHC	15	14	16	51	48	52
δ-BHC	17	18	26	56	57	87
NAC	3	21	18	11	69	60
Diethofencarb	26	24	15	85	75	49
MPP	29	8	8	97	27	28
Chlorpyrifos	27	22	18	88	72	61
Tebufenpyrad	10	16	14	36	53	47
Pyridaben	7	14	17	23	46	55

LC–MS has now emerged as an excellent alternative technique for simultaneous analysis of these compounds. Anastassiades et al. [55] first established dispersive SPE for the determination of pesticides in vegetables and fruits by GC–MS. Lehotay et al. [56–58] developed the analysis method of pesticides by GC–MS and LC–MS–MS. Posyniak et al. [59] recently applied the procedure to analyze sulfonamides in chicken by LC with fluorescence detection. LC–MS with a single quadrupole has also been widely reported to determine pesticides. However, the combination extraction procedure/determination technique has not been reported yet.

The purpose of this experiment was to establish a dispersive SPE method for the simultaneous determination of carbamates and OPPs in fruits and vegetables by LC–MS with a single quadrupole instead of the triple one. The system was optimized based on the chromatographic resolution and sensitivity of MS. Method validation was presented in terms of recovery and precision from spiked "neat" sample matrices. It confirmed that LC–MS can be used as an excellent alternative technique for the identification and determination of these compounds, and then it was applied to monitor real crop samples from various resources collected in the local marketplace.

18.5.1 Recording of Observations

A Shimadzu LC–MS 2010A system was employed. The chromatographic separation was performed on a Shimadzu Shim-Pack VP-ODS (150 × 2.0 mm i.d., 5 µm). The mobile phase was methanol/water (containing 0.2% formic acid) and a gradient program was used (total flow rate: 0.2 mL min^{-1}).

Sample preparation is the same as that described in Experiment I.

Linear dynamic range, precision, recovery, selectivity, and uncertainty for the analytical methodology were evaluated. Linearity was determined by calibration curves created with concentrations of 0.01, 0.02, 0.05, 0.10, 0.20, and 0.50 µg mL^{-1} by mixture standard solutions. For sample matrix testing, tomato, apple, cabbage, and carrot obtained from a supermarket were spiked and tested for recovery, with precision at 0.05 mg kg^{-1} for each pesticide with five replicates. Before spiked testing, the blank samples were analyzed. If contaminated, the recoveries were calculated by subtraction of the amount of blank samples.

TABLE 18.6
Comparison of Recovery (%) Achieved by GPC–GC–MS and GC–MS

	Potato			Cabbage			Carrot			Cucumber			Apple			Orange			Rice		
Pesticide	GPC–GC/MS Mean (%)	RSD (%)	GC/MS Mean (%)	GPC–GC/MS Mean (%)	RSD (%)	GC/MS Mean (%)	GPC–GC/MS Mean (%)	RSD (%)	GC/MS Mean (%)	GPC–GC/MS Mean (%)	RSD (%)	GC/MS Mean (%)	GPC–GC/MS Mean (%)	RSD (%)	GC/MS Mean (%)	GPC–GC/MS Mean (%)	RSD (%)	GC/MS Mean (%)	GPC–GC/MS Mean (%)	RSD (%)	GC/MS Mean (%)
1 Acephate	73	0.3	194	74	1.1	205	70	1.3	220	59	1.3	75	57	3.5	116	55	2.9	108	60	5.7	108
2 Acetamiprid	71	10.7	89	91	2.8	100	99	4.4	108	88	4.1	106	86	9.8	88	99	11.6	101	92	3.6	198
3 Acrinathrin	115	5.7	85	135	5.4	113	188	2.8	88	110	4.5	104	134	1.4	116	121	2.3	120	111	7.6	114
4 Bendiocarb	86	0.3	71	87	2.1	94	92	3.0	75	100	1.3	97	87	3.5	107	106	3.5	105	101	1.9	111
5 Benfuresate	94	1.0	75	94	0.9	77	102	0.8	77	108	1.3	71	96	1.8	78	111	2.4	114	97	3.2	87
6 α-BHC	97	1.5	76	97	1.0	79	102	2.4	79	113	3.8	73	104	2.4	56	114	5.0	107	118	2.5	69
7 β-BHC	107	0.3	80	106	3.2	83	112	4.1	83	103	1.1	90	120	1.7	99	137	3.5	113	104	1.3	89
8 γ-BHC	99	0.9	76	101	2.6	77	106	1.6	76	112	2.4	84	108	2.0	63	116	4.5	114	105	2.9	72
9 δ-BHC	105	1.1	66	106	2.7	82	108	0.7	75	110	1.2	68	115	1.2	53	111	4.3	76	110	2.0	56
10 Bitertanol	125	2.2	95	119	2.4	91	100	7.5	114	118	1.1	189	109	1.3	115	111	7.8	130	117	2.5	199
11 BPMC	98	1.7	79	97	1.3	86	100	1.4	74	104	1.4	93	109	1.3	79	119	1.4	119	111	1.9	83
12 Butylate	110	1.2	68	111	3.4	69	110	2.7	69	100	1.9	64	90	0.5	57	112	4.0	93	106	5.5	60
13 Cadusafos	104	0.7	74	102	2.3	77	108	2.8	80	110	2.1	92	118	3.3	81	112	6.7	97	116	1.6	84
14 Captan	91	3.0	69	66	1.4	ND	77	2.4	51	110	5.1	37	101	0.4	61	145	6.3	86	26	5.6	25
15 Chinomethionat	56	1.8	49	55	2.1	49	55	2.3	48	16	5.3	43	19	10.6	44	35	4.2	53	22	9.9	41
16 Chlorobenzilate	114	1.5	86	111	3.0	84	117	1.8	92	115	5.0	129	127	4.3	116	115	3.8	148	148	4.6	143
17 Chlorpropham	112	1.9	72	113	1.2	77	115	3.5	76	118	1.7	77	109	1.3	72	117	0.8	110	117	1.7	77
18 Chlorpyrifos	97	0.8	70	95	3.3	71	100	4.0	73	113	0.9	88	114	1.3	82	111	5.3	120	116	3.2	98
19 α-CVP	103	4.1	91	104	3.5	95	107	1.7	94	115	2.2	119	112	2.2	116	105	4.2	160	115	2.8	117
20 β-CVP	104	0.5	91	100	2.7	95	106	2.2	94	116	2.2	119	112	2.2	119	105	4.2	153	117	3.6	130
21 Cyfluthrin	92	7.2	67	101	2.6	79	124	7.0	75	101	0.9	105	96	4.9	110	115	7.4	95	112	1.3	116
22 Cyhalothrin	112	2.5	86	121	2.9	112	118	2.5	110	115	3.5	113	99	0.5	107	154	1.3	118	150	1.4	132
23 Cypermethrin	106	4.6	55	104	3.6	85	103	1.8	79	112	4.8	91	124	2.7	79	111	2.5	91	117	2.8	98
24 Cyproconazole	106	4.3	83	105	4.3	84	117	4.1	91	100	3.3	125	113	1.1	109	102	4.8	39	155	4.3	134
25 DCBP	104	6.0	78	108	4.7	82	120	1.7	80	191	6.9	88	181	0.8	81	159	8.1	114	197	7.8	95

No.	Name																					
26	p,p'-DDD	105	0.3	82	104	1.9	87	107	2.0	89	110	1.8	103	119	1.1	85	113	5.2	118	120	7.1	114
27	p,p'-DDE	100	4.0	82	97	1.6	82	95	0.8	81	113	0.3	85	109	2.6	80	120	5.3	115	113	4.8	87
28	DDVP	88	0.8	72	88	1.7	78	91	1.3	73	98	2.1	74	85	1.1	67	98	4.3	99	105	0.8	64
29	Deltamethrin	109	3.3	78	95	6.1	81	111	4.3	108	64	2.5	62	62	0.7	92	63	1.6	82	44	0.6	31
30	Diazinon	106	0.7	76	104	3.4	78	109	2.2	79	121	0.7	86	111	1.4	83	119	3.8	116	120	3.2	86
31	Dichlofluanid	72	5.6	63	62	2.0	58	50	3.9	61	37	0.2	47	61	8.4	58	56	7.6	81	20	5.5	44
32	Diethofencarb	117	3.1	86	106	4.9	84	115	2.1	89	114	2.1	102	99	2.7	102	115	3.1	152	132	0.9	128
33	Difenoconazole	123	4.1	95	115	8.9	87	137	8.1	117	163	3.0	143	113	1.8	111	112	5.0	142	112	4.8	144
34	Difolatan	90	4.2	ND	58	7.0	ND	112	5.3	ND	65	5.6	45	68	5.6	96	87	3.8	53	19	5.5	46
35	Dimethipin	77	1.3	84	79	3.5	89	79	1.7	73	72	3.7	85	54	1.5	85	71	7.4	112	72	2.3	73
36	(Z)-Dimethylvinphos	97	1.6	94	94	2.1	99	102	2.0	97	112	4.7	112	104	0.6	116	115	3.4	164	113	2.4	116
37	EDDP	107	0.9	132	107	1.3	143	118	2.8	128	94	2.3	143	119	2.0	136	116	1.8	180	115	2.3	169
38	EPN	110	0.9	95	112	2.9	97	138	4.7	113	94	4.8	158	116	1.5	103	97	6.3	113	112	4.8	183
39	EPTC	99	1.0	69	97	1.5	69	101	1.7	70	120	0.2	60	111	1.1	50	118	2.9	84	116	1.3	60
40	Esprocarb	99	1.3	77	98	1.9	79	103	1.4	82	117	0.7	91	111	3.0	85	117	1.2	116	116	2.3	94
41	Ethiofencarb	94	5.1	105	92	2.5	121	91	3.3	98	72	7.4	113	95	6.4	115	88	6.8	177	77	3.7	112
42	Ethoprophos	99	0.5	78	100	1.9	78	103	3.0	82	112	1.4	84	113	2.0	85	114	0.8	118	107	1.8	79
43	Etrimfos	103	0.6	74	101	2.3	75	107	3.4	78	119	1.5	103	119	1.2	93	110	6.1	132	114	2.5	95
44	Fenarimol	100	1.3	85	99	3.0	85	111	3.0	95	108	1.7	132	118	1.2	110	112	5.9	141	112	4.3	143
45	Fensulfothion	99	1.5	89	102	1.8	97	119	1.4	122	110	5.7	136	101	1.6	110	78	6.2	110	111	6.0	140
46	Fenvalerate	101	6.5	74	101	2.2	105	104	2.1	89	113	3.4	81	115	2.0	113	119	1.0	111	112	2.0	115
47	Flucythrinate	114	2.8	93	120	2.8	111	120	3.8	110	153	5.8	116	110	4.2	119	118	5.1	129	117	6.3	116
48	Flusilazole	93	1.2	80	92	2.5	85	97	2.1	87	116	5.0	115	98	4.7	107	117	1.8	120	111	3.3	117
49	Flutolanil	113	2.6	85	110	1.8	88	110	5.4	104	116	6.4	120	106	0.6	115	114	6.8	139	157	1.0	130
50	Fluvalinate	142	5.9	82	155	0.5	133	152	1.6	75	145	4.0	112	135	2.1	127	151	3.7	113	162	4.8	129
51	Fosthiazate	108	0.7	127	105	3.6	127	118	3.9	129	103	3.3	110	106	7.3	147	105	5.9	193	90	4.2	92
52	Halfenprox	94	1.1	92	97	1.6	99	103	3.7	115	158	1.0	120	106	1.5	120	117	2.0	130	120	2.6	139
53	Imibenconazole	157	8.1	88	156	9.2	99	180	5.9	113	119	3.2	110	110	2.6	75	115	6.3	107	105	5.1	104
54	Iprodione	103	0.8	138	104	2.6	156	112	1.7	91	104	1.1	111	120	0.7	117	106	4.3	97	119	1.5	136
55	Isofenphos Oxon	120	2.8	118	117	2.8	112	88	6.0	110	102	3.2	191	116	3.8	167	105	4.0	178	139	1.4	195
56	Isophenphos	116	1.5	80	114	2.7	83	120	1.2	88	118	2.5	118	115	0.8	112	118	2.1	135	144	1.7	118
57	Isoprocarb	95	0.6	80	96	2.1	87	96	2.1	69	112	1.3	88	100	0.8	81	113	3.5	114	115	0.7	83

(continued)

TABLE 18.6 (continued)
Comparison of Recovery (%) Achieved by GPC–GC–MS and GC–MS

	Potato			Cabbage			Carrot			Cucumber			Apple			Orange			Rice		
	GPC–GC/MS		GC/MS	GPC–GC/MS		GC/MS	GPC–GC/MS		GC/MS	GPC–GC/MS		GC/MS	GPC–GC/MS		GC/MS	GPC–GC/MS		GC/MS	GPC–GC/MS		GC/MS
Pesticide	Mean (%)	RSD (%)	Mean (%)	Mean (%)	RSD (%)	Mean (%)	Mean (%)	RSD (%)	Mean (%)	Mean (%)	RSD (%)	Mean (%)	Mean (%)	RSD (%)	Mean (%)	Mean (%)	RSD (%)	Mean (%)	Mean (%)	RSD (%)	Mean (%)
58 Lenacil	109	0.6	91	107	2.1	92	119	2.4	104	111	7.4	124	112	2.4	118	117	2.9	134	168	2.9	143
59 Malathion	97	0.8	83	95	1.3	91	104	2.1	90	116	3.1	113	110	1.4	106	98	4.3	132	119	0.7	112
60 Mefenacet	104	1.3	93	108	1.7	95	116	3.4	107	96	3.2	125	93	3.2	115	96	7.6	134	99	4.7	144
61 MEP	100	2.1	81	102	1.5	74	112	3.7	94	113	2.5	120	119	0.6	119	87	6.4	155	114	6.4	125
62 Mepronil	109	2.0	85	110	2.3	85	119	1.6	93	112	0.8	120	119	2.2	116	116	2.1	129	153	3.6	134
63 Methamidophos	68	1.2	71	72	2.0	84	71	0.8	68	61	9.3	109	57	5.9	110	56	3.6	111	48	5.0	103
64 Methiocarb	112	2.1	90	114	3.0	126	118	2.1	80	115	1.5	134	103	2.2	113	115	2.8	105	117	2.2	109
65 Methyl-parathion	104	1.4	85	120	3.5	82	115	4.6	92	119	1.4	108	118	2.7	115	84	2.1	151	114	2.0	120
66 Metolachlor	102	0.3	80	99	2.5	80	107	2.2	86	107	1.4	118	116	0.6	100	106	8.3	130	118	2.4	117
67 MPP	63	1.1	76	57	1.8	81	56	1.7	83	149	1.3	95	108	2.5	89	118	2.2	112	97	5.6	88
68 Myclobutanil	91	2.1	84	92	1.6	84	98	2.2	93	91	1.8	110	110	1.9	111	93	6.2	112	118	0.5	128
69 NAC	99	0.9	78	106	2.6	124	109	5.2	78	113	1.5	115	98	0.6	120	110	6.9	146	111	3.7	111
70 Paclobutrazol	106	0.1	84	102	2.9	83	112	4.1	92	106	6.2	118	115	0.9	110	108	3.7	150	120	7.8	81
71 PAP	98	0.8	78	95	2.2	80	106	3.2	89	119	1.3	111	118	2.8	103	102	5.6	125	112	6.2	114
72 Parathion	114	1.9	83	108	3.8	77	120	3.2	96	119	1.1	136	84	1.6	118	101	5.6	118	117	4.1	135
73 Pendimethalin	110	0.5	85	108	3.0	79	120	2.7	96	93	3.5	129	119	1.0	114	102	6.5	154	117	2.2	138
74 Permethrin	113	0.4	92	111	2.6	95	115	2.6	98	117	0.5	108	111	3.4	112	116	5.4	109	114	3.6	119
75 Phosalone	111	0.5	87	116	2.3	95	120	2.6	103	105	4.1	131	94	2.0	115	146	4.4	115	113	3.0	151
76 Pirimicarb	92	0.5	76	89	1.8	78	97	2.6	80	107	2.1	82	97	0.6	79	107	6.5	79	110	2.7	89
77 Pirimiphos-methyl	103	0.5	76	102	3.0	82	108	1.4	84	111	2.2	103	118	1.2	98	114	4.8	131	118	2.1	112
78 Pretilachlor	114	0.5	88	113	1.8	83	118	7.8	101	117	1.8	134	119	0.2	120	120	0.2	154	118	3.4	146
79 Propiconazole	97	0.6	85	98	4.2	86	105	2.9	92	94	1.2	143	94	2.5	113	94	3.9	115	110	6.2	136
80 Prothiofos	103	0.8	80	101	1.2	83	86	4.0	85	109	1.6	104	120	0.3	97	111	6.4	130	119	5.3	114
81 Pyraclofos	108	1.7	141	107	1.4	157	191	5.5	161	104	0.3	164	109	3.8	151	108	5.4	135	108	3.4	199
82 Pyridaben	120	1.1	89	121	1.4	96	83	1.6	103	118	2.0	116	118	2.5	117	118	1.9	113	117	5.4	130

No.	Pesticide																				
83	Pyrifenox	97	0.9	79	94	2.3	81	102	2.7	85	98	3.1	119	85	0.6	100	99	7.1	138	117	2.0	130
84	Pyrimidufen	82	1.3	94	85	2.8	101	86	7.0	107	114	2.3	120	111	3.3	125	113	3.2	138	120	2.2	132
85	Pyriproxyfen	99	0.8	87	103	1.2	89	106	2.2	93	109	1.1	119	90	3.2	117	110	5.4	120	116	0.9	130
86	Quinalphos	98	0.7	80	98	2.1	84	104	2.0	89	118	1.7	114	111	3.3	93	102	6.8	117	115	7.0	102
87	Silafluofen	88	0.1	90	119	1.1	92	118	1.9	92	176	2.6	119	137	2.1	115	143	2.9	120	147	1.5	128
88	Tebuconazole	109	2.7	82	102	0.5	84	117	3.0	93	135	4.2	132	112	1.8	111	106	4.5	133	113	7.8	136
89	Tebufenpyrad	112	1.8	89	108	2.2	91	119	2.1	92	114	2.7	114	114	1.0	111	115	3.4	129	118	2.1	130
90	Tefluthrin	118	1.8	76	116	2.9	79	117	3.1	80	186	2.9	86	207	2.2	78	168	4.0	114	190	3.0	79
91	Terbufos	120	0.9	75	122	2.4	78	120	1.3	81	156	0.5	111	109	1.3	101	116	1.8	129	112	1.2	108
92	Thenylchlor	100	0.4	88	101	1.7	85	109	3.5	98	110	2.5	117	112	1.6	95	113	5.4	119	115	6.1	129
93	Thiobencarb	96	0.6	77	97	1.1	78	101	2.3	81	118	1.9	95	111	1.9	75	115	1.3	105	116	1.2	86
94	Thiometon	103	1.2	74	107	1.8	75	107	2.6	79	150	2.8	93	100	3.1	84	104	2.5	120	110	4.1	88
95	Tolclofos-methyl	97	0.9	76	97	1.5	80	101	2.5	82	117	1.8	91	106	2.2	84	118	1.6	118	106	2.9	88
96	Triadimenol	102	1.4	79	97	3.5	82	109	5.1	91	110	2.2	86	112	3.7	114	112	7.8	116	117	2.1	117
97	Tricyclazole	94	3.2	94	100	2.4	100	104	4.6	111	105	5.4	110	111	1.9	84	101	7.0	92	109	7.5	108

Mean: average recovery; RSD: relative standard deviation; ND: not detected.

18.5.2 Computation of Results

18.5.2.1 Optimized LC–MS Method

For selecting the type of ionization probe (either ESI or APCI) and the ionization mode (positive or negative), flow-injection testing of individual standard solutions was performed using the carrier solutions of instrumental autotuning as mobile phases, that is, acetonitrile/water (90/10, v/v) for ESI; and methanol/water (50/50, v/v) for APCI, respectively. The results demonstrated higher responses in positive mode than in negative mode for both ESI and APCI. In addition, the signal responses were 10–20 times higher by ESI than APCI for all tested pesticides. In order to confirm the analytical conditions, results from ESI and APCI were further compared using the same mobile phase (methanol/water gradient program), and the same result was observed. ESI positive mode proved to be the most appropriate ionization mode for analysis of 14 pesticide mixture solution and was thus selected for the next experiment.

To improve the chromatographic resolution and ionization efficiency of MS, analytical conditions, such as the LC gradient program, mobile-phase composition, and flow rate of drying gas, were adjusted and optimized. The signal intensity of MS was found to be strongly influenced by the mobile-phase composition. The influence of the organic modifier on the signal response was compared. Methanol and acetonitrile as modifiers in the mobile phase were tested with gradient elution. Chromatographic resolution did not change dramatically, and the MS signal for most pesticides decreased by a factor of 5–10 when acetonitrile/water was compared with that of methanol/water. This is most likely due to the fact that acetonitrile is a weaker proton donor than methanol (both in aqueous phase and in gaseous phase as well). Methanol was suitable for obtaining high intensity of all carbamates since it is liable to provide hydrogen to the radical ion of carbamates. Consequently, ionization efficiency (in positive mode) produces better results with methanol as mobile phase.

For optimal LC separation, the more the formic acid that was added to the mobile phase, the better the resolution that was obtained for pesticides, especially for solute pairs such as pirimiphos-methyl and etrimfos, which were difficult to be separated. The addition of formic acid increased the signal of MH^+ ions; however, the signal-to-noise ratio (S/N) of analytes decreased since the baseline noise increased when formic acid was more than 0.2% (v/v) in the mobile phase for MS detection. The phenomenon agrees with previous results that analytes could be efficiently ionized when the mobile phase contains only trace amounts of acid. Regarding MS sensitivity and optimal mobile-phase composition for separation, a mixture of methanol and water with the addition of 0.2% formic acid was chosen as the eluting solvent.

During the process of optimization, it was found that the flow rate of drying gas also played an important role in MS sensitivity. Different flow rates ranging from 0.01 to 0.06 MPa were then investigated. At low flow rates such as 0.02 MPa, the noise of the baseline was relatively high while S/N decreased. By increasing the flow rate, baseline noise decreased, and S/N increased until the gas flow reached 0.04 MPa. By increasing the flow rate of drying gas to 0.05 MPa, the signal intensity decreased. This was probably caused by the volatilization of pesticides at the high flow rate of drying gas. Therefore, 0.04 MPa drying gas flow appeared suitable for the application.

After optimization, LODs were obtained by injection of the standard mixture and calculated with S/N ≥ 3 in SIM mode. Pesticides can be detected at the level of 0.5–10 ng mL^{-1} depending on the type of the analytes that could meet requirements of residue analysis.

18.5.2.2 Validation of Method

Calibration curves were established using six different concentrations by LC–MS in the selected ion mode, followed by extraction of one or two signals of the more abundant ions acquired in full-scan mode. Excellent linearity through the range of 0.01–0.5 μg mL^{-1} with correlation coefficients from 0.9950 to 0.9999 was obtained. For the tomato, lack of interfering peaks and low background noise were shown in the blank sample (Figure 18.5a). Chromatograms obtained from LC–MS analysis

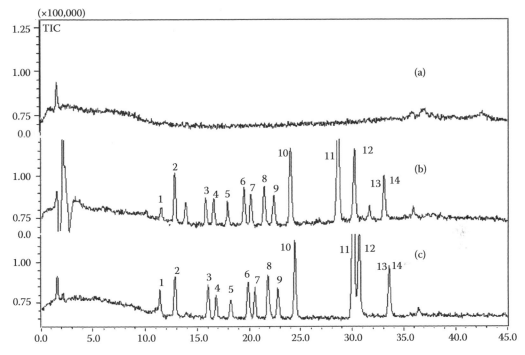

FIGURE 18.5 The chromatograms of pesticide mixture (a) unspiked tomato, (b) spiked tomato (0.05 mg kg^{-1}), and (c) standard solution (0.05 µg mL^{-1}). 1, Methiocarb-sulfone; 2, aldicarb; 3, carbaryl; 4, ethiofencarb; 5, isoprocarb; 6, methidathion; 7, azinphos-methyl; 8, baycarb; 9, methiocarb; 10, malathion; 11, pirimiphos-methyl; 12, etrimfos; 13, pyraclofos; 14, phosalone.

tomato spiked with 0.05 mg kg^{-1} and 0.05 µg mL^{-1} standard solutions are respectively illustrated in Figure 18.5b and c. Compared with the two chromatograms, it is easy to observe an obvious trend that pirimiphos-methyl and etrimfos were separated completely in that tomato sample, whereas only partial separation was achieved by injection of the standard solution. It is thought that pirimiphos-methyl was protonated in the acid tomato matrix, which resulted in decreasing the retention on the column. With regard to other types of samples matrices, there is some concern that target pesticides coeluted with other components at about the same retention time originating from the matrix itself (see chromatogram of blank real samples in Figure 18.6). As a result, and with the aid of selected ion chromatogram data and cochromatography of each pesticide standard, this enabled the selective and positive identification of peaks of interest. No interfering peaks from endogenous compounds of matrices were found in the retention time range of the target pesticides.

Due to the fact that maximum residue limits (MRLs) of most carbamates and OPP pesticides are more than or equal 0.05 mg kg^{-1} as shown in Table 18.7, the precision and accuracy of the above-mentioned method are validated by four spiked samples at 0.05 mg kg^{-1} (Table 18.7). The low recoveries of pirimiphos-methyl may have resulted from the hydrophilic structure, which led to considerable solubility in the water phase. Methiocarb-sulfone and aldicarb showed relatively low recovery and high RSD. There are two reasons: one is that these compounds are liable to degrade during the extraction process; on the other hand, many polar compounds from the matrices exhibited weak retention on the column, resulting in the suppression of some pesticides with similar retention behavior. As reported by Jansson et al. [60], the matrix effect is very compound dependent, probably due to the coeluting matrix components, which might interact with the target pesticide within the ionization interface. In addition, the high mean recovery of azinphos-methyl and phosalone could be partly explained by the lower ionization efficiency of the compounds containing an acryl group in working solutions of pure methanol than those ionized from the matrices containing water. Finally, for bulb vegetables and fruits, most pesticides gave the highest

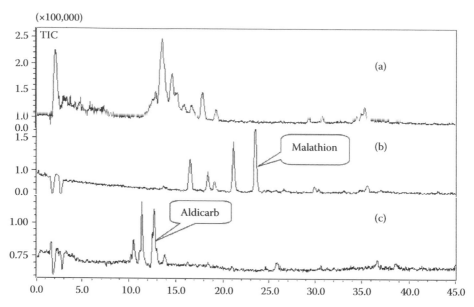

FIGURE 18.6 The chromatograms of real samples without spiked (a) apple; (b) cabbage, and (c) peach.

TABLE 18.7
Recovery and RSD of the Pesticides in Different Samples Spiked with 0.05 mg kg^{-1} ($n = 5$) and MRLs Established by Japan (a) and EU (b)

Pesticides	Cabbage Recovery (%)	RSD (%)	Tomato Recovery (%)	RSD (%)	Carrot Recovery (%)	RSD (%)	Apple Recovery (%)	RSD (%)	MRLs (mg kg^{-1}) Values	Matrixes
Methiocarb-sulfone	72	2.8	100	5.5	73	6.2	87.01	5.7	0.05[a]	Lettuce
Aldicarb	91	5.1	89	3.3	88	2.8	93	2.4	0.05[a]	Grape
Carbaryl	90	3.3	101	2.4	101	2.4	120	6.5	0.10[a]	Potato
Ethiofencarb	89	2.1	103	1.9	97	1.7	113	3.9	0.50[a]	Potato
Isoprocarb	86	4.0	79	4.7	85	1.7	103	5.9	0.05[b]	Pear
Methidathion	81	1.4	109	2.2	94	2.7	115	1.5	0.30[b]	Pear
Azinphos-methyl	105	0.3	116	1.7	95	1.7	120	2.0	0.50[b]	Apple
Baycarb	90	3.5	98	2.4	93	2.9	104	2.2	0.30[a]	Peach
Methiocarb	102	2.3	104	1.2	100	1.4	117	4.3	0.05[a]	Cabbage
Malathion	90	2.6	99	3.0	94	2.6	107	3.2	0.50[b]	Apple
Pirimiphos-methyl	68	0.8	58	0.3	58	1.0	71	1.4	0.05[b]	Apple
Etrimfos	93	1.9	99	2.5	98	1.0	114	2.2	0.05[a]	Cauliflower
Pyraclofos	81	1.9	94	1.2	96	1.8	112	4.2	0.05[a]	Potato
Phosalone	110	2.3	111	3.6	86	3.4	119	3.3	2.00[b]	Apple

recovery. These effects demonstrate a differential matrix affinity for pesticides as suppression and enhancement for one specific combination of pesticide and matrix. On the whole, the recoveries and RSDs were not influenced adversely by the kind of sample, and the method could serve as a quantitative method to identify and determine the pesticides in vegetables and fruits with reliable results at MRLs.

18.5.3 Applied to Real Samples

About 25 representative samples were collected from local markets, including root vegetables (carrot and potato), leafy vegetables (lettuce, cabbage, and spinach), bulb vegetables (onion, pumpkin-squash, and eggplant), fruit vegetables (cucumber and tomato), bean vegetables (kidney bean legume), and pome fruits (apple, melon, and peach). In these fruits and vegetables, the pesticides studied were monitored by Japan or EU, which established the corresponding MRLs as shown in Table 18.7. Table 18.8 illustrates the distribution and concentration of the main pesticide residues determined by LC–MS in the real samples. Some pesticides were also detected in one or two samples at different concentration levels (e.g., 12.66 mg kg^{-1} methidathion in one potato). Other pesticides (carbaryl, baycarb, methiocarb, pyraclofos, and etrimfos) were not found in any samples. The concentrations found in the samples except for two peaches were always lower than MRLs (see Table 18.7). Representative chromatograms of actual market apple, cabbage, and peach samples are shown in Figure 18.6. On the whole, 70% of the samples contained one or more pesticide residues. More than 30% of the samples contained multiresidues. In the worst case, there were 10 different pesticides found in a potato sample, but all concentrations found were below MRLs. Ostensibly, root and leafy vegetables are more susceptible to contamination than other samples compared with the soil food sources mentioned here . Azinphos-methyl and malathion widely existed in almost all types of fruits and vegetables, which indicated that these two pesticides are often used. Other pesticides existed in only one or two types of fruits and vegetables, which may be explained by the specific use of pesticides. For example, aldicarb residues surpassed the MRLs in two peaches, while it was not found in other samples. The reason may lie in the fact that aldicarb was the most frequently used pesticide for peach orchards (above-ground crop dusting or spraying).

TABLE 18.8
Distribution and Amount of Main Pesticide Residues in All Kinds of Fruits and Vegetables

Residues	Root (%)	Leafy (%)	Fruit (%)	Pome (%)	Bulb (%)	Total (%)	Amount (mg kg^{-1})
Azinphos-methyl	60	60	20	20	20	36	6.02–73.9
Malathion	40	16	—	20	20	36	2.13–223
Phosalone	60	—	—	—	—	28	0.52–88.6
Pirimiphos-methyl	40	—	20	—	—	12	0.54–6.54
Isoprocarb	—	60	—	—	—	12	9.87–24.3
Aldicarb	—	—	—	40	—	8	130–173
Methiocarb-sulfone	20	—	—	—	—	4	23.5

"—" means no residues.

Abbreviations

GC–EI-MS	gas chromatography–electron ionization–mass spectrometry
GC–PCI–MS–MS	gas chromatography–positive-ion chemical ionization–multiple-stage mass spectrometry
GC–NCI–MS	gas chromatography–negative-ion chemical ionization–mass spectrometry

LV–DMI–GC–TOF-MS	large volume-difficult matrix introduction–gas chromatography– time-of-flight mass spectrometry
GPC–GC–EI-MS	gel permeation chromatography–gas chromatograph–electron ionization–mass spectrometry
GC × GC–TOF-MS	two-dimensional gas chromatography–time-of-flight mass spectrometry
Thermal desorption-PTV–GC–EI-MS	thermal desorption-programmed temperature vaporization–gas chromatography–electron ionization–mass spectrometry
LC–ESI–MS	liquid chromatography–electrospray ionization–mass spectrometry
LC–IT–MS–MS	liquid chromatography–ion trap–multiple-stage mass spectrometry
LC–APCI-MS–MS	liquid chromatography–atmospheric pressure chemical ionization– multiple-stage mass spectrometry
LC–API–MS–MS	liquid chromatography–atmospheric pressure ionization– multiple-stage mass spectrometry
LC–TOF-MS	liquid chromatography–time-of-flight mass spectrometry
LC–QTOF-MS	liquid chromatography–quadrupole time-of-flight-mass spectrometry

REFERENCES

1. A Sanusi, V Guillet, and M Montury. Advanced method using microwaves and solid-phase microextraction coupled with gas chromatography–mass spectrometry for the determination of pyrethroid residues in strawberries. *J Chromatogr A* 1046: 35–40, 2004.
2. H Obana, K Akutsu, M Okihashi, and S Hori. Multiresidue analysis of pesticides in vegetables and fruits using two-layered column with graphitized carbon and water absorbent polymer. *Analyst* 126: 1529–1534, 2001.
3. A Agüera, M Contreras, J Crespob, and AR Fernández-Alba. Multiresidue method for the analysis of multiclass pesticides in agricultural products by gas chromatography–tandem mass spectrometry. *Analyst* 127: 347–354, 2002.
4. K Patel, RJ Fussell, DM Goodall, and BJ Keely. Analysis of pesticide residues in lettuce by large volume–difficult matrix introduction–gas chromatography–time of flight–mass spectrometry (LV-DMI-GC-TOF-MS). *Analyst* 128: 1228–1231, 2003.
5. E Soboleva, K Ahadb, and A Ambrusa. Applicability of some mass spectrometric criteria for the confirmation of pesticide residues. *Analyst* 129: 1123–1129, 2004.
6. JLM Vidala, FJ Arrebolaa, and M Mateu-Sa'nchez. Application of gas chromatography–tandem mass spectrometry to the analysis of pesticides in fruits and vegetables. *J Chromatogr A* 959: 203–213, 2002.
7. FJ Arrebola, JLM Vidal, M Mateu-Sánchez, and FJ Álvarez-Castellón. Determination of 81 multiclass pesticides in fresh foodstuffs by a single injection analysis using gas chromatography–chemical ionization and electron ionization tandem mass spectrometry. *Anal Chim Acta* 484: 167–180, 2003.
8. LB Liu, Y Hashi, YP Qin, HX Zhou, and JM Lin. Development of automated online gel permeation chromatography–gas chromatograph mass spectrometry for measuring multiresidual pesticides in agricultural products. *J Chromatogr B* 845: 61–68, 2007.
9. JLF Moreno, FJA Li'ebanas, AG Frenich, and JLM Vidal. Evaluation of different sample treatments for determining pesticide residues in fat vegetable matrices like avocado by low-pressure gas chromatography–tandem mass spectrometry. *J Chromatogr A* 1111: 97–105, 2006.
10. J Zrostl'ıková, J Hajšlová, and T Cajka. Evaluation of two-dimensional gas chromatography–time-of-flight mass spectrometry for the determination of multiple pesticide residues in fruit. *J Chromatogr A* 1019: 173–186, 2003.
11. S Walorczyk and B Gnusowski. Fast and sensitive determination of pesticide residues in vegetables using low-pressure gas chromatography with a triple quadrupole mass spectrometer. *J Chromatogr A* 1128: 236–243, 2006.
12. T Cajka and J Hajslova. Gas chromatography–high-resolution time-of-flight mass spectrometry in pesticide residue analysis: Advantages and limitations. *J Chromatogr A* 1058: 251–261, 2004.

13. DA Lambropoulou and TA Albanis. Headspace solid-phase microextraction in combination with gas chromatography–mass spectrometry for the rapid screening of organophosphorus insecticide residues in strawberries and cherries. *J Chromatogr A* 993: 197–203, 2003.

14. EM Kristenson, EGJ Haverkate, CJ Slooten, L Ramos, RJJ Vreuls, and UT Brinkman. Miniaturized automated matrix solid-phase dispersion extraction of pesticides in fruit followed by gas chromatographic–mass spectrometric analysis. *J Chromatogr A* 917: 277–286, 2001.

15. S Song, X Ma, and C Li. Multi-residue determination method of pesticides in leek by gel permeation chromatography and solid phase extraction followed by gas chromatography with mass spectrometric detector. *Food Control* 18: 448–453, 2007.

16. D Štajnbaher and L Zupancic-Kralj. Multiresidue method for determination of 90 pesticides in fresh fruits and vegetables using solid-phase extraction and gas chromatography–mass spectrometry. *J Chromatogr A* 1015: 185–198, 2003.

17. JLM Vidal, IJ Arrebola, and M Mateu-Sdnchez. Multi-residue method for determination of pesticides in vegetable samples by GC–MS–MS. *Chromatographia* 56: 475–481, 2002.

18. P Sandra, B Tienpontb, and F Davida. Multi-residue screening of pesticides in vegetables, fruits and baby food by stir bar sorptive extraction–thermal desorption–capillary gas chromatography–mass spectrometry. *J Chromatogr A* 1000: 299–309, 2003.

19. AG Frenich, MJ Gonzalez-Rodriguez, FJ Arrebola, and JLM Vidal. Potentiality of gas chromatography-triple quadrupole mass spectrometry in vanguard and rearguard methods of pesticide residues in vegetables. *Anal Chem* 77: 4640–4648, 2005.

20. FJ Arrebola, JLM Vidal, MJ Gonzalez-Rodrıgueza, A Garrido-Frenich, and NS Morito. Reduction of analysis time in gas chromatography: Application of low-pressure gas chromatography–tandem mass spectrometry to the determination of pesticide residues in vegetables. *J Chromatogr A* 1005: 131–141, 2003.

21. M Kirchnera, E Matisova, R Otrekala, A Hercegova, and J Zeeuwb. Search on ruggedness of fast gas chromatography–mass spectrometry in pesticide residues analysis. *J Chromatogr A* 1084: 63–70, 2005.

22. S Koning, G Lachb, M Linkerhagnerb, R Loschera, PH Tablack, and UAT Brinkman. Trace-level determination of pesticides in food using difficult matrix introduction–gas chromatography–time-of-flight mass spectrometry. *J Chromatogr A* 1008: 247–252, 2003.

23. Y Pico, C Blasco, and G Font. Environmental and food applications of LC–tandem mass spectrometry in pesticide-residue analysis: An overview. *Mass Spectrom Rev* 23: 45–85, 2004.

24. I Ferrer, JF Garcia-Reyes, and A Fernandez-Alba. Identification and quantitation of pesticides in vegetables by liquid chromatography time-of-flight mass spectrometry. *Trends Anal Chem* 24: 671–682, 2005.

25. C Soler and Y Pico. Recent trends in liquid chromatography–tandem mass spectrometry to determine pesticides and their metabolites in food. *Trends Anal Chem* 26: 103–115, 2007.

26. S Lacorte and AR Fernandez-Alba. Time of flight mass spectrometry applied to the liquid chromatographic analysis of pesticides in water and food. *Mass Spectrom Rev* 25: 866–880, 2006.

27. Y Pico, G Font, JC Molto, and J Manes. Pesticide residue determination in fruit and vegetables by liquid chromatography–mass spectrometry. *J Chromatogr A* 882: 153–173, 2000.

28. S Bogialli, R Curini, AD Corcia, M Nazzari, and D Tamburro. A simple and rapid assay for analyzing residues of carbamate insecticides in vegetables and fruits: Hot water extraction followed by liquid chromatography–mass spectrometry. *J Agric Food Chem* 52: 665–671, 2004.

29. GL Hall, J Engebretson, MJ Hengel, and T Shibamoto. Analysis of methoxyfenozide residues in fruits, vegetables, and mint by liquid chromatography–tandem mass spectrometry (LC–MS/MS). *J Agric Food Chem* 52: 672–676, 2004.

30. C Blasco, G Font, and Y Pico. Analysis of pesticides in fruits by pressurized liquid extraction and liquid chromatography–ion trap–triple stage mass spectrometry. *J Chromatogr A* 1098: 37–43, 2005.

31. S Grimalt, JV Sancho, OSJ Pozo, JM GarciA-Baudin, ML Fernandez-Cruz, and FL Hernandez. Analytical study of trichlorfon residues in kaki fruit and cauliflower samples by liquid chromatography–electrospray tandem mass spectrometry. *J Agric Food Chem* 54: 1188–1195, 2006.

32. R Romero-Gonzalez, E Pastor-Montoro, JL Martinez-Vidal, and A Garrido-Frenich. Application of hollow fiber supported liquid membrane extraction to the simultaneous determination of pesticide residues in vegetables by liquid chromatography/mass spectrometry. *Rapid Commun Mass Spectrom* 20: 2701–2708, 2006.

33. A Sannino, L Bolzoni, and M Bandini. Application of liquid chromatography with electrospray tandem mass spectrometry to the determination of a new generation of pesticides in processed fruits and vegetables. *J Chromatogr A* 1036: 161–169, 2004.

34. AD Muccio, P Fidente, DA Barbini, R Dommarcoa, S Seccia, and P Morrica. Application of solid-phase extraction and liquid chromatography–mass spectrometry to the determination of neonicotinoid pesticide residues in fruit and vegetables. *J Chromatogr A* 1108: 1–6, 2006.

35. C Soler, KJ James, and Y Pico. Capabilities of different liquid chromatography tandem mass spectrometry systems in determining pesticide residues in food: Application to estimate their daily intake. *J Chromatogr A* 1157: 73–84, 2007.

36. C Soler, J Manes, and Y Pico. Comparison of liquid chromatography using triple quadrupole and quadrupole ion trap mass analyzers to determine pesticide residues in oranges. *J Chromatogr A* 1067: 115–125, 2005.

37. M Hiemstra and A Kok. Comprehensive multi-residue method for the target analysis of pesticides in crops using liquid chromatography–tandem mass spectrometry. *J Chromatogr A* 1154: 3–25, 2007.

38. OJ Pozo, JM Marin, JV Sancho, and F Hernandez. Determination of abamectin and azadirachtin residues in orange samples by liquid chromatography–electrospray tandem mass spectrometry. *J Chromatogr A* 992: 133–140, 2003.

39. M Takino, K Yamaguchi, and T Nakahara. Determination of carbamate pesticide residues in vegetables and fruits by liquid chromatography–atmospheric pressure photoionization–mass spectrometry and atmospheric pressure chemical ionization–mass spectrometry. *J Agric Food Chem* 52: 727–735, 2004.

40. M Fernandez, Y Pico, and J Manes. Determination of carbamate residues in fruits and vegetables by matrix solid-phase dispersion and liquid chromatography–mass spectrometry. *J Chromatogr A* 871: 43–56, 2000.

41. H Obana, M Okihashi, K Akutsu, Y Kitagawa, and Shinjiro Hori. Determination of neonicotinoid pesticide residues in vegetables and fruits with solid phase extraction and liquid chromatography mass spectrometry. *J Agric Food Chem* 51: 2501–2505, 2003.

42. HGJ Mol, RCJ Dam, and OM Steijger. Determination of polar organophosphorus pesticides in vegetables and fruits using liquid chromatography with tandem mass spectrometry: Selection of extraction solvent. *J Chromatogr A* 1015: 119–127, 2003.

43. A Sannino and M Bandini. Determination of seven benzoylphenylurea insecticides in processed fruit and vegetables using high-performance liquid chromatography/tandem mass spectrometry. *Rapid Commun Mass Spectrom* 19: 2729–2733, 2005.

44. T Chen and G Chen. Identification and quantitation of pyrethroid pesticide residues in vegetables by solid-phase extraction and liquid chromatography/electrospray ionization ion trap mass spectrometry. *Rapid Commun Mass Spectrom* 21: 1848–1854, 2007.

45. C Blasco, G Font, and Y Picó. Multiple-stage mass spectrometric analysis of six pesticides in oranges by liquid chromatography–atmospheric pressure chemical ionization–ion trap mass spectrometry. *J Chromatogr A* 1043: 231–238, 2004.

46. D Ortelli, P Edder, and C Corvi. Multiresidue analysis of 74 pesticides in fruits and vegetables by liquid chromatography–electrospray–tandem mass spectrometry. *Anal Chim Acta* 520: 33–45, 2004.

47. F Hernandez, OJ Pozo, JV Sancho, L Bijlsma, M Barreda, and E Pitarch. Multiresidue liquid chromatography tandem mass spectrometry determination of 52 non gas chromatography-amenable pesticides and metabolites in different food commodities. *J Chromatogr A* 1109: 242–252, 2006.

48. MJ Taylor, K Hunter, KB Hunter, D Lindsay, and SL Bouhellec. Multi-residue method for rapid screening and confirmation of pesticides in crude extracts of fruits and vegetables using isocratic liquid chromatography with electrospray tandem mass spectrometry. *J Chromatogr A* 982: 225–236, 2002.

49. I Ferrer, JF Garcia-Reyes, M Mezcua, EM Thurman, and AR Fernandez-Alba. Multi-residue pesticide analysis in fruits and vegetables by liquid chromatography–time-of-flight mass spectrometry. *J Chromatogr A* 1082: 81–90, 2005.

50. JLM Vidal, AG Frenich, TL Lopez, IM Salvador, LH Hassani, and MH Benajiba. Selection of a representative matrix for calibration in multianalyte determination of pesticides in vegetables by liquid chromatography–electrospray tandem mass spectrometry. *Chromatographia* 61: 127–137, 2005.

51. M Liu, Y Hashi, Y Song, and JM Lin. Simultaneous determination of carbamate and organophosphorus pesticides in fruits and vegetables by liquid chromatography–mass spectrometry. *J Chromatogr A* 1097: 183–187, 2005.

52. T Goto, Y Ito, S Yamadaa, H Matsumoto, H Oka, and H Nagase. The high throughput analysis of N-methyl carbamate pesticides in fruits and vegetables by liquid chromatography electrospray ionization tandem mass spectrometry using a short column. *Anal Chim Acta* 555: 225–232, 2006.

53. CC Leandro, P Hancock, RJ Fussell, and BJ Keely. Ultra-performance liquid chromatography for the determination of pesticide residues in foods by tandem quadrupole mass spectrometry with polarity switching. *J Chromatogr A* 1144: 161–169, 2007.
54. S Grimalt, OJ Pozo, J V Sancho, and F Hernandez. Use of liquid chromatography coupled to quadrupole time-of-flight mass spectrometry to investigate pesticide residues in fruits. *Anal Chem* 79: 2833–2843, 2007.
55. M Anastassiades, SJ Lehotay, D Stajnbaher, and FJ Schenck. Fast and easy multiresidue method employing acetonitrile extraction/partitioning and "dispersive solid-phase extraction" for the determination of pesticide residues in produce. *J AOAC Int* 86: 412–431, 2003.
56. SJ Lehotay, AD Kok, M Hiemstra, and PV Bodegraven. Validation of a fast and easy method for the determination of residues from 229 pesticides in fruits and vegetables using gas and liquid chromatography and mass spectrometric detection. *J AOAC Int* 88: 595–614, 2005.
57. SJ Lehotay, K Mastovska, and SJ Yun. Evaluation of two fast and easy methods for pesticide residue analysis in fatty food matrixes. *J AOAC Int* 88: 630–638, 2005.
58. SJ Lehotay, K Mastovska, and AR Lightfield. Use of buffering and other means to improve results of problematic pesticides in a fast and easy method for residue analysis of fruits and vegetables. *J AOAC Int* 88: 615–629, 2005.
59. A Posyniak, J Zmudzki, and K Mitrowska. Dispersive solid-phase extraction for the determination of sulfonamides in chicken muscle by liquid chromatography. *J Chromatogr A* 1087: 259–264, 2005.
60. C Jansson, T Pihlström, BG Österdahl, and KE Markides. A new multi-residue method for analysis of pesticide residues in fruit and vegetables using liquid chromatography with tandem mass spectrometric detection. *J Chromatogr A* 1023: 93–104, 2004.

19 Pesticides in Fish and Wildlife

Leo M. L. Nollet

CONTENTS

19.1 INTRODUCTION

Pesticides are analyzed and quantified in fish and wildlife for various reasons. Such reasons may be monitoring of the wildlife and the environment, risk assessment, effects of pesticides on fauna, and estimation of pesticides use in the surrounding environment of the animals.

Effects of historical use of pesticides or continuing application of pesticides can be monitored by analyzing samples of fish and wildlife and their surrounding environment. A great number of pesticides have endocrine disrupting properties on fish and other wildlife. So, it is necessary to know the real concentrations of pesticides in them.

The consumption of fish and wildlife by other animals and humans also necessitates the estimation of pesticides concentration in them.

19.2 ANALYSIS

The analysis of pesticides in fish and wildlife has the same logic as in other matrices. The analysis starts with an extraction and is followed by a cleanup and separation step. If necessary, the analytes, in this case pesticides, are preconcentrated or enriched. The final steps consist of separation and detection.

The great majority of papers on pesticides in fish and wildlife deal with organochlorines (OCs). In almost all discussed methods, OC pesticides are determined by gas chromatography (GC) coupled with an electron capture detector (ECD). In some applications, the detection is by mass spectrometry (MS).

For further details on analytical methodology of pesticides, the reader is directed to Chapter 20.

19.3 DIFFERENT WILDLIFE MATRICES

Ten organochlorine pesticides in different biological matrices were analyzed by Volz and Johnston [1]. Sample preparation is done by acetonitrile extraction, followed by a solid-phase extraction cleanup using C18-Florisil cartridges in tandem. The pesticides were quantified by GC with an ECD.

The limit of detection for this method ranged from 1.1 to 2.6 ng/g. The mean recovery and standard deviation for the 10 pesticides in fortified deer muscle was 94.7% ± 7.9%. Recoveries for individual analytes ranged from 83.6% to 105%. This method was further applied successfully to quantify these organochlorine analytes in insects, bird eggs, calf liver, beef brain, boar, deer, elk, alligator, mussels, oysters, clams, crab, mahi-mahi, and tobacco.

The same authors [2] developed a gas chromatographic method for the analysis of 10 organochlorine pesticides in 0.5 mL of wildlife whole blood. Sample preparation involved an ethyl ether and hexane extraction, followed by a silica solid-phase extraction cleanup. The pesticides were quantified by GC/electron capture detection. The limit of detection for this method ranged from 1.1 to 5.2 µg/L. The mean and standard deviation for the recovery of 10 pesticides was 97.9% ± 5.5%.

A gas chromatographic method for the analysis of nine organochlorine pesticides in wildlife urine was developed by Petty et al. [3]. Reversed-phase solid-phase extraction was utilized to extract the organochlorine pesticides. The pesticides were recovered by elution with hexane–ethyl ether (1:1) and separated by GC with electron capture detection. Method detection limits (MDLs) ranged from 1.4 to 2.7 µg/L. Mean recoveries for all pesticides were 90.6%.

Soil sediments from two depths and tissue samples from eight species of aquatic animals were collected on or near Yazoo National Wildlife Refuge, Mississippi, and analyzed for organochlorine pesticide residues [4]. Residues of 12 OCs were found in most animal samples, and 0.0–4.6 mg/kg of five compounds were detected in soil sediments. With the exception of mosquitofish (*Gambusia affinis*), residues were similarly distributed in soil and animal samples among different watercourses within the watershed.

Biomagnification of organochlorine pesticide residues was evident from soil sediments to mosquitofish, a lower secondary consumer and forage fish, to spotted gar (*Lepisosteus oculatus*), a tertiary consumer. Residues in larger secondary consumers such as carp (*Cyprinus carpio*) and smallmouth buffalo (*Ictiobus bubalus*) and tertiary consumers such as water snakes (*Nerodia* spp.) and cottonmouths (*Agkistrodon piscivorus*) demonstrated no clear patterns of accumulation.

Sediment and biota samples were collected from Msimbazi and Kizinga rivers and from the coastal marine environment of Dar es Salaam (Tanzania) during both dry and wet seasons [5]. The samples were analyzed for various organochlorine pesticide residues using GC–ECD and GC–MS. Dieldrin, *p,p'*-DDT (dichlorodiphenyltrichloroethane), *p,p'*-DDE (dichlorodiphenyldichloroethylene), *p, p'*-DDD (dichlorodiphenyldichloroethane), *o,p'*-DDT, and γ-HCH (hexachlorocyclohexane) were detected at significantly higher concentrations than the MDLs. Recoveries of pesticide residues ranged from 86.5% to 120% in sediments and from 62% to 102% in biota. The average concentrations of total DDT in sediments for the two seasons were almost the same. Biota samples showed significant difference in levels of residues depending on the mode of feeding and age of analyzed biota. *p,p'*-DDT to total DDT ratios in all matrices indicated its recent use.

Sediment, mussel, and seawater samples were collected thrice during 2001–2003 at nine sampling stations along the mid-Black Sea coast of Turkey [6]. The samples were analyzed with GC-ECD for the contents of various OCs. DDT and its metabolites were detected at concentrations significantly above the detection limits. The highest concentrations of DDT metabolites measured in the sediment and mussel samples were 35.9 and 14.0 ng/g wet weight, respectively. Considerable levels of aldrin, dieldrin, endrin, heptachlor epoxide, lindane, endosulfan sulfate, and hexachlorobenzene (HCB) were also detected in the sediment, mussel, or seawater samples.

A GC/ion trap MS method was developed for the analysis of OCs in coral samples, which were extracted with accelerated solvent extraction (ASE) and cleaned up on a sulfuric acid-modified silica gel column [7]. The optimal ASE conditions were found to be 100°C and 2000 psi, with a mixture of acetone and methylene chloride (1:1, v/v). The pesticides included hexachlorocyclohexane (HCHs) isomers, specifically, α-, β-, γ-, and δ-HCH isomers, heptachlor, and HCB, *o,p'*-, *p,p'*-DDT, *o,p'*-, *p,p'*-DDE, and *o,p'*-, *p,p'*-DDD. Average recoveries of OCs ranged from 82% to 102%, with relative standard deviations of 3%–6%, at a level of 10 ng/g and from 50% to 68%, with relative standard deviations of 13%–19% at a level of 2 ng/g. The developed method was applied for the analysis of OCs in coral samples collected from Tern Island and Bikini Atoll in the Pacific Ocean. The concentrations of HCB were 7–26 pg/g dry weight in the samples from Bikini Atoll and 3–45 pg/g in those from Tern Island, and heptachlor concentrations were 208–2200 and 44–104 pg/g in the coral samples from Bikini Atoll and Tern Island, respectively. Σ HCH (sum of α-, β-, γ-, and δ-HCH) were 8–82 pg/g in Bikini Atoll coral and 86–629 pg/g in Tern island coral, and Σ DDT (sum of o,p'-, p,p'-DDD, o,p'-, p,p'-DDE, and o,p'-, p,p'-DDT) were 80–212 pg/g in Bikini Atoll coral and 593–3165 pg/g in Tern Island coral.

Moths were collected with a light trap from 15 sites in the Baltimore, Maryland–Washington, DC area and analyzed by GC–ECD for OC residues [8]. On an average, sampled moths contained 0.33 ppm heptachlor-chlordane compounds, 0.25 ppm DDE, and 0.11 ppm dieldrin. There were large differences in the concentrations detected in different species. Concentrations were especially high in moths whose larvae were cutworms, and were virtually absent from moths whose larvae fed on tree leaves.

Tissue samples from 56 bird and 11 mammal species of different trophic levels, collected from 1994 to 1995 from the Urbino–Pesaro area in the Marche region of central Italy, were analyzed by Alleva et al. [9] for the presence of organochlorine compounds (polychlorinated biphenyls – PCBs and p,p'-DDE) and heavy metals. Results revealed interspecies differences in pollutant residue concentrations. Polychlorinated biphenyls and p,p'-DDE were found in all bird and mammal species analyzed (bird- or fish-eating birds), and insectivorous mammals showed the highest level of these contaminants.

A solid-phase extraction method with an Oasis® hydrophilic–lipophilic-balanced cartridge was developed using 8 M urea to desorb and extract OCs and PCBs from avian serum for analysis by capillary GC with electron capture detection [10]. Recoveries for OCs ranged from 75% to 101%.

BHC and α, β, γ, and δ isomers of HCH were determined in 16 samples of surface and ground waters and mussels in the middle Black Sea region, Turkey, and the concentrations of PCB and OCs were analyzed in the eastern Aegean Sea water and fish samples [11]. Thirteen OC pesticides were determined in water, sediment, fish, and water birds in the Göksu delta. Thirteen OC pesticides were analyzed in the Sarçyar Dam Lake, in Sakarya basin, five lakes in central Anatolia, and the Meriç Delta in water sediment and fish samples. Some of these pesticides were found in almost all of the samples.

19.4 MAMMALS

PCBs, HCHs, DDT compounds (DDTs), and hexachlorobenzene (HCB) were measured in eight species of terrestrial mammals and 10 species of birds inhabiting Chubu region, Japan. In view of feeding habits, the contamination levels of OCs were found to be higher in omnivorous mammals than in herbivorous ones, and in fish-eating ones and raptores than in omnivorous birds [12].

One-fourth of the estimated population (about 100) of ocelots (*Felix pardalis*) in the Lower Rio Grande Valley in Texas, USA, was sampled to evaluate the impacts of chlorinated pesticides, PCBs, and trace elements on the population [13]. Hair was collected from 32 ocelots trapped between 1986 and 1992, and blood was collected from 20 ocelots trapped between 1993 and 1997. Few blood samples were obtained from individuals recaptured two or three times. Tissue samples from four road-killed ocelots were also analyzed. DDE, PCBs, and Hg were some of the most common contaminants detected in hair and blood. Mean DDE concentrations in plasma ranged from 0.005 µg/g wet weight to 0.153 mg/g wet weight. Concentrations of DDE did not increase significantly with age, although the highest concentrations of DDE were found in older animals. Overall, the concentrations of DDE were low and have been found at levels that currently do not pose any threat to health or survival of the ocelot.

Kayser et al. [14] were looking for reasons why the common hamster is declining. This may be due to ingestion of some persistent organochlorines.

Σ DDT or total DDT concentrations in the whole blood of pups of Steller sea lions (*Eumetopias jubatus*) from western Alaska ranged from 0.18 to 11 ng/g wet weight with a mean of 1.6 ± 0.23 ng/g wet weight [15]. In Russia, Σ DDT in the whole blood of pups ranged from undetectable to 26 ng/g wet weight with a mean of 3.3 ± 0.36 ng/g wet weight. Average OC concentrations were significantly higher in the blood of Russian animals when compared with western Alaska, and in both areas females had higher concentrations than males. Male pups from western Alaska had significantly lower levels of Σ DDT when compared with male pups from Russia. These data indicate that Steller sea lion pups have measurable concentrations of these synthetic chemicals. The analysis method was a combination of HPLC with Cosmosil PYE columns and PDA (photo diode array) [15].

Organochlorine pesticides were analyzed in captive giant and red panda tissues from China. The total concentrations of OCs in tissues ranged from 16.3 to 888 ng/g lipid weight. *p,p'*-DDE and β-HCH were the major OC contaminants [16].

Levels of PCBs, OCs, and polybrominated diphenyl ethers (PBDEs) in the cerebral cortex of river otters (*Lontra canadensis*) trapped from Ontario and Nova Scotia between 2002 and 2004 were measured. The mean concentration of total OCs was 21.2 ± 3.7 ng/g lipid weight, and hexachlorobenzene (32.6% of total) and DDE (28.1%) accounted for the majority [17]. The method used was capillary GC–MS in SIM (single ion monitoring) mode.

During summer, a grizzly bear (*Ursus arctos horribilis*) in the Greater Yellowstone Ecosystem (USA) can excavate and consume millions of army cutworm moths (*Euxoa auxiliaris*) (ACMs) [18]. ACMs are agricultural pests and may contain pesticides being possibly toxic to bears. This study investigated if ACMs contain and transport pesticides to bear foraging sites and, if so, if these levels could be toxic to bears.

ACMs were screened for 32 pesticides with GC with electron capture detection. Later, some ACMs were analyzed with GC–MS/MS. On both occasions, ACMs contained trace or undetectable levels of pesticides.

Blubber was analyzed for a wide range of contaminants from five subadult and eight adult male ringed seals sampled in 2004, namely, for PCBs, hexachlorobenzene, toxaphenes, chlordanes, DDE, and PBDEs [19]. Contaminant levels were compared with previously sampled animals from the same area, as well as to data from literature for other arctic wildlife species from a wide variety of locations. Ringed seals sampled in 2004 showed 50%–90% lower levels of PCBs and chlorinated pesticides when compared with animals sampled in 1996 of similar age (14 subadults and 7 adult males), indicating that the decline of chlorinated contaminants observed during the 1990s in a variety of arctic wildlife species is continuing into the 21st century.

Kidney, liver, and bone samples were taken from 19 wolves (*Canis lupus*) collected from two locations in the Yukon Territory, Canada [20]. Liver samples pooled by age and sex were analyzed for 22 organochlorine pesticides. Whereas most organochlorines were not present at detectable levels in wolf liver, some chlorobenzenes and dieldrin were present at low levels.

Naccari et al. [21] investigated the levels of contamination by OCs and PCBs in some organs and tissues of wild boars, utilized as biological indicators, from various areas of Calabria, Italy. Quantitative determinations of organochlorines were carried out using GC–ECD and confirmed with GC–MS in 154 samples from wild boars (heart, liver, lung, kidney, muscle tissue, and spleen) during the hunting season from 2000 to 2002. Low residual levels of DDE were found in eight samples and of DDT in four samples.

For the first time, Kannan et al. [22] studied concentrations of OCs, PCBs, and PBDEs in tissues of Irrawaddy dolphins collected from Chilika Lake, India, to understand the status of contamination. DDT and its metabolites were the predominant contaminants found in Irrawaddy dolphins; the highest concentration found was 10,000 ng/g lipid weight in blubber. HCHs were the second most prevalent contaminants in dolphin tissues. Concentrations of PCBs, chlordanes, hexachlorobenzene, *tris*(4-chlorophenyl)methane, and *tris*(4-chlorophenyl)methanol were in the ranges of few ng/g to few hundreds ng/g on a lipid-weight basis.

Rocky Mountain Arsenal (RMA) National Wildlife Area, near Denver, Colorado, USA, is a Superfund site contaminated by past military and industrial uses, including pesticide manufacturing [23]. Bats captured while foraging at RMA had measurable quantities of dieldrin and DDE in masticated insect samples from stomach contents and significantly higher concentrations of dieldrin, DDE, DDT, and mercury (juveniles) in carcasses than big brown bats ($n = 26$) sampled at a reference area (RA) 80 km to the north. Concentrations of dieldrin and DDE in brains of bats captured while foraging at RMA were also greater than those in bats from the reference area, but not high enough to suggest mortality. Maximum concentrations of DDE, DDT, and cyclodienes in brains of big brown bats were found in adult males from RMA. Guano from the two closest known roosts had significantly higher concentrations of dieldrin, DDE, and mercury than guano from two roosts at the reference area. Dieldrin concentrations in carcasses of bats from RMA were highest in juveniles, followed by adult males and adult females. DDE concentrations in carcasses were lowest in adult females at both sites and highest in adult males at RMA.

Indo-Pacific humpback dolphins (*Sousa chinensis*) may accumulate significant amounts of various pollutants, PCBs, OCs, PAHs, and petroleum hydrocarbons (PHCs) [24]. Blubber samples were collected from five free-ranging living Indo-Pacific humpback dolphins from Honk Hong, four stranded dolphins from Xiamen, Fujian Province, and one stranded specimen from Zhuhai, Guangdong Province, China. Organochlorines were quantified by GC–µECD with a DB-5 capillary column. HCHs (α, β, γ, and δ isomers), HCB, heptachlor, heptachlor epoxide, aldrin, dieldrin, endrin, kepone, chlordanes (CHLs—*cis*-chlordane, *trans*-chlordane, *cis*-nonachlor, *trans*-nonachlor, and oxychlordane) and DDT (sum of p,p'-DDT, o,p-DDT, p,p'-DDD, o,p-DDD, p,p'-DDE, and o,p-DDE) were measured. Recoveries were 95% ± 7.5% for DDTs, 100% ± 4.7% for CHLs, and 94% ± 5.9% for HCB. For total chlordanes, male dolphins had concentrations ranging from 276.1 to 488.7 ng/g, and concentrations of the two female specimens were 108.0 and 105.8 ng/g.

19.5 FISH

Levels of mercury and selected pesticides were determined in the muscle tissue of fish obtained from different regions in the state of Utah [25]. Only small amounts of pesticides were found in fish. A majority of the tissues contained appreciable levels of PCBs.

Bairdiella icistia (bairdiella), *Cynoscion xanthulus* (orangemouth corvina), and *Oreochromis* spp. (tilapia) were sampled from two river mouths and two near shore areas of the Salton Sea California, USA [26]. Muscle tissues were analyzed by GC–ECD for a complete suite of 14 trace metals and 53 pesticides. p,p'-DDE accounted for 94% of the total DDT metabolites. Total DDTs ranged between 17.1 and 239.0.

Fish were collected in late-1995 from 34 National Contaminant Biomonitoring Program (NCBP) stations and 13 National Water Quality Assessment Program (NAWQA) stations in the Mississippi River Basin (MRB) and in late-1996 from a reference site in West Virginia (USA) [27]. Four composite samples, each comprising (nominally) 10 adult common carp (*C. carpio*) or black bass (*Micropterus* spp.) of the same sex, were collected from each site and analyzed for organochlorine chemical residues by GC with electron capture detection. At the NCBP stations, which are located on relatively large rivers, concentrations of organochlorine chemical residues were generally lower than when previously sampled in the mid-1980s. Residues derived from DDT (primarily p,p'-DDE) were detected at all sites (including the reference site); however, only traces (≤ 0.02 µg/g) of the parent insecticide (p,p'-DDT) were present, which indicated continued weathering of residual DDT from previous use. Nevertheless, concentrations of DDT (as p,p'-DDE) in fish from the cotton-farming regions of the lower MRB were great enough to constitute a hazard to fish-eating wildlife and were especially high at the NAWQA sites on the lower-order rivers and streams of the Mississippi embayment. Mirex was detected at only two sites, both in Louisiana, and toxaphene was found exclusively in the lower MRB. Most cyclodiene pesticides (dieldrin, chlordane, and heptachlor epoxide) were more widespread, but concentrations were lower than in the 1980s except at a site on the Mississippi River near Memphis, TN. Concentrations were also somewhat elevated at sites in the Corn Belt. Endrin was detected exclusively at the Memphis site.

Liver concentrations of 30 OC pesticides/pesticide metabolites and total PCBs were measured and compared with *Dasyatis sabina* collected from four central Florida lakes of the St. Johns River: Lake George, Lake Harney, Lake Jesup, and Lake Monroe (USA) [28].

In 1996, OCs were measured in flooded soils and in black crappie, and in brown bullhead catfish and largemouth bass from different sites in central Florida, USA [29]. Concentrations of total residual OCs found in the flooded soils included dieldrin (385 ± 241 μg/kg), sum of DDT, DDD, and DDE (7,173 ± 1,710 μg/kg), and toxaphene (39,444 ± 11,284 μg/kg). Sum of chlordane residuals reached 1,766 ± 1,037 μg/kg. OCs in muscle tissue were below the US Food and Drug Administration action limits for human consumption. For three-year-old bass mean concentrations of chlordane residuals, DDT residuals, and dieldrin were 15–17 times higher in ovary tissue and 76–80 times higher in fat tissue when compared with muscle tissue. Mean toxaphene levels in bass ovary and fat tissues were 9 and 39 times higher, respectively, than in muscle tissues. Tissue OC concentrations were consistent, with site OCs, regardless of fish species.

Organochlorine pesticide and total PCB concentrations were measured in largemouth bass from the Tombigbee River near a former DDT manufacturing facility at McIntosh, Alabama (USA) by GC–EDC [30]. Evaluation of mean p,p'- and o,p'-DDT isomer concentrations and o,p'- versus p,p'-isomer proportions in McIntosh bass indicated that DDT is moving off-site from the facility and into the Tombigbee River. Concentrations of p,p'-DDT isomers in McIntosh bass remained unchanged from 1974 to 2004 and were four times greater than contemporary concentrations from a national program. Whereas concentrations of DDT and most other organochlorine chemicals in fish have generally declined in the US since their ban, concentrations of DDT in fish from McIntosh remain elevated and represent a threat to wildlife.

Common carp (*C. carpio*), black bass (*Micropterus* spp.), and channel catfish (*Ictalurus punctatus*) were collected from 14 sites in the Colorado River Basin, USA (CRB) to document spatial trends in accumulative contaminants, health indicators, and reproductive biomarkers [31]. Organochlorine residues (analyzed by GC–ECD), 2,3,7,8-tetrachlorodibenzo-p-dioxin-like activity (TCDD–EQ), and elemental contaminants were measured in composite samples of whole fish, grouped by species and gender, from each site. Pesticide concentrations were greatest in fish from agricultural areas in the Lower Colorado River and Gila River. Concentrations of p,p'-DDE were relatively high in fish from the Gila River at Arlington, Arizona (>1.0 μg/g wet weight) and Phoenix, Arizona (>0.5 μg/g wet weight). Concentrations of other formerly used pesticides including toxaphene, total chlordanes, and dieldrin were also the highest at these two sites, but did not exceed toxicity thresholds. Currently used pesticides such as dacthal, endosulfan, γ-HCH, and methoxychlor were also the highest in fish from the Gila River downstream of Phoenix.

In the study of Sapozhnikova et al. [32], organochlorine and organophosphorous pesticides in sediments and fish tissues in the Salton Sea, USA, were determined and the relative ecological risk of these compounds was evaluated. Sediment samples were taken during 2000–2001 and fish tissues (*Tilapia mossambique, Cynoscion xanthulu*) were collected in May 2001. All samples were analyzed for 12 chlorinated pesticides and 6 organophosphorus pesticides. Σ DDT observed in sediments ranged from 10 to 40 ng/g dry weight. DDT/DDD ratios in sediments and fish tissues of the northern Sea in 2001 indicated recent DDT exposure. Lindane, dieldrin, and DDE detected in sediments exceeded probable effect levels established for freshwater ecosystems. In fish liver, concentrations of endrin and Σ DDT exceeded threshold effect level established for invertebrates. Σ DDT concentrations detected in fish tissues were higher than threshold concentrations for the protection of wildlife consumers of aquatic biota. DDE concentrations in fish muscle tissues were above the 50 ng/g concentration threshold for the protection of predatory birds. Dimethoate, diazinon, malathion, chlorpyrifos, and disulfoton varied from ≤0.15 to 9.5 ng/g dry weight in sediments and from

≤0.1 to 80.3 ng/g wet weight in fish tissues. Disulfoton was found in relatively high concentrations (up to 80.3 ng/g) in all organs from *Tilapia* and *Corvina*.

The method used was GC (DB-5MS fused silica capillary column) equipped with an ECD.

Soil sediments from two depths and tissue samples from eight species of aquatic animals were collected on or near Yazoo National Wildlife Refuge, Mississippi, USA, and analyzed for organo-chlorine pesticide residues by GC–ECD [4]. Residues of 12 OCs were found in most animal samples, and 0.0–4.6 mg/kg of five compounds were detected in soil sediments. With the exception of mosquitofish (*Gambusia affinis*), residues were similarly distributed in soil and animal samples among different watercourses within the watershed.

Biomagnification of organochlorine pesticide residues was evident from soil sediments to mosquitofish, a lower secondary consumer and forage fish, to spotted gar (*Lepisosteus oculatus*), a tertiary consumer. Residues in larger secondary consumers such as carp (*C. carpio*) and smallmouth buffalo (*Ictiobus bubalus*) and tertiary consumers such as water snakes (*Nerodia* spp.) and cottonmouths (*Agkistrodon piscivorus*) demonstrated no clear patterns of accumulation.

The US Fish and Wildlife Service analyzed residues of OCs in 315 composite samples of whole fish collected in 1980–1981 from 107 stations nationwide as part of the National Pesticide Monitoring Program (NPMP) [33]. The mean concentrations of total DDT and all *p,p′*-homologs except *p,p′*-DDT showed significant but small declines relative to mean concentrations before 1978–1979. The mean concentration of *p,p′*-DDT did not change. The most persistent DDT homolog, *p,p′*-DDE, continued to constitute about 70% of total DDT residues.

Residues of other organochlorines (e.g., mirex, pentachloroanisole [PCA], benzene hexachloride [BHC] isomers, endrin, heptachlor, HCB, dacthal [DCPA], and methoxychlor) were either found in relatively few (<25%) stations sampled in 1980 1981 or were characterized by relatively low concentrations.

Fishes of the Great Lakes contain hazardous chemicals such as synthetic halogenated hydrocarbons and metals [34]. These fish can move from the lakes into the Great Lakes tributaries of Michigan. Concentrations of Hg, total PCBs, 2,3,7,8-tetrachlorodibenzo-*p*-dioxin equivalents, total DDT complex, aldrin, endrin, dieldrin, heptachlor, heptachlor epoxide, lindane, hexachlorobenzene, *cis*-chlordane, oxychlordane, endosulfan-I, methoxychlor, *trans*-chlordane, and *trans*-nonachlor were determined in composite samples of fishes from above and below Michigan hydroelectric dams, which separate the fishes that have access to the Great Lakes from fishes that do not. Mean concentrations of total PCBs, TCDD-EQ, DDT, and most of the other pesticides were greater in composite samples of six species of fishes from below than above the dams on the Au Sable, Manistee, and Muskegon Rivers.

Sharks are fish particularly threatened by anthropogenic pollution because of their tendency to bioaccumulate and biomagnify environmental contaminants. Gelsleichter et al. [35] examined concentrations of 29 OC pesticides in the bonnethead shark (*Sphyrna tiburo*). Quantifiable levels of 22 OCs were detected via GC and MS in liver of 95 *S. tiburo* from four estuaries on Florida's Gulf coast: Apalachicola Bay, Tampa Bay, Florida Bay, and Charlotte Harbor. In general, OC concentrations were significantly higher in Apalachicola Bay, Tampa Bay, and Charlotte Harbor *S. tiburo* in relation to the Florida Bay population. Pesticide concentrations did not appear to significantly increase with growth or age in *S. tiburo*, suggesting limited potential for OC bioaccumulation in this species when compared with other sharks for which contaminant data are available.

Samples of bluegills (*Lepomis macrochirus*) and common carp (*C. carpio*) collected from the San Joaquin River and two tributaries (Merced River and Salt Slough) in California, USA, were analyzed for 21 organochlorine chemical residues by GC to determine if pesticide contamination was confined to downstream sites exposed to irrigated agriculture, or if nonirrigated upstream sites were also contaminated [36]. Residues of *p,p′*-DDE were detected in all samples of both species. Six other contaminants were also present in both species at one or more of the collection sites: chlordane (*cis*-chlordane + *trans*-nonachlor), *p,p′*-DDD, *o,p′*-DDT, *p,p′*-DDT, DCPA (dimethyl tetrachloroterephthalate), and dieldrin. Concentrations of most of these residues were generally higher in carp than in bluegills; residues of other compounds were found only in carp: α-BHC, Aroclor® 1260, and toxaphene. Concentrations of most organochlorines in fish increased from upstream to downstream. In carp, concentrations of two residues – Σ DDT (p,p'-DDD + p,p'-DDE + p,p'-DDT; 1.43–2.21 mg/kg wet weight) and toxaphene (3.12 mg/kg wet weight) – approached the highest levels reported by the National Pesticide Monitoring Program for fish from other intensively farmed watersheds of the United States in 1980–1981, and surpassed criteria for whole-body residue concentrations recommended by the National Academy of Sciences and National Academy of Engineers for the protection of piscivorous wildlife.

Concentrations of OCs, p,p'-DDT, p,p'-DDD, p,p'-DDE, α-HCH, and γ-HCH were measured in the sediments, water, and burbot (*Lota iota* L.) (whole liver and liver lipids) of eight Russian Arctic rivers near their outflows to the Arctic Ocean between 1988 and 1994 [37]. Concentrations of Σ DDT up to 70 ng/g wet weight and Σ HCH up to 18 ng/g wet weight were found in burbot livers.

The method utilized was GC with an ECD using peak identification on two packed columns of different stationary phases and polarities (SE-30 and XE-60) [37].

Concentrations of dioxins (PCDD/PCDFs), PCBs, metals, metalloids, pesticides, and antimicrobial residues were gathered for the edible portion of Australian wild and farmed southern bluefin tuna (*Thunnus maccoyii*) [38]. In 2004, wild caught ($n = 5$) and farmed ($n = 26$) southern bluefin tuna were collected in Australia. No detectable residues of any pesticide or antimicrobial compounds were found. Analysis was by GC (HP-5973 column)–MS with PTV (programmed temperature vaporization) injection [38].

OCs in tissues and organs of silver carp (*Hypophthalmichthys molitrix*) from Guanting Reservoir, China, were investigated [39]. A total of 16 OCs were measured and the concentrations were in the range of 1.61–69.01 ng/g wet weight for total OCs, 0.16–0.75 ng/g wet weight for HCB, 0.75–26.80 ng/g wet weight for Σ HCH (sum of α-, β-, γ-, and δ-HCH) and 0.68–35.94 ng/g wet weight for Σ DDT (sum of p,p'-DDE, p,p'-DDD, o,p'-DDT, and p,p'-DDT). The mean concentrations of total OCs, HCB, Σ HCH, and Σ DDT were 18.04, 0.96, 7.14, and 9.28 ng/g wet weight, respectively. Among the organochlorine pesticides, β-HCH and p,p'-DDE were the most dominant compounds in tissues and organs with average concentrations of 4.42 and 8.14 ng/g, respectively.

Contamination by persistent OCs, such as DDTs, hexachlorocyclohexane isomers (HCHs), chlordane compounds (CHLs), HCB, and PCBs were examined in sediments, soils, fishes, crustaceans, birds, and aquaculture feed from Lake Tai, Hangzhou Bay, and in the vicinity of Shanghai city in China, in 2000 and 2001, by GC–ECD (DB1 column) and GC–MS [40]. OCs were detected in all samples, and DDT and its metabolites were the predominant contaminants in most sediments, soils, and biota. Concentrations of p,p'-DDT and ratio of p,p'-DDT to Σ DDTs were significantly higher in marine fishes than those in freshwater fishes.

Ribeiro et al. [41] measured concentrations of OCs, PAHs, and heavy metals, and their effects in the eel *Anguilla anguilla* from three locations in the Camargue Reserve in southern France. Livers and spleens were analyzed for histopathological, chemical, and organosomatic effects. Livers and muscles were sampled for metabolic parameters and persistent organic pollutant analysis. OC pesticides were extracted from lipids of muscles and livers and analyzed by GC. High concentrations of contaminants were found in eel tissues. La Capeliere had the highest OC and PAH concentrations. High pesticide and PAH concentrations and lesions in eels from the Camargue reserve demonstrated the contamination of the area.

In the study of Mazet et al. [42] samples of 10 species of fish (Drôme River, Rhône-Alpes region, France) were analyzed for concentrations of OCs, PCBs, and heavy metals (Pb, Cd, and Cu). Quantitative determination of OCs and PCBs compounds was performed GC–ECD. Samples contained detectable concentrations of lindane.

Twenty-nine specimens of a cichlid fish *Sarotherodon alcalicus grahami* were collected from Lake Nakuru, Kenya, between September and October 1990 and samples of liver, kidney, muscle, brain, and fat were removed for the analysis of organochlorine pesticides [43]. Fat was extracted and the concentration of three lindane (BHC/HCH) isomers (α, β, and γ), aldrin, heptachlor, heptachlor epoxide, endrin, dieldrin, DDD, DDE, and DDT was determined. No residues of o,p'-DDD, p,p'-DDD, aldrin, endrin, and dieldrin were detected. The highest residue concentration detected was 0.062 mg/kg of p,p'-DDT. The mean pesticide residue concentration levels were generally low. The $[p,p'$-DDT$]/[p,p'$-DDE$]$ ratio of 1.22 indicated that the residues of the parent DDT compound exist in the Lake Nakuru ecosystem.

Water, sediment, red swamp crayfish (*Procambarus clarkii*), and black bass (*Micropterus salmoides*) from Lake Naivasha (Kenya) were analyzed for selected organochlorine and organophosphorus pesticide residues [44]. The mean p,p'-DDT, o,p'-DDT, and p,p'-DDE residue levels recorded in black bass (28.3 (±30.0), 34.2 (±54.0), and 16.1 (±16.1) µg/kg, respectively) and crayfish (4.6 (±5.1), 3.2 (±2.8), and 1.4 (±1.1) µg/kg, respectively) were higher than what was previously recorded. This was an indication of recent usage of technical DDT in the lake's catchment. Levels of p,p'-DDT, higher than those of p,p'-DDE further accented this. Mean lindane, dieldrin, β-endosulfan, and aldrin concentrations in black bass were 100.5, 34.6, 21.6, and 16.7 µg/kg, respectively. The same residues were detected at lower concentrations in crayfish at 2.0, 2.0, 2.0, and 1.9 µg/kg, respectively. The higher fat content (3.7% ± 2.7% SD) in black bass (compared with 0.6% ± 0.3% in crayfish) accounted for the significantly higher residue concentrations in black bass. Organophosphate pesticides were the most commonly used pesticides in the lake's catchment, but none was detected in any of the samples.

Takazawa et al. [45] determined the concentrations and residue patterns of 20 persistent OCs, including HCHs, hexachlorobenzene, DDTs, chlordane-related compounds (CHLs), mirex, dieldrin, endrin, and aldrin, in muscle of rainbow trout from Lake Mashu, Japan. Total concentrations of OCs varied from 1.0 to 132 ng/g lipid weight. α-HCH was the most prevalent OC contaminant in the fish muscle.

Nile tilapia (*Oreochromis niloticus*) and Nile perch (*Lates niloticus*) samples were collected and screened for residues of 64 organochlorine, organophosphorus, carbamate, and pyrethroid pesticides [46]. The residue levels in the fish fillet were up to 0.003, 0.03, and 0.2 mg/kg fresh weight (0.7, 3.8, and 42 mg/kg lipid weight) of fenitrothion, DDT, and endosulfan, respectively. Mean levels within sites were up to 0.002, 0.02, and 0.1 mg/kg fresh weight (0.5, 0.5, and 16 mg/kg lipid weight), respectively. The detection of higher levels of p,p'-DDT than the degradation products (p,p'-DDD and p,p'-DDE), and higher levels of endosulfan isomers (α and β) than the sulfate, in fish samples, implied recent exposure of fish to DDT and endosulfan, respectively. Generally, most of the fish samples had residue levels above the average MDLs, but were within the calculated ADI.

19.6 BIRDS

GC was used to quantify residues of 14 persistent chlorinated hydrocarbon pollutants in whole blood, clotted blood, heart, kidney, liver, and muscle samples from African white backed (*Pseudogyps africanus*), Cape griffon (*Gyps coprotheres*), and Lappet faced (*Torgos tracheliotos*) vultures from different localities in South Africa [47]. The levels of pesticides measured in whole blood samples of live specimens were compared between nestlings from two natural breeding colonies, adults from a wildlife area and birds held in captivity. Statistically significant ($P < 0.05$) differences between populations were detected in geometric means calculated for γ-BHC (lindane), α(*cis*)-chlordane, and α-endosulfan. Five of the organochlorine contaminants, γ-BHC, α-chlordane, dieldrin, β-endosulfan, and heptachlor epoxide, showed significant variations in concentrations detected in the clotted blood, organs, and muscles excised from vulture carcasses.

Concentrations of DDE and mercury in nestling tissues of prothonotary warblers varied considerably across the USA [48]. Mean concentration of DDE was greater in eggs than all other tissues, with individual samples ranging from 0.24 to 8.12 µg/kg. In general, concentrations of DDT in soil were effective in describing the variation of contaminants in adipose samples.

Liver and muscle samples from seven species of aquatic and terrestrial predatory birds from Flanders (Belgium) were analyzed for PCBs, PBDEs, and organochlorine pesticides by capillary

GC (HT-8 column) and MS (ECNI—electron capture negative ion) [49]. Sparrowhawks had the highest levels of hexachlorobenzene, DDTs, and PBDEs. In contrast, kestrels (*Falco tinnunculus*) had relatively low levels of most of the measured organochlorines.

In a wildlife area, two buzzards (*Buteo buteo*) and one red kite (*Milvus milvus*) were found dead. In two cases, mice (*Apodemus sylvaticus*) were present in the pharynx or stomach, and a dead fox (*Vulpes vulpes*) was found near a dead buzzard [50]. A subsequent toxicological analysis revealed high residues of carbofuran, parathion, and paraoxon in organs of the dead raptors and a fox. The validity of the toxicological analysis for carbamates and organophosphates using extraction and a modified GC/MS procedure was demonstrated.

Concentrations of the principal organochlorine insecticides were determined in eggs and freshly dead chicks of the Squacco heron (*Ardeola ralloides*), Little egret (*Egretta garzetta*), and Night heron (*Nycticorax nycticorax*), as well as in frogs (*Rana* sp.), the main heron prey [51]. Material was collected from the wetlands of the Thermaikos Gulf (Macedonia, northern Greece) in 1992 and 1993. Residues of the organochlorine pesticides α-BHC, β-BHC, lindane, *p,p'*-DDD, *p,p'*-DDE, heptachlor, and dieldrin were found in the eggs, chicks, and prey of the herons. α-BHC, β-BHC, and lindane had the highest concentration in the night heron and the lowest in the little egret.

Variation in the pesticide contents in the different heron species was attributed to different feeding habits, the exception being the occurrence of dieldrin in eggs alone and *p,p'*-DDE as a remnant of past spraying.

Wildlife contamination studies found high levels of DDT and associated metabolites in bird eggs from Canadian orchard sites during the early 1990s. The study of Harris et al. [52] investigated local dietary uptake of DDT and geographic variability in tissue concentrations in the same orchards. Organochlorine pesticides and PCBs were measured in soil, earthworm, robin egg, and robin nestling samples collected from fruit orchards and reference sites. High average DDE (soil: 5.2 mg/kg; earthworm: 52 mg/kg; robin egg: 484 mg/kg dry weight) and DDT (soil: 9.2 mg/kg; earthworm: 21 mg/kg; robin egg: 73 mg/kg dry weight) concentrations in Okanagan (British Columbia) samples confirmed that previously recorded contamination was common in the region. Concentrations detected in Simcoe, Ontario, orchards were not as high, but were still significantly elevated relative to the levels in soils and robins from reference areas. Low concentrations of DDT and DDTr (DDT-related impurities) in robin eggs collected from nests in nearby nonorchard and post-DDT orchard habitats suggested that the local sources were in orchards. Persistence of DDT in orchard food chains is quite likely due to the combination of retarded degradation rates for DDT in soil and its extensive use historically.

Persistent organochlorine OCs, such as DDT and its metabolites, HCB, α-, β-, and γ-hexachlorocyclohexane isomers, together with PCBs were determined in tail feathers from 35 birds belonging to 15 species, all originating from the southwest of Iran (Khuzestan, coast of the Persian Gulf) and kept in museum collections by GC–ECD (DB-5 column) [53]. The patterns of OCs in birds varied depending on their migratory behavior. Resident birds contained higher median PCB concentrations (<LOQ-151 ng/g feather) than HCHs, DDTs, and HCB. Locally migrating birds had higher median concentrations of HCHs (19–83 ng/g feather). In contrast, long-distance migrants had lower concentrations of HCB and HCHs.

OCs residues were determined by GC–ECD (DB-608 column) in tissues of five Indian white-backed vultures and two of their eggs collected from different locations in India [54]. *p,p'*-DDE ranged between 0.002 μg/g in muscle of vulture from Mudumalai and 7.30 μg/g in liver of vulture from Delhi. Relatively higher levels of *p,p'*-DDT and its metabolites were documented in the bird from Delhi than other places. Dieldrin was 0.003 and 0.015 μg/g, whereas *p,p'*-DDE was 2.46 and 3.26 μg/g in egg one and two, respectively.

The great horned owl (*Bubo virginianus*) was used in a study of PCBs and DDT exposures at two regions of the Kalamazoo River Superfund Site (KRSS), Kalamazoo, Michigan, USA (upper, lower) and an upstream (RA) [55]. The study examined risks of total DDTs (sum of DDT, DDE, and DDD; Σ DDT) by measuring concentrations in eggs and nestling blood plasma. Egg Σ DDT concentrations were as great as 4.2×10^2 and 5.0×10^3 ng Σ DDT/g wet weight at the RA and combined KRSS, respectively.

Levels of mercury and selected pesticides were determined in muscle tissue of chukars, pheasants, and waterfowl collected from various regions within the state of Utah, USA [56]. None of the chukars or pheasants and only 2% of the waterfowl inspected contained dieldrin concentrations above the FDA tolerance level of 0.3 ppm. None of the chukars or pheasants contained levels of DDT + DDE above 5.0 ppm FDA tolerance level.

Although it has been documented that wildlife in the Rio Grande Valley (RGV) contain increased concentrations of OCs, particularly DDE, little has been published on residues of toxaphene throughout this major North American watershed [57]. In this study, 28 liver composites from adult swallows (*Petrochelidon* spp.) collected along the Rio Grande from 1999 to 2000 were analyzed for toxaphene residues using GC–electron-capture negative ionization–MS. Estimated total toxaphene concentrations ranged from 12 to 260 ng/g wet weight, and were highest in samples from the lower RGV near Llano Grande Lake in Hidalgo and Cameron counties (Texas). Toxaphene congener profiles were relatively invariant throughout the watershed and were dominated by 2,2,5-endo,6-exo,8,8,9,10-octachlorobornane with lesser amounts of several other Cl7–Cl9 compounds, many of which remain unidentified.

Braune and Noble [58] sampled 12 species of shorebirds from four locations across Canada to assess their exposure to PCBs, organochlorine pesticides, as well as four trace elements (Hg, Se, Cd, and As). OC analysis was by GC–ECD. Σ PCB and Σ DDT followed by Σ CHL were most frequently found above trace levels in the shorebird carcasses. In general, the plover species (American golden, semipalmated, black-bellied) appear to be the most contaminated with organochlorines, whereas Hudsonian and marbled godwits appear to be the least contaminated.

Eggs of double-crested and pelagic cormorants were collected between 1970 and 2002 from colonies in the Strait of Georgia, British Columbia, Canada, and assayed for concentrations of OC pesticides and PCBs [59]. Double-crested cormorant eggs from the early 1970s contained up to 4.1 mg/kg p,p'-DDE and 12.5 mg/kg Σ PCBs. Corresponding values for pelagic cormorant eggs were 1.5 mg/kg p,p'-DDE and 3.9 mg/kg Σ PCBs. Egg tissue concentrations of the dominant OC pesticides and Σ PCBs dropped mainly during the 1970s, with minor declines thereafter. The data suggest that contaminant levels in cormorants have now stabilized at low levels throughout the resident population. Small but significant latitudinal gradients in several OC pesticides and PCBs indicated that areas of the southern strait were more contaminated than areas of the less populated northern strait. Interspecific differences in contamination may indicate that pelagic cormorants have a reduced capacity to metabolize chlordanes, DDT, and PCBs when compared with double-crested cormorants. Alternatively, the two species may have more divergent prey bases than what was previously thought.

The levels of organochlorine compounds in eggs of water birds from the colony on Tai Lake in China were studied [60]. The eggs were collected in 2000 and belonged to the following species: 65 samples of black-crowned night heron (*Nycticorax nycticorax*), 36 samples of little egret (*E. garzetta*), 26 samples of cattle egret (*Bubulcus ibis*) from 13 clutches, and 43 samples of Chinese pond heron (*Ardeola bacchus*) from 17 clutches. DDT and its derivates (DDE and DDD), HCH and its isomers (α-HCH, β-HCH, γ-HCH, and δ-HCH), heptachlor, heptachlor epoxide, aldrin, dieldrin, endrin, endrin aldchyde, α-endosulfan, β-endosulfan, and endosulfan sulfate were determined by GC. The data showed that DDE was present at the highest levels in all the samples, followed by β-HCH. The mean levels of DDE among the water bird species were as follows: black-crowned night heron (5464.26 ng/g, dry weight) > Chinese pond heron (2791.12 ng/g, dry weight) > little egret (1979.97 ng/g, dry weight) > cattle egret (660.11 ng/g, dry weight). DDT and its metabolites accounted for 90% of the total organochlorines, except that it was only 73% for cattle egret. The differences of the residue among the bird species were statistically significant and could be attributed to their variations in prey and habitat.

OCs and PCBs residue levels were determined in 53 unhatched eggs from greater flamingos (*Phoenicopterus ruber*) [61]. Eggs were collected in 1996 from the National Park of Doñana

(Guadalquivir marshes, Southwest Spain), immediately after one breeding colony abandoned the nesting site due to predator attacks. *p,p′*-DDE was the OC residue found at higher concentrations with a geometric mean of 721 ng/g wet weight. Residues of other pesticides, including some hexachlorocyclohexane isomers, hexachlorobenzene, aldrin, heptachlor, and heptachlor epoxide were detected at much lower concentrations.

19.7 REPTILES

The accumulation of metals and OCs in caudal scutes of crocodiles from Belize and Costa Rica were examined [62]. Scutes from Morelet's crocodiles (*Crocodylus moreletii*) from two sites in northern Belize were analyzed for metals, and scutes from American crocodiles (*C. acutus*) from one site in Costa Rica were analyzed for metals and OC pesticides. American crocodile scutes from Costa Rica contained multiple OC pesticides, including endrin, methoxychlor, *p,p′*-DDE, and *p,p′*-DDT. Mean OC concentrations varied in relation to those previously reported in crocodilian scutes from other localities in North, Central, and South America. OC concentrations in American crocodile scutes were generally higher than those previously reported for other Costa Rican wildlife. OCs were analyzed by GC (DB-5 column)–ECD.

Eggs of eastern spiny softshell turtles (*Apalone spiniferus spiniferus*) were monitored from three populations, located at Thames River, Rondeau Provincial Park, and the Long Point National Wildlife Area, in southern Ontario, Canada in 1998 [63]. Organochlorine pesticides, PCBs, dibenzo-*p*-dioxins, and furans from eggs from the same nests were measured. Contaminant concentrations in eggs were similar among sites. There was no correlation between hatching success, parasitism, and depredation rates, or the proportion of hatchlings that were males with total PCBs or individual pesticides, but there was a positive correlation between egg viability with concentrations of total PCBs in eggs, and with five pesticides.

Sediment from a wetland adjacent to an industrial wastewater treatment plant in Sumgayit (Republic of Azerbaija) contained concentrations of total PAHs, total PCBs, aldrin, biphenyl, chlordane, DDT, mercury, β-endosulfan, heptachlor, α-HCH, γ-HCH, and several individual PAH congeners that were elevated relative to published sediment quality guidelines [64]. Chemical analyses of tissues from European pond turtles (*Emys orbicularis*) showed increased levels of many of the same chemicals including aldrin, chlordane, heptachlor, α-HCH, total PCBs, total PAHs, and mercury, compared with reference turtles. In addition, turtle tissues contained elevated levels of DDD, HCB, and pentachlorobenzene that were not elevated in the sediment sample. Some differences were observed in contaminant levels between European pond turtles and Caspian turtles (*Mauremys caspica*) taken from the ponds in Sumgayit.

19.8 AMPHIBIANS

Organochlorine pesticide concentrations in sediment and amphibians from playas in cropland and grassland watersheds in the Southern High Plains, USA, were determined [65]. Heptachlor, α- and β-BHC, γ-chlordane, and dieldrin were detected in sediment or tissue samples, typically at or below 1 ng/g dry weight. However, mean DDT and DDE reached 6.3 and 2.4 ng/g in tissues, respectively.

Green frogs were collected from seven southern Ontario (Canada) locations and analyzed for chlorinated organic chemicals [66]. At Hillman Marsh, a wildlife reserve in an agricultural area, green frogs accumulated significantly greater amounts of highly chlorinated PCBs than green frogs from all other collection sites. At Ancaster, DDE accumulated in green frogs to a significantly greater extent than at all other sites. This was attributed to the presence of agriculture at Ancaster and the historic use of DDT in agriculture.

Seven adult green frogs (*Rana clamitans*) were collected from three sites adjacent to intensive agriculture in the lower Fraser River Valley, British Columbia, Canada [67]. Detection was by

CGC–MS in SIM mode. The highest mean concentrations of chemicals were p,p'-DDE at 0.313 µg/g lipid weight. On a lipid weight basis, both p,p'-DDE and PCB concentrations varied by almost an order of magnitude among sites.

Artificial water reservoirs are important for fauna in arid–semiarid regions, because they provide suitable habitats for species that depend on water, such as amphibians. Jofré et al. [68] evaluated OCs contaminant levels in anurans from an artificial lake (Embalse La Florida) in a semiarid region of the Midwest Argentina. OCs were detected in all individuals. Levels ranged from 2.34 ± 0.62 ng/g wet mass of heptachlors to 9.76 ± 1.76 ng/g wet mass of hexachlorocyclohexanes. The distribution pattern of OCP was Σ HCH > Σ DDT > endosulfan > Σ chlordane > metoxichlor > Σ aldrin > Σ heptachlor.

REFERENCES

1. Volz, SA and Johnston, JJ. Solid phase extraction/gas chromatography/electron capture detector method for the determination of organochlorine pesticides in wildlife and wildlife food sources. *Journal of Separation Science*, 25 (3), 119–124, 2002.
2. Volz, SA, Johnston, JJ, and Griffin, DL. Solid phase extraction gas chromatography/electron capture detector method for the determination of organochlorine pesticides in wildlife whole blood. *Journal of Agricultural and Food Chemistry*, 49 (6), 2741–2745, 2001.
3. Petty, EE, Johnston, JJ, and Volz, SA. Solid-phase extraction method for the quantitative analysis of organochlorine pesticides in wildlife urine. *Journal of Chromatographic Science*, 35 (9), 430–434, 1997.
4. Ford, WM and Hill, EP. Organochlorine pesticides in soil sediments and aquatic animals in the upper steele bayou watershed of Mississippi. *Archives of Environmental Contamination and Toxicology*, 20 (2), 161–167, 1991.
5. Mwevura, H, Othman, OC, and Mhehe, GL. Organochlorine pesticide residues in sediments and biota from the coastal area of Dar es Salaam city, Tanzania. *Marine Pollution Bulletin*, 45 (1–12), 262–267, 2002 (special issue).
6. Ozkoc, HB, Bakan, G, and Ariman, S. Distribution and bioaccumulation of organochlorine pesticides along the Black Sea coast. *Environmental Geochemistry and Health*, 29 (1), 59–68, 2007.
7. Wang, D, Miao, X, and Li, QX. Analysis of organochlorine pesticides in coral (*Porites evermanni*) samples using accelerated solvent extraction and gas chromatography/ion trap mass spectrometry. *Archives of Environmental Contamination and Toxicology*, 54 (2), 211–218, 2008.
8. Beyer, WN and Kaiser, TE. Organochlorine pesticide residues in moths from the Baltimore, MD-Washington, DC area. *Environmental Monitoring and Assessment*, 4 (2), 129–137, 1984.
9. Alleva, E, Francia, N, Pandolfi, M, De Marinis, AM, Chiarotti, F, and Santucci, D. Organochlorine and heavy-metal contaminants in wild mammals and birds of Urbino-Pesaro province, Italy: an analytic overview for potential bioindicators. *Archives of Environmental Contamination and Toxicology*, 51 (1), 123–134, 2006.
10. Sundberg, SE, Ellington, JJ, and Evans, JJ. A simple and fast extraction method for organochlorine pesticides and polychlorinated biphenyls in small volumes of avian serum. *Journal of Chromatography B*, 831 (1–2), 99–104, 2005.
11. Kolankaya, D. Organochlorine pesticide residues and their toxic effects on the environment and organisms in Turkey. *International Journal of Environmental Analytical Chemistry*, 86 (1–2), 147–160, 2006.
12. Hoshi, H, Minamoto, N, Iwata, H, Shiraki, K, Tatsukawa, R, Tanabe, S, Fujita, S, Hirai, K, and Kinjo, T. Organochlorine pesticides and polychlorinated biphenyl congeners in wild terrestrial mammals and birds from Chubu region, Japan: Interspecies comparison of the residue levels and compositions. *Chemosphere*, 36 (15), 3211–3221, 1998.
13. Mora, MA, Laack, LL, Lee, MC, Sericano, J, Presley, R, Gardinali, PR, Gamble, LR, Robertson, S, and Frank, D. Environmental contaminants in blood, hair, and tissues of ocelots from the Lower Rio Grande Valley, Texas, 1986–1997. *Environmental Monitoring and Assessment*, 64 (2), 477–492, 2000.
14. Kayser, A, Voigt, F, and Stubbe, M. First results on the concentrations of some persistent organochlorines in the common hamster *Cricetus cricetus* (L.) in Saxony-Anhalt. *Bulletin of Environmental Contamination and Toxicology*, 67 (5), 712–720, 2001.
15. Myers, MJ, Ylitalo, GM, Krahn, MM, Boyd, D, Calkins, D, Burkanov, V, and Atkinson, S. Organochlorine contaminants in endangered Steller sea lion pups (*Eumetopias jubatus*) from western Alaska and the Russian Far East. *Science of The Total Environment*, 396 (1), 60–69, 2008.

16. Hu, GC, Luo, XJ, Dai, JY, Zhang, XL, Wu, H, Zhang, CL, Guo, W, Xu, MQ, Mai, BX, and Wei, FW. Brominated flame retardants, polychlorinated biphenyls, and organochlorine pesticides in captive giant panda (*Ailuropoda melanoleuca*) and red panda (*Ailurus fulgens*) from China. *Environmental Science & Technology*, 42 (13), 4704–4709, 2008.

17. Basu, N, Scheuhammer, AM, and O'Brien, M. Polychlorinated biphenyls, organochlorinated pesticides, and polybrominated diphenyl ethers in the cerebral cortex of wild river otters (*Lontra canadensis*). *Environmental Pollution*, 149 (1), 25–30, 2007.

18. Robison, HL, Schwartz, CC, Petty, JD, Brussard, PF, Robison, AF, Hillary, L, Schwartz, CC, Petty, JD, and Brussard, PF. Assessment of pesticide residues in army cutworm moths (*Euxoa auxiliaris*) from the Greater Yellowstone Ecosystem and their potential consequences to foraging grizzly bears (*Ursus arctos horribilis*). *Chemosphere*, 64 (10), 1704–1712, 2006.

19. van Bavel, B, Helgason, LB, Lydersen C, and Kovacs, KM. Biomarker responses and decreasing contaminant levels in ringed seals (*Pusa hispida*) from Svalbard, Norway. *Journal of Toxicology and Environmental Health*, 71 (15), 1009–1018, 2008.

20. Gamberg, M and Braune, BM. Contaminant residue levels in arctic wolves (*Canis lupus*) from the Yukon Territory, Canada. *Science of the Total Environment*, 243–244 (15), 329–338, 1999.

21. Naccari, F, Giofrè, F, Licata, P, Martino, D, Calò, M, and Parisi, N. Organochlorine pesticides and PCBs in wild boars from Calabria (Italy). *Environmental Monitoring and Assessment*, 96 (1–3), 191–202, 2004.

22. Kannan, K, Ramu, K, Kajiwara, N, Sinha, RK, and Tanabe, S. Organochlorine pesticides, polychlorinated biphenyls, and polybrominated diphenyl ethers in Irrawaddy Dolphins from India. *Archives of Environmental Contamination and Toxicology*, 49 (3), 415–420, 2005.

23. O'Shea, TJ, Everette, AL, and Ellison, LE. Cyclodiene insecticide, DDE, DDT, arsenic, and mercury contamination of big brown bats (*Eptesicus fuscus*) foraging at a Colorado superfund site. *Archives of Environmental Contamination and Toxicology*, 40 (1), 112–120, 2001.

24. Leung, CCM, Jefferson, TA, Hung, SK, Zheng, GJ, Yeung, LWY, Richardson, BJ, and Lam, PKS. Petroleum hydrocarbons, polycyclic aromatic hydrocarbons, organochlorine pesticides and polychlorinated biphenyls in tissues of Indo-Pacific humpback dolphins from south China waters. *Marine Pollution Bulletin*, 50 (2), 1713–1719, 2005.

25. Smith, FA, Sharma, RP, Lynn, RI, and Lowl, JB. Mercury and selected pesticide levels in fish and wildlife of Utah: I. levels of mercury, DDT, DDE, dieldrin and PCB in fish. *Bulletin of Environmental Contamination and Toxicology*, 12 (2), 218–223, 1974.

26. Riedel, R, Schlenk, D, Frank, D, and Costa-Pierce, B. Analyses of organic and inorganic contaminants in Salton Sea fish. *Marine Pollution Bulletin*, 44 (5), 403–411, 2002.

27. Schmitt, CJ. Organochlorine chemical residues in fish from the Mississippi River basin. *Archives of Environmental Contamination and Toxicology*, 43 (1), 81–97, 2002.

28. Gelsleichter, J, Walsh, CJ, Szabo, NJ, and Rasmussen, LEL. Organochlorine concentrations, reproductive physiology, and immune function in unique populations of freshwater Atlantic stingrays (*Dasyatis sabina*) from Florida's St. Johns Rive. *Chemosphere*, 63 (9), 1506–1522, 2006.

29. Marburger, J, Johnson, WE, Gross, TS, Douglas, DR, and Di, J. Residual organochlorine pesticides in soils and fish from wetland restoration areas in central Florida, USA. *Wetlands*, 22 (4), 7–711, 2002.

30. Hinck, JE, Norstrom, RJ, Orazio, CE, Schmitt, CJ, and Tillitt, DE. Persistence of organochlorine chemical residues in fish from the Tombigbee River (Alabama, USA): Continuing risk to wildlife from a former DDT manufacturing facility. *Environmental Pollution*, 157 (2), 582–591, 2009.

31. Hinck, JE, Blazer, VS, Denslow, ND, Echols, KR, Gross, TS, May, TW, Anderson, PJ, Coyle, JJ, and Tillitt, DE. Chemical contaminants, health indicators, and reproductive biomarker responses in fish from the Colorado River and its tributaries. *Science of the Total Environment*, 378 (3), 376–402, 2007.

32. Sapozhnikova, Y, Bawardi, O, and Schlenk, D. Pesticides and PCBs in sediments and fish from the Salton Sea, California, USA. *Chemosphere*, 55 (6), 797–809, 2004.

33. Schmitt, CJ, Zajicek, JL, and Ribick, MA. National Pesticide Monitoring Program: Residues of organochlorine chemicals in freshwater fish, 1980–1981. *Archives of Environmental. Contamination and Toxicology*, 14 (2), 225–260, 1985.

34. Giesy, JP, Verbrugge, DA, Othout, RA, Bowerman, WW, Mora, MA, Jones, PD, Newsted, JL, Vandervoort, C, Heaton, SN, Aulerich, RJ, Bursian, SJ, Ludwig, JP, Ludwig, M, Dawson, GA, Kubiak, TJ, Best, DA, and Tillitt, DE. Contaminants in fishes from Great Lakes-influenced sections and above dams of three Michigan rivers. I: Concentrations of organochlorine insecticides, polychlorinated biphenyls, dioxin equivalents, and mercury. *Archives of Environmental Contamination and Toxicology*, 27 (2), 202–212, 1994.

35. Gelsleichter, J, Manire, CA, Szabo, NJ, Cortés, E, Carlson, J, and Lombardi-Carlson, L. Organochlorine concentrations in bonnethead sharks (*Sphyrna tiburo*) from four Florida estuaries. *Archives of Environmental Contamination and Toxicology*, 48 (4), 474–483, 2005.

36. Saiki, MK and Schmitt, CJ. Organochlorine chemical residues in bluegills and common carp from the irrigated San Joaquin Valley Floor, California. *Archives of Environmental Contamination and Toxicology*, 15 (4), 357–366, 1986.

37. Zhulidov, AV, Robarts, RD, Headley, JV, Liber, K, Zhulidov, DA, Zhulidova, OV, and Pavlov, DF. Levels of DDT and hexachlorocyclohexane in burbot (*Lota iota* L.) from Russian Arctic rivers. *Science of the Total Environment*, 292 (3), 231–246, 2002.

38. Padula, DJ, Daughtry, BJ, and Nowak, BF. Dioxins, PCBs, metals, metalloids, pesticides and antimicrobial residues in wild and farmed Australian southern bluefin tuna (*Thunnus maccoyii*). *Chemosphere*, 72 (1), 34–44, 2008.

39. Sun, YZ, Wang, XT, Li, XH, and Xu, XB. Distribution of persistent organochlorine pesticides in tissue/organ of silver carp (*Hypophthalmichthys molitrix*) from Guanting Reservoir, China. *Journal of Environmental Sciences-China*, 17 (5), 722–726, 2005.

40. Nakata, H, Hirakawa, Y, Kawazoe, M, Nakabo, T, Arizono, K, Abe, SI, Kitano, T, Shimada, H, Watanabe, L, Li, WH, and Ding, XC. Concentrations and compositions of organochlorine contaminants in sediments, soils, crustaceans, fishes and birds collected from Lake Tai, Hangzhou Bay and Shanghai city region, China. *Environmental Pollution*, 133 (3), 415–429, 2005.

41. Ribeiro, CAO, Vollaire, Y, Sanchez-Chardi, A, and Roche, H. Bioaccumulation and the effects of organochlorine pesticides, PAH and heavy metals in the Eel (*Anguilla anguilla*) at the Camargue Nature Reserve, France. *Aquatic Toxicology*, 74 (1), 53–69, 2005.

42. Mazet, A, Keck, G, and Berney, P. Concentrations of PCBs, organochlorine pesticides and heavy metals (lead, cadmium, and copper) in fish from the Drôme river: Potential effects on otters (*Lutra lutra*). *Chemosphere*, 61 (6), 810–816, 2005.

43. Kairu, JK. Organochlorine pesticide and metal residues in a cichlid fish, Tilapia, Sarotherodon (= Tilapia) alcalicus grahami Boulenger from Lake Nakuru, Kenya. *International Journal of Salt Lake Research*, 8 (3), 253–266, 2004.

44. Gitahi, SM, Harper, DM, Muchiri, SM, Tole, MP, and Ng'ang'a, RN. Organochlorine and organophosphorus pesticide concentrations in water, sediment, and selected organisms in Lake Naivasha (Kenya). *Hydrobiologia*, 488 (1–3), 123–128, 2002.

45. Takazawa, Y, Tanaka, A, and Shibata, Y. Organochlorine pesticides in muscle of rainbow trout from a remote Japanese lake and their potential risk on human health. *Water, Air, and Soil Pollution*, 187 (1–4), 31–40, 2008.

46. Henry, L and Kishimba, MA. Pesticide residues in Nile tilapia (*Oreochromis niloticus*) and Nile perch (*Lates niloticus*) from Southern Lake Victoria, Tanzania. *Environmental Pollution*, 140 (2), 348–354, 2005.

47. van Wyk, E, Bouwman, H, van der Bank, H, Verdoorn, GH, and Hofmann, D. Persistent organochlorine pesticides detected in blood and tissue samples of vultures from different localities in South Africa. *Comparative Biochemistry and Physiology C-Toxicology & Pharmacology*, 129 (3), 243–264, 2001.

48. Reynolds, KD, Rainwater, TR, Scollon, EJ, Sathe, SS, Adair, BM, Dixon, KR, Cobb, GP, and McMurry, ST. Accumulation of DDT and mercury in prothonotary warblers (*Protonotaria citrea*) foraging in a heterogeneously contaminated environment. *Environmental Toxicology and Chemistry*, 20 (12), 2903–2909, 2001.

49. Jaspers, VLB, Covaci, A, Voorspoels, S, Dauwe, T, Eens, M, and Schepens, P. Brominated flame retardants and organochlorine pollutants in aquatic and terrestrial predatory birds of Belgium: Levels, patterns, tissue distribution and condition factors. *Environmental Pollution*, 139 (2), 340–352, 2006.

50. Meiser, H. Gas chromatography/mass spectrometry procedure for carbamate and organophosphate insecticide detection in the organs of poisoned wild birds. *Tierarztliche Umschau*, 59 (12), 703, 2004.

51. Albanis, TA, Hela, D, Papakostas, G, and Goutner, V. Concentration and bioaccumulation of organochlorine pesticide residues in herons and their prey in wetlands of Thermaikos Gulf, Macedonia, Greece. *Science of the Total Environment*, 182 (1–3), 11–19, 1996.

52. Harris, ML, Wilson, LK, Elliott, JE, Bishop, CA, Tomlin, AD, and Henning, KV. Transfer of DDT and metabolites from fruit orchard soils to American robins (*Turdus migratorius*) twenty years after agricultural use of DDT in Canada. *Archives of Environmental Contamination and Toxicology*, 39 (2), 205–220, 2000.

53. Behrooz, RD, Esmaili-Sari, A, Ghasempouri, SM, Bahramifar, N, and Hosseini, SM. Organochlorine pesticide and polychlorinated biphenyl in feathers of resident and migratory birds of South-West Iran. *Environment International*, 35 (2), 285–290, 2009.

54. Muralidharan, S, Dhananjayan, V, Risebrough, R, Prakash, V, Jayakumar, R, and Bloom, PH. Persistent organochlorine pesticide residues in tissues and eggs of white-backed vulture, *Gyps bengalensis* from different locations in India. *Bulletin of Environmental Contamination and Toxicology*, 81 (6), 561–565, 2008.

55. Strause, KD, Zwiernik, MJ, Im, SH, Bradley, PW, Moseley, PP, Kay, DP, Park, CS, Jones, PD, Blankenship, AL, Newsted, JL, and Giesy, JP. Risk assessment of great horned owls (*Bubo virginianus*) exposed to polychlorinated biphenyls and DDT along the Kalamazoo River, Michigan, USA. *Environmental Toxicology and Chemistry*, 26 (7), 1386–1398, 2007.
56. Smith, FA, Sharma, RP, Lynn, RI, and Low, JB. Mercury and selected pesticide levels in fish and wildlife of Utah: II. levels of mercury, DDT, DDE, dieldrin and PCB in chukars, pheasants and waterfowl. *Bulletin of Environmental Contamination and Toxicology*, 12 (2), 153–157, 1974,
57. Maruya, KA, Smalling, KL, und Mora, MA. Residues of toxaphene in insectivorous birds (*Petrochelidon* spp.) from the Rio Grande, Texas. *Archives of Environmental Contamination and Toxicology*, 48 (4), 567–574, 2005.
58. Braune, BM and Noble, DG. Environmental contaminants in Canadian shorebirds. Environmental Monitoring and Assessment, 148 (1–4), 185–204, 2009.
59. Harris, ML, Wilson, LK, and Elliott, JE. An Assessment of PCBs and OC pesticides in eggs of double-crested (*Phalacrocorax auritus*) and pelagic (*P. pelagicus*) cormorants from the West Coast of Canada, 1970 to 2002. *Ecotoxicology*, 14 (6), 607–625, 2005.
60. Dong, YH, Wang, H, An, Q, Ruiz, X, Fasola, M, and Zhang, YM. Residues of organochlorinated pesticides in eggs of water birds from Tai Lake in China. *Environmental Geochemistry and Health*, 26 (2), 259–268, 2004.
61. Guitart, R, Clavero, R, Mateo, R, and Mañez, M. Levels of persistent organochlorine residues in eggs of greater flamingos from the Guadalquivir Marshes (Doñana), Spain. *Journal of Environmental Science and Health, Part B*, 40 (5), 753–760, 2005.
62. Rainwater, TR, Wu, TH, Finger, AG, Canas, JE, Yu, L, Reynolds, KD, Coimbatore, G, Barr, B, Platt, SG, Cobb, GP, Anderson, TA, and McMurry, ST. Metals and organochlorine pesticides in caudal scutes of crocodiles from Belize and Costa Rica. *Science of the Total Environment*, 373 (1), 146–156, 2007.
63. de Solla, SR, Fletcher, ML, and Bishop, CA. Relative contributions of organochlorine contaminants, parasitism, and predation to reproductive success of eastern spiny softshell turtles (*Apalone spiniferus spiniferus*) from southern Ontario, Canada. *Ecotoxicology*, 12 (1–4), 261–270, 2003.
64. Swartz, CD, Donnelly, KC, Islamzadeh, A, Rowe, GT, Rogers, WJ, Palatnikov, GM, Mekhtiev, AA, Kasimov, M, Mcdonald, TJ, Wickliffe, JK, Presley, BJ, and Bickham, JW. Chemical contaminants and their effects in fish and wildlife from the industrial zone of Sumgayit, Republic of Azerbaija. *Ecotoxicology*, 12 (6), 509–521, 2003.
65. Venne, LS, Anderson, TA, Zhang, B, Smith, LM, and McMurry, ST. Organochlorine pesticide concentrations in sediment and amphibian tissue in playa wetlands in the Southern High Plains, USA. *Bulletin of Environmental Contamination and Toxicology*, 80 (6), 497–501, 2008.
66. Russell, RW, Gillan, KA, and Haffner, GD. Polychlorinated biphenyls and chlorinated pesticides in southern Ontario, Canada, green frogs. *Environmental Toxicology and Chemistry*, 16 (11), 2258–2263, 1997.
67. Loveridge, AR, Bishop, CA, Elliott, JE, and Kennedy, CJ. Polychlorinated biphenyls and organochlorine pesticides bioaccumulated in green frogs, *Rana clamitans*, from the Lower Fraser Valley, British Columbia, Canada. *Bulletin of Environmental Contamination and Toxicology*, 79 (3), 315–318, 2007.
68. Jofré, MB, Antón, RI, and Caviedes-Vidal, E. Organochlorine contamination in anuran amphibians of an artificial lake in the semiarid Midwest of Argentina. *Archives of Environmental Contamination and Toxicology*, 55 (3), 471–480, 2008.

20 Determination of Pesticides in Human Blood and Urine by High-Performance Liquid Chromatography

Haseeb Ahmad Khan

CONTENTS

20.1   INTRODUCTION

20.1.1   HUMAN DEPENDENCY ON PESTICIDES

Pesticides are among the most widely used chemicals in the world. They are primarily used in agriculture to increase crop yield, in household to kill various pests, and in health sector to combat disease vectors. The worldwide pesticide use is about 5 billion pounds each year, of which about 1.2 billion pounds is used in the United States alone. The three most commonly used pesticide types are herbicides, insecticides, and fungicides. Herbicides are chemicals with the capacity to kill plants selectively or nonselectively, representing nearly half of all pesticides employed. Herbicides shared 44% of the total pesticide use in 2001 in the United States [1]. About 430 million pounds of herbicides and plant growth regulators are used annually in the U.S. agriculture, with another 80 million pounds being used for home and garden applications, and an additional 40 million pounds used in industrial, commercial, and government applications [1]. Four of the five most abundantly used pesticides in agricultural and residential setting in the United States are herbicides [1]. Although the use of herbicides in agriculture has been questioned many times, it has been proven that the rational application of herbicides results in a steady and sufficient flow of food products of high quality. Herbicides protect plants against undue competition from weeds, and enhance the nutritional quality of foods. However, the intensive application of herbicides has resulted in the contamination of atmosphere, water, crops, and food products. Being toxic, herbicides represent environmental risk as well as human health hazard [2]. The main classes of herbicides include bipiridilium compounds, triazine derivatives, chlorophenoxyacid derivatives, urea derivatives, and sulfonylureas. Insecticides are widely used in agriculture to protect crops, in the household to control insect pests, and in public health to control diseases caused by insect vectors or intermediate hosts. The persistence of organochlorine pesticides has led to their replacement by other, more readily degradable and less-persistent pesticides, such as organophosphates, carbamates, and pyrethroids. Pyrethroids account for more than 25% of the worldwide insecticide market, and this percentage has increased substantially over the last few years as a result of the U.S. Environmental Protection Agency's restrictions on household and agricultural use of organophosphates. Pyrethroid pesticides possess high insecticidal potency, slow development of pest resistance, and relatively low toxicity in mammals. The fungicides vinclozolin and iprodione are widely used in agriculture. These pesticides are dicarboximide fungicides containing the common moiety 3,5-dichloroaniline (3,5-DCA).

20.1.2   HEALTH HAZARDS OF SOME COMMON PESTICIDES

Herbicides have received particular attention by health scientists because of their general toxicity, carcinogenicity, and neurotoxicity [3–6]. Although the standard use of paraquat (PQ) does not pose a risk due to its inactivation by natural components of soil, the accidental or intentional ingestion of PQ can result in severe clinical situations [7]. Death usually occurs within 2 days of ingestion of 2 mg/kg by general organ failure or several weeks later for lower doses due to progressive and irreversible pulmonary fibrosis [8,9]. Sulfonylurea herbicides are much more potent than other classes of herbicides; however, they are categorized as safe for workers and consumers. Chlorophenoxy herbicides such as 2,4-D, 2,4,5-T, and MCPA are used to control broad-leaved weeds and at high application rates for total vegetation control. Ingestion of chlorophenoxy herbicides can cause nausea, vomiting, abdominal pain, confusion, coma, metabolic acidosis, convulsions, and renal damage [10,11] and may also result in fatalities [12,13]. Ioxynil, a benzonitrile, has herbicidal properties similar to those of chlorophenoxy herbicides. The principal action of benzonitriles is to uncouple oxidative phosphorylation and the symptoms of acute poisoning are fatigue, excessive sweating, thirst, anxiety, tachycardia, and hyperventilation [14]. Fatalities with ioxynil poisoning have also been reported [15]. Phenylurea herbicides such as diuron have been widely used since their discovery in 1950. Diuron is considered as a highly toxic, persistent priority substance with a half-life of 300 days when applied to the soil [16]. Experimental studies have indicated that diuron is carcinogenic to rodents [17].

Both organophosphate and carbamate insecticides are potent cholinesterase inhibitors; however, the latter are of shorter duration with a reversible toxic action. It is estimated that 99% of all deaths from pesticide poisoning occur in developing countries where organophosphate insecticides are extensively used in agriculture, with little protection for the communities and individuals thus exposed [18]. A hospital-based survey in Japan revealed organophosphate insecticide as the most frequent inducer of clinical cases (36%), followed by bipyridylium herbicide (20%), and carbamate insecticide (6%) [19]. There are numerous reports on human health hazards associated with organophosphate insecticides [20–22]. Among the carbamate insecticides, the Environmental Protection Agency has classified aldicarb in the highest toxicity category, and has defined a strict control for its delivery and use. Ragoucy-Sengler et al. [23] have reviewed the aldicarb poisoning circumstances associated with clinical and analytical findings. Among the 39 cases of aldicarb poisoning, 31 were symptomatic with muscarinic signs (20 cases), digestive (15 cases), neurological (8 cases), and nicotinic signs (6 cases) [24]. Carbofuran and its major metabolites can cross the placental barrier and produce serious effects on the maternal–placental–fetal unit. Carbofuran's toxicity can be potentiated by simultaneous exposure with other cholinesterase inhibitors [25]. Typical signs and symptoms of acute poisoning of laboratory animals and humans by pyrethroids include salivation, hyperexcitability, and choreoathetosis. Immature rats are much more susceptible to acute neurotoxicity of pyrethroids than adults owing to inefficient metabolic detoxification of the parent compound. At present, there is a concern that deltamethrin and possibly other pyrethroids, like certain organophosphates, may exhibit the potential to cause developmental neurotoxicity in infants and children [26]. It has been suggested that low-level exposures to dicarboximide fungicides may be associated with adverse health effects such as endocrine disruption. Ethylenebisdithiocarbamate (EBDC) fungicides are also an important class of organic fungicides that exhibit a high degree of carcinogenicity, mutagenicity, and neurotoxicity.

20.1.3 IMPORTANCE OF PESTICIDES ANALYSIS IN BIOLOGICAL FLUIDS

Although most people are not occupationally exposed to pesticides, nearly everyone has some level of exposure resulting from food, air, water, or dermal contact. Owing to the potential for widespread exposure to herbicides in both occupational and environmental settings, health effects associated with herbicide exposures or determinants of herbicide exposures have been the focus of several studies, primarily evaluating occupational exposures [27–29]. There are also several reports on the misuse of pesticides for suicidal attempts or accidental ingestion [24]. Pesticide self-poisonings account for about one-third of the world's suicides [30].

The effective methods for the determination of pesticides in biological fluids are necessary to monitor pesticides in human body. Both blood and urine are very complex and multicomponent mixtures, as a large number of compounds ingested or formed in the body by catabolic/anabolic pathways are circulated in blood and excreted in urine. A plethora of reports are available on the analysis of herbicides in environmental samples such as water, soil, and vegetation, as have been reviewed by Cserhati et al. [31]. However, fewer attempts have been made to develop methods for estimation of herbicides in body fluids and tissues. Similarly, many high-performance liquid chromatography (HPLC)-based methods have been reported for the determination of fungicides in water [32,33], soil [34], food, and beverages [35,36], whereas studies on their estimation in blood and urine are scarce [37]. Hence, the safe application of pesticides in practice requires convenient methods for their determination not only in environmental samples but also in biological fluids.

20.1.4 ANALYSIS OF PESTICIDES IN BIOLOGICAL SAMPLES USING HPLC

20.1.4.1 Herbicides

HPLC is the method of choice for the analysis of herbicides because of its high sensitivity and compatibility for both thermoliable and nonvolatile herbicides. Taylor et al. [38] have reported a quantitative method for the analysis of PQ in plasma and urine using HPLC with ultraviolet (UV) detection at 260 nm. The herbicide was extracted from urine or plasma sample (1 mL) using C18 solid-phase extraction (SPE) followed by chromatographic separation on a Zorbax RX-Silica column. The mobile phase consisted of 96% sodium chloride (5 g/L) and 4% acetonitrile (pH 2.2) pumped at a flow rate of 1.0 mL/min. The accuracy and imprecision of the method over the linear range (0.1–5.0 mg/L) were 94.7%–104.9% and <12.2%, respectively. The limit of quantitation (LOQ) for both matrices was 0.1 mg/L. The absolute recovery of PQ from plasma and urine was found to be 79.9% ± 5.3% and 88.2% ± 5.3%, respectively. This method in conjunction with a qualitative urine PQ screen has been validated for clinical studies [38]. An HPLC method for the quantitation of PQ in urine was also applied to serum [39]. Sample preparation consisted of ion-pair extraction on disposable cartridges of end-capped octadecyl silica. The extracted PQ was quantitated by HPLC using 1,1′-diethyl-4,4′-dipyridyl dichloride as an internal standard (IS). The LOQ was found to be 0.025 μg/mL. This technique was devoid of interferences from muscle relaxants (pancuronium bromide and vecuronium bromide) and anticoagulants (heparin and K_2EDTA) [39]. Lee et al. [40] have developed a new ion-pair HPLC method with column-switching for the determination of PQ in human serum. The diluted serum sample was injected onto a precolumn packed with LiChroprep RP-8 (25–40 μm), and polar serum components were washed out by 3% acetonitrile in 0.05 M phosphate buffer (pH 2.0) containing 5 mM sodium octanesulfonate. After valve switching to inject position, concentrated compounds were eluted in the back-flush mode and separated on an Inertsil ODS-2 column with 17% acetonitrile in 0.05 M phosphate buffer (pH 2.0) containing 10 mM sodium octanesulfonate. The total analysis time per sample was about 30 min and the mean recovery was 98.5% ± 2.8%, with a linear range of 0.1–100 μg/mL [40]. Arys et al. [41] have quantified PQ in a victim of massive ingestion of this herbicide. Sample preparation involved a protein-precipitation step using trichloroacetic acid (necessary only for blood and tissue homogenate), followed by a chemical reduction with sodium borohydride of the fully ionized PQ to a diene, which is amenable to solvent extraction. The extract was subjected to HPLC with diode array detection. Quantitative results were obtained for all postmortem matrices available: blood: 5.05 mg/L, urine: 6.00 mg/L, stomach contents: 17.2 g/L, liver: 4.86 mg/kg, and kidney: 80.6 mg/kg [41].

A simple, sensitive, reliable, and economical method for the simultaneous determination of PQ and diquat in human biological materials using HPLC has been reported [42]. The herbicides were extracted from the autopsy sample with a Sep-Pak C18 cartridge and subjected to HPLC with the IS, L-tyrosine. Paraquat and diquat were clearly separated on the octadecylsilica column with a mobile phase of 0.5% potassium bromide in 5% methanol solution, containing triethylamine (1 mL/L). Two UV wavelengths were selected, 256 nm for PQ as well as the IS, and 310 nm for diquat. The calibration curves were linear in the concentration range 0.1–10 μg/g, and the limit of detection (LOD) was 0.05 μg/g [42]. Lee et al. [43] have determined PQ and diquat in human whole blood and urine by HPLC-MS/MS. The herbicides were extracted with Sep-Pak C18 cartridges from whole blood and urine samples containing ethyl PQ as an IS. The separation of herbicides was carried out using ion-pair chromatography with heptafluorobutyric acid in 20 mM ammonium acetate and acetonitrile gradient elution for successful coupling with MS. The recoveries of PQ and diquat were 80.8%–95.4% for whole blood and 84.2%–96.7% for urine. The calibration curves showed excellent linearity in the range of 25–400 ng/mL of whole blood and urine. The LOD was 10 ng/mL for PQ and 5 ng/mL for diquat in both body fluids [43]. A simple HPLC method has been described to quantify diquat in biological fluids permitting the separation and quantification of diquat from blood, bile, urine, liver, and kidney [44]. The procedure does not require special pretreatment of the samples prior to analysis and offers excellent recovery (95%–105%). This method was applied for the quantification of diquat in toxicological samples from diquat treated rats [44]. Recently, a rapid and sensitive HPLC method for the simultaneous determination of PQ and diquat in human serum has been developed by Hara et al. [45]. After deproteinization of the serum with 10% trichloroacetic acid, the samples were separated on a reversed-phase column, and subsequently reduced to their radicals with alkaline sodium hydrosulfite solution and monitored with a UV detector at 391 nm. This method permitted the reliable quantification of PQ over linear ranges of 50 ng to 10 μg/mL and 100 ng to 10 μg/mL for diquat in human serum. This technique was also utilized to determine the PQ and diquat serum levels in a patient who had ingested herbicide formulation containing PQ and diquat [45].

Hori et al. [46] have developed a new HPLC method for the quantification of glufosinate in human serum and urine. The p-nitrobenzoyl derivative of glufosinate was produced quantitatively over 10 min at room temperature, and was isolated from biological specimens by reversed-phase chromatography using Inertsil Ph-3 and detected by UV absorption at 273 nm. The LOD of the method has been reported to be 0.005 μg/mL, and the recovery rate was at least 93.8%. The method was also applied for the analysis of glufosinate in serum samples from patients intoxicated by ingestion of glufosinate. This technique is also applicable for glyphosat, which possesses a chemical structure similar to glufosinate, and has been suggested to be of great use for the determination of these two compounds [46]. Dickow et al. [47] have developed a method for simultaneous determination of 2,4-dichlorophenoxyacetic acid (2,4-D) and 2-methyl-4-chlorophenoxyacetic acid (MCPA) in canine plasma and urine. The extracted herbicides were derivatized with 9-anthrylmethane (ADAM) for the analysis by reversed-phase HPLC with fluorescence detection. Precision and accuracy were within the accepted limits of 15% and 85%–115%, respectively, for both analytes in plasma and urine. Calibration curves for 2,4-D and MCPA in plasma were linear between 0.50 and 5.0 mg/L and 5.0 and 100 mg/L, respectively, and in urine they were linear between 5.0 and 70.0 mg/L and 10.0 and 70.0 mg/L, respectively. The LOD was found to be 62.5 ng/mL for both 2,4-D and MCPA [47]. A comparative evaluation of HPLC and capillary electrophoresis (CE) for the determination of 2,4-dichlorophenoxypropionic acid (dichlorprop, 2,4-DP) in the case of herbicide intoxication has been performed [48]. Body fluids and tissues obtained at autopsy were analyzed for 2,4-DP by HPLC and CE. The concentrations of 2,4-DP in cardiac blood, stomach contents, bile, liver, spleen, kidney, and brain found by both methods were very similar [48].

Han et al. [49] have described a method for the determination of diuron and chlortoluron in beef and beef products with HPLC. The sample was extracted with a mixture of acetonitrile and methanol (50:50, v/v). After filtration, the filtrate was defatted with petroleum ether, and then water was added and further extracted with chloroform. The chloroform was evaporated in a rotary evaporator at 45°C. The residue was dissolved in acetonitrile–methanol (50:50, v/v) mixture, poured into an alumina column and eluted with the same mixture. The eluate was collected for HPLC analysis. The analytical column was Selectosil C18 (5 μm, 250 mm × 4.6 mm i.d.), mobile phase was methanol–water (60:40, v/v), and detection wavelength was 245 nm. The LOD for diuron and chlortoluron were 0.4 and 0.5 ng, respectively. Recoveries were 87.34%–87.64% for diuron and 88.78%–91.94% for chlortoluron [49].

20.1.4.2 Insecticides

Most of the new classes of insecticides are biodegradable; they are quickly catabolized in the body and excreted in urine. Estimation of their specific metabolites in blood or urine is commonly employed to assess the exposure level of parent insecticides. Montesano et al. [50] have reported an HPLC method for the determination of urine-specific biomarkers of various insecticides. Oneto et al. [51] have determined urinary p-nitrophenyl sulfate, p-nitrophenyl glucuronide, and free p-nitrophenol using reversed-phase HPLC with a C18 column on day 3 in a fatal case of acute parathion ingestion. *p*-Nitrophenyl sulfate amounted to about 81% of the total conjugates excreted. The excreted *p*-nitrophenol was equivalent to 76 mg parathion (lethal dose in humans between 20 and 100 mg). No changes in the concentrations of *p*-nitrophenol or its conjugates were noticed in the urine samples stored frozen over a 1 year period [51]. Bar et al. [52] developed a rapid, selective, and high-throughput method for quantifying *p*-nitrophenol, a biomarker of methyl parathion exposure using isotope dilution HPLC-tandem mass spectrometry. Smith et al. [53] have reported an HPLC-UV method for the simultaneous analysis of urinary 2-methyl-3-phenylbenzoic acid (MPA), a metabolite of bifenthrin and 3-phenoxybenzoic acid (3-PBA), a metabolite of several other common pyrethroid insecticides with a detection limit of 2.5 ng/mL. This method revealed that MPA ranged from 1.8 to 31.9 μg/g creatinine and PBA from 1.3 to 30.0 μg/g in the urine of pest control workers [53]. Yao et al. [54] have used HPLC method for the determination of three metabolites of deltamethrin namely, dibromovinyl-dimethylcyclopropane carboxylic acid, 3-phenoxybenzyl-hydroxy-ethyl acetate, and 3-phenoxyl-benzoic acid in the urine of spray men and one suicidal case.

Ageda et al. [55] investigated the stability of 14 organophosphorus insecticides in fresh blood (Table 20.1). Methyl phosphate types (dichlorvos) decomposed most rapidly followed by methyl thiophosphate types (fenitrothion and cyanophos) and methyl dithiophosphate types (methidathion, dimethoate, and thiometon). Methyl thiophosphate types decomposed faster than ethyl thiophosphate types (isoxathion and diazinon). Of the five methyl dithiophosphate type insecticides (malathion, phenthoate, methidathion, dimethoate, and thiometon), the compounds with a carboxylic ester bond (malathion and phenthoate) decomposed faster than the others. Temperature had a great effect on the decomposition of organophosphorus insecticides in blood; compounds left standing at

TABLE 20.1
Residual Levels of Organophosphorus Insecticides in Fresh Blood Incubated at Three Different Temperatures over 24 h

Insecticide	Residual Level After 24 h (%)		
	37°C	Room Temperature	4°C
Dichlorvos	0 (1 h)	0 (2 h)	0 (12 h)
Malathion	0 (8 h)	0	56
Trichlorfon	0 (8 h)	0	86
Phenthoate	14	39	91
Fenitrothion	18	72	85
Cyanophos	26	64	101
Methidathion	37	79	95
Dimethoate	57	83	93
Thiometon	62	92	100
Isoxanthion	65	91	92
EPN	72	89	104
Acephate	78	88	102
Diazinon	79	94	100
Sulprofos	109	107	114

Source: Reproduced from Ageda, S. et al., *Leg. Med.*, 8, 144, 2006. With permission.

The residual level is the concentration as a percentage of the concentration measured immediately after the addition of each organophosphorus compound, assumed to be 100%. Values in parentheses indicate the time at which the compounds could not be detected.

37°C decomposed faster than those at 4°C. Thus, in cases of suspected organophosphate poisoning, it should be considered that the blood concentration of the compound might decrease during the postmortem interval [55]. Meyer et al. [56] have developed a specific method for the quantification of fenthion in postmortem matrices using SPE combined with HPLC-diode array detector (HPLC-DAD). Fenitrothion was selected as the IS. The analytes are desorbed with 5 mL of dichloromethane, and aliquots of the extract are subjected to HPLC analysis using gradient elution with a mixture of methanol and water (10:90 to 90:10, v/v) containing 0.0125 M NaOH on an Aluspher RP-Select B column monitoring at 250 nm. This method was applied to a suicidal case involving unsuspected acute intoxication with fenthion and the blood concentration was found to be 3.8 µg/mL [56]. Cho et al. [57] have developed a simple and rapid method for measuring 11 organophosphorous insecticides (dichlorvos, methidathion, salithion, malathion, fenitrothion, fenthion, parathion, diazinon, ethylthiometon, O-ethyl O-(4-nitrophenyl)-phenylphosphonothioate, and chlorpyrifos) and one metabolite (3-methyl-4-nitrophenol) of fenitrothion in serum and urine of acute poisoning patients by HPLC-DAD. Biological sample was deproteinized using acetonitrile prior to injection into C18 column using acetonitrile–water as a mobile phase. The detection limits in serum and urine ranged from 0.05 to 6.8 µg/mL at a wavelength of 230 nm [57]. Jadhav et al. [58] have performed quantitative analysis of malathion using HPLC in six cases of suspected poisoning. Various body tissues and fluids including lungs, liver, kidneys, spleen, brain, heart, blood, muscles, urine, and gastric contents of all the cases were analyzed, and malathion was found positive. Sharma et al. [59] reported a method for rapid quantitative analysis of organophosphorous and carbamate pesticides using HPLC. Good separation was obtained among organophosphorus pesticides (methyl parathion, malathion, phosphomidon, monocrotophos, dichlorvos, and quinalphos) and carbamates (carbaryl and baygon) with a detection limit of 100 ng for all the pesticides. Recovery studies were made in the blood, lung, and liver and were found to be 85%–97% with reproducibility at greater than 95%. The method was suggested to be useful for the analysis of biological samples for the presence of organophosphorus and carbamate pesticides in poisoning cases [59].

Ramagiri et al. [60] performed stability study for propoxur in whole blood and urine samples stored over varying periods (0–60 days) at four different temperature conditions while applying SPE with a weak cation exchange cartridge for sample purification and HPLC-photodiode array detector (HPLC-DAD) for quantitation. Propoxur was spiked at two different concentration levels (10 and 100 µg/L) in both blood and urine samples. After 60 days storage of blood and urine samples at 25°C and 4°C, the decrease in concentration of propoxur was found to be 95% and 60%, respectively. The stability of propoxur was inversely proportional with temperature, and the pesticide was found to be stable below −20°C. The time-dependent decrease of propoxur in urine and blood samples was suggested to be of considerable significance in forensic toxicology [60]. Proenca et al. [61] analyzed aldicarb in a suicidal case using HPLC with a postcolumn derivatization with o-phthaldialdehyde and 2-mercaptoethanol and fluorescence detection at excitation and emission wavelengths of 339 and 445 nm. The toxic concentrations of aldicarb in the postmortem samples were found to be: blood (6.2 µg/mL), stomach (48.9 µg/g), liver (0.80 µg/g), kidney (8.10 µg/g), heart (6.70 µg/g), and urine (17.50 µg/mL) [61]. Nisse et al. [24] applied HPLC-DAD for the estimation of aldicarb in one of the deaths following aldicarb ingestion and reported its levels in blood (6.04 µg/mL), urines (1.88 µg/mL), and gastric contents (3.98 µg/mL). Tracqui et al. [62] assayed aldicarb by HPLC in 21 blood and 8 urine samples, successively taken during hospitalization of a nonfatal case of aldicarb poisoning. Blood aldicarb level was 3.11 µg/mL at the time of hospitalization and peaked 3.5 h later (3.22 µg/mL) and then followed a two-slope decay with a terminal half-life of about 20 h. Aldicarb was detected in all the urine samples with a peak level of 6.95 µg/mL at 31.5 h after hospitalization and was still present at the time of discharge [62]. Ichinoki et al. [63] used automatic on-line column enrichment technique followed by reversed-phase HPLC with photometric detection for the determination of carbaryl and propanil in human serum (ng/mL level) and urine (µg/mL level). The serum was filtered through a membrane filter (0.45 µm) and an aliquot of 0.1 mL of the filtrate was diluted with water up to 1 mL. The solution of 0.8 mL was directly injected to automatic HPLC without any preparation. Urine was incubated with bcta-glucuronidase/arylsulfate for 16 h at 37°C. The resultant solution was filtered through a membrane filter, and the filtrate was analyzed by the similar manner as serum [63]. Carbaryl and propanil in the sample solution were concentrated on an ODS mini-column and then separated using an ODS analytical column (Cosmosil 5 C18-MS) with acetonitrile/water (30:70, v/v) and detected with a UV detector at 220 and 210 nm for carbaryl and propanil, respectively [63]. Hori et al. [64] have determined carbaryl, propanil, and dichloroaniline in human serum using HPLC-UV detection. The LOQ in the serum were found to be 0.005 µg/mL for propanil and dichloroaniline and 0.001 µg/mL for carbaryl. When the three compounds were added to serum at concentrations ranging from 0.1 to 10.0 µg/mL, their recovery rates ranged between 91.1% and 101.9% [64]. A simple and sensitive HPLC method with fluorescence detection has been reported for simultaneous analysis of carbaryl and 1-naphthol in whole blood [65]. Spiked blood samples containing an IS were hemolyzed and extracted with ethyl acetate. After centrifugation, the extractant was evaporated to dryness, reconstituted and subjected to HPLC. A simple chemical hydrolysis study of carbaryl was also included to illustrate the effectiveness of the extraction procedure and assay [65]. Kim et al. [66] have developed and validated a rapid, sensitive, and selective HPLC method for the determination of pyrethroid insecticide deltamethrin in plasma and tissues. The LOD for deltamethrin was found to be 0.01 µg/mL for plasma. The method performances were linear over the concentration range of 0.01–20.0 µg/mL with accompanied recovery range of 93%–103% of deltamethrin in plasma [66]. The presence of lindane in blood was confirmed by HPLC [67].

20.1.4.3 Fungicides

Debbarh et al. [68] have described an HPLC assay for the determination of mancozeb in urine. The fungicide is derivatized with 1,2-benzenedithiol to yield a cyclocondensation product, 1,3-benzodithiole-2-thione, which is quantitated by reversed-phase HPLC at 365 nm using a microBondapak C18 column. The mobile phase is methanol/water (70:30, v/v). The assay has been reported to be linear from 0.25 to 100 µg/mL and the LOD and LOQ have been found as 0.1 and 0.25 µg/mL, respectively [68]. Dhananjeyan et al. [69] have developed a specific HPLC method for the simultaneous detection and determination of vinclozolin and its three degradation products. A baseline separation of vinclozolin and its degradation products was found with symmetrical peak shapes on an XTerra MS C18 column using 10 mM ammonium bicarbonate at pH 9.2 and acetonitrile as mobile phase. A linear calibration curve was obtained across a range from 5 to 200 µmol. for vinclozolin. Greater than 90% recoveries of vinclozolin from biofluids including plasma, serum, and urine have been obtained in a single step with a single solvent [69].

However, many investigators have used HPLC for the determination of various metabolites of fungicides as suitable biomarkers of parent compound's exposure in humans. Lindh et al. [70] have reported an HPLC/MS/MS method for the analysis of 3,5-DCA as a biomarker of exposure to fungicides vinclozolin and iprodione in human urine. The urine samples were treated by basic hydrolysis to degrade the fungicides, their metabolites, and conjugates to 3,5-DCA. Extraction was done using toluene and derivatization with pentafluoropropionic anhydride and analyzed using selected reaction monitoring (SRM) in the negative ion mode. The LOD was determined to be 0.1 ng/mL. The metabolites in urine were found to be stable during storage at −20°C. The dose was recovered as 3,5-DCA, between 78% and 107%, in the urine after 200 µg single oral exposure in volunteers [70]. A sensitive and reliable method to assess occupational exposure to vinclozolin based on biomonitoring principles has been reported [71]. The conditions for pretreating the human urinary samples are chosen to completely degrade vinclozolin metabolites, containing the intact 3,5-DCA moiety, into this amine by means of basic hydrolysis. After addition of 3,4-DCA as an IS and steam distillation and extraction, the analysis is carried out by HPLC and electrochemical detection. The LOQ of this method is 5 µg 3,5-DCA/L urine [71].

Ethylene thiourea (ETU) is a metabolite of EBDCs and is regarded as the best indicator of exposure to these fungicides. Sottani et al. [72] have developed rapid and selective HPLC/MS/MS method for the determination of the ETU in human urine. The purification system is based on the use of a Fluorosil phase of a BondElut column followed by a liquid–liquid extraction resulting in the mean extracted recoveries of more than 85%. The assay is linear over the range 0–50 μg/L with LOD and LOQ of 0.5 μg/L and 1.5 μg/L, respectively. The assay was applied to quantify ETU in human urine from growers who were regularly exposed to fungicides; their ETU urine levels varied between 1.9 and 8.2 μg/L [72]. Another simple and rapid high-performance liquid chromatographic method has been reported for the determination of ETU in biological fluids[37]. Samples were chromatographed on a Lichrosorb RP8 column after extraction with dichloromethane. The mobile phase was a mixture of hexane/isopropyl alcohol/ethyl alcohol (93:6:1 v/v) containing 0.6 mL/L butylamine and the detection was done with a UV detector set at 243 nm. The LOD was found to be 20 ng/mL in plasma, 25 ng/mL in 0.9% saline, and 0.5 ng/mL in urine [37]. El Balkhi et al. [73] optimized and validated an HPLC method for the determination of ETU in human urine. Urine samples were extracted by SPE using Extrelut and analyzed using HPLC-DAD set at 231 nm. The analyses were carried out using a mobile phase of 0.01 M phosphate buffer (pH 4.5) on a C18 Uptisphere NEC-5-20 column. The IS used was 4-pyridinecarboxylic acid hydrazide. The LOQ of this method was at 1 μg/L [73].

20.2 DETERMINATION OF HERBICIDES

20.2.1 Determination of Chlorophenoxy and Benzonitrile Herbicides in Blood

20.2.1.1 Materials and Methods

20.2.1.1.1 Herbicides
2,4-Dichlorophenoxyaceitc acid (2,4-D), 4-(2,4-dichlorophenoxy)butyric acid (2,4-DB), 2,4,5-trichlorophenoxyacetic acid (2,4,5-T), 2-(2,4-dichlorophenoxy)propionic acid (DCPP or dichlorprop), and 2-(4-chloro-2-methylphenoxy)propionic acid (MCPP or mecoprop) were obtained from FBC (Hauxton, U.K.). Ioxynil was purchased from May & Baker (Dagenham, U.K.). Standard solutions of herbicides are prepared in the range of 0.1–0.4 g/L from methanolic stock solution (5 g/L).

20.2.1.1.2 Internal Standard
2-(2,4,5-trichlorophenoxy)propionic acid (2,4,5-TP or fenoprop) (1 g/L solution in 50% methanol in Tris buffer).

20.2.1.1.3 Buffers and Solutions
Phosphate buffer: 67 mM (pH 7.4).
Tris buffer: 20 mM (pH 9.6).
Methanolic hydrochloric acid: 2 mL concentrated HCl/L (23 mM).

20.2.1.1.4 Mobile Phase
Aqueous potassium dihydrogen orthophosphate (0.05 M, pH 3.5):acetonitrile (3:1).

20.2.1.1.5 Standard and Sample Preparation
Standard solutions of chlorophenoxy herbicides prepared in deionized water or in 50% aqueous methanol showed some decomposition after 5–6 weeks at room temperature. This problem can be overcome by preparing solutions in Tris buffer:methanol (1:1) resulting in at least 6 months stability of calibration solutions at 2°C–8°C. The spiked equine serum matrices are stable for at least a year at −20°C [74].

An aliquot (100 μL) of whole blood or plasma/serum or standard is vortex mixed with 20 μL of IS solution. Methanolic hydrochloric acid (200 μL) is added to the tube and the contents were vortex mixed for 30 s followed by centrifugation at 10,000 g for 2 min. The supernatant is used for the analysis of herbicides as reported by Flanagan and Ruprah [74].

20.2.1.1.6   Chromatography

The HPLC system composed of isocratic pump, UV-visible detector, and a stainless steel column (250 × 5 mm i.d.) packed with Spherisorb S5 Phenyl (Hichrom, Reading, U.K.) is used at an optimal temperature. The flow rate of mobile phase is 1.8 mL/min and the injection volume is 10–20 μL. The analytes are detected by UV absorption at 240 nm [74].

20.2.1.2   Results and Discussion

The chromatogram for the separation of various chlorophenoxy and benzonitrile herbicides is shown in Figure 20.1. The retention times (RTs) and the assay precision are given in Table 20.2. The results of plasma:whole blood ratio of herbicides (Table 20.3) indicate that these herbicides are largely distributed in plasma; hence, the findings of whole blood analyses must be multiplied by a factor of ≈2 to facilitate comparison with plasma/serum data.

   This method is free from the interferences of endogenous compounds, drugs, or other pesticides [74]. In this method, 2,4,5-TP has been used as IS; however, if the presence of 2,4,5-TP is antici-pated in the sample then MCPP can be used as an alternative to 2,4,5-TP for IS. Flushing the HPLC

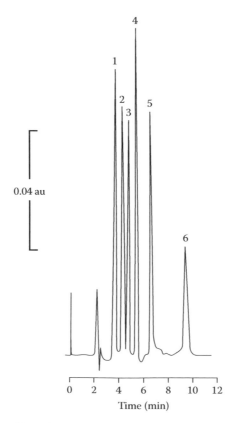

FIGURE 20.1   HPLC of some chlorophenoxy and benzonitrile herbicides. Peak identities are (1) 2,4-D, (2) DCPP, (3) MCPP, (4) 2,4,5-TP, (5) ioxynil, and (6) 2,4-DB. (Reproduced from Flanagan, R.J. and Ruprah, M., *Clin. Chem.*, 35, 1342, 1989. With permission.)

TABLE 20.2
RT and Intra- and Interassay Precision for the Determination of Chlorophenoxy and Benzonitrile Herbicides (N = 10)

Herbicide	RT Relative to 2,4,5-TP	Intra-Assay CV (%)	Interassay CV (%)
2,4-D	0.75	3.5	6.5
DCPP	0.84	2.9	7.2
2,4,5-T	0.86	4.1	7.5
MCPP	0.93	3.2	5.7
Ioxynil	1.16	2.9	4.3
2,4-DB	1.53	—	—

Source: Reproduced from Flanagan, R.J. and Ruprah, M., *Clin. Chem.*, 35, 1342, 1989. With permission.

TABLE 20.3
Whole Blood versus Plasma Levels of Chlorophenoxy and Benzonitrile Herbicides Analyzed by HPLC

Herbicide	Concentration in Whole Blood (mg/L)	Concentration Measured (mg/L) Whole Blood	Plasma	Plasma/Blood Ratio
2,4-D	400	380	750	1.97
DCPP	400	390	720	1.85
2,4,5-T	400	390	770	1.97
MCPP	400	390	720	1.85
Ioxynil	100	95	190	2.00
2,4-DB	400	400	640	1.60

Source: Reproduced from Flanagan, R.J. and Ruprah, M., *Clin. Chem.*, 35, 1342, 1989. With permission.

column with pure acetonitrile or methanol after use helps to maintain column efficiency. Acidified methanol has been found to be more efficient than methanol alone in precipitating plasma proteins; this also helps to prolong column life. The LOD of this method is 20 mg/L for chlorophenoxy herbicides and 10 mg/L for ioxynil [74].

20.2.2 DETERMINATION OF BIPYRIDINIUM HERBICIDES IN PLASMA AND SERUM

20.2.2.1 Materials and Methods

20.2.2.1.1 Herbicide
1,1'-Dimethyl-4,4'-bipyridinium dichloride (paraquat dichloride) was purchased from Sigma Chemical Company (St. Louis, MO).

20.2.2.1.2 Internal Standard
Diethyl PQ diiodide was kindly supplied by Dr. Bruce Woollen from Zeneca Pharmaceuticals (Macclesfield, U.K.).

20.2.2.1.3 Solutions
Methanolic perchloric acid: 6% Perchloric acid (v/v) in methanol.

20.2.2.1.4   Mobile Phase
A solution containing 3 mM 1-octanesulfonic acid and 100 mM orthophosphoric acid in 900 mL of distilled water; the pH of this solution is adjusted at 3.0 with the addition of diethylamine. Acetonitrile is then added to give a 10% (v/v) proportion [75].

20.2.2.1.5   Standard and Sample Preparation
A stock solution of PQ (1 mg/mL) is prepared in distilled water (DW). For the calibration curve, a further dilution is made to 50 µg/mL in DW. The diethyl PQ solution (210 µg/mL) is freshly prepared in DW. Plasma, serum, and water standards with known amounts of PQ (0, 0.5, 1.0, 2.5, 5.0, and 10.0 µg/mL) are prepared. To 200 µL of different standards or samples, 5 µL of IS and 45 µL of 6% methanolic perchloric acid are added. The contents are vortex mixed, incubated for 10 min at −18°C, and then centrifuged at 2000 *g* for 10 min. The supernatants are used for HPLC analysis [75].

20.2.2.1.6   Chromatography
The HPLC system composed of isocratic pump, UV-visible detector, a NovaPak C18 column (150 × 4.5 mm, 5 µm particle size; Waters Corporation, Milford, MA) and a C18 precolumn (100 × 4.6 mm, 10 µm particle size; HPLC Technologies, Cheshire, U.K.) is used at an optimum temperature. The flow rate of mobile phase is kept at 0.8 mL/min and the analytes are detected by UV absorption at 258 nm [75].

20.2.2.2   Results and Discussion

This is a simple, rapid, and sensitive HPLC method for the determination of PQ in human plasma and serum [75]. The sample treatment prior to analysis consists only of a protein-precipitation step by methanolic perchloric acid. The representative chromatograms (Figure 20.2) are free from any interference between PQ, IS, and peaks from plasma. The RTs of PQ and IS are 6.5 and 9.5 min, respectively. Similar behavior is observed with serum samples. The calibration curves are linear

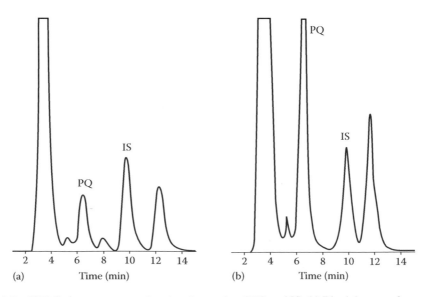

FIGURE 20.2   HPLC chromatograms showing the peaks of PQ and IS. (a) Blank human plasma with 1 µg/mL of PQ and 1.9 µg/mL of IS added. (b) Plasma of a patient of PQ poisoning with 1.9 µg/mL of IS added. (Reproduced from Paixao, P. et al., *J. Chromatogr. B*, 775, 109, 2002. With permission.)

TABLE 20.4
Recovery, Precision, and Accuracy of the Determination of Paraquat in Plasma and Serum (N = 6)

Paraquat Added (μg/mL)	Paraquat Found (μg/mL)		Recovery (%)		CV (%)	
	Serum	Plasma	Serum	Plasma	Serum	Plasma
0.50	0.59 ± 0.03	0.52 ± 0.04	100.06	98.81	4.75	7.25
1.00	1.10 ± 0.05	1.02 ± 0.06	100.70	103.03	4.36	5.98
2.50	2.40 ± 0.06	2.46 ± 0.12	91.69	98.87	2.50	4.68
5.00	4.80 ± 0.15	4.98 ± 0.18	93.10	98.48	3.08	3.56
10.00	10.09 ± 0.18	10.07 ± 0.31	98.70	100.74	1.77	3.05

Source: Reproduced from Paixao, P. et al., *J. Chromatogr. B*, 775, 109, 2002. With permission.

over the study concentration range and can be extrapolated to at least 50 μg/mL. The LOD and LOQ of this method are around 0.1 and 0.4 μg/mL, respectively, for both plasma and serum [75]. The interday precision and accuracy are presented in Table 20.4.

20.2.3 DETERMINATION OF PHENYLUREA HERBICIDES IN URINE

20.2.3.1 Materials and Methods

20.2.3.1.1 Herbicides
3-(3,4-Dichlorophenyl)-1,1-dimethylurea (diuron) and 3-(3,4-dichlorophenyl)-1-methoxy-1-methylurea (linuron).

20.2.3.1.2 Solvents
Acetonitrile, dichloromethane, methanol, *n*-hexane, and *n*-heptane—all of analytical grade.

20.2.3.1.3 Binders
Poly-methyloctylsiloxane (PMOS) and poly-methyloctadecylsiloxane (PMODS) from Petrarch Silanes and Silicones.

Silica for SPE: Irregular particles of 40–63 μm, pore size 10 nm (Fluka).

Silica for HPLC stationary phase: Spherical particles of 5 μm, pore size 8 nm (Sphrisorb, Phase Separations).

20.2.3.1.4 Mobile Phase
Acetonitrile:water (40:60)

20.2.3.1.5 Standard and Sample Preparation
Stock solutions of herbicides (100 μg/mL) are prepared in methanol. The solutions used to construct calibration curves and to spike urine samples are prepared in mobile phase and stored at 4°C. The standard solutions of herbicides may range between 20 and 1000 μg/L.

The urine samples are kept frozen at −20°C until analyzed. Prior to analysis, the urine is thawed and shaken to homogenize. For recovery analysis, urine samples (2 mL) are fortified with herbicides (40, 80, and 160 μg/L). The samples are basified by the addition of 200 μL of ammonium hydroxide (pH ≈ 9) and diluted with 4 mL of acetonitrile. Deproteinization is carried out by centrifugation (3000 *g* for 5 min). An aliquot (3 mL) of the supernatant is diluted with 20 mL of DW and percolated through SPE cartridge under vacuum at a flow rate of 3 mL/min. The SPE cartridge is prepared using silica and PMODS as reported earlier [76]. Before sample application, the SPE cartridge is conditioned with 10 mL of methanol and equilibrated with 5 mL of DW. After passing the sample, the cartridge is washed with 5 mL of DW and the sorbent bed is dried under vacuum for 3 min. The analytes are eluted with 3 mL of dichloromethane. The solvent is evaporated to dryness under a stream of nitrogen, and the residue is dissolved in 200 μL of acetonitrile for injection (10 μL) into HPLC column [76].

20.2.3.1.6 Chromatography

The HPLC system is composed of isocratic pump, injector with 10 μL loop, and UV-visible detector. The HPLC column is prepared using silica and PMOS as reported earlier [76]. The chromatography is performed at ambient temperature. The flow rate of mobile phase is kept at 0.8 mL/min and the UV detection of herbicides is performed at 254 nm [76].

20.2.3.2 Results and Discussion

A good separation of analytes (diuron and linuron) has been obtained from spiked urine samples; the peaks are well resolved and free from any matrix interference (Figure 20.3). Table 20.5 shows the recovery and the intra- and interassay precision of the method. Recoveries obtained by triplicate analysis of urine spiked with herbicides at three levels of fortification are given. The results of LOD and LOQ before and after preconcentration are presented in Table 20.6. The main advantages of this procedure to obtain new SPE and HPLC materials are good performance, lower cost, simplicity, and reduction of toxic residues [76].

20.2.4 DETERMINATION OF SULFONYLUREA HERBICIDES IN URINE

20.2.4.1 Materials and Methods

20.2.4.1.1 Herbicides

Bensulfuron methyl (99%), chlorsulfuron (98%), ethametsulfuron methyl (98%), halosulfuron methyl (99%), metsulfuron methyl (99%), primisulfuron methyl (99%), prosulfuron (98%), rimsulfuron (99%), sulfometuron methyl (98%), sulfosulfuron (98%), triasulfuron (98%), and triflusulfuron methyl (98%) were purchased from Chem Service (West Chester, PA). Foramsulfuron (97%), mesosulfuron methyl (98%), nicosulfuron (97%), oxasulfuron (97%), and thifensulfuron methyl (98%) were purchased from EQ Laboratories (Augsburg, Germany).

20.2.4.1.2 Internal Standards

Stable, isotopically labeled ($^{13}C_3$)-ethametsulfuron methyl (98%) was a generous gift from DuPont Corporation (Wilmington, DE). $^{13}C_6$-labeled 3-PBA was purchased from Cambridge Isotope Laboratories, Inc. (Andover, MA).

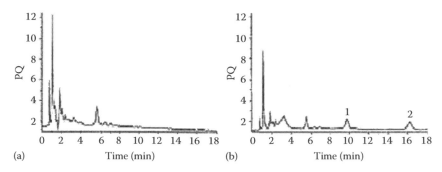

FIGURE 20.3 HPLC chromatograms of (a) blank urine and (b) urine spiked with herbicides. Peak identity is (1) diuron and (2) linuron. (Reproduced from Pozzebon, J.M. et al., *J. Chromatogr. A*, 987, 381, 2003. With permission.)

TABLE 20.5
Intra- and Interassay Precision for HPLC Determination of Diuron and Linuron (N = 3)

			Precision, RSD (%)	
Herbicide	Addition (µg/L)	Recovery (%)	Intra-Assay	Interassay
Diuron	40	95	1.80	
	80	99	0.38	1.6
	160	103	1.40	
Linuron	40	85	1.50	
	80	98	0.78	1.0
	160	100	0.37	

Source: Reproduced from Pozzebon, J.M. et al., *J. Chromatogr. A*, 987, 381, 2003. With permission.

TABLE 20.6
LOD and LOQ of Diuron and Linuron in Matrix-Matched Standards (N = 3)

	LOD (µg/L)		LOQ (µg/L)	
	Preconcentration		Preconcentration	
Herbicide	Zerofold	Fivefold	Zerofold	Fivefold
Diuron	14	2.8	40	8
Linuron	22	4.5	60	12

Source: Reproduced from Pozzebon, J.M. et al., *J. Chromatogr. A*, 987, 381, 2003. With permission.

20.2.4.1.3 SPE Cartridges
Oasis HLB 3 cc SPE cartridges (Waters Corporation, Milford, MA).

20.2.4.1.4 Mobile Phase
A (water with 0.1% acetic acid) + B (acetonitrile with 0.1% acetic acid).

20.2.4.1.5 Sample Preparation
Urine is collected from multiple donors, combined and mixed overnight at 4°C. After pressure filtering with a 0.2 µm filter capsule to remove bioparticulates, the urine is spiked with herbicide to yield an approximate concentration of 0, 3, and 10 µg/L. Urine (2 mL) is spiked with 25 µL of IS, resulting in a concentration of about 6 µg/L. After adding 1.5 mL of acetate buffer (pH 5), the sample tubes are briefly vortex mixed before being extracted with SPE cartridges. The SPE cartridges are first conditioned with 1 mL of methanol followed by 1 mL water. After the samples have been loaded onto the cartridges, they are washed with 1 mL of 5% methanol in water. Samples are eluted with 2 mL of 100% methanol into 20 mL conical tubes. Sample extracts are concentrated to dryness at 40°C using 10 psi of nitrogen for 30 min. After concentration, residual sulfonylurea herbicides are rinsed from the walls of the test tubes by adding 0.35 mL of 100% methanol to each tube and vortex mixing for 5 s. After concentrating for 7 min in addition, samples are reconstituted with 50 µL of acetonitrile, briefly vortex mixed, and transferred to autoinjection vials for analysis [77].

20.2.4.1.6 Chromatography

The analysis has been performed on an HP 1100 liquid chromatograph (Agilent Technologies, Palo Alto, CA), with a chilled (10°C) autosampler, interfaced to a Sciex API4000 triple quadrupole mass spectrometer (Applied Biosystems/MDS Sciex, Foster City, CA) as reported earlier [77]. The column is Synergi Polar–RP–80A column, 4 μm, 100 mm × 4.6 mm (Phenomenex, Torrance, CA) being held at constant temperature of 35°C. The injection volume is 10 μL and the flow rate of mobile phase is 1 mL/min. Initial mobile phase conditions are 55% A and 45% B. The analytes are separated using a gradient elution. From 0 to 6.5 min, B is increased to 64%. From 6.5 to 6.6 min B is increased to 100% and held until 7.5 min. From 7.5 to 7.6 min, B is decreased back to the starting conditions for a 3.9 min column equilibration period. Analyte RTs are between 2 and 6.5 min. The MS is operated in the SRM mode using negative turboionspray (TIS) atmospheric pressure ionization (API). The TIS heater temperature has been 650°C and the gas pressures of the collision-activated dissociation (CAD), nebulizer, and heater gases are 7, 18, and 15 psi, respectively. Zero air is used for CAD, nebulizer, and heater gases [77]. Nitrogen is used as curtain gas at 35 psi. The RTs, elution order, and ion masses are listed in Table 20.7.

20.2.4.2 Results and Discussion

Chromatographic RTs of the sulfonylurea herbicides are less than 7 min (Table 20.7). All analytes had chromatographic RTs that deviated for no more than 2 s from the mean value. A total ion current chromatogram with RTs, mass spectrometer acquisition periods, and elution order is shown in Figure 20.4. The LOD ranged from 0.05 to 0.10 μg/L with an average LOD of 0.06 μg/L (Table 20.8). Extraction efficiencies of the SPE cartridges ranged from 79% to 97%.

TABLE 20.7
HPLC Elution Order, RTs, and Mass Analyzer Information for Determination of Sulfonylurea Herbicides

		Mass		
			Product Ion	
Herbicide	HPLC RT (min)	Precursor Ion	Quantitation Ion	Confirmation Ion
Foramsulfuron	2.18	451	296	268
Nicosulfuron	2.21	409	154	228
Oxasulfuron	2.88	405	182	122
Thifensulfuron methyl	2.91	386	139	246
Metsulfuron methyl	3.08	380	139	182
Sulfometuron methyl	3.33	363	182	122
Chlorsulfuron	3.38	356	139	190
Ethametsulfuron methyl	3.46	409	168	182
Rimsulfuron	3.46	430	186	179
Triasulfuron	3.47	400	139	198
Mesosulfuron methyl	3.60	502	347	267
Sulfosulfuron	4.13	469	288	154
Bensulfuron methyl	4.41	409	254	154
Prosulfuron	5.04	418	139	252
Halosulfuron	5.42	433	252	154
Triflusuluron methyl	6.12	491	236	196
Primisulfuron methyl	6.24	467	226	176

Source: Reproduced from Baker, S.E. et al., *Anal. Bioanal. Chem.*, 383, 963, 2005. With permission.

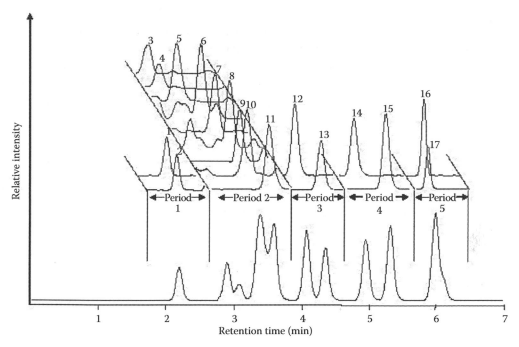

FIGURE 20.4 Individual ion chromatograms (upper) and total ion chromatogram (lower) showing the elution order of sulfonylurea herbicides. Peak identities are given in Table 20.6. (Reproduced from Baker, S.E. et al., *Anal. Bioanal. Chem.*, 383, 963, 2005. With permission.)

TABLE 20.8
LOD, Extraction Efficiency of SPE, and Recovery of Herbicides from Urine (N = 10)

Herbicide	LOD (μg/L)	SPE Extraction Efficiency		Total Recovery	
		%	SD	%	SD
Foramsulfuron	0.06	83	3.8	53	6.9
Nicosulfuron	0.10	89	3.2	59	7.2
Oxasulfuron	0.06	90	5.0	70	6.0
Thifensulfuron methyl	0.08	97	9.7	82	9.2
Metsulfuron methyl	0.05	91	5.3	76	5.8
Sulfometuron methyl	0.05	88	5.8	71	5.1
Chlorsulfuron	0.05	90	6.9	79	4.3
Ethametsulfuron methyl	0.10	90	7.6	70	6.1
Rimsulfuron	0.06	90	5.8	75	4.4
Triasulfuron	0.07	97	6.2	81	6.1
Mesosulfuron methyl	0.05	86	4.8	71	9.0
Sulfosulfuron	0.05	85	7.0	75	5.4
Bensulfuron methyl	0.05	85	6.8	76	6.2
Prosulfuron	0.05	87	6.2	78	5.9
Halosulfuron	0.07	84	6.9	80	5.5
Triflusuluron methyl	0.05	83	6.3	76	6.4
Primisulfuron methyl	0.06	79	5.7	72	6.3

Source: Reproduced from Baker, S.E. et al., *Anal. Bioanal. Chem.*, 383, 963, 2005. With permission.

Total recoveries using the method, which included all losses during sample preparation, ranged from 53% to 82%. Analyte concentration did not affect extraction efficiencies or total recoveries. The difference between the total recovery and the extraction efficiency, which reflects the loss of analyte in the evaporation and reconstitution steps, averaged about 15%. However, for two compounds, foramsulfuron and nicosulfuron, the difference was more than 30%. Because sulfonylurea herbicides are generally thermally labile and prone to rapid decay, their degradation properties, especially in urine and solvent, are of particular concern [77]. After periodically monitoring the degradation of all analytes in both urine and solvent matrices at $-10°C$, $6°C$, $23°C$, and 37°C, half-lives have been calculated for each analyte at 23°C and 37°C and the number of days required for 10% degradation have been estimated at $-10°C$, 6°C, and 23°C. Chemical degradation of sulfonylurea herbicides in acetonitrile and urine have been monitored over 250 days. Estimated days for 10% and 50% degradation in urine and acetonitrile ranged from 0.7 to >318 days [77]. Dadgar et al. [78] have recommended storing standards used as the reference standard for stability tests at temperatures lower than $-70°C$.

20.3 DETERMINATION OF INSECTICIDES

20.3.1 DETERMINATION OF ORGANOPHOSPHORUS INSECTICIDES IN SERUM

20.3.1.1 Materials and Methods

20.3.1.1.1 Insecticides
Acephate, methidathion, dichlorvos, fenthion, EPN, diazinon, phenthoate, malathion, fenitrothion, and cyanophos were obtained from Wako Pure Chemical Industries (Osaka, Japan). The stock solutions of each insecticide (1 mg/mL) are prepared in methanol. All the stock solutions were stored at $-20°C$ in the dark when not in use. The working solutions are prepared by diluting the stock solution with methanol [79].

20.3.1.1.2 Internal Standard
Diazinon-d10 and fenitrothion-d6 were obtained from Hayashi Pure Chemical Industrial, Co., Ltd. (Osaka, Japan). The working solution for the ISs, fenitrothion-d6 (3 µg/mL), and diazinon-d10 (50 µg/mL) are prepared by diluting an aliquot of respective stock solution with methanol.

20.3.1.1.3 Mobile Phase
Gradient mobile phase with (A) 10 mM ammonium formate in water and (B) methanol was used. The gradient solutions were filtered through a 0.45 µm Omnipore membrane filter before use.

20.3.1.1.4 Standard and Sample Preparation
Serum calibration standards of insecticides (0.1, 0.25, 0.5, 1, 2, 5, and 8 µg/mL) are prepared by spiking the working standard solutions into a pool of drug-free human serum (200 µL) followed by the addition of IS (5 µL). Diazinon-d10 is used for acephate, methidathion, dichlorvos, fenthion, EPN, diazinon, phenthoate, and malathion, and fenitrothion-d6 is used as IS for fenitrothion and cyanophos. For all samples, 200 µL of acetonitrile is added to the mixture. The resulting mixture is vortex mixed for 1 min and then centrifuged at 3000 g for 5 min. The supernatant is filtered through a 0.45 µm Millex-LH filter prior to injection in HPLC column [79].

20.3.1.1.5 Chromatography
The HPLC system composed of a pump (LC-10A; Shimadzu, Kyoto, Japan), a detector (SPD-10A; Shimadzu), an XTerra MS C18 stainless steel cartridge column (2.1 × 100 mm, 3.5 µm; Waters, Milford, MA), and an XTerra MS C18 guard column (2.1 × 20 mm, 3.5 µm; Waters) at 50°C. The elution gradient was 0% B to 100% B (0–3 min), 100% B (3–9.5 min), and 100% B to 0% B (9.5–10 min) at a flow rate of 0.3 mL/min. The mass spectrometer was a triple quadrupole QP8000α (Shimadzu) equipped with an APCI interface operating in the positive or negative

TABLE 20.9
RTs and Ions Observed for Various Organophosphate Insecticides

Insecticide	RT (min)	Ion Observed (m/z)	
		Positive Ion Mode	Negative Ion Mode
Acephate	3.72	143, 184	—
Methidathion	5.42	144.9, 338.2	157, 286.85
Dichlorvos	5.17	221, 338	—
Fenthion	5.85	279	262.95
EPN	5.99	294	293.85
Diazinon	5.88	305	—
Phenthoate	5.77	321, 247.9	—
Malathion	5.55	331	—
Diazinon-d10 (IS)	5.88	315	—
Fenitrothion	5.61	—	262
Cyanophos	5.33	—	228
Fenitrothion-d6 (IS)	5.61	—	265

Source: Reproduced from Inoue, S. et al., *J. Pharm. Biomed. Anal.*, 44, 258, 2007. With permission.

mode. The parameters of the interface and detector are optimized in single MS full scan mode (*m/z* 50–500) with direct injections of 10 µL of a 10 ng/µL standard solution. Nitrogen is used as a nebulizer gas with a flow rate of 2.5 L/min, APCI probe temperature at 400°C, CDL temperature at 250°C, and the detector voltage of 2.3 V [79].

20.3.1.2 Results and Discussion

This is a simple screening procedure for 10 organophosphorus insecticides in urine. To simplify the process, a small quantity (200 µL) of acetonitrile is used for extraction without further evaporation to dryness because some pesticides, such as dichlorvos are thermally unstable. The total extraction procedure takes less than 10 min and does not require the use of a large amount of organic solvents. Although three freeze–thaw cycles and storage (at 4°C for 7 days and at −30°C for 4 weeks) before analysis had little effect on the quantification; storage at room temperature for 24 h caused the decomposition of few compounds (dichlorvos and malathion). The RTs and ions for the identification of respective insecticides are given in Table 20.9. For quantification, molecular target ions of the 10 insecticides are used in the positive and negative selected ion monitoring (SIM) modes as shown in chromatograms (Figures 20.5 and 20.6). The method is free from matrix interferences. For all the compounds, the calibration curves are linear up to 8 µg/mL and the mean results within 20% of the expected concentration. The LODs of insecticides in serum range from 0.125 to 1 µg/mL and the LOQs range from 0.25 to 1.25 µg/mL (Table 20.10). This method has been successfully applied to a case of acute poisoning; the determined serum concentrations of acephate and fenitrothion have been found to be 7.2 and 4.5 µg/mL, respectively [79].

20.3.2 Determination of Aldicarb in Blood and Urine

20.3.2.1 Materials and Methods

20.3.2.1.1 Insecticide

Aldicarb standard was obtained from Riedel-de Haen. Stock solution (1 mg/mL) was prepared in acetonitrile and stored at 4°C.

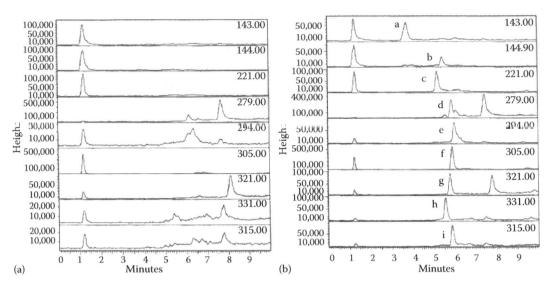

FIGURE 20.5 LC-MS chromatograms obtained in the positive ion mode. Selected ion chromatograms obtained by the analysis of (a) blank serum and (b) blank serum spiked with organophosphate insecticides (each 7.5 μg/mL). Peaks: a = acephate, b = methidathion, c = dichlorvos, d = fenthion, e = EPN, f = diazinon, g = phenthoate, h = malathion, and i = diazinon-d10 (IS). (Reproduced from Inoue, S. et al., *J. Pharm. Biomed. Anal.*, 44, 258, 2007. With permission.)

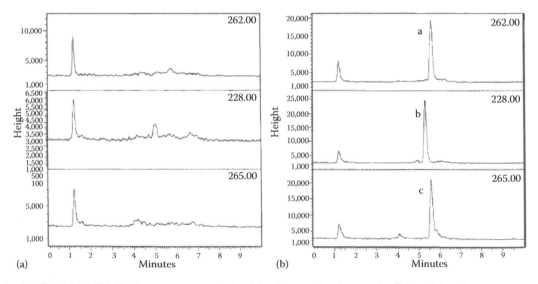

FIGURE 20.6 LC-MS chromatograms obtained in the negative ion mode. Selected ion chromatograms obtained by the analysis of (a) blank serum and (b) blank serum spiked with organophosphate insecticides (each 7.5 μg/mL). Peaks: a = fenitrothion, b = cyanophos, and c = fenitrothion-d6 (IS). (Reproduced from Inoue, S. et al., *J. Pharm. Biomed. Anal.*, 44, 258, 2007. With permission.)

20.3.2.1.2 Mobile Phase

The mobile phase is a nonlinear gradient of water/methanol/acetonitrile. The mobile phase is filtered with a 0.20 μm filter and degassed in an ultrasonic bath for 15 min just before use.

20.3.2.1.3 Standard and Sample Preparation

Standard working solutions are prepared from stock solution at concentrations of 5–50 μg/mL diluted with water, which has been acidified to pH 3.0 with concentrated HCl. Control and calibration

TABLE 20.10
Sensitivity and Linearity Parameters for Various Organophosphate Insecticides

Insecticide	LOD (μg/mL)	LOQ (μg/mL)	Linearity (μg/mL)
Acephate	0.250	0.375	0.375–8.000
Methidathion	0.500	0.625	0.625–8.000
Dichlorvos	0.500	0.625	0.625–8.000
Fenthion	1.000	1.250	1.250–8.000
EPN	0.375	0.500	0.500–8.000
Diazinon	0.125	0.250	0.250–8.000
Phenthoate	0.250	0.375	0.375–8.000
Malathion	0.250	0.375	0.375–8.000
Fenitrothion	0.125	0.250	0.250–8.000
Cyanophos	0.125	0.250	0.250–8.000

Source: Reproduced from Inoue, S. et al., *J. Pharm. Biomed. Anal.*, 44, 258, 2007. With permission.

samples are prepared by spiking drug-free blood samples with standard solutions. The samples are extracted three times with dichloromethane and filtered by anhydrous sodium sulfate. Supernatants are evaporated to dryness under a slow stream of nitrogen, at 40°C. The dried extracts are reconstituted with 250 μL water (pH 3) and an aliquot (10 μL) is injected into the HPLC system [61].

20.3.2.1.4 Chromatography

The HPLC system used is a Waters Carbamate Analysis System including a Model 600E Multisolvent Delivery System Controller, Fluid Handling Unit with integral postcolumn reaction system at 80°C, and a 7125 Rheodyne injector with a 10 μL loop. The flow rate of mobile phase is 1.5 mL/min. The separation is performed on a Waters carbamate analysis column, C18 (3.9 × 150 mm, 4 μm) at 25°C and the carbamate is detected using scanning fluorescence model 470 (Waters) detector set at $\lambda_{ex} = 339$ nm and $\lambda_{em} = 445$ nm [61].

20.3.2.2   Results and Discussion

Calibration curve of aldicarb in blood samples has been found to be linear over the concentration range of 0–50 μg/mL, with a correlation coefficient of 0.9998. The LOD of aldicarb is 1 ng/mL. This method has been used for toxicological analysis of the postmortem samples of aldicarb poisoning case and revealed the concentrations of aldicarb in blood and urine to be 6.2 and 17.5 μg/mL, respectively [61]. Most carbamate pesticides are thermally labile. Owing to their thermal instability and high polarity, HPLC methods are preferred over GC methods.

20.3.3   Determination of Pyrethroid Insecticide Deltamethrin in Plasma

20.3.3.1   Materials and Methods

20.3.3.1.1   Insecticide

Standard deltamethrin (DLM) was kindly provided by Bayer Crop Science AG (Monheim, Germany). A stock solution of deltamethrin is prepared in acetonitrile at a concentration of 1.0 mg/mL. The stock solution is stored at −20°C; though deltamethrin is believed to be stable at least 6 months at room temperature.

20.3.3.1.2   Mobile Phase

The mobile phase is a mixture (80:20, v/v) containing acetonitrile and sulfuric acid (1%, v/v).

20.3.3.1.3 Standard and Sample Preparation

Working standard solutions with concentrations of 0.05, 0.1, 0.5, 1.25, 1.5, 2.5, 3.75, 5.0, 10.0, 15.0, 25.0, and 100 µg/mL are prepared by diluting the stock solution with acetonitrile. Mixtures of appropriate volume (30 µL) of a working solution and plasma matrix (120 µL) are prepared for calibration standards. The final calibration standards are 0.01, 0.02, 0.1, 0.25, 0.3, 0.5, 0.75, 1.0, 2.0, 3.0, 5.0, and 20.0 µg deltamethrin/mL [66]. Plasma sample (65 µL) is added to 130 µL of acetonitrile in a microcentrifuge tube and vortex mixed for 30 s, followed by centrifugation at 13,000 rpm for 5 min. The clear supernatant is injected onto the column.

20.3.3.1.4 Chromatography

A Shimadzu HPLC (Shimadzu, Canby, OR) consisted of a pump (LC-10AT), degasser (DGU-14A), autosampler (SIL-HT), and detector (SPD-10AV). The analytical column was an Ultracarb 5 ODS (20) column (250 × 4.6 mm; 5 µm particle) (Phenomenex, Torrance, CA), and the guard column was a Phenomenex fusion RP 4 × 3 mm (Torrance, California). The flow rate of mobile phase is set at 1.0 mL/min and the eluate is monitored at 230 nm [66]. Under these chromatographic conditions, DLM was eluted at approximately 14.5 min.

20.3.3.2 Results and Discussion

This HPLC method requires a very simple extraction protocol; the direct injection of an aliquot of supernatant into the column without taking the samples through an evaporation process render the procedure rapid and sensitive for quantitation of deltamethrin in blood.

The RT for deltamethrin is 14.5 min with baseline resolution and there are no interfering or coeluting peaks with similar RTs in the chromatograms of blank biological samples (Figure 20.7). The calibration curve for plasma shows good linearity over the range from 0.01 to 20.0 µg/mL. The LOD and LOQ are found to be 0.01 and 0.05 µg/mL, respectively [66]. The intraday precision and accuracy of this method are given in Table 20.11. The absolute recoveries of deltamethrin from spiked plasma in the range of 93%–103% have been found to be higher than those (83%–89%) observed earlier for an HPLC method possibly owing to the evaporation and reconstitution steps used in that report [80].

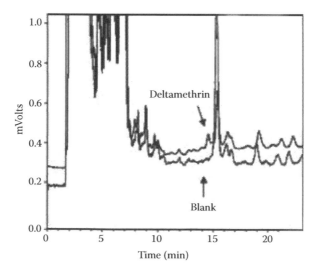

FIGURE 20.7　Chromatograms of blank and deltamethrin (0.05 µg/mL)-spiked plasma. (Reproduced from Kim, K.B. et al., *J. Chromatogr. B*, 834, 141, 2006. With permission.)

TABLE 20.11

Intraday Precision and Accuracy of Deltamethrin Analysis

Concentration of Deltamethrin (μg/mL)		Accuracy (% Error)	Precision (% RSD)
Added	Found		
4.00	4.03 ± 0.29	4.7	7.2
1.50	1.57 ± 0.10	5.0	6.4
0.20	0.19 ± 0.00	4.8	1.6
0.10	0.09 ± 0.01	7.4	5.5
0.05	0.047 ± 0.004	9.1	8.2

Source: Reproduced from Kim, K.B. et al., *J. Chromatogr. B*, 834, 141, 2006. With permission.

20.4 DETERMINATION OF FUNGICIDES

20.4.1 DETERMINATION OF FUNGICIDES IPRODIONE AND VINCLOZOLIN IN URINE

20.4.1.1 Materials and Methods

20.4.1.1.1 Fungicides

The stock solutions of iprodione and vinclozolin are prepared in acetonitrile at a concentration of 1 mg/mL. These solutions can be stored at −20°C for 1 month without any appreciable decomposition [81].

20.4.1.1.2 Internal Standard

Procymidone is used as IS. The stock solution is prepared in the same way as for fungicides. This solution is also stable for 1 month at −20°C.

20.4.1.1.3 Mobile Phase

The mobile phase consisted of a mixture of acetonitrile and water (60:40, v/v).

20.4.1.1.4 Standard and Sample Preparation

Standard solutions of fungicides are prepared by diluting the stock solutions with blank urine in the range 10–1000 ng/mL. For each solution, the internal standard is added at a constant level of 100 μL of a 100 μg/mL acetonitrile solution. These standards are treated concurrently in the same manner as the samples to be analyzed [81]. Urine samples are collected and kept frozen at −20°C until use. Prior to analysis, the urine samples are thawed, shaken for homogenization, and the required volume is sampled as quickly as possible to avoid sedimentation of any solids.

20.4.1.1.5 Extraction of Fungicides

Samples are thawed just before the extraction procedure, thoroughly agitated, and centrifuged at 1000 g for 10 min. The Isolute cartridges are placed in a luer that fitted the top of the Supelco vacuum manifold. A vacuum of 250–500 Torr is applied to the manifold to carry out various steps of extraction. Prior to the introduction of urine samples, the cartridge is rinsed with 5 mL of chloroform–methanol mixture (9:1, v/v) followed by 5 mL of methanol to desorb any organic impurities from it and to wet the silica packing. Urine sample (20 mL) is mixed with IS (100 μL) and passed through the cartridge followed by 10 mL of water. The fungicides are eluted with 3 mL of chloroform–methanol mixture (9:1, v/v). The eluate is centrifuged, transferred to a new tube, and evaporated to dryness with a nitrogen stream under vacuum, reconstituted with 500 μL of mobile phase and filtered with 0.22 μm pore filter [81].

20.4.1.1.6 Chromatography

The HPLC system composed of a model 515 pump, a model 996 DAD and an analytical 250 × 4.6 mm i.d. reversed-phase Spherisorb ODS2 5 μm column (Waters, Milford, MA). The mobile phase is delivered at a flow rate of 1.0 mL/min and the injection volume is 20 μL. The column eluate is monitored at 220 nm [81].

20.4.1.2 Results and Discussion

Typical chromatograms of a blank human urine and urine sample spiked with iprodione, vinclozolin, and IS are shown in Figure 20.8. The elution peaks do not show any interferences deriving from other human urine components and are characterized by RTs of 8.8 (iprodione), 9.2 (IS), and 10.5 (vinclozolin) min. Figure 20.9 shows the chromatograms of human urine samples obtained from greenhouse operators. This chromatographic method is free from interferences by urine matrix and results in a fair resolution between vinclozolin and iprodione peaks. The accuracy of method based on the recovery of fungicides from spiked urine samples is given in Table 20.12. The LOD and LOQ of fungicides in urine are 10 and 50 ng/mL, respectively. The SPE procedure eliminates endogenous interference, which is frequently present in the biological sample. The filtration of extracts before

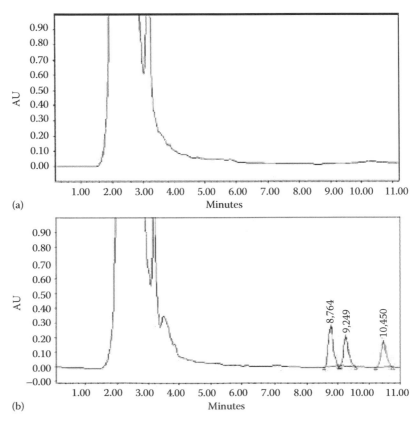

FIGURE 20.8 Typical chromatograms of human urine samples after extraction: (a) blank human urine and (b) blank human urine spiked with iprodione (50 ng/mL), IS (30 ng/mL), and vinclozolin (200 ng/mL). (Reproduced from Carlucci, G. et al., *J. Chromatogr. B*, 828, 108, 2005. With permission.)

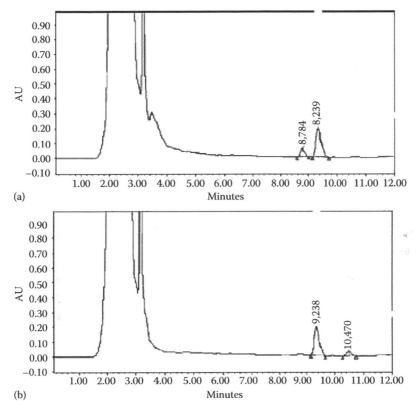

FIGURE 20.9 Chromatograms of human urine samples of two greenhouse operators after extraction: (a) after applications of iprodione and (b) vinclozolin, respectively. The concentration calculated are 42.5 ng/mL for iprodione and 64.4 ng/mL for vinclozolin. (Reproduced from Carlucci, G. et al., *J. Chromatogr. B*, 828, 108, 2005. With permission.)

TABLE 20.12
Accuracy of HPLC Method for Determination of Fungicides in Human Urine (N = 5)

Amount Added (ng/mL)	Vinclozolin		Iprodione	
	Amount Found (Mean ± SD)	Accuracy (%)	Amount Found (Mean ± SD)	Accuracy (%)
50	48.6 ± 0.5	−2.8	49.3 ± 1.2	−1.4
100	489.1 ± 0.9	−2.1	497.8 ± 0.4	−0.4
1000	997.6 ± 0.8	−0.2	996.8 ± 0.8	−0.3

Source: Reproduced from Carlucci, G. et al., *J. Chromatogr. B*, 828, 108, 2005. With permission.

injection onto chromatographic column avoids rapid obstruction of the precolumn and increases its life. The main advantage of this method is the rapid SPE procedure for the preparation of the biological sample, which gives high extraction yields and is convenient for the determination of iprodione and vinclozolin in urine samples with a typical assay time of about 12 min [81].

REFERENCES

1. Kiely, T., Donaldson, D., and Grube, A. 2004. *Pesticide Industry Sales and Usage: 2000 and 2001 Market Estimates*, U.S. EPA, Washington, DC.
2. Manahan, S.E. 2000. *Environmental Chemistry*, 7th Edition, CRC Press, Boca Raton, FL.
3. Mackay, D.W., Shiu, Y.K., and Ma, C. 1997. *Illustrated Handbook of Physico-Chemical Properties and Environmental Fate for Organic Chemicals*, Vol. 5, Pesticides Chemicals, NY.
4. Johnson, E.S. 1990. Association between soft tissue sarcomas, malignant lymphomas, and phenoxy herbicides/chlorophenols: Evidence from occupational cohort studies. *Fundam Appl Toxicol* 14: 219–234.
5. Filipov, N.M., Stewart, M.A., Carr, R.L., and Sistrunk, S.C. 2007. Dopaminergic toxicity of the herbicide atrazine in rat striatal slices. *Toxicology* 232: 68–78.
6. Dinis-Oliveira, R.J., Remião, F., Carmo, H., Duarte, J.A., Sánchez Navarro, A., Bastos, M.L. and Carvalho, F. 2006. Paraquat exposure as an etiological factor of Parkinson's disease. *Neurotoxicology* 27: 1110–1122.
7. Houze, P., Baud, F.J., Mouy, R., Bismuth, C., Bourdon, R., and Scherrmann, J.M. 1990. Toxicokinetics of paraquat in humans. *Hum Exp Toxicol* 9: 5–12.
8. Leroy, J.P., Volant, A., Guedes, Y., and Briere, J. 1982. Massive poisoning by paraquat with early death (19 h). Apropos of a case with ultrastructure study. *Ann Pathol* 2: 332–335.
9. Okonek, S., Baldamus, C.A., Hofmann, A., Schuster, C.J., Bechstein, P.B., and Zoller, B. 1979. Two survivors of severe paraquat intoxication by "continuous hemoperfusion". *Klin Wochenschr* 57: 957–959.
10. Berwick, B. 1970. 2,4-dichlorophenoxyacetic acid poisoning in man. *J Am Med Assoc* 214: 1114–1117.
11. Wells, W.D.E., Wright, N., and Yeoman, W.B. 1981. Clinical features and management of poisoning with 2,4-D and mecoprop. *Clin Toxicol* 18: 273–276.
12. Nielsen, K., Kaempe, B., and Jensen-Holm, J. 1965. Fatal poisoning in man by 2,4-dichlorophenoxyacetic acid (2,4-D): Detection of the agent in forensic material. *Acta Pharmacol Toxicol* 22: 224–234.
13. Kancir, C.B., Andersen, C., and Olesen, A.S. 1988. Marked hypocalcaemia in a fatal poisoning with chlorinated phenoxy acid derivatives. *Clin Toxicol* 26: 257–264.
14. Conso, F., Neel, P., Pouzoulet, C., Efthymiou, M.L., Gervais, P., and Gaultier, M. 1977. Acute toxicity in man of halogenated derivatives of hydroxybenzonitrile (ioxynil, bromoxynil). *Arch Mal Prof* 38: 674–677.
15. Dickey, W., McAleer, J.A., and Callender, M.E. 1988. Delayed sudden death after ingestion of MCPP and ioxynil: An unusual presentation of hormonal weedkiller intoxication. *Postgrad Med J* 64: 681–682.
16. Malato, S., Caceres, J., Fernandez-Alba, A.R., Piedra, L., Hernando, M.D., Aguera, A., and Vial, J. 2003. Photocatalytic treatment of diuron by solar photocatalysis: Evaluation of main intermediates and toxicity. *Environ Sci Technol* 37: 2516–2524.
17. Antony, M., Shukla, Y. and Mehrota, N.K. 1989. Tumor activity of a herbicide Diuron on mouse skin. *Cancer Lett* 30: 125–128.
18. De Silva, H.J., Samarawickrema, N.A., and Wickremasinghe, A.R. 2006. Toxicity due to organophosphorus compounds: What about chronic exposure? *Trans R Soc Trop Med Hyg* 100: 803–806.
19. Nagami, H., Nishigaki, Y., Matsushima, S., Matsushita, T., Asanuma, S., Yajima, N., Usuda, M., and Hirosawa, M. 2005. Hospital-based survey of pesticide poisoning in Japan, 1998–2002. *Int J Occup Environ Health* 11: 180–184.
20. Lee, W.J., Alavanja, M.C., Hoppin, J.A., Rusiecki, J.A., Kamel, F., Blair, A., and Sandler, D.P. 2007. Mortality among pesticide applicators exposed to chlorpyrifos in the Agricultural Health Study. *Environ Health Perspect* 115: 528–534.
21. Walker, B. Jr. and Nidiry, J. 2002. Current concepts: Organophosphate toxicity. *Inhal Toxicol* 14: 975–990.
22. Kwong, T.C. 2002. Organophosphate pesticides: Biochemistry and clinical toxicology. *Ther Drug Monit* 24: 144–149.
23. Ragoucy-Sengler, C., Tracqui, A., Chavonnet, A., Daijardin, J.B., Simonetti, M., Kintz, P., and Pileire, B. 2000. Aldicarb poisoning. *Hum Exp Toxicol* 19: 657–662.
24. Nisse, P., Deveaux, M., Tellart, A.S., Dherbecourt, V., Peucelle, D., and Mathieu-Nolf, M. 2002. Aldicarb poisoning: Review of cases in the North of France between 1998 and 2001. *Acta Clin Belg* (Suppl 1): 12–15.
25. Gupta, R.C. 1994. Carbofuran toxicity. *J Toxicol Environ Health* 43: 383–418.
26. Shafer, T.J., Meyer, D.A., and Crofton, K.M. 2005. Developmental neurotoxicity of pyrethroid insecticides: Critical review and future research needs. *Environ Health Perspect* 113: 123–136.

27. Hines, C.J., Deddens, J.A., Striley, C.A., Biagini, R.E., Shoemaker, D.A., Brown, K.K., MacKenzie, B.A. and Hull, R.D. 2003. Biological monitoring for selected herbicide biomarkers in the urine of exposed custom applicators: Application of mixed-effect models. *Ann Occup Hyg* 47: 503–517.
28. Alavanja, M.C., Dosemeci, M., Samanic, C., Lubin, J., Lynch, C.F., Knott, C., Barker, J., Hoppin, J.A., Sandler, D.P., Coble, J., Thomas, K. and Blair, A. 2004. Pesticides and lung cancer risk in the agricultural health study cohort. *Am J Epidemiol* 160: 876–885.
29. Garry, V.F., Tarone, R.E., Kirsch, I.R., Abdallah, J.M., Lombardi, D.P., Long, L.K., Burroughs, B.L., Barr, D.B. and Kesner, J.S. 2001. Biomarker correlations of urinary 2,4-D levels in foresters: Genomic instability and endocrine disruption. *Environ Health Perspect* 109: 495–500.
30. Gunnell, D., Eddleston, M., Phillips, M.R. and Konradsen, F. 2007. The global distribution of fatal pesticide self-poisoning: Systematic review. *BMC Public Health* 7: 357.
31. Cserhati, T., Forgacs, E., Deyl, Z., Miksik, I. and Eckhardt, A. 2004. Chromatographic determination of herbicide residues in various matrices. *Biomed Chromatogr* 18: 350–359.
32. López Monzón, A., Vega Moreno, D., Torres Padrón, M.E., Sosa Ferrera, Z., and Santana Rodríguez, J.J. 2007. Solid-phase microextraction of benzimidazole fungicides in environmental liquid samples and HPLC-fluorescence determination. *Anal Bioanal Chem* 387: 1957–1963.
33. Halko, R., Sanz, C.P., Ferrera, Z.S. and Rodriguez, J.J. 2004. Determination of benzimidazole fungicides by HPLC with fluorescence detection after micellar extraction. *Chromatographia* 60: 151–156.
34. Halko, R., Sanz, C.P., Ferrera, Z.S., and Rodriguez, J.J. 2006. Determination of benzimidazole fungicides in soil samples using microwave-assisted micellar extraction and liquid chromatography with fluorescence detection. *JAOAC Int* 89: 1403–1409.
35. Baggiani, C., Baravalle, P., Giraudi, G., and Tozzi, C. 2007. Molecularly imprinted solid-phase extraction method for the high-performance liquid chromatographic analysis of fungicide pyrimethanil in wine. *J Chromatogr A* 1141: 158–164.
36. De Melo, A.S., Correia, M., Herbert, P., Santos, L., and Alves, A. 2005. Screening of grapes and wine for azoxystrobin, kresoxim-methyl and trifloxystrobin fungicides by HPLC with diode array detection. *Food Addit Contam.* 22: 549–556.
37. Debbarh, I. and Moore, N. 2002. A simple method for the determination of ethylene-thiourea (ETU) in biological samples. *J Anal Toxicol* 26: 216–221.
38. Taylor, P.J., Salm, P., and Pillans, P.I. 2001. A detection scheme for paraquat poisoning: Validation and a five-year experience in Australia. *J Anal Toxicol* 25: 456–460.
39. Croes, K., Martens, F., and Desmet, K. 1993. Quantitation of paraquat in serum by HPLC. *J Anal Toxicol* 17: 310–312.
40. Lee, H.S., Kim, K., Kim, J.H., Do, K.S., and Lee, S.K. 1998. On-line sample preparation of paraquat in human serum samples using high-performance liquid chromatography with column switching. *J Chromatogr B* 716: 371–374.
41. Arys, K., Van Bocxlaer, J., Clauwaert, K., Lambert, W., Piette, M., Van Peteghem, C., and De Leenheer, A. 2000. Quantitative determination of paraquat in a fatal intoxication by HPLC-DAD following chemical reduction with sodium borohydride. *J Anal Toxicol* 24: 116–121.
42. Ito, S., Nagata, T., Kudo, K., Kimura, K., and Imamura, T. 1993. Simultaneous determination of paraquat and diquat in human tissues by high-performance liquid chromatography. *J Chromatogr* 617: 119–123.
43. Lee, X.P., Kumazawa, T., Fujishiro, M., Hasegawa, C., Arinobu, T., Seno, H., Ishii, A., and Sato, K. 2004. Determination of paraquat and diquat in human body fluids by high-performance liquid chromatography/tandem mass spectrometry. *J Mass Spectrom* 39: 1147–1152.
44. Madhu, C., Gregus, Z., and Klaassen, C.D. 1995. Simple method for analysis of diquat in biological fluids and tissues by high-performance liquid chromatography. *J Chromatogr B* 674: 193–196.
45. Hara, S., Sasaki, N., Takase, D., Shiotsuka, S., Ogata, K., Futagami, K., and Tamura, K. 2007. Rapid and sensitive HPLC method for the simultaneous determination of paraquat and diquat in human serum. *Anal Sci* 23: 523–526.
46. Hori, Y., Fujisawa, M., Shimada, K., Sato, M., Kikuchi, M., Honda, M., and Hirose, Y. 2002. Quantitative determination of glufosinate in biological samples by liquid chromatography with ultraviolet detection after p-nitrobenzoyl derivatization. *J Chromatogr B* 767: 255–262.
47. Dickow, L.M., Gerken, D.F., Sams, R.A., and Ashcraft, S.M. Simultaneous determination of 2,4-D and MCPA in canine plasma and urine by HPLC with fluorescence detection using 9-anthryldiazomethane (ADAM). *J Anal Toxicol* 25: 35–39.
48. West, A., Frost, M., and Kohler, H. 1997. Comparison of HPLC and CE for the analysis of dichlorprop in a case of intoxication. *Int J Legal Med* 110: 251–253.

49. Han, H., Li, J., Cao, S., and Huang, H. 1998. The determination of diuron and chlortoluron residues in beef and beef products by high performance liquid chromatography. *Se Pu* 16: 367–368.

50. Montesano, M.A., Olsson, A.O., Kuklenyik, P., Needham, L.L., Bradman, A.S., and Barr, D.B. 2007. Method for determination of acephate, methamidophos, omethoate, dimethoate, ethylenethiourea and propylenethiourea in human urine using high-performance liquid chromatography-atmospheric pressure chemical ionization tandem mass spectrometry. *J Expo Sci Environ Epidemiol* 17: 321–330.

51. Oneto, M.L., Basack, S.B., and Kesten, E.M. 1995. Total and conjugated urinary paranitrophenol after an acute parathion ingestion. *Sci Justice* 35: 207–211.

52. Barr, D.B., Turner, W.E., DiPietro, E., McClure, P.C., Baker, S.E., Barr, J.R., Gehle, K., Grissom, R.E. Jr., Bravo, R., Driskell, W.J., Patterson, D.G. Jr., Hill, R.H. Jr., Needham, L.L., Pirkle, J.L., and Sampson, E.J. 2002. Measurement of *p*-nitrophenol in the urine of residents whose homes were contaminated with methyl parathion. *Environ Health Perspect* 110: 1085–1091.

53. Smith, P.A., Thompson, M.J., and Edwards, J.W. 2002. Estimating occupational exposure to the pyrethroid termiticide bifenthrin by measuring metabolites in urine. *J Chromatogr B* 778: 113–120.

54. Yao, P.P., Li, Y.W., Ding, Y.Z., and He, F. 1992. Biological monitoring of deltamethrin in sprayers by HPLC method. *J Hyg Epidemiol Microbiol Immunol* 36: 31–36.

55. Ageda, S., Fuke, C., Ihama, Y., and Miyazaki, T. 2006. The stability of organophosphorus insecticides in fresh blood. *Leg Med* 8: 144–149.

56. Meyer, E., Borrey, D., Lambert, W., Van Peteghem, C., Piette, M., and De Leenheer, A. 1998. Analysis of fenthion in postmortem samples by HPLC with diode-array detection and GC-MS using solid-phase extraction. *J Anal Toxicol* 22: 248–252.

57. Cho, Y., Matsuoka, N., and Kamiya, A. 1997. Determination of organophosphorous pesticides in biological samples of acute poisoning by HPLC with diode-array detector. *Chem Pharm Bull* 45: 737–740.

58. Jadhav, R.K., Sharma, V.K., Rao, G.J., Saraf, A.K., and Chandra, H. 1992. Distribution of malathion in body tissues and fluids. *Forensic Sci Int* 52: 223–229.

59. Sharma, V.K., Jadhav, R.K., Rao, G.J., Saraf, A.K., and Chandra, H. 1990. High performance liquid chromatographic method for the analysis of organophosphorus and carbamate pesticides. *Forensic Sci Int* 48: 21–25.

60. Ramagiri, S., Kosanam, H., and Sai Prakash, P.K. 2006. Stability study of propoxur (Baygon) in whole blood and urine stored at varying temperature conditions. *J Anal Toxicol* 30: 313–316.

61. Proenca, P., Teixeira, H., de Mendonca, M.C., Castanheira, F., Marques, E.P., Corte-Real, F., and Nuno Vieira, D. 2004. Aldicarb poisoning: One case report. *Forensic Sci Int* 146: S79–S81.

62. Tracqui, A., Flesch, F., Sauder, P., Raul, J.S., Geraut, A., Ludes, B., and Jaeger, A. 2001. Repeated measurements of aldicarb in blood and urine in a case of nonfatal poisoning. *Hum Exp Toxicol* 20: 657–660.

63. Ichinoki, S., Takido, N., Fujii, Y., Morita, T., Ieiri, I., Otsubo, K., and Saitoh, N. 2003. Simultaneous determination of carbaryl and propanil in human serum and urine by on-line column-switching technique followed by automatic reversed-phase HPLC. *Chudoku Kenkyu* 16: 171–178.

64. Hori, Y., Nakajima, M., Fujisawa, M., Shimada, K., Hirota, T., and Yoshioka, T. 2002. Simultaneous determination of propanil, carbaryl and 3,4-dichloroaniline in human serum by HPLC with UV detector following solid phase extraction. *Yakugaku Zasshi* 122: 247–251.

65. DeBerardinis, M. Jr. and Wargin, W.A. 1982. High-performance liquid chromatographic determination of carbaryl and 1-naphthol in biological fluids. *J Chromatogr* 246: 89–94.

66. Kim, K.B., Bartlett, M.G., Anand, S.S., Bruckner, J.V., and Kim, H.J. 2006. Rapid determination of the synthetic pyrethroid insecticide, deltamethrin, in rat plasma and tissues by HPLC. *J Chromatogr B* 834: 141–148.

67. Seth, V., Ahmad, R.S., Suke, S.G., Pasha, S.T., Bhattacharya, A., and Banerjee, B.D. 2005. Lindane-induced immunological alterations in human poisoning cases. *Clin Biochem* 38: 678–680.

68. Debbarh, I., Titier, K., Deridet, E., and Moore, N. 2004. Identification and quantitation by high-performance liquid chromatography of mancozeb following derivatization by 1,2-benzenedithiol. *J Anal Toxicol* 28: 41–45.

69. Dhananjeyan, M.R., Erhardt, P.W., and Corbitt, C. 2006. Simultaneous determination of vinclozolin and detection of its degradation products in mouse plasma, serum and urine, and from rabbit bile, by high-performance liquid chromatography. *J Chromatogr A* 1115: 8–18.

70. Lindh, C.H., Littorin, M., Amilon, A., and Jönsson, B.A. 2007. Analysis of 3,5-dichloroaniline as a biomarker of vinclozolin and iprodione in human urine using liquid chromatography/triple quadrupole mass spectrometry. *Rapid Commun Mass Spectrom* 21: 536–542.

71. Will, W. 1995. Determination of vinclozolin metabolites in human urine by high-performance liquid chromatography and electrochemical detection. *Fresen J Anal Chem* 353: 215–218.

72. Sottani, C., Bettinelli, M., Lorena Fiorentino, M., and Minoia, C. 2003. Analytical method for the quantitative determination of urinary ethylenethiourea by liquid chromatography/ electrospray ionization tandem mass spectrometry. *Rapid Commun Mass Spectrom.* 17: 2253–2259.

73. El Balkhi, S., Sandouk, P., and Galliot-Guilley, M. 2005. Determination of ethylene thiourea in urine by HPLC-DAD. *J Anal Toxicol* 29: 229–233.

74. Flanagan, R.J. and Ruprah, M. 1989. HPLC measurement of chlorophenoxy herbicides, bromoxynil, and ioxynil, in biological specimens to aid diagnosis of acute poisoning. *Clin Chem* 35: 1342–1347.

75. Paixao, P., Costa, P., Bugalho, T., Fidalgo, C., and Pereira, L.M. 2002. Simple method for determination of paraquat in plasma and serum of human patients by high-performance liquid chromatography. *J Chromatogr B* 775: 109–113.

76. Pozzebon, J.M., Queiroz, S.C.N., Melo, L.F.C., Kapor, M.A., and Jardim, I.C.S. 2003. Application of new high-performance liquid chromatography and solid-phase extraction materials to the analysis of pesticides in human urine. *J Chromatogr A* 987: 381–387.

77. Baker, S.E., Olsson, A.O., Needham, L.L., and Barr, D.B. 2005. High-performance liquid chromatography-tandem mass spectrometry method for quantifying sulfonylurea herbicides in human urine: Reconsidering the validation process. *Anal Bioanal Chem* 383: 963–976.

78. Dadgar, D., Burnett, P., Choc, M.G., Gallicano, K., and Hooper, J.W. 1995. Application issues in bioanalytical method validation, sample analysis and data reporting. *J Pharmaceut Biomed* 13: 89–97.

79. Inoue, S., Saito, T., Mase, H., Suzuki, Y., Takazawa, K., Yamamoto, I., and Inokuchi, S. 2007. Rapid simultaneous determination for organophosphorus pesticides in human serum by LC-MS. *J Pharm Biomed Anal* 44: 258–264.

80. Ding, Y., White, C.A., Bruckner, J.V., and Bartlett, M.G. 2004. Determination of deltamethrin and its metabolites, 3-phenoxybenzoic acid and 3-phenoxybenzyl alcohol, in maternal plasma, amniotic fluid, and placental and fetal tissues by HPLC. *J Liq Chromatogr* 27: 1875–1892.

81. Carlucci, G., Di Pasquale, D., Ruggieri, F., and Mazzeo, P. 2005. Determination and validation of a simple high-performance liquid chromatographic method for simultaneous assay of iprodione and vinclozolin in human urine. *J Chromatogr B* 828: 108–112.

21 Analysis of Pesticide Residues in Animal Feed

Pramod Singh, Balbir K. Wadhwa,
and Aruna Chhabra

CONTENTS

21.1   INTRODUCTION

World over, chemical pesticide compounds viz., organochlorine (OC), organophosphorus (OP), organocarbamate (OCm), and pyrethroid (Ptd) contribute significantly in agricultural production, safer storage of food commodities, and public health security. These compounds remained the key factors, especially for the pest control technology in crop production. This unequaled value of pesticides brought an upsurge in synthesis and production of a large number of similar compounds. New technologies such as natural pesticides for crop protection and genomic interventions for the development of pest-resistant plant varieties have also been initiated, but their contribution is marginal. Synthetic pesticide compounds are chemically different and exhibit a variable degree of persistence in the environment. The OC compounds are nonpolar, most persistent, and bioaccumulative, whereas OP and OCm compounds are acute toxic and less persistent in nature. The later compounds are readily decomposed by physicochemical and enzymatic processes.

Residues of pesticide are now almost ubiquitous and a major concern for public health. The contamination of human foodstuffs of animal origin with pesticide residues primarily occurs due to the consumption of polluted feeds and fodders by the farm animals. By and large, the extent of pesticide residue in animal-based food products is dependent on its carryover ratio, which in turn depends on the diverse nature of biochemical activities during the digestion and metabolism in the animals [1,2].

The increased use of pesticides in crop protection increases the possibility of feed contamination [3]. Animal feedstuffs, such as cereal grains, brans, oilseed cakes/meals, green fodders, straws, stovers (kadabies), and agroindustrial by-products of cereals, oil seeds, pulses, vegetables and fruits, and so on, contain varying levels of different pesticides [4,5]. Their contamination may be due to (a) direct application on the field crops, (b) plant uptake of residues from soils, (c) application for seed treatment, (d) application for safer storage, (e) accidental or unintentional mixing of feeds with pesticides, and (f) storage of feeds in contaminated sacks that were previously used for packaging of pesticides [1]. The green fodders, however, are most likely contaminated due to uptake of residues from the soil and less likely due to direct application on the crops for protection against pests.

With increasing human health concern and advancement in the field of medical research, the maximum residue limits (MRL) of pesticide residues in different commodities are being reduced further. Most countries have laid permissible limits of pesticide residues in different ingredients of animal feeds through legislations. Under this situation, the pesticide analysis of animal feeds provides a safeguard [6].

Keeping in view the great diversity of pesticides and nature of feed samples, several types of methodologies have been developed during 40 years. Last decade alone witnessed the advent of various multiresidue analytical procedures with increasing number of pesticides. The pesticide analytical procedure for animal feeds involves special techniques for extraction and purification of samples in combination with classical separation and detection techniques. The essentials and use of gas and/or liquid chromatographic techniques with specific detectors for pesticide analyses have already been covered to a great extent in previous chapters of this book. In this chapter more emphasis is laid on the preparation and purification of feed sample.

21.2 PESTICIDE RESIDUE ANALYSIS IN FEEDS

The analysis of pesticide residues in different types of feed ingredients has been a time-consuming and cumbersome process. An enormous range of pesticides and the varied nature of animal feed ingredient samples for single or multiresidue extraction and cleanup procedures have encouraged a continuous development and refinement of these procedures. Several biological and chemical techniques viz., bioassay, immunoassay, thin-layer chromatography (TLC), column chromatography, and capillary electrophoresis can be used efficiently for the determination of pesticides, but gas chromatography (GC) and liquid chromatography (LC) are most popular. During recent years capillary GC with different sensitive and selective detectors including mass spectrometric (MS) detection has been used predominantly. LC with detectors such as ultraviolet (UV), photodiode array (PDA), fluorescence, and MS have also been used for such determinations [7].

Remarkable improvements have taken place in technologies of GC and LC for separation of pesticides and various detectors, especially MS in recent times. Table 21.1 lists an overview of different pesticide analytical techniques used for feeds or various commodities that can be used as animal feed. Sample preparation and separation-cum-detection are the two basic steps involved in the pesticide residue analysis of animal feeds.

21.2.1 SAMPLE PREPARATION

The preparation of the sample is a prerequisite to facilitate the introduction of analyte into the device that is capable of its separation and detection. It is usually achieved by extraction of the sample with a variety of combinations of organic solvents [8]. The residues of pesticide in feed ingredients are usually present in small quantities. The extraction step allows quantitative transfer of this small quantity from a relatively large volume of sample (which is either in solid, semisolid, or liquid state) into a liquid organic phase. The classical procedures involve homogenization of sample and extraction with polar or nonpolar organic solvents followed by liquid–liquid partitioning with other organic solvents of limited water-holding capacity. However, the method of extraction and the type of solvent are dependent on the nature of the sample and the chemical properties of the pesticide residues [9]. Certain cleanup steps are also required before the separation and detection of pesticides by gas and/or liquid chromatographic techniques.

TABLE 21.1
Methods for Multiresidue Pesticide Analysis of Animal Feeds or Commodities Which Can Be Used as Animal Feeds

Nature of Feed Sample	Extraction Method/Solvent	Partition/Cleanup Step	Instrument Conditions	Pesticides	Ref.
Concentrate feeds, fodders, and straws	MeCN with sodium sulfate	C18 SPE	LC–UVD	OC and OP	[2]
Cereal grains and other feedstuffs	MeCN	Dispersive SPE	GC with tandem quadruple MS	OC, OP, etc.	[6]
Rice	DCM	Florisil SPE		OC, OP, OCm, Ptd	[8]
Wheat middling	MeCN–water (65:35)	Florisil column	HPLC with hv-EC (photolysis-electrochemical detection)	OP	[10]
Fruits and vegetables	MeCN with NaCl and MgSO$_4$ as dehydrant	DSI	GC–MS	OC, OP, OCm, etc.	[13]
Postharvest plants used as animal feeds	DCM	—	Low-pressure GC/MS–MS	OC, OP, and Ptd	[14]
Cotton seed	Soxhet extraction (diethyl ether)	Silica gel column	GC–NPD	OP	[16]
Feed concentrates	Soxhet extraction; Acetone:hexane (1:1)	Partitioned into aqueous MeCN again into hexane for OC and DCM for OP	GC–ECD/NPD	OC, OP	[19]
Green fodder	MeCN	Partitioned into hexane for OC and DCM for OP	GC–ECD/NPD	OC, OP	[19]
Poultry feed	Soxhet extraction acetone–hexane (1:1)	Silica gel column	GC–ECD	OC	[20]
Vegetables	MeCN with anhydrous sodium sulfate	ENVI-Carb SPE	GC-MS and HPLC-fluorescence detector	OP, Ptd	[22]
Fruits and vegetables	EtAc	—	LC-MS/MS	OC and OP	[28]
Fish feeds	n-Hexane	ENVI Carb SPE	GC–ECD and MS	OC	[29]
Fruits and vegetables	MeCN with NaCl	C$_{18}$-SPE and aminopropyl SPE in tandem	GC–MS	OC, OP, Ptd	[30]
Animal feeds	ASE	Silica gel, GPC	GC–MS	OC	[34]
Animal feed concentrate	Microwave assisted, hexane–acetone (50:50)	SPE with three sorbents (alumina/ENVI™-Florisil®, ENVI-Carb and ENVI-Carb II/PSA)	GC/ECD and MS	OC	[35]
Cabbage	EtAc	Direct injection		OP	[43]
Soybean	—	—	LC-TOF MS	OC, OP, etc.	[47]

(continued)

TABLE 21.1 (continued)
Methods for Multiresidue Pesticide Analysis of Animal Feeds or Commodities Which Can Be Used as Animal Feeds

Nature of Feed Sample	Extraction Method/Solvent	Partition/Cleanup Step	Instrument Conditions	Pesticides	Ref.
Concentrate feed			HPLC–PDA GC–MS	OC and OP	[50]
Cereal grains and flour	MeCN–water (2:1), partitioned onto the petroleum ether	GPC and florisil column	GC–ECD	Ptd	[54]
Cereals, fruits, and vegetables	MeCN–water(2:1) for cereals and MeCN for fruits and vegetables, partitioned onto the hexane	Florisil	GC–ECD	Ptd	[55]
Plant-based foodstuffs	—	GPC	GC with atomic emission detection (AED)	OC, OP, etc.	[56]
Corn, oats, and wheat	MeCN–water	Graphitized carbon black and anion exchange SPE	GC with Hall electroconductivity and flame photometer detectors	OC and OP	[57]
Leaves (Ginkgo)	ASE	SPE	GC–FPD	OP	[58]
Barley	Acetone	GPC	GC–FPD	OC, OP	[59]
Roots	ASE	ENVI Carb SPE	GC–FPD	OP	[60]
Maize, oats, wheat, rice, and barley	ASE with MeCN and Celite	SPE (Envi-18 and by tandem Envicarb and Sep-Pak NH2 cartridges)	GC–MS and LC–MS/MS	OC, OP, etc.	[61]
Vegetable	EtAc	Dispersive SPE (using primary-secondary amine and graphitized carbon black)	GC–MS and LC–MS/MS	OC, OP	[62]
Wheat flour and vegetables	MeCN–water	Dispersive-SPE	GC–MS/MS	OC, OP	[63]
Malt, spent grains	Water–hexane for malt and spent grains partitioned on to the DCM	—	GC–ECD	OP, dinitroanilines, triazoles, etc.	[64]
Fruits and vegetables	MeCN with methanol and MgSO4 with sodium acetate as dehydrant	MgSO4 with primary and secondary amino sorbent	GC–MS and LC–MS/MS	OC, OP, OCm, etc.	[65]
Herbs (*Origanum syriacum*)	Soxhet extraction followed by MeCN–petroleum ether partition	Florisil	—	OC, etc.	[66]
Animal feeds	EtAc	GPC, PSA-SPE	Two-dimensional GC with TOF-MS	OC and OP, etc.	[67]
Fruits and vegetables	Acetone–EtAc–hexane (10:80:10) and acetone and hexane (5:95)	RP C18 SPE	GC–FPD	OP and OCm	[68]

21.2.1.1 Extraction

For extraction of pesticides from high-moisture-containing samples, most analytical methods still employ liquid–liquid partition. In case of low-moisture samples, a hydration step or aqueous polar organic solvent is generally used for extraction [2,6,10–12]. In the extraction step for high-moisture samples such as forages, fruits, and vegetables, frozen conditions are also used to check the enzymatic degradation. This is achieved by the use of either dry ice [13] or freezing mixtures. To dry the moist sample, lyophilization is also preferred by some workers [14].

In the practice of sample preparation, very often drying of extracted solvent is required and the presence of moisture, even in traces, hampers efficient evaporation of the organic solvent. To eliminate this moisture from the sample extract usually anhydrous sodium sulfate is used [15]; nevertheless, anhydrous magnesium sulfate provides better dehydration efficiency [13].

Classical Soxhlet extraction procedure with diethyl ether [16], acetonitrile [17], *n*-hexane [18], or acetone–hexane (1:1) mixture [19,20] and conventional liquid–liquid extraction (LLE) are routinely used for extracting pesticides from solid and liquid samples, respectively. Cyclic steam distillation process has also been used for the extraction of organochlorine pesticides (OCP) from animal feed samples [21].

Among the organic solvents used for the pesticide extraction from moist sample matrices, acetonitrile (MeCN) remains the most preferred [2,6,22]. Acetone [23], 2-propanol [24], acetone and hexane [25], ethyl acetate (EtAc; [26–28]), and dichloromethane (DCM; 15) are also among the preferred solvents for multipesticide extraction. The use of DCM for the multiresidue pesticide extraction from postharvest plant materials that are used as animal feeds [14] and rice [8] has been preferred over other organic solvents. For fatty matrices such as fish feeds, *n*-hexane was used efficiently for extraction of chlorinated pesticides [29]. After homogenization of samples with 2-propanol, the "salting out principal" was also used for transferring the pesticide residues into petroleum ether phase by addition of sodium chloride [24].

For extraction of residues of organophosphorus pesticides (OPP) from rice grain samples, acetone [23] and for extraction of multipesticide residues of OC and OP from a variety of agricultural products, EtAc [27] were specifically used. For extraction of multipesticide residues form animal feeds and fodders [2], vegetables [22,30], and fruits [30], MeCN was a privileged organic solvent because it minimized the coextraction of lipids from sample matrices.

Different combinations of solvents are used for pesticide extraction. A mixture of MeCN–methanol–acetone (3:1:1, v/v) was used [31]. For fruits and vegetables, a 40 g sample with 150 mL of a mixture of methylene chloride and EtAc (1:1, v/v) and 25 g sodium sulfate was macerated at 4000 rpm for 4 min. The extract was then cleaned up by passive diffusion using cyclohexane [15].

In recent years, there has been an increasing demand for new extraction techniques that are amenable to automation with shortened extraction times and reduced organic solvent consumption. This has incorporated new technologies to develop and use a procedure, which minimizes environmental concerns, time, labor, and exposure of laboratory personnel to toxic chemicals. This is also advantageous in reducing sample preparation costs.

Supercritical fluid extraction (SFE) technique is a good alternative for the organic solvent extraction of pesticide residues from nonfatty samples. SFE protocols extract the pesticide residues more selectively and eliminate postextraction cleanup steps, but the polarity range of the approach is compromised as these were generally developed for a single class of pesticides [32]. Accelerated solvent extraction (ASE; trade name Dionex), also known as pressurized liquid extraction (PLE) uses small volumes of extracting organic solvents. This technology has been used for extraction of pesticides from a varied range of sample matrices [33]. In an ASE (approved under method 3545A of EPA) based multiresidue extraction method for OC residues, a small quantity of fine-ground animal feed sample (1.0 g) is blended with a sorbent (Isolute; International Sorbent Technology, Ltd., UK) using a mortar and pestle. The sample is then placed into the extraction cell of the ASE system [34].

A number of other techniques such as microwave-assisted extraction (MAE; [35]) or matrix solid-phase dispersion (MSPD) have been used for solid-state samples [36]. Direct sample introduction (DSI) involves the placement of a little amount of sample material or liquid extract into a disposable micro vial. For GC analysis, it greatly minimizes the sample preparation and yet provides a good analytical approach for complex matrices [37]. For extracting analytes from liquid matrices, solid-phase extraction (SPE), solid-phase microextraction (SPME), stir-bar sorptive extraction (SBSE), and supported liquid membranes (SLM) are progressively replacing LLE techniques [35,38].

A multiresidue method appeared in the *Pesticide Analytical Manual* (PAM) in 1985 [39], where MeCN and acetone were used for extraction from fatty and nonfatty samples, respectively. It was followed by liquid–liquid partitioning and cleanup by florisil column. The Swedish National Food Administration developed a method using extraction by acetone followed by n-hexane/methylene chloride partitioning. The major disadvantage was that the extremely polar OPP were recovered poorly by this method. The change of acetone to EtAc as solvent and maceration in the presence of sodium sulfate provided better recoveries. The OPP can be isolated from oil and milk by mixing them with acetone and MeCN and passing through a column containing activated charcoal. Extraction can also be done with MeCN and the lipids can be removed by zinc acetate treatment [40].

21.2.1.2   Cleanup

Despite advances in the sensitivity of analytical instrumentation for the end-point determination of pesticide residues in feed samples, a pretreatment is usually required to extract and isolate them from the complex sample matrix. In most cases, although the pesticide components of interest are isolated from the sample, several contaminants as well as part of the matrix coextracted interfere in the determination step of the analysis too. Consequently, further refinement of the extract is obligatory before the determination [7]. This step is called as cleanup, and it greatly aims at the purification of the target pesticides.

Various cleanup procedures viz., steam distillation, oxidation, saponification, partition, sweep codistillation, adsorption chromatography, and so on, have been used in different types of sample matrices for multipesticide residue analyses. Liqiud–liquid partitions, SPE, gel permeation chromatography (GPC), and so on have been introduced to ease the sample purification.

Adsorption chromatography in diverse forms such as column chromatography, SPE, and SPME has been used widely for sample cleanup. The classical column chromatography using silica gel is still being used for cleanup in the determination of OC pesticides [20]. However, recent work indicates that the application of SPE cartridges is on the increase. The use of SPE not only reduces the extraction or processing time but also minimizes the wastage of organic solvents [22]. Different types of SPE cartridges viz., reverse phase (RP), aminopropyl, florisil, ENVI-Carb, and so on, have been used for the purification of different samples. The use of C_{18} SPE cartridges provides better results than LLE for cleanup after extraction of rice with acetone for 47 OP residues [23]. For removal of plant pigments and associated impurities, two different types of SPEs can also be used in tandem as an aminopropyl cartridge is placed beneath a C_{18} cartridge [30]. Use of ENVI-Carb SPE cartridges (containing nonporous, graphitized carbon material) provides 70%–100% recoveries of different pesticides from vegetables [22]. For fatty sample matrices, GPC is highly suitable for cleanup [41].

OC pesticide residues from fat-rich feeds may be extracted by n-hexane and subsequently purified with the help of sulfuric acid and ENVI-Carb SPE cartridges [29]. The C_{18} SPE (1g) cartridge proved better than liquid–liquid cleanup of rice samples for analysis of OPP on GC [23]. However, some workers not only evaded the cleanup step to enhance the recoveries but they also used sophisticated MS technologies for detection [42,43]. In some cases after exchanging the extracting solvent to acetone, it can directly be injected onto the GC fitted with an MS detector [27].

21.2.2 Separation and Detection

The estimation of pesticide residues can be performed with the help of any technique viz., microbiological, enzymatic, colorimetric, spectrophotometric, radiometric, and so on. Previously, enzyme inhibition tests were aimed to analyze the OPP, which were later replaced by colorimetric methods. In due course of time because of lesser sensitivity these methods became redundant. TLC and other methods were also not much used due to a similar reason.

Chromatographic techniques such as high-performance liquid chromatography (HPLC) and gas–liquid chromatography (GLC) are used for the fractionation/separation and detection of pesticide residues. Compounds of varying polarity such as metabolites of pesticides can be analyzed in a single run [2]. The wider wavelength variation in UV detection for monitoring of different pesticides also encouraged the use of HPLC by many workers [44]. A simple postcolumn photoconductivity detector (PCD) was attached with HPLC and used for multiresidue pesticide analysis in soil and feed samples. This method was capable of simultaneous determination of 15 pesticides [45]. In another method, components were separated in isocratic mode on an RP C_{18} column with MeCN–water (50:50, v/v) as mobile phase and UV detector at a wavelength of 200 nm [46]. Thus, the use of HPLC appeared to be more advantageous than GC for the determination of thermolabile and highly polar pesticides [47]. In GC furthermore, the use of a large number of columns was also problematic and sample cleanup is usually less problematic for HPLC [40]. The advent of ultraperformance liquid chromatography (UPLC) greatly reduces the time of multiresidue pesticide analysis [48].

A multiresidue method was developed for the analysis of 17 OPP (including some metabolites) by the HPLC system. It employed C_{18} HL 90 (250 × 4.6 mm i.d., 5 μm particle size) analytical column (maintained at 40°C temperature), PDA detector (190–350 nm scanning), and step gradient mobile phase. The initial mobile phase composition of MeCN–water was 55:45 (v/v) at a flow of 1.00 mL/min, which was held constant from 0.00 to 11.00 min. At 11.10 min, the MeCN was increased to 65%. From 11.10 to 34.00 min it was linearly increased to 70% [49]. An isocratic mobile phase (MeCN–water, 80:20 v/v) was used in the other method for the separation and determination of 21 components (11 OCP, 9 OPP, and 1 carbamate) on an ODS column (250 × 4.6 mm i.d.) at a flow of 0.5 mL/min, 40°C column temperature, and UV detection at 200 nm. The recoveries for different components ranged from 79.33% to 106% [50]. Recently, a gradient program on binary HPLC was developed for the determination of 28 OP and OC pesticides. It uses MeCN and water as mobile phases and RP C_{18} column (250 × 4.6 mm i.d.), and detection was done at 200nm wavelength [2].

The development of element selective or specific detectors and the substitution of packed columns with capillary columns made GC a very successful analytical tool for pesticide analysis. Capillary GC in combination with selective detectors, mainly nitrogen phosphorus detectors (NPD), electron capture detectors (ECD), flame photometric detectors (FPD), or electrolytic conductivity detectors (ELCD), has been exploited frequently. The use of a selective detector in GC partially reduces the need for cleanup. But at the same time, a high degree of selectivity, which is generally considered as the strength of the above-mentioned detectors and others, is also a weakness. Each of these detects only a narrow range of analytes. For analyzing GC amenable multiclass, multiresidue pesticides, several GC injections are required when these selective detectors are used. Therefore, the use of a selective detector also gave rise to a problem, as it was unable to detect all the residues in a single run.

MS detection linked to either GC [37] or HPLC is an efficient way of determining the specificity and accuracy in the pesticide residue analysis [7]. The determination of OC pesticides was also tried using neutron activation analysis (NAA), but it could account only for 1.6% of the total extractable organic chlorine element [51]. Analysis of the pesticides is also carried out by low-pressure gas chromatography (LP–GC) with mass spectrometry in tandem (MS–MS) mode [14]. More recent work indicates an increased use of UPLC coupled with different types of MS detectors. The major advantages of UPLC over HPLC are the speed of analysis, the narrower peaks (giving increased signal-to-noise ratio), and improved confirmation for the targeted pesticides in the analyses [48,52,53].

21.2.3 RECENT METHODS

For multipesticide residue analyses of feed samples, a hydration step with acetone as solvent was applied [50]. A 30.0 g sample aliquot was homogenized with 98.5 mL water, 198.5 mL acetone, and 35 g sodium chloride at 5000 rpm for 2 min. Subsequently, 100 mL mixture of EtAc and cyclohexane (1:1) were added, and contents were homogenized for 1 min. The resultant organic phase (200 mL) was then dehydrated over 100 g sodium sulfate and dried using a rotatory evaporator. This residue was reconstituted with 15 mL mixture of EtAc and cyclohexane (1:1) for GPC cleanup. The GPC employed an isocratic pump, Bio Beads SX-3 packed in a 450 × 20 mm i.d. column, an eluent of EtAc, and cyclohexane (1:1) at 2.5 mL/min flow rate. First 100 mL elution was discarded and a subsequent similar volume was collected, evaporated, and reconstituted with toluene for determination by GC-ECD/NPD and/or MS.

Another multiresidue method was developed for analysis of 28 pesticides on binary HPLC with UV detection at 200 nm. Three organic solvents (viz., EtAc, acetone, and MeCN) were tried without employing a hydration step for extraction of OC and OP pesticides from feeds. A 10.0 g sample was homogenized with 100 mL of organic solvent for 5 min at high speed, and after addition of 10 g sodium sulfate, the contents were homogenized for 2 min. The organic phases were dehydrated by anhydrous sodium sulfate and a 50 mL portion was evaporated to dryness. The residues were reconstituted with MeCN and cleaned up by an SPE (Discovery; C18Lt, 3 mL, 500 mg) cartridge. The SPE cartridges were conditioned with a sequence of methanol and MeCN. The pesticides were recovered with methanol, evaporated, and reconstituted with MeCN before the analysis by HPLC. The percent recoveries of different pesticides with MeCN extraction were better than other solvents. This method is advantageous over other methods because it could analyze the residues of pesticides that are most prevalent or highly persistent in Indian conditions. A standard chromatogram shows the separation of a mixture of these pesticides (Figure 21.1). It includes 18 OPP viz., acephate, chlorpyriphos, chlorpyriphos-methyl, diazinone, dichlorvos, dicrotophos, dimethoate, fenitrothion, malaxon, malathion, monocrotophos, paraoxon-ethyl, parathion-methyl, phorate, phosphamidon, profenofos, quinalphos, and tetrachlorvinphos; and 10 OCP viz., aldrin, diendrin, endosulfan-α, endrin, heptachlor, lindane, 2,4-DDE, 2,4-DDT, 4,4-DDD, and 4,4-DDT [2].

Recently, a multiresidue method for the determination of 450 pesticide residues in nonfatty liquid samples using double-cartridge SPE, GC–MS, and LC–MS–MS was developed [38]. Samples were first diluted with water–acetone and then extracted with portions of DCM. The extracts were concentrated and cleaned up with graphitized carbon black and aminopropyl SPE cartridges stacked in tandem. Pesticides were eluted with MeCN–toluene, and elutes were concentrated. For 383 pesticides, the elute was extracted, twice with hexane, and an internal standard solution was added

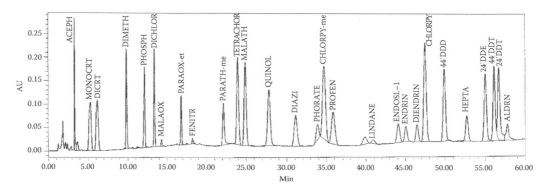

FIGURE 21.1 Chromatogram showing separation of a mixture of various pesticides on HPLC system. (From Singh, P., Organophosphorus pesticide residues in animal feeds and their excretion in milk, PhD dissertation, National Dairy Research Institute, Karnal, Haryana, India, 2004, pp. 56–76.)

before GC–MS determination. For 67 pesticides, extraction was performed with methanol prior to LC–MS–MS determination. The limit of detection for the method was between 1.0 and 300 µg/kg, depending on each pesticide analyte. At the three fortification levels of 2.0–3000 µg/kg, the average recoveries ranged between 59% and 123%, among which 413 pesticides (92%) had recovery rates of 70%–120% and 35 pesticides (8%) had recovery rates of 59%–70%.

21.3 CONCLUSION

Extraordinary progress took place in the extraction and cleanup procedures for multiresidue pesticide analyses of animal feed samples. An initial higher cost involvement still dictates the use of many classical methods for sample preparation in several laboratories. For multiresidue pesticide analyses, a modern generalized scheme and strategy involve sampling, homogenization, extraction, cleanup (for fatty samples), or no-cleanup (for nonfatty samples), separation by GC or LC, detection by specific and/or MS detectors, and data handling. Moreover, for extraction and cleanup, speed and robustness are the main criteria, and sensitivity and selectivity are important for the detection. Overall, the precision and cost effectiveness rule the pesticide analyses of feed samples. Recently, however rapid extraction and cleanup procedures have been put into use in many dedicated laboratories across the world. A thumping progress in the separation technologies such as GC and LC (HPLC and UPLC) and detection technologies involving MS has been favoring a simplification of multiresidue pesticide analyses, where the sample without much processing is injected directly. However, analytical techniques such as GC or LC with various types of columns and specific detectors involve great costs and more attention in standardization of methodology.

REFERENCES

1. IDF. 1997. Pesticides (Chapter 7). *Monograph on Residues and Contaminants in Milk and Milk Products.* IDF special issue 9701, pp. 56–64. Brussels: IDF General Secretariat.
2. Singh, P. 2004. Organophosphorus pesticide residues in animal feeds and their excretion in milk. PhD dissertation, pp. 56–76. Karnal, Haryana, India: National Dairy Research Institute (DU).
3. Lovell, R.A., D.G. McGghesney, and W.D. Price. 1996. Organohalogen and organophosphorus pesticides in mixed feed rations: Findings from FDA's domestic surveillance during fiscal years 1989–1994. *J. AOAC Int.* 79: 544–548.
4. Chhabra, A. and P. Singh. 2005. Antinutritional factors and contaminants in animal feeds and their detoxification: A review. *Ind. J. Anim. Sci.* 75: 101–112.
5. Kan, C.A. and G.A.L. Meijer. 2007. The risk of contamination of food with toxic substances present in animal feed. *Anim. Feed Sci. Technol.* 133: 84–108.
6. Walorczyk, S. 2007. Development of a multi-residue screening method for the determination of pesticides in cereals and dry animal feed using gas chromatography-triple quadrupole tandem mass spectrometry. *J. Chromatogr. A* 1165: 200–212.
7. Valverde, A. 2000. Chromatographic pesticide residue analysis. *J. AOAC Int.* 83: 679.
8. Pengyan, L.I.U., L.I.U. Qingxue, M.A. Yusong, L.I.U. Jinwei, and J.I.A. Xuan. 2006. Analysis of pesticide multiresidues in rice by gas chromatography-mass spectrometry coupled with solid phase extraction. *Chin. J. Chromatogr.* 24: 228–234.
9. PAM. 1999. Multiclass MRMS: Concepts and application. In: *Pesticide Analytical Manual Volume I: Multiresidue Methods,* 3rd edition 1994 (revised 1999), ed. C.M. Makovi and B.M. McMahon, Section 301:1–8. USFDA-2905a (6/92) transmittal No. 94–1 (1/94).
10. Ding, X.D. and I.S. Krull. 1984. Trace analysis of organothiophosphate agricultural chemicals by high performance liquid chromatography-photolysis-electrochemical detection. *J. Agric. Food Chem.* 32: 622–628.
11. Suri, K.S. and B.S. Joia. 1996. Persistence of chlorpyriphos in soil and its terminal residues in wheat. *Pest. Res. J.* 8: 186–190.
12. Chinniah, C., S. Kuttalam, and A. Regupathy. 1998. Harvest time residues of lindane and chlorpyriphos in/on paddy. *Pest. Res. J.* 10: 91–94.
13. Lehotay, S.J. 2000. Analysis of pesticide residues in mixed fruits and vegetables by direct sample introduction/gas chromatography/tandem mass spectroscopy. *J. AOAC Int.* 83: 680–697.

14. Garrido-Frenich, A., F.J. Arrebola, M.J. González-Rodríguez, J.L.M. Vidal, and N.M. Díez. 2003. Rapid pesticide analysis, in post-harvest plants used as animal feed, by low-pressure gas chromatography–tandem mass spectrometry. *Anal. Bioanal. Chem.* 377: 1038–1046.
15. Ahmad, N., G. Bugueno, L. Guo, and R. Marolt. 1999. Determination of organochlorine and organophosphate pesticide residues in fruits, vegetables and sediments. *J. Environ. Sci. Health B* 34: 829–848.
16. Kumar, K. and A. Regupathy. 1999. Residue of quinalphos from AF and EC formulations on cotton. *Pest. Res. J.* 11: 44:46.
17. Singh, R., H. Singh, and T.S. Kathpal. 1998. Harvest time residues of lindane, and chlorpyriphos in mustard (*Brassica juncea*) and sunflower (*Helianthus annus* L.) seeds. *Pest. Res. J.* 10: 219–223.
18. Bhatnagar, A. and A. Gupta. 1998. Lindane, chlorpyriphos and quinalphos residue in safflower seed and oil. *Pest. Res. J.* 10: 127–128.
19. Kang, B.K., B. Singh, K.K. Chahal, and R.S. Battu. 2002. Contamination of feed concentrate and green fodder with pesticide residues. *Pest. Res. J.* 14: 308–312.
20. Aulakh, R.S., J.P.S. Gill, J.S. Bedi, J.K. Sharma, B.S. Joia, and H.W. Ockerman. 2006. Organochlorine pesticide residues in poultry feed, chicken muscle and eggs at a poultry farm in Punjab, India. *J. Sci. Food Agric.* 86: 741–744.
21. Ober, A.G., I.S. Maria, and J.D. Carmi. 1987. Organochlorine pesticide residues in animal feeds by cyclic steam distillation. *Bull. Enviorn. Contam. Toxicol.* 38: 404–408.
22. Barwick, V.L., S.L.R. Ellison, S.J. Lacy, C.R. Mussell, and C.L. Lucking. 1999. Evaluation of solid phase extraction procedure for the determination of pesticide residues in foodstuffs. *J. Sci. Food Agric.* 79: 1190–1196.
23. Chu, C., S.S. Wong, and G.C. Li. 2000. Determination of organophosphate pesticide residues in rice by multiresidue analysis. *J. Food Drug Anal.* 8: 63–73.
24. Ramesh, A. and M. Balasubramanian. 1999. The impact of household preparations on the residues of pesticides in selected agricultural food commodities available in India. *J. AOAC Int.* 82: 725–737.
25. Surendranath, B., V. Unnikrishnan, C.N. Preeja, and M.K. Ramamurthy. 2000. A study of the transfer of organochlorine pesticide residues from the feed of the cattle into their milk. *Pest. Res. J.* 12: 68–73.
26. Pihlstrom, T. and B.G. Osterdahl. 1999. Analysis of pesticide residues in fruit and vegetables after cleanup with solid phase extraction using ENV + (polystyrene-divinybenzene) cartridges. *J. Agric. Food Chem.* 47: 2549–2552.
27. Aguera, A., M. Contreras, J. Crespo, and A.R. Fernandez-Alba. 2002. Multiresidue method for the analysis of multiclass pesticides in agricultural products by GC-Tandem MS. *Analyst* 127: 347–354.
28. Frenich, A.G., J.L.M. Vidal, T.L. López, S.C. Aguado, and I.M. Salvador. 2004. Monitoring multi-class pesticide residues in fresh fruits and vegetables by liquid chromatography with tandem mass spectrometry. *J. Chromatogr. A* 1048: 199–206.
29. Nardelli, V., C. Palermo, and D. Centonze. 2004. Rapid multiresidue extraction method of organochlorinated pesticides from fish feed. *J. Chromatogr. A* 1034: 33–40.
30. Fillion, J., F. Sauve, and J. Selvyn. 2000. Multiresidue method for determination of residues of 254 pesticides in fruits and vegetables by gas chromatography/mass spectrometry and liquid chromatography with fluorescence detection. *J. AOAC Int.* 83: 698–713.
31. Schenk, F.J. and J. Casanova. 1999. Rapid screening for organochlorine and organophosphorus pesticides in milk using C18 and graphitized black solid phase extraction cleanup. *J. Environ. Sci. Health B* 34: 349–362.
32. Lehotay, S.J. 1997. Supercritical fluid extraction of pesticides in foods. *J. Chromatogr. A* 785: 289–312.
33. Adou, K., W.R. Bontoyan, and P.J. Sweeney. 2001. Multiresidue method for the analysis of pesticide residues in fruits and vegetables by accelerated solvent extraction and capillary gas chromatography. *J. Agric. Food Chem.* 49: 4153–4160.
34. Chen, S., M. Gfrerer, E. Lankmayr, X. Quan, and F. Yang. 2003. Optimization of accelerated solvent extraction for the determination of chlorinated pesticides from animal feed. *Chromatographia* 58: 631–636.
35. Iglesias-García I., M. Barriada-Pereira, M.J. González-Castro, S. Muniategui-Lorenzo, P. López-Mahía, and D. Prada-Rodríguez. 2008. Development of an analytical method based on microwave-assisted extraction and solid phase extraction cleanup for the determination of organochlorine pesticides in animal feed. *Anal. Bioanal. Chem.* 391: 745–752.
36. Tekel, J. and S. Hatrik. 1996. Pesticide residue analysis in plant material by chromatographic methods: Clean-up and selective detectors. *J. Chromatogr. A* 754: 397–410.

37. Lehotay, S.J. 2000. Analysis of pesticide residues in mixed fruits and vegetables by direct sample introduction/gas chromatography/tandem mass spectroscopy. *J. AOAC Int.* 83: 680–697.

38. Pang, G.F., C.L. Fan, Y.M. Liu, Y.Z. Cao, J.J. Zhang, B.L. Fu, X.M. Li, Z.Y. Li, and Y.P. Wu. 2006. Multi-residue method for the determination of 450 pesticide residues in honey, fruit juice and wine by double-cartridge solid-phase extraction/gas chromatography-mass spectrometry and liquid chromatography-tandem mass spectrometry. *Food Addit. Contam.* 23: 777–810.

39. Pahadia, S., V. Sharma, V. Sharma, and B.K. Wadhwa. 2005. Evolution of multiresidue analysis of pesticides in dairy and fatty foods—A review. *Indian J. Dairy Sci.* 58: 1–5.

40. Manes, J., G. Font, Y. Pico, and J.C. Malto. 1996. Residues of organophosphates in food (Chapter 33). In: *Handbook of Food Analysis. Vol. 2: Residues and Other Food Component Analysis*, ed. L.M.L. Nollet, pp. 1437–1460. New York: Marcel Decker.

41. Chamberlin, S.J. 1990. Determination of multi-pesticide residues in cereals, cereal products and animal feed using gel permeation chromatography. *Analyst* 115: 1161–1165.

42. Faud, A.T., H. Mahler, T. Oliver, and D. Thomas. 2001. An analytical method for the rapid screening of organophosphate pesticides in human samples and foodstuffs. *Forensic Sci. Int.* 121: 126–133.

43. Mol, H.G., R.C.J. van Dam, and O.M. Steijger. 2003. Determination of polar organophosphorus pesticides in vegetables and fruits using liquid chromatography with tandem mass spectrometry: Selection of extraction solvent. *J. Chromatogr. A* 1015: 119–127.

44. Pryde, A. and M.T. Gilbert. 1979. Pesticide, carcinogens and industrial pollutants. In: *Applications of High Performance Lliquid Chromatography*, eds. A. Pryde and M.T. Gilbert, p. 171. New York: Chapman Hall.

45. Miles, C.J. and M. Zhou. 1990. Multiresidue pesticide determination with a simple photoconductivity HPLC detector. *J. Agric. Food Chem.* 38: 986–989.

46. Cabras, P., C. Tuberoso, M. Melis and M.G. Martini. 1992. Multiresidue method for pesticide determination in high performance liquid chromatography. *J. Agric. Food Chem.* 40: 817–819.

47. Chu, X., W. Yong, Y. Ling, H. Yao, J. Zweigenbaum, and Y. Fang. 2007. Screening method for multi herbicide and insecticide residues in soybeans using high performance liquid chromatography-time-of-flight mass spectrometry. *Se Pu.* 25: 907–916.

48. Leandro, C.C., P. Hancock, R.J. Fussell, and B.J. Keely. 2006. Comparison of ultra-performance liquid chromatography and high performance liquid chromatography for the determination of priority pesticides in baby foods by tandem quadrupole mass spectrometry. *J. Chromatogr. A* 1103 (1): 94–101.

49. Iorger, B.P. and J.S. Smith. 1993. Multiresidue method for extraction and detection of organophopsphate pesticides and their primary and secondary metabolites from beef tissues using HPLC. *J. Agric. Food Chem.* 41: 303–307.

50. Sharma, V. 2003. Multiresidue analysis of pesticides (organophosphates and organocarbamates) in milk and infant foods. PhD dissertation, pp. 59–115. Karnal, Haryana, India: National Dairy Research Institute.

51. Zhong, W., D. Xu, Z. Chai, and X. Mao. 2004. Neutron activation analysis of extractable organohalogens in milk from China. *J. Radioanal. Nucl. Chem.* 259: 485–488.

52. Pozo, O.J., M. Barreda, J.V. Sancho, F. Hernandez, J. Ll. Lliberia, M.A. Cortes, and B. Bago. 2007. Multiresidue pesticide analysis of fruits by ultra-performance liquid chromatography tandem mass spectrometry. *Anal. Bioanal. Chem.* 389: 1765–1771.

53. Kovalczuk, T., O. Lacina, M. Jech, J. Poustka, and J. Hajslová. 2008. Novel approach to fast determination of multiple pesticide residues using ultraperformance liquid chromatography-tandem mass spectrometry (UPLC-MS/MS). *Food Addit. Contam.* 25: 444–457.

54. Dicke, W., H.D. Ocker, and H.P. Thier. 1988. Residue analysis of pyrethroid insecticides in cereal grains, milled fractions and bread. *Z. Lebensm Unters Forsch.* 186: 125–129.

55. Pang, G.F., Y.Z. Chao, C.L. Fan, J.J. Zhang, X.M. Li, and T.S. Zhao. 1995. Modification of AOAC multiresidue method for determination of synthetic pyrethroid residues in fruits, vegetables, and grains. Part I: Acetonitrile extraction system and optimization of florisil cleanup and gas chromatography. *J. AOAC Int.* 78: 1481–1488.

56. Hans-Jürgen, S. and M. Linkerhägner. 1996. Pesticide residue analysis in foodstuffs applying capillary gas chromatography with atomic emission detection state-of-the-art use of modified multimethod S19 of the Deutsche Forschungsgemeinschaft and automated large-volume injection with programmed-temperature vaporization and solvent venting. *J. Chromatogr. A* 750: 369–390.

57. Schenk, F.J. and V. Howard-King. 2000. Determination of organochlorine and organophosphorus pesticide residues in low moisture, nonfatty products using a solid phase extraction clean up and gas chromatography. *J. Environ. Sci. Health B.* 35: 1–12.
58. Yi, X. and Y. Lu. 2005. Multiresidue determination of organophosphorus pesticides in ginkgo leaves by accelerated solvent extraction and gas chromatography with flame photometric detection. *J. AOAC Int.* 88: 729–735.
59. Díez, C.W.A., P. Zommer, P. Marinero, and J. Atienza. 2006. Comparison of an acetonitrile extraction/ partitioning and "dispersive solid-phase extraction" method with classical multi-residue methods for the extraction of herbicide residues in barley samples. *J. Chromatogr. A* 1131: 11–23.
60. Yi. X., Q. Hua, and Y. Lu. 2006. Determination of organophosphorus pesticide residues in the roots of *Platycodon grandiflorum* by solid-phase extraction and gas chromatography with flame photometric detection. *J. AOAC Int.* 89: 225–231.
61. Pang, G.F., Y.M. Liu, C.L. Fan, J.J. Zhang, Y.Z. Cao, X.M. Li, Z.Y. Li, Y.P. Wu, and T.T. Guo. 2006b. Simultaneous determination of 405 pesticide residues in grain by accelerated solvent extraction then gas chromatography-mass spectrometry or liquid chromatography-tandem mass spectrometry. *Anal. Bioanal. Chem.* 384: 1366–1408.
62. Mol, H.G., A. Rooseboom, R. van Dam, M. Roding, K. Arondeus, and S. Sunarto. 2007. Modification and re-validation of the ethyl acetate-based multi-residue method for pesticides in produce. *Anal. Bioanal. Chem.* 389: 1715–54.
63. Paya, P., M. Anastassiades, D. Mack, I. Sigalova, B. Tasdelen, J. Oliva, and A. Barba. 2007. Analysis of pesticide residues using the Quick Easy Cheap Effective Rugged and Safe (QuEChERS) pesticide multiresidue method in combination with gas and liquid chromatography and tandem mass spectrometric detection. *Anal. Bioanal. Chem.* 389: 1697–714.
64. Vela, N., G. Pérez, G. Navarro, and S. Navarro. 2007. Gas chromatographic determination of pesticide residues in malt, spent grains, wort, and beer with electron capture detection and mass spectrometry. *J. AOAC Int.* 90: 544–549.
65. Lehotay, S.J. 2007. Determination of pesticide residues in foods by acetonitrile extraction and partitioning with magnesium sulfate: Collaborative study. *J. AOAC Int.* 90: 485–520.
66. Hajjo, R.M., F.U. Afifi, and A.H. Battah. 2007. Multiresidue pesticide analysis of the medicinal plant *Origanum syriacum. Food Addit. Contam.* 24: 274–279.
67. van der Lee, M.K., G. van der Weg, W.A. Traag, and H.G. Mol. 2008. Qualitative screening and quantitative determination of pesticides and contaminants in animal feed using comprehensive two-dimensional gas chromatography with time of flight mass spectrometry. *J. Chromatogr. A* 1186: 325–339.
68. Lal, A., G. Tan, and M. Chai. 2008. Multiresidue analysis of pesticides in fruits and vegetables using solid-phase extraction and gas chromatographic methods. *Anal. Sci.* 24: 231–236.

22 Analysis of Pesticide Residues in Soils

Balbir K. Wadhwa, Pramod Singh, Sumit Arora, and Vivek Sharma

CONTENTS

22.1 INTRODUCTION

Owing to the great efficacy of a variety of pesticide compounds against the pests, their use had been escalating both in agriculture and public health sectors. Approximately, 940 million pounds of active ingredients are applied annually to croplands [1]. Pesticides are chemical substances that are used to kill or control pests. This term includes different insecticides, herbicides, fungicides, bactericides, rodenticides, disinfectants, and repellants; however, very often we thoughtlessly refer it only to insecticides. The pervasive uses of pesticide instigated pollution of the environment and a great concern for human health. Although some pesticides, especially the more persistent organochlorines, have been restricted/banned, they are still used in some countries to control the pests in agricultural production and to control vector-borne diseases. According to WHO and FAO, the contamination of various food commodities, drinking water, and ground water with different pesticides has been recognized all over the world.

Soil is a complex mixture of different components: (a) abiotic such as solids, liquids, and gases, and (b) biotic like soil invertebrate animals, algae, and microorganisms, viz., bacteria and fungi. It provides the life-support system for roots of growing plants and other flora and fauna. The symbiotic and dynamic nature of abiotic and biotic components is responsible for soil fertility. Soil quality also encompasses the impacts that soil use and management can have on water and air quality, and on human and animal health. The presence and bioavailability of pesticides in soil can adversely impact human and animal health, and beneficial plants and soil organisms. Soil to an extent acts as a reservoir of pesticides in the environment [2,3].

When a pesticide enters soil, some of it will stick on to soil particles, particularly organic matter through a process called sorption, and some will dissolve and mix with the water between soil particles, called the soil water. As more water enters the soil through rain or irrigation, the sorbed pesticide molecules may get detached from soil particles through a process called desorption. The solubility of a pesticide and its sorption on soil are inversely related, i.e., increased solubility results in less sorption. The residues of pesticide may be toxic and may disturb the biotic components [4–6], organic matter, and thus the fertility of the soil [7,8].

The significance of soil as a reservoir of several types of pesticide compounds has been well recognized [3], and their analysis for pesticide residues could be useful in ensuring the food and health security of mankind, and the soil quality. Several methodologies, viz. liquid or gas chromatographic analysis, have been developed for individual or multipesticide determinations in soil samples. Basic steps of separation and detection for the soil pesticide residue analysis remained similar to those of other sample matrices like food, vegetable, beverages, etc. Such techniques have been dealt comprehensively in other chapters of this book. In this chapter, all the steps have been embarked and a greater emphasis has been laid, however, on the sample preparation for pesticide analysis in the soil.

22.2 GENERAL ASPECTS OF PESTICIDE RESIDUES IN SOILS

The pesticides are principally applied on the aerial vegetation cover. They are also applied to the soil [9,10] for protection against common insects, like termites, and production of underground crops, such as potato tubers, etc. The presence of residues of various pesticides in soils has been reported across the world [11–14] and India [3,15,16]. However, more persistent organochlorine pesticides (OCP) have been reported frequently [2,3,15].

After the application of pesticides on croplands, an array of processes may occur. Pesticides may be (a) taken up by plants or ingested by animals, insects, worms, or microorganisms in the soil and subsequently metabolized, (b) moved downward in the soil and either adhered to particles or dissolved, (c) vaporized and enter the atmosphere, or break down via microbial and chemical pathways, (d) leached out of the root zone or washed off the surface of land by rain or irrigation water. The evaporation of water at the ground surface can lead to upward flow of pesticides as well.

Degradation of pesticides by chemical pathways is usually partial. However, soil microorganisms can completely break down many pesticides to carbon dioxide, water, and other inorganic constituents. Exposure to sunlight degrades several pesticides to varying extents. The organic matter content represents the quality of various soils in terms of microbiological activity, and plays an important role in pesticide degradation. The biological activity of these substances may also have environmental significance. Because microbial population decreases rapidly below the root zone, pesticides leached beyond this depth are less likely to be degraded. However, some pesticides will continue to degrade by chemical reactions after they have left the root zone. Retention of pesticide residue refers to the ability of the soil to hold a pesticide in place, not allowing it to be transported. Adsorption is the primary process of soil retention of a pesticide, and is defined as the accumulation of a pesticide on the soil particle surfaces. Pesticide adsorption to soil depends on both the chemical properties of the pesticide (i.e., water solubility, polarity) and properties of the soil (i.e., organic matter and clay contents, pH, surface charge characteristics, permeability). Organic matter is the most important soil property in controlling the movement of pesticide residues.

22.3 ANALYSIS OF PESTICIDE RESIDUES IN SOILS

The present situation greatly influenced the development and refinement of procedures for soil pesticide residue analyses, as a large number of pesticide compounds are already available, and with period their number is increasing further. Special and extreme care and proper sampling are most important for this analysis [17]. Erroneous sampling could compromise the entire analyses, and therefore basic principles for true representative sampling must be adhered to [18].

The analysis of pesticide residues in soil is complicated, primarily because of high and variable level of interfering components of organic matter such as humic and fulvic acids. Different molecules of pesticide residues also tend to adsorb on to the soil particles very strongly and make the estimates of percent recoveries doubtful. The most common method for the isolation of these residues from soil is extraction with organic solvents, followed by cleanup procedures to remove interferences prior to analysis. Blending the soils with organic solvents is generally used for the isolation of pesticides. The organic solvents can be filtered or decanted to remove the particulate materials. Anhydrous sodium sulfate is commonly used as a dehydrant.

Soxhlet extraction technique being simple and economical is used for dried soil samples, and often, good recoveries are obtained by this method. But the thermal degradation of analytes is obvious, and in addition (a) the time-consuming cleanup procedure with great chances of analyte losses and (b) the high consumption of high purity, expensive organic solvents, which pose a burden to the environment, thus offsetting the benefits of pesticide residue analysis, make this method unpopular [19].

Several types of biological and chemical techniques, viz. bioassay, enzyme assay, immunoassay, radio assay, thin layer chromatography (TLC), column chromatography, and capillary electrophoresis (CE) can be used for the determination of pesticides. The methodologies of gas chromatography (GC) and liquid chromatography (LC) are the most popular. Recent trends indicate that capillary column GC with different sensitive and selective detectors including mass spectrometer has been used predominantly. For many of the pesticide compounds (which may not be amenable to GC methodology), LC with detectors like ultraviolet, photodiode array, fluorescence, and MS has been an obvious choice [20]. Table 22.1 gives an impression about the pesticide residue analysis of soil.

22.3.1 Sample Preparation

For pesticide analysis using a particular type of instrument, soil preparation is obligatory. It is performed by extraction with several organic solvents, individually or in combinations. The extraction allows quantitative transfer of soil pesticides into a liquid organic phase. Traditional procedures involve mixing of sample with polar or nonpolar organic solvents followed by liquid–liquid partitioning (LLP) with other organic solvents of limited water-holding capacity. The choice of the method of extraction and the type of solvent is dependent on the nature of the sample and chemical properties of the pesticide residues. Soil behaves as an adsorbent and retains pesticides on its surface. These are eluted by a solvent mixture that should be chosen for appropriate polarity characteristics [21]. Various impurities are also extracted along with desired components and these interfere with the proper analysis. Therefore, certain cleanup steps are also required prior to the separation and detection by gas or liquid chromatographic techniques. However, some methodologies evaded the cleanup steps.

22.3.1.1 Extraction Procedures

For the extraction of pesticides, usually air drying of soil samples is employed [16]. As in the case of other high-moisture samples, viz. forages, fruits, and vegetables, so far none of the studies have indicated the application of cool conditions while extracting the pesticide residues from moist soil samples. To restrain the enzymatic degradation of pesticides in moist soil samples, frozen conditions may be used. These conditions may be achieved by the use of either dry ice [17,22] or freezing mixtures. Recently, a method based on ultrasonic-assisted extraction has been put forward, where the moist soil sample without drying can be used for multipesticide residue analysis [23]. In the majority of cases, 50 g [16,24–26] soil was taken for the analysis. However, lesser quantities of soil samples (20 g [27] and 10 g [28]) were also used in some cases.

Review of literature suggests that acetone remained the most preferred extracting solvent [10,29–32] for the isolation of pesticides from soil samples. Mixtures of acetone with hexane [9,16,25,33–35], benzene [36], dichloromethane (DCM; [24,32]), and petroleum ether (PE; [37]) were also favored. A variety of other solvents, viz. hexane [16,38], ethyl acetate (EtAc [39,40]), DCM [32],

TABLE 22.1
Overview of Various Methods Used for the Pesticide Residue Analysis in Soil

Sl. No.	Particulars of Pesticide Residues	Extraction Procedure	Cleanup Technique	Instrumentation Details	Ref.
1	Dimethoate, monocrotophos, triazophos, deltamethrin, cypermethrin, and endosulfan	Extraction with acetone	—	GC-ECD	[10]
2	Atrazine	SFE with carbon dioxide	—	—	[13]
3	22 OCP with their degradation products and 16 EPA-PAH	PLE	—	GC-MS	[14]
4	15 OCP and 6 SP	Acetone–hexane	DSI	GC-ECD	[16]
5	Pesticides (hexachlorocyclohexane, cyclodiene, DDTs, chlordane, hexachlorobenzene, heptachlor, endrin aldehyde, heptachlor epoxide, dicofol, acetochlor, alachlor, metolachlor, chlorpyrifos, nitrofen, trifluralin, cypermethrin, fenvalerate, and deltamethrin)	UAE	SPE	GC	[23]
6	Insecticides (atrazine, propham, chlorpropham, diflubenzuron, cypermethrin, and tetramethrin)	Ultrasonication with acetone	—	Quantitative TLC on RP-18 plates	[29]
7	Dimethoate and phosphamidon	Acetone	—	TLC	[30]
8	55 pesticides (OCP, OPP, carbryl, etc.)	Acetone	LLE with 2 × 50 mL DCM	LC-MS/MS using TurboIonSp	[31]
9	140 PCBs and pesticides	Extracted sequentially using mechanical shaker with acetone, 1:1 acetone–DCM, and DCM. The extracts were combined to give solutions in 1:1 acetone–DCM before concentration by evaporation.	—	GC-TOFMS	[32]
10	OPP (dichlorvos, trichlorphon, and isofenphos)	Blending with acetone–benzene (19:1, v/v) and charcoal	—	GC with alkali flame ionization detection	[36]
11	15 OCP (α-HCH, β-HCH, γ-HCH, δ-HCH, hexachlorobenzene, heptachlor, aldrin, α-chlordane, γ-chlordane, dieldrin, endrin, o,p'-DDE, o,p'-DDT, p,p'-DDT, and p,p'-DDD)	Sonication with mixture of acetone and PE (50:50,v/v)	LLE	GC-ion trap MS/MS	[37]
12	Pesticides (trifluralin, metolachlor, chlorpyrifos, and triadimefon)	MAEP with MeCN–water and n-hexane	—	GC-ECD	[38]
13	9 pesticides (OCP, OPP, and SP)	Sonication extraction using Acetone–EtAc	—	GC-MS	[39]

No.	Pesticides	Extraction	Cleanup	Detection	Reference
14	Multiresidue herbicides and insecticides	Sonication with EtAc	—	GC-MS (EI)	[40]
15	Pesticide and acaricide—azinphos-methyl	Methanol extraction and alkaline hydrolysis to its main metabolite anthranilic acid	—	Spectrofluorimetric determination	[41]
16	Herbicide—sulfosulfuron	Extraction with MeCN-1M ammonium carbonate	LLE with DCM	HPLC (RP-8, 250 × 4 mm column; UV detector)	[43]
17	46 Pesticides (triazines, OCP, OPP, OCm anilides, anilines, and amides)	Extraction with methanol–water	SPE membrane workstation on Empore C_{18} disks	GC-MS/IT	[44]
18	10 Pesticides (OCP, OPP, and SP)	EtAc with ascorbic acid solution (pH 2.15)	—	GC-AED	[45]
19	Multiresidues of OCP, OPP, and SP	SFE	C-18 SPF cartridges	GC-ECD	[49]
	OCP, OPP, and SP insecticides and triazine and acetanilide herbicides (lindane, dieldrin, endosulfan, endosulfan sulfate, 4,4'-DDE, 4,4'-DDD, atrazine, desethylatrazine, alachlor, dimethoate, chlorpyrifos, pendimethalin, procymidone, and chlorfenvinphos)	USE	—	GC-MS	[50]
20	11 SP insecticides (tetramethrin, bifenthrin, phenothrin, λ-cyhalothrin, permethrin, cyfluthrin, cypermethrin, flucythrinate, esfenvalerate, fluvalinate, and deltamethrin)	MAE with toluene–water	Florisil	GC-MS/IT(EI and NCI)	[51]
21	8 OPP	MAME using nonionic surfactants polyoxyethylene 10 lauryl ether (POLE) and oligoethylene glycol monoalkyl ether (Genapol X-080)	—	LC-PDA	[52]
22	Insecticide (pirimicarb) and fungicide (azoxystrobin)	—	—	CE	[61]
23	Multipesticides (alachlor, atrazine, captan, carbofuran, chlorpyrifos, chlorsulfuron, chlorthal-dimethyl, cypermethrin, 2,4-D, diuron, glyphosate, malathion, methomyl, methyl arsonic acid, mocap, norflurazon, oxyfluorfen, paraquat, temik, and trifluralin)	—	—	Direct surface analyses by static SIMS	[64]
24	Eight different pesticides	—	—	GC-MS	[66]
25	DDT	—	—	ELISA	[67]
26	Six pesticides (acetochlor, atrazine, diazinon, carbendazim, imidacloprid, and isoproturon)	Hot water percolation apparatus	—	HPLC with UV detection	[68]

(continued)

TABLE 22.1 (continued)
Overview of Various Methods Used for the Pesticide Residue Analysis in Soil

Sl. No.	Particulars of Pesticide Residues	Extraction Procedure	Cleanup Technique	Instrumentation Details	Ref.
27	Herbicides—cyclohexanedione oxime (alloxydim, clethodim, sethoxydim) and their sulfoxide metabolites, sulfonylurea and imidazolinone herbicides (chlorsulfuron, metsulfuron-methyl, thifensulfuronmethyl, triasulfuron, imazethapyr, imazamox) Rodenticides—coumarin (bromoadiolone and warfarin) and indandione (chlorophacinone and pindone)	LLE	C18-SPE	LC-ES/MS and LC-ES/MS/MS	[69]
28	Insecticide—monosultap	Extraction with water. Hydrolysis under alkaline condition (monosultap was converted to nereistoxin)	LLP	GC-FPD(S)	[70]

TLC, thin layer chromatography; GC-MS(IT) EI & NCI, gas chromatography-ion trap mass spectrometry electron impact and negative chemical ionization; GC-MS/IT, gas chromatography-ion trap mass spectrometry; GC-TOFMS, gas chromatography-time of flight mass spectrometry; CE, capillary electrochromatography, capillary electrophoresis; GC-ECD, gas chromatography-electron capture detection; GC-FPD-S, gas chromatography-flame photometric detector with S mode; GC-AED, gas chromatography-atomic emission detection; LC-ES/MS, liquid chromatography-electron spray mass spectrometry; LC-PDA, liquid chromatography-photodiode array detection; ELISA, enzyme-linked sorbent assay; MAE, microwave-assisted extraction; MAEP, microwave-assisted extraction and partitioning method; MAME, microwave-assisted micellar extraction; UAE, ultrasonic-assisted extraction; USE, ultrasonic extraction; PLE, pressurized liquid extraction; SPE, solid-phase extraction; LPME, liquid-phase microextraction; LLE, liquid–liquid extraction; LLP, liquid–liquid partition; SFE, supercritical fluid extraction; DSI, direct sample injection; OCP, organochlorine pesticides; OPP, organophosphorus pesticides; SP, synthetic pyrethroid compounds; OCm, organocarbamate pesticides; PCB, polychlorobiphenyls; PAH, polycyclic aromatic hydrocarbon; EtAc, ethyl acetate; DCM, dichloromethane; MeCN, acetonitrile; PE, petroleum ether; EPA, environment protection agency.

and methanol [41] were well used. Combinations of acetonitrile (MeCN) and water [26,38,42], MeCN and aqueous solution of ammonium carbonate [43], methanol and water [44], methanol and buffer [27], and EtAc and aqueous solution of ascorbic acid [45] were also used for specific analyses. Soxhlet extraction procedures are still preferred for simpler reasons of convenience and economy [34].

After the initial step of extraction, liquid–liquid extraction (LLE) or partitioning is an obvious choice for the removal of interfering substances [9]. The process of LLE is now less commonly used and newer techniques are overtaking the cleanup of the sample. After homogenization of samples with proper organic solvent, "salting out principal" was also employed to facilitate the partitioning of analytes from the extracting solutions to other organic solvents of limited water-holding capacity by the addition of sodium chloride [9,46].

In the traditional practice of sample preparation, drying of extract is frequently required and the presence of even traces of moisture obstructs the effective evaporation of organic solvent. Use of anhydrous sodium sulfate [9,24,47,48] as dehydrating agent is helpful in quick drying of organic solvents; however, anhydrous magnesium sulfate can provide better dehydration efficiency [22].

Onus has been for new sample preparation techniques, which are acquiescent to automation with reduced duration for the extraction with smaller volumes of organic solvent. This necessity incorporated new technologies in the progress and utilization of procedures, which minimize environmental concerns, time, labor, exposure of laboratory personnel to toxic chemicals, and preparation costs. Several extraction techniques were initiated for easing the sample preparation and analyses of pesticide residues in soil. Some of them include supercritical fluid extraction (SFE [13,49]), ultrasonic extraction [23,29,40,50], microwave-assisted extraction (MAE [51]), microwave-assisted micellar extraction (MAME [52]), and MAE and partition method [38]. SFE technique is a good alternative for the organic solvent extraction of pesticide residues from nonfatty samples. These protocols extract the pesticide residues more selectively and eliminate postextraction cleanup steps, but polarity range of the approach is compromised as these were generally developed for single class of pesticides [53]. Accelerated solvent extraction (ASE; Dionex trade name), also known as pressurized liquid extraction (PLE), uses small volumes of extracting organic solvents. This technology has been used for the extraction of pesticides from a varied range of sample matrices [54]. Direct sample introduction (DSI) involves the placement of a little amount of sample material or liquid extract into a disposable microvial, and for GC analysis, it greatly minimizes the sample preparation and yet provides a good analytical approach for complex matrices [22].

22.3.1.2   Sample Cleanup

Despite the improvements in the sensitivity of instrumentation, a pretreatment is usually required to isolate the pesticides from the complex sample matrix. Along with the pesticide components of interest, several contaminants as well as part of the matrix also get coextracted. These interfere in the separation and detection of pesticide residues. Hence, further refinement of the extract is a requisite [20].

For a variety of samples, different cleanup procedures, viz. steam distillation, oxidation, saponification, partition, sweep codistillation, adsorption chromatography, etc. have been used. LLP, solid-phase extraction (SPE), gel permeation chromatography (GPC), etc. have been introduced to ease the sample purification. The principle of adsorption chromatography has been well utilized for sample cleanup. Different forms of adsorption chromatography such as column chromatography, SPE, and solid-phase microextraction (SPME) are widely used. Traditional column chromatography using alumina [24], silica gel [55], and Florisil [42] are still being used. Dialysis tube [24] was also used for cleanup in the determination of different pesticides.

Recent work, however, indicated that the application of SPE cartridges is on the raise. Different types of SPE cartridges viz., reverse phase [48], aminopropyl, Florisil, ENVI-Carb, etc. have been used for the purification of different samples. A combination of classical extraction of soil by an organic solvent and SPME was applied for the isolation of pesticides from soil [19].

22.3.2 SEPARATION AND DETECTION

Pesticide residues can be determined with the help of any technique, viz. microbiological, enzymatic, colorimetric, spectrophotometric, radiometric, etc. Previously, enzyme inhibition tests were aimed to analyze some pesticides, especially the organophosphorus pesticides (OPP), which were later replaced by colorimetric methods. In due course of time, these methods became redundant. TLC methods were also not used frequently. The introduction of high-caliber analytical columns made some of the instrument techniques, viz. GC [16,19,28], HPLC [25,26,56], ultraperformance liquid chromatography (UPLC [57–59]), and CE [60,61] very much adaptable for pesticide analysis. Mass spectroscopic detection linked to either GC [19,37] or HPLC is a good approach for specificity and accuracy in the pesticide residue analysis [20]. The determination of organochlorine (OC) pesticides was also attempted using neutron activation analysis (NAA), but its poor accuracy [62] did not favor its use for the analysis.

Analysis of the pesticides is also carried out by low-pressure gas chromatography (LP-GC) with mass spectrometry in tandem (MS-MS) mode [63] and capillary GC with atomic emission detection (AED; 45). More recent work indicates an increased use of UPLC coupled with different types of MS detectors. The major advantages of UPLC over HPLC are the speed of analysis, the narrower peaks (giving increased signal-to-noise ratio), and improved confirmation for the targeted pesticides in the analyses [57–59].

Direct surface analyses by static secondary ion mass spectrometry (SIMS) has been used satisfactorily for the determination of pesticides, viz. alachlor, atrazine, captan, carbofuran, chlorpyrifos, chlorsulfuron, chlorthal-dimethyl, cypermethrin, 2,4-D, diuron, glyphosate, malathion, methomyl, methyl arsonic acid, mocap, norflurazon, oxyfluorfen, paraquat, temik, and trifluralin. The advantage of direct surface analysis over conventional pesticide analysis methods is the elimination of sample pretreatment including extraction, which streamlines the analysis substantially; total analysis time for SIMS analysis was about 10 min/sample [64].

CE has been proved a suitable microseparation technique for the analysis of a wide variety of chiral and achiral pesticides. It was also revealed that by combining selective precolumn derivatization schemes, sensitive detection methods (e.g., laser-induced fluorescence detection), and trace enrichment techniques, CE is capable of determining pesticides satisfactorily at trace levels [61].

A property of humic acid (HA) in soil relates to the binding and transport of pesticides. An attempt was made to use this unique property for pesticide monitoring. In this procedure, HA was immobilized on a support and a novel chromatography column was prepared. Then, analyses of some herbicides and rodenticides were attempted on this column. It was observed that HA had a lower affinity for neutral pesticides than polar compounds. Furthermore, it was established that the HA column was stable during an extended period of time, indicating that the HA column could soon become very attractive to determine the risk assessment of pesticides [65].

22.3.3 CONTEMPORARY METHODS OF ANALYSIS

Some fundamental steps of soil pesticide residue analysis from the literature are delineated under this section. For the isolation of chlorpyrifos (an OPP), 50 g of soil sample was shaken with 100 mL mixture of acetone and hexane (1:9) on wrist action mechanical shaker for 1 h. The filtrate was passed through anhydrous sodium sulfate and exchanged to 10 mL hexane before its analysis on GC [33]. Other workers extracted 10 g soil sample thrice with 50 mL acetone and partitioned onto 50 mL hexane thrice [28]. For the extraction of endosulfan, 50 g soil was mixed with 0.5 g each of activated charcoal and Florisil and 10 g anhydrous sodium sulfate. The mixture was shaken with 100 mL acetone–hexane (1:9) mixture for 1 h [35]. For the isolation of chlorpyrifos and endosulfan together, 100 g soil was mixed with 0.5 g activated charcoal and 10 g anhydrous sodium sulfate, and extracted by Soxhlet apparatus with n-hexane and acetone (9:1) mixture [34].

For the extraction of multiresidues of OCP and OPP from high-moisture soil samples, LLP was used. A 50 g soil (river sediment) sample was shaken with 150 mL mixture of DCM and acetone (4:1) for 4 h on a mechanical shaker. The extraction was repeated with 50 mL DCM after the addition of 6 mL aqueous *o*-phosphoric acid solution (1 mL *o*-phosphoric acid and 5 mL water) for 20 min. The combined DCM extract was exchanged to 10 mL *n*-hexane before its purification [24].

In another case, for determination of several OCPs, 50 g soil was thoroughly mixed with 150 mL mixture of MeCN and water (2:1) and kept overnight. The extract was filtered through Whatman No. 1 paper. The filtrate was diluted with 600 mL of 5% aqueous sodium chloride solution in 1 L separating funnel and partitioned twice with 100 mL *n*-hexane [42]. The residues of 15 OCP and 6 synthetic pyrethroids (SP) were extracted successfully from 50 g air-dried sediment (soil) by 150 mL hexane–acetone (4:1) mixture [16].

Trace amounts of pesticides in soil were also determined by liquid-phase microextraction (LPME) coupled to gas chromatography-mass spectrometry (GC-MS). The technique involved the use of a small amount (3 µL) of organic solvent impregnated in a hollow fiber membrane, which was attached to the needle of a conventional GC syringe. Various aspects of this procedure such as organic solvent selection, extraction time, movement pattern of plunger, concentrations of HA and salt, and the proportion of organic solvent in the soil sample, were optimized. Limits of detection (LODs) were between 0.05 and 0.1 µg/g with GC-MS analysis under selected-ion monitoring (SIM). Also, this method provided good precision ranging from 6% to 13% and the relative standard deviations were lower than 10% for most target pesticides. This process was completed within 4 min [66].

For the extraction of carbofuran insecticide, 20 g soil was extracted with 100 mL of 9:1 mixture of methanol and a buffer of pH 8.0. The insecticide was partitioned thrice with 50 mL DCM. After concentration, silica gel column cleanup was performed prior to analysis on GC-electron capture detection (ECD). The residues of quinalphos and phorate were, however, extracted with 10% aqueous acetone and partitioned thrice with 50 mL DCM. After concentration, silica gel column cleanup was performed prior to analysis on GC-thermal ionization detector (TID; 27). The imidacloprid is a lately introduced insecticide in the Indian province of Punjab. From 50 g soil samples, it was extracted by shaking thrice with 50 mL mixture of 70% aqueous MeCN and subsequently partitioned thrice onto 75 mL DCM. After exchanging with 5 mL MeCN, the sample was analyzed by isocratic HPLC using MeCN–water (7:3) mobile phase at 0.5 mL/min flow rate and 270 nm detection wavelength [26].

Recently, simultaneous determination of 11 SP insecticides, viz. tetramethrin, bifenthrin, phenothrin, λ-cyhalothrin, permethrin, cyfluthrin, cypermethrin, flucythrinate, esfenvalerate, fluvalinate, and deltamethrin in soil was carried out by gas chromatography-ion trap-mass spectrometry (GC-IT-MS), by means of two different ionization modes: electron impact (EI) and negative chemical ionization (NCI) and three data acquisition procedures: full scan, SIM, and MS/MS. The soil samples were treated with toluene–water, extracted by MAE for 9 min at 700 W, and subsequently purified by Florisil. Clean soil samples were spiked with SP at a spiking level of 10, 25, and 50 ng/g. Methane gas was used as ionization gas in the NCI mode. Owing to high selectivity and sensitivity, MS/MS acquisition in EI mode of ionization provided best results [51].

Herbicides are chemically different from other compounds and hence require specific solvent mixtures for extraction from the soil samples. Sulfosulfuron was extracted thrice by 80 mL mixture of acetonitrile (MeCN) and 1 M ammonium carbonate (9:1). The combined filtrate was condensed to 20 mL, diluted with saline water and subsequently partitioned thrice onto 50 mL DCM. The composite DCM layer was dried over anhydrous sodium sulfate and exchanged to MeCN before analysis by HPLC [43]. Trifluralin was extracted with acidic methanol (2% HCl in methanol, w/v), diluted 4–5 times with 10% aqueous sodium chloride solution, and partitioned thrice onto 30, 20, and 20 mL hexane. It was dried over anhydrous sodium sulfate and concentrated to 5–10 mL before further cleanup and analysis on GC [47]. Fenazaquin, an acaricide, is applied for the protection of fruit crops. For its isolation, 50 g soil was mixed with 0.3 g each of activated charcoal and Florisil and 10 g anhydrous sodium sulfate. Through column chromatography, it was eluted with 100 mL of acetone–hexane (1:9) mixture [25].

22.4 CONCLUSION

The existence of pesticide residues in our environment greatly affects the quality of life. Amongst other environmental components, soil is the largest reservoir of pesticides. The analysis for pesticide residues in soil would be useful in assuring the food and health security of mankind and soil quality as well. The pesticide analysis is tricky; moreover, the increase in the number of pesticides has made it more problematic. Consequently, onus has been shifted toward multipesticide residue analyses. Still, individual compound-based analysis finds the way, especially for newer pesticide compounds. Owing to great variation in chemical properties, all the pesticides cannot be analyzed by a single method, and we need to use diversified extraction, cleanup, and instrumental techniques. Despite the fact that at the time of processing, pesticide gets degraded by chemical and biochemical agents on exposure to elevated temperature, air drying is commonly used. It is suggested that suitable techniques such as vacuum drying under low temperature or lyophilization may be employed for the drying of soil sample, prior to the extraction. Traditional "Soxhlet" extraction procedure is being used, basically because of its simplicity and cost, although it should not be used for thermolabile compounds.

The process of LLE is now rarely used and newer techniques are overtaking the cleanup of the sample. The requisite has been for new sample preparation techniques, which are acquiescent to automation with reduced duration for extraction with smaller volumes of organic solvent. This necessity incorporated new technologies in the development and use of procedures, which minimized environmental concerns, time, labor, exposure of laboratory personnel to toxic chemicals, and preparation costs. The use of SPE not only reduces the extraction or processing time but also minimizes the wastage of organic solvents. Recently, LPME coupled with GC-MS has been regarded as a fast (within 4 min) and accurate method to determine trace amounts of pesticides in soil. Various separation cum detection devices, viz. GC, HPLC, UPLC, and CE are being used with different types of columns and detectors, and tremendous progress has been achieved. Overall, there is a collective need for cost-effective and simple sample preparation and analytical methodologies for soil multipesticide residues, which may apply globally for ensuring the quality of soil and ultimately the human life.

REFERENCES

1. Aspelin, A.A. and A.H. Gruber. 2000. *Pesticide Industry, Sales and Usage: 1996 and 1997 Market Estimates*. Office of Pesticide Programs, U.S. Environment Protection Agency, Washington, DC.
2. Kumari, B., R. Singh, V.K. Madan, R. Kumar, and T.S. Kathpal. 1996. DDT and HCH compounds in soils, ponds and drinking water of Haryana, India. *Bull. Environ. Contam. Toxicol.* 57(5): 787–793.
3. Awasthi, M.D., D. Sharma, and A.K. Ahuja. 2002. Monitoring of horticulture ecosystem: Orchard soil and water bodies from pesticide residues around North Bangalore. *Pest. Res. J.* 14(2): 286–291.
4. Logan, T.J. 2000. Chapter 6: Soils and environmental quality. In: *Handbook of Soil Science*, ME Sumner (editor in chief). CRC Press, Washington, DC, pp. G155–G169.
5. Frampton, G.K., S. Jänsch, J.J.S. Fordsmand, J. Römbke, and P.J. Van den Brink. 2006. Effects of pesticides on soil invertebrates in laboratory studies: A review and analysis using species sensitivity distributions. *Environ. Toxicol. Chem.* 25: 2480–2489.
6. Sharma, N. and Savita. 2007. Isoproturon persistence in field soil and its impact on microbial population under north western Himalayan conditions. *Pest. Res. J.* 19(1): 116–118.
7. Anderson, J.P.E. and K.H. Donsch. 1980. Influence of selected pesticides on the microbial degradation of ^{14}C-triallate and ^{14}C-diallate in soil. *Arch. Environ. Contam. Toxicol.* 9: 115–123.
8. Seven, D.J. and G. Ballard. 1990. Risk/benefit and regulations. In: *Pesticides in the Soil Environment: Process, Impact and Modeling*, H.H. Cheng (ed). Soil Science Society of America, Madison, WI, pp. 467–491.

9. Suri, K.S. and B.S. Joia. 1996. Persistence of chlorpyriphos in soil and its terminal residues in wheat. *Pest. Res. J.* 8(2): 186–190.

10. Vig, K., D.K. Singh, H.C. Agarwal, A.K. Dhawan, and P. Dureja. 2001. Insecticide residues in cotton crop soil. *J Environ. Sci. Health B*. 36(4): 421–434.

11. Sirvporn, S., P. Soraya, H. Poonsook, M. Sripan, and C. Aree. 1995. Accumulation of pesticides in soil, water, sediment and fish in pummelo orchard under IPM project; GTZ, pp. 142–153. Department of Agriculture, Bangkok, Thailand; Agricultural Toxic Substances Div. 1. Technical Conference of Agricultural Toxic Substances Division, 224 p. Kan prachum wichakan kong watthu mi phit kan kaset khrang thi 1. Bangkok, Thailand, 1995.

12. Bin, Z.T., R. Yong, W.H. Fu, Y.G. Yi, and X.Y. Sheng. 2005. Content and compositions of organochlorinated pesticides in soil of Dongguan City. *China Environ. Sci.* 25(Suppl.): 89–93.

13. Castelo, G.T., P.A. Augusto, and D. Barbosa. 2005. Removal of pesticides from soil by supercritical extraction—A preliminary study. *Chem. Eng. J.* 111(2/3): 167–171.

14. Hildebrandt, A., S. Lacorte, and D. Barceló. 2008. Occurrence and fate of organochlorinated pesticides and PAH in agricultural soils from the Ebro river basin. *Arch. Environ. Contam. Toxicol.* December 4, 2008. <http://www.ncbi.nlm.nih.gov/pubmed/19052798?>

15. Gupta, H.C.L. 2001. Persistence, resistance and tolerance of insecticides. In: *Insecticides: Toxicity and Uses*. Agrotech Publishing Academy, Udaipur, India, pp. 221–264.

16. Kathpal, T.S., S. Rani, B. Kumari, and G. Prasad. 2004. Magnitude of pesticide contamination of sediment and water of Keoladeo national park lake, Bharatpur. *Pest. Res. J.* 16(2): 75–77.

17. Saxton, G.N. and B. Engel. 2007. Fipronil insecticide and soil-sample handling techniques of state regulatory agencies. *Environ. Foren.* 8(3): 283–288.

18. James, D.W. and K.L. Wells. 1990. Soil sample collection and handling: Techniques based on source and degree of field availability. In: *Soil Testing and Plant Analysis*, 3rd edn., R.L. Westermond (ed.). Soil Science Society of America, Madison, WI, pp. 25–44.

19. Prosen, H. and L. Zupencic-Kralj. 1998. Use of solid-phase microextraction in analysis of pesticides in soil. *Acta Chim. Slov.* 45: 1–17.

20. Valverde, A. 2000. Chromatographic pesticide residue analysis. *J. AOAC Int.* 83: 679.

21. Mangani, F., G. Crescentlni, and F. Bruner. 1981. Sample enrichment for determination of chlorinated pesticides in water and soil by chromatographic extraction. *Anal. Chem.* 53: 1627–1632.

22. Lehotay, S.J. 2000. Analysis of pesticide residues in mixed fruits and vegetables by direct sample introduction/gas chromatography/tandem mass spectroscopy. *J. AOAC Int.* 83: 680–697.

23. Xue, N., F. Li, H. Hou, and B. Li. 2008. Occurrence of endocrine-disrupting pesticide residues in wetland sediments from Beijing, China. *Environ. Toxicol. Chem.* 27(5):1055–1062.

24. Ahmad, N., G. Bugueno, L. Guo, and R. Marolt. 1999. Determination of organochlorine and organophosphate pesticide residues in fruits, vegetables and sediments. *J. Environ. Sci. Health B*. 34: 829–848.

25. Sharma, I.D., J.K. Dubey, and S.K. Patyal. 2006. Persistence of fenazaquin in apple fruits and soil. *Pest. Res. J.* 18(1): 79–81.

26. Kang, B.K., J. Gagan, B. Singh, R.S. Battu, and H.K. Cheema. 2007. Persistence of imidacloprid in paddy and soil. *Pest. Res. J.* 19(2): 237–238.

27. Bhuwaneswari, K. and A. Reghupahy. 2006. Insecticide residues in soil of Nilgiris District. *Pest. Res. J.* 18(2): 225–227.

28. George, T., S.N. Beevi, and G. Priya. 2007. Persistence of chlorpyriphos in acidic soils. *Pest. Res. J.* 19(1): 113–115.

29. Babic, S., M. Petrovic, and M.K. Telan Macan. 1998. Ultrasonic solvent extraction of pesticides from soil *J. Chromatogr. A* 823 (1–2): 3–9.

30. Kaur, K. and R.K. Garg. 2003. Results of detection on the persistence of insecticides (dimethoate and phosphamidon) from soil and paper substrates by thin layer chromatography. *Anil Aggrawal's Internet J. Forensic Med. Toxicol.* 4(2), http://www.geradts.com/anil/ij/vol_004_no_002/papers/paper003.html

31. Anonymous. 2004. Fast multi-residue pesticide analysis in soil and vegetable samples. Publication 114AP30-01, Applied Biosystems 850, Foster City, CA.

32. Anonymous. 2008. Rapid analysis of SVOCS, PCBs and pesticides in crude soil extracts by GC-TOFMS. Form No. 203-821-253 4/08-REV1. LECO Corporation, St. Joseph, MI.

33. Awasthi, M.D. and N.B. Prakash. 1998. Persistence of chlorpyriphos in soil irrigated with saline water. *Pest. Res. J.* 10(1): 112–116.

34. Diwan, K., B.A. Patel, M.F. Raj, P.G. Shah, J.A. Patel, and B.K. Patel. 2002. Dissipation of chlorpyriphos and endosulfan in soil and their residues in wheat. *Pest. Res. J.* 14(1): 107–112.

35. Raikwar, M.K., S.K. Nag, T. Banerjee, and N.K. Shah. 2003. Persistence behaviour of endosufan in fodder maize. *Pest. Res. J.* 15(2): 186–190.

36. Holland, P.T. 1977. Routine methods for analysis of organophosphorus and carbamate insecticides in soil and ryegrass. *Pest. Sci.* 8(4): 354–358.

37. Zhang, P., X. Hu, Y. Wang, and T. Sun. 2008. Simultaneous determination of 15 organochlorine pesticide residues in soil by GC/MS/MS. In: *Bioinformatics and Biomedical Engineering, ICBBE 2008*. The 2nd International Conference, 16–18 May, 2008, Shanghai, China, pp. 4113–4116.

38. Fuentes, E., M.E. Báez, and D. Reyes. 2006. Microwave assisted extraction through an aqueous medium and simultaneous cleanup by partition on hexane for determining pesticides in agricultural soils by gas chromatography: A critical study. *Anal. Chim. Acta.* 578(2): 122–130.

39. Castro, J., C. Sanchez Brunete, and J.L. Tadeo. 2001. Multiresidue analysis of insecticides in soil by gas chromatography with electron-capture detection and confirmation by gas chromatography-mass spectrometry. *J. Chromatogr.* 918(2): 371–380.

40. Sánchez Brunete, C., B. Albero, and J.L. Tadeo. 2004. Multiresidue determination of pesticides in soil by gas chromatography-mass spectrometry detection. *J. Agric. Food Chem.* 52(6), 1445–1451.

41. García Sánchez, F. and A.A. Gallardo. 1992. Spectrofluorimetric determination of the insecticide azinphos-methyl in cultivated soils following generation of a fluorophore by hydrolysis. *Analyst* 117: 195–198.

42. Surendra Nath, B., M.A. Usha, V. Unnikrishnan, and S.R. Gawali. 2000. Organochlorine pesticide residues in fodder, soil and water from dairy farms. *Ind. J. Dairy Biosci.* 11: 127–130.

43. Saha, S., N.T. Yaduraju, and G. Kulshrestha. 2003. Residue studies and efficacy of sulfosulfuron in wheat crop. *Pest. Res. J.* 15(2): 173–175.

44. Bao, M.L., F. Pantani, K. Barbieri, D. Burrini, and O. Griffini. 1996. Multi-residue pesticide analysis in soil by solid-phase disk extraction and gas chromatography/ion-trap mass spectrometry. *Int. J. Environ. Anal. Chem.* 64(4): 233–245.

45. Vinas, P., N. Campillo, I.L. Garcia, N. Aguinaga, and M.H. Cordoba. 2003. Capillary gas chromatography with atomic emission detection for pesticide analysis in soil samples. *J. Agric. Food Chem.* 51(13): 3704–3708.

46. Ramesh, A. and M. Balasubramanian. 1999. The impact of household preparations on the residues of pesticides in selected agricultural food commodities available in India. *J. AOAC Int.* 82: 725–737.

47. Goyal, V., V.K. Madan, and V.K. Fogat. 2003. Persistence of Trifluralin in soils under intermittent and continuous pending conditions. *Pest. Res. J.* 15(2): 181–184.

48. Singh, P. and A. Chhabra. 2008. Influence of dietary activated charcoal on carryover of monocrotophos in goat milk. *Ind. J. Dairy Sci.* 61(2): 127–135.

49. Rissato, S.R., S.M. Galhiane, B.M. Apon, and M.S.P. Arruda. 2005. Multiresidue analysis of pesticides in soil by supercritical fluid extraction/gas chromatography with electron-capture detection and confirmation by gas chromatography-mass spectrometry. *J. Agric. Food Chem.* 53(1): 62–69.

50. Gonçalves, C. and M.F. Alpendurada. 2005. Assessment of pesticide contamination in soil samples from an intensive horticulture area, using ultrasonic extraction and gas chromatography-mass spectrometry. *Talanta* 65(5): 1179–1189.

51. Esteve-Turrillas, F.A., A. Pastor, and M. de la Guardia. 2006. Comparison of different mass spectrometric detection techniques in the gas chromatographic analysis of pyrethroid insecticide residues in soil after microwave-assisted extraction. *Anal. Bioanal. Chem.* 384(3): 801–809.

52. Padron-Sanz, C., R. Halko, Z. Sosa-Ferrera, and J.J. Santana-Rodriguez. 2005. Combination of microwave assisted micellar extraction and liquid chromatography for the determination of organophosphorous pesticides in soil samples. *J. Chromatogr. A* 1078(1/2): 13–21.

53. Lehotay, S.J. 1997. Supercritical fluid extraction of pesticides in foods. *J. Chromatogr. A* 785: 289–312.

54. Adou, K., W.R. Bontoyan, and P.J. Sweeney. 2001. Multiresidue method for the analysis of pesticide residues in fruits and vegetables by accelerated solvent extraction and capillary gas chromatography. *J. Agric. Food Chem.* 49: 4153–4160.

55. Ober, A.G., I.S. Maria, and J.D. Carmi. 1987. Organochlorine pesticide residues in animal feeds by cyclic steam distillation. *Bull. Environ. Contam. Toxicol.* 38: 404–408.

56. Singh, P. 2004. Organophosphorus pesticide residues in animal feeds and their excretion in milk, PhD dissertation, National Dairy Research Institute (DU), Karnal, Haryana, India, pp. 56–76.

57. Leandro, C.C., P. Hancock, R.J. Fussell, and B.J. Keely. 2006. Comparison of ultra-performance liquid chromatography and high performance liquid chromatography for the determination of priority pesticides in baby foods by tandem quadrupole mass spectrometry. *J. Chromatogr. A* 1103(1): 94–101.

58. Pozo, O.J., M. Barreda, J.V. Sancho, F. Hernandez, J.L. Lliberia, M.A. Cortes, and B. Bago. 2007. Multiresidue pesticide analysis of fruits by ultra-performance liquid chromatography tandem mass spectrometry. *Anal. Bioanal. Chem.* 389: 1765–1771.
59. Kovalczuk, T., O. Lacina, M. Jech, J. Poustka, and J. Hajslová. 2008. Novel approach to fast determination of multiple pesticide residues using ultraperformance liquid chromatography-tandem mass spectrometry (UPLC-MS/MS). *Food Addit. Contam.* 25: 444–457.
60. El Rassi, Z. 1997. Capillary electrophoresis of pesticides. *Electrophoresis* 18(12–13): 2465–2481.
61. Cooper, P.A., K.M. Jessop, and F. Moffatt. 2000. Capillary electrochromatography for pesticide analysis: Effects of environmental matrices. *Electrophoresis* 21(8): 1574–1579.
62. Zhong, W., D. Xu, Z. Chai, and X. Mao. 2004. Neutron activation analysis of extractable organohalogens in milk from China. *J. Radioanal. Nucl. Chem.* 259: 485–488.
63. Garrido-Frenich, A., F.J. Arrebola, M.J. González-Rodríguez, J.L.M. Vidal, and N.M. Díez. 2003. Rapid pesticide analysis, in post-harvest plants used as animal feed, by low-pressure gas chromatography–tandem mass spectrometry. *Anal. Bioanal. Chem.* 377: 1038–1046.
64. Ingram, J.C., G.S. Groenewold, A.D. Appelhans, J.E. Delmore, J.E. Olson, and D.L. Miller. 1997. Direct surface analysis of pesticides on soil, leaves, grass, and stainless steel by static secondary ion mass spectrometry. *Environ. Sci. Technol.* 31(2): 402–408.
65. Andre, C., T.T. Truong, J.F. Robert, M. Thomassin, and Y.C. Guillaume. 2005. Construction and evaluation of a humic acid column: Implication for pesticide risk assessment. *Anal. Chem. (Washington)*, 77(13): 4201–4206.
66. Li, H. and H.K. Lee. 2004. Determination of pesticides in soil by liquid-phase microextraction and gas chromatography-mass spectrometry. *J. Chromatogr. A.* 1038(1/2): 37–42.
67. Maestroni, B.M., J.H. Skerritt, I.G. Ferris, and A. Ambrus. 2001. Analysis of DDT residues in soil by ELISA: An international interlaboratory study. *J. AOAC Int.* 84(1): 134–142.
68. Konda, L.N., G. Fuleky, and G. Morovjan. 2002. Subcritical water extraction to evaluate desorption behavior of organic pesticides in soil. *J. Agric. Food Chem.*, 50(8): 2338–2343.
69. Koskinen, W.C. and L.J. Marek. 2005. LC/MS/MS multiresidue analysis of polar pesticides in soil and water [Abstract]. Canadian Society of Chemistry Conference, Canada, p. 7.
70. Tao, C.J., J.Y. Hu, and J.Z. Li. 2007. Determination of insecticide monosultap residues in tomato and soil by capillary gas chromatography with flame photometric detection. *Can. J. Anal. Sci. Spectr.* 52 (5): 296–304.

Index